Adaptive Radiation Therapy

IMAGING IN MEDICAL DIAGNOSIS AND THERAPY
William R. Hendee, Series Editor

Adaptive
Radiation Therapy

Edited by
X. Allen Li

CRC Press
Taylor & Francis Group
Boca Raton   London   New York

CRC Press is an imprint of the
Taylor & Francis Group, an **informa** business

A TAYLOR & FRANCIS BOOK

CRC Press
Taylor & Francis Group
6000 Broken Sound Parkway NW, Suite 300
Boca Raton, FL 33487-2742

First issued in paperback 2020

© 2011 by Taylor and Francis Group, LLC
CRC Press is an imprint of Taylor & Francis Group, an Informa business

No claim to original U.S. Government works

ISBN-13: 978-0-367-57700-1 (pbk)
ISBN-13: 978-1-4398-1634-9 (hbk)

Visit the Taylor & Francis Web site at
http://www.taylorandfrancis.com

and the CRC Press Web site at
http://www.crcpress.com

To my wife, Rong,
to my children, Vincent and Hana,
and to my parents, for their love, patience, and inspiration.

Contents

SECTION I   Treatment Planning

SECTION II   Treatment Delivery

SECTION III Clinical Applications

Series Preface

Advances in science and technology related to medical imaging and radiation therapy, since their inception over a century ago, are more profound and rapid than ever before. Further, the disciplines are increasingly cross-linked as imaging methods are being more widely used to plan, guide, monitor, and assess treatments in radiation therapy. Today, the technologies of medical imaging and radiation therapy are so complex and computer-driven that it is difficult for the persons responsible for their clinical use (physicians and technologists) to know exactly what is happening at the point-of-care, when a patient is being examined or treated. The persons best equipped to understand the technologies and their applications are medical physicists, and these individuals are assuming greater responsibilities in the clinical arena to ensure that what is intended for the patient is actually delivered in a safe and effective manner.

However, the growing responsibilities of medical physicists in the clinical arenas of medical imaging and radiation therapy are not without their challenges. Most medical physicists are knowledgeable in either radiation therapy or medical imaging, and are experts in one or a small number of areas within their discipline. They sustain their expertise in these areas by reading scientific articles and attending various scientific conferences. In contrast, their responsibilities increasingly extend beyond their specific areas of expertise. To meet these responsibilities, medical physicists periodically must refresh their knowledge of advances in medical imaging or radiation therapy, and they must be prepared to function at the intersection of these two fields. The manner in which these objectives are achieved remains a challenge.

At the 2007 annual meeting of the American Association of Physicists in Medicine in Minneapolis, this challenge was the topic of conversation during a lunch hosted by the publishers Taylor & Francis, involving a group of senior medical physicists (Arthur L. Boyer, Joseph O. Deasy, C.-M. Charlie Ma, Todd A. Pawlicki, Ervin B. Podgorsak, Elke Reitzel, Anthony B. Wolbarst, and Ellen D. Yorke). The conclusion of this discussion was that a book series should be launched under the Taylor & Francis banner, with each volume in the series addressing a rapidly advancing area of medical imaging or radiation therapy of importance to medical physicists. The aim of each volume would be to provide medical physicists with the information needed to understand the technologies advancing at a rapid rate and to apply them to safe and effective delivery of patient care.

Each volume in the series is edited by one or more individuals with recognized expertise in the technological area encompassed by the book. The editors are responsible for selecting the authors of individual chapters and ensuring that the chapters are comprehensive and intelligible to someone without such expertise. The enthusiasm of the volume editors and chapter authors has been gratifying and reinforces the conclusion of the Minneapolis luncheon that this series of books addresses a major need of medical physicists.

Imaging in Medical Diagnosis and Therapy would not have been possible without the encouragement and support of the series manager, Luna Han of Taylor & Francis Publishers. The editors and authors, and most of all I, are indebted to her steady guidance during the entire project.

William Hendee
Series Editor
Rochester, Minnesota

Foreword

Recent national focus on "personalized" cancer treatment may strike many radiation oncologists and medical physicists, who are responsible for the administration of modern radiation therapy to cancer patients, as familiar territory. The quest to tailor radiation treatment to the *specific* needs of individual patients has been actively pursued for more than a century, since the very inception of radiation therapy. This investigational process is reaching its zenith.

Pioneering radiation therapists grasped the conceptual imperative of providing full radiation doses to tumor target areas, while protecting vulnerable normal tissues. Many of them no doubt also envisioned a future that has only recently arrived. Improved understanding of tumor and normal tissue radiation responses and a long technological evolution in treatment delivery machines, imaging methodology, and radiation treatment planning algorithms were first necessary. With those advances, the ideal of providing every patient with individualized radiation therapy that is maximally accurate as well as precise is within reach.

Dr. Li and his colleagues are to be commended for assembling within a single volume a compendium of recent vast technological progress that advances us toward this paramount goal. Elucidating its initial application in the context of disease-specific clinical models in which radiation therapy plays an important role increases our appreciation of its impact in human terms. Significant additional work lies ahead to fully optimize radiation therapy delivery beyond what is achievable today, but the path forward is clearly visible in these chapters. We look ahead to a future in which day-to-day adaptation of the radiation plan to *intra* as well as *inter* fractional biological and anatomical changes in tumor and normal tissues is routine.

J. Frank Wilson MD, FACR, FASTRO

Preface

Adaptive radiation therapy (ART) is a state-of-the-art approach that uses a feedback process to account for patient-specific anatomic and/or biological changes, thus delivering highly individualized radiation therapy for cancer patients. Basic components of ART include (1) detection of anatomic and biological changes, often facilitated by multimodality images, (2) treatment plan optimization to account for the patient-specific spatial morphological and biological changes with consideration of radiation responses, and (3) technologies to precisely deliver the optimized plan to the patient. Interventions of ART may consist of both online and offline approaches. Accumulated clinical data have demonstrated the need for ART in clinical settings, assisted by the wide application of intensity modulated RT (IMRT) and image-guided RT (IGRT). The technology and methodology for ART have advanced significantly in the last few years and are moving rapidly into the clinic. This book describes these technological and methodological advances as well as initial clinical experiences using ART for selected anatomic sites.

The contents have been divided into three main sections. Section I begins with the radiobiological basis for ART (Chapter 1), followed by a detailed discussion dealing with the use of morphological (Chapter 2) and biological (Chapter 3) images and biomarkers (Chapter 4) to detect patient-specific anatomic and biological information for ART planning. The

methodology and technology to design and optimize treatment plans in light of available radiation response relationships are discussed in Chapters 5 through 7.

Section II looks at the technologies and methodologies used to accurately deliver the planned treatment to the patient. These technologies and methodologies include the delivery of IMRT and IGRT (Chapters 8 through 10), intervention methodologies of ART (Chapters 11 and 12), management of intrafraction variations with respiratory motion in particular (Chapters 13 and 14), and the quality assurance needed to ensure the safe delivery of ART (Chapter 15).

Section III presents clinical examples of ART applications in several common cancer types or anatomic sites including central nervous system (Chapter 16), head and neck (Chapter 17), breast (Chapter 18), lung (Chapter 19), liver (Chapter 20), prostate (Chapter 21), gynecological cancers (Chapter 22), and soft-tissue sarcoma (Chapter 23).

This book is intended primarily for medical physicists and radiation oncologists although it should also benefit dosimetrists and radiation therapists. It is meant to serve as both a reference for those who have been working in the field for some years and an educational text for those who are entering the field.

Acknowledgments

The editor is indebted to Dr. William Hendee for seeding the concept of this volume and to Dr. J. Frank Wilson for providing endless encouragement and guidance. Sincere thanks to all the authors and co-authors for their excellent contributions to the volume; to the members of the medical physics group and the faculty in the Department of Radiation Oncology at the Medical College of Wisconsin for their continuous support; to Jessica Kotowicz for her editorial and administrative assistance; and to Luna Han and the staff at Taylor & Francis for making this book a reality. Special thanks to Drs. David Rogers, Charlie Ma, Peter Raaphorst, Lee Gerig, James Chu, and Cedric Yu for their inspiration throughout my career as a medical physicist.

Editor

X. Allen Li, PhD, DABMP, FAAPM, is professor and chief physicist in the Department of Radiation Oncology at the Medical College of Wisconsin. He received his PhD from Concordia University in Canada and a masters degree in physics from Yunnan University in China. He has served in faculty positions at the Ottawa Cancer Center, University of Ottawa, Rush-Presbyterian-St. Luke's Medical Center in Chicago, and in the Department of Radiation Oncology at the University of Maryland in Baltimore. Dr. Li is the recipient of over 20 research grants and has published over 120 peer-reviewed papers. His research interests include adaptive radiation therapy, biological modeling for radiation treatment planning, and biological/functional imaging for radiotherapy planning and delivery.

Contributors

Justus Adamson, PhD
Department of Radiation Oncology
Duke University Medical Center
Durham, North Carolina

Ergun Ahunbay, PhD
Department of Radiation Oncology
Medical College of Wisconsin
Milwaukee, Wisconsin

James Balter, PhD
Department of Radiation Oncology
University of Michigan
Ann Arbor, Michigan

Julie A. Bradley, MD
Department of Radiation Oncology
Medical College of Wisconsin
Milwaukee, Wisconsin

Daliang Cao, PhD
Swedish Cancer Institute
Seattle, Washington

Yue Cao, PhD
Department of Radiation Oncology
   and Radiology
University of Michigan
Ann Arbor, Michigan

David J. Carlson, PhD
Department of Therapeutic Radiology
Yale University School of Medicine
New Haven, Connecticut

Laura I. Cerviño, PhD
Department of Radiation Oncology
University of California, San Diego
La Jolla, California

Zheng Chang, PhD
Department of Radiation Oncology
Duke University Medical Center
Durham, North Carolina

Kihwan Choi, PhD
Department of Electrical Engineering
Stanford University
Stanford, California

Jeffrey M. Craft, MD, PhD
Department of Radiation Oncology
Washington University
St. Louis, Missouri

Laura A. Dawson, MD
Department of Radiation Oncology
Princess Margaret Hospital
University of Toronto
Toronto, Ontario, Canada

Joseph O. Deasy, PhD
Division of Bioinformatics and
   Outcomes Research
Department of Radiation Oncology and
   Mallinckrodt Institute of Radiology
Washington University School of
   Medicine
St. Louis, Missouri

Thomas F. DeLaney, MD
Department of Radiation Oncology
Harvard Medical School
Boston, Massachusetts

Lei Dong, PhD
The University of Texas M.D. Anderson
   Cancer Center
Houston, Texas

Eric Donnelly, MD
Department of Radiation Oncology
Northwestern University
Chicago, Illinois

Beth Erickson, MD
Department of Radiation Oncology
Medical College of Wisconsin
Milwaukee, Wisconsin

Anthony Fyles, MD
Department of Radiation Oncology
Princess Margaret Hospital
Toronto, Ontario, Canada

Jeho Jeong, MS
Nuclear Science and Engineering
   Institute
University of Missouri
Columbia, Missouri

Robert Jeraj, PhD
Department of Medical Physics
University of Wisconsin
Madison, Wisconsin

Steve B. Jiang, PhD
Department of Radiation Oncology
University of California, San Diego
La Jolla, California

Jian-Yue Jin, PhD
Department of Radiation Oncology
Henry Ford Hospital System
Detroit, Michigan

Kristofer Kainz, PhD
Department of Radiation Oncology
Medical College of Wisconsin
Milwaukee, Wisconsin

Paul Keall, PhD
Department of Radiation Oncology
Stanford University
Stanford, California

Chad Lee, PhD
CK Solutions
Edmond, Oklahoma

X. Allen Li, PhD
Department of Radiation Oncology
Medical College of Wisconsin
Milwaukee, Wisconsin

Jun Lian, PhD
Department of Radiation Oncology
University of North Carolina
Chapel Hill, North Carolina

Karen Lim, MBBS
Department of Radiation Oncology
Princess Margaret Hospital
Toronto, Ontario, Canada

C-M Charlie Ma, PhD
Radiation Oncology Department
Fox Chase Cancer Center
Philadelphia, Pennsylvania

Lawrence B. Marks, MD
Department of Radiation Oncology
University of North Carolina
Chapel Hill, North Carolina

Dan McShan, PhD
Department of Radiation Oncology
University of Michigan
Ann Arbor, Michigan

Michael Milosevic, MD
Department of Radiation Oncology
Princess Margaret Hospital
Toronto, Ontario, Canada

Doug Miller, MD
Department of Radiation Oncology
Washington University
St. Louis, Missouri

Issam El Naqa, PhD
Department of Radiation Oncology
Washington University
St. Louis, Missouri

Simeon Nill, PhD
Department of Medical Physics in
   Radiation Therapy
DKFZ
Heidelberg, Germany

Matt Nyflot, MSc
Department of Medical Physics
University of Wisconsin
Madison, Wisconsin

Jennifer O'Daniel, PhD
Department of Radiation Oncology
Duke University Medical Center
Durham, North Carolina

Uwe Oelfke, PhD
Department of Medical Physics in
   Radiation Therapy
DKFZ
Heidelberg, Germany

Jung Hun Oh, PhD
Department of Radiation Oncology
Washington University
St. Louis, Missouri

Joo Han Park, MSc
School of Health Sciences
Purdue University
West Lafayette, Indiana

Jianguo Qian, PhD
Department of Radiation Oncology
   and Molecular Imaging Program
   at Stanford (MIPS)
Stanford University
Stanford, California

David L. Schwartz, MD
The University of Texas M.D. Anderson
   Cancer Center
Houston, Texas

Vladimir A. Semenenko, PhD
Department of Radiation Oncology
Medical College of Wisconsin
Milwaukee, Wisconsin

David Shepard, PhD
Swedish Cancer Institute
Seattle, Washington

William Small, MD
Department of Radiation Oncology
Northwestern University
Chicago, Illinois

James Stewart, MSc
Department of Radiation Oncology
Princess Margaret Hospital
Toronto, Ontario, Canada

Robert D. Stewart, PhD
School of Health Sciences
Purdue University
West Lafayette, Indiana

Tae-Suk Suh, PhD
Research Institute of Biomedical
   Engineering
The Catholic University of Korea
Seoul, South Korea

An Tai, PhD
Department of Radiation Oncology
Medical College of Wisconsin
Milwaukee, Wisconsin

Dian Wang, MD, PhD
Department of Radiation Oncology
Medical College of Wisconsin
Milwaukee, Wisconsin

Julia White, MD
Department of Radiation Oncology
Medical College of Wisconsin
Milwaukee, Wisconsin

Jianzhou Wu, PhD
Swedish Cancer Institute
Seattle, Washington

Qiuwen Wu, PhD
Department of Radiation Oncology
Duke University Medical Center
Durham, North Carolina

Qingrong Jackie Wu, PhD
Department of Radiation Oncology
Duke University Medical Center
Durham, North Carolina

Lei Xing, PhD
Department of Radiation Oncology
   and Molecular Imaging Program
   at Stanford (MIPS)
Stanford University
Stanford, California

Fang-Fang Yin, PhD
Department of Radiation Oncology
Duke University Medical Center
Durham, North Carolina

Cedric X. Yu, DSc
Department of Radiation Oncology
University of Maryland School of
   Medicine
Baltimore, Maryland

Treatment Planning

I

Radiobiological Principles for Adaptive Radiotherapy

Joseph O. Deasy
*Washington University
School of Medicine*

Jeho Jeong
University of Missouri

1.1 Introduction

This chapter introduces radiobiological issues relevant to adaptive radiotherapy. The goal is to briefly describe the basic radiobiological factors that enter into deciding whether modifying a planned course of therapy would likely benefit a patient, either in improving the chances of local control or in reducing the risk of complications, or not.

In this chapter, we take adaptation to include both changes in treatment based on customization to an individual's biology and modifications in dose delivery in response to tissue changes, delivery inaccuracies, uncertainties, or outright delivery errors observed during treatment. We discuss the potential effect of dose "cold spots" either inside a target volume or at the target edge, the effect of dose nonuniformity caused by delivery errors, the rationale for boosting tumors, and the rationale for modifying plans for some patients during a course of therapy to reduce the risk of a complication. We also briefly discuss several other types of adaptation, including adaptation to observable spatial variations across a tumor, in particular, for positron emission tomography (PET)-F-18-fluorodeoxyglucose (FDG) uptake; adaption to a patient's inherent radiosensitivity; and adaptation based on treatment response.

Although fractionation effects are important in general, and tumors vary in sensitivity to fraction size effects (Steel 2002), we choose to make the calculations independent of the quadratic fraction size effect in order to focus on the central effects; however, if a change in fraction size is contemplated as an adaptive strategy, this issue should be addressed. We also do not address the actual process of generating radiobiologically optimal or robust treatment plans, which is covered in Chapter 6. Other recent reviews are available that focus either on normal tissue volume effects (QUANTEC 2010) or tumor response modeling (Deasy and Fowler 2005; Moiseenko, Deasy, and Van Dyk 2005).

1.2 Clinically Meaningful Changes in Tumor Dose Levels: What Is the Threshold?

A fundamental issue is at what point a change in dose level is of a clinically meaningful magnitude. We do not state the question as follows: What accuracy is desirable? Instead, we ask: What level of dose change really matters with respect to response, and should therefore potentially be a focus for adaptive interventions?

This "radiobiological action threshold" should certainly be as large as the typical uncertainties in actual doses delivered. Because dose variations on the order of 1%–3% are considered within the realm of current variations on quality assurance procedures, we consider 2%–3% to be the cutoff for a clinically meaningful change in dose rates. The current accuracy of typical intensity modulated radiation therapy (IMRT) delivery may be much less than what is commonly believed. To support clinical trial research, the Radiological Physics Center (RPC) in Houston, Texas, conducts tests of radiotherapy delivery accuracy by requiring clinics to plan and deliver treatments to standard head and neck and lung phantoms. The success rate, under rather liberal pass criteria (at least 90% of points within ±7% of the planned dose), is only 70%–80% (Ibbott et al. 2008). Although these failures may partially have to do with a physicist performing the treatment planning and delivery (as opposed to working dosimetrists/therapists), the origins of these frequent delivery inaccuracies remain obscure and may indicate that even chasing absolute dose variations of 3% (with IMRT delivery) is elusive. More generally, the implications of the outcomes of these unspecified problems remain unknown.

The radiobiological action threshold must also account for the position on the dose-response curve: If tumor response is very high (e.g., early stage prostate cancer or breast cancer), then a 3% change in dose level may cause a 1% or less change in response probability. However, near the midpoint of dose-response curves, even a 1% change in dose can have quite a sharp effect on local control. For a typical tumor dose-response curve (Okunieff et al. 1995), the corresponding absolute reduction in tumor control probability (TCP) is about 0.02. This is the steepest part of the dose-response curve. On the other hand, if local control is very poor (less than about 20%), then the subtraction or addition of 1%–2% of dose does little to change the overall response. We thus observe that whether a treatment plan should be changed or not depends on the rate of success under normal circumstances, corrections being most critical for disease/stage/treatment combinations with responses from 30%–70%.

From yet another point of view, any determination of the radiobiological action threshold should consider the relative imprecision of *optimal dose* as determined from randomized clinical trials. For example, the current Radiation Therapy Oncology Group (RTOG) trial number 0617 (external beam for non-small cell lung cancer) compares 60 Gy to 74 Gy, a 20% difference in dose.

Taking these observations together, although changes in dose of even 1% are known to change the dose response, variability in delivery and the clinical precision of optimal dose imply that the radiobiological action threshold is probably at least 2%–3% in dose unless one is operating at the steepest part of the dose-response curve.

1.3 Idea of the Planning Target Volume: Should It Be the Basis for Treatment Planning in the Future?

A key clinical question, implicitly faced every time a conventional treatment plan is reviewed, is what cold spots should be allowed in a gross target volume (GTV), a clinical target volume (CTV) possibly containing occult disease, or a planning target volume (PTV). The motivation behind the PTV margin concept (ICRU 1999) is to ensure adequate dose to disease by planning full dose to the entire PTV. However, despite its usefulness as a standard, from a radiobiological view the concept is inherently flawed for several reasons, in particular, the following:

- The margin is usually selected so that there is negligible risk of tissue identified within a target volume actually being shifted during delivery to the edge of the PTV volume, yet a full-dose level is required there. The probability of a systematic shift throughout a course of therapy decreases smoothly with the magnitude of the shift, whereas the PTV concept implicitly assumes the probability of a large shift (to the edge of the PTV) is just as big as the probability of a small shift (well inside the PTV).
- Despite the fact that radiotherapy is clearly a trade-off between normal tissue toxicity and disease control, the PTV concept implicitly assumes that normal tissue volume effects are not important, even if small cold spots in the PTV can significantly reduce toxicity at negligible risk of failure to control the disease.
- On purely dosimetric grounds, it can be argued that the PTV concept is problematic. Any treatment delivery device will have an essentially Gaussian spread in dose near field edges (Bortfeld, Oelfke, and Nill 2000); the tail of this Gaussian will, by physical necessity, invade the PTV (due to a lack of scattered secondary electrons). Requiring that this tail essentially reach near full-dose levels at the edge of the PTV necessitates the placement of the 50% point (or isosurface) of the Gaussian at a distance at least two to three times the Gaussian's standard deviation away from the edge of the PTV. This amounts to a huge increase in the normal tissue volume irradiated. In practice, planners routinely accept some type of cold spot on the edge of the PTV.

We will not discuss in detail the interesting and important topic of designing treatment plans to maximize disease control at satisfactory toxicity levels, or how to make robust plans that can withstand treatment motion or setup errors. Indeed, several groups have worked on methods that either allow for cold spots in the target volume explicitly (such as the Memorial Sloan-Kettering IMRT prostate trial [Zelefsky et al. 2006]) or do it implicitly using planning methods that can develop robust plans to withstand setup and immobilization uncertainties (Chan, Tsitsiklis, and Bortfeld 2010; McShan et al. 2006; Baum et al. 2006; Gordon and Siebers 2009). We expect an eventual move from current treatment planning to a practice that more explicitly includes such uncertainties.

Despite these caveats, the PTV concept is a foundation of conventional treatment planning. Consequently, one focus of this chapter is on what it means to have a "reasonable" cold spot in the PTV or the GTV within the current clinical paradigm.

1.4 Tumor Response: When Are Cold Spots in the Target Volume Too Cold or Too Large?

In order to understand the potential effect of GTV cold spots, we utilize an idealized model of a tumor that is completely homogeneous. Although clearly untrue, this simple model gives an "upper limit" on the effect of cold spots—one would expect tumor non-uniformity to reduce the effect of dose cold spots (Deasy 2001). We consider cold spots in the GTV and then the PTV, in turn. In the context of adaptive radiotherapy, we are actually referring to the best estimate of the GTV's dose-volume histogram (DVH) based on the accumulated "true dose." Of course, knowing the "true DVH" of the full tumor could be made difficult by many effects, such as tumor shrinkage, delivery uncertainties, imaging imprecision, and target contouring, which are discussed more fully in Chapters 2, 3, 9, 10, and 13.

In order to isolate the effect of cold spots (and later, hot spots), we use the cell-kill-based concept of equivalent uniform dose (EUD). Niemierko (1997) introduced the cell-kill-based EUD (hereafter denoted as cEUD to distinguish it from the generalized EUD, or gEUD) to capture the effect of dose variations. Two key advantages of working with the cEUD concept rather than a full-blown TCP model are (1) cEUD is independent of assumptions regarding clonogen density and (2) it is highly insensitive to the precise cell-kill parameter used. Here, we follow Niemierko's lead and use the parameter SF_2 (defined as the level of survival after a dose of 2 Gy). The main equation (Niemierko 1997) is

$$cEUD = 2 \times \ln \left((1/N_{voxels}) \sum_{i=1}^{N_{voxels}} SF_2^{D_i/2} \right) / \ln(SF_2) \qquad (1.1)$$

where the number of equal-volume tumor voxels is N_{voxels}, and the ith voxel's dose is denoted as D_i. Although actual dose-response curves are composed of patients with slightly varying radiosensitivities, the effect of a cold spot should move the cEUD down by the same amount regardless of patient radiosensitivity; hence, cEUD provides a useful investigative tool despite its simplicity.

In order to understand the effect of cold spots, we first consider the local control curve for a fraction of a tumor considered in isolation [voxel control probability (VCP) curves]. The VCP curves, compared to local control curves for the entire tumor, have 50% values at lower doses (because less dose is required

to control fewer clonogens), as shown in Figure 1.1. Also, if the actual TCP is anywhere greater than even a few percent, then the probability of controlling any particular tumor voxel is quite high, because the overall TCP curve is the multiplicative product of the VCPs of all the voxels. Thus, although the overall TCP may be, say, 50%, the corresponding operating point on a VCP curve is much higher. For a uniform dose (ignoring patient heterogeneity), $\log_{10}(TCP) = N_{voxels} \times \log_{10}(VCP)$, where N_{voxels} is the number of voxels and the typical values of VCP will be close to 0.9998 or even 0.99998 (depending on the number of voxels). Because these S-shaped VCP curves operate at such high operating points they are necessarily flatter, as a function of dose, than the TCP curves, as can be seen in Figure 1.1. It follows that the effect of dose changes on small cold spots (admittedly much larger than a single voxel) will nevertheless be much less than the potential effect of the same dose variation on the entire tumor.

To further examine the effects of small cold spots, Figure 1.2 shows how cold a uniform cold spot of varying fractional tumor volume would be to cause a 1%, 2%, or 3% drop in the estimated cEUD. This calculation assumes $SF_2 = 0.5$, and a reference dose prescription of 70 Gy. These results are very insensitive to those particular parameter values. It is assumed that the dose distribution consists of a cold spot of a given dose; the rest of the tumor is assumed to be irradiated uniformly to the prescription dose. The curve for a 1% drop in cEUD shows that a 10% volume corresponds to a cold spot dose of 95% of the prescription dose. This might be taken as a reasonable goal for the (accumulated) GTV D95 ("the D95 should receive 95% of the prescription dose"), as D95 is the median value of the coldest 10% in a DVH. From this

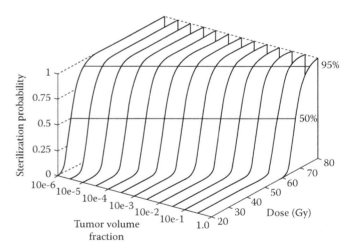

FIGURE 1.1 Theoretical curves showing the dose required to control a fraction of an idealized, biologically homogeneous tumor. The curves are known as voxel control probability (VCP) curves. One can note how the dose required to control the disease moves lower as the fractional volume decreases. (Reproduced from Moiseenko, V., et al. 2005. Radiobiological modeling for treatment planning. In *The Modern Technology of Radiation Oncology: A Compendium for Medical Physicists and Radiation Oncologists*, ed. J. Van Dyk, 85–220. Madison, WI: Medical Physics Publishing. With permission.)

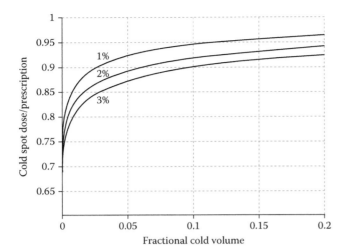

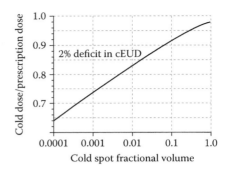

FIGURE 1.3 The magnitude of dose deficit causing a given drop in cEUD as a function of fractional volume: same as Figure 1.2 but for a logarithmic progression of cold spot volumes. The allowed dose deficit goes down linearly with the logarithm of the cold spot volume. For a gross target volume of 100 cc, a minimum dose to the coldest 1 cc could be about 10% less than the median dose of the tumor, with less than 1% effect on the cEUD.

FIGURE 1.2 The magnitude of dose deficit causing a given drop in cEUD as a function of fractional volume: This calculation assumes the level of survival after a dose of 2 Gy (SF_2) = 0.5, and a reference dose prescription of 70 Gy. However, the shape of these curves is relatively insensitive to those particular values. It is assumed that the dose distribution consists of a cold spot of a given dose, and the rest of the tumor is assumed to be irradiated uniformly to the prescription dose. The curve for a 1% drop in cEUD shows that a volume of 10% corresponds to a cold spot dose of 0.95. Uniform clonogen sensitivity and density is assumed. These assumptions are nearly always violated, as is discussed in Section 1.4, and this makes actual tumor response more insensitive to cold spot dose deficits. cEUD = cell-kill-based equivalent uniform dose.

discussion, it would be justifiable to adopt the 2% or even the 3% curve as a guide. The 2% cEUD-effect curve supports a target GTV D95 of 92%, which is a deficit larger than what most clinicians are willing to tolerate in a planned GTV. In order for these calculations to be applicable in the clinic, at least 90% of the tumor should receive at least the prescription dose, which is usually true in practice. Similar calculations relying on full TCP models have been published (Deasy 1997; Goitein, Niemierko, and Okunieff 1996; Tome and Fowler 2000). Again, we emphasize that we are talking about the accumulated true dose, subject to the effects of tumor shrinkage, delivery uncertainties, etc.

Figure 1.3 further shows the effect on cEUD of very small cold spots, nearly down to the (unrealistic) size of a few voxels. The use of the log scale reveals the nearly linear dependence between the allowed dose deficit and the cold spot volume.

Unfortunately, there is currently a general lack of information regarding detailed dose-volume parameters and their relationship to tumor response, both in humans and in human-tumor xenografts. Exceptions include a study by Terahara et al. (1999) that showed both the cEUD and the minimum dose correlated with the local control in very slow-growing chordoma tumors treated with proton therapy. Other dosimetric parameters (Vx or Dx values) also may have correlated with the outcome but were not considered in detail. In another TCP study of isolated non-small lung tumors from Washington University (Hope et al. 2005; El Naqa et al. 2010), the dosimetric parameter best related to improved tumor response was V75—the volume receiving at least 75 Gy. The best threshold

for correlation with local control was a V75 value between 20% and 40%. This finding may seem surprising, but it is perhaps consistent with a picture of tumor response that is not driven by small cold spots and might be affected by tumor heterogeneity.

Now, we turn our attention to PTV. First note that location matters. The location of any cold spot is important. Some dose deficit at the edge of the PTV is inevitable unless the beam portals are made very wide (to the detriment of the normal tissues). The implication is that dose at the edge of the PTV can be lower than the dose at the edge of the GTV by a significant amount if the cold spot really is on the edge of the PTV. Thus, conventional PTV plan criteria such as "the D95 should be at least 95% of the prescription dose" is almost certainly overkill, figuratively and literally. Hence, allowing for a D95 as low as 90% is certainly justifiable from a radiobiological point of view if the D90 reaches the prescription dose and most of the cold dose is near the edge of the PTV.

1.5 Tumor Dose Interfraction Fluctuations: When Do They Matter?

In this section, we briefly discuss the potential impact of fluctuations in dose due to patient motion during delivery. Shortly after the introduction of IMRT, Cedric Yu et al. observed that the interaction of breathing motion and (IMRT) dose delivery, which uses a series of open beam portals, could lead to large dose deficits (Yu, Jaffray, and Wong 1998). Bortfeld et al. subsequently argued that the errors due to breathing motion average out adequately over a typical course of radiotherapy (i.e., about 30 dose fractions) and the final effect would be insignificant (Bortfeld, Jiang, and Rietzel 2004; Bortfeld et al. 2002). As reviewed by Webb (2006), most publications conclude that there is an effective "smoothing out" of the dose distribution when anything like 30 fractions is used. However, in the most extensive experimental examination of statistical uncertainties in tumor delivery, Berbeco, Pope, and Jiang (2006) reported, for 1 IMRT plan, that the coldest 1 cc would typically be 6% less than the median target dose even when the voxel standard deviation was as low as 0.5%. This unexpected result remains unexplained. Theoretically, the dose smearing appealed to by Bortfeld and others would be expected to smooth out fluctuations. In addition to the fact that patient breathing changes over the course of therapy, we note that the relative orientation of the beams to the pattern of motion might be important when considering dose fluctuations.*

As a computational guide, we again use the cEUD model (also used by Duan et al. [2006] to model results from a motion phantom.) We compute the degradation in the final cEUD with the simple assumption that each fraction is delivered with an effective number of "degrees of freedom" (to roughly correspond to the number of effectively independent dose regions in a dose distribution) and a given coefficient of variation (defined as the standard deviation divided by the mean) of the dose values around the prescription dose. The results, as a function of the number of fractions delivered, are shown in Figure 1.4. There is little degradation in cEUD for relatively modest fluctuations in dose (coefficient of variation, CV = 5%), for five fractions or more, and only modest degradation if the CV is increased to 10%, which likely comprises the upper limit of noise levels that could be expected. The error bars represent the standard deviation of the resulting spread in individual simulated patient

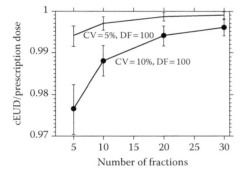

FIGURE 1.4 The hypothetical effect of statistical fluctuations in dose delivery on the resulting cEUD. The effective number of degrees of freedom was set equal to 100 in the results shown. The error bars are not uncertainties in the mean values, but rather one standard deviation estimates of the range of the final patient cEUD values over a fractionated course of therapy. For each point, 1000 hypothetical treatments were simulated with dose values drawn from Gaussian random number generation around the prescription dose (70 Gy) with coefficients of variation as stated. Small fluctuations (CV = 5%) do not result in a large effect on cEUD, even with a modest number of fractions. Only relatively noisy deliveries (CV = 10%) given over no more than about 5–10 fractions could result in significant degradation of cEUD (>2%).

---

* It is easy to imagine that if one particular beam's dose distribution is spread out due to motion, whereas another beam's distribution is less spread out, then the resulting distributions may not "fit together" as designed by the optimization algorithm, potentially creating hot and cold spots. This could happen if some beams are not orthogonal to some of the motions. Usually, only the case of cranio-caudal motion and coplanar transverse beams is considered.

cEUD values. Again using our 2% degradation as a standard, only the noisy delivery at five fractions or less would be of concern. Similar results are obtained when the degrees of freedom are smaller, although then the spread in interpatient results is smaller (not shown in Figure 1.4).

These results support a cautious attitude toward the use of IMRT when synchronization effects might induce significant dose fluctuations, unless a significant number of fractions is used (perhaps 10 or more). Again, the "blurring effect" may not hold if there is significant motion that is not orthogonal to all the beams.

1.6 Tumor Response: Value of Partial Tumor Boosts

In the era of IMRT, it is time to reconsider the wisdom of continuing the conventional planning goal of homogeneous target doses, a common planning goal that runs counter to sound radiobiology and results in a mathematical disadvantage in the optimization phase because it decreases the search space. Clinically, insisting on flat dose distributions restricts the ability of IMRT to sculpt out small dose "indentations" in order to reduce dose to nearby critical structures. Another reason it is appropriate to reconsider the desirability of homogeneous target doses is that state-of-the-art IMRT planning systems are poised to take advantage of increased flexibility regarding the allowance of target hot spots. The potential ability of partial tumor boosts to increase TCP has previously been discussed by others (Deasy 1997; Goitein, Niemierko, and Okunieff 1996; Deasy 2001; Tome and Fowler 2000). Our discussion is consistent with those publications, which contain extensive numerical results.

The schematic advantage of boosting a significant fraction of a tumor is depicted in Figure 1.5. Suppose a small part of a CTV or a GTV must be kept to a relatively low dose in order to protect a nearby normal structure. A strong partial tumor

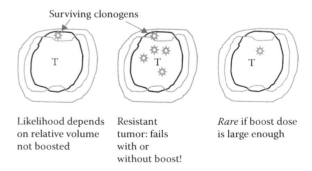

FIGURE 1.5 The potential advantage of partial tumor boosts. Contours depicting a PTV (outer contour, gray); gross tumor (black contour, denoted "T"); and a boost region (inside the PTV, in gray) are shown. A partial tumor boost would make it unlikely that a tumor recurrence would involve surviving clonogens only in the boost region, not in the cold spot. The advantage of eliminating this category of failure increases if the tumor has nonuniform radioresistance, because the most resistant part is then likely in the larger (boost) volume.

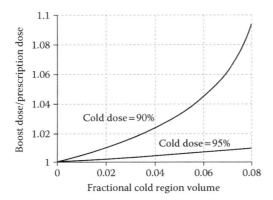

FIGURE 1.6 Calculations of the dose required to effectively cancel the negative effect of a cold spot. The figure shows whether a boost to the rest of the tumor can compensate for a cold spot. The parameters used here are the same as those used in Figure 1.2. The reduction in cEUD due to small dose deficits can be compensated quite well. If the cold dose is 95% of the prescription dose, then the negative effect on cEUD can be compensated by adding a small boost dose to the rest of the tumor, as indicated. If the cold dose is much lower, for example, 90%, then compensation quickly becomes difficult, and above a cold spot volume of about 8%, compensation is impossible. Thus, compensation for cold spots of modest volumes is readily achievable.

boost would reduce the probability of tumor recurring only in the boost volume (and not the cold spot). Of course, some tumors may be so resistant they are likely to recur in both the cold region and the boost region. Also, if the cold spot is large enough and cold enough, it could dominate response, although this would have to be a large cold spot indeed. The impact on TCP from partial tumor boosts may be substantial (up to an increase of 0.2 on a scale of 0–1), as previous publications show (Goitein, Niemierko, and Okunieff 1996; Tome and Fowler 2000; Deasy 1997).

Although our simple model assumes tumor homogeneity, in fact the opposite is true: tumor radiosensitivity is almost always effectively nonuniform, in which case the impact of a boost is expected to increase because it is more likely that the boost will hit the most resistant part of the tumor (Deasy 2001). This is true even when no information is available concerning the location of the most resistant tumor regions. In Section 1.11, we discuss data supporting the idea that FDG-PET avid regions may be resistant, in which case partial tumor boosts may be even more effective.

Partial tumor boosting can also be used to make up for a small cold spot. Figure 1.6 shows calculations of the dose required to effectively cancel the negative effect (on cEUD) of a cold spot of varying volume and 90% or 95% of the prescription dose. As in Figure 1.2, we assume that SF_2 equals 0.5 and the prescription dose is 70 Gy.

Chapet et al. (2005) performed a treatment planning study on the use of nonuniform lung tumor dose distributions when the esophagus is in the PTV. Treatment plan optimizations were performed using the gEUD function to model tumor response (the relationship between gEUD and cEUD was studied by Zhou et al. [2004]). They concluded that boosting a maximum target

volume fraction could plausibly lead to higher tumor response rates without increasing esophagitis toxicity.

There is already substantial clinical experience indicating that nonuniform dose distributions can be effective, for example, in stereotactic or brachytherapy treatments. Of course, allowing the dose to rise without bound in some target volumes containing nerves or blood vessels might lead to increased toxicity. However, the current paradigm of "worshipping at the altar of flat dose distribution" is counterproductive. Routinely allowing tumor doses to rise by 10%–15% above prescription values during treatment planning is likely to increase local control, and the rise is usually well-tolerated unless radiosensitive structures are contained in the boost volume.

1.7 Subclinical Disease Response: More Forgiving than Gross Disease Response

In this section, we briefly discuss the dose response (and dose volume response) of regional subclinical disease. Unseen subclinical disease is undoubtedly a source of real failures for many treatment sites, including lung, prostate, head and neck, and breast. For dose-volume effects, relevant points include the following:

- Occult disease, compared to gross disease, has a more shallow dose-response curve, likely due to a logarithmic-like size distribution of underlying foci (Withers, Peters, and Taylor 1995; Withers and Suwinski 1998). Although 55 Gy in 30 weekday fractions would normally mop up all occult disease, if some modest dose depression is necessary due to the presence of a sensitive normal structure, the response rate may still be high.
- Another reason that the impact of small volume cold (but not too cold) spots is likely to be less for occult disease compared to the impact on gross disease is that most of the CTV typically will not contain occult disease.
- Here again, location matters. Chao, Blanco, and Dempsey (2003) have surveyed pathology data showing how the probability of occult disease falls off exponentially, as a function of increasing distance from the primary tumor edge for breast and lung carcinomas. Of course, occult disease in lymphatics is a separate issue.

Because occult disease is likely to be even less sensitive to cold spots than gross disease, we again note that conventional measures of PTV plan goodness (when the PTV surrounds CTV and no gross disease) are probably more aggressive than warranted if they entail increased toxicity.

Again, this is not a declaration that dose variations do not matter for subclinical disease. As a modeling investigation into the effect of cold spots on subclinical disease, consider the TCP calculations for plans shown in Figures 1.7 and 1.8. These postsurgical breast cancer patients were treated using high-dose-rate afterloader brachytherapy at Washington University.

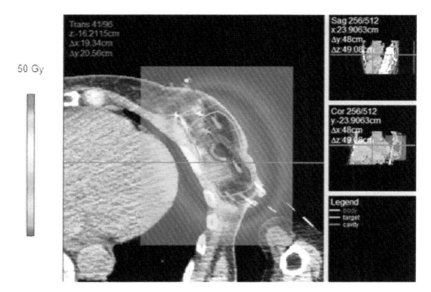

FIGURE 1.7 (See color insert following page 204.) Example of the estimated impact of dose variations on subclinical disease control: a brachytherapy (high-dose-rate afterloading) case to treat breast carcinoma (50 Gy dose in 10 fractions). The estimated regional control rate is 93.9%, based on the assumption that 40% of patients have at least one subclinical microscopic focus (cf. Figure 1.8).

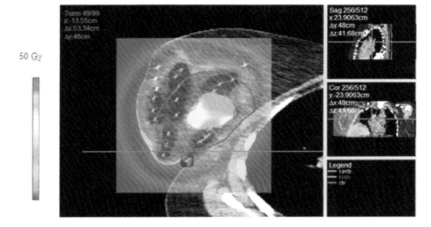

FIGURE 1.8 (See color insert following page 204.) Example of the estimated impact of dose variations on subclinical disease control: a brachytherapy (high-dose-rate afterloading) case to treat breast carcinoma (50 Gy dose in 10 fractions). The estimated regional control rate is 84.2%, based on the assumption that 40% of patients have at least one subclinical microscopic focus (cf. Figure 1.7).

After loading the plans into our research treatment planning system (CERR) (Deasy, Blanco, and Clark 2003), we calculated the TCP by assuming a given rate of regional metastases and using the logarithmic fall-off function from Chao et al. (2003) from the surgical margin, and a linear dose-response curve that is 0 at 0 dose and 1 at 55 Gy (Deasy, Powell, and Zoberi 2006). A Poisson distribution of the number of foci was assumed, using the average rate of patients with at least 1 microscopic focus (taken to be 40%). The computed TCP values show that the locations of cold spots and heterogeneity of the dose distribution could definitely affect the probability of regional relapse. In particular, cold spots near the surgical margin are definitely more important than those away from the margin, an effect that is usually ignored.

Despite all this, *failure to adequately irradiate occult disease may well be a prime cause of many treatment failures*, especially when close margins are used. Engels et al. support this point; they report that biochemical disease-free survival after external beam prostate radiotherapy decreased from 91% to 58% when margins were shrunk from the original 6–10 mm down to 3–5 mm based on seed localization ($p = .02$; Engels et al. 2009). Also for external beam prostate therapy, a study by de Crevoisier et al. (2005) and a Dutch trial (Al-Mamgani et al. 2008) both report that large rectal cross sections on simulation planning scans (due to gas or fecal matter) correlate with a large loss of local control. A subsequent analysis by Witte et al. ascribes this as most likely due to the presence of occult disease outside the prostate that was inadequately irradiated (Witte et al. 2010).

1.8 Critical Structure Response: Serial End Points Are Sensitive to Delivery Fluctuations

Dose response for normal tissues can be roughly divided into two categories with respect to dose-volume effects: "serial" end points, mostly sensitive to high doses, and "parallel" end points, mostly sensitive to mean doses. It almost immediately follows that serial end points are more likely to be significantly affected by systematic or interfraction errors. For example, cases of late rectal bleeding may increase for patients whose actual delivered high-dose region overlaps the rectum more than expected from the treatment plan. This possibility is schematically depicted in Figure 1.9, which shows a hypothetical wide probability distribution for the delivered gEUD. The NTCP model parameters used were similar to those reported for late rectal bleeding. The implication from this figure is that when population-based NTCP values are low, a significant fraction of complications may be cases where the delivery is worse than the plan. The standard use of daily image guidance would likely reduce the unexpected overirradiation of organs, attributed to poor setup precision. Theoretically, there may still be an important subgroup of patients, however, whose accumulated dose distributions significantly exceed treatment plan goals as seen late in a course of radiotherapy; the projected actual gEUD probably needs to exceed the planned gEUD by at least 10% to make an intervention worthwhile. This is likely to still leave the physician with a trade-off between reduction of target and normal tissue irradiation. Most normal tissue late toxicities are not life threatening

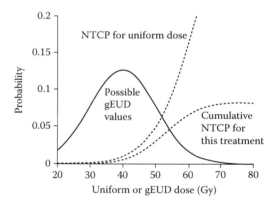

FIGURE 1.9 Observed complications can potentially be dominated by cases in which delivery is worse than the plan. This figure schematically shows how an uncertainty in the delivered gEUD translates into an increase in NTCP for that specific patient. Substantial uncertainty in the actual, delivered gEUD value may arise from anatomic variations, delivery inaccuracies, or setup inaccuracies, resulting in the Gaussian spread in potential gEUD values, as shown. Although the NTCP probability would be expected to be very low if the delivered gEUD value is close to the delivered one, the potential for much worse (higher) gEUD values significantly increases the potential for a complication. The "cumulative" curve across potential gEUD values shows the effect of the potential high values of gEUD, resulting in a best estimate of NTCP in this hypothetical case of about 8%.

and, therefore, have to be considered secondary to the goal of local control.

Nonetheless, one could imagine adaptive radiotherapy protocols that invoke increased imaging or other methods to increase the precision of target volume setup, especially for those patients whose sensitive normal tissue structures are projected to be close to tolerance levels. Such strategies, of course, would rely on the ability to accumulate an accurate map of dose values registered to patient anatomy. Potential strategies of adapting plans based on normal tissue tolerance are also challenged by the typically shallow dose-volume-response curves for nearly all end points studied to date. Despite this, as the precision of normal tissue complication probability (NTCP) models increases, for example, for late rectal bleeding, one could imagine that patients whose accumulated dose distributions are particularly poor could have altered treatment regiments in the last few weeks of therapy.

1.9 Critical Structure Response: Parallel End Points Can Withstand Patient Motion

In contrast to the so-called serial end points, complications that are thought to be related to a more "parallel" or global response of the organ or tissue structure in question tend to be less sensitive (but not completely insensitive) to patient setup errors. End points that fall into this category include radiation pneumonitis (Marks et al. 2010), radiation-induced liver disease (Dawson et al. 2002), and parotid function (Deasy et al. 2010). The reason for the reduced sensitivity is twofold: (1) systematic setup errors would need to be a significant fraction (>10%) of the entire organ width to have an effect, which is unusual; and (2) the associated end points tend to track the mean organ dose, which is not sensitive to random setup errors or breathing motion. Hence, these end points are not attractive for adaptive radiotherapy techniques, such as dose accumulation. An important caveat is that what now appears to be a parallel/mean-dose end point, may, upon further investigation, be found to be more sensitive to high doses.*

1.10 Adaptation to Patient's Inherent Radiosensitivity

In addition to the impact of delivery uncertainties, it has long been established that patients differ in their ability to repair normal tissue radiation damage (Safwat et al. 2002). This is illustrated in Figure 1.10, which shows a hypothetical normal tissue complication probability curve and the contributions to the curve from the most sensitive third, most resistant third, and middle third of the patient distribution. As recently reviewed by Andreassen and Alsner (2009), there are approximately 60 published studies that test potential correlations between normal tissue genomic variations and in vitro or clinical toxicity rates.

---

* A point often made by Randy Ten Haken.

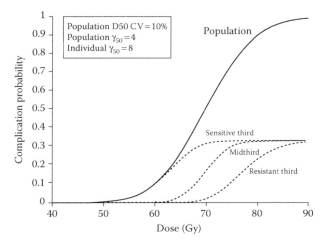

FIGURE 1.10 The effect of patient heterogeneity on normal tissue dose-response curves. This figure shows a hypothetical NTCP dose-response curve comprised of patients of varying radiobiological sensitivities, along with the dose-response curves of three subgroups—the most sensitive third, most resistant third, and the middle third. It is assumed that an individual patient's dose-response curve is also S-shaped (logistic functions were used), but steeper. The absolute percent increase in response divided by the relative increase in dose (γ_{50}) was set to 8 for individuals. The patient population coefficient of variation was set to 10%, resulting in a patient population γ_{50} of 4. Note that if NTCP is low, that is, less than about 20%, then in this model, NTCP is dominated by patients who fall within the most sensitive third of the patient distribution.

Although in this chapter we do not discuss these results in detail (see Chapter 4), we do note that many significant results have been reported for radiation repair genes, including genes associated with fibrosis and tissue remodeling. For example, single letter differences in the deoxyribonucleic acid (DNA) code (single nucleotide polymorphisms [SNPs]) in transforming growth factor beta-1 (TGFbeta1) have been statistically significantly associated with radiotherapy toxicity in approximately 7 out of 14 studies reviewed by Andreassen and Alsner (2009). Another gene commonly tested is the DNA repair gene XRCC1. Out of approximately 15 published studies, 4 studies find XRCC1 SNPs to be significantly associated with radiation toxicity. It is unlikely that 4 positive studies would arise out of 15 studies if XRCC1 had no actual correlation with toxicity. The typical odds ratios for these genetic variances are in the range of 1.1–1.3 and are usually not larger than 1.5. The fact that larger odds ratios have not been found may be due to the fact that radiation repair is a highly complex and multifactorial process involving many genes and pathways (Ismail et al. 2004; Bentzen 2006). Thus, highly predictive models for potential "genomic dose modifying factor" will probably require the identification of many relevant SNPs. We are perhaps 5–10 years away from the introduction of such a genomic dose modifying factor into routine clinical use, although it does appear likely, to us at least, that such a factor will eventually become a standard part of radiotherapy practice.

1.11 Tumor Heterogeneity and the Rationale for Boosting F-18-Fluorodeoxyglucose-Positron Emission Tomography-Avid Regions

Much of the discussion in this chapter involves the potential impact of dose cold spots or dose boosts on tumor response. Out of necessity, we include a short discussion about factors affecting tumor response. Excellent reviews on the topic are available (Steel 2002; Hall and Giaccia 2005).

Tumors arise through genetic changes to normally functioning cells that are passed to progeny. A hallmark of cancer, however, is the widespread number of changes to the tumor genome (mutations, deletions, and reorderings) compared to the cell line of origin. These changes partially arise from a phenomenon called "genomic instability," which is related to the loss of the tumor cell's ability to monitor and limit DNA misrepair through mechanisms such as apoptosis and cell-cycle arrest.

Tumors are biologically heterogeneous (Marusyk and Polyak 2010; Brown and Wilson 2004), partly due to cellular adaptation and partly due to limitations in local oxygen and nutrients. Cellular hypoxia, proliferation, and glycolytic metabolism are all thought to affect treatment response.

Hypoxia arises from uncontrolled tumor cell growth (Höckel and Vaupel 2001a, 2001b). As tumor cell growth exceeds the capacity of the local blood supply, despite the ongoing creation of new blood vessels, some cells are pushed far enough away from the nearest capillary to become oxygen and nutrient deficient, thus forming a microscopic hypoxic region. At larger distances from any capillaries, cell death (necrosis) often occurs. Many studies have reported that tumor hypoxia, measured either with a probe or with imaging compounds, is a negative predictive factor for local control and survival (Dehdashti et al. 2008; Dehdashti et al. 2003; Rajendran et al. 2006). This is probably not only because hypoxic cells are more resistant to radiotherapy and chemotherapy (Brown and Wilson 2004; Brown 2002; Brown 1999) but also because hypoxia promotes malignant progression by stimulating angiogenesis, proliferation, and metastasis (Vaupel and Mayer 2007; Höckel and Vaupel 2001a). The radiation dose necessary to achieve the same level of cell kill for hypoxic conditions compared to normoxic conditions (known as the *oxygen enhancement effect*) is about three times higher in vitro (Hall and Giaccia 2005), although this factor may be very different in situ because chronic hypoxia is stressful and changes gene expression. Nonetheless, this implies that more aggressive therapy directed toward macroscopic hypoxic regions may improve local control. Several planning studies predict that therapeutic response may be improved by selectively boosting the dose to the hypoxic region identified by PET imaging (Popple, Ove, and Shen 2002; Thorwarth et al. 2007). Hypoxia-inducible factor 1 (HIF-1) is likely a key cause of malignant progression under hypoxia; therefore, HIF-1 is a possible drug target (Semenza 2003). The level of hypoxia is difficult to predict using other tumor parameters such as size, grade, and level of necrosis. Noninvasive PET compounds with hypoxia-specific markers have been developed (Wang et al. 2009; Carlin et al. 2009; Krohn, Link, and Mason 2008; Varia et al. 1998; Scigliano et al. 2008). Local oxygen levels can fluctuate throughout the tumor, and several studies show that the resulting intermittent hypoxia increases the progression toward malignancy (Cárdenas-Navia et al. 2008; Graeber et al. 1996; Cairns, Kalliomaki, and Hill 2001; Martinive et al. 2006).

Tumor cell proliferation generally decreases with the distance from the nearest capillary. It is sometimes forgotten that the level of proliferation is reduced in hypoxic cells (Evans et al. 2001; Kennedy et al. 1997). The fraction of proliferative cells has been reported as a negative prognostic factor (Valera et al. 2005; Hoos et al. 2001; Viberti et al. 1997; Müller et al. 1996; Ishida et al. 1993; Larsson et al. 1993). The level of proliferation can be visualized by immunohistochemical staining with Ki-67 or in situ by F-18-tagged fluorinated thymidine (FLT)-PET imaging (Shields et al. 1998; Yamamoto et al. 2007; Buck et al. 2003; Chen et al. 2005; Been et al. 2004). It had been thought that accelerated hyperfractionated radiotherapy (Kim and Tannock 2005) would reduce the effect of tumor repopulation during a course of radiotherapy. However, clinical studies report the opposite result: tumors with the least pretreatment proliferation benefitted the most from treatment acceleration (Wilson et al. 2006; Sakata et al. 2000).

Most tumors demonstrate increased glucose uptake compared to normal tissues, as a result of upregulated glycolysis (Gatenby and Gillies 2004). Enhanced glycolysis might partially be a result of hypoxia, because hypoxic cells produce energy through glycolysis, without oxygen. However, tumor cells also demonstrate increased glycolysis even in the presence of oxygen (the "Warburg effect"; Warburg 1956). This aerobic glycolysis is thought to be caused by a number of genetic (or epigenetic) changes in a malignant tumor (Heiden, Cantley, and Thompson 2009; Kim and Dang 2006; Dang and Semenza 1999). The pattern of glucose uptake in a tumor varies spatially, as can be seen in FDG-PET imaging. This imaging has become an important clinical tool for cancer detection, staging, and monitoring of response after therapy. Although there are contrary results (Agarwal et al. 2009; Vesselle et al. 2007), many clinical studies have shown that increased FDG-PET image intensity, usually measured as a standardized uptake value (SUV), is a significant negative predictor of prognosis (Xie et al. 2009; Chen et al. 2009; van Baardwijk et al. 2007; Kidd et al. 2007; Xue et al. 2006; Sasaki et al. 2005; Cerfolio et al. 2005; Borst et al. 2005; Downey et al. 2004; Jeong et al. 2002; Dhital et al. 2000; Vansteenkiste et al. 1999; Ahuja et al. 1998; Casali et al. 2009). Also, the FDG uptake pattern after therapy correlates with tumor response and survival (van Loon et al. 2009; Schwarz et al. 2007; Mac Manus et al. 2005; Grigsby et al. 2004; Mac Manus et al. 2003; Grigsby et al. 2003; Kong et al. 2007).

The underlying mechanism of increased tumor glycolysis is still unclear. However, many studies using PET imaging or immunostaining have correlated FDG uptake with other physiological parameters, such as hypoxia, proliferation, blood flow, histology, and differentiation (Yamamoto et al. 2007; Bruechner et al. 2009; Vesselle et al. 2008; Dierckx and Van De Wiele 2008; Buchmann et al. 2008; Kelly et al. 2007; Zimny et al. 2006; Rajendran et al. 2004; Pugachev et al. 2005; Hara, Bansal, and DeGrado 2006; Vesselle et al. 2000).

Many studies have reported a correlation between microscopic FDG uptake and hypoxia (van Baardwijk et al. 2007; Kelly et al. 2007; Zimny et al. 2006; Rajendran et al. 2004; Pugachev et al. 2005). Other factors reported to affect glycolysis uptake include lactate, a by-product of glycolysis in hypoxic cells used as a fuel for oxidative metabolism in normoxic cells. The inhibition of lactate-fueled respiration is another target for intervention (Sonveaux et al. 2008; Semenza 2008).

The precise meaning of increased FDG regarding the dose required to sterilize a voxel is complicated by the fact that FDG is related to multiple factors such as blood flow and microscopic hypoxia, and may be related to changes in the fraction of cells in the cell cycle, and even intrinsic radioresistance. What does seem certain is that for standard fractionation regimes, high SUV values imply a higher dose is required for tumor eradication.

Publications by the Ann Arbor group at the University of Michigan (Kong et al. 2007), and the Maastro clinic group (Petit et al. 2009) indicate that, at least for non-small cell lung cancer patients, the likely location of failure on posttreatment FDG-PET scans appears to preferentially be the PET high-SUV region determined in pretreatment scans. Previously, Chao et al. (2001), Ling et al. (2000), and others suggested that IMRT delivery could nonuniformly boost a "biological target volume." These new results regarding lung cancer PET SUV values and the location of treatment failure motivate the use of boosts for high-SUV regions. The strategy of boosting FDG-PET-avid regions is under study (Petit et al. 2009; Feng et al. 2009; Vanderstraeten et al. 2006; Das et al. 2004). Future studies will hopefully supply similar data on the site of treatment failure regarding hypoxia imaging markers.

1.12 Adaptation Based on Treatment Response: Regression-Guided Radiotherapy

Tumor regression is commonly observed for many sites, for example, non-small cell lung tumors (Seibert et al. 2007; Kupelian et al. 2005). Moreover, regression itself has been reported to correlate with response (Li et al. 2007). The potential then exists to modify the treatment based on this information. Indeed, the conventional "shrinking field" boost already represents treatment adaptation. Monitoring treatment regression during radiotherapy could also allow for a strategy in which treatment could be modified or intensified for tumors that do not follow a favorable regression trajectory. This would probably require at least two weeks of observation. Unfortunately, this treatment modification or intensification would probably come at the cost of increased risk of toxicity.

1.13 Adaptation Based on Treatment Response: Response Guided by F-18-Fluorodeoxyglucose-Positron Emission Tomography Changes

FDG-PET changes during a course of radiotherapy have been shown to be predictive of tumor response, for radiotherapy alone (Kong et al. 2007; Erdi et al. 2000) or radiotherapy with targeted drugs (Sunaga et al. 2008). Erdi et al. (2000) showed that patients dramatically differed in their uptake of FDG during a course of radiotherapy for lung cancer, potentially providing a source of data for adapting treatment. Another recent study on 15 non-small-cell lung cancer patients reported by Kong et al. (2007) demonstrate a strong correlation between FDG-PET peak intensity images taken during a course of radiotherapy (at 45 Gy) and a follow-up FDG-PET image taken 3 months later. This also supports the idea that intratreatment PET could be used to modify lung cancer radiotherapy.

1.14 Summary

This chapter briefly focuses on radiobiological issues of relevance to potential adaptive radiotherapy strategies. Of course, all the arguments discussed here are limited by uncertainties in the accumulating true dose. We have argued that, for several reasons, small cold spots, not too large and not too cold, even in the GTV and more so in the PTV can probably be tolerated better than currently believed, and they could be used in some cases to reduce damage to nearby normal tissues. Perhaps more importantly, it may be time for the concept of the PTV to give way to concepts that more fully take into account the underlying geometrical uncertainties, as others have proposed. Nonetheless, actually hitting the tumor and any occult disease with an effective dose will always be the most important principle in radiation oncology.

We have taken a broad view that adaptation will ultimately encompass biological, imaging, and physical variables measured over time. The use of "one-size fits all" dose prescriptions should probably be discarded in favor of customized plans that could, for example, treat patients to a common level of complication risk, a strategy introduced by Lawrence, Kessler, and Robertson (1996). The application of radiobiological principles to adaptive radiotherapy is replete with uncertainties and associated caveats. Despite this, partial tumor boosts have significant advantages, and they become even more attractive if guided by FDG-PET images. As always, scientifically sound studies will be required to test these ideas.

Acknowledgments

This chapter benefits greatly from past conversations and debates with many colleagues, including Randy Ten Haken, Larry Marks, Soeren Bentzen, Marcus Alber, Vitali Moiseenko, Andrzej Niemierko, Herman Suit, Micheal Goitein, Andy Jackson, Jack Fowler, Allen Nahum, Giovanni Gagliardi, Philippe Lambin, Joe Roti Roti, and my long-time collaborator, Issam El Naqa. This work was partially supported by a grant from TomoTherapy, Inc., Wisconsin, and by National Institutes of Health (NIH) R01 grant CA85181.

References

Agarwal, M., G. Brahmanday, S. K. Bajaj, K. P. Ravikrishnan, and C. Y. Wong. 2009. Revisiting the prognostic value of preoperative 18F-fluoro-2-deoxyglucose (18F-FDG) positron emission tomography (PET) in early-stage (I & II) non-small cell lung cancers (NSCLC). *Eur J Nucl Med Mol Imaging* (e-pub ahead of print, November 2009) 37(4):691–8.

Ahuja, V., R. E. Coleman, J. Herndon, and E. R. Patz Jr. 1998. The prognostic significance of fluorodeoxyglucose positron emission tomography imaging for patients with nonsmall cell lung carcinoma. *Cancer* 83(5):918–24.

Al-Mamgani, A., W. L. van Putten, W. D. Heemsbergen, et al. 2008. Update of Dutch multicenter dose-escalation trial of radiotherapy for localized prostate cancer. *Int J Radiat Oncol Biol Phys* 72(4):980–8.

Andreassen, C. N., and J. Alsner. 2009. Genetic variants and normal tissue toxicity after radiotherapy: a systematic review. *Radiother Oncol* 92(3):299–309.

Baum, C., M. Alber, M. Birkner, and F. Nusslin. 2006. Robust treatment planning for intensity modulated radiotherapy of prostate cancer based on coverage probabilities. *Radiother Oncol* 78(1):27–35.

Been, L. B., A. J. Suurmeijer, D. C. Cobben, P. L. Jager, H. J. Hoekstra, and P. H. Elsinga. 2004. [18F]FLT-PET in oncology: Current status and opportunities. *Eur J Nucl Med Mol Imaging* 31(12):1659–72.

Bentzen, S. M. 2006. Preventing or reducing late side effects of radiation therapy: Radiobiology meets molecular pathology. *Nat Rev Cancer* 6(9):702–13.

Berbeco, R. I., C. J. Pope, and S. B. Jiang. 2006. Measurement of the interplay effect in lung IMRT treatment using EDR2 films. *J Appl Clin Med Phys* 7(4):33–42.

Borst, G. R., J. S. Belderbos, R. Boellaard, et al. 2005. Standardised FDG uptake: A prognostic factor for inoperable non-small cell lung cancer. *Eur J Cancer* 41(11):1533–41.

Bortfeld, T., S. B. Jiang, and E. Rietzel. 2004. Effects of motion on the total dose distribution. *Semin Radiat Oncol* 14(1):41–51.

Bortfeld, T., K. Jokivarski, M. Goitein, J. Kung, and S. B. Jiang. 2002. Effects of intra-fraction motion on IMRT dose delivery: Statistical analysis and simulation. *Phys Med Biol* 47(13):2203–20.

Bortfeld, T., U. Oelfke, and S. Nill. 2000. What is the optimum leaf width of a multileaf collimator? *Med Phys* 27(11):2494–502.

Brown, J. M. 1999. The hypoxic cell: A target for selective cancer therapy—eighteenth Bruce F. Cain Memorial Award lecture. *Cancer Res* 59(23):5863–70.

Brown, J. M. 2002. Tumor microenvironment and the response to anticancer therapy. *Cancer Biol Ther* 1(5):453–8.

Brown, J. M., and W. R. Wilson. 2004. Exploiting tumour hypoxia in cancer treatment. *Nat Rev Cancer* 4(6):437–47.

Bruechner, K., R. Bergmann, A. Santiago, et al. 2009. Comparison of [18F]FDG uptake and distribution with hypoxia and proliferation in FaDu human squamous cell carcinoma (hSCC) xenografts after single dose irradiation. *Int J Radiat Biol* 85(9):772–80.

Buchmann, I., U. Haberkorn, I. Schmidtmann, et al. 2008. Influence of cell proportions and proliferation rates on FDG uptake in squamous-cell esophageal carcinoma: A PET study. *Cancer Biother Radiopharm* 23(2):172–80.

Buck, A. K., G. Halter, H. Schirrmeister, et al. 2003. Imaging proliferation in lung tumors with PET: 18F-FLT versus 18F-FDG. *J Nucl Med* 44(9):1426–31.

Cairns, R. A., T. Kalliomaki, and R. P. Hill. 2001. Acute (cyclic) hypoxia enhances spontaneous metastasis of KHT murine tumors. *Cancer Res* 61(24):8903–8.

Cárdenas-Navia, L. I., D. Mace, R. A. Richardson, D. F. Wilson, S. Shan, and M. W. Dewhirst. 2008. The pervasive presence of fluctuating oxygenation in tumors. *Cancer Res* 68(14):5812–9.

Carlin, S., A. Pugachev, X. Sun, et al. 2009. In vivo characterization of a reporter gene system for imaging hypoxia-induced gene expression. *Nucl Med Biol* 36(7):821–31.

Casali, C., M. Cucca, G. Rossi, et al. 2009. The variation of prognostic significance of maximum standardized uptake value of [18F]-fluoro-2-deoxy-glucose positron emission tomography in different histological subtypes and pathological stages of surgically resected non-small cell lung carcinoma. *Lung Cancer* 69(2):187–93.

Cerfolio, R. J., A. S. Bryant, B. Ohja, and A. A. Bartolucci. 2005. The maximum standardized uptake values on positron emission tomography of a non-small cell lung cancer predict stage, recurrence, and survival. *J Thorac Cardiovasc Surg* 130(1):151–9.

Chan, T. C., J. N. Tsitsiklis, and T. Bortfeld. 2010. Optimal margin and edge-enhanced intensity maps in the presence of motion and uncertainty. *Phys Med Biol* 55(2):515–33.

Chao, K. S., A. I. Blanco, and J. F. Dempsey. 2003. A conceptual model integrating spatial information to assess target volume coverage for IMRT treatment planning. *Int J Radiat Oncol Biol Phys* 56(5):1438–49.

Chao, K. S., W. R. Bosch, S. Mutic, et al. 2001. A novel approach to overcome hypoxic tumor resistance: Cu-ATSM-guided intensity-modulated radiation therapy. *Int J Radiat Oncol Biol Phys* 49(4):1171–82.

Chapet, O., E. Thomas, M. L. Kessler, B. A. Fraass, and R. K. Ten Haken. 2005. Esophagus sparing with IMRT in lung tumor irradiation: An EUD-based optimization technique. *Int J Radiat Oncol Biol Phys* 63(1):179–87.

Chen, W., T. Cloughesy, N. Kamdar, et al. 2005. Imaging proliferation in brain tumors with 18F-FLT PET: Comparison with 18F-FDG. *J Nucl Med* 46(6):945–52.

Chen, J. C., T. W. Huang, Y. L. Cheng, et al. 2009. Prognostic value or 18-FDG uptake in early stage NSCLC. *Thorac Cardiovasc Surg* 57(7):413–6.

Dang, C. V., and G. L. Semenza. 1999. Oncogenic alterations of metabolism. *Trends Biochem Sci* 24(2):68–72.

Das, S. K., M. M. Miften, S. Zhou, et al. 2004. Feasibility of optimizing the dose distribution in lung tumors using fluorine-18-fluorodeoxyglucose positron emission tomography and single photon emission computed tomography guided dose prescriptions. *Med Phys* 31(6):1452–61.

Dawson, L. A., D. Normolle, J. M. Balter, C. J. McGinn, T. S. Lawrence, and R. K. Ten Haken. 2002. Analysis of radiation-induced liver disease using the Lyman NTCP model. *Int J Radiat Oncol Biol Phys* 53(4):810–21.

Deasy, J. O. 1997. Tumor control probability models for nonuniform dose distributions. In *Fifth International Conference on Dose, Time and Fractionation in Radiation Oncology: Volume and Kinetics in Tumor Control and Normal Tissue Complications,* ed. J. Fowler, D. Herbert, and B. Paliwal. Madison, WI: Medical Physics Publishing.

Deasy, J. O. 2001. Partial tumor boosts: Even more attractive than theory predicts? *Int J Radiat Oncol Biol Phys* 51(1):279–80.

Deasy, J. O., A. I. Blanco, and V. H. Clark. 2003. CERR: A computational environment for radiotherapy research. *Med Phys* 30:979–85.

Deasy, J. O., and J. F. Fowler. 2005. The radiobiology of intensity modulated radiation therapy. In *Intensity Modulated Radiation Therapy: A Clinical Perspective,* ed. A. J. Mundt and J. Roeske, 53–74. Hamilton, ON: BC Decker.

Deasy, J., V. Moiseenko, L. Marks, K. Chao, J. Nam, A. Eisburch. 2010. Radiation therapy dose-volume effects on salivary gland function. *Int J Radiat Oncol Biol Phys* 76(01): S58–S63.

Deasy, J. O., S. N. Powell, and I. Zoberi. 2006. The predicted effect of HDR dose heterogeneity on local/regional control for post-surgical patients: Location matters. *Brachytherapy* 5:105.

de Crevoisier, R., S. L. Tucker, L. Dong, et al. 2005. Increased risk of biochemical and local failure in patients with distended rectum on the planning CT for prostate cancer radiotherapy. *Int J Radiat Oncol Biol Phys* 62(4):965–73.

Dehdashti, F., P. W. Grigsby, J. S. Lewis, R. Laforest, B. A. Siegel, and M. J. Welch. 2008. Assessing tumor hypoxia in cervical cancer by PET with 60Cu- labeled diacetyl-bis (N4-methylthiosemicarbazone). *J Nucl Med* 49(2):201–5.

Dehdashti, F., P. W. Grigsby, M. A. Mintun, J. S. Lewis, B. A. Siegel, and M. J. Welch. 2003. Assessing tumor hypoxia in cervical cancer by positron emission tomography with 60Cu-ATSM: Relationship to therapeutic response—A preliminary report. *Int J Radiat Oncol Biol Phys* 55(5):1233–8.

Dhital, K., C. A. Saunders, P. T. Seet, M. J. O'Doherty, and J. Dussk. 2000. [18F]Fluorodeoxyglucose positron emission tomography and its prognostic value in lung cancer. *Eur J Cardiothorac Surg* 18(4):425–8.

Dierckx, R. A., and C. Van De Wiele. 2008. FDG uptake, a surrogate of tumour hypoxia? *Eur J Nucl Med Mol Imaging* 35(8):1544–9.

Downey, R. J., T. Akhurst, M. Gonen, et al. 2004. Preoperative F-18 fluorodeoxyglucose-positron emission tomography maximal standardized uptake value predicts survival after lung cancer resection. *J Clin Oncol* 22(16):3255–60.

Duan, J., S. Shen, J. B. Fiveash, R. A. Popple, and I. A. Brezovich. 2006. Dosimetric and radiobiological impact of dose fractionation on respiratory motion induced IMRT delivery errors: a volumetric dose measurement study. *Med Phys* 33(5):1380–7.

El Naqa, I., J. O. Deasy, Y. Mu, et al. 2010. Datamining approaches for modeling tumor control probability. *Acta Oncol.*

Engels, B., G. Soete, D. Verellen, and G. Storme. 2009. Conformal arc radiotherapy for prostate cancer: Increased biochemical failure in patients with distended rectum on the planning computed tomogram despite image guidance by implanted markers. *Int J Radiat Oncol Biol Phys* 74(2):388–91.

Erdi, Y. E., H. Macapinlac, K. E. Rosenzweig, et al. 2000. Use of PET to monitor the response of lung cancer to radiation treatment. *Eur J Nucl Med* 27(7):861–6.

Evans, S. M., S. M. Hahn, D. P. Magarelli, and C. J. Koch. 2001. Hypoxic heterogeneity in human tumors: EF5 binding, vasculature, necrosis, and proliferation. *Am J Clin Oncol* 24(5):467–72.

Feng, M., F. M. Kong, M. Gross, S. Fernando, J. A. Hayman, and R. K. Tan Haken. 2009. Using fluorodeoxyglucose positron pmission tomography to assess tumor volume during radiotherapy for non-small-cell lung cancer and its potential impact on adaptive dose escalation and normal tissue sparing. *Int J Radiat Oncol Biol Phys* 73(4):1228–34.

Gatenby, R. A., and R. J. Gillies. 2004. Why do cancers have high aerobic glycolysis? *Nat Rev Cancer* 4(11):891–9.

Goitein, M., A. Niemierko, and P. Okunieff. 1996. The probability of controlling an inhomogeneously irradiated tumour: A stratagem for improving tumour control through partial tumour boosting. In *19th L H Gray Conference: Quantitative Imaging in Oncology*, ed. K. Faulkner. London: British Institute of Radiology.

Gordon, J. J., and J. V. Siebers. 2009. Coverage-based treatment planning: Optimizing the IMRT PTV to meet a CTV coverage criterion. *Med Phys* 36(3):961–73.

Graeber, T. G., C. Osmanian, T. Jacks, et al. 1996. Hypoxia-mediated selection of cells with diminished apoptotic potential in solid tumours. *Nature* 379(6560):88–91.

Grigsby, P. W., B. A. Siegel, F. Dehdashti, and D. G. Mutch. 2003. Posttherapy surveillance monitoring of cervical cancer by FDG-PET. *Int J Radiat Oncol Biol Phys* 55(4):907–13.

Grigsby, P. W., B. A. Siegel, F. Dehdashti, J. Rader, and I. Zoberi. 2004. Posttherapy [18F] fluorodeoxyglucose positron emission tomography in carcinoma of the cervix: Response and outcome. *J Clin Oncol* 22(11):2167–71.

Hall, E. J., and A. J. Giaccia. 2005. *Radiobiology for the Radiologist.* 6th ed. Philadelphia, PA: Lippincott Williams & Wilkins.

Hara, T., A. Bansal, and T. R. DeGrado. 2006. Effect of hypoxia on the uptake of [methyl-3H]choline, [1-14C] acetate and [18F]FDG in cultured prostate cancer cells. *Nucl Med Biol* 33(8):977–84.

Heiden, M. G. V., L. C. Cantley, and C. B. Thompson. 2009. Understanding the Warburg effect: The metabolic requirements of cell proliferation. *Science* 324(5930):1029–33.

Höckel, M., and P. Vaupel. 2001a. Biological consequences of tumor hypoxia. *Semin Oncol* 28(2 Suppl. 8):36–41.

Höckel, M., and P. Vaupel. 2001b. Tumor hypoxia: Definitions and current clinical, biologic, and molecular aspects. *J Natl Cancer Inst* 93(4):266–76.

Hoos, A., A. Stojadnovic, S. Mastorides, et al. 2001. High Ki-67 proliferative index predicts disease specific survival in patients with high-risk soft tissue sarcomas. *Cancer* 92(4):869–74.

Hope, A. J., P. E. Lindsay, I. El Naqa, J. D. Bradley, M. Vivic, and J. O. Deasy. 2005. Clinical, dosimetric, and location-related factors to predict local control in non-small cell lung cancer. *Inter J Radiat Oncol Biol Phys* 63(2):S231.

Ibbott, G. S., D. Followill, H. A. Molineu, J. R. Lowenstein, P. E. Alvarez, and J. E. Roll. 2008. Challenges in credentialing institutions and participants in advanced technology multi-institutional clinical trials. *Int J Radiat Oncol Biol Phys* 71(1 Suppl):S71–5.

ICRU, Report 62. 1999. *Prescribing, Recording and Reporting Photon Beam Therapy (Supplement to ICRU Report 50).* Bethesda, MD: International Commission on Radiation Units and Measurements.

Ishida, T., S. Kaneko, K. Akazawa, M. Tateishi, K. Sugio, and K. Sugimachi. 1993. Proliferating cell nuclear antigen expression and argyrophilic nucleolar organizer regions as factors influencing prognosis of surgically treated lung cancer patients. *Cancer Res* 53(20):5000–3.

Ismail, S. M., T. Bucholz, M. Story, W. A. Brock, and C. W. Stevens. 2004. Radiosensitivity is predicted by DNA end-binding complex density, but not by nuclear levels of band components. *Radiother Oncol* 72(3):325–32.

Jeong, H. J., J. J. Min, J. M. Park, et al. 2002. Determination of the prognostic value of [18F]fluorodeoxyglucose uptake by using positron emission tomography in patients with non-small cell lung cancer. *Nucl Med Commun* 23(9):865–70.

Kelly, C. J., K. Smallbone, T. Roose, et al. 2007. A model to investigate the feasibility of FDG as a surrogate marker of hypoxia. In *ISBI 2007: Proceedings of the 2007 IEEE International Symposium on Biomedical Imaging: From Nano to Macro.* Washington, DC.

Kennedy, A. S., J. A. Raleigh, G. M. Perez, et al. 1997. Proliferation and hypoxia in human squamous cell carcinoma of the cervix: First report of combined immunohistochemical assays. *Int J Radiat Oncol Biol Phys* 37(4):897–905.

Kidd, E. A., B. A. Siegel, F. Dehdashti, and P. W. Grigsby. 2007. The standardized uptake value for F-18 fluorodeoxyglucose is a sensitive predictive biomarker for cervical cancer treatment response and survival. *Cancer* 110(8):1738–44.

Kim, J. W., and C. V. Dang. 2006. Cancer's molecular sweet tooth and the Warburg effect. *Cancer Res* 66(18):8927–30.

Kim, J. J., and I. F. Tannock. 2005. Repopulation of cancer cells during therapy: An important cause of treatment failure. *Nat Rev Cancer* 5(7):516–25.

Kong, F. M., K. A. Frey, L. E. Quint, et al. 2007. A pilot study of [18F] fluorodeoxyglucose positron emission tomography scans during and after radiation-based therapy in patients with non small-cell lung cancer. *J Clin Oncol* 25(21):3116–23.

Krohn, K. A., J. M. Link, and R. P. Mason. 2008. Molecular imaging of hypoxia. *J Nucl Med* 49(Suppl 2):129S–48.

Kupelian, P. A., C. Ramsey, S. K. Meeks, et al. 2005. Serial megavoltage CT imaging during external beam radiotherapy for non-small-cell lung cancer: observations on tumor regression during treatment. *Int J Radiat Oncol Biol Phys* 63(4):1024–8.

Larsson, P., G. Roos, R. Stenling, and B. Ljungberg. 1993. Tumor-cell proliferation and prognosis in renal-cell carcinoma. *Int J Cancer* 55(4):566–70.

Lawrence, T. S., M. L. Kessler, and J. M. Robertson. 1996. D conformal radiation therapy in upper gastrointestinal cancer. In *3-D Conformal Radiotherapy,* ed. J. L. Meyer and J. A. Purdy, 221–8. Basel: Karger.

Li, J., S. M. Bentzen, M. Renschler, and M. P. Mehta. 2007. Regression after whole-brain radiation therapy for brain metastases correlates with survival and improved neurocognitive function. *J Clin Oncol* 25(10):1260–6.

Ling, C. C., J. Humm, S. Larson, et al. 2000. Towards multidimensional radiotherapy (MD-CRT): Biological imaging and biological conformality. *Int J Radiat Oncol Biol Phys* 47(3):551–60.

Mac Manus, M. P., R. J. Hicks, J. P. Matthews, et al. 2003. Positron emission tomography is superior to computed tomography scanning for response-assessment after radical radiotherapy or chemoradiotherapy in patients with non-small-cell lung cancer. *J Clin Oncol* 21(7):1285–92.

Mac Manus, M. P., R. J. Hicks, J. P. Matthews, A. Wirth, D. Rischin, and D. L. Ball. 2005. Metabolic (FDG-PET) response after radical radiotherapy/chemoradiotherapy for non-small cell lung cancer correlates with patterns of failure. *Lung Cancer* 49(1):95–108.

Marks, L., S. Bentzen, J. Deasy, et al. 2010. Radiation dose volume effects in the lung. *Int J Radiat Oncol Biol Phys* 76(01):S70–6.

Martinive, P., F. Defresene, C. Bouzin, et al. 2006. Preconditioning of the tumor vasculature and tumor cells by intermittent hypoxia: Implications for anticancer therapies. *Cancer Res* 66(24):11736–44.

Marusyk, A., and P. Polyak. 2010. Tumor heterogeneity: Causes and consequences. *Biochim Biophys Acta* 1805(1):105–17.

McShan, D. L., M. L. Kessler, K. Vineberg, and B. A. Fraass. 2006. Inverse plan optimization accounting for random geometric uncertainties with a multiple instance geometry approximation (MIGA). *Med Phys* 33(5):1510–21.

Moiseenko, V., J. O. Deasy, and J. Van Dyk. 2005. Radiobiological modeling for treatment planning. In *The Modern Technology of Radiation Oncology: A Compendium for Medical Physicists and Radiation Oncologists,* ed. J. Van Dyk, 85–220. Madison, WI: Medical Physics Publishing.

Müller, W., A. Schneiders, S. Meier, et al. 1996. Immunohistochemical study on the prognostic value of MIB-1 in gastric carcinoma. *Br J Cancer* 74(5):759–65.

Niemierko, A. 1997. Reporting and analyzing dose distributions: A concept of equivalent uniform dose. *Med Phys* 24(1):103–10.

Okunieff, P., D. Morgan, A. Niemierko, and H. D. Suit. 1995. Radiation dose-response of human tumors. *Int J Radiat Oncol Biol Phys* 32(4):1227–37.

Petit, S. F., H. J. Aerts, J. G. van Loon, et al. 2009. Metabolic control probability in tumour subvolumes or how to guide tumour dose redistribution in non-small cell lung cancer (NSCLC): An exploratory clinical study. *Radiother Oncol* 91(3):393–8.

Popple, R. A., R. Ove, and S. Shen. 2002. Tumor control probability for selective boosting of hypoxic subvolumes, including the effect of reoxygenation. *Int J Radiat Oncol Biol Phys* 54(3):921–7.

Pugachev, A., S. Ruan, S. Carlin, et al. 2005. Dependence of FDG uptake on tumor microenvironment. *Int J Radiat Oncol Biol Phys* 62(2):545–53.

QUANTEC (Quantitative Analyses of Normal Tissue Effects). 2010. *Int J Radiat Oncol Biol Phys* (special issue) 76(01):S1–S160.

Rajendran, J. G., D. A. Mankoff, F. O'Sullivan, et al. 2004. Hypoxia and glucose metabolism in malignant tumors: Evaluation by [18F]fluoromisonidazole and [18F]]fluorodeoxyglucose positron emission tomography imaging. *Clin Cancer Res* 10(7):2245–52.

Rajendran, J. G., D. L. Schwartz, J. O'Sullivan, et al. 2006. Tumor hypoxia imaging with [F-18] fluoromisonidazole positron emission tomography in head and neck cancer. *Clin Cancer Res* 12(18):5435–41.

Safwat, A., S. M. Bentzen, I. Turesson, and J. H. Hendry. 2002. Deterministic rather than stochastic factors explain most of the variation in the expression of skin telangiectasia after radiotherapy. *Int J Radiat Oncol Biol Phys* 52(1):198–204.

Sakata, K., A. Oouchi, H. Nagakura, et al. 2000. Accelerated radiotherapy for T1, 2 glottic carcinoma: Analysis of results with KI-67 index. *Int J Radiat Oncol Biol Phys* 47(1):81–8.

Sasaki, R., R. Komaki, H. Macapinlac, et al. 2005. [18F]fluorodeoxyglucose uptake by positron emission tomography predicts outcome of non-small-cell lung cancer. *J Clin Oncol* 23(6):1136–43.

Schwarz, J. K., B. A. Siegel, F. Dehdashti, and P. W. Grigsby. 2007. Association of posttherapy positron emission tomography with tumor response and survival in cervical carcinoma. *JAMA* 298(19):2289–95.

Scigliano, S., S. Pinel, S. Poussier, et al. 2008. Measurement of hypoxia using invasive oxygen-sensitive electrode, pimonidazole binding and 18F-FDG uptake in anaemic or erythropoietin-treated mice bearing human glioma xenografts. *Int J Oncol* 32(1):69–77.

Seibert, R. M., C. R. Ramsey, J. W. Hines, et al. 2007. A model for predicting lung cancer response to therapy. *Int J Radiat Oncol Biol Phys* 67(2):601–9.

Semenza, G. L. 2003. Targeting HIF-1 for cancer therapy. *Nat Rev Cancer* 3(10):721–32.

Semenza, G. L. 2008. Tumor metabolism: Cancer cells give and take lactate. *J Clin Invest* 118(12):3835–7.

Shields, A. F., J. R. Grierson, B. M. Dohmen, et al. 1998. Imaging proliferation in vivo with [F-18]FLT and positron emission tomography. *Nat Med* 4(11):1334–6.

Sonveaux, P., F. Vegran, T. Schroeder, et al. 2008. Targeting lactate-fueled respiration selectively kills hypoxic tumor cells in mice. *J Clin Invest* 118(12):3930–42.

Steel, G. G. 2002. *Basic Clinical Radiobiology.* London: Hodder Arnold Publications.

Sunaga, N., N. Oriuchi, K. Kaira, et al. 2008. Usefulness of FDG-PET for early prediction of the response to gefitinib in non-small cell lung cancer. *Lung Cancer* 59(2):203–10.

Terahara, A., A. Niemierko, M. Goitein, et al. 1999. Analysis of the relationship between tumor dose inhomogeneity and local control in patients with skull base chordoma. *Int J Radiat Oncol Biol Phys* 45(2):351–8.

Thorwarth, D., S. M. Eschmann, F. Paulsen, and M. Alber. 2007. Hypoxia dose painting by numbers: A planning study. *Int J Radiat Oncol Biol Phys* 68(1):291–300.

Tome, W. A., and J. F. Fowler. 2000. Selective boosting of tumor subvolumes. *Int J Radiat Oncol Biol Phys* 48(2):593–9.

Valera, V., N. Yokoyama, B. Walter, H. Okamoto, T. Suda, and K. Hatakeyama. 2005. Clinical significance of Ki-67 proliferation index in disease progression and prognosis of patients with resected colorectal carcinoma. *Br J Surg* 92(8):1002–7.

van Baardwijk, A., C. Dooms, R. J. van Suylen, et al. 2007. The maximum uptake of 18F-deoxyglucose on positron emission tomography scan correlates with survival, hypoxia inducible factor-1α and GLUT-1 in non-small cell lung cancer. *Eur J Cancer* 43(9):1392–8.

Vanderstraeten, B., W. Duthoy, W. De Gersem, W. De Neve, and H. Thierens. 2006. [18F]fluoro-deoxy-glucose positron emission tomography ([18F]FDG-PET) voxel intensity-based intensity-modulated radiation therapy (IMRT) for head and neck cancer. *Radiother Oncol* 79(3):249–58.

van Loon, J., J. Grutters, R. Wanders, et al. 2009. Follow-up with 18FDG-PET-CT after radical radiotherapy with or without chemotherapy allows the detection of potentially curable progressive disease in non-small cell lung cancer patients: A prospective study. *Eur J Cancer* 45(4):588–95.

Vansteenkiste, J. F., S. G. Stroobants, P. J. Dupont, et al. 1999. Prognostic importance of the standardized uptake value on 18F-fluoro- 2-deoxy-glucose-positron emission tomography scan in non-small-cell lung cancer: An analysis of 125 cases. *J Clin Oncol* 17(10):3201–6.

Varia, M. A., D. P. Calkins-Adams, L. H. Rinker, et al. 1998. Pimonidazole: A novel hypoxia marker for complementary study of tumor hypoxia and cell proliferation in cervical carcinoma. *Gynecol Oncol* 71(2):270–7.

Vaupel, P., and A. Mayer. 2007. Hypoxia in cancer: Significance and impact on clinical outcome. *Cancer Metastasis Rev* 26(2):225–39.

Vesselle, H., J. D. Freeman, L. Wiens, et al. 2007. Fluorodeoxyglucose uptake of primary non-small cell lung cancer at positron emission tomography: New contrary data on prognostic role. *Clin Cancer Res* 13(11):3255–63.

Vesselle, H., A. Salskov, E. Turcotte, et al. 2008. Relationship between non-small cell lung cancer FDG uptake at PET, tumor histology, and Ki-67 proliferation index. *J Thorac Oncol* 3(9):971–8.

Vesselle, H., R. A. Schmidt, J. M. Pugsley, et al. 2000. Lung cancer proliferation correlates with [F-18]fluorodeoxyglucose uptake by positron emission tomography. *Clin Cancer Res* 6(10):3837–44.

Viberti, L., M. Papotti, G. C. Abbona, et al. 1997. Value of Ki-67 immunostaining in preoperative biopsies of carcinomas of the lung. *Hum Pathol* 28(2):189–92.

Wang, W., J. C. Georgi, S. A. Nehmeh, et al. 2009. Evaluation of a compartmental model for estimating tumor hypoxia via FMISO dynamic PET imaging. *Phys Med Biol* 54(10):3083–99.

Warburg, O. 1956. On the origin of cancer cells. *Science* 123(3191):309–14.

Webb, S. 2006. Motion effects in (intensity modulated) radiation therapy: A review. *Phys Med Biol* 51(13):R403–25.

Wilson, G. D., M. I. Saunders, S. Dische, et al. 2006. Pretreatment proliferation and the outcome of conventional and accelerated radiotherapy. *Eur J Cancer* 42(3):363–71.

Withers, H. R., L. J. Peters, and J. M. Taylor. 1995. Dose-response relationship for radiation therapy of subclinical disease. *Int J Radiat Oncol Biol Phys* 31(2):353–9.

Withers, H. R., and R. Suwinski. 1998. Radiation dose response for subclinical metastases. *Semin Radiat Oncol* 8(3):224–8.

Witte, M. G., W. D. Heemsbergen, R. Bohoslavsky, et al. 2010. Relating dose outside the prostate with freedom from failure in the Dutch Trial 68 Gy vs. 78 Gy. *Int J Radiat Oncol Biol Phys* 77(1):131–8.

Xie, P., J. B. Yue, H. X. Zhao, et al. 2009. Prognostic value of (18)F-FDG PET-CT metabolic index for nasopharyngeal carcinoma. *J Cancer Res Clin Oncol* (epub ahead of print), November 20, 2009.

Xue, F., L. L. Lin, F. Dehdashti, T. R. Miller, B. A. Siegel, and P. W. Grigsby. 2006. F-18 fluorodeoxyglucose uptake in primary cervical cancer as an indicator of prognosis after radiation therapy. *Gynecol Oncol* 101(1):147–51.

Yamamoto, Y., Y. Nishiyama, S. Ishikawa, et al. 2007. Correlation of 18F-FLT and 18F-FDG uptake on PET with Ki-67 immunohistochemistry in non-small cell lung cancer. *Eur J Nucl Med Mol Imaging* 34(10):1610–6.

Yu, C. X., D. A. Jaffray, and J. W. Wong. 1998. The effects of intrafraction organ motion on the delivery of dynamic intensity modulation. *Phys Med Biol* 43(1):91–104.

Zelefsky, M. J., H. Chan, M. Hunt, Y. Yamada, A. M. Shippy, and H. Amols. 2006. Long-term outcome of high dose intensity modulated radiation therapy for patients with clinically localized prostate cancer. *J Urol* 176(4 Pt 1):1415–9.

Zhou, S. M., S. Das, Z. Wang, and L. B. Marks. 2004. Relationship between the generalized equivalent uniform dose formulation and the Poisson statistics-based tumor control probability model. *Med Phys* 31(9):2606–9.

Zimny, M., B. Gagel, E. DiMartino, et al. 2006. FDG—A marker of tumour hypoxia? A comparison with [18F] fluoromisonidazole and pO2-polarography in metastatic head and neck cancer. *Eur J Nucl Med Mol Imaging* 33(12):1426–31.

2

Three- and Four-Dimensional Morphological Imaging for Adaptive Radiation Therapy Planning

Lei Xing
Stanford University

Jianguo Qian
Stanford University

Kihwan Choi
Stanford University

Tae-Suk Suh
The Catholic University of Korea

2.1 Introduction

Onboard volumetric imaging is developing to meet the increased needs for accurate patient setup in modern radiation therapy (RT) techniques such as intensity-modulated RT (IMRT) and volumetric-modulated arc therapy (VMAT). Although exquisite dose distribution can be planned, how to deliver the dose to the right volume at the right time has become increasingly important. The development of modern therapy techniques has not occurred very meaningfully and it could even be harmful if there is no effective way of ensuring patient setup and accurate delivery (Xing et al. 2006; Timmerman and Xing 2009).

In current practice, two-dimensional (2D) digitally reconstructed radiography (DRR) is used to guide patient positioning and tumor target localization. In general, the dimensionality of the image guidance tool should commeasure with that of planning and dose delivery. The 2D image-based approach is appropriate if there were no internal organ motions. The advantage of volumetric imaging is that it provides accurate displacement information of every voxel even in the presence of organ motion and/or deformation. Additionally, volumetric imaging provides a basis for one to model the multidimensional organ motion and to adaptively modify the treatment plan so as to compensate for any interfractional anatomy change of the target as well as the sensitive structures. In this sense, onboard volumetric imaging is fast becoming an indispensable part of modern RT.

Although it is theoretically valuable, currently available cone-beam computed tomography (CBCT) imaging is far from satisfactory. There are a number of problems that hinder the maximal utilization of this technique. Briefly, the following three factors adversely influence the quality and widespread application of CBCT: (1) photon scattering, (2) motion artifacts, and (3) imaging dose. Other factors such as metal artifacts may also diminish the value of onboard CBCT imaging. These artifacts not only deteriorate the image quality but also make accurate CBCT-based dose calculation impossible. Much of current CBCT research is focused on these three areas. In this chapter, we summarize the current method of CBCT imaging and reconstruction techniques, and highlight some of the recent progress in three-dimensional (3D) and four-dimensional (4D) morphological imaging and their applications in adaptive RT.

2.2 Onboard CBCT

Both kilovoltage (kV) and megavoltage (MV) flat-panel imagers integrated with a linear accelerator (linac) have become available for therapy guidance. The former typically consists of a kV-source and amorphous silicon flat-panel detector combination mounted on the drum of a linac (Jaffray et al. 2002), with the kV imaging axis orthogonal to that of the MV therapy beam. The system provides online 3D or even 4D (Sonke, Zijp Remeijer, and van Herk 2005; Lu et al. 2007; Berbeco et al. 2007; Ball 2006) patient anatomy data that is valuable for patient setup and, more importantly, adaptive replanning (de la Zerda, Armbruster, and Xing 2007; Hong et al. 2007; Aldridge, Reckwerdt, and Mackie 1999; Court et al. 2005; Welsh et al. 2006; Wu, Liang, and Yan 2006).

Presently, onboard morphological imaging systems are primarily used for guiding patient setup by using rigid 3D–3D registration technique (Oldham et al. 2005; Masi et al. 2008; Li et al. 2008; Chang et al. 2007). Although CBCT images can clearly reveal setup error and readily detect rotational errors, the rigid-registration-based patient setup procedure falls short in the presence of organ deformation or relative displacement of the involved organs. Vendors of major linac manufacturers are working on the development of deformable registration for improved patient setup. In reality, RT in the presence of organ deformation is a multidimensional problem, which cannot be solved completely by translation and rotation of the patient or by deformable image registration. When deformable registration is used, there are a number of options to achieve the registration depending on whether the primary aim is to match soft-tissue detail to compensate for organ motion or to align 3D bony structures. The multiple choices result from the fact that the dimensionality of patient data is much greater than that of patient setup, suggesting that deformable registration is not the ultimate in volumetric image-guided RT. Nevertheless, the technique improves the current approach since it partially takes into account organ deformation by achieving the closest overlay match possible between the planning and CBCT data sets according to the clinical objective and serves as an interim solution until a better, integrated approach becomes available. The true value of onboard volumetric imaging lies in its ability to provide a patient's on-treatment geometric model for dose reconstruction and adaptive replanning. This will be discussed in Section 2.8.

2.3 CBCT Image Reconstruction

Generally, the cone beam–filtered backprojection (FBP) algorithm is an appealing approach for practical CBCT image reconstruction since it can make a reconstruction much more computationally efficient and easy to implement in a multiprocessor parallel computing structure. In CBCT literature, a variety of FBP algorithms are popularly used for image reconstruction from projection data. An FBP-type algorithm, originally proposed by Feldkamp, Davis, and Kress (FDK; Feldkamp, Davis, and Kress

1984), and its derivatives (Grass, Kohler, and Proksa 2000; Grass, Kohler, and Proksa 2001; Kohler, Proksa, and Grass 2001; Turbell 2001) are widely used for CBCT reconstruction. Although only theoretically approximate, the algorithm and its derivatives represent important progress in CBCT imaging. All commercially available onboard imaging systems use FDK or similar algorithms to reconstruct images from collected projection data. A vast number of studies have been published in the literature on various aspects of the algorithm, and we refer the readers to some textbooks (Buzug 2008; Kak and Slaney 1988) and the references therein for detailed information. In this chapter, we focus on some of the current issues associated with CBCT imaging.

2.4 Scatter Correction in CBCT

In a single-slice CT, the X-ray beam and the detector spans a narrow plane. Most of the scattered photons fall outside the plane. In a cone-beam or multidetector CT, the X-ray is collimated to a wide beam. The possibility of a scattered photon getting mixed with the primary photons is therefore much higher in the second case. The projection images from a cone-beam or multidetector CT are therefore much noisier than those from a single-slice CT; the uncertainty in the reconstruction is correspondingly higher, which causes higher noise in the reconstructed image. Indeed, the quality of current CBCT reconstruction is still far from optimal due to the high scatter-to-primary ratio caused by the increased exposed volume in the cone-beam geometry.

Various methods of scatter reduction and strategies for motion artifacts removal are being investigated (Zhu et al. 2010). Some of the popular approaches perform scatter suppression during the acquisition of projection data based on the incident angle difference between primary photons and scatter photons (e.g., the antiscatter grid method and the air gap method; Boone et al. 2000; Siewerdsen et al. 2004; Kyriakou and Kalender 2007). Another approach involves the proper modeling of scatter photons and the postprocessing of scatter-contaminated projection images (Kyriakou, Riedel, and Kalender 2006). Zhu et al. recently proposed a simple patient setup and scatter removal protocol for RT applications (Zhu et al. 2009a; Zhu, Wang, and Xing 2009b; Zhu et al. 2010). In their approach, a sheet of lead strips is inserted between the X-ray source and the patient to extract the patient-specific scatter profile, which is then employed to correct subsequent CBCT scans with respect to potential patient setup variation. The procedure is outlined in Figure 2.1. The CBCT images of an anthropomorphic phantom with and without scatter removal are displayed in Figure 2.2. Improvement of image quality makes it possible to use CBCT for accurate dose computation and replanning. The approach proposed by Zhu et al. is clinically useful but involves the use of scatter template derived from the first day of treatment. Given that internal anatomy may change, which alters the scatter distribution, the extraction of scatter template may have to be done more frequently. This issue was recently solved, and it is now possible to obtain scatter-free CBCT images with only a single gantry rotation occurring with any change in clinical workflow (Choi and Xing 2010).

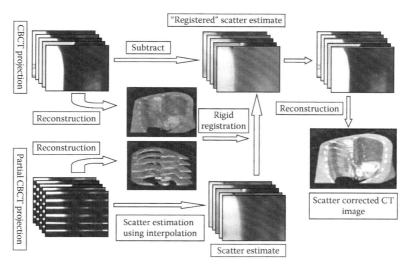

FIGURE 2.1 Workflow of the scatter correction using prior scatter measurement from partially blocked CBCT. (Reprinted from Zhu, L., et al. 2009a. *Med Phys* 36:2258–68. With permission.)

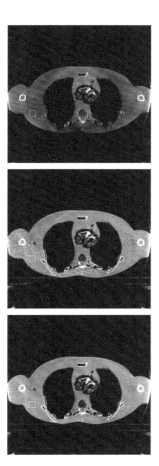

FIGURE 2.2 Axial views of the reconstructed anthropomorphic phantom. Top: no scatter correction; middle: measurement-based scatter correction using a partially blocked cone-beam computed tomography as a prescan; and bottom: scatter corrected using the proposed method. (Reprinted from Zhu, L., et al. 2009b. *Med Phys* 36:741–52. With permission.)

2.5 CBCT Motion Artifacts Removal

A few research groups (Sonke, Zijp Remeijer, and van Herk 2005; Lu et al. 2007; Li, Xing, et al. 2006; Li, Koong, and Xing 2007) have investigated strategies to acquire 4D CBCT images based on phase binning of the CBCT projection data. The phase-binned projections are reconstructed using either the conventional Feldkamp algorithm or a more advanced method to yield 4D CBCT images. Li and Xing (2007) studied several factors that are important to the clinical implementation of this technique, such as scanning time, number of projections, and radiation dose, and proposed an optimal 4D CBCT acquisition protocol for an individual breathing pattern. Figure 2.3 shows a 4D CBCT image (one phase) of a phantom with reduced breathing artifacts compared to its 3D counterpart. They also investigated a motion compensation method for slow-gantry-rotation CBCT scanning by incorporating into the reconstructed image a patient-specific motion model (Li, Schreibmann, et al. 2006), which is derived from 4D treatment-planning CT images of the same patient via deformable registration. It is demonstrated that the algorithm can reduce the motion artifacts locally and restore the tumor size and shape, which may thereby improve the accuracy of target localization and patient positioning. Four-dimensional CBCT is important for future 4D adaptive RT as it allows one to derive a patient's on-treatment 4D model.

The 4D CBCT image is commonly obtained by respiratory phase binning of projections, followed by independent reconstructions of the rebinned data in each phase bin. Due to the significantly reduced number of projections per reconstruction, the quality of 4D CBCT images is often degraded by view-aliasing artifacts seen in the axial view. Acquisitions using multiple gantry rotations or slow gantry rotation can increase the number of projections and substantially improve the 4D images. However, the extra cost of the scan time may set fundamental limits on applications of such images in clinics. Improving the trade-off

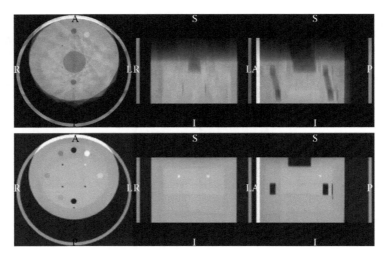

FIGURE 2.3   The 3D cone-beam computed tomography scan of the phantom with (top) and without (bottom) motion (second mode) with 80-mA tube current. (Reprinted from Li, T., L. Xing, et al. 2006. *Med Phys* 33:3825–33. With permission.)

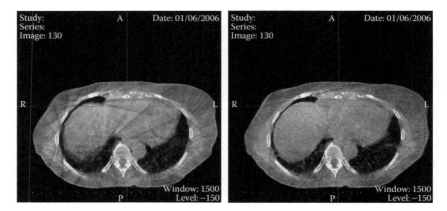

FIGURE 2.4   The 4D cone-beam computed tomography images of a liver patient: the left image shows an example of the peak inspiration phase, and the right shows the corresponding image after the proposed enhancement. (Reprinted from Li, T., et al. 2007. *Med Phys* 34:3688–95. With permission.)

between image quality and scan time is the key to making 4D onboard imaging practical and more useful. Li, Koong, and Xing (2007) improved the method by correlating the projection data from different phases; their approach was similar to the 4D positron emission tomography (PET) image enhancement strategy proposed a few years ago. Their method correlates the data in different phase bins and integrates the internal motion into the 4D CBCT image formulation. The central idea is that, given the projection data at two angles and two phases (assume that the angular and temporal separations are relatively small), it is possible, to a certain degree, to derive the projection at an intermediate phase and angle through spatial–temporal interpolation. In other words, the reconstruction of a phase will consider not only the projections corresponding to that phase but also those corresponding to other phases. As a result, the number of gantry rotations needed to obtain a decent 4D CBCT image will be greatly reduced. The enhanced 4D CBCT images for a liver patient are shown in Figure 2.4 using the proposed approach.

2.6 Metal Artifacts Removal in CBCT Imaging

The presence of metals in patients, such as dental fillings, hip prostheses, implanted markers, applicators used in brachytherapy, and surgical chips, may cause streaking artifacts in the diagnostic X-ray CT scan, and this has long been recognized as a problem that limits various applications of CT imaging (Robertson et al. 1997; Kalender, Hebel, and Ebersberger 1987; De Man et al. 1999; Williamson et al. 2002). There are two types of techniques for metal artifacts removal. One is to first identify the metal-contaminated region in the projection space and then replace the contaminants using different interpolation schemes, such as linear interpolation (Kalender, Hebel, and Ebersberger 1987), multiresolution interpolation (Zhao et al. 2000), or variational method (Zhang et al. 2007), based on the uncontaminated projection data. The CT images are then reconstructed from the completed projection data by analytical FBP-type algorithms. In reality, the identification of a metal object in the projection is not always an easy thing to do, especially when the object lies behind a high-density structure such as the bone. Alternatively, metal artifacts are reduced using model-based iterative reconstruction algorithms (De Man et al. 1999; Williamson et al. 2002; Wang et al. 1996). Iterative image reconstruction is advantageous in modeling the image formation and incorporating a priori knowledge. A major shortcoming of existing iterative algorithms is that knowledge about the shape and location, and sometimes even the attenuation coefficients (Williamson et al. 2002), of the metal objects is required to ensure the effective removal of the metal artifacts.

Wang and Xing (2010) developed an effective image intensity gradient threshold-based method for auto-identifying the shape and location of metallic objects in the image space. Instead of developing a full-fledged iterative algorithm to obtain the tissue and metal density distributions simultaneously, which usually does not completely remove the artifacts due to the incompleteness of the projection data, Wang and Xing proceed in two separate but related steps. First, they obtain a binary image of the patient reconstructed in such a way that the high density metals possess a "density" 1 and the remaining regions 0 with all anatomic detail ignored. An edge-preserving iterative algorithm (Fessler 1994; Wang, Li, and Xing 2009) is then applied to obtain the binary image by effectively utilizing the enormous difference in attenuation coefficients of metals and the surrounding tissues

and bones. Mathematically, reconstruction of such an image is much more manageable because there is no missing data issue for this problem. That is, the projection data at hand is sufficient to define the metal boundary within the patient. The obtained geometric information of the metal objects is by itself valuable for many clinical applications, such as the localization of implanted seeds and intracavitary applicators (Lerma and Williamson 2002) for accurate dose calculation and planning in brachytherapy. Figure 2.5 illustrates the results of inserting the 10 ball bearings (BBs) in the head phantom. Figure 2.5a shows one slice of the image reconstructed by the FDK algorithm. Strong artifacts caused by the BBs are observed in Figure 2.5a, which blur the image and make the identification of the BBs difficult. Figure 2.5b shows the image reconstructed by the proposed algorithm, and it is seen that the 10 BBs can be clearly identified and the structure of the head is presented as background. The reconstructed BBs show different sizes in Figure 2.5b because the inserted bolus and BBs are not perfectly perpendicular to the z axis and not all the centers of the BBs are within the reconstructed slice. Moreover, the information about the shape and location of the metal objects can be incorporated into iterative image reconstruction algorithms to more reliably reconstruct the images with substantially reduced artifacts. Results of the head phantom embedded with a triangularly shaped brass object are shown in Figure 2.6. Figure 2.6a shows the FDK-reconstructed image in which severe artifacts are

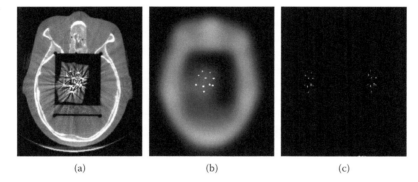

(a)　　　　　　(b)　　　　　　(c)

FIGURE 2.5 One slice of the image of the head phantom with 10 BBs: (a) image reconstructed by the FDK algorithm; (b) the metal-only image reconstructed by the proposed algorithm; and (c) the binary image extracted from (b). (Reprinted from Wang, J., and L. Xing. 2010. *J Xray Sci Technol*. With permission.)

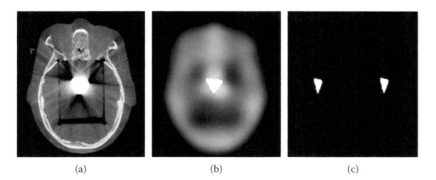

(a)　　　　　　(b)　　　　　　(c)

FIGURE 2.6 One slice of the image of the head phantom embedded with a brass object: (a) image reconstructed by the FDK algorithm; (b) the metal-only image reconstructed by the proposed algorithm; and (c) the binary image extracted from (b).

seen surrounding the brass object. Figure 2.6b shows the image reconstructed by the proposed gradient-controlled penalized weighted least squares (PWLS) algorithm, and Figure 2.6c shows the binary image extracted from Figure 2.6b through thresholding. The nonstraight edge in the reconstructed image caused by the uneven boundaries of the brass object is clearly seen.

2.7 Radiation Dose Reduction

Although onboard volumetric imaging offers welcome information regarding on-treatment patient anatomy, there is critical concern over the risk associated with excessive radiation dose when imaging is done repeatedly (Brenner et al. 2007; Islam et al. 2006; Murphy et al. 2007). The risk is invisible, long term, and cumulative—every scan compounds the dose and the risk. The 2006 report on the Biological Effects of Ionizing Radiation (BEIR) provides a framework for estimating the lifetime risk of cancer incidence from radiation exposure using the most current data on the health effects of radiation. In general, the risk is significantly modulated by the polymorphisms of genes involved in deoxyribonucleic acid (DNA) damage and repair (such as the BRCA1–BRCA2 mutation). It has been reported that the dose delivered to the patient is more than 3 cGy for central tissue and about 5 cGy for most of the peripheral tissues from a kV-CBCT scan with current clinical protocols (Wen et al. 2007). When a patient is imaged daily, this amounts to an imaging dose of more than 100 cGy to the region inside the field of view during a treatment course with a conventional fractionation scheme. The risk is exacerbated by the frequent use of other modern X-ray imaging modalities such as 4D simulation CT and fluoroscopic imaging in modern radiation oncology clinics. Given that the radiological dose is directly and linearly related to risk and based on the "as low as reasonably achievable" (ALARA) principle, the unwanted kV-CBCT dose must be minimized in order for the patient to truly benefit from modern image guidance technology (Murphy et al. 2007).

Much effort has been devoted to reducing the imaging dose in CBCT scanning. A strategy for CBCT dose reduction is to image the patient with lower milliamperes (mAs) and then recover the quality of the resultant images through the use of a statistical analysis–based noise removal technique (Wang et al. 2008). An iterative image reconstruction algorithm based on a PWLS principle has been developed to incorporate the noise spectrum into the reconstruction calculation and to effectively suppress the adverse effect of lowering the mAs (Wang, Yu, and De Man 2008) The PWLS principle consists of two terms: a weighted least squares (WLS) term that models the measurement data, and a penalty (P) term that encourages image smoothness of reconstructed images. The WLS criterion is formulated in such a way that the measured projection data with a lower contrast-to-noise ratio (CNR) will contribute less to the estimation of attenuation map. The CBCT images are reconstructed by minimizing the PWLS objective function. With this technique, it has been shown that high-quality CBCT images can be obtained with an imaging dose lesser by an order of magnitude while preserving the edges of various structures and the spatial resolution.

2.8 Closing the Loop of IMRT/VMAT Process: CBCT-Based Dose Reconstruction

An important application of onboard CBCT is to provide an on-treatment patient model for retrospective reconstruction. Yang et al. (2007) and Yoo and Yin (2006) evaluated the accuracy of CBCT-based dose calculation of existing CBCT systems. Researchers Chen et al. (2006) and van Elmpt et al. (2009) carried out dose reconstruction using the pretreatment MV-CBCT of patients and the energy fluence maps converted from treatment-time portal images. Retrospective IMRT dose reconstruction methods based on onboard kV-CBCT scanning and multileaf collimator (MLC) log-files as well as measured leaf sequences using electronic portal imaging devices (EPIDs) have been reported by Lee et al. (Lee, Le, and Xing 2008; Lee, Mao, and Xing 2008). The dosimetric impact of retrospective dose reconstruction taking into consideration a patient's geometric changes over time, residual setup errors, and the inherent delivery errors associated with the MLC for three head and neck (HN) patients has been assessed. A similar technique for retrospectively reconstructing the actual dose delivered in VMAT, a new clinical modality for conformal RT, has been reported by Qian et al. (2010). This pragmatic technique based on the pretreatment CBCT, which supplies the most updated patient model, and dynamic log-files that record the actual leaf positions, gantry angles, and cumulative monitoring units (MUs) during dose delivery, provides an invaluable tool for future online or offline adaptive VMAT. In this work, CBCT was performed before the dose delivery and the system's log-files were retrieved after the delivery. The actual delivery status at a control point, including MLC leaf positions, gantry angles, and cumulative doses, was reconstituted. The compiled data was used to replace the control points in the original digital imaging and communications in medicine (DICOM) radiotherapy plan using in-house software. This reconstituted plan was then imported into the Eclipse treatment planning system, and dosage was reconstructed on the corresponding CBCT. The workflow of VMAT dose reconstruction based on Varian's RapidArc radiotherapy system is shown in Figure 2.7.

Figure 2.8 presents the dose distributions of four situations for a pelvis phantom study: the original VMAT plan on a planning CT (origin.pCT), and reconstituted plans on CBCTs (recon.CBCT) of the phantom positioned with different setups. Figures 2.8a through d display the dose contributions in the same transverse slice of the pelvis phantom for each case. When the phantom was set up as planned, the high-dose levels are properly focused on the target as the pCT-based original plan. However, considerable shift in the posterior–anterior (PA) direction of high-dose levels is observed when moderate errors were introduced in the phantom setup. For larger setup errors, the high-dose region significantly deviated from the target and covered its periphery. These observations are also reflected in the final dose volume histograms (DVHs). Figure 2.8g depicts

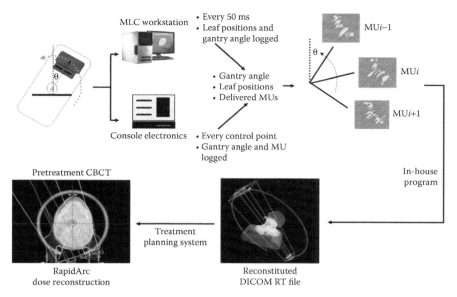

FIGURE 2.7 The workflow for using pretreatment CBCT, MLC, and delivery dynamic log-files for volumetric-modulated arc therapy dose reconstruction. (Reprinted from Qian, J., et al. 2010. *Phys Med Biol* 55:3597–610. With permission.)

the DVHs of the hypothetical target from the pCT-based original plan and the CBCT-based reconstituted plans. Compared with the origin.pCT, the DVH of the error-free setup case (recon. CBCT1) demonstrates an increase of 1.3% in dose for the entire target. This small discrepancy may be attributed to the inaccuracies of the dose delivery system and the density calibration of CBCT and pCT. When the moderate (recon.CBCT2) and large (recon.CBCT3) setup errors were introduced, the DVHs of the two resultant plans showed significant compromises in the target dose coverage as expected. Compared with the recon.CBCT1 plan, the minimum target dose is reduced from 94.8% to 76.1% and 47.3%, respectively. The mean target dose decreases and the respective differences are 0.8% and 4.6% for the two plans with moderate and large setup errors.

For the four dose distributions, the right–left (RL) and anterior–posterior (AP) dose profiles are plotted in Figures 2.8e and f, respectively. The dose profiles of the recon.CBCT2 and recon. CBCT3 plans show degraded gradient behaviors at the target edges as compared to the recon.CBCT1 plan. It is observed that the edges of the dose profiles of recon.CBCT2 are shifted in the AP direction. Similarly, the edges of the dose profiles of recon.CBCT3 are shifted in both the LR and AP directions. The extent of the shift is consistent with the intentionally introduced setup errors: 2 mm in the AP direction for recon.CBCT2, −3 mm in the RL and 5 mm in the AP directions for recon.CBCT3. There is no discernable shift of the dose profile along either direction for the error-free setup case.

The works by Lee et al. and Qian et al. on IMRT/VMAT dose reconstruction mark the first time in radiation oncology history when the dose actually delivered to patients was examined in a rigorous manner, thus closing the loop of the RT process (Lee, Le, and Xing 2008; Lee, Mao, and Xing 2008; Qian et al. 2010). These works provide practical platforms for implementing a workflow in reconstructing the IMRT/VMAT delivered dose

and the dosimetric information needed to adaptively modify the treatment plan, if indicated, based on the accumulated dose given to a patient. This maneuver affords an objective dosimetric basis for making the clinical decision on whether a replanning or reoptimization is necessary during the course of treatment.

2.9 Recent Progress in CBCT Image Reconstruction Using Sparse Projection Data with Compressed Sensing Theory

The classical Shannon–Nyquist sampling theorem specifies that to avoid the loss of information when capturing a signal, one must sample at least two times faster than the signal bandwidth. In many applications, including digital imaging and video cameras, the Nyquist rate is so high that too many samples result, making samples compressible for efficient storage or transmission. In other applications, such as medical imaging systems and high-speed analog-to-digital converters, increasing the sampling rate is either impractical or too expensive. A good example of this is 4D CBCT and 4D PET imaging, where the sampling per phase is rather limited because increasing the sampling rate is very costly in a clinical environment. Compressed sensing (CS) has recently emerged as a promising method for obtaining superresolved signals and images from far fewer data/measurements than what is usually considered necessary (Donoho 2006; Candes, Romberg, and Tao 2006). Briefly, CS is a technique for acquiring and reconstructing a signal that is known to be sparse or compressible. A mathematical manifestation of a sparse signal contains many coefficients close to or equal to zero, when represented in some appropriate transform domain, such as the Fourier domain,

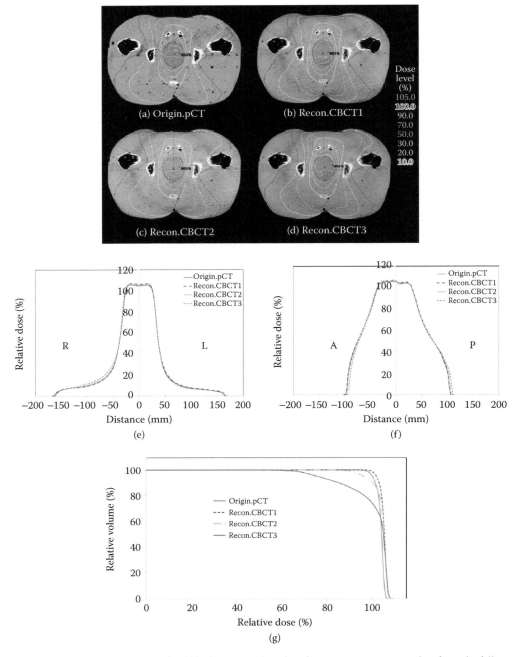

FIGURE 2.8   (See color insert following page 204 for (a).) The VMAT dose distributions on a transverse slice from the following: (a) the original plan on the pCT and (b–d) the reconstituted plan on the CBCTs (CBCT1, CBCT2, CBCT3) of the pelvis phantom set up with three predefined errors, respectively. Dose profiles are shown along (e) RL direction and (f) AP direction for the pCT-based original plan and the CBCT-based reconstructions. (g) The DVHs of the hypothetical target for the four calculations: origin.pCT, recon.CBCT1, recon.CBCT2, and recon.CBCT3 are also shown. (Reprinted from Qian, J., et al. 2010. *Phys Med Biol* 55:3597–610. With permission.)

total-variation (TV) norm, and wavelet domain. Effective utilization of this type of prior knowledge of the system can potentially reduce the required number of measurement samples determined by the Shannon–Nyquist theorem.

Several methods are under investigation for tomographic image reconstruction from sparse samples. When FBP or FDK (Feldkamp, Davis, and Kress 1984) algorithms are applied to undersampled projection data, the quality of the resultant images degrades dramatically due to incomplete information in the Fourier domain. For cone-beam geometry, obtaining Fourier-domain samples from projection data is less straightforward. Approximation algorithms commoly used (e.g., Fourier rebinning) interpolate the projection data and may result in some distortion. Heuristic iterative algorithms, such as projection on convex sets (POCS; Sidky and Pan 2008; Wang, Yu, and De Man 2008; Chen, Tang, and Leng 2008), random search, and first-order methods, are developed to find solutions. Computationally, although the standard second-order methods work well, it is necessary to solve a large system of linear equations in order to compute the Newton steps. A large number of first-order methods are available to tackle the problem of CS. Choi et al. have applied this technique to CBCT image reconstruction from a set of highly undersampled and noisy CBCT projection measurements, and they have shown that high-quality CBCT images are attainable under the condition of sparse and even noisy projection data (Choi et al. 2010). They formulated the CS problem as a TV norm minimization with a quadratic inequality constraint, and mitigated the manual parameter selection in previous approaches by enabling the physical interpretation of data. The use of first-order methods reduces significantly the use of computer memory by iterative forward-projection and backprojection to handle large-scale problems. The method is summarized as follows:

The CS theory–based CBCT recovery of Choi et al. (2010) starts from a cost function in the image domain:

$$\Phi(x) = (y - Px)^T \Sigma^{-1} (y - Px) \qquad (2.1)$$

where y is the vector of sinogram data, and x is the vector of attenuation coefficients to be reconstructed. Operator P represents the system or projection matrix. The symbol T denotes the transpose operator, and thus, P^T is the backprojection matrix. The matrix Σ is a diagonal matrix with i th element of σ_i^2, that is, an estimate of the variance of noise of the scatter-corrected line integral at detector bin i, which can be calculated from the measured projection data (Wang et al. 2008 and Wang et al. 2009). For the standard least squares formulation, setting $A = \Sigma^{-1/2} P$ and $b = \Sigma^{-1/2} \hat{y}$, we can rewrite Equation 2.1 as

$$\Phi(x) = \| Ax - b \|_{\ell_2}^2 \qquad (2.2)$$

Introducing the tolerance level of measurement inconsistency ε, we can formulate the image reconstruction problem as a quadratically constrained problem:

$$\begin{aligned} \text{minimize} \quad & f(x) \\ \text{subject to} \quad & \| Ax - b \|_{\ell_2} \le \varepsilon \end{aligned} \qquad (2.3)$$

where f is an ℓ_1-norm-related regularization function depending on prior assumptions about the image x. The quadratic constraint here can be interpreted as follows: the Euclidian distance between detection and estimation is not greater than ε. The Euclidian distance ε quantifies the tolerable uncertainty level of the noisy projection measurements. Among the many possible candidates for applying the CS penalty function, we select the 3D TV of the reconstructed image, that is, $f(x) = \| x \|_{\mathrm{TV}}$, defined by

$$\| x \|_{\mathrm{TV}} := \sum_{i,j,k} \left\| \nabla x[i,j,k] \right\|_{\ell_2}$$

to form the objective function, where $\nabla x[i,j,k] \in R^3$ is the difference vector at each position (i,j,k) of the object image.

Problems of the form of Equation 2.3 can be solved using a variety of algorithms, including interior point methods (Kim et al. 2007), projected gradient methods (Saunders and Kim 2002), homotopy methods (Hale, Yin, and Zhang 2008), Bregman iterative regularization algorithms (Yin et al. 2008), and a first-order method based on Nesterov's algorithm (Becker, Bobin, and Candes 2009). A first-order method developed by Nesterov, which provides an accurate and efficient solution to large-scale CS reconstruction problems using a smoothing technique, seems to be quite efficient and suitable for dealing with the aforementioned problem because of the method's superior computational efficiency. In the calculations performed by Choi et al. (2010), the CBCT projection function call, rather than the matrix-vector product, is used. Storing a projection matrix requires excessive memory space; this can be problematic for large-scale CBCT imaging problems. Compared to conventional approaches, efficient CBCT forward- and backprojection functions can solve the problems associated with very large numbers of variables (voxels to be determined) and measurements (CBCT projection data) with more efficiency. The convergence speed of the proposed method and the existing POCS method is shown in Figure 2.9 in terms of the number of iterations.

Figures 2.10 and 2.11 show a representative slice of the head phantom images reconstructed by the low- and high-dose protocol using FDK and CS algorithms. The experimental CBCT projection data were acquired by an Acuity simulator (Varian Medical Systems, Palo Alto, California). The number of

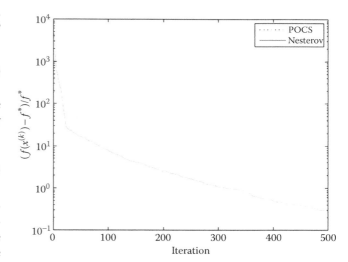

FIGURE 2.9 Convergence comparison between the POCS and Nesterov's algorithms in terms of the first 500 iterations. For fair comparison, the POCS algorithm uses a backtracking line search rather than a constant step size. The compared criterion f is unconstrained least absolute shrinkage and selection operator (LASSO) regression value, and f is the LASSO value with the digital phantom. (Reprinted from Choi, K., et al. 2010. *Med Phys* 37:5113–25. With permission.)

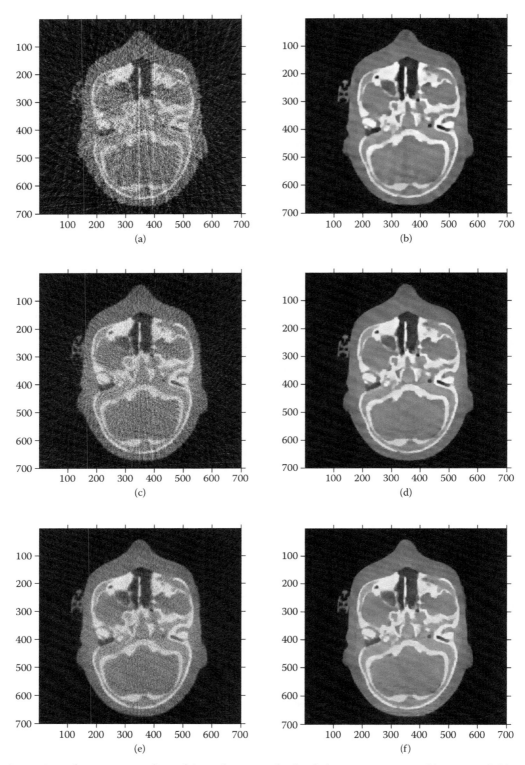

FIGURE 2.10 Comparison of representative slices of the anthropomorphic head phantom reconstructed by FDK and CS-WLS using 10-mA tube-current projection data: (a) FDK reconstruction using 56 projection views; (b) CS-WLS reconstruction using 56 projection views; (c) FDK reconstruction using 113 projection views; (d) CS-WLS reconstruction using 113 projection views; (e) FDK reconstruction using 226 projection views; (f) CS-WLS reconstruction using 226 projection views; (g) FDK reconstruction using 339 projection views; (h) CS-WLS reconstruction using 339 projection views; (i) FDK reconstruction using 678 projection views; and (j) CS-WLS reconstruction using 678 projection views. (Reprinted from Choi, K., et al. 2010. *Med Phys* 37:5113–25. With permission.)

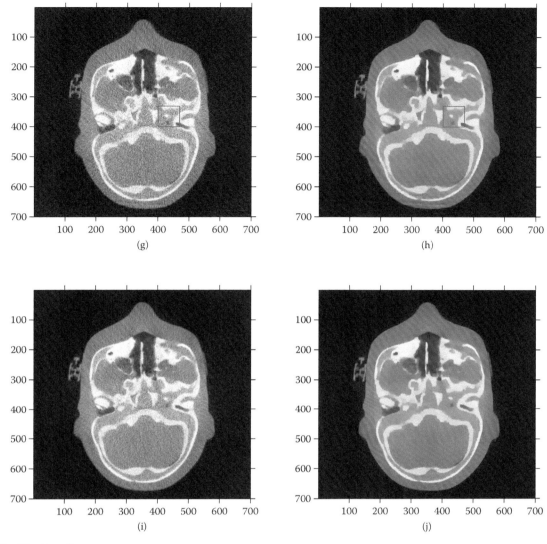

FIGURE 2.10 *(Continued)*

projections for a full 360-degree rotation is 680, and the total time for the acquisition is about 1 minute. The dimension of each acquired projection image is 397 × 298 mm² containing 1024 × 768 pixels. As can be seen from the figures, the CS techinque efficiently suppresses noise in the low-dose protocol resulting in images with sharper edges compared to the FDK reconstruction. The CS technique thus provides a useful method to effectively reduce the imaging dose with minimal compromise on the resultant image quality.

It is important to note that CS-based reconstruction favors piecewise constant solution (Chen, Tang, and Leng 2008; Donoho 2006; Candes, Romberg, and Tao 2006), and a phantom construct with no noise as the one used here may not reflect realistic clinical situations completely. This issue has been compensated in the presented anthropomorphic head phantom experiments with different dose levels. But the digital phantom study is useful because it shows the ideal performance of CS-based reconstruction and helps in comparing a proposed method with existing methods.

2.10 Summary

3D and 4D morphological imaging plays a pivotal role in modern adaptive RT. In the last decade, much progress has been made in integrating novel imaging techniques into the RT process and in improving the quality of onboard imaging systems. However, it is important to note that the 3D and 4D CBCT imaging techniques available currently are far from ideal, and further research from the imaging and therapeutic communities is necessary to maximally utilize the technical capacity of onboard imaging system. Looking forward, there are a number of important directions of research that should be pursued in order to make adaptive RT a routine clinical practice. The first, and perhaps the most difficult one because of

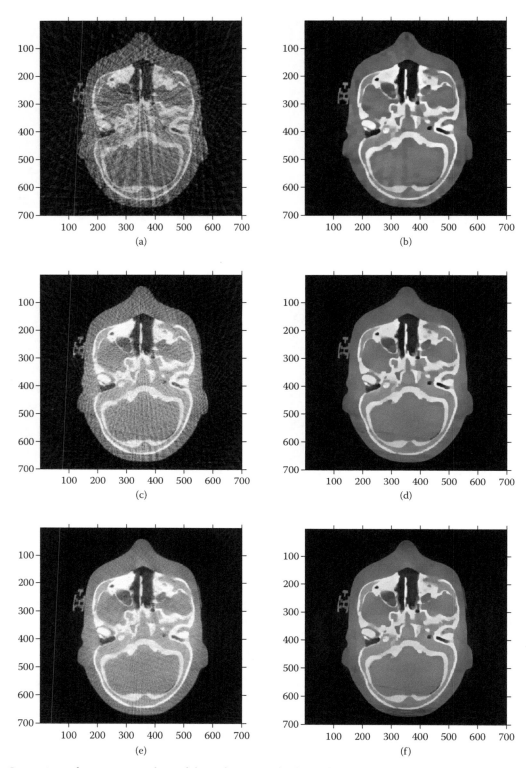

FIGURE 2.11 Comparison of representative slices of the anthropomorphic head phantom reconstructed by FDK and CS-WLS using 80-mA tube-current projection data: (a) FDK reconstruction using 56 projection views; (b) CS-WLS reconstruction using 56 projection views; (c) FDK reconstruction using 113 projection views; (d) CS-WLS reconstruction using 113 projection views; (e) FDK reconstruction using 226 projection views; (f) CS-WLS reconstruction using 226 projection views; (g) FDK reconstruction using 339 projection views; (h) CS-WLS reconstruction using 339 projection views; (i) FDK reconstruction using 678 projection views; and (j) CS-WLS reconstruction using 678 projection views. (Reprinted from Choi, K., et al. 2010. *Med Phys.* With permission.)

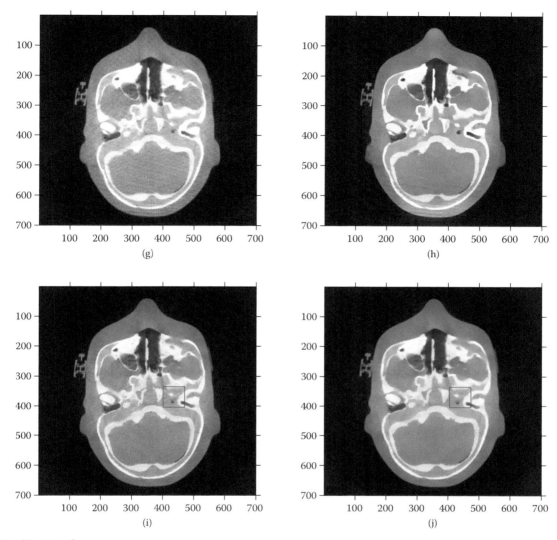

FIGURE 2.11 *(Continued)*

its multidisciplinary nature, is the development of molecular or biological imaging techniques and their effective integration into the RT process. Molecular imaging will allow one not only to delineate the boundary of the tumor based on biological characteristics but also to map out the biology distribution of cancer cells, affording a significant opportunity for biologically conformal radiation therapy (Yang and Xing 2005a; Yang and Xing 2005b) treatment in the future. Real-time image guidance and even real-time adaptive therapy is becoming increasingly possible. Along this line, a number of stereoscopic X-ray imaging techniques based on kV, hybrid kV/MV X-ray (Wiersma, Mao, and Xing 2008; Wiersma et al. 2009; Liu, Wiersma, and Xing 2010a; Liu, Wiersma, and Xing 2010b; Liu et al. 2008) systems (with or without other auxiliary external surrogate devices such as Real-time Position Management (RPM), have been proposed for real-time therapeutic guidance.

A general strategy of minimizing the patient imaging dose by the effective use of partial information of the system has also been proposed (Liu, Wiersma, and Xing 2010b). Finally, development of effective adaptive therapy workflow (de la Zerda, Armbruster, and Xing 2007) and quality assurance tools is also of paramount importance in the successful translation of state-of-the-art imaging and planning tools to widespread clinical use.

Acknowledgments

This project was supported in part by grants from the United States National Science Foundation (0854492), United States Department of Defense (PC080941), and Global R&D Resource Recruitment & Support Program, Korean Ministry of Education, Science and Technology.

References

Aldridge, J. S., P. J. Reckwerdt, and T. R. Mackie. 1999. A proposal for a standard electronic anthropomorphic phantom for radiotherapy. *Med Phys* 26:1901–3.

Ball, D. 2006. Is the way up the way forward? Radiotherapy dose escalation for non-small cell lung cancer. *J Thorac Oncol* 1:107–8.

Becker, S., J. Bobin, and E. J. Candes. 2009. NESTA: A fast and accurate first-order method for sparse recovery. Tech. Report, Caltech. Arxiv preprint arXiv:0904.3367.

Berbeco, R. I., F. Hacker, D. Ionascu, and H. J. Mamon. 2007. Clinical feasibility of using an EPID in CINE mode for image-guided verification of stereotactic body radiotherapy. *Inter J Radiat Oncol Biol Phys* 69:258–66.

Boone, J. M., K. K. Lindfors, V. N. Cooper 3rd, and J. A. Seibert. 2000. Scatter/primary in mammography: Comprehensive results. *Med Phys* 27:2408–16.

Brenner, D. J., and E. J. Hall. 2007. Computed tomography—an increasing source of radiation exposure. *NEJM* 357:2277–84.

Buzug, T. M. 2008. *Computed Tomography.* Berlin: Springer.

Candes, E. J., J. Romberg, and T. Tao. 2006. Robust uncertainty principles: Exact signal reconstruction from highly incomplete frequency information. *IEEE Trans Inf Theory* 52:489–509.

Chang, J., K. M. Yenice, A. Narayana, and P. H. Gutin. 2007. Accuracy and feasibility of cone-beam computed tomography for stereotactic radiosurgery setup. *Med Phys* 34:2077–84.

Chen, J., O. Morin, M. Aubin, M. K. Bucci, C. F. Chuang, and J. Pouliot. 2006. Dose-guided radiation therapy with megavoltage cone-beam CT. *Br J Radiol* 79 Spec No 1:S87–98.

Chen, G. H., J. Tang, and S. Leng. 2008. Prior image constrained compressed sensing (PICCS): A method to accurately reconstruct dynamic CT images from highly undersampled projection data sets. *Med Phys* 35:660–3.

Choi, K., J. Wang, L. Zhu, T. Suh, S. Boyd, and L. Xing. 2010b. Compressed sensing with a first-order method for cone-beam CT dose reduction. *Med Phys* 37:5113–25.

Choi, K., and L. Xing. 2010. Scatter correction in cone beam CT. *Med Phys.*

Court, L. E., L. Dong, A. K. Lee, et al. 2005. An automatic CT-guided adaptive radiation therapy technique by online modification of multileaf collimator leaf positions for prostate cancer. *Inter J Radiat Oncol Biol Phys* 62:154–63.

de la Zerda, A., B. Armbruster, and L. Xing. 2007. Formulating adaptive radiation therapy (ART) treatment planning into a closed-loop control framework. *Phys Med Biol* 52:4137–53.

De Man, B., J. Nuyts, P. Dupont, G. Marchal, and P. Suetens. 1999. Reduction of metal streak artifacts in X-ray computed tomography using a transmission maximum a posteriori algorithm. *IEEE Nuclear Science Symposium Conference Record* 2:850–4.

Donoho, D. L. 2006. Compressed sensing. *IEEE Trans Inf Theory* 52:1289–306.

Feldkamp, L. A., L. C. Davis, and J. W. Kress. 1984. Practical cone-beam algorithm. *J Opt Soc Am* A 1:612–9.

Fessler, J. A. 1994. Penalized weighted least-squares image-reconstruction for positron emission tomography. *IEEE Trans Med Imaging* 13:290–300.

Grass, M., T. Kohler, and R. Proksa. 2000. 3D cone-beam CT reconstruction for circular trajectories. *Phys Med Biol* 45:329–47.

Grass, M., T. Kohler, and R. Proksa. 2001. Angular weighted hybrid cone-beam CT reconstruction for circular trajectories. *Phys Med Biol* 46:1595–610.

Hale, E. T., W. T. Yin, and Y. Zhang. 2008. Fixed-point continuation for ℓ_1-minimization: Methodology and convergence. *SIAM J Opt* 19:1107–30.

Hong, T. S., J. S. Welsh, M. A. Ritter, et al. 2007. Megavoltage computed tomography: An emerging tool for image-guided radiotherapy. *Am J Clin Oncol* 30:617–23.

Islam, M. K., T. G. Purdie, B. D. Norrlinger, et al. 2006. Patient dose from kilovoltage cone beam computed tomography imaging in radiation therapy. *Med Phys* 33:1573–82.

Jaffray, D. A., J. H. Siewerdsen, J. W. Wong, A. A. Martinez. 2002. Flat-panel cone-beam computed tomography for image-guided radiation therapy. *Inter J Radiat Oncol Biol Phys* 53:1337–49.

Kak, A. C., and M. Slaney. 1988. *Principles of Computerized Tomographic Imaging.* Philadelphia: Siam.

Kalender, W. A., R. Hebel, and J. Ebersberger. 1987. Reduction of CT artifacts caused by metallic implants. *Radiology* 164:576–77.

Kim, S. J., K. Koh, M. Lustig, S. Boyd, and D. Gorinevsky. 2007. An interior-point method for large-scale l(1)-regularized least squares. *IEEE J Sel Top Signal Process* 1:606–17.

Kohler, T. H., R. Proksa, M. Grass. 2001. A fast and efficient method for sequential cone-beam tomography. *Med Phys* 28:2318–27.

Kyriakou, Y., and W. Kalender. 2007. Efficiency of antiscatter grids for flat-detector CT. *Phys Med Biol* 52:6275–93.

Kyriakou, Y., T. Riedel, and W. A. Kalender. 2006. Combining deterministic and Monte Carlo calculations for fast estimation of scatter intensities in CT. *Phys Med Biol* 51:4567–86.

Lee, L., Q. Le, and L. Xing. 2008. Retrospective IMRT dose reconstruction based on cone-beam computed tomography and the MLC positional log-file recorded during treatment. *Int J Radiat Oncol Biol Phys* 70:634–44.

Lee, L., W. Mao, L. Xing. 2008. The use of EPID-measured leaf sequence files for IMRT dose reconstruction in adaptive radiation therapy. *Med Phys* 35:5019–29.

Lerma, F. A., and J. F. Williamson. 2002. Accurate localization of intracavitary brachytherapy applicators from 3D CT imaging studies. *Med Phys* 29:325–33.

Li, T., A. Koong, and X. Xing. 2007. Enhanced 4D cone-beam computed tomography using an on-board imager. *Med Phys* 34:3688–95.

Li, T., E. Schreibmann, Y. Yang, and L. Xing. 2006. Motion correction for improved target localization with on-board cone-beam computed tomography. *Phys Med Biol* 51:253–67.

Li, T., and L. Xing. 2007. Optimizing 4D cone-beam CT acquisition protocol for external beam radiotherapy. *Int J Radiat Oncol Biol Phys* 67:1211–9.

Li, T., L. Xing, C. McGuinness, P. Munro, B. Loo, and A. Koong. 2006. Four-dimensional cone-beam CT using an on-board imager. *Med Phys* 33:3825–33.

Li, H., X. R. Zhu, L. Zhang, et al. 2008. Comparison of 2D radiographic images and 3D cone beam computed tomography for positioning head-and-neck radiotherapy patients. *Inter J Radiat Oncol Biol Phys* 71:916–25.

Liu, W., R. D. Wiersma, W. Mao, G. Luxton, and L. Xing. 2008. Real-time 3D internal fiducial marker tracking during arc radiotherapy by use of combined MV-kV imaging. *Phys Med Biol* 53:7197–213.

Liu, W., R. D. Wiersma, and L. Xing. 2010a. Optimized hybrid MV-kV imaging protocol for volumetric prostate arc therapy. *Int J Radiat Oncol Biol Phys* 78:595–604.

Liu, W., R. D. Wiersma, and L. Xing. 2010b. A failure detection strategy for real-time image guided prostate IMRT. *Int J Radiat Oncol Biol Phys*.

Lu, J., T. M. Guerrero, P. Munro, et al. 2007. Four-dimensional cone beam CT with adaptive gantry rotation and adaptive data sampling. *Med Phys* 34:3520–9.

Masi, L., F. Casamassima, C. Polli, C. Menichelli, I. Bonucci, and C. Cavedon. 2008. Cone beam CT image guidance for intracranial stereotactic treatments: Comparison with a frame guided set-up. *Int J Radiat Oncol Biol Phys* 71:926–33.

Murphy, M. J., J. Balter, S. Balter, et al. 2007. The management of imaging dose during image-guided radiotherapy: Report of the AAPM Task Group 75. *Med Phys* 34:4041–63.

Oldham, M., D. Letourneau, L. Watt, et al. 2005. Cone-beam-CT guided radiation therapy: A model for on-line application. *Radiother Oncol* 75:271 E271–8.

Qian, J., L. Lee, W. Liu, et al. 2010. Dose reconstruction for volumetric modulated arc therapy (VMAT) using cone-beam CT and dynamic log-files. *Phys Med Biol* 55:3597–610.

Robertson, D. D., J. Yuan, G. Wang, and M. W. Vannier. 1997. Total hip prosthesis metal-artifact suppression using iterative deblurring reconstruction. *J Comput Assist Tomogr* 21:293–8.

Saunders, M. A., and B. Kim. 2002. PDCO: Primal-dual interior method for convex objectives. Tech. Report, Stanford University. http://www.stanford.edu/group/SOL/software/pdco.html (accessed October 1, 2010).

Sidky, E. Y., and X. C. Pan. 2008. Image reconstruction in circular cone-beam computed tomography by constrained, total-variation minimization. *Phys Med Biol* 53:4777–807.

Siewerdsen, J. H., D. J. Moseley, B. Bakhtiar, S. Richard, and D. A. Jaffray. 2004. The influence of antiscatter grids on soft-tissue detectability in cone-beam computed tomography with flat-panel detectors. *Med Phys* 31:3506–20.

Sonke, J. J., L. P. Zijp Remeijer, and M. van Herk. 2005. Respiratory correlated cone beam CT. *Medl Phys* 32:1176–86.

Timmerman, R., and L. Xing. 2009. *Image Guided and Adaptive Radiation Therapy*. Baltimore: Lippincott Williams & Wilkins.

Turbell, H. 2001. Cone-beam reconstruction using filtered back-projection. Dissertation No. 672, Linkoping Studies in Science and Technology.

van Elmpt, W., S. Nijsten, S. Petit, B. Mijnheer, P. Lambin, and A. Dekker. 2009. 3D in vivo dosimetry using megavoltage cone-beam CT and EPID dosimetry. *Int J Radiat Oncol Biol Phys* 73:1580–7.

Wang, J., T. Li, Z. Liang, and L. Xing. 2008. Dose reduction for kilovotage cone-beam computed tomography in radiation therapy. *Phys Med Biol* 53:2897–909.

Wang, J., T. Li, and L. Xing. 2009. Iterative image reconstruction for CBCT using edge-preserving prior. *Med Phys* 36:252–60.

Wang, G., D. L. Snyder, J. A. O'Sullivan, and M. W. Vannier. 1996. Iterative deblurring for CT metal artifact reduction. *IEEE Trans Med Imaging* 15:657–64.

Wang, J., and L. Xing. 2010. Accurate determination of the shape and location of metal objects in x-ray computed tomography. *J Xray Sci Technol*.

Wang, G., H. Yu, and B. De Man. 2008. An outlook on x-ray CT research and development. *Med Phys* 35:1051–64.

Welsh, J. S., M. Lock, P. M. Harari, et al. 2006. Clinical implementation of adaptive helical tomotherapy: A unique approach to image-guided intensity modulated radiotherapy. *Technol Cancer Res Treat* 5:465–79.

Wen, N., H. Q. Guan, R. Hammoud, et al. 2007. Dose delivered from Varian's CBCT to patients receiving IMRT for prostate cancer. *Phys Med Biol* 52:2267–76.

Wiersma, R. D., W. Mao, and L. Xing. 2008. Combined kV and MV imaging for real-time tracking of implanted fiducial markers. *Med Phys* 35:1191–98.

Wiersma, R. D., N. Riaz, S. Dieterich, Y. Suh, and L. Xing. 2009. Use of MV and kV imager correlation for maintaining continuous real-time 3D internal marker tracking during beam interruptions. *Phys Med Biol* 54:89–103.

Williamson, J. F., B. R. Whiting, J. Benac, et al. 2002. Prospects for quantitative computed tomography imaging in the presence of foreign metal bodies using statistical image reconstruction. *Med Phys* 29:2404–18.

Wu, Q., J. Liang, and D. Yan. 2006. Application of dose compensation in image-guided radiotherapy of prostate cancer. *Phys Med Biol* 51:1405–19.

Xing, L., B. Thorndyke, E. Schreibmann, et al. 2006. Overview of image-guided radiation therapy. *Med Dosim* 31:91–112.

Yang, Y., E. Schreibmann, T. Li, C. Wang, and L. Xing. 2007. Evaluation of on-board kV cone beam CT (CBCT)-based dose calculation. *Phys Med Biol* 52:685–705.

Yang, Y., and L. Xing. 2005a. Towards biologically conformal radiation therapy (BCRT): Selective IMRT dose escalation under the guidance of spatial biology distribution. *Med Phys* 32:1473–84.

Yang, Y., and L. Xing. 2005b. Optimization of radiation dose-time-fractionation scheme with consideration of tumor specific biology. *Med Phys* 32:3666–77.

Yin, W., S. Osher, D. Goldfarb, and J. Darbon. 2008. Bregman iterative algorithms for ℓ_1-minimization with applications to compressed sensing. *SIAM J Imag Sci* 1:143–68.

Yoo, S., and F. F. Yin. 2006. Dosimetric feasibility of cone-beam CT-based treatment planning compared to CT-based treatment planning. *Int J Radiat Oncol Biol Phys* 66:1553–61.

Zhang, Y. B., L. F. Zhang, R. Zhu, A. K. Lee, M. Chambers, and L. Dong. 2007. Reducing metal artifacts in cone-beam CT images by preprocessing projection data. *Int J Radiat Oncol Biol Phys* 67:924–32.

Zhao, S. Y., D. D. Roberston, G. Wang, B. Whiting, and K. T. Bae. 2000. X-ray CT metal artifact reduction using wavelets: Arm application for imaging total hip prostheses. *IEEE Trans Med Imaging* 19:1238–47.

Zhu, L., J. Wang, Y. Xie, J. Starman, R. Fahrig, and L. Xing. 2008. A patient set-up protocol based on partially blocked cone-beam CT. *Med Phys.*

Zhu, L., J. Wang, and L. Xing. 2009. Noise suppression in scatter correction for cone-beam CT. *Med Phys* 36(3):741–52.

Biological Imaging for Adaptive Radiation Therapy

Robert Jeraj
University of Wisconsin

Matt Nyflot
University of Wisconsin

3.1 Introduction

Imaging is intimately interlaced with radiation oncology. X-rays and, more recently, computed tomography (CT; Baker 1975; New et al. 1975; Reich and Seidelmann 1975), are the workhorses of tumor localization, which is an essential component of the radiation therapy (RT) process. Magnetic resonance imaging (MRI; Hawkes et al. 1980; Crooks et al. 1983) is essential for tumor localization in certain anatomical regions, particularly brain, because of its improved soft-tissue contrast. Currently, CT and MRI are the primary modalities used in the treatment process for determining patient and tumor anatomy. However, because disease progression and treatment response at the molecular and cellular levels precedes visible structural changes to tissue, "biological imaging" is receiving increasing attention as a method to understand disease progression and treatment response. Biological imaging does not have a precise definition. It encompasses virtually all imaging modalities and can be divided by the type of imaging processes into *functional* and *molecular* imaging. Note that the terms biological, molecular, and functional imaging are often used interchangeably, due to a lack of consensus regarding their definitions.

Functional imaging refers to either imaging of physiologic processes, such as blood flow to an organ or diseased tissue, and visualizing ongoing biochemical and metabolic activities of normal and abnormal tissues, or using established pharmacologic methods to assess disease processes and develop new drugs. Functional imaging is typically performed with ultrasound, CT, and MRI.

Molecular imaging refers to imaging specific molecular interactions and pathways, which reflect physiological, biochemical, and pharmacological processes. The term molecular imaging was defined by the Commission on Molecular Imaging of the American College of Radiology as "the spatially localized and/or temporally resolved sensing of molecular and cellular processes in vivo." The main modalities for molecular imaging are positron emission tomography (PET), single photon emission tomography (SPECT), magnetic resonance spectroscopy (MRS), and optical imaging.

Applications of biological imaging in radiation oncology are numerous and can be divided into three categories (Figure 3.1):

1. Diagnosis and staging: Imaging is performed during the initial phases of treatment process to establish whether there is a tumor and if so, how advanced it is.
2. Target definition: Imaging is performed prior to RT in order to determine the extent of the tumor and the position of normal tissue.
3. Treatment assessment: Imaging is performed either during or after the treatment to establish the efficacy of treatment process, predict outcome, and potentially modify the therapy.

In radiation oncology (and oncology in general), emphasis on anatomical imaging, particularly the use of CT/MRI for diagnosis, CT for target definition, and CT for image guidance (image-guided radiotherapy [IGRT]) is still very high. The application of biological imaging, except ^{18}F-2-deoxy-2-fluoro-D-glucose (FDG)-PET imaging, for diagnosis and staging is very limited. The use of biological imaging modalities (e.g., non-FDG-PET, MRS, dynamic contrast-enhanced [DCE]-MRI, DCE-CT) is rare. In the future, however, one can expect much more significant and intensive use of biological imaging for all steps of the treatment process. In this chapter, we discuss various biological

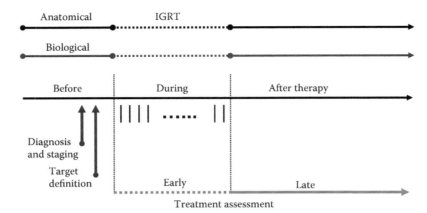

FIGURE 3.1 Biological imaging in the radiation oncology process. Imaging can be divided into three categories: (1) imaging for diagnosis and staging; (2) imaging for target definition; and (3) imaging for treatment assessment, which can be further divided into early treatment assessment (during therapy) and late treatment assessment (post-therapy). Traditionally, anatomical imaging, most often in the form of computed tomography, is used for diagnosis, staging, and late treatment assessment. Recently, anatomical image guidance is being increasingly used during the course of radiotherapy, which is widely known as image-guided radiotherapy. Similar to anatomical imaging, biological imaging can be used for all of the treatment steps.

imaging techniques and their applications with special focus on adaptive treatments and prospects for the future.

3.2 Biological Imaging

3.2.1 Biological Imaging Modalities

The main imaging modalities that enable biological imaging are PET, SPECT, MRS, and DCE imaging such as DCE-MRI and DCE-CT. Development of the first biological imaging modalities, SPECT (Kuhl and Edwards 1964; Kuhl et al. 1973) and PET (Phelps, Hoffman, et al., "Effect of positron," 1975; Phelps, Hoffman, et al., "Application of annihilation," 1975; Ter-Pogossian et al. 1975; Hoffmann et al. 1976), coincided with the development of computers that enabled tomographic image reconstruction. Development of MRS started around the same time (Moon and Richards 1972; Moon and Richards 1973; Hoult et al. 1974), which resulted in the first imaging of a tumor in early 1980s. In the late 1980s, a variety of functional MRI techniques were developed for imaging physiologic phenomena, such as blood flow, perfusion, or diffusion (Singer and Crooks 1983; Runge et al. 1985; Thomsen, Henriksen, and Ring 1987; Ogawa et al. 1990). For a list of hallmarks of cancer targeted by molecular and functional imaging, refer to Figure 3.2. However, it was not until the late 1990s that the use of biological imaging for clinical applications became more common, primarily due to the extremely rapid adoption of FDG-PET imaging.

When discussing biological imaging, it is important to remember that biological imaging is typically just a surrogate for underlying biological processes (Lewis and Welch 2001; Becherer et al. 2003; Haberkorn and Altmann 2003; Laking and Price 2003; Rehman and Jayson 2005). The correlation between the imaging information and actual biology is often limited because of the following factors:

- Complex uptake mechanisms
- Complicated pharmacokinetics
- Technological limitations of a particular imaging modality

For example, although certain imaging agents have been promoted as markers of specific tumor properties (e.g., ^{18}F-fluoromisonidazole [FMISO]-PET or blood oxygen level dependent [BOLD]-MRI as markers of hypoxia), this is in general a very simplified view. Most of these markers are not comparable with the much more specific biochemical markers used in cellular and molecular biology (e.g., TUNEL labeling for apoptosis). For example, even the uptake mechanisms of FDG, a surrogate marker of tumor metabolism, are not entirely known. The FDG uptake is related to a variety of mechanisms like microvasculature structure for delivering nutrients, overexpression of glucose membrane transporters Glut-1 and Glut-3 for the transportation of FDG into the cell, hexokinase activity for entering FDG into glycolysis, number of tumor cells/volume, proliferation rate (also reflected in necrosis), number of lymphocytes (not macrophages), and hypoxia-inducible factor (HIF)-1α for upregulating Glut-1 (Bos et al. 2002; de Geus-Oei et al. 2007). However, this complicated biological correlation and the still-incomplete understanding of its uptake mechanism did not prevent FDG from being extremely successful in the clinic and, by virtue of clinical use, also thoroughly clinically validated.

Biological imaging agent uptake mechanism and pharmacokinetics can be complicated, and a complex compartmental kinetic analysis might be necessary to extract the information of interest (Yamaguchi et al. 1986; Hawkins et al. 1992; Minn et al. 1993; Minn et al. 1995; Tsuyuguchi 1997; Mankoff et al. 1998; Mankoff et al. 1999; Mineura et al. 1999; Weber, Schwaiger, and Avril 2000; Lehtio et al. 2001; Wells et al. 2002; Sutinen et al. 2004; Tseng et al. 2004; Muzi et al. 2005). Although compartmental kinetic analysis generally improves correlation between the imaging data and the underlying biological processes, it is often

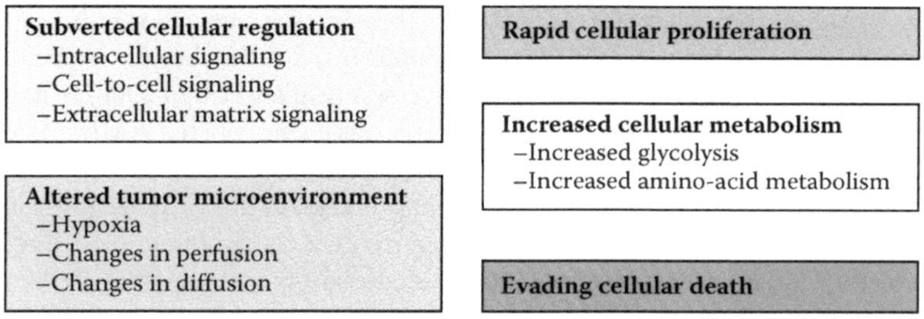

Subverted cellular regulation –Intracellular signaling –Cell-to-cell signaling –Extracellular matrix signaling	**Rapid cellular proliferation**
	Increased cellular metabolism –Increased glycolysis –Increased amino-acid metabolism
Altered tumor microenvironment –Hypoxia –Changes in perfusion –Changes in diffusion	**Evading cellular death**

FIGURE 3.2 Hallmarks of cancer targeted by molecular and functional imaging. Over 500 biological imaging agents have been developed to target these processes. Interestingly, only a handful of the agents are commonly imaged with biological imaging.

time consuming and relatively unstable. Therefore, it is usually avoided in clinical practice and more stable integral measures (e.g., standardized uptake value in PET) are used for biological image characterization.

Another crucial consideration is the technical limitations of imaging modalities, which can result in significant, often poorly characterized, imaging uncertainties. Biomedical imaging has been developed principally with diagnostic application in mind, where the primary goal is tumor detection and exact quantification is only secondary. When the aim is to characterize and quantify biological properties of tumors or quantify treatment response, it becomes critical that there are reliable numbers in the imaging data. Even a simple change of image acquisition parameters might lead to significant quantification uncertainties. For example, different image reconstruction algorithms can lead to differences in the reconstructed PET images in excess of 20% (Boellaard et al. 2001; Oda et al. 2001; Etchebehere et al. 2002; Lartizien et al. 2003; Jaskowiak et al. 2005). The problem of image quantification is especially important for radiation oncology applications, when treatment dose is to be correlated with the imaging data.

From the practical user standpoint, it is also important to know which imaging modality or imaging agent to use for the assessment of a given biological process (Cook 2003; Apisarnthanarax and Chao 2005). For example, in order to assess tumor perfusion, one has the option to choose among CT perfusion (DCE-CT), MRI perfusion (DCE-MRI), and PET perfusion (^{15}O-H$_2$O) imaging. What are the advantages, disadvantages, limitations, and associated uncertainties of each of these modalities? How do these modalities perform for different anatomical and tumor sites? Why pick one over the other? For example, DCE-CT has the advantages of a linear relationship between signal intensity and concentration of contrast agent, and ease of accessibility in clinics, including standardized analysis packages. However, limitations include large CT slice thicknesses, small fields of view on conventional hardware, nontrivial radiation doses, and relatively slow clearance of contrast agents. DCE-MRI has no radiation burden and has wide flexibility in field of view and resolution, but the relationship between signal intensity and contrast varies based on the scan protocol and analysis techniques are not standardized. ^{15}O-PET can be used over a larger field of view and the short tracer half-life easily accommodates short-term follow-up

imaging, but it requires onsite production of the tracer, which may severely limit the routine adoption of this technique into an adaptive radiotherapy context.

3.2.2 Biological Imaging Targets

3.2.2.1 Metabolism

By far the most common biological imaging modality is the imaging of cellular metabolism with ^{18}F-FDG PET. Most malignant tissues have increased ^{18}F-FDG uptake, which is associated with an increased rate of glycolysis and glucose transport. Warburg first described this fundamental aberration of malignant cells in the 1930s, and more recently, several research groups have described the specific cellular mechanisms associated with glucose uptake in malignant tissue (Bos et al. 2002; Avril 2004). The increase in ^{18}F-FDG uptake noted in malignant tissue is related in a complex manner to the proliferative activity of malignant tissue and to the number of viable tumor cells (Haberkorn et al. 1991; Higashi, Clavo, and Wahl 1993; Vesselle et al. 2000). For these reasons, investigators have postulated that alterations in ^{18}F-FDG uptake after treatment of cancer should reflect the cellular response to the treatment, likely including effects such as changes in the number of viable tumor cells and altered cellular proliferation. Changes in tumor glucose metabolism precede changes in tumor size. These changes in tumor metabolism reflect drug effects at a cellular level. For these reasons, FDG-PET enables the prediction of therapy response early in the course of treatment, as well as determines the viability of residual masses after completion of treatment for a variety of tumor types (Smith 1998; Avril and Weber 2005). However, one should not forget that a complex mix of different cellular processes determines the rate of glucose metabolism. The precise mechanism by which alterations in these cellular processes with cancer treatment lead to changes in ^{18}F-FDG uptake is incompletely understood and may be different for different tumor types and treatments.

3.2.2.2 Hypoxia

Another relevant measure of tumor microenvironment is hypoxia, historically defined as decreased oxygenation as a result of insufficient blood supply to a cell. Whereas classical perspectives of hypoxia focus on the cause, namely distance from tumor blood supply (chronic hypoxia) or transient changes in local perfusion (acute hypoxia), more recent discussions focus on hypoxia as a driver of malignant progression in tumors (Höckel et al. 1996; Bussink et al. 2003; Rofstad et al. 2007). Further, hypoxia has been identified as a predictive and prognostic indicator of poor treatment outcome (Ressel, Weiss, and Feyerabend 2001; Nordsmark et al. 2005). Several compounds have been synthesized for PET hypoxia imaging studies, of which FMISO (Koh et al. 1992; Koh et al. 1995; Rasey et al. 1996) and Cu-diacetyl-bis(N4-methylthiosemicarbazone) (CuATSM; Fujibayashi et al. 1997; Lewis et al. 1999; Takahashi et al. 2000; Lewis et al. 2001; Takahashi et al. 2001; Maurer et al. 2002; Dehdashti, Grigsby, et al. 2003; Dehdashti, Mintun, et al. 2003; Obata et al. 2003; Obata et al. 2005) have been extensively clinically validated. All these compounds diffuse into both normally oxygenated and hypoxic cells, but they are retained in substantially higher concentrations in hypoxic tissues. The compound FMISO appears suboptimal for assessing hypoxia because of low uptake in hypoxic cells, slow clearance from the normal tissues, and clinically conflicting results (Bentzen et al. 2002; Bentzen et al. 2003; Rajendran et al. 2003; Rajendran et al. 2004). Despite these difficulties, FMISO has been well-validated for clinical use (Gagel et al. 2004; Lawrentschuk et al. 2005). Similarly, the compound CuATSM has already proved to be an effective agent for this purpose. Clinically, CuATSM has shown promising results in prospective studies of nonsmall cell lung cancer (NSCLC) and cervical cancer patients (Dehdashti, Grigsby, et al. 2003; Dehdashti, Mintun, et al. 2003) with high discriminatory power in selecting patients who responded to the treatment as compared to the nonresponders.

3.2.2.3 Proliferation

Proliferative imaging is a more useful tool for overall treatment assessment, compared to metabolic imaging. Recently, the ^{18}F-labeled 3′-deoxy-3′-fluorothymidine (FLT) was proposed as a new marker for imaging tumor proliferation by PET (Shields et al. 1998; Grierson and Shields 2000; Krohn, Mankoff, and Eary 2001; Mier, Haberkorn, and Eisenhut 2002; Buck et al. 2003; Schwartz et al. 2003; Vesselle et al. 2003; Wagner et al. 2003). The uptake of FLT is correlated primarily with TK1 activity (Rasey et al. 2002; Schwartz et al. 2003), which shows an S-phase-regulated expression. Importantly, direct correlation between FLT uptake and proliferation Ki-67 labeling index from biopsy samples has been observed (Vesselle et al. 2002; Buck et al. 2003; Muzi et al. 2005), indicating the potential of FLT as a surrogate marker of tumor proliferation. In the context of RT, FLT shows a rapid decrease during the therapy (Everitt et al. 2009; Menda et al. 2009) and provides the optimal imaging window for treatment response assessment after 10–20 Gy, which is typically after 1–2 weeks of fractionated RT.

3.2.2.4 Angiogenesis

Imaging angiogenesis, the formation of new blood vessels, is an obvious corollary to the investigation of tumor vasculature. Most molecular techniques focus on imaging signaling the vascular endothelial growth factor (VEGF) pathway or $\alpha_v\beta_3$ integrins, a cell adhesion molecule located on immature blood vessels (Cai and Chen 2008). Angiogenic imaging of $\alpha_v\beta_3$ integrins has been reported in clinical and preclinical PET (Beer et al. 2005; Haubner 2006; Zhang et al. 2006; Beer et al. 2007). Some approaches to imaging the VEGF receptor (VEGFR) pathway, in particular, consist of radiolabeling VEGF antibodies such as bevacizumab (Nagengast et al. 2007). Due to the low uptake kinetics and long clearance times of antibody-based tracers, a better approach may be radiolabeling peptide-based tracers, which has been achieved in preclinical models with [^{64}Cu]DOTA-VEGF$_{121}$ (Cai et al. 2006; Cai and Chen 2008). Further research in molecular imaging of angiogenesis will be focused on improving uptake kinetics, specificity, and production yields.

3.2.2.5 Perfusion and Functional Vascular Characteristics

Dynamic CT or MR techniques could illuminate characteristics of tumor vasculature (such as "leakiness" of immature blood vessels) by providing measurements of blood perfusion, blood volume, and contrast enhancement (Miller et al. 2005; Cuenod et al. 2006; Provenzale 2007). These *in vivo* techniques could supplement molecular imaging data, indicating that antiangiogenic therapy reduces microvessel density and interstitial pressure (Lee et al. 2000; Tong et al. 2004; Willett et al. 2004). DCE-CT has been used to quantify perfusion in head and neck squamous cell carcinoma (HNSCC), including radiotherapy response (Hermans et al. 1999; Hermans et al. 2003), and to assess response to bevacizumab in patients with rectal cancer (Willett et al. 2004); however, the simplicity of the maximum-slope analysis used in such studies may limit their utility. More recently, improved convolution-based models were clinically implemented in head and neck studies (Bisdas, Medov, et al. 2008; Bisdas, Spicer, et al. 2008).

3.3 Biological Imaging in Radiotherapy

As mentioned in Section 3.1, applications of biological imaging in radiation oncology can be divided into three categories.

3.3.1 Diagnosis and Staging

Of all the biological imaging modalities, FDG-PET has become by far the most valuable tool for diagnosis and staging. In a landmark study by Hillner, Siegel, et al. ("Impact of positron emission tomography," 2008), the questionnaire data from the National Oncological PET Registry (NOPR) was analyzed to assess the correlation between FDG-PET and changes in patient management (Figure 3.3). Analysis of 22,975 studies revealed that FDG-PET significantly influenced patient management in 36.5% of cases (Hillner, Siegel, et al., "Impact of positron emission tomography," 2008). The authors concluded that FDG-PET was invaluable for diagnosis across all cancer presentations, even for sites traditionally regarded as nonavid for FDG uptake (e.g., prostate; Hillner, Siegel, et al., "Relationship between cancer type and

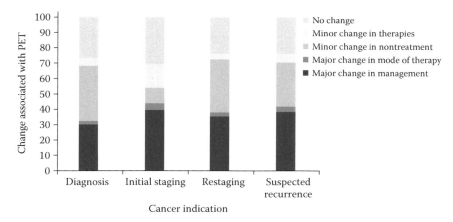

FIGURE 3.3 Change in intended patient management associated with ^{18}F-2-deoxy-2-fluro-D-glucose (FDG)-positron emission tomography (PET) stratified by cancer indication. Note that over 30% of patients experienced major change in management, often related to the identification of more advanced disease, with less than 30% of the patients unaffected. (Reprinted from Hillner, B. E., et al. 2008. *J Clin Oncol* 26(13):2155–61. With permission.)

impact of PET," 2008). In general, benefits of FDG-PET include improved detection of lymph nodes, microscopic infiltration, and distant metastases in many sites (Schwartz et al. 2005).

Although biological imaging might play the most prominent role in diagnostic staging, its prognostic value could also impact patient management. For instance, FMISO imaging has been shown to correlate with radiotherapy outcome in cancers of the head and neck (Rajendran, Hendrickson, et al. 2006; Thorwarth et al. 2006) and the lung (Eschmann et al. 2005). As treatment regimes increase in complexity to include combinations of radiotherapy, chemotherapy, and molecular therapies, imaging will guide patient stratification and treatment selection. In a study of the addition of tirapazamine (Tpz), a potential hypoxic cytotoxin, to conventional chemoradiation, Rischin et al. (2006) investigated the prognostic relationship of hypoxia imaging with [^{18}F]FMISO PET to locoregional failure (Figure 3.4). The authors found that patients with hypoxic tumors had lower rates of locoregional failure when receiving Tpz in addition to chemoradiation, hypothesizing that hypoxia imaging could be used as a predictive marker. Eventually, similar biological imaging techniques may be used to personalize cancer care by selecting therapies that target individual tumor phenotype.

3.3.2 Target Definition

Target definition is at the heart of RT and is a critical step in achieving satisfactory tumor control. To date, most target definition has been based on manual definition of CT volumes due to its high resolution and correlation with electron density. Although MRI is generally unsuitable for treatment planning due to its geometric distortion and nonlinearity with respect to electron density, it is a useful complement to CT in sites susceptible to density artifacts or requiring improved soft-tissue contrast (Khoo et al. 1997). Such sites include brain (Prabhakar et al. 2007), prostate (Debois et al. 1999; Chen et al. 2004), and head and neck (Rasch et al. 1997; Webster et al. 2009).

Biological imaging, especially FDG-PET, is increasingly being used to define tumor volume. In a study by Daisne et al. (2004), gross tumor volumes in patients with pharyngolaryngeal squamous cell carcinoma were defined via FDG-PET, CT, and MRI. When compared with pathological specimens, FDG-PET gross tumor volumes were more accurate than those defined by CT or MRI (Daisne et al. 2004; Grosu, Piert, et al. 2005), although substantial discrepancies existed. Regardless, the hope that FDG-PET imaging might become a new standard for target definition remains high (Gregoire et al. 2007; Ford et al. 2009). On the other hand, there are areas where FDG-PET imaging results in low signal-to-noise ratios (e.g., brain, due to high background uptake of FDG, or prostate, due to low FDG avidity); this encourages the use of alternative imaging techniques. PET tracers such as ^{11}C-acetate, ^{11}C-choline, ^{11}C-methionine, or ^{18}F-tyrosine are surrogates for other characteristics of tumor metabolism (Hara, Kosaka, and Kishi 1998; Weber et al. 2000) that may have clinical benefit (Grosu, Piert, et al. 2005; Grosu, Weber et al. 2005). Functional MR techniques may also aid in target definition in sites such as brain (Nelson et al. 2002; Pirzkall et al. 2004; Chang et al. 2008) and prostate (van Dorsten et al. 2004; Payne and Leach 2006). Although these alternative imaging biomarkers may have site-specific advantages over FDG-PET, they are not standards of care.

Image quantification is of paramount importance for applications such as target definition. Currently, the quantitative uncertainties of biological imaging modalities are poorly characterized and generally prohibitively high (Boellaard 2009). So far, most of the methods have been based on FDG standardized uptake value (SUV) threshold segmentation, either relative (e.g., $SUV_{40\%}$) or absolute (e.g., $SUV_{bm} = 2.5$; Nestle et al. 2005; Greco et al. 2008). Most of the studies, however, neglect the fact that threshold-based methods are highly sensitive to imaging uncertainties, such as PET image acquisition and image reconstruction methods (Figure 3.5), and studies recommending different threshold criteria for specific tumor types continue to appear (Hellwig et al. 2007;

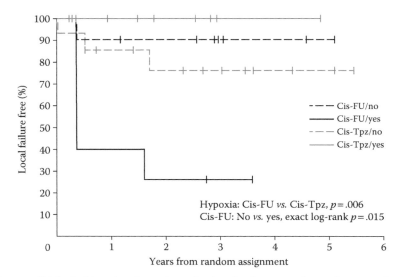

FIGURE 3.4 Example of the use of biological imaging for patient selection. In this particular study, it was shown that patients with hypoxic tumors had lower rates of locoregional failure when receiving tirapazamine (Tpz) in addition to chemoradiation, hypothesizing that hypoxia imaging could be used as a predictive marker. (Reprinted from Rischin, D., et al. 2006. *J Clin Oncol* 24(13):2098–104. With permission.)

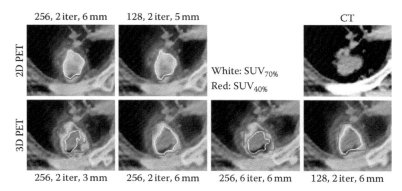

FIGURE 3.5 **(See color insert following page 204.)** Impact of different scan acquisition and image reconstruction parameters on target definition: An example is shown for the same patient with 2D and 3D FLT PET scan acquisition and a different set of reconstruction parameters (labels: reconstruction matrix, number of iterations, and postfilter width). Note the high discrepancies between the SUV contours ($SUV_{40\%}$), which would result in significantly different treatment volumes (volumes from 26 to 32 cc and SUV_{mean} from 3.1 to 4.1). The $SUV_{70\%}$ contours are even more different, indicating large discrepancies if one was to design a biologically conformal treatment plan (volumes from 4 to 11 cc and SUV_{max} from 3.8 to 5.3).

Hoisak et al. 2009; Han et al. 2010). Alternatives to thresholding methods include edge detection, watershed, and Bayesian methods (Geets et al. 2007; Li et al. 2008; Hatt et al. 2009), which can reduce the impact of different uncertainties. However, all of these methods lack rigorous clinical validation and have not been widely adopted.

The use of biological imaging, particularly PET, for target definition has been extended further to the identification of tumor subvolumes that are more or less radioresistant. Selective targeting of such regions with more or less radiation dose would lead to biological dose conformity, a process most often termed *dose painting* (Ling et al. 2000; Bentzen 2005). Most often, the regions considered for selective dose boosting are either highly metabolic

regions (Vanderstraeten et al. 2006; Gregoire and Haustermans 2007) or regions of increased tumor hypoxia (Chao et al. 2001; Rajendran, Schwartz, et al. 2006; Grosu et al. 2007; Thorwarth et al. 2007; Lee et al. 2008). The first clinical study with selective dose escalation to FDG-PET-avid tumor subvolumes has been reported (Madani et al. 2007), however, without much evidence that those are indeed the regions that would require the increased dose.

Dose painting represents a further challenge for imaging as determination of tumor heterogeneities assumes critical importance in this process. The recovery of heterogeneities is dependent on the inherent scanner detector resolution, which leads to partial-volume effects. The magnitude of the quantitative error

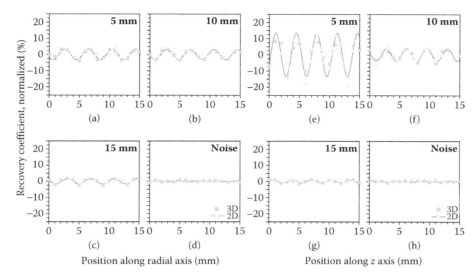

FIGURE 3.6 Partial-volume effects and their variability for a typical PET scanner. Plots show recovery coefficients (normalized by mean) as measured in 2D (□) and 3D (■) PET images of spherical volumes of radioactivity as the spheres were moved in 1-mm intervals along the axes of a PET scanner. Parts a, b, and c show variations in the recovery coefficients of the 5-mm, 10-mm, and 15-mm diameter spheres, at positions along the radial direction. Parts e, f, and g show position-dependent variations along the axial direction. However, parts d and h show the recovery coefficients of a stationary sphere, and are examples of the intrinsic noise in repeated measurements of recovery coefficients.

introduced by partial-volume effects is complex and dependent on object size, contrast, and image resolution. For example, for PET, the most commonly considered modality for dose painting, quantitative error increases as object size decreases, object contrast increases, or as the distribution becomes more nonuniform (sharper gradients occurring over shorter distances), and the error can easily increase beyond 40% even for relatively large heterogeneities of 1 cm. The impact of such imaging uncertainties on dose prescriptions could be prohibitive, especially in the setting of threshold-based prescriptions. Some of these errors can be mitigated by employing partial-volume correction methods (Teo et al. 2007; Kirov, Piao, and Schmidtlein 2008; Barbee et al. 2010). Uncertainties due to patient setup and tumor motion represent another challenge for dose painting target definition. These uncertainties could be seen as a "second-order" partial-volume effect, as they represent uncertainty in recovering heterogeneity due to variable object position with respect to the detector plane or the reconstruction matrix. As seen in Figure 3.6, the peak contrast-recovery occurs at radial and axial positions that coincide with the center of the image voxel, and the minimum contrast-recovery occurs when the object is positioned at the edges of the image voxel. These uncertainties can be substantial, in particular, for small objects where errors larger than 30% can occur due to changing the position of high-contrast spheres by one-half the voxel dimension (McCall et al. 2010).

Even if the technical issues can be successfully resolved, the most important questions remain: Which biological phenotypes should be selectively targeted, and how should they be combined in the overall dose painting strategy? As shown in Figure 3.7, considering the example of two phenotypes (hypoxia and proliferation), one can identify three regions—regions of

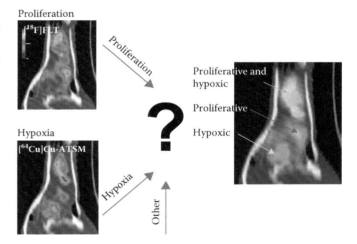

FIGURE 3.7 (See color insert following page 204.) How to combine biological phenotypes. This figure shows how two phenotypes determine three distinctive regions—regions of high hypoxia/low proliferation, low hypoxia/low proliferation, and high hypoxia/high proliferation. Each of these regions will likely respond differently to therapy and would require a different dose painting strategy.

high hypoxia/low proliferation, low hypoxia/low proliferation, and high hypoxia/high proliferation. Likely, each of these regions will respond differently to therapy. If we add other phenotypes, the number of possibilities further increases. The complexity of this problem calls for immense effort in improving understanding of the underlying radiobiology, as well as numerous clinical trials to determine the appropriate response relations.

3.3.3 Treatment Assessment

Assessment of therapy response early, before the onset of clinical symptoms, has been pursued for a long time. Imaging has been used extensively to monitor the effect of therapy on the tumor itself (Figure 3.8). Similarly, imaging can be used to assess treatment response of normal tissues (Jeraj et al. 2010).

For most solid tumors and many systemic diseases, CT is the standard imaging modality used to monitor and quantify treatment response. Objective response criteria were first defined by the World Health Organization (WHO) in 1979 (WHO 1979; Miller et al. 1981), and were subsequently included in Response Evaluation Criteria in Solid Tumors (RECIST) in order to integrate data from three-dimensional (3D) imaging technologies like CT and MR into the response guidelines. Recently, the RECIST guidelines were updated (RECIST 1.1); but the basic assessment methodology remains unchanged. Although RECIST 1.1 does allow for PET as an adjunct for the determination of progression, the report notes that the use of PET for response assessment "requires appropriate and rigorous clinical validation studies."

In the decade since the adoption and implementation of RECIST, various weaknesses and limitations of this response assessment methodology have come to light (Thiesse et al. 1997; Gwyther et al. 1999; Stroobants et al. 2003; Avril and Weber 2005; Jaffe 2006; Biganzoli et al. 2007) often resulting in a weak correlation between RECIST tumor response and patient outcome (Avril and Weber 2005; Schuetze et al. 2005). PET has been seen as one of the prime candidates to fill this gap. A remarkable 2003 (Stroobants et al. 2003) study examined the value of FDG-PET in the assessment of early tumor response in patients with gastrointestinal stromal tumors (soft-tissue sarcomas) treated with imatinib mesylate (Gleevec). PET response was strongly associated with a longer progression-free survival (92% vs. 12% after 1 year), closely correlated with subjective symptom control (Stroobants et al. 2003). In this particular case, the RECIST criteria were found to be of little use or even potentially misleading. Several other studies have shown the higher sensitivity and specificity of PET compared to CT for different sites and treatment regimens (Jerusalem et al. 1999; de Wit et al. 2001; Hueltenschmidt et al. 2001; Spaepen et al. 2001; Antoch et al. 2004; Hutchings et al. 2006).

In 1999, the European Organization for Research and Treatment of Cancer (EORTC) proposed initial recommendations for common measurement standards and criteria for the use of FDG-PET to assess treatment response. Note, however, that no recommendations for the use of FDG-PET in RT were provided because of the "lack of clinical data and potential confounding effects of radiation-induced inflammatory reactions." In 2005, the Cancer Imaging Program of the National Cancer Institute (NCI) recommended procedures for the analysis of PET scans acquired in diagnostic and therapeutic clinical trials. Again, they noted that "timing of scans after changes due to radiotherapy needs further investigation." Recently, a novel framework for PET Response Criteria in Solid Tumors (PERCIST) was proposed. The PERCIST guidelines propose significant changes from and additions to the EORTC recommendations in order to improve response assessment using PET. One of these changes includes the recommendation to use SUV_{peak} (rather than SUV_{mean} or SUV_{max}) values. The other important difference is PERCIST suggests that unique response thresholds for specific diseases and specific treatments should be established. In a large multicenter trial, Jacene et al. (2009) found that SUV_{max} had higher interobserver reproducibility than CT size, which supports PERCIST recommendations that SUV_{max} or SUV_{peak} should be implemented for treatment assessment.

FDG PET is the most commonly used imaging biomarker for treatment response, and its clinical implementation has been reviewed for many cancer presentations (Weber 2009). Although FDG-PET is widely used in the clinic to assess treatment response, it has certain disadvantages, including high avidity in areas of posttreatment inflammation and fibrosis and low avidity in tumors such as prostate. As a result, suggested time points for FDG follow-up scans are as late as 10–12 weeks following radiotherapy, although there is very little evidence that earlier time points would not be predictive of clinical response. The majority of the studies so far have investigated late response to RT and chemoRT. The results are mixed, ranging from excellent sensitivity, specificity, and accuracy of FDG-PET response for predicting pathologic response in NSCLC (90% sensitivity, 100% specificity, and 96% accuracy; Cerfolio et al. 2004) to mixed results in HNSCC, where postchemoRT seems to have a high negative predictive value (95%) but a low positive predictive value (50%; Schoder et al. 2009), and esophageal cancer, where in adenocarcinomas a negative FDG-PET postchemoRT has a high positive predictive value, whereas the response in other tumor types was inconclusive (Krause et al. 2009). The heterogeneity of the response indicates that FDG-PET has to be used carefully in the context of treatment response evaluation and that more specific studies are necessary to determine its overall value.

Much more significant than late treatment response assessment is early treatment response assessment, as the latter potentially provides means to alter the course of treatment. Such studies are rare at present time. In an interesting study

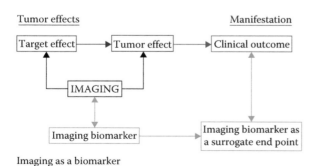

FIGURE 3.8 Role of imaging in treatment response assessment. Biological imaging can be used as a biomarker to probe the effects of therapy on the treatment target. If an imaging biomarker is related to the clinical outcome, it can become a surrogate end point.

by van Baardwijk et al. (2007) on a cohort of 22 patients with NSCLC, a large intraindividual heterogeneity in the evolution of SUV_{max} was noted. A nonsignificant increase in the first week, followed by a decrease in the second week, which persisted after radiotherapy, was observed. Different time trends for metabolic responders (no change during radiotherapy) and metabolic nonresponders (48% increase during the first week and 15% decrease in the second week) were determined (Figure 3.9).

Focusing on a later response to RT, Kong et al. (2007) found that the FDG-PET response after approximately 45 Gy correlated well with the FDG-PET activity and overall response 3 months after RT in a cohort of 15 patients with NSCLC. Although these studies indicate that there is a potential for using FDG-PET during RT, the confounding effects of post-RT inflammatory response need to be investigated further before determining its absolute value. On the other hand, FLT-PET, a surrogate for cellular proliferation, may be superior to FDG-PET. Yang et al. (2006) reported that C_3H/HeN mice bearing murine squamous cell carcinomas showed decreased FLT uptake 24 hours after receiving radiation, which corresponded with decreased fraction of cells in S phase, higher apoptosis, and reduced clonogenic cell survival. FDG-PET uptake was not significantly changed at the same time point, leading the authors to suggest that changes in cellular proliferation precede changes in cellular metabolism after RT (Yang et al. 2006). Recently, Menda et al. (2009) evaluated FLT-PET uptake of patients with HNSCC at baseline and after 10 Gy of radiotherapy and found significant decreases in FLT uptake as evaluated using compartmental kinetic analysis

and SUV measurement. The main disadvantages of FLT-PET for treatment assessment are its "investigatory new drug" Food and Drug Administration (FDA) status, limiting its use to research environments, and concerns that FLT cannot distinguish between reactive and metastatic lymph nodes due to FLT avidity for B lymphocytes, causing false-positive results (Troost et al. 2007). Clinical trials are ongoing, but the correlation of FLT with established immunohistochemical techniques such as Ki-67 scoring and staining of iododeoxyuridine and bromodeoxyuridine (Valente et al. 1994; Zackrisson et al. 2002) suggests it will be a valuable marker for radiotherapy outcome.

A variety of literature exists on the assessment of radiation response through perfusion imaging using dynamic CT, MR, or PET techniques. Although results should theoretically be similar between modalities, the lack of standardized imaging protocols and pharmacokinetic models makes comparisons difficult. It has been shown that DCE-CT predicts radiotherapy outcome in head and neck cancers. Hermans et al. (2003) reported that low rates of perfusion indicated poor outcome in a group of 105 patients, and Bisdas et al. (2009) reported similar changes in several related parameters in a smaller group, although differences in implementation prevent any direct comparison of the studies. Perfusion imaging through DCE-MRI has been shown to predict response in a variety of cancer presentations, although quantitative analyses are somewhat rare. In a review of 29 DCE-MRI studies, Zahra et al. (2007) indicate that high tumor enhancement at early time points and low tumor enhancement at follow-up time points are associated with local control in multiple sites.

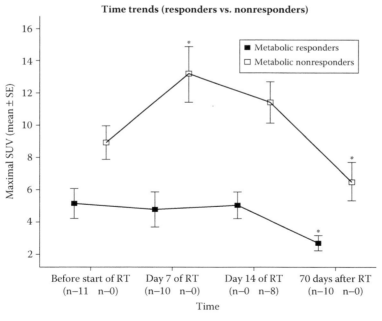

FIGURE 3.9 Difference in the FDG PET response between metabolic responders and nonresponders. The response trends between metabolic responders and nonresponders were significantly different during the course of RT. Whereas metabolic responders did not show a significant metabolic response during the course of radiotherapy, the response of metabolic nonresponders was significantly different. This persistence of FDG uptake underlines the problem of confounding inflammatory response to RT, which needs to be considered in the design of the treatment response assessment trials. (Reprinted from van Baardwijk, A., et al. 2007. *Radiother Oncol* 82(2):145–52. With permission.)

Change in K_{trans}, in particular, may indicate tumor response to chemoradiotherapy in colorectal (George et al. 2001) and cervical cancers (Zahra et al. 2009), and pretreatment cerebral blood volume was predictive of survival in glioma patients receiving radiotherapy (Cao et al. 2006). In contrast to the DCE-CT and DCE-MRI results, Lehtiö et al. utilized ^{15}O-PET in HNSCC patients and found that high blood flow was associated with poor local control following RT (Lehtiö et al. 2004). Although the aforementioned perfusion techniques are based on similar pharmacokinetic models, the quantitative impact of diverse imaging protocols, contrast agents, injection protocols, mathematical implementations, and physical properties of imaging modalities make comparisons between trials difficult. Two studies conducted on the interchangeability of DCE-CT and DCE-MRI in rectal and squamous cell carcinoma using proprietary analysis methods concluded that the two modalities deliver similar results (Bisdas, Medov, et al. 2008; Kierkels et al. 2009). However, going forward, there is a clear need for standardized protocols for imaging and analysis, validation of kinetic models, and further intermodality comparisons.

The MRS technique has shown promise in radiotherapy response assessment. When prostate MRS was performed after RT to identify residual malignancy compared with biopsy, sensitivities and specificities of 89% and 92% (Menard et al. 2001) and 89% and 82% (Coakley et al. 2004) were reported. Other promising results for radiotherapy assessment have been published for cancers of the breast (Merchant et al. 1995; Leach et al. 1998) and brain (Zeng et al. 2007; Laprie et al. 2008). Although MRS is undoubtedly a powerful tool, more work is required for its quantitative assessment, validation, and standardization. Additionally, although MRS allows measurement of multiple chemical signals, PET assessment of individually radiolabeled metabolites (e.g., methionine) may be preferable in some cases.

Another potential application of biological imaging in radiotherapy assessment is in normal tissue assessment. Many FDG-PET studies have been used to correlate clinical symptoms of radiation pneumonitis with metabolic uptake; they indicate strong patient variability in biological response to radiation (Hassaballa et al. 2005; Guerrero et al. 2007; Hart et al. 2008). Similarly, MRI studies demonstrate changes in kinetics of gadolinium enhancement (Muryama et al. 2004) or hyperpolarized ^{3}H (Ireland et al. 2007), which can be associated with different phases of radiation pneumonitis. Although SPECT lung perfusion imaging has been used to quantitatively relate changes in regional perfusion/ventilation (e.g., function) to the regional radiation dose, the association between the sum of these regional injuries (i.e., the integrated response) is not that well-correlated with corresponding changes in global lung function (e.g., as assessed by pulmonary function tests; Fan et al. 2001; Seppenwoolde et al. 2004; Zhang et al. 2004). Cardiac functional imaging may allow early detection of treatment-associated dysfunction, which is important as, in general, these changes do not manifest clinically for at least 10 years after treatment. Available data using SPECT myocardial perfusion imaging demonstrate perfusion defects in the irradiated left ventricle, which can be associated with wall motion abnormalities and reduced ejection fraction in patients treated for breast cancer (Darby et al. 2005; Marks et al. 2005). Available data on the functional imaging of other organs such as liver (Cao et al. 2007; Cao et al. 2008), brain (Zeng et al. 2007), prostate (Ireland et al. 2007), and bone marrow (Everitt et al. 2009) reveal correlations of changes measurable by a variety of functional imaging modalities with radiation dose.

Although most of the studies focus on investigating a response of a single biological phenotype to RT, the biology of the response is much more complex. This is further complicated for modern therapies, where RT is often combined with chemotherapy and targeted molecular therapies. An example of the complexity of PET response to a therapy regimen combining an antiangiogenic therapy (bevacizumab) with chemoradiation therapy is shown in Figure 3.10. In another study, Willett et al. (2009) utilized DCE-CT to evaluate the addition of bevacizumab to standard chemoradiotherapy regimens in patients with localized rectal cancer. Blood flow and permeability surface-area product decreased after bevacizumab monotherapy, and blood flow, blood volume, median transit time, and permeability surface decreased relative to baseline at the conclusion of chemoradiotherapy (Willett et al. 2009). Many more studies like this are needed to finally uncover the complexity of treatment response and get closer to the goal of personalized approach to treatment.

3.4 Conclusions

Biological imaging is quickly becoming an integral part of radiation oncology practice. As the flagship biological imaging method, FDG-PET has proven indispensable in tumor diagnosis and staging. Although it will continue to play the central role in cancer management, other biological imaging methods that target general tumor phenotypes like hypoxia, cellular proliferation, angiogenesis, or even more specific cellular processes will come to play increasingly important roles in future. Their roles will be multifaceted—to design the most optimal treatment strategy, better define the target volume, or estimate the expected treatment outcome. As the complexity of treatment management, which often includes chemotherapy or molecular targeted therapies in addition to RT, increases, the appropriate selection and use of biological imaging will become essential. The path to optimal selection is not straightforward, as it requires a comprehensive clinical validation of the particular imaging modality in addition to its biological validation. Because tumor biology is incredibly complex and biological imaging surrogates are imperfect, it is unreasonable to expect that such validations will be perfect. One of the most detrimental aspects of surrogacy is the limited quantitative accuracy of biological imaging modalities. Understanding these limitations is of paramount importance, although they are often neglected in the desire to push the field forward.

Target definition is at the heart of the RT process and will benefit significantly from expanded incorporation of biological

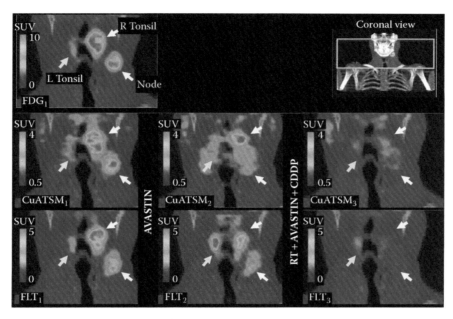

FIGURE 3.10 **(See color insert following page 204.)** Complexity of the positron emission tomography response to a combined radiation therapy/molecular targeted therapy. The behavior and magnitude of the response is variable and might depend on the pretreatment status of different biological phenotypes. For example, note that hypoxia (as indicated by the Cu-diacetyl-bis(N4-methylthiosemicarbazone) uptake) is reduced after avastin therapy in the nodal volume but remains high in the primary tumor.

imaging information. The efforts to incorporate biological imaging are concentrated in two areas: (1) to better define the tumor volume boundaries; and (2) to identify biological tumor subvolumes that would benefit from selective dose targeting or sparing, the process most often termed dose painting. Many methods utilize the established molecular imaging modalities without much consideration of inherent imaging uncertainties. Neglecting such uncertainties can easily lead to inappropriate attribution of imaging artifacts to biological phenomena, further complicating the already complex understanding of tumor biology. With significant efforts in the field of image quantification, one can be hopeful that quantitative imaging uncertainties will soon be better understood and kept under control, so that biological imaging will be able to show its full potential.

Treatment response studies have been primarily aimed at establishing biological imaging as biomarkers and surrogate end points of clinical outcomes. Although this is a worthwhile endeavor, treatment assessment also offers possibilities for treatment adaptation, particularly if done early during treatment. In order to pursue this goal, one has to fully understand the dynamics of treatment response, as well as the complexity and interaction of different biological phenotypes. This is becoming even more important as RTs are becoming more complex, often combined with either chemotherapy or molecular targeted therapies. Only when we understand the complexity of treatment response, allowing for better treatment design and potential treatment adaptation, can we fully explore the full potential of treatment assessment strategies.

References

Antoch, G., J. Kanja, S. Bauer, et al. 2004. Comparison of PET, CT, and dual-modality PET/CT imaging for monitoring of imatinib (STI571) therapy in patients with gastrointestinal stromal tumors. *J Nucl Med* 45(3):357–65.

Apisarnthanarax, S., and K. S. Chao. 2005. Current imaging paradigms in radiation oncology. *Radiat Res* 163(1):1–25.

Avril, N. 2004. GLUT1 Expression in tissue and 18F-FDG uptake. *J Nucl Med* 45(6):930–2.

Avril, N. E., and W. A. Weber. 2005. Monitoring response to treatment in patients utilizing PET. *Radiol Clin North Am* 43(1):189–204.

Baker, H. L. 1975. The impact of computed tomography on neuroradiologic practice. *Radiology* 116(3):637–40.

Barbee, D., R. Flynn, J. E. Holden, R. J. Nickles, and R. Jeraj. 2010. A method for partial volume correction of PET-imaged tumor heterogeneity using expectation maximization with a spatially varying point spread function. *Phys Med Biol* 55(1):221–36.

Becherer, A., G. Karanikas, M. Szabo, et al. 2003. Brain tumour imaging with PET: A comparison between [18F]fluoro-dopa and [11C]methionine. *Eur J Nuc Med Mol Imaging* 30(11):1561–7.

Beer, A. J., A. L. Grosu, J. Carlsen, et al. 2007. [18F]galacto-RGD positron emission tomography for imaging of {alpha}v{beta}3 expression on the neovasculature in patients with squamous cell carcinoma of the head and neck. *Clin Cancer Res* 13(22):6610–6.

Beer, A. J., R. Haubner, M. Goebel, et al. 2005. Biodistribution and pharmacokinetics of the {alpha}v{beta}3-selective tracer 18F-galacto-RGD in cancer patients. *J Nucl Med* 46(8):1333–41.

Bentzen, L., S. Keiding, M. R. Horsman, T. Gronroos, S. B. Hansen, and J. Overgaard. 2002. Assessment of hypoxia in experimental mice tumours by [18F]fluoromisonidazole PET and pO2 electrode measurements. Influence of tumour volume and carbogen breathing. *Acta Oncol* 41(3):304–12.

Bentzen, L., S. Keiding, M. Nordsmark, et al. 2003. Tumour oxygenation assessed by 18F-fluoromisonidazole PET and polarographic needle electrodes in human soft tissue tumours. *Radiother Oncol* 67(3):339–44.

Bentzen, S. M. 2005. Theragnostic imaging for radiation oncology: Dose-painting by numbers. *Lancet Oncol* 6(2):112–7.

Biganzoli, L., R. Coleman, A. Minisini, et al. 2007. A joined analysis of two European Organization for the Research and Treatment of Cancer (EORTC) studies to evaluate the role of pegylated liposomal doxorubicin (Caelyx(TM)) in the treatment of elderly patients with metastatic breast cancer. *Crit Rev Oncol Hematol* 61(1):84–9.

Bisdas, S., L. Medov, M. Baghi, et al. 2008. A comparison of tumor perfusion assessed by deconvolution-based analysis of dynamic contrast-enhanced CT and MR imaging in patients with squamous cell carcinoma of the upper aerodigestive tract. *Eur Radiol* 18(4):843–50.

Bisdas, S., S. A. Nguyen, S. K. Anand, G. Glavina, T. Day, and Z. Rumbolt. 2009. Outcome prediction after surgery and chemoradiation of squamous cell carcinoma in the oral cavity, oropharynx, and hypopharynx: Use of baseline perfusion CT microcirculatory parameters vs. tumor volume. *Int J Radiat Oncol Biol Phys* 73(5):1313–8.

Bisdas, S., K. Spicer, Z. Rumbolt, et al. 2008. Whole-tumor perfusion CT parameters and glucose metabolism measurements in head and neck squamous cell carcinomas: A pilot study using combined positron-emission tomography/CT imaging. *Am J Neuroradiol* 29(7):1376–81.

Boellaard, R. 2009. Standards for PET image acquisition and quantitative data analysis. *J Nucl Med* 50(S1):11S–20.

Boellaard, R., A. van Lingen, and A. A. Lammertsma. 2001. Experimental and clinical evaluation of iterative reconstruction (OSEM) in dynamic PET: Quantitative characteristics and effects on kinetic modeling. *J Nucl Med* 42(5):808–17.

Bos, R., J. J. van der Hoeven, E. van Der Wall, et al. 2002. Biologic correlates of (18)fluorodeoxyglucose uptake in human breast cancer measured by positron emission tomography. *J Clin Oncol* 20(2):379–87.

Buck, A. K., G. Halter, H. Schirrmeister, et al. 2003. Imaging proliferation in lung tumors with PET: 18F-FLT versus 18F-FDG. *J Nucl Med* 44(9):1426–31.

Bussink, J., J. H. Kaanders, A. J. van der Kogel, et al. 2003. Tumor hypoxia at the micro-regional level: Clinical relevance and predictive value of exogenous and endogenous hypoxic cell markers. *Radiother Oncol* 67:3–15.

Cai, W., and X. Chen. 2008. Multimodality molecular imaging of tumor angiogenesis. *J Nucl Med* 49(S2):113S–128.

Cai, W., K. Chen, K. A. Mohamedali, et al. 2006. PET of vascular endothelial growth factor receptor expression. *J Nucl Med* 47(12):2048–56.

Cao, Y., C. Pan, J. M. Balter, et al. 2008. Liver function after irradiation based on computed tomographic portal vein perfusion imaging. *Int J Radiat Oncol Biol Phys* 70(1):154–60.

Cao, Y., J. F. Platt, I. R. Francis, et al. 2007. The prediction of radiation-induced liver dysfunction using a local dose and regional venous perfusion model. *Med Phys* 34(2):604–12.

Cao, Y., C. I. Tsien, V. Nagesh, et al. 2006. Clinical investigation survival prediction in high-grade gliomas by MRI perfusion before and during early stage of RT. *Int J Radiat Oncol Biol Phys* 64(3):876–85.

Cerfolio, R. J., A. S. Bryant, T. S. Winkur, B. Ohja, and A. A. Bartolucci. 2004. Repeat FDG-PET after neoadjuvant therapy is a predictor of pathologic response in patients with non-small cell lung cancer. *Ann Thorac Surg* 78(6):1903–09.

Chang, J., S. B. Thankur, W. Huang, and A. Narayana. 2008. Magnetic resonance spectroscopy imaging (MRSI) and brain functional magnetic resonance imaging (fMRI) for radiotherapy treatment planning of glioma. *Technol Cancer Res and Treat* 7(5):349–62.

Chao, K. S., W. R. Bosch, S. Mutic, et al. 2001. A novel approach to overcome hypoxic tumor resistance: Cu-ATSM-guided intensity-modulated radiation therapy. *Int J Radiat Oncol Biol Phys* 49(4):1171–82.

Chen, L., R. A. Price Jr, T. B. Nguyen, et al. 2004. Dosimetric evaluation of MRI-based treatment planning for prostate cancer. *Phys Med Biol* 49(22):5157–70.

Coakley, F. V., H. S. Teh, A. Qayyum, et al. 2004. Endorectal MR imaging and MR spectroscopic imaging for locally recurrent prostate cancer after external beam radiation therapy: Preliminary experience. *Radiology* 233(2):441–8.

Cook, G. J. 2003. Oncological molecular imaging: Nuclear medicine techniques. *Br J Radiol* 76 Spec No 2:S152–8.

Crooks, L. E., D. A. Ortendahl, L. Kaufman, et al. 1983. Clinical efficiency of nuclear magnetic resonance imaging. *Radiology* 146(1):123–8.

Cuenod, C. A., L. Fournier, D. Balvay, and J. M. Guinebretiere. 2006. Tumor angiogenesis: Pathophysiology and implications for contrast-enhanced MRI and CT assessment. *Abdom Imaging* 31(2):188–93.

Daisne, J. F., T. Duprez, B. Weynand, et al. 2004. Tumor volume in pharyngolaryngeal squamous cell carcinoma: Comparison at CT, MR imaging, and FDG PET and validation with surgical specimen. *Radiology* 233(1):93–100.

Darby, S. C., P. McGale, C. W. Taylor, and R. Peto. 2005. Long-term mortality from heart disease and lung cancer after radiotherapy for early breast cancer: Prospective cohort study of about 300,000 women in US SEER cancer registries. *Lancet Oncol* 6(8):557–65.

Debois, M., R. Oyen, F. Maes, et al. 1999. The contribution of magnetic resonance imaging to the three-dimensional treatment planning of localized prostate cancer. *Int J Radiat Oncol Biol Phys* 45(4):857–65.

de Geus-Oei, L. F., J. H. van Krieken, R. P. Aliredjo, et al. 2007. Biological correlates of FDG uptake in non-small cell lung cancer. *Lung Cancer* 55(1):79–87.

Dehdashti, F., P. W. Grigsby, M. A. Mintun, J. S. Lewis, B. A. Siegel, and M. J. Welch. 2003. Assessing tumor hypoxia in cervical cancer by positron emission tomography with 60Cu-ATSM: Relationship to therapeutic response-a preliminary report. *Int J Radiat Oncol Biol Phys* 55(5):1233–8.

Dehdashti, F., M. A. Mintun, J. S. Lewis, et al. 2003. In vivo assessment of tumor hypoxia in lung cancer with 60Cu-ATSM. *Eur J Nucl Med Mol Imaging* 30(6):844–50.

de Wit, M., K. H. Bohuslavizki, R. Buchert, D. Bumann, M. Clausen, and D. K. Hossfeld. 2001. 18FDG-PET following treatment as valid predictor for disease-free survival in Hodgkin's lymphoma. *Ann Oncol* 12(1):29–37.

Eschmann, S. M., F. Paulsen, M. Reimold, et al. 2005. Prognostic impact of hypoxia imaging with 18F-Misonidazole PET in non-small cell lung cancer and head and neck cancer before radiotherapy. *J Nucl Med* 46(2):253–60.

Etchebehere, E. C., H. A. Macapinlac, M. Gonen, et al. 2002. Qualitative and quantitative comparison between images obtained with filtered back projection and iterative reconstruction in prostate cancer lesions of (18)F-FDG PET. *Q J Nucl Med* 46(2):122–30.

Everitt, S., R. J. Hicks, D. Ball, et al. 2009. Imaging cellular proliferation during chemo-radiotherapy: A pilot study of serial 18F-FLT positron emission tomography/computed tomography imaging for non-small-cell lung cancer. *Int J Radiat Oncol Biol Phys* 75(4):1098–104.

Fan, M., L. B. Marks, P. Lind, et al. 2001. Relating radiation-induced regional lung injury to changes in pulmonary function tests. *Int J Radiat Oncol Biol Phys* 51(2):311–7.

Ford, E. C., J. Herman, E. Yorke, and R. L. Wahl. 2009. 18F-FDG PET/CT for image-guided and intensity-modulated radiotherapy. *J Nucl Med* 50(10):1655–65.

Fujibayashi, Y., H. Taniuchi, Y. Yonekura, H. Ohtani, J. Konishi, and A. Yokoyama. 1997. Copper-62-ATSM: A new hypoxia imaging agent with high membrane permeability and low redox potential. *J Nucl Med* 38(7):1155–60.

Gagel, B., P. Reinartz, E. Dimartino, et al. 2004. pO2 Polarography versus positron emission tomography ([18F] fluoromisonidazole, [18F]-2-fluoro-2'-deoxyglucose). An appraisal of radiotherapeutically relevant hypoxia. *Strahlenther Onkol* 180(10):616–22.

Geets, X., J. A. Lee, A. Bol, M. Lonneux, and V. Gregoire. 2007. A gradient-based method for segmenting FDG-PET images: Methodology and validation. *Eur J Nucl Med Mol Imaging* 34(9):1427–38.

George, M. L., A. S. Dzik-Jurasz, A. R. Padhani, et al. 2001. Non-invasive methods of assessing angiogenesis and their value in predicting response to treatment in colorectal cancer. *Br J Surg* 88(12):1628–36.

Greco, C., S. A. Nehmeh, H. Schoder, et al. 2008. Evaluation of different methods of 18F-FDG-PET target volume delineation in the radiotherapy of head and neck cancer. *A J Clin Oncol* 31(5):439–45.

Gregoire, V., K. Haustermans, et al. 2007. PET-based treatment planning in radiotherapy: A new standard? *J Nucl Med* 48(Suppl 1): S68–77.

Grierson, J. R., and A. F. Shields. 2000. Radiosynthesis of 3'-deoxy-3'-[(18)F]fluorothymidine: [(18)F]FLT for imaging of cellular proliferation in vivo. *Nucl Med Biol* 27(2):143–56.

Grosu, A. L., M. Piert, W. A. Weber, et al. 2005. Positron emission tomography for radiation treatment planning. *Strahlenther Onkol* 181(8):483–99.

Grosu, A. L., M. Souvatzoglou, B. Roper, et al. 2007. Hypoxia imaging with FAZA-PET and theoretical considerations with regard to dose painting for individualization of radiotherapy in patients with head and neck cancer. *Int J Radiat Oncol Biol Phys* 69(2):541–51.

Grosu, A. L., W. A. Weber, M. Franz, et al. 2005. Reirradiation of recurrent high-grade gliomas using amino acid PET (SPECT)/CT/MRI image fusion to determine gross tumor volume for stereotactic fractionated radiotherapy. *Int J Radiat Oncol Biol Phys* 63(2):511–9.

Guerrero, T., V. Johnson, J. Hart, et al. 2007. Radiation pneumonitis: Local dose versus [18F]-Fluorodeoxyglucose uptake response in irradiated lung. *Int J Radiat Oncol Biol Phys* 68(4):1030–5.

Gwyther, S. J., M. S. Aapro, S. R. Hatty, P. E. Postmus, and I. E. Smith. 1999. Results of an independent oncology review board of pivotal clinical trials of gemcitabine in non-small cell lung cancer. *Anticancer Drugs* 10(8):693–8.

Haberkorn, U., and A.Altmann. 2003. *Functional Genomics and Radioisotope-Based Imaging Procedures: Molecular Diagnosis and Therapy of Cancer.* Basingstoke, ROYAUME-UNI. Boca Raton, FL: Taylor and Francis.

Haberkorn, U., L. G. Strauss, C. Reisser, et al. 1991. Glucose uptake, perfusion, and cell proliferation in head and neck tumors: Relation of positron emission tomography to flow cytometry. *J Nucl Med* 32(8):1548–55.

Han, D., J. Yu, Y. Yu, et al. 2010. Comparison of 18F-fluorothymidine and 18F-fluorodeoxyglucose PET/CT in delineating gross tumor volume by optimal threshold in patients with squamous cell carcinoma of thoracic esophagus. *Int J Radiat Oncol Biol Phys* 76(4):1235–41.

Hara, T., N. Kosaka, and H. Kishi. 1998. PET imaging of prostate cancer using carbon-11-choline. *J Nucl Med* 39(6):990–5.

Hart, J., M. R. McCurdy, M. Ezhil, W. Wei, M. Khan, D. Luo, R. F. Munden, V. E. Johnson, and T. M. Guerrero. 2008. Radiation pneumonitis: Correlation of toxicity with pulmonary metabolic radiation response. *Int J Radiat Oncol Biol Phys* 71(4):967–71.

Hassaballa, H. A., E. S. Cohen, A. J. Khan, A. Ali, P. Bonomi, and D. B. Rubin. 2005. Positron emission tomography demonstrates radiation-induced changes to nonirradiated lungs in lung cancer patients treated with radiation and chemotherapy. *Chest* 128(3):1448–52.

Hatt, M., C. Cheze le Rest, A. Turzo, C. Roux, and D. Visvikis. 2009. A fuzzy locally adaptive Bayesian segmentation approach for volume determination in PET. *IEEE Trans Med Imaging* 28(6):881–93.

Haubner, R. 2006. Alphavbeta3-integrin imaging: A new approach to characterise angiogenesis? *Eur J Nuc Med Mol Imaging* 33(Suppl 1):54–63.

Hawkes, R. C., G. N. Holland, W. S. Moore, and B. S. Worthington. 1980. Nuclear magnetic resonance (NMR) tomography of the brain: A preliminary clinical assessment with demonstration of pathology. *J Comput Assist Tomogr* 4(5):577–86.

Hawkins, R. A., Y. Choi, S. C. Huang, et al. 1992. Evaluation of the skeletal kinetics of fluorine-18-fluoride ion with PET. *J Nucl Med* 33(5):633–42.

Hellwig, D., T. P. Graeter, D. Ukena, et al. 2007. 18F-FDG PET for mediastinal staging of lung cancer: Which SUV threshold makes sense? *J Nucl Med* 48(11):1761–6.

Hermans, R. P., P. Lambin, A. Van der Goten, et al. 1999. Tumoural perfusion as measured by dynamic computed tomography in head and neck carcinoma. *Radiother Oncol* 53(2):105–11.

Hermans, R. M., M. Meijerink, W. Van den Bogaert, A. Rijnders, C. Weltens, and P. Lambin. 2003. Tumor perfusion rate determined noninvasively by dynamic computed tomography predicts outcome in head-and-neck cancer after radiotherapy. *Int J of Radiat Oncol Biol Phys* 57(5):1351–6.

Higashi, K., A. C. Clavo, and R. L. Wahl. 1993. Does FDG uptake measure proliferative activity of human cancer cells? In vitro comparison with DNA flow cytometry and tritiated thymidine uptake. *J Nucl Med* 34(3):414–9.

Hillner, B. E., B. A. Siegel, D. Liu, et al. 2008. Impact of positron emission tomography/computed tomography and positron emission tomography (PET) alone on expected management of patients with cancer: Initial results from the National Oncologic PET Registry. *J Clin Oncol* 26(13):2155–61.

Hillner, B. E., B. A. Siegel, A. F. Shields, et al. 2008. Relationship between cancer type and impact of PET and PET/CT on intended management: Findings of the National Oncologic PET Registry. *J Nucl Med* 49(12):1928–35.

Höckel, M., K. Schlenger, B. Aral, M. Mitze, U. Schaffer, and P. Vaupel. 1996. Association between tumor hypoxia and malignant progression in advanced cancer of the uterine cervix. *Cancer Res* 56(19):4509–15.

Hoffmann, E. J., M. E. Phelps, N. A. Mullani, C. S. Higgins, and M. M. Ter-Pogossian. 1976. Design and performance characteristics of a whole-body positron transaxial tomograph. *J Nucl Med* 17(6):493–502.

Hoisak, J., H. Keller, et al. 2009. TH-D-213A-01: An evaluation of FDG-PET uptake thresholds for head & neck target definition based on local regions of high inter-observer concordance. *Med Phys* 36:2820.

Hoult, D. I., S. J. Busby, D. G. Gadian, G. K. Radda, R. E. Richards, and P. J. Seeley. 1974. Observation of tissue metabolites using 31P nuclear magnetic resonance. *Nature* 252(5481):285–7.

Hueltenschmidt, B., M. L. Sautter-Bihl, O. Lang, et al. 2001. Whole body positron emission tomography in the treatment of Hodgkin disease. *Cancer* 91(2):302–10.

Hutchings, M., A. Loft, M. Hansen, et al. 2006. FDG-PET after two cycles of chemotherapy predicts treatment failure and progression-free survival in Hodgkin lymphoma. *Blood* 107(1):52–9.

Ireland, R. H., C. M. Bragg, M. McJury, et al. 2007. Feasibility of image registration and intensity-modulated radiotherapy planning with hyperpolarized helium-3 magnetic resonance imaging for non-small-cell lung cancer. *Int J Radiat Oncol Biol Phys* 68(1):273–81.

Jacene, H. A., S. Leboulleux, S. Baba, et al. 2009. Assessment of interobserver reproducibility in quantitative 18F-FDG PET and CT measurements of tumor response to therapy. *J Nucl Med* 50(11):1760–9.

Jaffe, C. C. 2006. Measures of response: RECIST, WHO, and new alternatives. *J Clin Oncol* 24(20):3245–51.

Jaskowiak, C. J., J. A. Bianco, S. B. Perlman, and J. P. Fine. 2005. Influence of reconstruction iterations on 18F-FDG PET/CT standardized uptake values. *J Nucl Med* 46(3):424–8.

Jeraj, R., Y. Cao, R. K. Ten Haken, C. Han, and L. Marks. 2010. Imaging for assessment of radiation-induced normal tissue effects. *Int J Radiat Oncol Biol Phys* 76(3, S1):S140–4.

Jerusalem, G., Y. Beguin, M. F. Fassotte, et al. 1999. Whole-body positron emission tomography using 18F-fluorodeoxyglucose for posttreatment evaluation in Hodgkin's disease and non-Hodgkin's lymphoma has higher diagnostic and prognostic value than classical computed tomography scan imaging. *Blood* 94(2):429–33.

Khoo, V. S., D. P. Dearnaley, D. J. Finnigan, A. Padhani, S. F. Tanner, and M. O. Leach. 1997. Magnetic resonance imaging (MRI): considerations and applications in radiotherapy treatment planning. *Radiother Oncol* 42(1):1–15.

Kierkels, R. G., W. H. Backes, M. H. Janseen, et al. 2009. Comparison between perfusion computed tomography and dynamic contrast-enhanced magnetic resonance imaging in rectal cancer. *Int J Radiat Oncol Biol Phys* 77(2):400–8.

Kirov, A., J. Z. Piao, and C. R. Schmidtlein. 2008. Partial volume effect correction in PET using regularized iterative deconvolution with variance control based on local topology. *Phys Med Biol* 53(10):2577.

Koh, W. J., K. S. Bergman, J. S. Rasey, et al. 1995. Evaluation of oxygenation status during fractionated radiotherapy in human non-small cell lung cancers using [F-18]fluoromisonidazole positron emission tomography. *Int J Radiat Oncol Biol Phys* 33(2):391–8.

Koh, W. J., J. S. Rasey, M. L. Evans, et al. 1992. Imaging of hypoxia in human tumors with [F-18]fluoromisonidazole. *Int J Radiat Oncol Biol Phys* 22(1):199–212.

Kong, F. M., K. A. Frey, L. E. Qunt, et al. 2007. A pilot study of [18F] fluorodeoxyglucose positron emission tomography scans during and after radiation-based therapy in patients with non small-cell lung cancer. *J Clin Oncol* 25(21):3116–23.

Krause, B. J., K. Herrmann, H. Wieder, and C. M. zum Buschenfelde. 2009. 18F-FDG PET and 18F-FDG PET/CT for assessing response to therapy in esophageal cancer. *J Nucl Med* 50(S1):S89–96.

Krohn, K. A., D. A. Mankoff, and J. F. Eary. 2001. Imaging cellular proliferation as a measure of response to therapy. *J Clin Pharmacol* 41(Suppl):S96–103.

Kuhl, D., and R. Edwards. 1964. Cylindrical and section radioisotope scanning of the liver and brain. *Radiology* 83:926–36.

Kuhl, D., R. Edwards, A. R. Ricci, and M. Reivich. 1973. Quantitative section scanning using orthogonal tangent correction. *J Nucl Med* 14(4):196–200.

Laking, G. R., and P. M. Price. 2003. Positron emission tomographic imaging of angiogenesis and vascular function. *Br J Radiol* 76(Suppl 1):S50–59.

Laprie, A., I. Catalaa, E. Cassol, et al. 2008. Proton magnetic resonance spectroscopic imaging in newly diagnosed glioblastoma: Predictive value for the site of postradiotherapy relapse in a prospective longitudinal study. *Int J Radiat Oncol Biol Phys* 70(3):773–81.

Lartizien, C., P. E. Kinahan, and R. Swensson, et al. 2003. Evaluating image reconstruction methods for tumor detection in 3-dimensional whole-body PET oncology imaging. *J Nucl Med* 44(2):276–90.

Lawrentschuk, N., A. M. Poon, S. S. Foo, et al. 2005. Assessing regional hypoxia in human renal tumors using 18F-fluoromisonidazole positron emission tomography. *BJU Int* 96(4):540–6.

Leach, M., M. Verrill, J. Glaholm, et al. 1998. Measurements of human breast cancer using magnetic resonance spectroscopy: A review of clinical measurements and a report of localized 31P measurements of response to treatment. *NMR Biomed* 11(7):314–40.

Lee, C. G., M. Heijn, E. di Tomaso, et al. 2000. Anti-vascular endothelial growth factor treatment augments tumor radiation response under normoxic or hypoxic conditions. *Cancer Res* 60(19):5565–70.

Lee, N. Y., J. G. Mechalakos, S. Nehmeh, et al. 2008. Fluorine-18-labeled fluoromisonidazole positron emission and computed tomography-guided intensity-modulated radiotherapy for head and neck cancer: A feasibility study. *Int J Radiat Oncol Biol Phys* 70(1):2–13.

Lehtiö, K., O. Eskola, T. Viljanen, et al. 2004. Imaging perfusion and hypoxia with PET to predict radiotherapy response in head-and-neck cancer. *Int J Radiat Oncol Biol Phys* 59(4):971–82.

Lehtiö, K., V. Oikonen, T. Gronroos, et al. 2001. Imaging of blood flow and hypoxia in head and neck cancer: Initial evaluation with [(15)O]H(2)O and [(18)F] fluoroerythronitroimidazole PET. *J Nucl Med* 42(11):1643–52.

Lewis, J. S., D. W. McCarthy, T. J. McCarthy, Y. Fujibayashi, and M. J. Welch. 1999. Evaluation of 64Cu-ATSM in vitro and in vivo in a hypoxic tumor model. *J Nucl Med* 40(1):177–83.

Lewis, J. S., T. L. Sharp, R. Laforest, Y. Fujibayashi, and M. J. Welch. 2001. Tumor uptake of copper-diacetyl-bis(N4-Methylthiosemicarbazone): Effect of changes in tissue oxygenation. *J Nucl Med* 42(4):655–61.

Lewis, J. S., and M. J. Welch. 2001. PET imaging of hypoxia. *Q J Nucl Med* 45(2):183–8.

Li, H., W. L. Thorstad, K. J. Biehl, et al. 2008. A novel PET tumor delineation method based on adaptive region-growing and dual-front active contours. *Med Phys* 35(8):3711–21.

Ling, C. C., J. Humm, S. Larson, et al. 2000. Towards multidimensional radiotherapy (MD-CRT): Biological imaging and biological conformality. *Int J Radiat Oncol Biol Phys* 47(3):551–60.

Madani, I., W. Duthoy, C. Derie, et al. 2007. Positron emission tomography-guided, focal-dose escalation using intensity-modulated radiotherapy for head and neck cancer. *Int J Radiat Oncol Biol Phys* 68(1):126–35.

Mankoff, D. A., A. F. Shields, M. M. Graham, J. M. Link, J. F. Eary, and K. A. Krohn. 1998. Kinetic analysis of 2-[carbon-11] thymidine PET imaging studies: Compartmental model and mathematical analysis. *J Nucl Med* 39(6):1043–55.

Mankoff, D. A., A. F. Shields, J. M. Link, et al. 1999. Kinetic analysis of 2-[11C]thymidine PET imaging studies: Validation studies. *J Nucl Med* 40(4):614–24.

Marks, L. B., X. Yu, R. G. Prosnitz, et al. 2005. The incidence and functional consequences of RT-associated cardiac perfusion defects. *Int J Radiat Oncol Biol Phys* 63(1):214–23.

Maurer, R. I., P. J. Blower, J. R. Dilworth, C. A. Reynolds, Y. Zheng, G. E. Mullen. 2002. Studies on the mechanism of hypoxic selectivity in copper bis(thiosemicarbazone) radiopharmaceuticals. *J Med Chem* 45(7):1420–31.

McCall, K., D. Barbee, et al. 2010. PET imaging for the quantification of biologically heterogeneous tumors: Measuring the effect of relative position on image-based quantification of dose painting targets. *Phys Med Biol* 55:2789.

Menard, C., I. C. Smith, R. L. Somorjai, et al. 2001. Magnetic resonance spectroscopy of the malignant prostate gland after radiotherapy: A histopathologic study of diagnostic validity. *Int J Radiat Oncol Biol Phys* 50(2):317–23.

Menda, Y., L. L. Boles Ponto, K. J. Dornfeld, et al. 2009. Kinetic analysis of 3′-deoxy-3′-18F-fluorothymidine (18F-FLT) in head and neck cancer patients before and early after initiation of chemoradiation therapy. *J Nucl Med* 50(7):1028–35.

Merchant, T. E., A. A. Alfieri, T. Glonek, and J. A. Koutcher. 1995. Comparison of relative changes in phosphatic metabolites and phospholipids after irradiation. *Radiat Res* 142(1):29–38.

Mier, W., U. Haberkorn, and M. Eisenhut. 2002. [18F]FLT; portrait of a proliferation marker. *Eur J Nucl Med Mol Imaging* 29(2):165–9.

Miller, A. B., B. Hoogstraten, M. Staquet, and A. Winkler. 1981. Reporting results of cancer treatment. *Cancer* 47(1):207–14.

Miller, J. C., H. H. Pien, D. Sahani, A. G. Sorensen, and J. H. Thrail. 2005. Imaging angiogenesis: Applications and potential for drug development. *J Natl Cancer Inst* 97(3):172–87.

Mineura, K., M. Shioya Kowada, T. Ogawa, J. Hatazawa, and K. Uemura. 1999. Blood flow and metabolism of oligodendrogliomas: A positron emission tomography study with kinetic analysis of 18F-fluorodeoxyglucose. *J Neurooncol* 43(1):49–57.

Minn, H., S. Leskinen-Kallio, P. Lindholm, et al. 1993. [18F]fluorodeoxyglucose uptake in tumors: Kinetic vs. steady-state methods with reference to plasma insulin. *J Comput Assist Tomogr* 17(1):115–23.

Minn, H., K. R. Zasadny, L. E. Quint, and R. L. Wahl. 1995. Lung cancer: Reproducibility of quantitative measurements for evaluating 2-[F-18]-fluoro-2-deoxy-D-glucose uptake at PET. *Radiology* 196(1):167–73.

Moon, R., and J. Richards. 1972. Conformation studies of various hemoglobins by natural-abundance 13C NMR spectroscopy. *Proc Natl Acad Sci USA* 69(8):2193–7.

Moon, R. B., and J. H. Richards. 1973. Determination of intracellular pH by 31P magnetic resonance. *J Biol Chem* 248(20):7276–8.

Muryama, S., T. Akamine, S. Sakai, et al. 2004. Risk factor of radiation pneumonitis: Assessment with velocity-encoded cine magnetic resonance imaging of pulmonary artery. *J Comput Assist Tomogr* 28(2):204–8.

Muzi, M., H. Vesselle, J. R. Grierson, et al. 2005. Kinetic analysis of 3′-deoxy-3′-fluorothymidine PET studies: Validation studies in patients with lung cancer. *J Nucl Med* 46(2):274–82.

Nagengast, W. B., E. G. de Vries, G. A. Hospers, et al. 2007. In vivo VEGF imaging with radiolabeled bevacizumab in a human ovarian tumor xenograft. *J Nucl Med* 48(8):1313–9.

Nelson, S., E. Graves, A. Pirzkall, et al. 2002. In vivo molecular imaging for planning radiation therapy of gliomas: An application of 1H MRSI. *J Magn Reson Imaging* 16(4):464–76.

Nestle, U., S. Kremp, A. Schaefer-Schuler, et al. 2005. Comparison of different methods for delineation of 18F-FDG PET-positive tissue for target volume definition in radiotherapy of patients with non-small cell lung cancer. *J Nucl Med* 46(8):1342–8.

New, P. F., W. R. Scott, J. A. Schnur, K. R. Davis, J. M. Taveras, and F. H. Hochberg. 1975. Computed tomography with the EMI scanner in the diagnosis of primary and metastatic intracranial neoplasms. *Radiology* 114(1):75–87.

Nordsmark, M., S. M. Bentzen, V. Rudat, et al. 2005. Prognostic value of tumor oxygenation in 397 head and neck tumors after primary radiation therapy: An international multicenter study. *Radiother Oncol* 77(1):18–24.

Obata, A., S. Kasamatsu, J. S. Lewis, et al. 2005. Basic characterization of (64)Cu-ATSM as a radiotherapy agent. *Nucl Med Biol* 32(1):21–8.

Obata, A., M. Yoshimoto, S. Kasamatsu, et al. 2003. Intra-tumoral distribution of (64)Cu-ATSM: A comparison study with FDG. *Nucl Med Biol* 30(5):529–34.

Oda, K., H. Toyama, K. Uemura, Y. Ikoma, Y. Kimura, and M. Senda. 2001. Comparison of parametric FBP and OS-EM reconstruction algorithm images for PET dynamic study. *Ann Nucl Med* 15(5):417–23.

Ogawa, S., T. M. Lee, A. R. Kay, and D. W. Tank. 1990. Brain magnetic resonance imaging with contrast dependent on blood oxygenation. *Proc Natl Acad Sci USA* 87(24):9868–72.

Payne, G. S., and M. O. Leach. 2006. Applications of magnetic resonance spectroscopy in radiotherapy treatment planning. *Br J Radiol* 79(S1):S16–26.

Phelps, M. E., E. J. Hoffman, S. C. Huang, and M. M. Ter-Pogossian. 1975. Effect of positron range on spatial resolution. *J Nucl Med* 16(7):649–52.

Phelps, M. E., E. J. Hoffman, N. A. Mullani, and M. M. Ter-Pogossian. 1975. Application of annihilation coincidence detection to transaxial reconstruction tomography. *J Nucl Med* 16(3):210–24.

Pirzkall, A., X. Li, J. Oh, et al. 2004. 3D MRSI for resected high-grade gliomas before RT: Tumor extent according to metabolic activity in relation to MRI. *Int J Radiat Oncol Biol Phys* 59(1):126–37.

Prabhakar, R., P. K. Julka, T. Ganesh, A. Munshi, R. C. Joshi, and G. K. Rath. 2007. Feasibility of using MRI alone for 3D radiation treatment planning in brain tumors. *Jpn J Clin Oncol* 37(6):405–11.

Provenzale, J. M. 2007. Imaging of angiogenesis: Clinical techniques and novel imaging methods. *Am J Roentgenol* 188(1):11–23.

Rajendran, J. G., K. R. Hendrickson, A. M. Spence, M. Muzi, K. A. Krohn, and D. A. Mankoff. 2006. Hypoxia imaging-directed radiation treatment planning. *Eur J Nuc Med Mol Imaging* 33(Suppl 1):44–53.

Rajendran, J. G., D. A. Mankoff, F. O'Sullivan, et al. 2004. Hypoxia and glucose metabolism in malignant tumors: Evaluation by [18F]fluoromisonidazole and [18F]fluorodeoxyglucose positron emission tomography imaging. *Clin Cancer Res* 10(7):2245–52.

Rajendran, J. G., D. L. Schwartz, J. O'Sullivan, et al. 2006. Tumor hypoxia imaging with [F-18]fluoromisonidazole positron emission tomography in head and neck cancer. *Clin Cancer Res* 12(18):5435–41.

Rajendran, J. G., D. C. Wilson, E. U. Conrad, et al. 2003. [(18)F]FMISO and [(18)F]FDG PET imaging in soft tissue sarcomas: Correlation of hypoxia, metabolism and VEGF expression. *Eur J Nucl Med Mol Imaging* 30(5):695–704.

Rasch, C., R. Keus, F. A. Pameijer, et al. 1997. The potential impact of CT-MRI matching on tumor volume delineation in advanced head and neck cancer. *Int J Radiat Oncol Biol Phys* 39(4):841–8.

Rasey, J. S., J. R. Grierson, L. W. Wiens, P. D. Kolb, and J. L. Schwartz. 2002. Validation of FLT uptake as a measure of thymidine kinase-1 activity in A549 carcinoma cells. *J Nucl Med* 43(9):1210–7.

Rasey, J. S., W. J. Koh, M. L. Evans, et al. 1996. Quantifying regional hypoxia in human tumors with positron emission tomography of [18F]fluoromisonidazole: A pretherapy study of 37 patients. *Int J Radiat Oncol Biol Phys* 36(2):417–28.

Rehman, S., and G. C. Jayson. 2005. Molecular Imaging of Antiangiogenic Agents. *Oncologist* 10(2):92–103.

Reich, N., and F. Seidelmann. 1975. Computed tomography using the EMI scanner: Part II. Intracranial pathology. *J Am Osteopath Assoc* 74(12):1133–8.

Ressel, A., C. Weiss, and T. Feyerabend. 2001. Tumor oxygenation after radiotherapy, chemotherapy, and/or hyperthermia predicts tumor free survival. *Int J Radiat Oncol Biology Phys* 49(4):1119–25.

Rischin, D., R. J. Hicks, R. Fisher, et al. 2006. Prognostic significance of [18F]-misonidazole positron emission tomography-detected tumor hypoxia in patients with advanced head and neck cancer randomly assigned to chemoradiation with or without tirapazamine: A substudy of Trans-Tasman Radiation Oncology Group study 98.02. *J Clin Oncol* 24(13):2098–104.

Rofstad, E. K., K. Galappathi, B. Mathiesen, and E. B. Ruud. 2007. Fluctuating and diffusion-limited hypoxia in hypoxia-induced metastasis. *Clin Cancer Res* 13(7):1971–8.

Runge, V. M., J. A. Clanton, A. C. Price, et al. 1985. The use of GD DTPA as a perfusion agent and marker of blood-brain barrier disruption. *Magn Reson Imaging* 3(1):43–55.

Schoder, H., M. Fury, N. Lee, and D. Kraus. 2009. PET monitoring of therapy response in head and neck squamous cell carcinoma. *J Nucl Med* 50(S1):S74–88.

Schuetze, S., J. Eary, et al. 2005. FDG PET but not RECIST agrees with histologic response of soft tissue sarcoma to neoadjuvant chemotherapy. *Proc Am Soc Clin Oncol* 23:817s [abstr 9005].

Schwartz, D. L., E. Ford, J. Rajendran, et al. 2005. FDG-PET/CT imaging for preradiotherapy staging of head-and-neck squamous cell carcinoma. *Int J Radiat Oncol Biol Phys* 61(1):129–36.

Schwartz, J. L., Y. Tamura, R. Jordan, J. R. Grierson, and K. A. Krohn. 2003. Monitoring tumor cell proliferation by targeting DNA synthetic processes with thymidine and thymidine analogs. *J Nucl Med* 44(12):2027–32.

Seppenwoolde, Y., K. De Jaeger, L. J. Boersma, J. S. Belderbos, and J. V. Lebeaque. 2004. Regional differences in lung radiosensitivity after radiotherapy for non-small-cell lung cancer. *Int J Radiat Oncol Biol Phys* 60(3):748–58.

Shields, A. F., J. R. Grierson, B. M. Dohmen, et al. 1998. Imaging proliferation in vivo with [F-18]FLT and positron emission tomography. *Nat Med* 4(11):1334–6.

Singer, J. R., and L. Crooks. 1983. Nuclear magnetic resonance blood flow measurements in the human brain. *Science* 221(4611):654–56.

Smith, T. A. 1998. FDG uptake, tumour characteristics and response to therapy: A review. *Nucl Med Commun* 19(2):97–105.

Spaepen, K., S. Stroobants, P. Dupont, et al. 2001. Prognostic value of positron emission tomography (PET) with fluorine-18 fluorodeoxyglucose ([18F]FDG) after first-line chemotherapy in non-Hodgkin's lymphoma: Is [18F]FDG-PET a valid alternative to conventional diagnostic methods? *J Clin Oncol* 19(2):414–9.

Stroobants, S., J. Goeminne, M. Seegers, et al. 2003. 18FDG-Positron emission tomography for the early prediction of response in advanced soft tissue sarcoma treated with imatinib mesylate (Glivec®). *Eur J Cancer* 39(14):2012–20.

Sutinen, E., M. Nurmi, A. Roivainen, et al. 2004. Kinetics of [(11)C]choline uptake in prostate cancer: A PET study. *Eur J Nucl Med Mol Imaging* 31(3):317–24.

Takahashi, N., Y. Fujibayashi, Y. Yonekura, et al. 2000. Evaluation of 62Cu labeled diacetyl-bis(N4-methylthiosemicarbazone) as a hypoxic tissue tracer in patients with lung cancer. *Ann Nucl Med* 14(5):323–8.

Takahashi, N., Y. Fujibayashi, Y. Yonekura, et al. 2001. Copper-62 ATSM as a hypoxic tissue tracer in myocardial ischemia. *Ann Nucl Med* 15(3):293–6.

Teo, B. K., Y. Seo, S. L. Bacharach, et al. 2007. Partial-volume correction in PET: Validation of an iterative postreconstruction method with phantom and patient data. *J Nucl Med* 48(5):802–10.

Ter-Pogossian, M. M., M. E. Phelps, E. J. Hoffman, and N. A. Mullani. 1975. A positron-emission transaxial tomograph for nuclear imaging (PETT). *Radiology* 114(1):89–98.

Thiesse, P., L. Ollivier, D. Di Stefano-Louineau, et al. 1997. Response rate accuracy in oncology trials: Reasons for interobserver variability. Groupe Francais d'Immunotherapie of the Federation Nationale des Centres de Lutte Contre le Cancer. *J Clin Oncol* 15(12):3507–14.

Thomsen, C., O. Henriksen, and P. Ring. 1987. In vivo measurement of water self diffusion in the human brain by magnetic resonance imaging. *Acta Radiol* 28(3):353–61.

Thorwarth, D., S. M. Eschmann, F. Paulsen, M. Alber, et al. 2007. Hypoxia dose painting by numbers: A planning study. *Int J Radiat Oncol Biol Phys* 68(1):291–300.

Thorwarth, D., S. M. Eschmann, F. Hozner, F. Paulsen, and M. Alber. 2006. Combined uptake of [18F]FDG and [18F]FMISO correlates with radiation therapy outcome in head-and-neck cancer patients. *Radiother Oncol* 80(2):151–6.

Tong, R., Y. Boucher, S. V. Kozin, F. Winkler, D. J. Hicklin, and R. K. Jain. 2004. Vascular normalization by vascular endothelial growth factor receptor 2 blockade induces a pressure gradient across the vasculature and improves drug penetration in tumors. *Cancer Res* 64(11):3731–6.

Troost, E. G., W. V. Vogel, M. A. Merkx, et al. 2007. 18F-FLT PET does not discriminate between reactive and metastatic lymph nodes in primary head and neck cancer patients. *J Nucl Med* 48(5):726–35.

Tseng, J., L. K. Dunnwald, E. K. Schubert, et al. 2004. 18F-FDG kinetics in locally advanced breast cancer: Correlation with tumor blood flow and changes in response to neoadjuvant chemotherapy. *J Nucl Med* 45(11):1829–37.

Tsuyuguchi, N. 1997. Kinetic analysis of glucose metabolism by FDG-PET versus proliferation index of Ki-67 in meningiomas—comparison with gliomas. *Osaka City Med J* 43(2):209–23.

Valente, G., R. Orecchia, S. Gandolfo, et al. 1994. Can Ki67 immunostaining predict response to radiotherapy in oral squamous cell carcinoma? *J Clin Pathol* 47(2):109–12.

van Baardwijk, A., G. Bosmans, M. van Kroonenburgh, et al. 2007. Time trends in the maximal uptake of FDG on PET scan during thoracic radiotherapy. A prospective study in locally advanced non-small cell lung cancer (NSCLC) patients. *Radiother Oncol* 82(2):145–52.

van Dorsten, F. A., M. van der Graaf, M. R. Engelbrecht, et al. 2004. Combined quantitative dynamic contrast-enhanced MR imaging and 1H MR spectroscopic imaging of human prostate cancer. *J Magn Reson Imaging* 20(2):279–87.

Vanderstraeten, B., W. Duthoy, W. De Gersem, W. De Neve, and H. Thierens. 2006. [18F]fluoro-deoxy-glucose positron emission tomography ([18F]FDG-PET) voxel intensity-based intensity-modulated radiation therapy (IMRT) for head and neck cancer. *Radiother Oncol* 79(3):249–58.

Vesselle, H., J. Grierson, M. Muzi, et al. 2002. In vivo validation of 3′deoxy-3′-[18F]fluorothymidine ([18F]FLT) as a proliferation imaging tracer in humans: Correlation of [18F] FLT uptake by positron emission tomography with Ki-67 immunohistochemistry and flow cytometry in human lung tumors. *Clin Cancer Res* 8(11):3315–23.

Vesselle, H., J. Grierson, L. M. Peterson, M. Muzi, D. A. Mankoff, K. A. Krohn, et al. 2003. 18F-Fluorothymidine radiation dosimetry in human PET imaging studies. *J Nucl Med* 44(9):1482–8.

Vesselle, H., R. A. Schmidt, J. M. Pugsley, et al. 2000. Lung cancer proliferation correlates with [F-18]fluorodeoxyglucose uptake by positron emission tomography. *Clin Cancer Res* 6(10):3837–44.

Wagner, M., U. Seitz, A. Buck, et al. 2003. 3′-[18F]fluoro-3′-deoxythymidine ([18F]-FLT) as positron emission tomography tracer for imaging proliferation in a murine B-Cell lymphoma model and in the human disease. *Cancer Res* 63(10):2681–7.

Weber, W. A. 2009. Assessing Tumor Response to Therapy. *J Nucl Med* 50(S1):S1–10.

Weber, W. A., M. Schwaiger, and N. Avril. 2000. Quantitative assessment of tumor metabolism using FDG-PET imaging. *Nucl Med Biol* 27(7):683–7.

Weber, W. A., H. J. Wester, A. L. Grosu, et al. 2000. O-(2-[18F]Fluoroethyl)-l-tyrosine and l-[methyl-11C]methionine uptake in brain tumors: Initial results of a comparative study. *Eur J Nucl Med* 27(5):542.

Webster, G. J., J. E. Kilgallon, K. F. Ho, C. G. Rowbottom, N. J. Slevin, and R. I. Mackay. 2009. A novel imaging technique for fusion of high-quality immobilised MR images of the head and neck with CT scans for radiotherapy target delineation. *Br J Radiol* 82(978):497–503.

Wells, J. M., D. A. Mankoff, M. Muzi, et al. 2002. Kinetic analysis of 2-[11C]thymidine PET imaging studies of malignant brain tumors: Compartmental model investigation and mathematical analysis. *Mol Imaging* 1(3):151–9.

WHO. 1979. *WHO Handbook for Reporting Results of Cancer Treatment.* Geneva: The Organization.

Willett, C. G., Y. Boucher, E. di Tomaso, et al. 2004. Direct evidence that the VEGF-specific antibody bevacizumab has antivascular effects in human rectal cancer. *Nat Med* 10(2):145–7.

Willett, C. G., D. G. Duda, E. di Tomaso, et al. 2009. Efficacy, safety, and biomarkers of neoadjuvant bevacizumab, radiation therapy, and fluorouracil in rectal cancer: A multidisciplinary phase II Study. *J Clin Oncol* 27(18):3020–6.

Yamaguchi T., H. Sasaki, T. Ogawa, K. Mineura, K. Uemura, I. Kanno, F. Shishido, M. Murakami, A. Inugami, S. Higano. 1986. Relation between tissue nature and (18F) fluorodeoxyglucose kinetics evaluated by dynamic positron emission tomography in human brain tumors. *Acta Radiol Suppl* 369:415–8.

Yang, Y. J., J. S. Ryu, S. Y. Kim, et al. 2006. Use of 3′-deoxy-3′-[18F]fluorothymidine PET to monitor early responses to radiation therapy in murine SCCVII tumors. *Eur J Nucl Med Mol Imaging* 33(4):412–9.

Zackrisson, B., P. Flygare, H. Gustafsson, B. Sjostrom, and G. D. Wilson. 2002. Cell kinetic changes in human squamous cell carcinomas during radiotherapy studied using the in vivo administration of two halogenated pyrimidines. *Eur J Cancer* 38(8):1100–6.

Zahra, M. A., K. G. Hollingsworth, E. Sala, D. J. Lomas, and L. T. Tan. 2007. Dynamic contrast-enhanced MRI as a predictor of tumor response to radiotherapy. *Lancet Oncol* 8:63–74.

Zahra, M. A., L. T. Tan, A. N. Priest, et al. 2009. Semiquantitative and quantitative dynamic contrast-enhanced magnetic resonance imaging measurements predict radiation response in cervix cancer. *Int J Radiat Oncol Biol Phys* 74(3):766–73.

Zeng, Q. S., C. F. Li, K. Zhang, H. Liu, X. S. Kang, and J. H. Zhen. 2007. Multivoxel 3D proton MR spectroscopy in the distinction of recurrent glioma from radiation injury. *J Neurooncol* 84(1):63–9.

Zhang, W. J., R. Zheng, L. J. Zhao, L. H. Wang, and S. Z. Chen. 2004. Utility of SPECT lung perfusion scans in assessing the early changes in pulmonary function after radiotherapy for patients with lung cancer. *Ai Zheng (Chinese Journal of Cancer)* 23(10):1180–4.

Zhang, X., Z. Xiong, Y. Wu, et al. 2006. Quantitative PET imaging of tumor integrin {alpha}v{beta}3 expression with 18F-FRGD2. *J Nucl Med* 47(1):113–21.

4

Biomarkers of Early Response for Adaptive Radiation Therapy

Issam El Naqa
Washington University

Jeffrey M. Craft
Washington University

Jung Hun Oh
Washington University

Joseph O. Deasy
Washington University
School of Medicine

4.1 Introduction

Biologically based measures ("biomarkers") have emerged as useful tools for disease diagnosis and prognosis, as well as for drug discovery and treatment selection in cancer (Ludwig and Weinstein 2005). The recent evolution in biotechnology (in particular high-throughput genomics and proteomics techniques) has led to the foundation of a new field in radiation oncology denoted "radiogenomics" (West et al. 2005). These new molecular techniques carry great potential to revolutionize the field of radiobiology (West, Elliott, and Burnet 2007; Bentzen 2008; Bentzen, Buffa, and Wilson 2008). Recent years have witnessed an extraordinary surge in the application of these techniques for investigating radiotherapy response in different cancer sites such as breast (Nimeus-Malmstrom et al. 2008), cervix (Klopp et al. 2008), and head and neck (Pramana et al. 2007).

Biomarkers can be selected based on candidate or global transcriptome (or proteome) screening of differentially expressed genes or proteins (West, Elliott, and Burnet 2007; Wouters 2008), or they could be chosen based on investigating evolutionary genetic variations (Andreassen and Alsner 2009). In either approach, the biomarkers would be selected to achieve a key component of radiation oncology research, namely, to predict at the time of treatment planning or during the course of fractionated radiation treatment, the probability of tumor eradication and normal tissue risks for the type of treatment being considered for that particular patient (Torres-Roca and Stevens 2008).

In this chapter, we provide an overview of the current status of biomarkers for predicting tumor response and normal tissue toxicities for patients who receive radiotherapy. Then, we present clinical examples of applying different biomarkers for predicting early tumor response and acute normal tissue toxicities in radiation oncology. Finally, we discuss potential applications and the challenging obstacles to the use of effective biomarkers in adaptive radiation therapy (ART).

4.2 Background

4.2.1 What Is a Biomarker?

A biomarker is defined as "a characteristic that is objectively measured and evaluated as an indicator of normal biological processes, pathological processes, or pharmacological responses to a therapeutic intervention" (Biomarkers Definitions Working Group 2001). Biomarkers can be categorized into exogenous or endogenous based on the biochemical source of the marker.

Exogenous biomarkers are based on introducing a foreign substance into the patient's body such as those used in molecular imaging by positron emission tomography. For instance, a synthesized fluorodeoxyglucose compound (a glucose analog*) has been used in radiotherapy clinical practice for tumor detection, staging/restaging, and volumetric target definition of many

---

* A "biological analog" acts in a similar way, that is, biologically in some (but usually not all) aspects.

different types of cancers (Mutic et al. 2003; Biersack, Bender, and Palmedo 2004; Bradley et al. 2004), and as a prognostic factor for predicting outcomes (Borst et al. 2005; Levine et al. 2006; Ben-Haim and Ell 2009; El Naqa et al. 2009). Conversely, endogenous biomarkers can further be classified as (1) *expression biomarkers*, measuring changes in gene expression (using corresponding RNA[*] levels) or protein levels, or (2) *genetic biomarkers*, based on variations, for tumors or normal tissues, in the underlying DNA genetic code. Measurements are typically based on tissue or fluid specimens, which are analyzed using molecular biology laboratory techniques. Endogenous biomarkers will be the subject of this chapter.

The biological world presents a bewildering array of chemical signals that might potentially be relevant to cancer prognosis and treatment predictions. In addition to 30,000 or so genes that are known to code for proteins, other regulatory mechanisms exist, and it is highly likely that more mechanisms are waiting to be discovered or elucidated. An understanding of how all these signals work together is currently far beyond our capabilities, although this is the ultimate goal of the rapidly expanding field of systems biology. Despite the many and large gaps in our understanding of biological networks, a vast amount of knowledge has been established, including many examples of measured biological assays that correlate with disease prognosis and treatment outcomes.

4.2.1.1 Expression Biomarkers

Expression biomarkers are the result of gene expression changes in tissues or body fluids due to the disease itself or the disease or normal tissues' response to treatment (Mayeux 2004). These can be further divided into single-parameter and bioarray biomarkers. For instance, prostate-specific antigen (PSA) levels in blood serum, measured using immunoassay techniques, is a successful example of a single-parameter biomarker that is widely used for screening prostate cancer patients. For example, PSA levels above 4 ng/mL are considered clinically suspicious (Brawer 2001). The role of PSA as a radiotherapy response biomarker has long been considered (Dundas, Porter, and Venner 1990; Ritter et al. 1992), and its usefulness has been confirmed in a multi-institutional study, where it was demonstrated that PSA nadir can predict biochemical and distant failures after radiotherapy for prostate cancer patients (Michael et al. 2006). Other promising single-parameter biomarkers that predict radiotherapy response include circulating tumor antigen[†] molecules such as carcinoembryonic antigen (Vider et al. 1974; Das et al. 2007) and cancer antigen (Bast et al. 1998; Prat et al. 2008).

Array-based expression biomarkers have witnessed a rapid increase in recent years due to the extraordinary advances in biotechnology and corresponding knowledge advances due to the Human Genome Project and its offshoots. These biomarkers follow from the central dogma of molecular biology articulated by Francis Crick (1970), in which biological information is expressed via the sequential transcription of the DNA genetic code into the RNA and subsequent translation of the RNA into proteins. These can be based on disease pathophysiology or pharmacogenetics studies or they can be extracted from several methods, such as high-throughput gene expression (aka transcriptomics; Svensson et al. 2006; Ogawa et al. 2007; Nuyten and van de Vijver 2008), resulting protein expressions (aka proteomics; Alaiya, Al-Mohanna, and Linder 2005; Wouters 2008), or metabolites[‡] (aka metabolomics; Tyburski et al. 2008; Spratlin, Serkova, and Eckhardt 2009) as further discussed in Sections 4.3.1.7 and 4.3.2.3.

4.2.1.2 Genetic Variant Markers

The inherent genetic variability of the human genome is an emerging resource for studying disposition to cancer and the variability of patient responses to therapeutic agents. These variations in the DNA sequences of humans, in particular single-nucleotide polymorphisms (SNPs) have strong potential to elucidate complex disease onset and response in cancer (Erichsen and Chanock 2004). A SNP is defined as the stable substitution of a single base with a minor allele frequency of 1%. SNPs are differentiated from other mutations by the fact that they generally do not have a phenotypic effect. It is estimated that there are around 10 million SNPs that could be organized into distinct haplotype blocks (sets of SNPs) based on their genetic association by a genetic measure known as linkage disequilibrium (Engle, Simpson, and Landers 2006). Methods based on the candidate gene approach and high-throughput (genome-wide associations) are currently heavily investigated to analyze the functional effect of SNPs in predicting response to radiotherapy (West, Elliott, and Burnet 2007; Alsner, Andreassen, and Overgaard 2008; Andreassen and Alsner 2009).

4.2.2 Molecular Basis of Radiotherapy Response

4.2.2.1 DNA Damage and Cell Death

Classical radiobiology has been defined by "the four R's"—cellular damage repair, cell-cycle redistribution, reoxygenation over a course of therapy, and cellular repopulation/division over a course of therapy (Hall and Giaccia 2006). It is believed that radiation-induced cellular lethality is primarily caused by DNA damage in the targeted cells. Two types of cell death have been linked to radiation: apoptosis and postmitotic cell death. However, tumor cell radiosensitivity is controlled by many factors (known and unknown) related to tumor DNA repair

---

[*] RNA stands for ribonucleic acid, and is the intermediate product of gene expression, after the process of gene transcription. The "final expression" is the resulting protein.

[†] An antigen is a molecule that a cell recognizes and can link to via cell-surface receptors.

[‡] Metabolites are molecules that are involved in the intracellular production of energy.

efficiency (e.g., homologous recombination or nonhomologous end joining), cell-cycle control (cells are most resistant during S phase), oxygen concentration (hypoxia), and the radiation dose rate (Hall and Giaccia 2006; Lehnert 2008).

4.2.2.2 Sensitivity vs. Resistance

The seminal work of Fertil and Malaise has shown that the survival of cell lines, given small doses of radiation in vitro, correlates with the perceived ability to cure human tumors from which the cell lines were derived (Fertil and Malaise 1985). Shortly afterward, a great deal of work was undertaken within the radiation biology community to understand whether tumor biopsies or normal tissues biopsies could be used to derive radiosensitivity measures *in vitro*, which could then be applied to predict patients' treatment outcomes. Several metrics for describing differential radiosensitivity, or conversely, radioresistance of tumor types were typically explored. These can include the shape of the initial portion of the radiation cell-survival curve, the survival fraction of cells for a 2 Gy irradiation (SF2), the initial slope of the survival curve, or the mean inactivation dose (West 1995). It was subsequently noticed that there is a link between radiosensitivity and radiation-produced DNA double-strand breaks (DSBs) measured in single-cell gel electrophoresis (comet assay; Marples et al. 1998). It has also been established that the density of a particular DNA end-binding protein complex involved in DNA DSB repair, postirradiation, shows a high correlation with SF2 for a variety of normal and malignant mammalian cells (Ismail et al. 2004). The protein complex density was more predictive than the individual densities of any of the contributing genes, implying that the proper functioning of this protein complex is not a simple function of the available constituent protein densities. Figure 4.1 illustrates the basic concept of DNA DSB repair via protein complexes.

4.2.3 Specimen Types

4.2.3.1 Tissue Specimens

Formalin fixation and tissue embedding in paraffin wax (FFPE) is a standard approach for tissue processing prior to histological examination. To analyze protein contents, FFPE samples are typically first deparaffinized using xylene, followed by an ethanol wash. The cross-linked proteins are digested by protease K prior to analysis (von Ahlfen et al. 2007).

4.2.3.2 Fluid Specimens

Fluid specimens of use include urine, saliva, seminal fluid, vaginal secretions, and peripheral blood (Brunzel 1994). Among these, peripheral blood specimens are (currently) by far the most widely used in oncology research. Typically, blood samples are drawn from the patient as whole blood, collected in tubes containing a coagulation inhibitor, such as EDTA. Serum (blood without red blood cells, white blood cells, or coagulating factors) is the most frequently analyzed blood specimen. However, the generation of serum is a time-consuming process and is associated with the activation of the coagulation cascade and complement immune system. Another extracted component, buffy coat, contains white blood cells and their DNA, which can be further analyzed. The improper collection and mishandling of samples is considered one of the most frequent pitfalls in biomarker research (Plebani and Carraro 1997).

4.2.4 Early vs. Late Toxic Effects of Radiotherapy

Radiation-induced injuries and toxicities involve a complex cascade of radiobiological processes that can begin within a few hours after irradiation and progress over weeks, months, and years (Emami et al. 1991; Fajardo, Berthrong, and Anderson 2001). According to the onset time of these toxicities, they are clinically

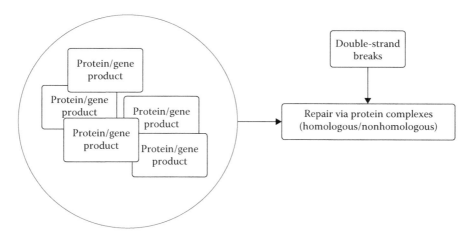

FIGURE 4.1 DNA repair is the result of a process that involves the recruitment of multiprotein complexes to the site of repair. This is a complicated process that involves the actions of many genes and the interactions of their protein products as well as other epigenetic factors that control post-transcriptional protein status. The most effective biomarkers are likely to incorporate multiple information sources concerning many aspects of intracellular damage processing.

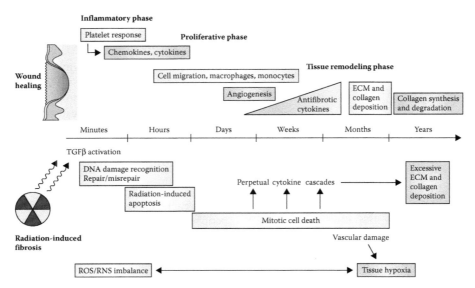

FIGURE 4.2 Normal wound healing and radiation-induced fibrosis progression timeline. TGF-β = transforming growth factor β; ECM = extracellular matrix; ROS = reactive oxygen species; RNS = reactive nitrogen species. (Reprinted from Bentzen, S.M. 2006. *Nat Rev Cancer* 6:702–713. With permission.)

classified into early and late effects. Early effects are typically transient and manifest during treatment or within a few weeks of the completion of a fractionated radiotherapy schedule. These effects include skin erythema, mucositis, esophagitis nausea, and diarrhea. Late effects are typically expressed after a latent period that can extend from months to years. These effects include radiation-induced inflammation and fibrosis, atrophy, vascular degeneration, and neural damage. The timeline for radiation-induced inflammation or fibrosis is depicted in Figure 4.2 (Bentzen 2006).

4.2.5 Toxicity Grading

Uncovering causal relationships between two complicated sets of variables is obviously challenging. A variety of end-point reporting systems are in use in radiotherapy, including widely used tools such as the radiotherapy oncology group (RTOG), late effects in normal tissues—subjective, objective, measured, analytical (LENT-SOMA), and the Common Toxicity Criteria Adverse Event (CTCAE) grading systems. The RTOG system groups several end points together to form a single-graded toxicity score for either early or late effects. As of version 3, the CTCAE has incorporated questions from the LENT-SOMA system. As an example, Table 4.1 compares RTOG and CTCAE v3.0 criteria for scoring of radiation-induced lung injury or radiation pneumonitis (RP).

4.3 Radioresponse Biomarkers

4.3.1 Biomarkers for Tumor Response

4.3.1.1 Clonogenic Assays

Clonogenic cell-survival assays have been considered the standard for judging cellular response to irradiation (Hall and Giaccia 2006). The resulting *in vitro* survival curve measurements could

TABLE 4.1 Comparison of Radiation Therapy Oncology Group and Common Terminology Criteria Adverse Events v3.0 Scoring of Radiation Pneumonitis

Grade	Radiation Therapy Oncology Group	Common Terminology Criteria for Adverse Events
1	Mild symptoms of dry cough or dyspnea on exertion	Asymptomatic, radiographic findings only
2	Persistent cough requiring narcotic, antitussive agents or dyspnea with minimal effort, but not at rest	Symptomatic, not interfering with activities of daily living
3	Severe cough unresponsive to narcotic antitussive agent or dyspnea at rest; clinical or radiological evidence of acute pneumonitis; intermittent oxygen or steroids may be required	Symptomatic, interfering with activities of daily living; oxygen indicated
4	Severe respiratory insufficiency; continuous oxygen or assisted ventilation	Life-threatening; ventilatory support indicated
5	Fatal	Fatal

be utilized to measure radioresponsiveness as described by the seminal work of Fertil and Malaise.

4.3.1.2 Oxygen Status Markers

Oxygen deprivation or hypoxia is an important phenomenon in solid tumors, which leads to genetic instability and induction of aggressive phenotypes (Bristow and Hill 2008). Tumors containing regions of acute and chronic hypoxia or anoxia can herald a negative clinical prognosis for cancer patients owing to local resistance and a greater tendency for systemic metastases. *In vitro* and *in vivo* experiments have demonstrated that at partial pressures of oxygen below 10 mm Hg, tumor cells become relatively resistant to radiotherapy. This is due to decreased fixation and increased repair of potentially lethal DNA DSBs produced by radiation-induced free radicals (Lehnert 2008). In a recent study, osteopontin in serum has been shown to be a good correlate of tumor hypoxia and a significant prognostic factor for relapse risk in lung cancer (Le et al. 2006). Similarly, a pro-angiogenic protein (vascular endothelial growth factors) has been correlated with both poor rectal tumor control (Zlobec et al. 2005) and worsening of radiotherapy outcomes in prostate cancer (Green et al. 2007).

4.3.1.3 Phenotypic Progression

Tumor progression biomarkers constitute a set of markers that are related to tumor invasiveness, which includes the matrix metalloproteinases (MMP) and tissue inhibitors of metalloproteinases (TIMP; Susskind et al. 2003). Interestingly, Vorotnikova, Tries, and Braunhut (2004) determined that inhibition of MMPs by TIMP was sufficient to block radiation-induced apoptosis. Unsal et al. (2007) showed that MMP-9 expression correlates with poor tumor response in patients with locally advanced rectal cancer undergoing preoperative chemoradiotherapy.

4.3.1.4 Proliferation Markers

Proliferation biomarkers include immortality and cell-cycle progression factors. A randomized study of postradiotherapy outcomes in a cohort of 304 head and neck cancer patients reported that the apoptosis gene (*BCL-2*) and epidermal growth factor receptor (EGFR) expressions increased the time to local and regional failures. Furthermore, increasing the cell-cycle cyclin D1 expression in these patients decreased the time to nodal failure (Ataman et al. 2004; Bentzen et al. 2005). In lung cancer, p53 expression and the apoptotic index (a correlate of *BCL-2*) were shown to be prognostic factors with regard to local control in patients with inoperable non-small cell lung cancer (NSCLC) treated with radiotherapy (Langendijk et al. 2000).

4.3.1.5 Growth Factors

The inverse relationship between EGFR expression levels and the effectiveness of radiotherapy to achieve tumor control has been long established, whereas ionization radiation activates EGFR signaling leading to radioresistance by inducing cell proliferation and enhanced DNA repair (Akimoto et al. 1999; Dent et al. 1999; Milas et al. 2004). Irradiation leads to EGFR tyrosine phosphorylation (a gene expression activation process) and increased proliferation during treatment through downstream activation of PI3K/AKT or RAF/MAPK pathways (Li, Dowbenko, and Lasky 2002).

4.3.1.6 Genetic Variants

The study of the role of genetic variation in radiosensitivity has witnessed high activity in recent years in radiation oncology, with focus not only on DNA damage and repair genes such as (TP53, XRCC1, XRCC3, LIG4, APEX) but also on cell-cycle checkpoint control (CDKN1A), response to oxidative stress (e.g., (SOD2)), and induction of apoptosis (BCL2); (West, Elliott, and Burnet 2007).

There are several ongoing SNPs genotyping initiatives in radiation oncology, including the pan-European GENEPI project (Baumann, Hölscher, and Begg 2003), the British RAPPER project (Burnet et al. 2006), the Japanese RadGenomics project (Iwakawa et al. 2006), and the US Gene-PARE project (Ho et al. 2006). As an example, it was shown recently that polymorphisms in repair genes might be involved in poor outcomes postradiotherapy in NSCLC (Yoon et al. 2005; Su et al. 2007).

4.3.1.7 High-Throughput Signatures

We will now discuss examples from high-throughput gene expression (genomics) using RNA microarrays and protein expression (proteomics) analysis using mass spectroscopy (MS). An RNA microarray is a multiplex technology that allows for the analysis of thousands of gene expressions from multiple samples at the same time, in which short nucleotides in the array hybridize to the sample, which is subsequently quantified by fluorescence (Schena et al. 1995). Recently, Klopp et al. (2008) used microarray profiling to identify a set of 58 genes using pretreatment biopsy samples that were differentially expressed in cervical cancer patients with and without recurrence.

Another interesting set of RNA markers include microRNAs (miRNAs), which are a family of small noncoding RNA molecules (~22 nucleotides), each of which can suppress the expression of hundreds of protein-coding gene (targets). Specific miRNAs can control related groups of pathways. Thus, miRNAs comprise a particularly powerful part of the cellular gene expression control system and are particularly attractive as potential biomarkers. We have developed a machine-learning algorithm for detecting miRNA targets (Wang and El Naqa 2008) and showed that miR-200 miRNA clusters could be used as prognostic marker in advanced ovarian cancer (Hu et al. 2009). In another example of the usefulness of miRNAs as biomarkers, several miRNA (miR-137, miR-32, miR-155, let-7a) expression levels have been correlated with poor survival and relapse in NSCLC (Yu et al. 2008).

MS is an analytical technique for determinining the molecular composition of a sample (Sparkman 2000). This tool is the main vehicle for large-scale protein profiling, also known as proteomics (Twyman 2004). Allal et al. (2004) applied proteomics to study radioresistance in rectal cancer. They identified tropomodulin, heat shock protein 42, beta-tubulin, annexin V, and calsenilin as radioresistive biomarkers, and keratin type I, notch 2 protein homolog, and DNA repair protein RAD51L3 as radiosensitive biomarkers. Zhu et al. (2009) applied proteomics to study tumor response in cervical carcinoma cancer. They found that increased expression of S100A9 and galectin-7 and decreased expression of NMP-238 and HSP-70 associated with significantly increased local response to concurrent chemoradiotherapy in cervical cancer.

4.3.2 Biomarkers for Early Normal Tissues Toxicities

Most of the research in the radiation biomarker field has focused primarily on tumor response. However, much recent work focuses on normal tissues (Okunieff et al. 2008). Research in this area has focused primarily on inflammatory cytokines, gene expressions, and genetic variations.

4.3.2.1 Cytokines

Cytokines is a family of signaling polypeptides, mainly proteins that are secreted by different immune cells that mediate inflammatory and immune reactions (Abbas, Lichtman, and Pillai 2007). Circulating cytokines such as transforming growth factor β1 (TGF-β1) and the interleukins have been shown to play an important role in radiation-related inflammatory responses. For instance, TGF-β1 has been widely related to play an important role in RP and fibrosis (Kong et al. 1996; Anscher, Kong, and Jirtle 1998; Kong et al. 2001; Anscher et al. 2003). However, there are conflicting reports concerning its role in RP in breast and lung cancers, possibly due to its dual role as a pro- and anti-inflammatory factor (Hill 2005). Another related set of markers is composed of variations in intercellular adhesion molecules (ICAM-1), which have been reported to occur during irradiation of the lung causing increase in the arrest of inflammatory cells (macrophages and neutrophils) in the capillaries (Hallahan, Geng, and Shyr 2002).

In several recent animal studies cytokines have been shown to act by mediating the activation and translocation of nuclear transcription factor kappa B (NF-κB), which has been identified as a good candidate for therapeutic intervention by inhibiting NF-κB activation via caffeic acid phenethyl ester (Linard et al. 2004; Chen et al. 2005; Jones et al. 2005). In addition, humoral factors including chemokines and other adhesion molecules such as MCP-1, MIP, and selectins (Johnston et al. 1998; Chen, Okunieff, and Ahrendt 2003) can further modulate the expression of the fibrotic cytokines bFGF and TGF-β (Johnston et al. 1996; Anscher, Kong, and Jirtle 1998; Kong et al. 2001; Chen et al. 2005), IL-1α (Chen et al. 2005; Hart et al. 2005), and IL-6 (Arpin et al. 2005; Chen et al. 2005).

4.3.2.2 Genetic Variants

As reviewed by Alsner et al. and West et al., normal tissue SNPs in multiple genes have been found to correlate with toxicity for several end points (West, Elliott, and Burnet 2007; Alsner, Andreassen, and Overgaard 2008). Andreassen et al. have reported on the presence of positive correlations between five types of SNPs (TGF-β1, SOD2, XRCC1, XRCC3, APEX) and the risk of radiosensitivity in cases of fibrosis and telangiectasia in breast cancer after radiotherapy (Andreassen et al. 2003). Particularly, it has been found that SNPs in TGF-β1, SOD2, and XRCC1 correlated with an increased risk of fibrosis. An SNP in the XRCC3 gene correlated with increased risk of subcutaneous fibrosis as well as telangiectasia. Burri et al. (2009) reported an association of SNPs in the SOD2, XRCC1, and XRCC3 genes with postradiotherapy adverse events for prostate cancer patients. Specifically, patients with a particular XRCC1 SNP were more likely to develop erectile dysfunction, whereas patients with a particular SOD2 SNP or a combination of the SOD2 SNP and an XRCC3 SNP exhibited a significant increase in the risk of late rectal bleeding.

4.3.2.3 High-Throughput Signatures

RNA expression profiles for studying early and late effects of radiation are still in their early stage of research (Kruse and Stewart 2007; Okunieff et al. 2008). For instance, Quamby et al. analyzed a cohort of six breast cancer patients with three patients diagnosed with fibrosis and three without fibrosis, in which they identified that mRNA transcripts relevant to TNFα, PDGF, and nerve growth factor were elevated in the fibrosis group (Quamby et al. 2002). Rieger et al. used microarrays to measure radiotherapy response in lymphoblastoid cells derived from 14 patients with acute radiation toxicity from different cancer sites (Rieger et al. 2004). The toxicities were scored using the RTOG criteria and included mainly skin reactions and some cases of pneumonitis, mucositis, diarrhea, and edema. A signature of 24 of 52 genes found by hierarchical clustering (see Figure 4.3) was correlated with radiation toxicity. The gene list included DNA repair, general stress response, the ubiquitin/proteasome protein degradation pathway, and apoptosis genes. In this case, MS is used as a tool for screening proteins that could be associated with normal tissues radiation response.

In a feasibility proteomics screening study of RP in locally advanced NSCLC patients, we selected a control and a disease case (Oh et al. 2009; Spencer et al. 2009). The control patient developed no adverse health conditions throughout a follow-up period of 14 months. The disease case selected for the study died due to a severe RP episode 1 month after the end of the treatment. In each case, a serum sample was drawn before the treatment and again at the last available follow-up, which was then submitted for liquid chromatography mass spectrometry analysis with the peptide features extracted. To separate radiation treatment effects from hypersensitivity effects, we developed a three-way strategy to screen relevant proteins in each case from precontrol, postcontrol, and disease samples as shown in Figures 4.4a through 4.4c.

The preliminary screening of this strategy yielded candidate biomarker proteins that were classified into three categories according to their potential functional roles in radiation treatment response, as shown in Figure 4.4d. Thirteen proteins that were identified were associated with general radiation response. These candidate proteins shared radiation-induced concentration changes in both the control patient and the patient who had developed RP. Another 39 proteins of interest changed over the course of treatment in the RP patients out of only 56 postradiotherapy candidates between the control and disease cases, indicating their potential roles in disease onset. Functional analysis of the RP group enriched the pathways related to inflammatory response, tissue injury, and cell-mediated and humoral immune response, which includes glycoproteins, complement factors, inter-alpha-trypsin inhibitors, and gelsolin/plasma kallikrein precursors. The radiation group enriched the pathways related to cell death, free radical scavenging, and immune response cross talk. A third set of 24 candidate proteins displayed concentration

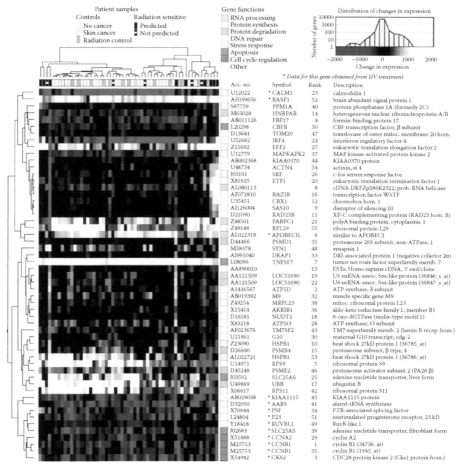

Patient samples

Controls — No cancer, Skin cancer, Radiation control

Radiation sensitive — Predicted, Not predicted

Gene functions — RNA processing, Protein synthesis, Protein degradation, DNA repair, Stress response, Apoptosis, Cell cycle regulation, Other

Distribution of changes in expression — Number of genes (10000, 1000, 100, 10, 1); Change in expression (−2000, −1000, 0, 1000, 2000)

* *Data for this gene obtained from UV treatment*

Acc. no.	Symbol	Rank	Description
U12022	* CALM1	23	calmodulin 1
AF039656	* BASP1	52	brain abundant signal protein 1
S87759	PPM1A	40	protein phosphatase 1A (formerly 2C)
M65028	HNRPAB	14	heterogeneous nuclear ribonucleoprotein A/B
AB011126	FBP17	4	formin-binding protein 17
L20298	CBFB	50	CBF transcription factor, β subunit
D13641	TOM20	47	translocase of outer mitoc. membrane 20 hom.
U52682	IRF4	24	interferon regulatory factor 4
Z11692	EEF2	27	eukaryotic translation elongation factor 2
U12779	MAPKAPK2	37	MAP kinase-activated protein kinase 2
AB002368	KIAA0370	44	KIAA0370 protein
U48734	ACTN4	54	actinin, α 4
J03161	SRF	26	c-fos serum response factor
X81625	ETF1	20	eukaryotic translation termination factor 1
AL080113		8	cDNA DKFZp586K2322; prob. RNA helicase
AF072810	BAZIB	16	transcription factor WSTF
U35451	CBX1	12	chromobox hom. 1
AI126004	SAS10	9	disrupter of silencing 10
D21090	RAD23B	11	XP-C complementing protein (RAD23 hom. B)
Z48501	PABPC1	21	polyA binding protein, cytoplasmic 1
Z49148	RPL29	55	ribosomal protein L29
AL022318	* APOBECIL	6	similar to APOBEC1
D44466	PSMD1	31	proteasome 26S subunit, non-ATPase, 1
M58378	SYN1	48	synapsin 1
AI991040	DRAP1	33	DR1-associated protein 1 (negative cofactor 2α)
L08096	TNFSF7	7	tumor necrosis factor superfamily memb. 7
AA890010		13	ESTs; Homo sapiens cDNA, 3' end/clone
AA121509	LOC51690	19	U6 snRNA-assoc. Sm-like protein (36846_s_at)
AA121509	LOC51690	22	U6 snRNA-assoc. Sm-like protein (36847_s_at)
A1436567	ATP5D	2	ATP synthase, δ subunit
AB019392	M9	32	muscle specific gene M9
Z49254	MRPL23	38	mitoc. ribosomal protein L23
X15414	AKRIB1	36	aldo-keto reductase family 1, member B1
D16581	NUDT1	18	8-oxo-dGTPase (nudix-type motif 1)
X83218	ATP5O	28	ATP synthase, O subunit
AF023676	TM7SF2	43	TM7 superfamily memb. 2 (lamin B recep. hom.)
U11861	G10	30	maternal G10 transcript; edg-2
Z23090	HSPB1	10	heat shock 27kD protein 1 (36785_at)
D26600	PSMB4	15	proteasome subunit, β type, 4
AL022721	HSPB1	53	heat shock 27kD protein 1 (36786_at)
U14971	RPS9	5	ribosomal protein S9
D45248	PSME2	46	proteasome activator subunit 2 (PA28 β)
J03592	SLC25A6	25	adenine nucleotide transporter, liver form
U4869	UBB	17	ubiquitin B
X06617	RPS11	42	ribosomal protein S11
AB029038	* KIAA1115	45	KIAA1115 protein
D32050	* AARS	41	alanyl-tRNA synthetase
X70944	* PSF	34	PTB-associated splicing factor
L24804	* P23	51	unstimulated progesterone receptor, 23 kD
Y18418	* RUVBL1	49	RuvB-like 1
J02683	* SLC25A5	39	adenine nucleotide transporter, fibroblast form
X51688	* CCNA2	29	cyclin A2
M25753	* CCNB1	1	cyclin B1 (34736_at)
M25753	* CCNB1	35	cyclin B1 (1945_at)
X54942	* CKS2	3	CDC28 protein kinase 2 (Cks1 protein hom.)

FIGURE 4.3 A heat map summary of hierarchical clustering results of microarray RNA expression for prediction of radiation toxicity. (From Rieger, K. E., et al. 2004. *Proc Natl Acad Sci U S A* 101:6635–40. With permission.)

changes due to radiation, but only in the control case. These candidates may be involved in protective mechanisms against excessive inflammation. However, these proteins and their exact roles in RP remain to be verified.

4.3.2.4 Clinical Examples

Here we present selected clinical examples of acute toxicities, in which biomarkers could play an important role in predicting and aiding adaptive treatment plans not only based on shrinkage in tumor volume as seen in the images but also based on expected acute adverse events that could be analyzed pretreatment or during treatment. The selected cases are included in Section 4.3.2.4.1.

4.3.2.4.1 Acute Esophagitis due to Chemoradiotherapy for Lung Cancer

Acute esophagitis is a common complication of concurrent chemoradiotherapy in NSCLC patients. Symptoms generally develop by week 3 or 4 of radiotherapy and can compromise the patient's fluid balance and nutritional status, resulting in

treatment interruptions and limited ability to complete the prescribed course (Bradley et al. 2004).

Models to predict the development of acute esophagitis based on dosimetric variables continue to be developed (Maguire et al. 1999; Werner-Wasik et al. 2000; Bradley et al. 2004; Kim et al. 2005; Watkins et al. 2009). No known independent biomarkers have been identified for prediction of the development of esophagitis as compared to pneumonitis (Chen, Okunieff, and Ahrendt 2003). However, in an animal study, Epperly et al. demonstrated that the administration of manganese superoxide dismutase-plasmid/liposome prior to a single fraction of radiation protected the mice from lethal esophagitis (Epperly et al. 2001). This was explained in terms of the resulting significant increase in peroxidized lipids and reduction in overall antioxidant levels, reduced thiols, and decreased glutathione. These could be regarded as potential biomarker candidates for esophagitis onset.

4.3.2.4.2 Severe Mucositis due to Head and Neck Radiotherapy

Head and neck cancer involves a group of pathologically related cancers that originate from the upper aerodigestive tracts (oral

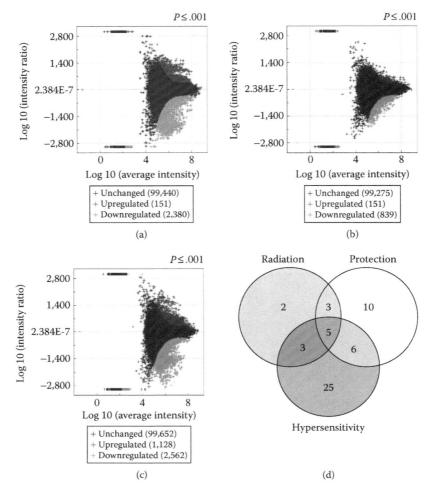

FIGURE 4.4 Feasibility proteomics screening of radiation pneumonitis data. First, categorization of upregulated and downregulated features within ratio data. The log ratio is plotted as a function of average intensity for (a) postcontrol to precontrol, (b) predisease to postdisease, and (c) predisease to precontrol. (d) Venn diagram summary of the corresponding proteins divided into radiation, protection, and hypersensitivity groups.

cavity, larynx, pharynx, and so on). Approximately 90% of these are classified as squamous cell carcinoma, which accounts for 4% of all malignancies (Jemal et al. 2008). Radiotherapy treatment may lead to mucositis, an early inflammation of the mucous membranes lining the digestive tract. This toxicity could further lead to swallowing dysfunction. In a study of 13 patients, Handschel et al. (1999) observed that an increased expression of adhesion molecules including endothelial ICAM-1 and E-selectin were associated with the severity of mucositis. The therapeutic addition of several growth factors were investigated in targeted studies, including granulocyte colony stimulating factor (Sprinzl et al. 2001; Mantovani et al. 2003) and the fibroblast growth factor-7 (Brizel et al. 2008). However, randomized clinical trials have yet to show the efficacy of these interventions (Rosenthal and Trotti 2009). Ki et al. (2009) showed that a change in the mean C-reactive protein levels in serum closely correlated with the progression of the mean grade of mucositis according to fraction number.

4.3.2.4.3 Diarrhea due to Prostate Cancer Radiotherapy

According to the RTOG criteria, side effects occurring within 120 days from the start of radiotherapy for prostate cancer patients are considered to be acute radiation morbidity (Michalski et al. 2000). These toxicities are scored according to the RTOG and/ or the European Organization for Research and Treatment of Cancer (Cox, Stetz, and Pajak 1995). Typically, these toxicities manifest themselves as acute bowel toxicities with different degrees of diarrhea severity. Dose–volume relations were reported in association with radiation-induced acute diarrhea in prostate (Fiorino et al. 2009) and rectal cancers (Robertson et al. in press). Hille et al. (2008) reported that fecal calprotectin and lactoferrin are correlated with acute proctitis. However, Wedlake et al. (2008) showed no significant correlation of calprotectin or lactoferrin with bowel toxicities when evaluating calprotectin or plasma citrulline for bowel toxicities. Nevertheless, these biomarkers are thought to potentially be valuable if evaluated on large homogeneous populations (West and Davidson 2009).

4.3.2.4.4 *Skin Reactions due to Breast Cancer Radiotherapy*

Breast cancer is possibly one of the most well-studied cancers in terms of radiation sensitivity genetics (Popanda et al. 2009). Skin reactions due to radiation treatment are experienced by up to 95% of patients and appear within the first 70 days following treatment (Archambeau, Pezner, and Wasserman 1995). Early effects are generally present in the first 70 days following treatment and can be graded according to increasing severity using a number of available skin scoring systems. Suga et al. (2007) identified genetic variations in CD44 to be associated with adverse skin reactions. Lincza et al. (2009) showed that plasma coagulation and fibrinolysis factors [prothrombin factor 1+2 (F1+2) and plasminogen activator inhibitor-1 (Pai-1)] correlate with skin reactions measured by skin reflectance method. Thus, skin reflectance measurement may be a particularly promising method to help identify relevant biomarkers.

4.4 Issues, Controversies, and Problems

4.4.1 Solutions and Recommendations

The field of radiation oncology is witnessing a tremendously changing landscape due to current advances in imaging and biotechnology. This is being manifested in the transition process from cellular predictive assays, which produced very little in terms of clinical applicability, into new promising biomarkers in the era of "omics." To achieve success in this transition process of developing new predictive assays and discovering relevant biomarkers for radiotherapy studies, Bentzen (2008) described what he referred to as "7 steps to heaven." These seven steps emphasize the following: (1) validating, (2) improving study reporting and data analysis, (3) ensuring clinical data quality, (4) performing quality assurance and interlaboratory comparisons, (5) distinguishing between statistical, biological, and clinical significance, (6) big is beautiful (large collaborative studies), but less is more (in the sense of testing a well-conceived specific hypothesis), and (7) being critical, creative, and remaining enthusiastic. We share the belief that these steps, if applied properly, may help eliminate many of the existing controversies in the field and aid in achieving faster progress in translating rapid biomarker discoveries into clinical practice.

It is also noted that the complexity of radiation-induced damage and signaling interactions may hinder the applicability of any individual biomarker by itself. This has been a long-standing issue in the field of oncology biomarkers with limited focus toward a single gene or protein discovery. This single gene (protein) "hunting" approach has yielded, in many cases, conflicting results even in well-studied sites such as breast cancer (Dent et al. 2003; Payne et al. 2008). This issue has been noted by West et al., "It is unlikely for single markers studies to be robust" (West, Elliott, and Burnet 2007, p. 471), and it was emphasized again in a recent review article in *Journal of Clinical Oncology* that stated, "The combination of several of molecular biomarkers might constitute a marker profile with increased predictive power over any one marker alone" (Riesterer, Milas, and Ang 2007, p. 4081). Therefore, investigating the interactions between these biomarkers as part of a more comprehensive approach becomes an essential task as it presents a new technical challenge for radiobiology modelers. Moreover, radiation response is known to be multifactorial and involves interactions among several physical parameters along with the biological processes as depicted in Figure 4.5. This further constitutes a new challenge of properly integrating dose–volume metrics with complex biological responses. Therefore, in this "omics" era, the study of this physics–biology interaction could be better denoted as "biophysiomics." In this context, biophysiomics would refer to large-scale analysis of dose–volume response in conjunction with different biological processes as understood from radiogenomics, radioproteomics, and epigenetic studies.

In our recent work, we have shown a complementary relationship between dose–volume metrics (mean lung dose) and center of mass in the superior–inferior direction and candidate biomarkers (interleukin-6 [IL-6] and angiotensin-converting enzyme [ACE]) extracted based on literature survey in modeling radiation-induced pneumonitis in a pilot study of 19 patients (Craft et al. 2009). Specifically, our original dose–volume model was improved by 89% when multiple biomarkers (IL-6 and ACE) were combined with dose–volume information on leave-one-out cross-validation analysis (Bradley et al. 2007). This indicates

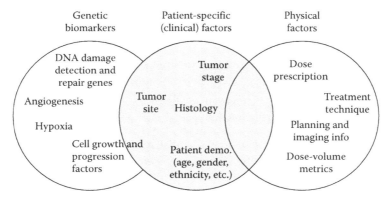

FIGURE 4.5 A schematic of the heterogeneous variable space for radiotherapy response. Note that there are several levels of interactions in this space; intracategory (e.g., biology–biology interactions), and intercategory (biology–physics interaction).

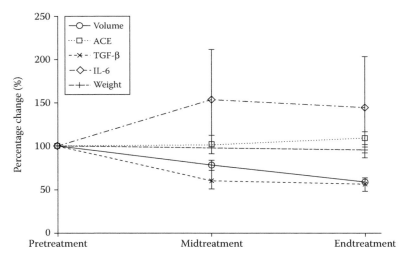

FIGURE 4.6   Changes in tumor volume and selected biomarkers during the course of radiotherapy measured at three time points (pretreatment, midtreatment, and end of treatment) for inoperable non-small cell lung cancer. Measurements are normalized with respect to pretreatment levels representing 100%. It is expected that changes in both tumor volume and biomarkers together will aid better adaptation of radiotherapy treatment to patients' expected response. TGF-β = transforming growth factor β; IL-6 = interleukin-6; ACE = angiotensin-converting enzyme.

the potential statistical power improvement in joint analysis of physical and biological determinants of radiation response.

4.5 Future Biomarker Research Directions

4.5.1 Adaptive Radiation Therapy and Biomarkers

Advances in 3D treatment planning and onboard imaging technology have recently enabled, at least partially, translating ART concepts into the clinical practice (Yan et al. 1997; Xing, Siebers, and Keall 2007; Tanyi and Fuss 2008; Godley et al. 2009; Yang et al. 2009). However, volumetric changes alone, despite their prognostic value (Tanyi and Fuss 2008), may not convey a comprehensive picture of tumor response and particularly the expected acute normal tissue toxicities. Therefore, biomarkers may provide a complementary role as a noninvasive tool for predicting the efficacy of the treatment in eradicating and providing an early warning against radiation-induced toxicity risks. In this scenario, imaging could be used to redefine the margins for replanning, and the extracted biomarkers could be used to convey a more accurate benefit–risk assessment.

Figure 4.6 shows an example of tumor volume changes during therapy and the corresponding changes in selected biomarkers that may aid in predicting early response and enable better adaptation of treatment plans in NSCLC patients receiving fractionated radiotherapy course.

4.6 Conclusions

Recent evolution in biotechnology has provided new opportunities for reshaping our understanding of radiotherapy response. However, its application in treatment planning and ART is still in

its infancy. Future directions should emphasize the development of new methods and techniques for personalizing the selection of cancer treatment, as well as for adapting treatment plans not only based on volumetric changes but also on biological changes as measured from relevant biomarkers. The intersection of physical and biological study mechanisms in predicting complex radiotherapy response is necessary to make significant progress toward the goal of personalized treatment planning and improved quality of life for radiotherapy patients.

Acknowledgments

This work was partially supported by NIH grants K25 CA128809 and R01 CA85181.

References

Abbas, A. K., A. H. Lichtman, and S. Pillai. 2007. *Cellular and Molecular Immunology.* Philadelphia, PA: Saunders Elsevier.

Akimoto, T., N. R. Hunter, L. Buchmiller, K. Mason, K. K. Ang, and L. Milas. 1999. Inverse relationship between epidermal growth factor receptor expression and radiocurability of murine carcinomas. *Clin Cancer Res* 5:2884–90.

Alaiya, A., M. Al-Mohanna, and S. Linder. 2005. Clinical cancer proteomics: promises and pitfalls. *J Proteome Res* 4:1213–22.

Allal, A. S., T. Kähne, A. K. Reverdin, H. Lippert, W. Schlegel, and M.-A. Reymond. 2004. Radioresistance-related proteins in rectal cancer. *Proteomics* 4:2261–9.

Alsner, J., C. N. Andreassen, and J. Overgaard. 2008. Genetic markers for prediction of normal tissue toxicity after radiotherapy. *Semin Radiat Oncol* 18:126–35.

Andreassen, C. N., and J. Alsner. 2009. Genetic variants and normal tissue toxicity after radiotherapy: A systematic review. *Radiother Oncol* 92:299–309.

Andreassen, C. N., J. Alsner, M. Overgaard, and J. Overgaard. 2003. Prediction of normal tissue radiosensitivity from polymorphisms in candidate genes. *Radiother Oncol* 69:127–35.

Anscher, M. S., F. M. Kong, and R. L. Jirtle. 1998. The relevance of transforming growth factor beta 1 in pulmonary injury after radiation therapy. *Lung Cancer* 19:109–20.

Anscher, M. S., L. B. Marks, T. D. Shafman, R. Clough, H. Huang, A. Tisch, et al. 2003. Risk of long-term complications after TFG-beta1-guided very-high-dose thoracic radiotherapy. *Int J Radiat Oncol Biol Phys* 56:988–95.

Archambeau, J. O., R. Pezner, and T. Wasserman. 1995. Pathophysiology of irradiated skin and breast. *Int J Radiat Oncol Biol Phys* 31:1171–85.

Arpin, D., D. Perol, J. Y. Blay, L. Falchero, L. Claude, S. Vuillermoz-Blas, et al. 2005. Early variations of circulating interleukin-6 and interleukin-10 levels during thoracic radiotherapy are predictive for radiation pneumonitis. *J Clin Oncol* 23:8748–56.

Ataman, O. U., S. M. Bentzen, G. D. Wilson, F. M. Daley, P. I. Richman, M. I. Saunders, et al. 2004. Molecular biomarkers and site of first recurrence after radiotherapy for head and neck cancer. *Eur J Cancer* 40:2734–41.

Bast, R. C., Jr., F. J. Xu, Y. H. Yu, S. Barnhill, Z. Zhang, and G. B. Mills. 1998. CA 125: The past and the future. *Int J Biol Markers* 13:179–87.

Baumann, M., T. Hölscher, and A. C. Begg. 2003. Towards genetic prediction of radiation responses: ESTRO's GENEPI project. *Radiother Oncol* 69:121–25.

Ben-Haim, S., and P. Ell. 2009. 18F-FDG PET and PET/CT in the evaluation of cancer treatment response. *J Nucl Med* 50:88–99.

Bentzen, S. M. 2006. Preventing or reducing late side effects of radiation therapy: Radiobiology meets molecular pathology. *Nat Rev Cancer* 6:702–713.

Bentzen, S. M. 2008. From cellular to high-throughput predictive assays in radiation oncology: Challenges and opportunities. *Semin Radiat Oncol* 18:75–88.

Bentzen, S. M., B. M. Atasoy, F. M. Daley, S. Dische, P. I. Richman, M. I. Saunders, et al. 2005. Epidermal growth factor receptor expression in pretreatment biopsies from head and neck squamous cell carcinoma as a predictive factor for a benefit from accelerated radiation therapy in a randomized controlled trial. *J Clin Oncol* 23:5560–7.

Bentzen, S. M., F. M. Buffa, and G. D. Wilson. 2008. Multiple biomarker tissue microarrays: Bioinformatics and practical approaches. *Cancer Metastasis Rev* 27:481–94.

Biersack, H. J., H. Bender, and H. Palmedo. 2004. FDG-PET in monitoring therapy of breast cancer. *Eur J Nucl Med Mol Imaging* 31(Suppl 1):S112–7.

Biomarkers Definitions Working Group. 2001. Biomarkers and surrogate endpoints: Preferred definitions and conceptual framework. *Clin Pharmacol Ther* 69:89–95.

Borst, G. R., J. S. Belderbos, R. Boellaard, E. F. Comans, K. De Jaeger, A. A. Lammertsma, et al. 2005. Standardised FDG uptake: A prognostic factor for inoperable non-small cell lung cancer. *Eur J Cancer* 41:1533–41.

Bradley, J. D., A. Hope, I. El Naqa, A. Apte, P. E. Lindsay, W. Bosch, et al. 2007. A nomogram to predict radiation pneumonitis, derived from a combined analysis of RTOG 9311 and institutional data. *Int J Radiat Oncol Biol Phys* 69:985–92.

Bradley, J., J. O. Deasy, S. Bentzen, and I. El Naqa. 2004. Dosimetric correlates for acute esophagitis in patients treated with radiotherapy for lung carcinoma. *Int J Radiat Oncol Biol Phys* 58:1106–13.

Bradley, J., W. L. Thorstad, S. Mutic, T. R. Miller, F. Dehdashti, B. A. Siegel, et al. 2004. Impact of FDG-PET on radiation therapy volume delineation in non-small-cell lung cancer. *Int J Radiat Oncol Biol Phys* 59:78–86.

Brawer, M. K. 2001. *Prostate Specific Antigen.* New York: Marcel Dekker.

Bristow, R. G., and R. P. Hill. 2008. Hypoxia and metabolism: Hypoxia, DNA repair and genetic instability. *Nat Rev Cancer* 8:180–92.

Brizel, D. M., B. A. Murphy, D. I. Rosenthal, K. J. Pandya, S. Gluck, H. E. Brizel, et al. 2008. Phase II study of palifermin and concurrent chemoradiation in head and neck squamous cell carcinoma. *J Clin Oncol* 26:2489–96.

Brunzel, N. A. 1994. *Fundamentals of Urine and Body Fluid Analysis.* Philadelphia, PA: Saunders.

Burnet, N. G., R. M. Elliott, A. Dunning, and C. M. L. West. 2006. Radiosensitivity, Radiogenomics and RAPPER. *Clin Oncol* 18:525–28.

Burri, R. J., R. G. Stock, J. A. Cesaretti, D. P. Atencio, S. Peters, C. A. Peters, et al. 2009. Association of single nucleotide polymorphisms in SOD2, XRCC1 and XRCC3 with susceptibility for the development of adverse effects resulting from radiotherapy for prostate cancer. *Radiat Res* 170:49–59.

Chen, M. F., P. C. Keng, P. Y. Lin, C. T. Yang, S. K. Liao, and W. C. Chen. 2005. Caffeic acid phenethyl ester decreases acute pneumonitis after irradiation in vitro and in vivo. *BMC Cancer* 5:158.

Chen, Y., O. Hyrien, J. Williams, P. Okunieff, T. Smudzin and P. Rubin. 2005. Interleukin (IL)-1A and IL-6: Applications to the predictive diagnostic testing of radiation pneumonitis. *Int J Radiat Oncol Biol Phys* 62:260–6.

Chen, Y., P. Okunieff, and S. A. Ahrendt. 2003. Translational research in lung cancer. *Semin Surg Oncol* 21:205–19.

Cox, J. D., J. Stetz, and T. F. Pajak. 1995. Toxicity criteria of the Radiation Therapy Oncology Group (RTOG) and the European Organization for Research and Treatment of Cancer (EORTC). *Int J Radiat Oncol Biol Phys* 31:1341–46.

Craft, J., S. Spencer, D. Almiron, J. D. Bradley, J. O. Deasy, and I. El Naqa. 2009. Integrating serum biomarkers and dose-volume metrics to predict radiation pneumonitis. *ASTRO 51st Annual Meeting*, Chicago, IL.

Crick, F. 1970. Central dogma of molecular biology. *Nature* 227:561–3.

Das, P., J. M. Skibber, M. A. Rodriguez-Bigas, B. W. Feig, G. J. Chang, R. A. Wolff, et al. 2007. Predictors of tumor response and downstaging in patients who receive preoperative chemoradiation for rectal cancer. *Cancer* 109:1750–55.

Dent, P., A. Yacoub, J. Contessa, R. Caron, G. Amorino, K. Valerie, et al. 2003. Stress and radiation-induced activation of multiple intracellular signaling pathways. *Radiat Res* 159:283–300.

Dent, P., D. B. Reardon, J. S. Park, G. Bowers, C. Logsdon, K. Valerie, et al. 1999. Radiation-induced release of transforming growth factor alpha activates the epidermal growth factor receptor and mitogen-activated protein kinase pathway in carcinoma cells, leading to increased proliferation and protection from radiation-induced cell death. *Mol Biol Cell* 10:2493–506.

Dundas, G. S., A. T. Porter, and P. M. Venner. 1990. Prostate-specific antigen. Monitoring the response of carcinoma of the prostate to radiotherapy with a new tumor marker. *Cancer* 66:45–8.

El Naqa, I., P. W. Grigsby, A. Apte, E. Kidd, E. Donnelly, D. Khullar, et al. 2009. Exploring feature-based approaches in PET images for predicting cancer treatment outcomes. *Pattern Recognit* 42:1162–71.

Emami, B., J. Lyman, A. Brown, L. Coia, M. Goitein, J. E. Munzenrider, et al. 1991. Tolerance of normal tissue to therapeutic irradiation. *Int J Radiat Oncol Biol Phys* 21:109–22.

Engle, L. J., C. L. Simpson, and J. E. Landers. 2006. Using high-throughput SNP technologies to study cancer. *Oncogene* 25:1594–601.

Epperly, M. W., V. E. Kagan, C. A. Sikora, J. E. Gretton, S. J. Defilippi, D. Bar-Sagi, et al. 2001. Manganese superoxide dismutase-plasmid/liposome (MnSOD-PL) administration protects mice from esophagitis associated with fractionated radiation. *Int J Cancer* 96:221–31.

Erichsen, H. C. and S. J. Chanock. 2004. SNPs in cancer research and treatment. *Br J Cancer* 90:747–51.

Fajardo, L. F., M. Berthrong, and R. E. Anderson. 2001. *Radiation Pathology.* New York: Oxford University Press.

Fertil B and E. Malaise. 1985. Intrinsic radiosensitivity of human cell lines is correlated with radioresponsiveness of human tumors: Analysis of 101 published survival curves. *Int J Radiat Oncol Biol Phys.* 11:1699–707.

Fiorino, C., F. Alongi, L. Perna, S. Broggi, G. M. Cattaneo, C. Cozzarini, et al. 2009. Dose-volume relationships for acute bowel toxicity in patients treated with pelvic nodal irradiation for prostate cancer. *Int J Radiat Oncol Biol Phys* 75:29–35.

Godley, A., E. Ahunbay, C. Peng, and X. A. Li 2009. Automated registration of large deformations for adaptive radiation therapy of prostate cancer. *Med Phys* 36:1433–41.

Green, M. M. L., C. T. Hiley, J. H. Shanks, I. C. Bottomley, C. M. L. West, R. A. Cowan, et al. 2007. Expression of vascular endothelial growth factor (VEGF) in locally invasive prostate cancer is prognostic for radiotherapy outcome. *Int J Radiat Oncol Biol Phys* 67:84–90.

Hall, E. J. and A. J. Giaccia 2006. *Radiobiology for the Radiologist.* Philadelphia, PA: Lippincott Williams & Wilkins.

Hallahan, D. E., L. Geng, and Y. Shyr. 2002. Effects of intercellular adhesion molecule 1 (ICAM-1) null mutation on radiation-induced pulmonary fibrosis and respiratory insufficiency in mice. *J Natl Cancer Inst* 94:733–41.

Handschel, J., F.-J. Prott, C. Sunderkötter, D. Metze, U. Meyer, and U. Joos. 1999. Irradiation induces increase of adhesion molecules and accumulation of [beta]2-integrin-expressing cells in humans. *Int J Radiat Oncol Biol Phys* 45:475–81.

Hart, J. P., G. Broadwater, Z. Rabbani, B. J. Moeller, R. Clough, D. Huang, et al. 2005. Cytokine profiling for prediction of symptomatic radiation-induced lung injury. *Int J Radiat Oncol Biol Phys* 63:1448–54.

Hill, R. P. 2005. Radiation effects on the respiratory system. *BJR Suppl* 27:75–81.

Hille, A., E. Schmidt-Giese, R. M. Hermann, M. K. Herrmann, M. Rave-Frank, M. Schirmer, et al. 2008. A prospective study of faecal calprotectin and lactoferrin in the monitoring of acute radiation proctitis in prostate cancer treatment. *Scand J Gastroenterol* 43:52–8.

Ho, A. Y., D. P. Atencio, S. Peters, R. G. Stock, S. C. Formenti, J. A. Cesaretti, et al. 2006. Genetic predictors of adverse radiotherapy effects: The Gene-PARE project. *Int J Radiat Oncol Biol Phys* 65:646–55.

Hu, X., D. M. Macdonald, P. C. Huettner, Z. Feng, I. M. El Naqa, J. K. Schwarz, et al. 2009. A miR-200 microRNA cluster as prognostic marker in advanced ovarian cancer. *Gynecol Oncol* 114:457–64.

Ismail, S. M., T. A. Buchholz, M. Story, W. A. Brock, and C. W. Stevens. 2004. Radiosensitivity is predicted by DNA end-binding complex density, but not by nuclear levels of band components. *Radiother Oncol* 72:325–32.

Iwakawa, M., S. Noda, S. Yamada, N. Yamamoto, Y. Miyazawa, H. Yamazaki, et al. 2006. Analysis of non-genetic risk factors for adverse skin reactions to radiotherapy among 284 breast cancer patients. *Breast Cancer* 13:300–307.

Jemal, A., R. Siegel, E. Ward, Y. Hao, J. Xu, T. Murray, et al. 2008. Cancer statistics, 2008. *CA Cancer J Clin* 58:71–96.

Johnston, C. J., B. Piedboeuf, P. Rubin, J. P. Williams, R. Baggs, and J. N. Finkelstein. 1996. Early and persistent alterations in the expression of interleukin-1 alpha, interleukin-1 beta and tumor necrosis factor alpha mRNA levels in fibrosis-resistant and sensitive mice after thoracic irradiation. *Radiat Res* 145:762–7.

Johnston, C. J., T. W. Wright, P. Rubin, and J. N. Finkelstein 1998. Alterations in the expression of chemokine mRNA levels in fibrosis-resistant and -sensitive mice after thoracic irradiation. *Exp Lung Res* 24:321–37.

Jones, M. R., B. T. Simms, M. M. Lupa, M. S. Kogan, and J. P. Mizgerd 2005. Lung NF-{kappa}B activation and neutrophil recruitment require IL-1 and TNF receptor signaling during pneumococcal pneumonia. *J Immunol* 175:7530–5.

Ki, Y., W. Kim, J. Nam, D. Kim, D. Park, and D. Kim 2009. C-reactive protein levels and radiation-induced mucositis in patients with head-and-neck cancer. *Int J Radiat Oncol Biol Phys* 75:393–8.

Kim, T. H., K. H. Cho, H. R. Pyo, J. S. Lee, J. Y. Han, J. I. Zo, et al. 2005. Dose-volumetric parameters of acute esophageal toxicity in patients with lung cancer treated with three-dimensional conformal radiotherapy. *Int J Radiat Oncol Biol Phys* 62:995–1002.

Klopp, A. H., A. Jhingran, L. Ramdas, M. D. Story, R. R. Broadus, K. H. Lu, et al. 2008. Gene expression changes in cervical squamous cell carcinoma after initiation of chemoradiation and correlation with clinical outcome. *Int J Radiat Oncol Biol Phys* 71:226–36.

Kong, F. M., M. S. Anscher, T. A. Sporn, M. K. Washington, R. Clough, M. H. Barcellos-Hoff, et al. 2001. Loss of heterozygosity at the mannose 6-phosphate insulin-like growth factor 2 receptor (M6P/IGF2R) locus predisposes patients to radiation-induced lung injury. *Int J Radiat Oncol Biol Phys* 49:35–41.

Kong, F.-M., M. K. Washington, R. L. Jirtle, and M. S. Anscher 1996. Plasma transforming growth factor-[beta]1 reflects disease status in patients with lung cancer after radiotherapy: A possible tumor marker. *Lung Cancer* 16:47–59.

Kruse, J. J. and F. A. Stewart. 2007. Gene expression arrays as a tool to unravel mechanisms of normal tissue radiation injury and prediction of response. *World J Gastroenterol* 13:2669–74.

Langendijk, H., E. Thunnissen, J. W. Arends, J. de Jong, G. ten Velde, R. Lamers, et al. 2000. Cell proliferation and apoptosis in stage III inoperable non-small cell lung carcinoma treated by radiotherapy. *Radiother Oncol* 56:197–207.

Le, Q.-T., E. Chen, A. Salim, H. Cao, C. S. Kong, R. Whyte, et al. 2006. An evaluation of tumor oxygenation and gene expression in patients with early stage non-small cell lung cancers. *Clin Cancer Res* 12:1507–14.

Lehnert, S. 2008. *Biomolecular Action of Ionizing Radiation.* New York: Taylor & Francis.

Levine, E. A., M. R. Farmer, P. Clark, G. Mishra, C. Ho, K. R. Geisinger, et al. 2006. Predictive value of 18-fluoro-deoxy-glucose-positron emission tomography (18F-FDG-PET) in the identification of responders to chemoradiation therapy for the treatment of locally advanced esophageal cancer. *Ann Surg* 243:472–8.

Li, Y., D. Dowbenko, and L. A. Lasky. 2002. AKT/PKB phosphorylation of p21Cip/WAF1 enhances protein stability of p21Cip/WAF1 and promotes cell survival. *J Biol Chem* 277:11352–61.

Linard, C., C. Marquette, J. Mathieu, A. Pennequin, D. Clarencon and D. Mathe. 2004. Acute induction of inflammatory cytokine expression after gamma-irradiation in the rat: Effect of an NF-kappaB inhibitor. *Int J Radiat Oncol Biol Phys* 58:427–34.

Lincz, L. F., S. A. Gupta, C. R. Wratten, J. Kilmurray, S. Nash, M. Seldon, et al. 2009. Thrombin generation as a predictor of radiotherapy induced skin erythema. *Radiother Oncol* 90:136–40.

Ludwig, J. A. and J. N. Weinstein 2005. Biomarkers in cancer staging, prognosis and treatment selection. *Nat Rev Cancer* 5:845–56.

Maguire, P. D., G. S. Sibley, S. M. Zhou, T. A. Jamieson, K. L. Light, P. A. Antoine, et al. 1999. Clinical and dosimetric predictors of radiation-induced esophageal toxicity. *Int J Radiat Oncol Biol Phys* 45:97–103.

Mantovani, G., E. Massa, G. Astara, V. Murgia, G. Gramignano, M. R. Lusso, et al. 2003. Phase II clinical trial of local use of GM-CSF for prevention and treatment of chemotherapy- and concomitant chemoradiotherapy-induced severe oral mucositis in advanced head and neck cancer patients: An evaluation of effectiveness, safety and costs. *Oncol Rep* 10:197–206.

Marples, B., D. Longhurst, A. M. Eastham, and C. M. West. 1998. The ratio of initial/residual DNA damage predicts intrinsic radiosensitivity in seven cervix carcinoma cell lines. *Br J Cancer* 77:1108–14.

Mayeux, R. 2004. Biomarkers: Potential uses and limitations. *NeuroRx* 1:18–8.

Michael, E. R., D. T. Howard, B. L. Larry, M. H. Eric, A. K. Patrick, A. M. Alvaro, et al. 2006. PSA nadir predicts biochemical and distant failures after external beam radiotherapy for prostate cancer: A multi-institutional analysis. *Int J Radiat Oncol Biol Phys* 64:1140–50.

Michalski, J. M., J. A. Purdy, K. Winter, M. Roach, S. Vijayakumar, H. M. Sandler, et al. 2000. Preliminary report of toxicity following 3D radiation therapy for prostate cancer on 3DOG/RTOG 9406. *Int J Radiat Oncol Biol Phys* 46:391–402.

Milas, L., Z. Fan, N. H. Andratschke, and K. K. Ang 2004. Epidermal growth factor receptor and tumor response to radiation: In vivo preclinical studies. *Int J Radiat Oncol Biol Phys* 58:966–71.

Mutic, S., R. S. Malyapa, P. W. Grigsby, F. Dehdashti, T. R. Miller, I. Zoberi, et al. 2003. PET-guided IMRT for cervical carcinoma with positive para-aortic lymph nodes-a dose-escalation treatment planning study. *Int J Radiat Oncol Biol Phys* 55:28–35.

Nimeus-Malmstrom, E., M. Krogh, P. Malmstrom, C. Strand, I. Fredriksson, P. Karlsson, et al. 2008. Gene expression profiling in primary breast cancer distinguishes patients developing local recurrence after breast-conservation surgery, with or without postoperative radiotherapy. *Breast Cancer Res* 10:R34.

Nuyten, D. S. and M. J. van de Vijver. 2008. Using microarray analysis as a prognostic and predictive tool in oncology: Focus on breast cancer and normal tissue toxicity. *Semin Radiat Oncol* 18:105–14.

Ogawa, K., S. Murayama and M. Mori 2007. Predicting the tumor response to radiotherapy using microarray analysis. *Oncol Rep* 18:1243–8.

Oh, J., S. Spencer, C. Lichti, R. Townsend, C. Craft, J. Deasy, et al. 2009. Discovery of blood biomarkers for radiation pneumonitis by proteomics analysis. *Translational Advances in Radiation Oncology and Cancer Imaging Symposium*, St. Louis, MO.

Okunieff, P., Y. Chen, D. Maguire, and A. Huser. 2008. Molecular markers of radiation-related normal tissue toxicity. *Cancer and Metastasis Reviews* 27:363–74.

Payne, S. J., R. L. Bowen, J. L. Jones, and C. A. Wells 2008. Predictive markers in breast cancer—the present. *Histopathology* 52:82–90.

Plebani, M. and P. Carraro. 1997. Mistakes in a stat laboratory: Types and frequency. *Clin Chem* 43:1348–51.

Popanda, O., J. U. Marquardt, J. Chang-Claude, and P. Schmezer. 2009. Genetic variation in normal tissue toxicity induced by ionizing radiation. *Mutat Res* 667:58–69.

Pramana, J., M. W. Van den Brekel, M. L. van Velthuysen, L. F. Wessels, D. S. Nuyten, I. Hofland, et al. 2007. Gene expression profiling to predict outcome after chemoradiation in head and neck cancer. *Int J Radiat Oncol Biol Phys* 69:1544–52.

Prat, A., M. Parera, S. Peralta, M. A. Perez-Benavente, A. Garcia, A. Gil-Moreno, et al. 2008. Nadir CA-125 concentration in the normal range as an independent prognostic factor for optimally treated advanced epithelial ovarian cancer. *Ann Oncol* 19:327–31.

Quarmby, S., C. West, B. Magee, A. Stewart, R. Hunter, and S. Kumar 2002. Differential expression of cytokine genes in fibroblasts derived from skin biopsies of patients who developed minimal or severe normal tissue damage after radiotherapy. *Radiat Res* 157:243–8.

Rieger, K. E., W. J. Hong, V. G. Tusher, J. Tang, R. Tibshirani, and G. Chu. 2004. Toxicity from radiation therapy associated with abnormal transcriptional responses to DNA damage. *Proc Natl Acad Sci U S A* 101:6635–40.

Riesterer, O., L. Milas, and K. K. Ang. 2007. Use of molecular biomarkers for predicting the response to radiotherapy with or without chemotherapy. *J Clin Oncol* 25:4075–83.

Ritter, M. A., E. M. Messing, T. G. Shanahan, S. Potts, R. J. Chappell, and T. J. Kinsella 1992. Prostate-specific antigen as a predictor of radiotherapy response and patterns of failure in localized prostate cancer. *J Clin Oncol* 10:1208–17.

Robertson, J. M., M. Söhn, and D. Yan. 2010. Predicting grade 3 acute diarrhea during radiation therapy for rectal cancer using a cutoff-dose logistic regression normal tissue complication probability model. *Int J Radiat Oncol Biol Phys* 77:66–72.

Rosenthal, D. I. and A. Trotti. 2009. Strategies for managing radiation-induced mucositis in head and neck cancer. *Semin Radiat Oncol* 19:29–34.

Schena, M., D. Shalon, R. W. Davis, and P. O. Brown. 1995. Quantitative monitoring of gene expression patterns with a complementary DNA microarray. *Science* 270:467–70.

Sparkman, O. D. 2000. *Mass Spectrometry Desk Reference*. Pittsburgh, PA: Global View Pub.

Spencer, S., D. A. Bonnin, J. Deasy, O., J. Bradley, D. and I. El Naqa. 2009. *Bioinformatics Methods for Learning Radiation-Induced Lung Inflammation from Heterogeneous Retrospective and Prospective Data*. New York: Hindawi Publishing Corporation.

Spratlin, J. L., N. J. Serkova, and S. G. Eckhardt 2009. Clinical applications of metabolomics in oncology: A review. *Clin Cancer Res* 15:431–40.

Sprinzl, G. M., O. Galvan, A. de Vries, H. Ulmer, A. R. Gunkel, P. Lukas, et al. 2001. Local application of granulocyte-macrophage colony stimulating factor (GM-CSF) for the treatment of oral mucositis. *Eur J Cancer* 37:2003–9.

Su, D., S. Ma, P. Liu, Z. Jiang, W. Lv, Y. Zhang, et al. 2007. Genetic polymorphisms and treatment response in advanced non-small cell lung cancer. *Lung Cancer* 56:281–8.

Suga, T., A. Ishikawa, M. Kohda, Y. Otsuka, S. Yamada, N. Yamamoto, et al. 2007. Haplotype-based analysis of genes associated with risk of adverse skin reactions after radiotherapy in breast cancer patients. *Int J Radiat Oncol Biol Phys* 69:685–93.

Susskind, H., M. H. Hymowitz, Y. H. Lau, H. L. Atkins, A. N. Hurewitz, E. S. Valentine, et al. 2003. Increased plasma levels of matrix metalloproteinase-9 and tissue inhibitor of metalloproteinase-1 in lung and breast cancer are altered during chest radiotherapy. *Int J Radiat Oncol Biol Phys* 56:1161–9.

Svensson, J. P., L. J. Stalpers, R. E. Esveldt-van Lange, N. A. Franken, J. Haveman, B. Klein, et al. 2006. Analysis of gene expression using gene sets discriminates cancer patients with and without late radiation toxicity. *PLoS Med* 3:e422.

Tanyi, J. A. and M. H. Fuss 2008. Volumetric image-guidance: Does routine usage prompt adaptive re-planning? An institutional review. *Acta Oncol* 47:1444–53.

Torres-Roca, J. F. and C. W. Stevens. 2008. Predicting response to clinical radiotherapy: Past, present, and future directions. *Cancer Control* 15:151–6.

Twyman, R. M. 2004. *Principles of Proteomics*. New York: BIOS Scientific Publishers.

Tyburski, J. B., A. D. Patterson, K. W. Krausz, J. Slavik, A. J. Fornace, Jr., F. J. Gonzalez, et al. 2008. Radiation metabolomics. 1. Identification of minimally invasive urine biomarkers for gamma-radiation exposure in mice. *Radiat Res* 170:1–14.

Unsal, D., A. Uner, N. Akyurek, P. Erpolat, A. Dursun, and Y. Pak. 2007. Matrix metalloproteinase-9 expression correlated with tumor response in patients with locally advanced rectal cancer undergoing preoperative chemoradiotherapy. *Int J Radiat Oncol Biol Phys* 67:196–203.

Vider, M., R. Kashmiri, L. Hunter, B. Moses, W. R. Meeker, J. F. Utley, et al. 1974. Carcinoembryonic antigen (CEA) monitoring in the management of radiotherapeutic patients. *Oncology* 30:257–72.

von Ahlfen, S., A. Missel, K. Bendrat, and M. Schlumpberger. 2007. Determinants of RNA Quality from FFPE Samples. *PLoS ONE* 2:e1261.

Vorotnikova, E., M. Tries, and S. Braunhut 2004. Retinoids and TIMP1 prevent radiation-induced apoptosis of capillary endothelial cells. *Radiat Res* 161:174–84.

Wang, X. and I. M. El Naqa. 2008. Prediction of both conserved and nonconserved microRNA targets in animals. *Bioinformatics* 24:325–32.

Watkins, J. M., A. E. Wahlquist, K. Shirai, E. Garrett-Mayer, E. G. Aguero, J. A. Fortney, et al. 2009. Factors associated with severe acute esophagitis from hyperfractionated radiotherapy with concurrent chemotherapy for limited-stage small-cell lung cancer. *Int J Radiat Oncol Biol Phys* 74:1108–13.

Wedlake, L., C. Mcgough, C. Hackett, K. Thomas, P. Blake, K. Harrington, et al. 2008. Can biological markers act as non-invasive, sensitive indicators of radiation-induced effects in the gastrointestinal mucosa? *Aliment Pharmacol Ther* 27:980–7.

Werner-Wasik, M., E. Pequignot, D. Leeper, W. Hauck, and W. Curran. 2000. Predictors of severe esophagitis include use of concurrent chemotherapy, but not the length of irradiated esophagus: A multivariate analysis of patients with lung cancer treated with nonoperative therapy. *Int J Radiat Oncol Biol Phys* 48:689–96.

West, C. M. L. 1995. Intrinsic radiosensitivity as a predictor of patient response to radiotherapy. *Br J Radiol* 68:827–37.

West, C. M. L. and S. E. Davidson. 2009. Measurement tools for gastrointestinal symptoms in radiation oncology. *Curr Opin Support Palliat Care* 3:36–40.

West, C. M. L., M. J. McKay, T. Hölscher, M. Baumann, I. J. Stratford, R. G. Bristow, et al. 2005. Molecular markers predicting radiotherapy response: Report and recommendations from an International Atomic Energy Agency technical meeting. *Int J Radiat Oncol Biol Phys* 62:1264–73.

West, C. M. L., R. M. Elliott, and N. G. Burnet. 2007. The genomics revolution and radiotherapy. *Clin Oncol* 19:470–480.

Wouters, B. G. 2008. Proteomics: Methodologies and applications in oncology. *Semin Radiat Oncol* 18:115–125.

Xing, L., J. Siebers and P. Keall. 2007. Computational challenges for image-guided radiation therapy: Framework and current research. *Semin Radiat Oncol* 17:245–57.

Yan, D., J. Wong, F. Vicini, J. Michalski, C. Pan, A. Frazier, et al. 1997. Adaptive modification of treatment planning to minimize the deleterious effects of treatment setup errors. *Int J Radiat Oncol Biol Phys* 38:197–206.

Yang, D., S. R. Chaudhari, S. M. Goddu, D. Pratt, D. Khullar, J. O. Deasy, et al. 2009. Deformable registration of abdominal kilovoltage treatment planning CT and tomotherapy daily megavoltage CT for treatment adaptation. *Med Phys* 36:329–38.

Yoon, S. M., Y. C. Hong, H. J. Park, J. E. Lee, S. Y. Kim, J. H. Kim, et al. 2005. The polymorphism and haplotypes of XRCC1 and survival of non-small-cell lung cancer after radiotherapy. *Int J Radiat Oncol Biol Phys* 63:885–91.

Yu, S. L., H. Y. Chen, G. C. Chang, C. Y. Chen, H. W. Chen, S. Singh, et al. 2008. MicroRNA signature predicts survival and relapse in lung cancer. *Cancer Cell* 13:48–57.

Zhu, H., H.-p. Pei, S. Zeng, J. Chen, L.-f. Shen, M.-z. Zhong, et al. 2009. Profiling protein markers associated with the sensitivity to concurrent chemoradiotherapy in human cervical carcinoma. *J Proteome Res* 8:3969–76.

Zlobec, I., R. Steele, N. Nigam, and C. C. Compton. 2005. A predictive model of rectal tumor response to preoperative radiotherapy using classification and regression tree methods. *Clin Cancer Res* 11:5440–3.

Treatment Plan Optimization for Adaptive Radiation Therapy

Qiuwen Wu
Duke University Medical Center

Justus Adamson
Duke University Medical Center

Qingrong Jackie Wu
Duke University Medical Center

5.1 Introduction

In current standard radiation therapy process, patient anatomy is represented by the snapshot of computed tomography (CT) images at the simulation for treatment planning. The planned radiation dose distribution, obtained from either three-dimensional (3D) conformal radiation therapy (3DCRT) or intensity-modulated radiation therapy (IMRT) can have submillimeter precision through computer-based treatment planning. The improvement in the quality assurance of the treatment delivery system, as described in Chapter 15 in this book, can also achieve similar accuracy in geometry of the collimation components and <1%–2% on the radiation output. However, patient anatomy during the the treatment course is not static; the changes can be in the orders of centimeters. These deviations in patient anatomy from the time of initial simulation to the time of treatment delivery should be minimized or accounted for, to ensure that the optimal planned dose distribution is delivered to the patient.

Recent advances in image-guided radiation therapy (IGRT), such as the integrated imaging devices on board the medical linear accelerators (Dawson and Jaffray 2007), provides many opportunities to measure some of these deviations such as interfractional patient setup and internal tumor or organ motions and subsequent corrections by changing patient positioning. The correct and optimal ways of using these new technologies can lead to reduced planning margins and in turn improve the therapeutic ratio. For those uncertainties that cannot be corrected such as organ deformation, and residual errors in the correction,

margins are still required during the treatment planning. Another useful technology is the incorporation of deformable registrations to track minute regions of tumor or organs in 3D space, or voxels, throughout the treatment course (Brock 2010). The use of deformable registration, coupled with fast dose calculations and capability of handling multiple data sets in modern treatment planning systems, facilitates the estimation of the delivered dose to the patient anatomy with reasonable accuracy.

Many studies have shown that the deviations of the patient anatomy from simulation to treatment are patient-specific, that is, there is no optimal one-size-fits-all solution. An individualized approach should be taken in the treatment planning to find the optimal management for each patient. However, these deviations are generally not known a priori during the initial treatment planning. Therefore, the treatment plan in principle needs to be modified as well during the treatment course when the changes in anatomy are measured. Adaptive radiation therapy (ART) therefore usually means that multiple treatment planning sessions are performed during the treatment course with feedback from the measurement on the patient, including setup, patient internal organ motion, and tumor changes. Simply put, the goal of the IGRT is to measure and correct the uncertainties in the treatment process, and the goal of the ART is to account for those uncertainties that cannot be corrected by IGRT in the treatment planning, in order to ensure that the optimal planned dose distribution is the same as the final delivered dose distribution. The field of the ART is rapidly changing. This chapter presents the methodologies of different ART planning currently developed and used by different investigators.

5.1.1 Geometry-Based vs. Dose-Based Adaptive Radiation Therapy

For individualized treatment planning, patient-specific uncertainties are necessary to construct the margins for the planning. However, these uncertainties are unknown before the treatment begins and are measured during the treatment course. Therefore, the margins used in the initial plan are most of the time based on the characteristics for the general population, which is usually larger than for a given individual patient. After a number of treatment fractions are given and the measured uncertainties are adequate and better than the initial single CT to represent the patient anatomy during the treatment course, a patient-specific target position can be determined, and the margin can be constructed and used for replanning (Yan et al. 2000). This patient-specific target is, in principle, more accurate in depicting the true target position, and the margin is also smaller than the population-based margin. When the new targets are used in the modified planning in ART, the target coverage should be improved, the irradiated volume should decrease, and the doses to organ at risks (OAR) should also decrease.

In general, there are two types of ART in varying complexities: geometry-based and dose-based. In geometry-based ART, the feedback from the treatment takes the form of geometric measurements, including setup errors and organ motions and so on (de la Zerda, Armbruster, and Xing 2007; Wu et al. 2002; Yan et al. 2005). They are taken into account in the replanning and decision-making process through the margin calculation. In contrast, the dose-based ART requires that the doses delivered in prior treatment fractions be included in the replanning (de la Zerda, Armbruster, and Xing 2007; Wu et al. 2002). The former is simple to implement, as the IMRT and inverse planning may not be necessary. The latter is in general more complex and more effective as well.

5.1.2 Online vs. Offline

Another common classification of ART is defined based on when the decision of replanning is made. If the decision of replanning is made while the patient is on the treatment table or in the treatment room, then it is online ART; if it is done when the patient is off the treatment table, then this is offline ART. Traditionally, replanning is performed offline for several reasons. Treatment planning is a computing and resource-intensive process, the target and organ definition can take a certain amount of time to complete; multiple personnel including physicians, dosimetrists, and physicists are involved in the planning process in different roles; dose calculation, plan optimization and evaluation, and prescription dose also take significant computing time. Typically, all these cannot be completed on time while a patient is still on the table during the allotted treatment time. If the treatment time is prolonged as a result of online replanning, this may not only affect the patient throughput, but would also be detrimental to the exact issue that ART is adopted to handle

in the first place, that is, longer treatment time induces larger changes in patient geometry and anatomy. Therefore, the online ART can only be performed in limited and simplified fashion for some tumor sites and protocols.

5.1.3 Replan vs. Correction

The simplest form of online ART is the online correction, in which the patient position is changed according to the "target of the day" based on treatment-day volumetric imaging to correct the setup errors and rigid organ motions. In more advanced implementation, the gantry and collimator angle can be adjusted as well to correct for the target rotation, the multileaf collimator shape can be changed to accommodate the changes in target shape (Court et al. 2005; Rijkhorst et al. 2007). As the research and development in this area are fast proliferating, more complex strategies are being investigated and implemented. For example, the benefit of online replanning based on the target shape of the day has been reported (Lerma et al. 2009; Schulze et al. 2009; Wu et al. 2008). To reduce the inverse planning time, the optimal plan from the initial planning can be chosen as a starting point for the replanning and so on.

5.1.4 Single vs. Multiple Plan

Another example of simplified ART is to classify the patient population into different subgroups using discriminant analysis based on motion magnitude and other features in the measurement of the day. Each group has different margins. During initial planning, multiple plans are generated. Then, on the treatment day, based on the imaging and measurement of that day, the most appropriate plan can be chosen from the plan library for the treatment delivery. The library can be as simple as two plans. The number of plans in the library depends on the complexity of the problem, which typically varies with the tumor sites, the required precision of the adaptation, and the implementation requirement.

5.1.5 Margins and Uncertainties

Whenever a plan is generated, the uncertainties of executing the plan must be estimated, and some of them may be unexpected. Therefore, a margin is usually reserved to account for these uncertainties. There are some recipes available to calculate the margin (Stroom and Heijmen 2002; van Herk, Remeijer, and Lebesque 2002); however, their use should be thoroughly investigated to make sure that the conditions from which these margin recipes were derived are met in the problems where they are applied to. For example, when some forms of image guidance, either online correction or offline correction, are used in the treatment, and the corrective action is taken, the uncertainties should be the residual errors, not the ones before the correction. In general, the residual error is smaller, and therefore, a smaller margin should be used in ART when combined with IGRT. The margin is in general a dosimetric concept; the margin investigations are complex and depend on many factors, including characteristics of the underlining uncertainties that it compensates, the clinical treatment protocols, the frequency of corrections in image guidance, the modality of the treatment, and the dose distributions. These will not be covered in this chapter.

5.2 Inverse Planning for Intensity-Modulated Radiation Therapy

5.2.1 Objective Functions

In the inverse planning for IMRT, an objective function is typically used. The goal of the objective function is to simplify the description of the desirable dose distributions (Wu and Mohan 2000). A simple objective function can be written like this:

$$f = \sum_{n=1}^{N} P_n^{T} \cdot f_n^{T} + \sum_{m=1}^{M} P_m^{OAR} \cdot f_m^{OAR} \qquad (5.1)$$

Here, f_n^{T} is the objective function for nth target volume, and P_n^{T} is the weight (i.e., penalty for deviating from the specified constraints) for that target. Similarly, f_m^{OAR} is the objective for mth OAR and P_m^{OAR} is the corresponding penalty factor. The penalties are adjusted to reflect the overall treatment objectives and the priorities of each structure; they depend on the treatment site and the location and size of the target.

In dose-based objectives, only the dose values are needed, for example, the prostate is prescribed at 72 Gy, and the rectum is limited to 60 Gy. The quadratic functions shown below (see Equations 5.2 and 5.3) are used by many systems:

$$f^{T} = \frac{1}{N_{T}} \sum_{i=1}^{N_{T}} \left(D_i - D_0^{T} \right)^2 \qquad (5.2)$$

$$f^{OAR} = \frac{1}{N_{OAR}} \sum_{i=1}^{N_{OAR}} H\left(D_i - D_0^{OAR} \right) \cdot \left(D_i - D_0^{OAR} \right)^2 \qquad (5.3)$$

where N_{T} and N_{OAR} are the number of voxels in a target volume and in an OAR, respectively. These voxels can be uniformly spaced or randomly chosen to allow a fixed number of voxels to represent a structure. The subscripts n and m have been dropped for clarity. D_i is the dose in the ith voxel, D_0^{T} is the prescription dose (72 Gy for prostate in this case), and D_0^{OAR} is the tolerance dose for the OAR (60 Gy for rectum in this case). $H\left(D_i - D_0^{OAR} \right)$ is the Heaviside function defined as

$$H\left(D_i - D_0^{OAR} \right) = \begin{cases} 1, D_i > D_0^{OAR} \\ 0, D_i \le D_0^{OAR} \end{cases} \qquad (5.4)$$

In other words, only the OAR points with a dose greater than the tolerance dose will contribute to its objective function. Dose at a point in space is the sum of the dose contributions from all rays or beamlets from all the beams, given by

$$D_i = \sum K_{ij} \cdot \omega_j \qquad (5.5)$$

where ω_j is the weight/intensity for jth ray and is the optimization variable. The K_{ij} is the contribution of the jth ray to the ith point. The K_{ij} depends on patient geometry and photon beam characteristics. The goal of the optimization is to minimize the objective function f, which has a minimum value of zero when all criteria are met.

The matrix K_{ij}, also called dose kernel, is largely a sparse matrix, because each ray or beamlet only deposits the dose along the ray and its close neighbors, the dose falls off rapidly with increasing distance from the ray. In fact, the dose calculation during the optimization can be approximated to be primary fluence only or analytic Gaussian forms (Wu 2004; Wu et al. 2003), which is usually accurate enough for many cases. The kernel can be updated periodically during the iterations. The exception is in the tumor site where large inhomogeneities occur, such as the lung where the kernel can be asymmetric.

Dose–volume-based criteria differ from dose-based criteria, in which not only the dose values, but also the volume is specified. A slight modification can be made to the dose-based objective functions to accommodate the dose–volume criteria, as suggested by Bortfeld, Stein, and Preiser (1997) and used by many other investigators.

Figure 5.1 is a simple schematic example of one OAR, illustrating how the optimization based on dose–volume objectives works. The dose–volume constraint is specified as $V (>D_1) < V_1$. In other words, the volume that receives a dose greater than D_1 should be less than V_1. To incorporate the constraint into the objective function, another dose value, D_2, is sought so that in

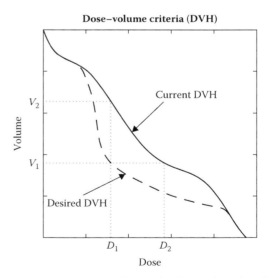

FIGURE 5.1 Optimization technique for dose–volume-based objectives. The dose–volume criteria are specified to constrain the volume that receives a dose greater than D_1 to less than V_1. This means that only the points with a dose between D_1 and D_2 are considered, and points with a dose above D_2 are ignored. (Reprinted from Wu, Q., and R. Mohan. 2000. *Med Phys* 27(4):1–11. With permission.)

the current dose–volume histogram (DVH) $V(D_2) = V_1$. The objective function may then be written as

$$f = \frac{1}{N}\left(p \cdot \sum_i H(D_2 - D_i) \cdot H(D_i - D_1) \cdot (D_i - D_1)^2 + \cdots \right) \quad (5.6)$$

That is, only the points with dose values between D_1 and D_2 contribute to the score. Therefore, they are the only ones that are penalized. For the target volumes, two types of dose–volume criteria may be specified to limit hot and cold spots. For instance, for the desired target dose of 80 Gy, it may be specified as V (>79 Gy) > 95% and V (>82 Gy) < 5%. In effect, the volume of target receiving 79 Gy or higher should be at least 95%, and the volume of the target receiving a dose greater than 82 Gy should be less than 5%. Dose-based criteria can be considered a subset of the dose–volume criteria in which the volume is set to an extreme value (0 or 100% as appropriate). Dose–volume criteria are more flexible for the optimization process and have greater control over the dose distributions. For example, referring to Figure 5.1, for an OAR, the dose–volume-based optimization process attempts to bring only the points between D_1 and D_2 into compliance with the constraint. As a result, the intensities of the rays passing through these dose points need to be adjusted. In contrast, the dose-based optimization process attempts to constrain all the points above D_1.

Because dose–volume criteria have many advantages, they are becoming the de facto standard in IMRT optimization. Dose-volume-based criteria are more tolerant than dose-based criteria. Dose-based optimization penalizes all the points above the dose limit, whereas dose–volume-based optimization penalizes only the subset of points within the lower end of the range of dose values above the dose limit. Dose–volume-based objectives are also intuitive and easy to use in treatment planning. In particular, dose–volume constraints can be placed directly on the DVH, a common tool for treatment plan evaluations, and changes made in the constraint parameters usually lead to predictable results. Some specific clinical outcomes are known to result from certain dose–volume end points, and these constraints can be incorporated directly into the treatment planning. An example is the fact that in prostate cancer patients receiving 3D conformal radiation treatment, late rectal bleeding has been correlated with the volume of rectum receiving doses of 46 and 77 Gy (Jackson et al. 2001; Skwarchuk et al. 2000). Further, in patients with lung cancer, radiation-induced pneumonitis has been associated with the mean dose of the lung and the volume of the lung receiving a dose of 20 Gy, the V_{20} (Graham et al. 1999; Hernando et al. 2001; Kwa et al. 1998). In patients with head and neck cancer, the salivary gland flow rate is highly sensitive to the mean radiation dose delivered; glands that receive a mean dose higher than 26 Gy threshold produce little saliva, with no improvement over time (Chao et al. 2001; Eisbruch et al. 1999).

The objectives described in this chapter are based on physical doses. Other types of objective functions such as biological objective functions based on tumor control probability, normal tissue complication probability, and equivalent uniform dose have also been used in the IMRT optimization. They can be found in Chapters 1 and 6 in this book.

5.2.2 Optimization Algorithms

Once the form of objective function has been defined and parameters for the objective function have been chosen, the next step is to find an optimal solution among many possible solutions, typically through iterative procedures. At each iteration, the intensity or weight for each beamlet is adjusted, the dose distributions are computed, and the objective function is evaluated. The objective function's value, also called *score*, is used to rank the competing solutions. Depending on the optimization algorithms used, further analysis of the objective function may be performed to determine the search direction and step size for the changes of the weight for each beamlet.

In gradient-based optimization algorithms, the derivatives of the objective function need to be computed, either analytically or numerically. This is also called the *deterministic* or *local-search* algorithm. Typically, the first-order derivative is computed, but some techniques need the second-order derivative as well. Generally speaking, the higher the order of the derivatives being computed, the faster the optimal solution can be found and the more efficient the optimization algorithm is. If the objective function is well-defined and its mathematical behavior is well-understood, as with the quadratic dose difference shown in Equations 5.1 through 5.3, analytical computation of the derivative is possible. Otherwise, the numerical calculation can be used, which is generally slower, requiring the objective functions to be evaluated many times. These derivatives will determine the direction of the changes of the beamlet weight and the step size at each iteration. Common algorithms are the steepest descent, the conjugate gradient, and the quasi-Newton's method (Bortfeld et al. 1994; Cho et al. 1998; Shepard et al. 2000; Spirou and Chui 1998; Wu and Mohan 2000; Xing et al. 1998). For example, in the steepest descent algorithm, the weight for one beamlet is adjusted according to the following:

$$\omega_j^{k+1} = \omega_j^k - \lambda \cdot \frac{\partial f}{\partial \omega_j} \quad (5.7)$$

Here, ω_j is the weight for jth beamlet, k is the iteration number, and $\frac{\partial f}{\partial \omega_j}$ is the derivative of the objective function on the weight for the jth beamlet. λ is a positive number, and its value is chosen so that the updated weights can lead to a minimum value of the objective function f afterward. Normally, a line-search algorithm is used to determine the value of λ. In general, only a limited number of iterations are required, which may be in the range of tens to a few hundred. The disadvantage is that it needs the information of the derivatives, and this limits the choices of the objective function formats. Another concern is that the objective function has to be of the convex nature, that is, there is only one global minimum; otherwise the search could potentially be trapped in the local minima.

Another type of optimization algorithm has a stochastic nature, represented by the *simulated annealing* (SA) algorithm and genetic algorithms (Ezzell 1996; Mageras and Mohan 1993; Morrill et al. 1995; Webb 1992). In this category, the computation of the derivatives of the objective function is not necessary. The direction of the change and the step size of the beamlet intensity are of "random" nature; the proposed change would be acceptable if these changes in the intensities would lead to a lower value of the objective function (assume the goal is to minimize the objective function). Typically, this would require many more iterations, in the order of tens of thousands, to find an acceptable solution, and it is not guaranteed that the solution found would be the optimal one. If there is no limit on the number of iterations and the computing time, then the solution can be optimal. Therefore, it generally takes longer for the SA algorithm to converge. However, there are many advantages of the SA algorithm. For example, if the objective function cannot be expressed in any analytic forms, its derivatives cannot be readily computed; this renders SA a viable optimization method. The most important and attractive feature of SA may be its capability to optimize an objective function with the potential existence of local minima. One common use of SA in radiation therapy optimization lies with beam angle optimization in which the local minima are known to exist (Djajaputra et al. 2003; Pugachev et al. 2001).

5.3 Adaptive Radiation Therapy Planning

5.3.1 Deformable Registration and Dose Accumulation

To accurately evaluate the cumulative dose distributions delivered to the clinical target volume (CTV), deformable organ registration (DOR) needs to be performed. In DOR, an organ is divided into many subvolumes, and the relative position of each subvolume varies from fraction to fraction. Assume v is the subvolume of an organ, V, that is, $v \in V$, it has a coordinate of $\vec{x}(v) \in \Re^3$ (Birkner et al. 2003). The displacement of the subvolume of the organ is subject to the constraints of the tissue elastic properties and the force exerted. Let $\vec{x}_{ref}(v)$ be the position of v in reference CT, then its position at fraction i becomes

$$\vec{x}_i(v) = \vec{x}_{ref}(v) + \vec{\delta}_i(v) + \mathrm{TM}_i(V) \qquad (5.8)$$

where $\vec{\delta}_i(v)$ is the displacement of the subvolume v, $\mathrm{TM}_i(V)$ is the rigid transformation matrix used in online image guidance and it affects the whole image set. If online image guidance is not performed, then $\mathrm{TM}_i(V) = 0$. Previously, tools have been developed to perform the mappings of the subvolumes based on biomechanical model of human organs and finite element analysis method (Brock et al. 2008; Liang and Yan 2003). The inputs to the tool are the surface information of the organ (i.e., contours) from

treatment and reference CTs, and the parameters describing elasticity properties of the organ. The outputs are the mappings of the subvolume, that is, $\vec{\delta}_i(v)$. Other types of DOR can be used as well (Lu et al. 2004; Wang et al. 2005). The tools can be expanded to compute the cumulative dose, which is the sum of the doses delivered in prior fractions. Assuming the dose matrix at fraction i is $d_i(\vec{x})$, then cumulative dose for subvolume v after k fractions is

$$\mathrm{CD}_k(v) = \sum_{i=1}^{k} d_i(\vec{x}(v)) \qquad (5.9)$$

5.3.2 Geometry-Adaptive Planning

The benefit of IGRT is that it can provide much information about patient anatomy during treatment, which can be used to take corrective actions to improve the treatment precision. However, it also poses challenges to ART planning. There are multiple CT images available that can be used in the planning; this greatly increases the complexity of the planning task. The inverse planning system must be modified to be able to handle multiple data sets, and a new technique needs to be adopted to improve the optimization algorithm. Here, we present a few examples.

In conventional inverse planning following standard target definitions (ICRU-50 1993; ICRU-62 1999), when margin is added to the CTV to form planning target volume (PTV) to handle the uncertainty, it represents a "hard boundary" around the CTV. The spatial probability distribution is completely ignored. Many investigators have explored the use of *probability density functions*, or *probability distribution functions*, or simply, pdf, to describe the target positions during treatment. A simple form is described here (Li and Xing 2000).

Given a set of beam configurations and fluence profiles, the corresponding dose distribution D_f is assumed to be fixed in the space and is not affected by the small random displacement of the internal structures, that is, a single data set is used for the optimization. This is valid for deep-seated tumors such as prostate cancer. The displacement causes a point in the target or in the sensitive structures to move to a series of spatial points in the radiation field D_f. We can then introduce a theoretical probability distribution, $P(i,j)$, to describe the probability for a point in a voxel indexed by i to move to another voxel j. The probability function is normalized according to $\sum_j P(i,j) = 1$. In this form, the average dose D_i received by a point in voxel i is given by

$$D_i = \sum_j P(i,j) \cdot D_f(j) \qquad (5.10)$$

Equation 5.10 relates the dose distributions in the patient for the cases with and without organ motion. Different organs can have different probability distributions. When $P(i,j) = \delta_{i,j}$, the two distributions D_i and D_j become identical. For a simple example, the random errors can be approximated by a 3D Gaussian probability distribution. For computational purposes, the distribution can be truncated when the distance was greater than 2–3σ. This is similar to a dose convolution or blurring. Other, more realistic probability distributions can similarly be incorporated into the algorithm.

Combining Equations 5.5 and 5.10,

$$D_i = \sum_j P(i,j) \cdot \sum_k K_{jk} \cdot \omega_k =$$
$$\sum_k \left(\sum_j P(i,j) \cdot K_{jk} \right) \cdot \omega_k = \sum_k K'_{ik} \cdot \omega_k \quad (5.11)$$

Here we define

$$K'_{ik} = \sum_j P(i,j) \cdot K_{jk} \quad (5.12)$$

Equation 5.11 is similar to Equation 5.5 except that the kernel K_{ij} is replaced by K'_{ik}. The new kernel matrix is not as sparse as the original K_{ij}, which increases the difficulties in the optimization algorithm when the derivatives are calculated.

Tests on a few examples have shown that this implementation of pdf led to an improved sparing of OAR for approximately the same target coverage when compared with the results obtained using simple margins only. The aforementioned technique works well for many sites where the tumors are deep-seated such as prostate cancer, in which the internal organ motion is much larger than the patient surface changes. However, this is not the case for other sites, such as superficial tumors like head and neck tumors, or sites with larger inhomogeneities where the dose convolution may not be valid. For these cases, the spatial dose distribution is not invariant from one data set to the other. A method called multiple instance geometry approximation (MIGA), developed by the University of Michigan, is an improvement over this (McShan et al. 2006; Yang et al. 2005). This method allows the use of multiple data sets (geometric instances) in the optimization and planning system.

The key difference is that the dose calculation procedure is repeated for each of the different geometrical instances. The geometrical shifts between the different data sets affect both the density grid used for heterogeneity corrections and the positions of the calculation points relative to the beam. For each data set, the same ray or beamlet description is used, and the calculations are again computed to the same set of points (but shifted or deformed) for each region. Equation 5.13 is used to determine the effective unit beamlet dose per calculation point:

$$K_{ij} = \sum_k W_k \cdot K_{ijk} \quad (5.13)$$

where the kernel from Equation 5.5, the dose to voxel point i from the jth ray or beamlet is replaced, by the sum over all geometrical instances k with weighting W_k. The weight W_k for each

instance represents the integral probability for a particular geometrical instance normalized so that $\sum_k W_k = 1.0$.

In MIGA optimization, the calculated results from each of the data sets are first retrieved, and the kernels for each point are summed using the probability weights from the motion file. In this way, different probability weightings specified in different motion files can be assessed without recalculating the point doses. The resulting point doses, which are the motion averaged dose to each point per unit intensity, are then used to determine the beamlet weights that result in the optimal cost function, just as is done in the static planning case. DVHs can be calculated using the dose points contained in each structure, just as in the non-MIGA case. The DVH analyzes the dose delivered to each point over the patient treatment course as approximated by the MIGA-weighted multiple instances.

A number of simple examples demonstrated that it is possible to use a limited number of instances, typically at least three in each direction, while still achieving a good representation of a Gaussian setup uncertainty distribution. Tests of using seven instances in a head and neck IMRT plan demonstrate significant improvements over what would typically occur if PTV expansions of the CTVs were used to compensate for the setup uncertainty.

5.3.3 Dose-Adaptive Planning

The cumulative dose distribution calculated from Equation 5.9 can be fed back to the treatment planning system for further evaluation. If it deviates significantly from the original goal, the dose deficit can be made up using dose compensation. Both 3DCRT and IMRT can be used for dose compensation (Wu, Liang, and Yan 2006). If 3DCRT is used, the underdosed region will be identified in the beam's-eye-view display window, so proper beam aperture can be designed to deliver boost dose to the underdosed region. However, IMRT may be more suitable for this purpose to incorporate previously delivered dose distributions into the treatment plan optimization for future treatment delivery. Several investigators have also addressed similar issues in the framework of adaptive plan optimization (Birkner et al. 2003; Lof, Lind, and Brahme 1998).

In dose compensation planning, the objective function for plan optimization in general can be written as

$$f(D_{Rx}(v), CD_k(v) + FD(v)) \quad (5.14)$$

Here $D_{Rx}(v)$ is the prescription dose, $CD_k(v)$ is the cumulative dose already delivered after k fractions from Equation 5.9, and $FD(v)$ is the dose to be delivered in future fractions, each term being a function for each subvolume. The type of the objective function can vary.

It is straightforward to incorporate fractional effect into the formula if the CTV dose is not uniform and the cumulative biological effect is significant. In this case, the $CD_k(v)$ is simply replaced by its corresponding biological equivalent, for example,

$$\text{NTCD}_k(v) = \sum_{i=1}^{k} d_i(\vec{x}(v)) \left(\frac{\alpha/\beta + d_i(\vec{x}(v))}{\alpha/\beta + d_f} \right) \qquad (5.15)$$

where $\text{NTCD}_k(v)$ is the normalized total dose equivalent to fractional prescription dose at d_f, which is usually 1.8 Gy, α/β is the biological parameter for CTV. More examples of biologically effective dose-based optimization can be found in Chapter 6 of this book.

Dose compensation frequency can vary in many ways, and effectiveness, workload, and efficiency should be balanced. One extreme is to do it at every fraction, and another is to perform only once at the end of the treatment course. Other options include once every few fractions such as weekly compensation. The dose compensation can also be triggered by predefined criteria in the offline dose analysis. For example, if the deviation of cumulative target dose from the original plan exceeds a threshold, the compensation will be performed. Daily compensation increases the workload dramatically but may only offer limited benefits considering the random nature of organ motion. Compensation at the end of the treatment course is relatively easy to implement but may not be beneficial for some patients. For example, the level of underdose at the end may be too large for the dose compensation to be completed in one fraction. In addition, even though dose compensation can correct the mistakes and residuals from previous fractions, it cannot compensate for the fraction where the boost dose is delivered. As a result, a large execution error could occur.

One example is shown in Figure 5.2. The minimum dose, D_{99}, of prostate cumulative dose after online image guidance is shown. For the ideal online correction using contour-based-registration (CBR) the dose compensation may not be necessary if we allow 2% underdose at the end of the treatment course. If the realistic online correction using image-based-registration (IBR) is applied, that is, with residual errors, then the underdose can

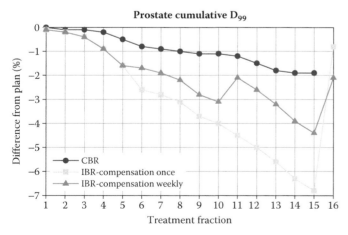

FIGURE 5.2 Reduction of D_{99} of the prostate cumulative dose as a function of treatment fractions for ideal online correction (CBR), realistic online correction (IBR) with residual errors. The dose compensation was applied to the latter in two frequencies: once at the end and once a week. (Reprinted from Wu, Q., et al. 2006. *Phys Med Biol* 51(6):1405–19. With permission.)

be expected to be as high as 7% at the end of the treatment course. The underdose monotonically decreases as a function of treatment fraction, indicating that the underdose is a gradual effect and cumulates from each fraction. The weekly compensation was performed at fractions 6, 11, and 16. At fraction 6, the additional dose to target was not able to reduce the overall dose difference; this was caused by the residual from online image guidance and deformation that occurred at fraction 6. However, there was still an observable improvement over the IBR correction alone. At fractions 11 and 16, such effects were small and the compensations were able to reduce the overall dose deficit. This shows that weekly compensation can reduce the execution errors by spreading boost doses into several fractions.

5.4 Examples of Adaptive Radiation Therapy Planning

In this section, we describe the benefits of different ART strategies in the clinical implementations for prostate and head and neck cancer.

5.4.1 Prostate

5.4.1.1 Offline Adaptive Radiation Therapy

In one study for prostate RT, the setup errors from 533 prostate cancer patients who underwent standard radiotherapy were analyzed by registering orthogonal megavoltage (MV) portal images with the digitally reconstructed radiograph (DRR) from initial planning CT (Wu et al. 2007). Data were collected from a few early fractions of treatment without any image guidance applied. The systematic errors $[\Sigma(\mu_i)]$ were found to be 2.6, 3.5, and 2.4 mm and the random errors were 1.5, 1.9, and 0.9 mm in left–right (LR), anterior–posterior (AP), and superior–inferior (SI) directions. The systematic error $\Sigma(\mu_i)$ measures the variations of mean setup error among the patients, and the random error $\text{RMS}(\sigma_i)$ measures the variations of setup among different treatment fractions. Ideally, they should be at the same magnitude, that is, variations of setup error between different patients and between different fractions for the same patient should be similar. In this case, the $\Sigma(\mu_i)$ is much larger than $\text{RMS}(\sigma_i)$, indicating that the magnitude of the systematic errors is larger than those of the random errors. To put it in another way, setup of the same patient on the treatment machine among different treatment fractions is more consistent than the setup among different patients. Elimination or reduction of the systematic errors is therefore of utmost importance. This is the main motivation of the offline ART for prostate cancer.

In offline ART, a population margin of 1 cm is used to expand the initial CTV to PTV for the first five fractions of radiation treatment, during which the patient-specific setup errors were measured and analyzed. The systematic error is eliminated in the modified planning through isocenter change. In addition, four additional helical CT (HCT) scans of the patient are performed, the CTV on each daily HCT is delineated, and the combination of these CTVs and the one from planning HCT is used to form the patient-specific internal target volume (ITV), which accounts for interfractional organ motion and shape change.

The ITV is typically a union or convex hull of the CTVs. A nonuniform margin, derived from the measurements of the setup errors from the first-week treatments to account for the residual of systematic setup error prediction and compensation of random setup error, is added to the ITV to form the patient-specific PTV in the modified planning for the remainder treatment fractions. During daily treatment, either the initial five treatments or the ones after the plan modification, the patient setup is not corrected; therefore, there is no negative effect in delivery efficiency, as the analyses and decision making are performed offline. The ART was originally performed using 3DCRT and the latter using IMRT. The ART has been shown to reduce the effective margins of the PTV and improve the local control on these patients (Brabbins et al. 2005; Martinez et al. 2001). There is a substantial margin reduction with the use of offline ART.

There are alternative methods for offline ART. For example, an average position of the prostate (CTV) can be found based on the center of mass (COM) or volume registration between CTVs (Nuver et al. 2007). An average shape of the CTV can also be constructed based on the isoprobability matrix of the occurrence of the prostate during the early fractions. Then, a margin can be added to form the modified PTV. Similarly, average shape of the OARs can be constructed in this way and included in the modified planning. The offline ART has also been adapted with the cone-beam CT (CBCT) instead of the HCT. Two approaches have been proposed (Nijkamp et al. 2008; Wang, Wu, and Yan 2010).

5.4.1.2 Online and Offline Adaptive Radiation Therapy

In the past few years, the IGRT devices have been increasingly available. The in-room on-board CBCT provides many opportunities for the image guidance. Currently, the image quality of CBCT is still inferior to the HCT in the pelvic region, making it difficult to distinguish the prostate gland from the surrounding soft tissues. However, with the implanted radiopaque markers as surrogate to the prostate gland, the online image guidance for prostate has become possible and widely used in many clinics. A lot of clinical data have been accumulated so that we now have better knowledge of the capability of the IGRT device and the characteristics of the uncertainties during the treatment process.

One immediate advantage of the online image guidance vs. the traditional offline ART for prostate is that the setup error and rigid organ motion (if implanted markers are used) are eliminated online; there is no distinction of systematic and random errors. This can drastically reduce the margins that are required for the treatment, and studies have been performed to discover the margins that can be reduced (Liang, Wu, and Yan 2009; Wu et al. 2006). The margins are still required to account for those errors or uncertainties that cannot be accounted for by the online correction, including residuals of online corrections, uncorrectable rotations and deformations, and intrafraction motion. The margin is much smaller than those required by the

offline ART. The margins also depend on the clinical protocols adopted.

An interesting question is whether the offline image guidance (ART replanning) is still necessary when online image guidance is used. One study analyzing the CBCT data from 16 patients who underwent an online image-guided hypofractionated prostate protocol using implanted markers was performed to investigate this issue (Lei and Wu 2010). A total of 592 CBCT images before (pre-CBCT) and after (post-CBCT) each treatment fraction was analyzed. The CBCT images were registered to planning CT based on marker localizations, and translational online correction was performed by couch shift based on pre-CBCT. Specifically, the rotational characteristics of the target were investigated in this study because it cannot be corrected online and additional margins are necessary. The results are summarized as follows:

Patient specificity: ANOVA shows that the rotation along LR and SI directions are patient-specific with $p < .05$; therefore, each patient has different rotation magnitudes.

Systematicness: There are (10, 10, 11) out of 16 patients who have rotations significantly different from 0 in LR, SI, and AP directions, respectively.

Persistence: Good correlations were observed between pre- and post-CBCTs for LR ($R^2 = 0.77$), AP ($R^2 = 0.66$), and SI ($R^2 = 0.36$). This means, if the interfractional rotation exists before the treatment, it is likely to remain through the treatment fraction and intrafractional rotation is relatively small.

Representativeness: The statistical analyses (Welch's test on the mean and Levene's test on variance) between the first 5 fraction rotations and the remaining 15 fractions show that the <mean> and <var> in the early 5 fractions can be used to predict the remaining fractions.

These results support the notion that during the online image guidance, the residuals (only rotation was investigated in this study) can be potentially corrected through offline replanning.

To further investigate quantitatively the benefits of the offline replanning in online image guidance protocols, a set of 412 HCT images from 25 patients were included in the study (Lei and Wu 2010), each patient has one planning CT and multiple CTs taken during the treatment course (23 patients have > 15 CTs). For each patient, contours of prostate, seminal vesicles (SVs), bladder, and rectum were delineated by experts. Both low-risk patients (CTV = Gross Tumor Volume (GTV) = prostate) and intermediate-risk patients (CTV = GTV + SVs = prostate + SVs) are simulated. The online image guidance was simulated by matching the COM of CTVs in treatment CT to that in planning CT through couch translation.

Here, the HCT was chosen instead of the CBCT because of higher image quality in the HCT that allows more accurate delineations of the prostate. The CBCT image quality is not adequate to perform this study. Therefore, in addition to the rotations, the effect of deformation is also included in this study.

New internal target volumes (ITV_n) for CTV are constructed based on the treatment image of the first five fractions, as follows:

$$ITV_n = \bigcup_{i=1}^{5} TM_i(CTV_i) \quad (5.16)$$

where TM_i is the translational transformation matrix used in online correction of the *i*th fraction.

We define the volume overlap index (OI_i) of *i*th treatment fraction for both ITV_n and CTV_0, that is, online corrected CTV_i volume that intersects with the volume of margin added to ITV_n or CTV_0, as follows:

$$OI_i = \frac{Volume\left((ITV_n + M) \bigcap TM_i(CTV_i)\right)}{Volume(CTV_i)}, i = 6, 7, ..., N_f \quad (5.17)$$

We can define the equivalent margin difference, ΔM, as a measure of benefit of ITV_n over CTV_0

Here the M_0 is defined as $Vol(ITV_n) = Vol(CTV_0 + M_0)$, that is, the equivalent margin for ITV_n and CTV_0 due to their volume difference. This is to remove the bias in the comparison. To achieve the same overlap index, OI_i, different margins need to be added. That is,

$$OI_i = OI(CTV_i, ITV_n + M_{ITV,i})$$
$$= OI(CTV_i, CTV_0 + M_{CTV_0,i}) \quad (5.18)$$

$$\Delta M_i = M_{CTV_0,i} - M_{ITV_0,i} - M_0 \quad (5.19)$$

Compared with CTV_0, which has an average volume of 50 cc for low-risk group patients and 67.9 cc for intermediate-risk group patients, the ITV_n has the volume of 61.2 and 88.3 cc, respectively. This corresponds to the equivalent margin M_0 of 1 mm for low-risk group and 1.3 mm for intermediate-risk group.

The margin benefit is a function of the OI values chosen. For OI = 0.99, the ΔM is 1.5 and 2.3 mm for low- and intermediate-group patients. We want to point out that, while the margin benefit may appear small, on the order of 1–2 mm, this is a significant reduction, considering that the margins have already been drastically reduced when online image guidance is used (10 mm → 3–5 mm).

5.4.1.3 Online Modification and Replanning

Several research groups have investigated the benefit of online planning or replanning for prostate cancer, even though there are many practical obstacles associated with the online planning that still need to be handled. The simplest method is to ignore these limitations and try to find the ultimate benefit of the online planning by moving the planning process to the online environment (Lerma et al. 2009; Schulze et al. 2009).

Wu et al. (2008) tried to use the preplan as a reference to guide the online replanning. The planning is performed according to the "anatomy of the day." Here, only the geometric adaptation is performed, that is, there is no dose adaptation. The unique aspect of this technique is that the original planned dose distribution from planning CT is used as the "goal" dose distribution for adaptation and to ensure the planning quality. There are several reasons for this arrangement. Due to the nature of the objective functions and optimization algorithms used in the conventional inverse planning, the achievable plan for a specific case is normally unknown; therefore, the constraints for targets and normal tissues as input to the optimization are usually stricter than the actual clinical goals. Planners iteratively adjust these constraints and associated weights during the optimization process. Therefore, it is not unusual for planners to perform several iterations in an attempt to find the "best" set of constraint parameters and dose distribution. This is not desirable for online reoptimization because of tight timing requirements. When the original planned dose distribution is chosen as the "goal" dose distribution for adaptation, at least the starting plan is clinically optimal and the adjustment of the constraints and weights during the optimization process can be avoided. The fluence maps are reoptimized via linear goal programming (LGP), and a plan solution can generally be achieved within a few minutes.

Online ART is a process of modifying the original IMRT plan just prior to the treatment to encompass interfractional changes in the patient's anatomy. The procedure involves four steps: (1) on-board 3D CT or CBCT images are acquired with the patient setup in the treatment position; (2) the daily image set is registered with the original planning CT, the "structures-of-interest" are identified/contoured, and their changes are analyzed; (3) the original IMRT plan is adapted (reoptimized) based on the changing patient anatomy; and (4) the adapted plan is delivered.

Deformable image registration is used to provide position variation information on each voxel in 3D. Daily on-board CBCT images and the planning CT images were registered using a thin plate spline (TPS) deformable registration algorithm. The TPS algorithm was selected for its flexibility in choosing deformation space and its computational efficiency, which was about 1 minute for this application. The deformable registration can be performed in parallel with contouring because they are independent tasks. The output from the deformable registration, a deformation map, is the voxel displacement in the *x*-, *y*- and *z*-directions from the planning CT images to the daily CBCT images. The new prescription dose distribution, or the "goal" dose distribution for the reoptimization algorithm, is generated by deforming dose distributions from the original plan onto the CBCT images. This is used to approximate the ideal dose distribution for the reoptimization process, which will match the "anatomy-of-the-day."

The reoptimization process is formulated as an LGP model that minimizes the weighted sum of deviations from the prescribed doses in the target volumes and overdosages in the nontarget volumes subject to constraints that control hot and cold spots. This approach is suitable for online plan reoptimization for several reasons. First, it is based on linear programming (LP) and can be solved quickly and efficiently. Second, a "globally optimal" solution is always guaranteed once the objective function is properly defined and accepted. Finally, this approach provides an efficient and clinically meaningful way to manage dose conformity through flexible control of hot spots, cold spots, and falloffs in the critical regions between the target and OARs. This LP-based model avoids the time-consuming trial-and-error adjustment of DVH constraints and weights that is characteristic of current commercial treatment planning systems.

For target volume, the following equation holds:

$$D_i - d_i^+ + d_i^- = D_i^P \qquad (5.20)$$

where D_i is the calculated dose at voxel i on treatment images or data set, D_i^P is the goal dose, that is, dose matrix from initial plan deformed on the treatment image. d_i^+ and d_i^- represent the over and underdoses for this voxel, $d_i^+ \geq 0$ and $d_i^- \geq 0$.

For OARs and normal tissues

$$D_i - d_i^+ \leq D_i^P \qquad (5.21)$$

Therefore, the underdose is not penalized in the reoptimization.

The objective is to minimize the following function through LP:

$$f = \sum_{i \in \text{Target}} w_{\text{T},i} \cdot \left(d_i^+ + d_i^- \right) + \sum_{i \in \text{OAR}} w_{\text{OAR},i} \cdot \left(d_i^+ \right) \qquad (5.22)$$

The technique was demonstrated on an example in which large variations exist in the shapes of rectum and bladder from day to day, shown in Figure 5.3. For this type of case, the online reoptimization is superior to other techniques such as online corrections.

For intermediate- and high-risk prostate cancer, both the prostate gland and seminal vesicles are included in the CTV. However, the internal motion patterns of these two organs vary, which can be a challenge for online correction only (Liang, Wu, and Yan 2009). Reoptimization of the IMRT plan is a valuable approach with large deformations, where other correction schemes can fail. Online ART process can be highly valuable with hypofractionated prostate IMRT treatment, as the accuracy of dose delivery is more stringent. Comparison with repositioning shows that the reoptimization plans are in general better than the patient correction technique, with plan quality close to that of the online planning from scratch (Thongphiew et al. 2009).

5.4.1.4 Intrafractional Motion Management

When prostate interfractional motion is managed by an online correction, remaining uncertainties include residual setup error and intrafractional motion, which are typically accounted for using population margins. However, similar to interfractional

motion, intrafractional uncertainties may also in principle be managed using either an online correction or an adaptive strategy. Online correction techniques for intrafraction motion include tracking and gating, the implementation of which may be possible using technologies capable of target localization during radiotherapy. Adaptive strategies would use intrafractional uncertainties measured during initial fractions to account for the intrafractional uncertainties of future fractions on a patient-specific basis. With this method, patients with relatively little intrafractional uncertainty would not be penalized by a population-wide approach that includes patients with large intra-fractional motion. Furthermore, while tracking and gating would require real-time localization, synchronization, and analysis, adaptive strategies may be implemented offline using retrospective measurements.

Figure 5.4 shows the intrafractional uncertainty in each axis after online CBCT correction for 30 prostate patients (a), and for three individual patients (b–d) taken from a recent study (Adamson and Wu 2009). It is clear from the figure that the individual uncertainties can be quite different from the population probability of motion, especially in the AP and SI axes. For some patients, the intrafractional uncertainty is much smaller than the population (b), while for others it is similar or larger (c), and for some there is a large systematic displacement even after online correction (d). Adaptive strategies for intrafraction motion management after online correction may benefit these patients if the intrafractional uncertainty can be predicted with sufficient accuracy in early fractions.

Adaptive management of intrafractional uncertainties after online correction consists of first estimating the mean (systematic component) and standard deviation (random component) of the target displacement from the expected position after online correction. This measurement can be performed over the first few fractions, and can be carried out using imaging of implanted fiducial markers, electromagnetic transponders, and so on. Then the systematic component is corrected in future fractions as part of the online correction. In essence, the patient is overcorrected during the online correction in anticipation of intrafractional drift. The random component can then be compensated using a patient-specific margin.

The most straightforward method for defining a patient-specific margin is to use a statistical criterion, which consists of defining a margin that will account for a given percentage of intrafractional motion. For example, a margin could be defined that would encompass 90% of the patient-specific intrafractional motion after online correction. Because only a few initial fractions are used to estimate the patient-specific mean and standard deviation, there will be some error in their calculation; however the margin can be designed to also account for this error.

One recent study investigated the feasibility of accounting for intrafractional uncertainties after online correction using an adaptive technique (Adamson and Wu 2009). The study compared an adaptive technique to population margins, both designed to account for 90% of motion. Adaptive margins

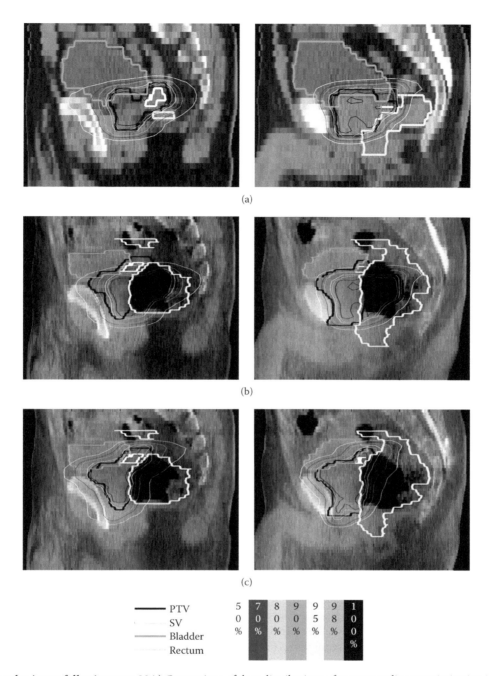

PTV
SV
Bladder
Rectum

50% | 70% | 80% | 90% | 95% | 98% | 100%

FIGURE 5.3 **(See color insert following page 204.)** Comparison of dose distributions of prostate online reoptimization in two sagittal slices: (a) initial optimized plan on planning computed tomography (CT), (b) initial plan applied to treatment CT, and (c) adapted plan on treatment CT. Note the large anatomy changes between planning CT and treatment CT. PTV = planning target volume; SV = seminal vesicles. (Reprinted from Wu, Q. J., et al. 2008. *Phys Med Biol* 53(3):673–91. With permission.)

applied after five fractions ranged from 1–3.2 mm, 2–7 mm, and 2–6.6 mm in RL, AP, and SI, respectively, while population margins were 1.7 mm, 4 mm, and 4.1 mm in RL, AP, and SI axes. The rather broad range of adaptive margins in AP and SI axes indicate that a one-size-fits-all approach may not be ideal.

At present, adaptive management of intrafractional uncertainties after online correction is still at an early stage of investigation and has only been applied for prostate. One important source of uncertainty that has not been dealt with is prostate deformation. Ideally, translational, rotational, and deformation uncertainties should all be addressed in an adaptive strategy; however, implementation becomes increasingly difficult with added complexity. Current adaptive strategies for intrafractional uncertainties after online correction deal only with translational uncertainties, and the patient-specific

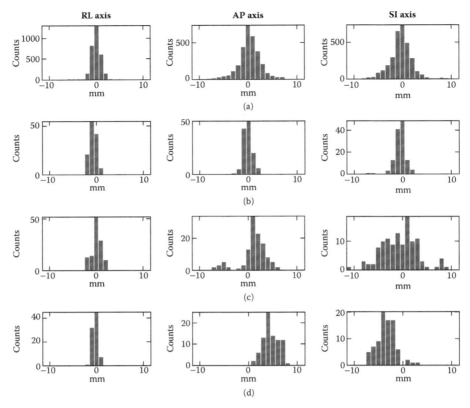

FIGURE 5.4 Prostate intrafractional uncertainty in the right–left (RL), anterior–posterior (AP), and superior–inferior (SI) axis after online cone-beam CT correction, for 30 patients (a), and for three individual patients (b–d).

margins defined must be increased to also account for deformation and rotational uncertainties.

5.4.2 Head and Neck Cancer

For head and neck cancer undergoing IMRT, the rigid setup error can be measured through the daily imaging: either orthogonal MV or kV projection images, or CBCTs. The setup error reported by one study using daily CBCT measurement and bony alignment for a group of 30 patients shows the systematic errors to be 1.3, 1.3, and 1.4 mm and random setup errors to be 2, 2.6, and 2.4 mm in the LR, AP, and SI directions (Worthy and Wu 2009). Both are relatively small compared to other tumor sites, primarily due to the use of thermoplastic masks in the immobilization of the head and neck region, the shallow depth of the tumor, and the use of offline correction protocol. The fact that the systematic error is smaller than random error suggests that the patient population-based margins should work well, even without patient-specific correction protocols. However, if the values from the standard recipes are adopted in planning, a margin in the order of 5 mm should be used to account for these errors (Stroom and Heijmen 2002; van Herk, Remeijer, and Lebesque 2002). Further direct dosimetric evaluation shows that this is not the case. In fact, only a 1.9-mm margin is necessary for CTV1 and a 1.5-mm margin for CTV2 (the

electively treated nodes) to compensate for rigid setup errors of this magnitude.

The overestimation of the published margins recipes for head and neck IMRT is due to differences in the tumor sites and the prescription criteria. In both the works of van Herk et al. (2002) and Stroom and Heijmen (2002), prostate cancer treatment was studied and margins were selected to ensure that the minimum dose (D_{min}) or dose to 99% of the CTV volume (D_{99}) was greater than or equal to 95% of the prescription dose. This criterion can be reasonably met in the treatment of prostate cancer, which is a deep-seated tumor, with a target volume that is generally of convex shape and is surrounded by few critical structures, mainly the bladder and rectum. For head and neck cancer, this criterion cannot be easily met often. This is because head and neck treatment is more complex due to the superficial location of the tumors and the usually concave shape of the target volume and the proximity of many critical and radiation-sensitive structures. Therefore, a more appropriate criterion is commonly used and margins were chosen based on D_{90} to ensure that 90% of the target volume receives the prescription dose. This is considerably relaxed compared to the minimum dose requirement for prostate and may contribute to the significantly smaller margins required. For this type of treatment, utilizing either offline or online image guidance, a single plan should suffice for the rigid setup error, and no replanning or adaptive planning is necessary.

However, there is another important issue that needs to be considered for head and neck IMRT. Significant anatomic and volumetric changes occur in head and neck cancer patients during fractionated radiotherapy (Barker et al. 2004; O'Daniel et al. 2007). The dose distribution for head and neck IMRT is usually highly conformal and complex. The actual dose can therefore be considerably different from those in the original plan. A recent publication investigated the dosimetric effect of the tumor shrinkage and the possible benefit of the adaptive planning (Wu et al. 2009). Eleven patients, each with one planning and six weekly HCTs, were included in the study. The IMRT plans were generated using simultaneous integrated boost technique. Weekly CTs were rigidly registered to planning CT before deformable registration was performed. Figure 5.5 shows the weekly volumetric changes of target and parotid glands. The following replanning strategies were investigated with different margins (0, 3, 5 mm): mid-course (one replan), every other week (two replans), and every week (six replans). Doses were accumulated on planning CT for various dose indices comparison for target and critical structures. The cumulative doses to targets were preserved even at zero margin. Doses to cord, brainstem, and mandible were unchanged. Significant increases in parotid doses were observed. Margin reduction from 5 to 0 mm leads to 22% improvement in parotid mean dose. The parotid sparing can be preserved with replanning. Because target doses are preserved throughout the treatment course, there is no need to perform dose-adaptive replanning to incorporate the previous dose distributions; therefore, simple replanning on updated CT scans should be adequate. More frequent replanning leads to better preservation, replanning more than once a week is unnecessary. Shrinkage does not result in significant

dosimetric difference in targets and critical structures except parotid glands, of which the mean dose increases by ~10%. The benefit of replanning is the improved sparing of parotid. The combination of replanning and reduced margin can lead up to a 30% difference in parotid dose.

5.5 Conclusions

There have been rapid developments in the past decade in the technology used in radiation therapy, such as image guidance capable linear accelerators and the deformable image registration algorithms. As they are increasingly used in many clinical protocols, a vast amount of patient images are collected and measurements during treatment are performed. These have greatly enhanced our knowledge of the previously known treatment "uncertainties." The uncertainty, if known and measurable, should be corrected if possible, as this has the largest impact on the treatment outcome. This is typically done with the IGRT. The residual uncertainties and those that cannot be corrected or are unavoidable should be compensated in the treatment planning through margins. This is achieved through ART. To put it simply, the goal of IGRT is to improve the radiation precision and minimize target miss. The goal of ART is to ensure that the delivered dose to the patient is the same as in prescription dose. The combination of ART and IGRT will lead to improved target coverage, reduced planning margin, improved sparing of organs at risks, and ultimately an improved therapeutic ratio.

References

Adamson, J., and Q. Wu. 2009. Prostate intrafraction motion assessed by simultaneous kV fluoroscopy at MV delivery II: Adaptive strategies. *Int J Radiat Oncol Biol Phys.* In press.

Barker Jr., J. L., A. S. Garden, K. K. Ang, et al. 2004. Quantification of volumetric and geometric changes occurring during fractionated radiotherapy for head-and-neck cancer using an integrated CT/linear accelerator system. *Int J Radiat Oncol Biol Phys* 59(4):960–70.

Birkner, M., D. Yan, M. Alber, J. Liang, and F. Nusslin. 2003. Adapting inverse planning to patient and organ geometrical variation: Algorithm and implementation. *Med Phys* 30(10):2822–31.

Bortfeld, T., D. L. Kahler, T. J. Waldron, and A. L. Boyer. 1994. X-ray field compensation with multileaf collimators. *Int J Radiat Oncol Biol Phys* 28:723–30.

Bortfeld, T., J. Stein, and K. Preiser. 1997. Clinically relevant intensity modulation optimization using physical criteria. In *XII International Conference on the Use of Computers in Radiation Therapy,* ed. D. D. Leavitt, G. Starkschall. Salt Lake City, Utah: Medical Physics Publishing.

Brabbins, D., A. Martinez, D. Yan, et al. 2005. A dose-escalation trial with the adaptive radiotherapy process as a delivery system in localized prostate cancer: Analysis of chronic toxicity. *Int J Radiat Oncol Biol Phys* 61(2):400–8.

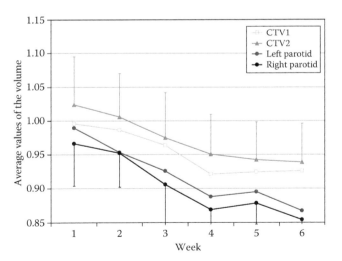

FIGURE 5.5 Volumetric changes of target and parotid gland during treatment course. Plotted on vertical (Y) axis are the average values of the volume from 11 patients normalized to those in planning computed tomography. CTV1 = clinical target volume expanded from gross tumor volume; CTV2 = clinical target volume for electively treated lymph nodes. (Reprinted from Wu, Q., et al. 2009. *Int J Radiat Oncol Biol Phys* 75(3):924–32. With permission.)

Brock, K. K. 2010. Results of a multi-institution deformable registration accuracy study (MIDRAS). *Int J Radiat Oncol Biol Phys* 76(2):583–96.

Brock, K. K., A. M. Nichol, C. Menard, et al. 2008. Accuracy and sensitivity of finite element model-based deformable registration of the prostate. *Med Phys* 35(9):4019–25.

Chao, K. S., J. O. Deasy, J. Markman, et al. 2001. A prospective study of salivary function sparing in patients with head-and-neck cancers receiving intensity-modulated or three-dimensional radiation therapy: Initial results. *Int J Radiat Oncol Biol Phys* 49(4):907–16.

Cho, P. S., S. Lee, R. J. Marks II., S. Oh, S. G. Sutlief, and M. H. Phillips. 1998. Optimization of intensity modulated beams with volume constraints using two methods: Cost function minimization and projections onto convex sets. *Med Phys* 25(4):435–43.

Court, L. E., L. Dong, A. K. Lee, et al. 2005. An automatic CT-guided adaptive radiation therapy technique by online modification of multileaf collimator leaf positions for prostate cancer. *Int J Radiat Oncol Biol Phys* 62(1):154–63.

Dawson, L. A., and D. A. Jaffray. 2007. Advances in image-guided radiation therapy. *J Clin Oncol* 25(8):938–46.

de la Zerda, A., B. Armbruster, and L. Xing. 2007. Formulating adaptive radiation therapy (ART) treatment planning into a closed-loop control framework. *Phys Med Biol* 52(14):4137–53.

Djajaputra, D., Q. Wu, Y. Wu, and R. Mohan. 2003. Algorithm and performance of a clinical IMRT beam-angle optimization system. *Phys Med Biol* 48(19):3191–212.

Eisbruch, A., R. K. Ten Haken, H. M. Kim, L. H. Marsh, and J. A. Ship. 1999. Dose, volume, and function relationships in parotid salivary glands following conformal and intensity-modulated irradiation of head and neck cancer. *Int J Radiat Oncol Biol Phys* 45(3):577–87.

Ezzell, G. A. 1996. Genetic and geometric optimization of three-dimensional radiation therapy treatment planning. *Med Phys* 23(3):293–305.

Graham, M. V., J. A. Purdy, B. Emami, et al. 1999. Clinical dose-volume histogram analysis for pneumonitis after 3D treatment for non-small cell lung cancer (NSCLC). [see comments]. *Int J Radiat Oncol Biol Phys* 45(2):323–9.

Hernando, M. L., L. B. Marks, G. C. Bentel, et al. 2001. Radiation-induced pulmonary toxicity: A dose-volume histogram analysis in 201 patients with lung cancer. *Int J Radiat Oncol Biol Phys* 51(3):650–9.

ICRU-50. 1993. *Prescribing, Recording and Reporting Photon Beam Therapy*. Bethesda, MD: International Commission on Radiation Units and Measurements.

ICRU-62. 1999. *Prescribing, Recording and Reporting Photon Beam Therapy (Supplement to ICRU Report 50)*. Bethesda, MD: International Commission on Radiation Units and Measurements.

Jackson, A., M. W. Skwarchuk, M. J. Zelefsky, et al. 2001. Late rectal bleeding after conformal radiotherapy of prostate cancer. II. Volume effects and dose-volume histograms. *Int J Radiat Oncol Biol Phys* 49(3):685–98.

Kwa, S. L., J. V. Lebesque, J. C. Theuws, et al. 1998. Radiation pneumonitis as a function of mean lung dose: an analysis of pooled data of 540 patients. *Int J Radiat Oncol Biol Phys* 42(1):1–9.

Lei, Y., and Q. Wu. 2010. A hybrid strategy of offline adaptive planning and online image guidance for prostate cancer radiotherapy. *Phys Med Biol* 55(8):2221–34.

Lerma, F. A., B. Liu, Z. Wang, et al. 2009. Role of image-guided patient repositioning and online planning in localized prostate cancer IMRT. *Radiother Oncol* 93(1):18–24.

Li, J. G., and L. Xing. 2000. Inverse planning incorporating organ motion. *Med Phys* 27(7):1573–8.

Liang, J., Q. Wu, and D. Yan. 2009. The role of seminal vesicle motion in target margin assessment for online image-guided radiotherapy for prostate cancer. *Int J Radiat Oncol Biol Phys* 73(3):935–43.

Liang, J., and D. Yan. 2003. Reducing uncertainties in volumetric image based deformable organ registration. *Med Phys* 30(8):2116–22.

Lof, J., B. K. Lind, and A. Brahme. 1998. An adaptive control algorithm for optimization of intensity modulated radiotherapy considering uncertainties in beam profiles, patient set-up and internal organ motion. *Phys Med Biol* 43(6):1605–28.

Lu, W., M. L. Chen, G. H. Olivera, K. J. Ruchala, and T. R. Mackie. 2004. Fast free-form deformable registration via calculus of variations. *Phys Med Biol* 49(14):3067–87.

Mageras, G. S., and R. Mohan. 1993. Application of fast simulated annealing to optimization of conformal radiation treatments. *Med Phys* 20(3):639–47.

Martinez, A. A., D. Yan, D. Lockman, et al. 2001. Improvement in dose escalation using the process of adaptive radiotherapy combined with three-dimensional conformal or intensity-modulated beams for prostate cancer. *Int J Radiat Oncol Biol Phys* 50(5):1226–34.

McShan, D. L., M. L. Kessler, K. Vineberg, and B. A. Fraass. 2006. Inverse plan optimization accounting for random geometric uncertainties with a multiple instance geometry approximation (MIGA). *Med Phys* 33(5):1510–21.

Morrill, S. M., K. S. Lam, R. G. Lane, M. Langer, and II. Rosen. 1995. Very fast simulated reannealing in radiation therapy treatment plan optimization. *Int J Radiat Oncol Biol Phys* 31(1):79–88.

Nijkamp, J., F. J. Pos, T. T. Nuver, et al. 2008. Adaptive radiotherapy for prostate cancer using kilovoltage cone-beam computed tomography: First clinical results. *Int J Radiat Oncol Biol Phys* 70(1):5–82.

Nuver, T. T., M. S. Hoogeman, P. Remeijer, M. van Herk, and J. V. Lebesque. 2007. An adaptive off-line procedure for radiotherapy of prostate cancer. *Int J Radiat Oncol Biol Phys* 67(5):1559–67.

O'Daniel, J. C., A. S. Garden, D. L. Schwartz, et al. 2007. Parotid gland dose in intensity-modulated radiotherapy for head and neck cancer: Is what you plan what you get? *Int J Radiat Oncol Biol Phys* 69(4):1290–6.

Pugachev, A., J. G. Li, A. L. Boyer, et al. 2001. Role of beam orientation optimization in intensity-modulated radiation therapy. *Int J Radiat Oncol Biol Phys* 50(2):551–60.

Rijkhorst, E. J., M. van Herk, J. V. Lebesque, and J. J. Sonke. 2007. Strategy for online correction of rotational organ motion for intensity-modulated radiotherapy of prostate cancer. *Int J Radiat Oncol Biol Phys* 69(5):1608–17.

Schulze, D., J. Liang, D. Yan, and T. Zhang. 2009. Comparison of various online IGRT strategies: The benefits of online treatment plan re-optimization. *Radiother Oncol* 90(3):367–76.

Shepard, D. M., G. H. Olivera, P. J. Reckwerdt, and T. R. Mackie. 2000. Iterative approaches to dose optimization in tomotherapy. *Phys Med Biol* 45(1):69–90.

Skwarchuk, M. W., A. Jackson, M. J. Zelefsky, et al. 2000. Late rectal toxicity after conformal radiotherapy of prostate cancer (I): Multivariate analysis and dose-response. *Int J Radiat Oncol Biol Phys* 47(1):103–13.

Spirou, S. V., and C. S. Chui. 1998. A gradient inverse planning algorithm with dose-volume constraints. *Med Phys* 25(3):321–33.

Stroom, J. C., and B. J. Heijmen. 2002. Geometrical uncertainties, radiotherapy planning margins, and the ICRU-62 report. *Radiother Oncol* 64(1):75–83.

Thongphiew, D., Q. J. Wu, W. R. Lee, et al. 2009. Comparison of online IGRT techniques for prostate IMRT treatment: Adaptive vs repositioning correction. *Med Phys* 36(5):1651–62.

van Herk, M., P. Remeijer, and J. V. Lebesque. 2002. Inclusion of geometric uncertainties in treatment plan evaluation. *Int J Radiat Oncol Biol Phys* 52(5):407–22.

Wang, H., L. Dong, M. F. Lii, et al. 2005. Implementation and validation of a three-dimensional deformable registration algorithm for targeted prostate cancer radiotherapy. *Int J Radiat Oncol Biol Phys* 61(3):725–35.

Wang, W., Q. Wu, and D. Yan. 2010. Quantitative evaluation of cone-beam computed tomography in target volume definition for offline image-guided radiation therapy of prostate cancer. *Radiother Oncol* 94(1):71–5.

Webb, S. 1992. Optimization by simulated annealing of three-dimensional, conformal treatment planning for radiation fields defined by a multileaf collimator: II. Inclusion of two-dimensional modulation of the x-ray intensity. *Phys Med Biol* 37(8):1689–704.

Worthy, D., and Q. Wu. 2009. Dosimetric margin assessment for rigid setup error by CBCT for HN-IMRT. *Med Phys* 36(6):2498.

Wu, Q. 2004. A dose calculation method including scatter for IMRT optimization. *Phys Med Biol* 49(19):4611–21.

Wu, Q., Y. Chi, P. Y. Chen, D. J. Krauss, D. Yan, and A. Martinez. 2009. Adaptive replanning strategies accounting for shrinkage in head and neck IMRT. *Int J Radiat Oncol Biol Phys* 75(3):924–32.

Wu, Q., D. Djajaputra, M. Lauterbach, Y. Wu, and R. Mohan. 2003. A fast dose calculation method based on table lookup for IMRT optimization. *Phys Med Biol* 48(12):N159–66.

Wu, Q., G. Ivaldi, J. Liang, D. Lockman, D. Yan, and A. Martinez. 2006. Geometric and dosimetric evaluations of an online image-guidance strategy for 3D-CRT of prostate cancer. *Int J Radiat Oncol Biol Phys* 64(5):1596–609.

Wu, C., R. Jeraj, G. H. Olivera, and T. R. Mackie. 2002. Reoptimization in adaptive radiotherapy. *Phys Med Biol* 47(17):3181–95.

Wu, Q., J. Liang, and D. Yan. 2006. Application of dose compensation in image guided radiotherapy of prostate cancer. *Phys Med Biol* 51(6):1405–19.

Wu, Q., D. Lockman, J. Wong, and D. Yan. 2007. Effect of the first day correction on systematic setup error reduction. *Med Phys* 34(5):1789–96.

Wu, Q., and R. Mohan. 2000. Algorithms and functionality of an intensity modulated radiotherapy optimization system. *Med Phys* 27(4):1–11.

Wu, Q. J., D. Thongphiew, Z. Wang, et al. 2008. Online re-optimization of prostate IMRT plans for adaptive radiation therapy. *Phys Med Biol* 53(3):673–91.

Xing, L., R. J. Hamilton, D. Spelbring, C. A. Pelizzari, G. T. Chen, and A. L. Boyer. 1998. Fast iterative algorithms for three-dimensional inverse treatment planning. *Med Phys* 25(10):845–9.

Yan, D., D. Lockman, D. Brabbins, L. Tyburski, and A. Martinez. 2000. An offline strategy for constructing a patient-specific planning target volume in adaptive treatment process for prostate cancer. *Int J Radiat Oncol Biol Phys* 48(1):289–302.

Yan, D., D. Lockman, A. Martinez, et al. 2005. Computed tomography guided management of interfractional patient variation. *Semin Radiat Oncol* 15(3):168–79.

Yang, J., G. S. Mageras, S. V. Spirou, et al. 2005. A new method of incorporating systematic uncertainties in intensity-modulated radiotherapy optimization. *Med Phys* 32(8):567–79.

Treatment Planning Using Biologically Based Models

Vladimir A. Semenenko
Medical College of Wisconsin

X. Allen Li
Medical College of Wisconsin

6.1 Rationale for the Use of Biologically Based Models in the Treatment Planning Process

In radiation treatment planning, there is continuing interest in replacing empirical methods and dose-volume (DV)-based surrogate measures of treatment effectiveness with quantities that provide more intuitive descriptions of treatment outcomes. Recent progress in the application of biologically based models in radiation therapy (RT) is the subject of this chapter. The current section highlights conventional approaches in treatment planning and describes how biologically based models can help enhance the current methods. In Section 6.2, desired properties of biologically based models are discussed using examples of existing formalisms. Section 6.3 provides a historical overview of applications of biologically based models in research and clinical settings. Section 6.4 provides a short description of recent developments in implementing biologically based models in commercial treatment planning software. The chapter concludes with remarks about the current and future role of biologically based models in treatment planning (Section 6.5).

In the literature related to the application of biologically based models in RT, treatment planning is often used in a narrow sense to mean formulation of objective functions for inverse plan optimization.[*] In this chapter, we will use the term *treatment planning* in a broader sense to also include the selection of an optimal fractionation schedule and assessment of plan quality. Specifically, discussion will be broken down to the following four main applications of biologically based models in treatment planning:

1. Optimization of dose distributions
2. Evaluation and ranking of treatment plans
3. Isoeffect calculations
4. Optimization of prescription dose and fractionation

Since the advent of intensity-modulated RT (IMRT) and the closely related inverse planning techniques, objective functions for *optimization of dose distributions* in tumors and organs at risk (OARs) have been typically designed using DV-based (i.e., physical) criteria. Optimization based on a single DV constraint may produce a number of dose distributions characterized by DV histograms (DVHs) pivoted around the constraint (Figure 6.1). It is intuitive that not all possible dose distributions

---

[*] We will use the term *optimization* to mean finding the optimum of some objective function subject to specified physical constraints rather than finding the best radiotherapy treatment plan for a patient. For further discussion on this point, the readers are referred to Rowbottom, Webb, and Oldham (1999).

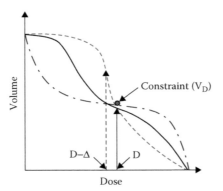

FIGURE 6.1 Limitations of dose-volume constraints for plan optimization and evaluation.

that meet the constraint would lead to the same biological outcome. Additional control over the shape of a DVH can be gained by specifying more than one DV constraint, but this approach increases the solution space and overall complexity of the optimization problem. The inverse planning techniques could benefit from objective functions that work on a selected region of a DVH or the entire DV domain rather than a single point. Biological indices of tumor and normal tissue response, such as tumor control probability (TCP), normal tissue complication probability (NTCP), and equivalent uniform dose (EUD), are typically calculated from a DVH (or directly from the dose grid) and, therefore, take the entire volumetric dose distribution into account. An emphasis on different regions within the DV domain could be placed by proper selection of model parameters. Another drawback of physical constraints is that the solution space is nonconvex and may result in multiple local minima (Deasy 1997). Nonconvex criteria may make the optimization problem hard to solve and require less efficient global optimization algorithms. It has been shown that several frequently used biologically based models can be cast as convex optimization criteria (Choi and Deasy 2002; Hoffmann et al. 2008), which makes such cost functions a more mathematically attractive choice. On the other hand, Wu and Mohan (2002) and Llacer et al. (2003) argue that the problem of multiple minima has no significant impact on finding clinically acceptable solutions using gradient-based optimization techniques.

DV criteria are also widely used in the current clinical practice of *evaluation and ranking of treatment plans*. Typically, the same DV constraint is used for both plan optimization and plan evaluation. If Figure 6.1 represents DVHs for alternative plans, suppose that, due to an uncertainty in identifying the correct threshold for the DV evaluation criterion V_D, a dose D – Δ represents a more relevant threshold for the considered end point. A dose distribution represented by the dashed line would then be clearly inferior to the other two, although the three DVHs appear to be equivalent according to the V_D criterion. Therefore, a more sensitive test that involves more than one point on a DVH is needed to distinguish between the dose distributions shown in Figure 6.1. Also, the current approach to plan evaluation using DV points implies a binary response model for clinical outcomes,

that is, a complication either occurs or does not occur depending on whether a DVH passes below or above the evaluation criterion. If the criterion is not met, there is no estimate of how grave the consequences of accepting such a plan for treatment could be. Hence, there is a need to move toward continuous risk models that attempt to quantify possible outcomes. The TCP, NTCP, and EUD models all have the potential to address the shortcomings of DV-based criteria. These models are calculated based on the entire volumetric dose distribution (and therefore can discriminate between alternative plans with intersecting DVHs, provided that the model parameters correctly describe organ tolerance as a function of irradiated volume, i.e., the "volume effect"), and they provide quantitative estimates of risk or benefit.

Isoeffect calculations are used to convert total dose, dose per fraction, and overall treatment time between two or more fractionation regimens so that these regimens result in a similar clinical outcome (i.e., tumor control or complication rate). The DVHs based on physical dose rescale with the change in prescription dose, but remain the same regardless of changes in dose per fraction. If a plan under consideration has been designed for one fractionation scheme and the plan evaluation criteria have been established for a different fractionation scheme, either the evaluation criteria or plan DVHs must be transformed using isoeffect relationships to obtain meaningful comparisons. The transformation of physical dose to biologically effective dose (BED) normalized to a standard fraction size facilitates a more consistent reporting of data between studies employing different fractionation schedules, and this is sometimes performed in research settings. However, the routine application of biological DVH corrections is yet to make its way to clinical practice.

Commonly accepted prescription doses and fractionation schedules for major disease types have been established over many years in clinical trials and in the daily practice of RT. Yet, as values of population-based radiobiological parameters come to be determined more accurately, new fractionation schemes that may increase the therapeutic gain are likely to be devised. Furthermore, recent developments in molecular imaging methods may one day produce three-dimensional (3D) maps of various biological parameters, such as intrinsic radiosensitivity, proliferation capacity, and hypoxic status, which will allow the *optimization of prescription dose and fractionation* based on individual tumor and normal tissue characteristics. Models will play an essential role in relating dose distribution and patient-specific radiobiological parameters to an expected biological outcome. Biologically based models could be combined into optimization scores that directly reflect the goal of RT in achieving tumor control while maximally sparing surrounding OARs.

6.2 Desired Model Properties

In this section, we highlight the desired properties of biologically based models with respect to each of the four applications outlined in Section 6.1. Various properties are illustrated using existing models as examples. A list of the most common biologically based models that have been applied in treatment planning is given in Table 6.1. In this and subsequent sections, models will be referred to according to the number assigned to them in Table 6.1. Readers seeking detailed information on each model are referred to the works of Yorke (2001) and Nahum and Kutcher (2007) and the references provided in the table.

6.2.1 Optimization of Dose Distributions

In order to efficiently steer the optimization process, biologically based models need to be able to take into account the volume dependence of an organ or tissue tolerance to radiation, that is, the volume effect. According to a hypothesis by Withers, Taylor, and Maciejewski (1988), the magnitude of the volume effect is related to the organization of functional subunits (FSUs) within the organ. These subunits are structurally or functionally discrete regions of an organ capable of being regenerated from surviving clonogenic cells following irradiation. In the two limiting cases, FSUs may be arranged in series or in parallel. Tissues with serially arranged FSUs exhibit threshold-like responses, where the threshold is defined by a dose required to inactivate a single FSU. There is little dependence of the tolerance dose on the irradiated volume. On the other hand, tissues with FSUs arranged in parallel exhibit graded responses to the increase in irradiated volume. A complication occurs when a critical number of

FSUs, referred to as "functional reserve," is inactivated. Many organs do not fall exactly into one of these categories, rather they exhibit mixed serial or parallel FSU organization. A tumor can be thought of as a tissue with very significant volume effect, in which every FSU, composed of a single clonogenic cell, must be inactivated in order to achieve tumor control.

Organs responding in a serial fashion tend to be sensitive to high doses or "hot spots," even if those are localized to small volumes. Organs with parallel architecture can tolerate hot spots in relatively small volumes, but may be damaged by intermediate or even low doses when those are delivered to a large volume. An objective function should be able to preferentially target a given domain of the dose distribution depending on the organ's FSU architecture. This task is usually accomplished using physical constraints. Although DV constraints do not have a strong biological basis, they nonetheless imply the presence of a certain volume effect. For example, a volume exposed to doses in excess of 20 Gy (V_{20}) is commonly used for the optimization and evaluation of dose distributions in the lung because pulmonary complications are related to the lung volume exposed to intermediate and low doses; maximum dose is used for the spinal cord because the cord represents a typical serial organ, which is sensitive to hot spots. However, a problem with DV-based optimization, which has already been outlined in Section 6.1, is that

TABLE 6.1 Frequently Used Biologically Based Models

Model	Formulas and Parameter Description	Comments	References
1. LQ cell-survival model	$S = \exp[-(\alpha D + G\beta D^2)]$, $G = \dfrac{2}{D^2} \int\limits_{-\infty}^{\infty} \dot{D}(t)\,dt \int\limits_{-\infty}^{t} e^{-\lambda(t-t')}\dot{D}(t')\,dt'$ α and β are linear and quadratic coefficients, D is the total dose, G is the generalized Lea-Catcheside dose-protraction factor, $\dot{D}(t)$ is the dose rate function, and λ is the first-order repair rate constant.	Premised on the idea that Poisson-distributed lethal events are produced by single-track (αD term) and double-track ($G\beta D^2$ term) mechanisms. Factor G accounts for the effects of protracting dose delivery.	Fowler 1989; Sachs and Brenner 1998
2. BED	$\text{BED} = D\left(1 + \dfrac{d}{\alpha\beta}\right)$ $D = nd$ is the total dose for a fractionated regimen, n is the number of fractions, d is the dose per fraction, and α/β is the ratio of parameters of the LQ model that defines the response of a tissue to fractionation.	Represents an isoeffective total dose given in infinitesimally small fractions ($d \to 0$). Two radiotherapeutic regimens with the same BED for a given outcome (tumor control or complication incidence) are considered to be equally effective. Used in isoeffect calculations for slow-growing tumors and late-responding tissues.	Barendsen 1982; Fowler 1989
3. BED including cell repopulation	$\text{BED} = D\left(1 + \dfrac{d}{\alpha\beta}\right) - \dfrac{\gamma}{\alpha}(T - T_k)$ D, d, and α/β have the same meaning as defined for the BED; γ is the rate of cell proliferation, T is the overall treatment time, and T_k is the time at which proliferation begins after the start of treatment.	Used in isoeffect calculations for fast-growing tumors and early responding tissues.	Fowler 1989
4. LQ equivalent dose in 2-Gy fractions (LQED$_2$)	$\text{LQED}_2 = D\,\dfrac{\alpha/\beta + d}{\alpha/\beta + d_{\text{ref}}}$ D, d, and α/β have the same meaning as defined for the BED, and $d_{\text{ref}} = 2$ Gy is the reference dose per fraction.	Used to convert dose delivered with nonconventional fractionation to the dose that is isoeffective when delivered with standard fractionation of 2 Gy per fraction. Has also been called "normalized total dose" (NTD), "normalized isoeffective dose" (NID), and "equivalent dose in 2-Gy fractions" (EQD$_2$).	Withers, Thames, and Peters 1983; Wheldon et al. 1998

(Continued)

TABLE 6.1   Frequently Used Biologically Based Models (*Continued*)

Model	Formulas and Parameter Description	Comments	References
5. LKB NTCP model[a]	$NTCP = \dfrac{1}{\sqrt{2\pi}} \int_{-\infty}^{t} e^{-\frac{x^2}{2}}\, dx,$ $t = \dfrac{D_{ref} - TD_{50}(V_{eff})}{m\,TD_{50}(V_{eff})} = \dfrac{D_{eff} - TD_{50}(1)}{m\,TD_{50}(1)},$ $TD_{50}(V_{eff}) = TD_{50}(1)V_{eff}^{-n},$ $V_{eff} = \sum_i v_i(D_i/D_{ref})^{1/n}, D_{eff} = \left(\sum_i v_i D_i^{1/n}\right)^n$ D_{ref} is an arbitrary reference dose usually chosen to be the maximum dose in the distribution, $TD_{50}(1)$ is the uniform dose given to the entire organ that results in 50% complication risk, m is a measure of the slope of the sigmoid curve, and n is the volume effect parameter.	The original model was designed to describe complication probabilities for uniformly irradiated whole or partial organ volumes. Effective volume DVH reduction algorithm converts a nonuniform DVH to a biologically equivalent single-step DVH according to which either partial organ volume V_{eff} is uniformly irradiated to dose D_{ref} or the whole organ is uniformly irradiated to dose D_{eff}. NTCP does not depend on the choice of D_{ref}. 3 parameters: $TD_{50}(1)$, m, and n.	Lyman 1985; Kutcher and Burman 1989; Appendix A in Luxton, Keall, and King 2008
6. Relative seriality NTCP model[a]	$NTCP = \left\{1 - \prod_i [1 - P(D_i)^s]^{v_i}\right\}^{1/s}$ $P(D_i)$ is the complication probability for the entire organ uniformly irradiated to dose D_i and s is the relative seriality parameter that describes the organization of FSUs within the organ. $P(D_i) = \exp\left(-\exp\left[e\gamma - \alpha D_i - \beta D_i^2/n\right]\right),$ $\alpha = \dfrac{e\gamma - \ln\ln 2}{\left(1 + \dfrac{\bar{d}}{\alpha/\beta}\right)D_{50}}, \beta = \dfrac{e\gamma - \ln\ln 2}{(\bar{d} + \alpha/\beta)D_{50}}$ D_{50} is the dose that results in a response probability of 50%, γ is the maximum normalized value of the dose-response gradient, n is the number of fractions, α/β is the ratio of parameters of the LQ model, and $\bar{d}$ is the dose per fraction for which D_{50} and γ were determined (usually $\bar{d} \approx 2$ Gy).	Describes response of an organ composed of a mixture of FSUs arranged in series and in parallel. Relative contribution of each type of architecture is described by the relative seriality parameter: $s = 1$ for a fully serial organ and $s = 0$ for a fully parallel organ. DVH reduction algorithm is not required. 4 parameters: D_{50}, γ, s, and α/β.	Källman, Ågren, and Brahme 1992; Lind et al. 1999
7. Parallel architecture NTCP model[a]	$NTCP = \dfrac{1}{\sigma_v\sqrt{2\pi}} \int_0^{f_{dam}} \exp\left[-\dfrac{(v - v_{50})^2}{2\sigma_v^2}\right]dv,$ $f_{dam} = \sum_i p(D_i)v_i, \; p(D_i) = \dfrac{1}{1 + (D_{50}/D_i)^k}$ v_{50} and σ_v are the mean and standard deviation of a functional reserve distribution, f_{dam} is the fraction of FSUs destroyed by radiation, $p(D_i)$ is the probability of destroying a single FSU following a uniform irradiation with dose D_i, D_{50} is the dose at which 50% of FSUs are damaged, and k determines the rate at which the probability of damaging the FSUs increases with dose.	Hypothesizes that a complication occurs when the fraction of damaged FSUs exceeds some critical value, the functional reserve. Functional reserves are normally distributed in a patient population. DVH reduction algorithm is not required. 4 parameters: v_{50}, σ_v, D_{50}, and k.	Yorke et al. 1993; Jackson, Kutcher, and Yorke 1993; Jackson et al. 1995
8. LQ-Poisson TCP model[a]	$TCP = \prod_i [P(D_i)]^{v_i}$ $P(D_i)$ is the tumor control probability for the entire organ uniformly irradiated to dose D_i.	$P(D_i)$ is calculated according to the same equations as those used for the relative seriality model. 3 parameters: D_{50}, γ, and α/β.	Källman, Ågren, and Brahme 1992; Lind et al. 1999
9. LQ-Poisson TCP model with interpatient heterogeneity[a]	$TCP = \dfrac{1}{\sigma_\alpha\sqrt{2\pi}} \int_0^\infty TCP(\alpha)\exp\left[-\dfrac{(\alpha - \bar{\alpha})^2}{2\sigma_\alpha^2}\right]d\alpha,$ $TCP(\alpha) = \prod_i \exp\left[-\rho v_i \exp(-\alpha D_i - \beta D_i^2/n)\right]$ ρ is the density of clonogenic cells, α and β are the coefficients of the LQ model, n is the number of fractions, $\bar{\alpha}$ and σ_α are the mean and standard deviation of the distribution of α.	Assumes that the linear coefficient of the LQ model is distributed normally among the patients' tumors. 4 parameters: ρ, $\bar{\alpha}$, σ_α, and β.	Webb and Nahum 1993; Nahum and Sanchez-Nieto 2001

TABLE 6.1 Frequently Used Biologically Based Models

Model	Formulas and Parameter Description	Comments	References
10. Complication-free tumor control probability (P_+)	$P_+ = P_B - P_I + \delta P_I (1 - P_B)$ $P_B = \prod_j \text{TCP}_j$ is the probability of getting benefit from controlling the jth tumor, $P_I = 1 - \prod_k (1 - \text{NTCP}_k)$ is the probability of causing injury to the kth critical organ, and δ is the fraction of patients with statistically independent tumor and normal tissue responses.	Provides a composite index of plan quality that represents the probability of achieving tumor control without causing severe injury to normal tissues.	Källman, Ågren, and Brahme 1992
11. EUD model[a]	$\text{EUD} = D_{\text{ref}} \dfrac{\ln\left[\sum_i v_i (SF_2)^{D_i/D_{\text{ref}}}\right]}{\ln(SF_2)}$ $D_{\text{ref}} = 2$ Gy is the reference dose and SF_2 is the surviving fraction at the reference dose.	Represents a uniform dose that, if delivered over the same number of fractions as the nonuniform dose distribution of interest, would result in the same clonogenic survival. Applicable only to tumors. 1 parameter: SF_2.	Niemierko 1997
12. gEUD model[a]	$\text{gEUD} = \left(\sum_i v_i D_i^a\right)^{1/a}$ a is a tissue-specific parameter that describes the magnitude of the volume effect.	Applicable to tumors and OARs. gEUD is equivalent to D_{eff} in the LKB model with $a = 1/n$. 1 parameter: a.	Niemierko 1999

[a] v_i is the fractional organ volume receiving a dose D_i, and i refers to an index of a DVH bin.

a DVH tends to bend around the DV constraint. The bent DVH most likely does not resemble dose distributions in a data set from which the DV constraint had been derived and, therefore, may result in unexpected (higher or lower) toxicity (Figure 6.1).

A good objective function should be able to optimize a *region* of a DVH around the relevant DV constraint rather than a *single point*. A generalized EUD (gEUD) equation proposed by Niemierko (1999; model 12) has proven to be a popular choice for an optimization objective because it is calculated based on the entire volumetric dose distribution and the emphasis on different DV regions can be placed by modifying a single parameter a. The mathematical properties of the gEUD equation are such that for $a \to -\infty$, gEUD approaches the minimum dose; for $a \to +\infty$, it approaches the maximum dose; and when $a = 1$, gEUD is equal to the mean dose. Generally, negative a values are an appropriate choice for target volumes (gEUD is influenced by cold spots), positive a values are used for structures with predominantly serial architecture (gEUD is influenced by hot spots), and values of a around unity are used for structures with parallel architecture. The absolute values of the parameter a are not very important as long as they are able to steer the optimization algorithm in the desired direction (Choi and Deasy 2002). The major criticism of the gEUD function is that the power-law relationship between a tolerance dose and the organ volume, on which the function is based, has limited radiobiological basis.

More advanced outcome models, such as TCP and NTCP models, can also be used in plan optimization. It is highly attractive to construct overall objective scores in terms of maximizing TCP for all targets and minimizing NTCP for all critical organs (e.g., model 10), which will more closely represent the goal of RT compared to the optimization based on surrogate DV parameters that correlate with TCP and NTCP. However, commonly used TCP and NTCP models require at least three parameters (models 5–9),

which increases uncertainties in absolute values of model predictions and complexity of the optimization problem. Langer, Morrill, and Lane (1998) suggest that absolute errors in NTCP and TCP estimates may result in the wrong ranking of plans, forcing the optimization algorithm in the wrong direction. We believe that the models used for plan optimization should capture the general trends in organ behavior with respect to the volume effect, and yet be minimally parameterized in order to preserve their ability to efficiently steer the optimization process.

6.2.2 Evaluation and Ranking of Treatment Plans

The main requirement for the use of biologically based models in plan evaluation and ranking is prediction accuracy. Two factors that impact a model's predictive power should be recognized: (1) how accurately the model captures the biological mechanisms underlying tumor or normal tissue responses to radiation and (2) how well the characteristics of a patient cohort from which parameter estimates had been derived agree with the characteristics of the patient group to which they are applied.

The concepts of TCP and NTCP are well-suited for plan evaluation because they have a clear clinical interpretation. A large number of TCP and NTCP models have been proposed in the literature. Of these, mechanistic models that attempt to link radiation response to survival of clonogenic cells are generally preferred over empirical models, which are aimed at mere data fitting. With empirical models, results are applicable only to conditions very similar to those used in the data analysis, and any extrapolation should be done with great caution. Regarding mechanism-based TCP and NTCP models, a trade-off always exists between the number and complexity of phenomena included in the model and the number of adjustable parameters. Most models that remain in current use (models 5–9) strike a compromise between these two factors. It should be noted that the Lyman–Kutcher–Burman (LKB) model (model 5), despite being widely used, is the least sound in terms of its mechanistic basis.

When it comes to ranking alternative treatment plans using TCP and NTCP models, the choice is clear if one plan offers lower NTCP or higher TCP, given that all other biological metrics remain approximately the same. However, the situation is not straightforward if one has to choose between a plan with low NTCP and relatively low TCP, and a plan with higher NTCP and higher TCP. The answer might depend on a physician's professional judgment or a patient's informed decision to accept or not to accept a higher risk of complications for a chance of cure (Amols et al. 1997). This dilemma can be approached mathematically by constructing a formula that combines all TCP and NTCP predictions into a single score that reflects the overall quality of a plan. The P_+ index (model 10) has been used extensively for both optimization and evaluation of treatment plans (e.g., Källman, Lind, and Brahme 1992; Söderström and Brahme 1993; Lind et al. 1999; Mavroidis et al. 2007). Other figures of merit have also been proposed for this purpose (Amols et al. 1997; Brenner and Sachs 1999).

The importance of appropriate parameter estimates cannot be emphasized enough. The most advanced TCP or NTCP model can yield predictions that are of little clinical value when used with parameter estimates that had been obtained using noncomparable fractionation schedules, organ definitions, dosimetry, adjuvant therapies, etc. Factors that affect the quality of predictions could be more easily matched if TCP/NTCP parameter estimates are derived from data originating in the same institution. However, not many radiotherapy centers can afford the resources and expertise required to perform such analyses. Most institutions interested in exploring TCP/NTCP models for plan evaluation have to resort to using parameters reported in the literature. Ideally, parameter estimates for different tumor control (i.e., local, regional, and distant control) and normal tissue end points (grade $\geq$ 1, 2, etc.; early vs. late toxicity) are needed, but are rarely available, especially from a single study.

Apart from possible differences in patient characteristics, large variability in end point definitions and TCP/NTCP models used for the analyses exists between different studies. Radiation oncology practitioners are often confronted with the problem of choosing the most reliable published studies to use. Several investigators have attempted meta-analyses of NTCP data sets collected at different institutions (e.g., Kwa et al. 1998; Seppenwoolde et al. 2003; Rancati et al. 2004; Bradley et al. 2007; Semenenko and Li 2008). The expectation from such efforts is that the derived parameter estimates are more likely to be applicable to the practice of radiation oncology at a wide variety of institutions. The Quantitative Analysis of Normal Tissue Effects in the Clinic (QUANTEC) initiative was created in 2007 to summarize the current knowledge of DV dependencies of normal tissue complications from external beam radiotherapy (EBRT) and provide quantitative guidance for clinical treatment planning. QUANTEC analyses for 16 organs or tissues have been published in a special issue of the *International Journal of Radiation Oncology Biology Physics* (Marks, Ten Haken, and Martel 2010). If TCP and NTCP models are to make their way to wide clinical practice, similar efforts to summarize TCP data may soon need to be undertaken.

6.2.3 Isoeffect Calculations

The BED formalism (models 2 and 3) originating from the linear-quadratic (LQ) model (model 1) has secured its position as a tool of choice for isoeffect calculations in RT because of its simplicity and applicability to a wide range of clinically relevant doses. For late-responding normal tissue effects and some slowly growing tumors, where repopulation effects can be ignored, only one parameter is needed for BED calculations (model 2): the ratio of linear and quadratic coefficients in the LQ model, α/β (Withers, Thames, and Peters 1983). Values of α/β for many animal and human tissues and tumors have been reported (Fowler 1989; Bentzen and Joiner 2009), and generic values of 10 Gy for tumors and early responding tissues and 3 Gy for late-responding tissues are often assumed. Some notable exceptions to this rule of thumb include the low α/β ratios for malignant melanoma (< 1 Gy; Bentzen et al. 1989), prostate cancer (1–3 Gy; Brenner and Hall 1999; Wang, Guerrero, and Li 2003; Williams et al. 2007), and breast cancer (4–5 Gy; START Trialists' Group 2008; Whelan, Kim, and Sussman 2008). For most tumors and early responding tissues, BED is modified by the addition of a repopulation term (model 3). Although the introduction of this term into the LQ model was controversial in the beginning (Tucker and Travis 1990; Van Dyk et al. 1990), the modified formalism has remained in use. Two additional parameters (γ/α and T_k) are required for isoeffect calculations involving the time factor. These parameters are not well-explored, except for head and neck cancers (Roberts and Hendry 1999), causing such isoeffect calculations to be subject to greater uncertainties (Tucker 1999). Further efforts to investigate alternative formulations of the time factor (e.g., Newcomb et al. 1993) and to obtain accurate estimates of relevant parameters are needed.

To be useful for isoeffect calculations, TCP and NTCP models must be sensitive to changes in fractionation schedule. This is the case for some (e.g., models 6, 8, and 9), but not all (e.g., models 5 and 7), models. Strictly speaking, models that do not include fractionation effects can be used only if the fraction size in a patient group from which the model parameters had been derived is the same as in the group to which they are applied. Fractionation sensitivity could be added to such models by converting physical doses to biologically equivalent doses given in 2-Gy fractions using an LQ-derived formalism ($LQED_2$; model 4). This process constitutes a nonlinear transformation of a dose scale in a DVH. If parameter estimation is performed with LQ-corrected DVHs, such parameters could be used to obtain TCP/NTCP predictions for treatments employing a different dose per fraction, as long as all physical DVHs are also corrected for heterogeneity of fraction size using the same α/β value.

The validity of the LQ model for large doses per fraction has been questioned earlier with regard to radiosurgery data (Marks 1995; Hall and Brenner 1995). Due to a widespread use of stereotactic body RT (SBRT), the debate has been resumed recently (Brenner 2008; Kirkpatrick, Meyer, and Marks 2008; Fowler 2008b; Kirkpatrick, Brenner, and Orton 2009; Courdi et al. 2010). The general consensus appears to be that the LQ model is applicable to single doses up to 10 Gy and may still be valid up to 15–20 Gy. More detailed binary misrepair models of cell survival, which reduce to the LQ model in the limit of low doses (Brenner et al. 1998), can be used above 20 Gy. Additional work is required to either validate the LQ model in the dose range between 10 and 20 Gy or develop alternative approaches. Several new formalisms have been proposed to describe modifications of the LQ model applicable to large fractional doses (Guerrero and Li 2004; Park et al. 2008; Kavanagh and Newman 2008; Astrahan 2008; Tomé 2008; McKenna and Ahmad 2009); but it is too early to tell whether any of these models have the potential to make an impact in the clinic.

Concerns about the validity of the LQ model at low doses have also been raised. Some cell types are known to exhibit a hyper-radiosensitivity (HRS) below 0.3 Gy followed by an increased radioresistance (IRR) in the 0.3–1.0 Gy dose range (Joiner et al. 2001). The two phenomena, if they occur *in vivo*, could lead to the underestimation of cell killing at doses < 1 Gy by the LQ model. Although DVH corrections for fraction size heterogeneity may be modified to accommodate the HRS/IRR phenomenon (Honoré and Bentzen 2006), the accuracy of this procedure has not been thoroughly investigated.

6.2.4 Optimization of Prescription Dose and Fractionation

Although isoeffect calculations and the selection of optimal fractionation schemes are closely related and are currently performed using the same underlying formalism (model 1), important differences exist between these two applications of biologically based models. In isoeffect calculations, one starts with an existing fractionation regimen of known effectiveness to produce a new fractionation scheme that is approximately as effective as the original one but is different in fraction size, number of fractions, or overall treatment time. For optimizing prescription dose, one does not need to have prior clinical knowledge of an effective treatment, rather one attempts to design a fractionation regimen that achieves a reasonable compromise between tumor cell killing and adverse normal tissue effects based on mathematical models and parameters that describe these processes.

Candidate models for the selection of optimal fractionation schedules should include major radiobiological mechanisms that determine the efficacy of RT. These mechanisms are known as the five Rs: radiosensitivity, repair, repopulation, redistribution, and reoxygenation (Withers 1975; Steel, McMillan, and Peacock 1989). The LQ model of cell survival (model 1) currently provides the most consistent formalism to describe these processes. Variations in the intrinsic *radiosensitivity* of different tissues are modeled

using absolute values of the parameters α and β. The capacity for sublethal damage *repair* is modeled using the ratio α/β and the dose protraction factor G (Sachs and Brenner 1998). *Repopulation* is included using the time factor discussed in Section 6.2.3. The description of repopulation for tumors may be further enhanced by including the effects of cell loss (Jones and Dale 1995). Although not commonly used, extensions to incorporate *redistribution* (Brenner et al. 1995; Zaider, Wuu, and Minerbo 1996) and *reoxygenation* (Wouters and Brown 1997; Carlson, Stewart, and Semenenko 2006) have also been proposed.

Accurate estimates of model parameters to describe the involved radiobiological processes are essential. Whereas for isoeffect calculations the number of adjustable parameters can be reduced by the use of parameter ratios, estimates of absolute values of model parameters are generally needed to optimize dose and dose per fraction. The great promise of this approach lies in the ability to optimize prescription dose for each patient using tumor-specific and normal tissue-specific rather than population-based radiobiological parameters. Although the theoretical framework already exists, introduction of such methods in the clinic will not be possible until breakthroughs are made in predictive assays and imaging technology to obtain spatial and temporal distribution of relevant radiobiological parameters. For additional discussion on this topic, the readers are referred to reviews by Brahme (2001) and Stewart and Li (2007).

6.3 Overview of Biologically Based Approaches in Treatment Planning

Sections 6.3.1–6.3.4 provide an overview of the research efforts made within the past 20 years in each of the four categories identified in Section 6.1. This review is by no means comprehensive, but is intended to describe major research directions in the field of biologically based treatment planning. For a historical excursion into the evolution of the treatment planning techniques, including the application of biologically based models, the reader is referred to an excellent review by Orton et al. (2008).

6.3.1 Optimization of Dose Distributions

A sharp increase in the efforts to incorporate biological metrics into the treatment planning process occurred in the early 1990s. This was fueled by several factors, including the intensifying of interest in inverse treatment planning, more widespread use of DVHs (Drzymala et al. 1991), publication of normal tissue toxicity data by Emami et al. (1991) and parameter estimates by Burman et al. (1991), and formulation of several TCP and NTCP models (models 5–9). In the following years, several research groups investigated the possibility of using cost functions constructed from biological indices for the optimization of treatment plans (Källman, Lind, and Brahme 1992; Mohan et al. 1992; Niemierko, Urie, and Goitein 1992; Söderström and Brahme 1993; Wang et al. 1995; De Gersem et al. 1999; Stavrev et al. 2003), and reported that the method had the potential to generate plans that are at least as good as those produced by an experienced planner, that is, better normal tissue sparing with same or improved tumor coverage. It was also realized that additional physical constraints had to be imposed in order to obtain plans that are compatible with conventional clinical judgment, the most important being the constraint on dose inhomogeneity in target volumes (Mohan et al. 1992). Despite the potential benefits of TCP/NTCP-based optimization outlined in these studies, it was widely recognized that much additional work was needed to increase confidence in model predictions (Bortfeld et al. 1996; Mohan and Wang 1996).

The idea of a single metric for summarizing 3D dose distributions for tumors was initially discussed by Brahme (1984) and presented in the form of the EUD concept by Niemierko (1997). The EUD concept was later extended to OARs (Niemierko 1999). A formula mathematically equivalent to Niemierko's gEUD equation (model 12) has also appeared in the work of Mohan et al. (1992). Jones and Hoban (2002) investigated a possibility of optimizing dose distributions based on the cell-survival-based formulation of EUD (model 11), again pointing out the inability of biologically based cost functions to limit large dose gradients in target volumes, but it is the gEUD that has drawn considerable attention from the proponents of biologically based optimization. The mathematical properties of the gEUD, as applied to IMRT inverse treatment planning, have been investigated by Choi and Deasy (2002) and Ólafsson, Jeraj, and Wright (2005). A first practical implementation was developed by Wu et al. (2002), who reported that the gEUD-based objective function had the potential to explore larger solution spaces, improving the chances of identifying new, favorable solutions that would otherwise be missed by DV-based optimization scores. The problem of inhomogeneous target dose distributions was solved by adding a second gEUD constraint that treated target volumes as "virtual OARs" (Wu et al. 2002). In subsequent studies, several hybrid methods that combine EUD-based and DV-based cost functions have been proposed (Wu, Djajaputra, et al. 2003; Thieke et al. 2003; Yang and Xing 2004; Chvetsov, Dempsey, and Palta 2007; Bortfeld et al. 2008; Hartmann and Bogner 2008; Das 2009) to address the limitations of the EUD-only optimization approach and facilitate a smooth transition from the DV-based inverse planning paradigm. In the approach used by investigators at the University of Michigan, objective functions based on the LKB model (model 5) for normal tissues were combined with gEUD-based objective functions for targets (Thomas et al. 2005; Chapet, Thomas et al. 2005; Spalding et al. 2007). An overall consensus emerged that incorporating gEUD-based cost functions into inverse treatment planning algorithms often leads to an improved OAR sparing with comparable target coverage, or may help to escalate the dose to target volumes while maintaining the dose to the OARs within acceptable limits. The popularity of gEUD in inverse treatment planning studies is attributed to the fact that it offers a compromise between purely biological indices, such as TCP and NTCP, and traditional DV metrics. Because it does not attempt

to predict the actual biological response and requires only one parameter, the behavior of gEUD in optimization problems is less ambiguous than that of TCP and NTCP.

6.3.2 Evaluation and Ranking of Treatment Plans

The TCP, NTCP, and EUD models all have found multiple applications in RT research. They are often used to supplement dosimetric quantities in the treatment planning comparisons of different EBRT techniques (e.g., Luxton, Hancock, and Boyer 2004; Mavroidis et al. 2007). Also, the models are indispensable when direct dosimetric comparisons cannot be performed, such as among EBRT, brachytherapy, and combined EBRT/brachytherapy schedules (e.g., King, DiPetrillo, and Wazer 2000; Wang and Li 2003; Bovi et al. 2007; Fatyga et al. 2009) or between sequential and concomitant boosts (e.g., Li et al. 2005; Guerrero et al. 2005; Pieters et al. 2008). To promote the use of TCP/NTCP models, several freely available tools for plan evaluation have been created, for example, BIOPLAN (Sanchez-Nieto and Nahum 2000), TCP_NTCP_CALC (Warkentin et al. 2004), CalcNTCP (Khan 2007), and DORES (Tsougos et al. 2009).

Despite their universal acceptance in research settings, biologically based models have not been used for routine plan evaluation in the clinic. The progress in that area is impeded by insufficient confidence in model predictions. Langer, Morrill, and Lane (1998) have challenged the popular belief that accurate estimates of TCP and NTCP are not essential in order to use these quantities for plan ranking. Using an example of a frequently employed score, the authors showed that plan rankings may be reversed even when the relative probabilities remain unchanged. Zaider and Amols (1999) have cautioned against extrapolating NTCP predictions to DV domains and complication levels not used in the initial testing of the model. Moiseenko, Battista, and Van Dyk (2000) and Muren et al. (2001) observed high sensitivity of NTCP predictions and plan rankings to the choice of an NTCP model and input parameters. One important limitation of most NTCP/TCP models (models 5–9) is that calculations are performed based on DVHs, in which all spatial information is lost. Several NTCP models have been proposed to take into account information about the location and distribution of cold and hot spots within an OAR (Stavreva et al. 2001; Thames et al. 2004; Tucker et al. 2006; Huang et al. 2010), but these models have not yet been widely adopted.

Concerns have also been raised about the predictive power of TCP models. Levegrün et al. (2000) investigated correlations of clinically assessed local tumor control with predictions from four TCP models. Unexpectedly, weak correlations of tumor control were observed with all quantities that are sensitive to cold spots, including TCP model predictions and minimum dose in the planning target volume (PTV). However, the authors reported a strong correlation with the mean PTV dose. This result was explained by the fact that cold spots, which are usually located on a periphery of the PTV, may not coincide with the location of clonogenic cells and, therefore, would be inconsequential for tumor control. This study highlighted problems with choosing a proper target volume for TCP calculations.

The importance of having accurate parameter estimates for TCP and NTCP models has already been emphasized in Section 6.2.2. The modern knowledge of normal tissue tolerance is summarized in a seminal report by Emami et al. (1991). The authors compiled tolerance dose values for uniform whole- and partial-organ irradiation of 28 critical structures based on a review of the literature and personal experience. In an accompanying article, Burman et al. (1991) provided parameter estimates for the LKB model (model 5) based on the tolerance data summarized by Emami et al. (1991). In the following years, many investigators compiled new clinical data sets and provided updated parameter estimates for various NTCP models. For most recent summaries of DV data and NTCP model parameters for 16 major OARs, the readers are referred to the QUANTEC papers (Marks, Ten Haken, and Martel 2010). Although TCP data have not yet been summarized in a clinically useful manner, parameter estimates for various forms of the Poisson statistics–based models (models 8 and 9) have been obtained for head and neck tumors (Roberts and Hendry 1993; Wu et al. 1997), breast tumors (Brenner 1993; Guerrero and Li 2003), malignant melanoma (Brenner 1993), squamous cell carcinoma of the respiratory and upper digestive tracts (Brenner 1993), cervical carcinoma (Buffa et al. 2001), prostate cancer (Brenner and Hall 1999; Levegrün et al. 2001; Wang, Guerrero, and Li 2003; Nahum et al. 2003), brain tumors (Qi et al. 2006), rectal cancer (Suwinski et al. 2007), and liver cancer (Tai et al. 2008). Okunieff et al. (1995) have collected and analyzed dose-response data for local control of 28 tumors treated with adjuvant intent and 62 tumors treated with curative intent using an empirical sigmoid TCP model. In order to facilitate analyses of new clinical data, several tools for outcome modeling have been made publically available: DREES (El Naqa et al. 2006), EUCLID (Gayou, Parda, and Miften 2007), and computational platform for outcome analysis (Liu et al. 2009).

6.3.3 Isoeffect Calculations

Isoeffect calculations have traditionally been performed using empirical relationships, such as Strandqvist's power-law relationship (Strandqvist 1944), nominal standard dose (NSD; Ellis 1969), cumulative radiation effect (Kirk, Gray, and Watson 1971), and time dose fractionation (Orton and Ellis 1973). However, significant limitations of such empirical approaches, particularly the most widely used NSD formalism, became evident over time (reviewed by Barendsen [1982]). Following the demise of the NSD concept, the mechanistically based LQ model (Sachs and Brenner 1998) became a *de facto* standard for performing isoeffect calculations in RT.

Lee et al. (1995) proposed that treatment planning systems (TPSs) have the capability to convert physical dose distributions into BED distributions and demonstrated the practical application of this technique using two clinical cases. Wheldon et al. (1998) suggested that a transformation of DVHs always be performed using $LQED_2$ (model 4) to account for the variation in fraction size throughout the different parts of an OAR or target volume. However, only a few institutions routinely correct DVHs for fractionation when performing outcome analyses (e.g., Oetzel et al. 1995; Kwa et al. 1998; Seppenwoolde et al. 2003; Willner et al. 2003; Dawson et al. 2002; Chapet, Kong, et al. 2005; Belderbos et al. 2005). Comparisons of model parameters derived from data sets involving substantially different fraction sizes should not be attempted without such normalization (Ten Haken, Lawrence, and Dawson 2006). Park et al. (2005) discuss the implications of not accounting for fractionation when using the gEUD index and propose a modification of the gEUD formula to include this effect.

The recent renewal of interest in hypofractionation has created a need to design OAR constraints for altered RT regimens based on existing constraints obtained with conventional fractionation. Such isoeffect calculations have been performed using the BED formalism for clinical trials of SBRT of lung tumors (Timmerman et al. 2006). Dawson, Eccles, and Craig (2006) used the LKB model (model 5) to set the prescription dose for liver SBRT based on the estimated risk of radiation-induced liver disease. In order to use the parameter estimates obtained in hyperfractionation studies for hypofractionated treatments, these authors transformed DVHs for SBRT plans using the fraction size and α/β ratio that had been used to derive the model parameters (Dawson et al. 2002). Song et al. (2005) used the same approach to guide dose prescription in lung SBRT, but the correction for fraction size differences was apparently not performed.

Interestingly, several investigators have proposed the use of BED formalism to quantify the biological effect of adding chemotherapy (Kasibhatla, Kirkpatrick, and Brizel 2007; Fowler 2008a; Plataniotis and Dale 2008; Lee and Eisbruch 2009) or hyperthermia (Plataniotis and Dale 2009) to RT.

6.3.4 Optimization of Prescription Dose and Fractionation

Jones, Tan, and Dale (1995) presented an analytical method for deriving optimal dose per fraction using the LQ model (model 1) with the time factor. Mehta et al. (2001) designed a dose-per-fraction escalation strategy for nonsmall cell lung cancer (NSCLC) based on maximizing the estimated TCP for a given level of late normal tissue effects, which has since been implemented in a clinical trial (Adkison et al. 2008). Radiobiological modeling has been used to derive alternative fractionation schedules for prostate cancer (Fowler et al. 2003; Li et al. 2008), head and neck cancer (Fowler 2007), and arteriovenous malformations (Qi et al. 2007).

These studies have been performed using population-based estimates of radiobiological parameters for each cancer type. An even greater promise of this method lies in using patient-specific parameters. The potential benefit of modifying dose prescription based on individual normal tissue radiosensitivity has been investigated in several theoretical studies (Tucker et al. 1996; Bentzen 1997; MacKay and Hendry 1999; Sanchez-Nieto, Nahum, and Dearnaley 2001; Guirado and Ruiz de Almodóvar 2003). All but one analysis suggest that significant therapeutic gains may be achievable. Modeling assumptions used in the study by Bentzen (1997), which reported the lack of therapeutic benefit, have been debated by MacKay et al. (1998) and Bentzen and Tucker (1998). Sanchez-Nieto, Nahum, and Dearnaley (2001) reported that individualizing dose prescription based on variations in patient anatomy alone may bring considerable gains to TCP, which may be further increased when the variability in normal tissue radiosensitivity is also taken into account.

Several dose escalation studies have been designed and executed using the idea of individualizing prescription dose based on a maximum acceptable level of NTCP for a dose-limiting complication. Ten Haken et al. (1993) proposed a dose escalation approach for tumors limited by OARs exhibiting a large volume effect (i.e., lung, liver) based on segregating patients according to the effective volume (model 5). This strategy has been implemented in clinical trials for hepatic tumors (McGinn et al. 1998) and NSCLC (Kong et al. 2006) conducted at the University of Michigan. Researchers at the Memorial Sloan-Kettering Cancer Center, New York, used the LKB model (model 5) and later the fraction of damaged FSUs, a by-product of the parallel architecture model (model 7), to assign patients into different prescription dose groups in the dose escalation trial for NSCLC (Rosenzweig et al. 2005). Several NSCLC trials to prescribe dose based on normal tissue constraints were designed by stratifying patients according to the V_{20} for the lung (Wu, Jiang, et al. 2003; Bradley et al. 2005) or the mean lung dose (Belderbos et al. 2006; van Baardwijk et al. 2010). Although these quantities are physical rather than biological, they have been shown to closely correlate with the incidence of radiation pneumonitis (Graham et al. 1999; Kwa et al. 1998), and are regarded as a close surrogate for biological response of the lung to radiation. Song et al. (2005) and Dawson, Eccles, and Craig (2006) used the LKB model (model 5) to select prescription doses for SBRT of lung and liver tumors, respectively.

Further individualization of RT is currently impeded by the lack of reliable predictive assays of normal tissue and tumor radiosensitivity. Molecular and functional imaging techniques capable of providing spatial distribution of biological parameters relevant to RT present the greatest potential when combined with IMRT delivery (Ling et al. 2000; Bentzen 2005). Although additional research efforts and clinical trials are needed before this potential can be realized, some theoretical frameworks for deriving optimal nonuniform dose distributions in tumors have already been described (e.g., Yang and Xing 2005; Chen et al. 2007; South, Partridge, and Evans 2008) and preclinical studies initiated (e.g., Søvik et al. 2007; Christian et al. 2009). Temporal changes in the distribution of biological parameters can be used as a basis for biologically adapted RT (Søvik, Malinen, and Olsen 2010).

6.4 Initial Experience with Commercial Biologically Based Treatment Planning Systems

In the past, biologically based models were used primarily in large radiotherapy centers that had the expertise and resources to develop in-house tools, and subsequently the biological modeling has not been able to enter wide clinical practice. Recently, commercial products employing some of the biologically based methods discussed in Section 6.3 have become available. Sections 6.4.1 through 6.4.3 provide a brief overview of biologically based tools implemented in three commercial TPSs.

6.4.1 CMS Monaco

Monaco (Elekta CMS Software, Maryland Heights, Missouri) is an IMRT-only TPS that has been specifically designed to utilize biologically based cost functions for inverse treatment planning and a Monte Carlo–based dose calculation engine. Monaco (these comments refer to versions 2.03 and lower) offers three biological cost functions: Poisson statistics cell kill model, serial complication model, and parallel complication model. These functions are based on a formalism developed at the University of Tübingen (Alber and Nüsslin 1999; Alber 2001). As their names imply, the Poisson cell kill model is used to create optimization objectives for target volumes, and the serial and parallel models are used to create constraints for OARs with different FSU architectures. For each function, the 3D dose distribution in a structure of interest is reduced to a single index that reflects a putative biological response of the structure to radiation. This index is referred to as an "isoeffect." Isoeffects for Poisson, serial, and parallel cost functions are conceptually identical to the cell-killing-based EUD (model 11), gEUD (model 12), and f_{dam} (model 7), respectively. Values specified by the user as optimization goals are referred to as "isoconstraints." Following each iteration, isoeffects are recomputed and compared with isoconstraints to determine whether the user-specified criteria have been met.

Monaco also provides five conventional, DV-based cost functions referred to as quadratic overdose penalty, quadratic underdosage penalty, overdose DVH constraint, underdose DVH constraint, and maximum dose constraint. The Poisson cell kill model is a mandatory cost function for targets. In the case of multiple target volumes, this model must be specified for at least one target. Because the Poisson cell kill model does not include a mechanism to control hot spots, a physical cost function, either the quadratic overdose penalty or the maximum dose constraint, should be added to create optimization goals for target volumes. Although the software does not require secondary constraints for target volumes, the optimization algorithm encounters convergence problems and the resulting target dose distributions are characterized by clinically unacceptable heterogeneity if such constraints are not used. In our experience, it was always possible to design acceptable plans using the three biological cost functions and the quadratic overdose penalty constraint (Semenenko et al. 2008).

Monaco is based on a concept of constrained optimization. That is, the serial and parallel cost functions and all physical cost functions are treated as hard constraints. All optimization goals specified using these cost functions must be met by the TPS. The Poisson cell kill model is only an objective. As a result, there are no weights to specify, that is, effectively, the Poisson cell kill model is assigned a very small weight and all other cost functions are assigned very large weights. Because target coverage is only an objective, achieving this objective may often be compromised by constraints on the exposure of nearby OARs or on hot spots in target volumes. A sensitivity analysis tool (Alber, Birkner, and Nüsslin 2002) is provided to help the planner identify the limiting constraints. Desired target coverage could then be obtained by relaxing (increasing) isoconstraint values for the most restrictive optimization criteria.

We have compared plans designed in Monaco using biological cost functions with plans created with a conventional TPS using DV-based optimization criteria (Semenenko et al. 2008). The Monaco plans were generally characterized by better OAR sparing while maintaining equivalent target coverage. However, less uniform target dose distributions were obtained in each case. In a study by Grofsmid et al. (2010), the dosimetric accuracy of the Monte Carlo algorithm implemented in Monaco has been tested and found to be clinically acceptable.

6.4.2 Philips Pinnacle3

Pinnacle[3] (Philips Healthcare, Andover, Massachusetts) offers biological optimization features incorporated into its P[3]IMRT inverse treatment planning module. Three biological objective functions (version 8.0h), denoted Min EUD, Max EUD, and Target EUD, have been developed by RaySearch Laboratories AB (Stockholm, Sweden; Hårdemark et al. 2004). These may be used alone or in any combination with traditional, DV-based cost functions (Min dose, Max dose, Uniform dose, Min DVH, Max DVH, and Uniformity). For target volumes, it is advised to supplement EUD-based objectives with DV-based objectives to better control dose uniformity (Hårdemark et al. 2004). The Min EUD, Max EUD, and Target EUD cost functions are designed to penalize for too low EUD, too high EUD, or any deviation from the desired EUD, respectively. The EUD is computed according to the gEUD formula (model 12) and therefore requires a single volume parameter a. The Min EUD cost function with negative a values or Target EUD objective with $a \approx 1$ are logical choices for targets, and the Max EUD cost function with $a \geq 1$ ($a \approx 1$ for parallel structures and $a > 1$ for serial structures) is a logical choice for OARs. However, no limits on values of the volume parameter or desired EUD are imposed by the software. Pinnacle[3] employs the traditional unconstrained optimization approach, that is, target and OAR cost functions contribute to the overall optimization score in proportion to user-specified weights. The three EUD-based cost functions can be used as either an objective or a constraint.

Pinnacle[3] also provides two separate tools for plan evaluation. NTCP and TCP estimates can be obtained using the NTCP/TCP Editor. NTCP is computed according to the LKB model (model 5). A customizable library of parameter estimates with default values based on the data provided by Burman et al. (1991) is available. TCP is estimated using an empirical sigmoid curve corresponding to the cumulative distribution function of the normal distribution. A database of TCP model parameters is not provided.

Users licensed for biological evaluation may take advantage of an enhanced plan evaluation tool, the Biological Response panel, which includes the following features:

- A customizable library of complication- and tumor-stage-specific parameter values accompanied by literature references
- NTCP and TCP calculations for individual structures
- Composite estimates of NTCP, TCP, and the probability of complication-free tumor control for the entire plan
- Plots representing changes in all NTCP and TCP estimates as the prescription dose is scaled between 0% and 200%
- Side-by-side comparisons of competing treatment plans

Models and parameter estimates implemented in the Biological Response panel are based on the work conducted at the Karolinska Institute and Stockholm University (Källman, Ågren, and Brahme 1992; Lind et al. 1999). NTCP is calculated using the relative seriality model (model 6). Although the source of provided model inputs is not readily available in the open literature, it appears that the parameter estimates were obtained by fitting the relative seriality model to the tolerance data of Emami et al. (1991). TCP is calculated using the LQ-Poisson model (model 8). Parameter estimates provided for TCP calculations are compiled from old literature dating back to the 1960s, with the most recent report dating to 1993. Both NTCP and TCP models include the α/β ratio as one of the input parameters and, therefore, are sensitive to the choice of fractionation schedule. Computation of composite TCP/NTCP estimates is based on the P_+ formalism (model 10). For simplicity, completely correlated tumor and normal tissue responses ($\delta = 0$) are assumed when calculating the probability of complication-free tumor control.

Widesott et al. (2008) and Qi, Semenenko, and Li et al. (2009) evaluated the capabilities of the gEUD-based approach implemented in Pinnacle[3] and reported better OAR sparing compared to DV-based optimization. The biological optimization implemented in Pinnacle[3] does not necessarily lead to increased dose heterogeneity in target volumes (Qi, Semenenko, and Li et al. 2009). Therefore, the loss of target dose uniformity observed in Monaco (Semenenko et al. 2008) is likely related to a particular implementation of the optimization algorithm. Widesott et al. (2008) also reported a substantial reduction in the time required to obtain a clinically acceptable plan with the use of the gEUD-based cost functions.

6.4.3 Varian Eclipse*

The biological optimization and evaluation algorithms implemented in Eclipse (Varian Medical Systems, Palo Alto, California) have also been developed by RaySearch Laboratories AB (Stockholm, Sweden). Currently (version 10.0), all biologically based functionality is accessible though an external software application. Because Eclipse and Pinnacle³ biological tools are based on the same initial set of ideas, we will frequently refer to the information presented in Section 6.4.2 in an effort to highlight the similarities and differences between the two TPSs.

In Eclipse, all cost functions are divided into physical and biological. The physical functions include Min Dose, Max Dose, Min DVH, Max DVH, Min EUD, Max EUD, Target EUD, Target Conformance, and Uniformity Constraint. The Min EUD, Max EUD, and Target EUD functions are the familiar gEUD-based cost functions from Pinnacle³, which, interestingly, are listed among physical functions in Eclipse. The idea of biologically based optimization has been taken a step further in Eclipse. Dose distributions may be optimized using three TCP/NTCP models listed under biological cost functions. These are termed TCP Poisson-LQ, NTCP Poisson-LQ, and NTCP Lyman. The TCP Poisson-LQ (model 8) and the NTCP Poisson-LQ (model 6) are the same models that are employed for plan evaluation in the Biological Response panel in Pinnacle³. The NTCP Lyman is the LKB model (model 5), which, in contrast to the same model in the NTCP/TCP Editor in Pinnacle³, is calculated based on LQ-corrected DVHs (model 4), and therefore takes an extra parameter α/β. All TCP and NTCP models used in Eclipse are thus sensitive to fractionation schedules. Another difference from Pinnacle³ is that in Eclipse, all three models include optional parameters to describe biphasic sublethal damage repair, that is, halftimes for slow and fast repair components and a fraction of damage with the slow repair. Additionally, the TCP Poisson-LQ model allows the specification of two parameters to describe clonogen repopulation, that is, the potential doubling time and the time when the repopulation starts. The biological functions allow the user to specify the weight (constraint bound percentage) used in the calculation of the cost function. Physical functions are treated as constraints that must be met by the TPS.

A treatment plan generated in Eclipse may be evaluated using the same NTCP and TCP models available in the optimization phase. Similar to Pinnacle³, the effect of changes in the prescription dose (0%–250%) on NTCP and TCP estimates is presented graphically. Additional features in Eclipse not available in Pinnacle³ include a tool for the evaluation of the effect of changes in fractionation schedule (i.e., twice daily vs. once daily, different number of fractions) and a plot of LQ-scaled DVHs. The normalization is performed to a standard fraction size of 2 Gy using the α/β ratio specified for the NTCP or TCP model. The same library of tissue-specific parameter values for NTCP and

TCP models as implemented in Pinnacle³ is provided in Eclipse. The library may be edited to include user-specified end points and model inputs.

6.5 Future Development of Biologically Based Treatment Planning

Application of biologically based models in RT has come a long way in the past 20 years. For the first time, models have become available in commercial treatment planning software. We believe that these products will be eventually used widely in clinical practice. At the present time, uncertainties in models and parameter estimates still create a great impediment to the dissemination of biologically based methods. As attractive as it may be to both optimize and evaluate treatment plans based on TCP and NTCP indices (just like most plans are currently both optimized and evaluated based on DV constraints), the absolute values of TCP and NTCP cannot currently be trusted. Despite the recent QUANTEC efforts to summarize current knowledge of normal tissue complications, much work to improve the standards of collecting and reporting toxicity data lies ahead (Bentzen et al. 2010). Similar efforts are needed to increase the reliability of tumor control data. Until further improvements to increase the predictive power of TCP and NTCP estimates are achieved, DV-based evaluation criteria verified in clinical practice should always be used as part of *plan evaluation*. With the support of TPS manufacturers who should keep pace with the latest developments (e.g., the QUANTEC results), additional model-based plan evaluation tools should help clinicians to eventually bridge the gap between the DV constraints and biological metrics.

In the meantime, biologically based models have matured enough to be used in *plan optimization*. Optimization objectives based on simpler functions, such as the gEUD, appear to be a logical first step in introducing biologically based models in the clinic. The gEUD-based objective functions have several features that make them attractive over both DV- and TCP/NTCP-based objective functions: gEUD belongs to a familiar dose domain, only one parameter needs to be specified, quantities such as the mean organ dose can be directly optimized, and a single cost function may replace multiple DV constraints. The simpler models would still encourage the treatment planners to learn basic concepts, for example, the serial vs. parallel tissue architecture theory; yet they would not be as overwhelming as TCP/NTCP-based cost functions whose behavior is harder to understand because of the larger number of parameters. Initial testing of commercial TPSs employing gEUD-based optimization objectives (Semenenko et al. 2008; Widesott et al. 2008; Qi, Semenenko, and Li et al. 2009) has revealed that plans offering additional OAR sparing can be generated using fewer trials, which may be attributed to the fact that gEUD-based cost functions are capable of exploring a larger solution space and finding additional solutions that result in redistributing the dose away from the most critical OARs. It is doubtful whether TCP/NTCP-based objective functions can bring any additional

---

* The authors would like to thank Charles Mayo for providing information about the biological tools in Eclipse.

benefits at this time. However, a strong rationale for switching to TCP/NTCP optimization in the future still remains so that the optimized TCP and NTCP values could be directly used to evaluate the plan. By the time TCP and NTCP are ready to enter clinical practice, a generation of treatment planners will be trained in the basic principles of biologically based treatment planning and will be prepared to embrace these more powerful tools.

References

Adkison, J. B., D. Khuntia, S. M. Bentzen, et al. 2008. Dose escalated, hypofractionated radiotherapy using helical tomotherapy for inoperable non-small cell lung cancer: Preliminary results of a risk-stratified phase I dose escalation study. *Technol Cancer Res Treat* 7:441–7.

Alber, M. 2001. A concept for the optimization of radiotherapy. PhD diss., University of Tübingen, Tübingen, Germany.

Alber, M., M. Birkner, and F. Nüsslin. 2002. Tools for the analysis of dose optimization: II. Sensitivity analysis. *Phys Med Biol* 47:N265–70.

Alber, M., and F. Nüsslin. 1999. An objective function for radiation treatment optimization based on local biological measures. *Phys Med Biol* 44:479–93.

Amols, H. I., M. Zaider, M. K. Hayes, and P. B. Schiff. 1997. Physician/patient-driven risk assignment in radiation oncology: Reality or fancy? *Int J Radiat Oncol Biol Phys* 38:455–61.

Astrahan, M. 2008. Some implications of linear-quadratic-linear radiation dose-response with regard to hypofractionation. *Med Phys* 35:4161–72.

Barendsen, G. W. 1982. Dose fractionation, dose rate and iso-effect relationships for normal tissue responses. *Int J Radiat Oncol Biol Phys* 8:1981–97.

Belderbos, J. S., W. D. Heemsbergen, K. De Jaeger, P. Baas, and J. V. Lebesque. 2006. Final results of a Phase I/II dose escalation trial in non-small-cell lung cancer using three-dimensional conformal radiotherapy. *Int J Radiat Oncol Biol Phys* 66:126–34.

Belderbos, J., W. Heemsbergen, M. Hoogeman, K. Pengel, M. Rossi, and J. Lebesque. 2005. Acute esophageal toxicity in non-small cell lung cancer patients after high dose conformal radiotherapy. *Radiother Oncol* 75:157–64.

Bentzen, S. M. 1997. Potential clinical impact of normal-tissue intrinsic radiosensitivity testing. *Radiother Oncol* 43:121–31.

Bentzen, S. M. 2005. Theragnostic imaging for radiation oncology: Dose-painting by numbers. *Lancet Oncol* 6:112–7.

Bentzen, S. M., L. S. Constine, J. O. Deasy, et al. 2010. Quantitative analyses of normal tissue effects in the clinic (QUANTEC): An introduction to the scientific issues. *Int J Radiat Oncol Biol Phys Suppl* 76:S3–9.

Bentzen, S. M., and M. C. Joiner. 2009. The linear-quadratic approach in clinical practice. In *Basic clinical radiobiology.* 4th ed. Eds. M. C. Joiner and A. J. van der Kogel, 120–34. London: Hodder Arnold.

Bentzen, S. M., J. Overgaard, H. D. Thames, et al. 1989. Clinical radiobiology of malignant melanoma. *Radiother Oncol* 16:169–82.

Bentzen, S. M., and S. L. Tucker. 1998. Individualization of radiotherapy dose prescriptions by means of an in vitro radiosensitivity assay. *Radiother Oncol* 46:216–8.

Bortfeld, T., D. Craft, J. F. Dempsey, T. Halabi, and H. E. Romeijn. 2008. Evaluating target cold spots by the use of tail EUDs. *Int J Radiat Oncol Biol Phys* 71:880–9.

Bortfeld, T., W. Schlegel, C. Dykstra, S. Levegrün, and K. Preiser. 1996. Physical vs. biological objectives for treatment plan optimization. *Radiother Oncol* 40:185–7.

Bovi, J., X. S. Qi, J. White, and X. A. Li. 2007. Comparison of three accelerated partial breast irradiation techniques: Treatment effectiveness based upon biological models. *Radiother Oncol* 84:226–32.

Bradley, J., M. V. Graham, K. Winter, et al. 2005. Toxicity and outcome results of RTOG 9311: A Phase I-II dose-escalation study using three-dimensional conformal radiotherapy in patients with inoperable non-small-cell lung carcinoma. *Int J Radiat Oncol Biol Phys* 61:318–28.

Bradley, J. D., A. Hope, I. El Naqa, et al. 2007. A nomogram to predict radiation pneumonitis, derived from a combined analysis of RTOG 9311 and institutional data. *Int J Radiat Oncol Biol Phys* 69:985–92.

Brahme, A. 1984. Dosimetric precision requirements in radiation therapy. *Acta Radiol Oncol* 23:379–91.

Brahme, A. 2001. Individualizing cancer treatment: Biological optimization models in treatment planning and delivery. *Int J Radiat Oncol Biol Phys* 49:327–37.

Brenner, D. J. 1993. Dose, volume, and tumor-control predictions in radiotherapy. *Int J Radiat Oncol Biol Phys* 26:171–9.

Brenner, D. J. 2008. The linear-quadratic model is an appropriate methodology for determining isoeffective doses at large doses per fraction. *Semin Radiat Oncol* 18:234–9.

Brenner, D. J., and E. J. Hall. 1999. Fractionation and protraction for radiotherapy of prostate carcinoma. *Int J Radiat Oncol Biol Phys* 43:1095–101.

Brenner, D. J., L. R. Hlatky, P. J. Hahnfeldt, E. J. Hall, and R. K. Sachs. 1995. A convenient extension of the linear-quadratic model to include redistribution and reoxygenation. *Int J Radiat Oncol Biol Phys* 32:379–90.

Brenner, D. J., L. R. Hlatky, P. J. Hahnfeldt, Y. Huang, and R. K. Sachs. 1998. The linear-quadratic model and most other common radiobiological models result in similar predictions of time-dose relationships. *Radiat Res* 150:83–91.

Brenner, D. J., and R. K. Sachs. 1999. A more robust biologically based ranking criterion for treatment plans. *Int J Radiat Oncol Biol Phys* 43:697–8.

Buffa, F. M., S. E. Davidson, R. D. Hunter, A. E. Nahum, and C. M. West. 2001. Incorporating biologic measurements (SF$_2$, CFE) into a tumor control probability model increases their prognostic significance: A study in cervical carcinoma treated with radiation therapy. *Int J Radiat Oncol Biol Phys* 50:1113–22.

Burman, C., G. J. Kutcher, B. Emami, and M. Goitein. 1991. Fitting of normal tissue tolerance data to an analytic function. *Int J Radiat Oncol Biol Phys* 21:123–35.

Carlson, D. J., R. D. Stewart, and V. A. Semenenko. 2006. Effects of oxygen on intrinsic radiation sensitivity: A test of the relationship between aerobic and hypoxic linear-quadratic (LQ) model parameters. *Med Phys* 33:3105–15.

Chapet, O., F. M. Kong, J. S. Lee, J. A. Hayman, and R. K. Ten Haken. 2005. Normal tissue complication probability modeling for acute esophagitis in patients treated with conformal radiation therapy for non-small cell lung cancer. *Radiother Oncol* 77:176–81.

Chapet, O., E. Thomas, M. L. Kessler, B. A. Fraass, and R. K. Ten Haken. 2005. Esophagus sparing with IMRT in lung tumor irradiation: An EUD-based optimization technique. *Int J Radiat Oncol Biol Phys* 63:179–87.

Chen, G. P., E. Ahunbay, C. Schultz, and X. A. Li. 2007. Development of an inverse optimization package to plan nonuniform dose distributions based on spatially inhomogeneous radiosensitivity extracted from biological images. *Med Phys* 34:1198–205.

Choi, B., and J. O. Deasy. 2002. The generalized equivalent uniform dose function as a basis for intensity-modulated treatment planning. *Phys Med Biol* 47:3579–89.

Christian, N., J. A. Lee, A. Bol, M. De Bast, B. Jordan, and V. Grégoire. 2009. The limitation of PET imaging for biological adaptive-IMRT assessed in animal models. *Radiother Oncol* 91:101–6.

Chvetsov, A. V., J. F. Dempsey, and J. R. Palta. 2007. Optimization of equivalent uniform dose using the L-curve criterion. *Phys Med Biol* 52:5973–84.

Courdi, A. 2010. High doses per fraction and the linear-quadratic model. *Radiother Oncol* 94:121–2; author reply 122–3.

Das, S. 2009. A role for biological optimization within the current treatment planning paradigm. *Med Phys* 36:4672–82.

Dawson, L. A., C. Eccles, and T. Craig. 2006. Individualized image guided iso-NTCP based liver cancer SBRT. *Acta Oncol* 45:856–64.

Dawson, L. A., D. Normolle, J. M. Balter, C. J. McGinn, T. S. Lawrence, and R. K. Ten Haken. 2002. Analysis of radiation-induced liver disease using the Lyman NTCP model. *Int J Radiat Oncol Biol Phys* 53:810–21. Erratum in: 2002. *Int J Radiat Oncol Biol Phys* 53:1422.

De Gersem, W. R., S. Derycke, C. O. Colle, C. De Wagter, and W. J. De Neve. 1999. Inhomogeneous target-dose distributions: A dimension more for optimization? *Int J Radiat Oncol Biol Phys* 44:461–8.

Deasy, J. O. 1997. Multiple local minima in radiotherapy optimization problems with dose-volume constraints. *Med Phys* 24:1157–61.

Drzymala, R. E., R. Mohan, L. Brewster, et al. 1991. Dose-volume histograms. *Int J Radiat Oncol Biol Phys* 21:71–8.

Ellis, F. 1969. Dose, time and fractionation: A clinical hypothesis. *Clin Radiol* 20:1–7.

El Naqa, I., G. Suneja, P. E. Lindsay, et al. 2006. Dose response explorer: An integrated open-source tool for exploring and modelling radiotherapy dose-volume outcome relationships. *Phys Med Biol* 51:5719–35.

Emami, B., J. Lyman, A. Brown, et al. 1991. Tolerance of normal tissue to therapeutic irradiation. *Int J Radiat Oncol Biol Phys* 21:109–22.

Fatyga, M., J. F. Williamson, N. Dogan, et al. 2009. A comparison of HDR brachytherapy and IMRT techniques for dose escalation in prostate cancer: A radiobiological modeling study. *Med Phys* 36:3995–4006.

Fowler, J. F. 1989. The linear-quadratic formula and progress in fractionated radiotherapy. *Br J Radiol* 62:679–94.

Fowler, J. F. 2007. Is there an optimum overall time for head and neck radiotherapy? A review, with new modelling. *Clin Oncol (R Coll Radiol)* 19:8–22. Erratum in: 2008. *Clin Oncol (R Coll Radiol)* 20:124–6.

Fowler, J. F. 2008a. Correction to Kasibhatla et al. How much radiation is the chemotherapy worth in advanced head and neck cancer? (*Int J Radiat Oncol Biol Phys* 2007;68:1491–5). *Int J Radiat Oncol Biol Phys* 71:326–9.

Fowler, J. F. 2008b. Linear quadratics is alive and well: In regards to Park et al. (*Int J Radiat Oncol Biol Phys* 2008;70:847–52). *Int J Radiat Oncol Biol Phys* 72:957; author reply 958.

Fowler, J. F., M. A. Ritter, R. J. Chappell, and D. J. Brenner. 2003. What hypofractionated protocols should be tested for prostate cancer? *Int J Radiat Oncol Biol Phys* 56:1093–104.

Gayou, O., D. S. Parda, and M. Miften. 2007. EUCLID: An outcome analysis tool for high-dimensional clinical studies. *Phys Med Biol* 52:1705–19.

Graham, M. V., J. A. Purdy, B. Emami, et al. 1999. Clinical dose-volume histogram analysis for pneumonitis after 3D treatment for non-small cell lung cancer (NSCLC). *Int J Radiat Oncol Biol Phys* 45:323–9.

Grofsmid, D., M. Dirkx, H. Marijnissen, E. Woudstra, and B. Heijmen. 2010. Dosimetric validation of a commercial Monte Carlo based IMRT planning system. *Med Phys* 37:540–9.

Guerrero, M., and X. A. Li. 2003. Analysis of a large number of clinical studies for breast cancer radiotherapy: Estimation of radiobiological parameters for treatment planning. *Phys Med Biol* 48:3307–26.

Guerrero, M., and X. A. Li. 2004. Extending the linear-quadratic model for large fraction doses pertinent to stereotactic radiotherapy. *Phys Med Biol* 49:4825–35.

Guerrero, M., X. A. Li, L. Ma, J. Linder, C. Deyoung, and B. Erickson. 2005. Simultaneous integrated intensity-modulated radiotherapy boost for locally advanced gynecological cancer: Radiobiological and dosimetric considerations. *Int J Radiat Oncol Biol Phys* 62:933–9.

Guirado, D., and J. M. Ruiz de Almodóvar. 2003. Prediction of normal tissue response and individualization of doses in radiotherapy. *Phys Med Biol* 48:3213–23.

Hall, E. J., and D. J. Brenner. 1995. In response to Dr. Marks. *Int J Radiat Oncol Biol Phys* 32:275–6.

Hårdemark, B., A. Liander, H. Rehbinder, J. Löf, and D. Robinson 2004. P³IMRT. Biological optimization and EUD. Pinnacle³ White Paper No. 4535 983 02482, Philips Medical Systems.

Hartmann, M., and L. Bogner. 2008. Investigation of intensity-modulated radiotherapy optimization with gEUD-based objectives by means of simulated annealing. *Med Phys* 35:2041–9.

Hoffmann, A. L., D. den Hertog, A. Y. Siem, J. H. Kaanders, and H. Huizenga. 2008. Convex reformulation of biologically-based multi-criteria intensity-modulated radiation therapy optimization including fractionation effects. *Phys Med Biol* 53:6345–62.

Honoré, H. B., and S. M. Bentzen. 2006. A modelling study of the potential influence of low dose hypersensitivity on radiation treatment planning. *Radiother Oncol* 79:115–21.

Huang, Y., M. Joiner, B. Zhao, Y. Liao, and J. Burmeister. 2010. Dose convolution filter: Incorporating spatial dose information into tissue response modeling. *Med Phys* 37:1068–74.

Jackson, A., G. J. Kutcher, and E. D. Yorke. 1993. Probability of radiation-induced complications for normal tissues with parallel architecture subject to non-uniform irradiation. *Med Phys* 20:613–25.

Jackson, A., R. K. Ten Haken, J. M. Robertson, M. L. Kessler, G. J. Kutcher, and T. S. Lawrence. 1995. Analysis of clinical complication data for radiation hepatitis using a parallel architecture model. *Int J Radiat Oncol Biol Phys* 31:883–91.

Joiner, M. C., B. Marples, P. Lambin, S. C. Short, and I. Turesson. 2001. Low-dose hypersensitivity: Current status and possible mechanisms. *Int J Radiat Oncol Biol Phys* 49:379–89.

Jones, B., and R. G. Dale. 1995. Cell loss factors and the linear-quadratic model. *Radiother Oncol* 37:136–9.

Jones, L., and P. Hoban. 2002. A comparison of physically and radiobiologically based optimization for IMRT. *Med Phys* 29:1447–55.

Jones, B., L. T. Tan, and R. G. Dale. 1995. Derivation of the optimum dose per fraction from the linear quadratic model. *Br J Radiol* 68:894–902.

Källman, P., A. Ågren, and A. Brahme. 1992. Tumour and normal tissue responses to fractionated non-uniform dose delivery. *Int J Radiat Biol* 62:249–62.

Källman, P., B. K. Lind, and A. Brahme. 1992. An algorithm for maximizing the probability of complication-free tumour control in radiation therapy. *Phys Med Biol* 37:871–90.

Kasibhatla, M., J. P. Kirkpatrick, and D. M. Brizel. 2007. How much radiation is the chemotherapy worth in advanced head and neck cancer? *Int J Radiat Oncol Biol Phys* 68:1491–5.

Kavanagh, B. D., and F. Newman. 2008. Toward a unified survival curve: In regards to Park et al. (*Int J Radiat Oncol Biol Phys* 2008;70:847–52) and Krueger et al. (*Int J Radiat Oncol Biol Phys* 2007;69:1262–71). *Int J Radiat Oncol Biol Phys* 71:958–9.

Khan, H. A. 2007. CalcNTCP: A simple tool for computation of normal tissue complication probability (NTCP) associated with cancer radiotherapy. *Int J Radiat Biol* 83:717–20.

King, C. R., T. A. DiPetrillo, and D. E. Wazer. 2000. Optimal radiotherapy for prostate cancer: Predictions for conventional external beam, IMRT, and brachytherapy from radiobiologic models. *Int J Radiat Oncol Biol Phys* 46:165–72.

Kirk, J., W. M. Gray, and E. R. Watson. 1971. Cumulative radiation effect: I. Fractionated treatment regimes. *Clin Radiol* 22:145–55.

Kirkpatrick, J. P., D. J. Brenner, and C. G. Orton. 2009. Point/counterpoint: The linear-quadratic model is inappropriate to model high dose per fraction effects in radiosurgery. *Med Phys* 36:3381–4.

Kirkpatrick, J. P., J. J. Meyer, and L. B. Marks. 2008. The linear-quadratic model is inappropriate to model high dose per fraction effects in radiosurgery. *Semin Radiat Oncol* 18:240–3.

Kong, F. M., J. A. Hayman, K. A. Griffith, et al. 2006. Final toxicity results of a radiation-dose escalation study in patients with non-small-cell lung cancer (NSCLC): Predictors for radiation pneumonitis and fibrosis. *Int J Radiat Oncol Biol Phys* 65:1075–86.

Kutcher, G. J., and C. Burman. 1989. Calculation of complication probability factors for non-uniform normal tissue irradiation: The effective volume method. *Int J Radiat Oncol Biol Phys* 16:1623–30.

Kwa, S. L., J. V. Lebesque, J. C. Theuws, et al. 1998. Radiation pneumonitis as a function of mean lung dose: An analysis of pooled data of 540 patients. *Int J Radiat Oncol Biol Phys* 42:1–9.

Langer, M., S. S. Morrill, and R. Lane. 1998. A test of the claim that plan rankings are determined by relative complication and tumor-control probabilities. *Int J Radiat Oncol Biol Phys* 41:451–7.

Lee, I. H., and A. Eisbruch. 2009. Mucositis versus tumor control: The therapeutic index of adding chemotherapy to irradiation of head and neck cancer. *Int J Radiat Oncol Biol Phys* 75:1060–3.

Lee, S. P., M. Y. Leu, J. B. Smathers, W. H. McBride, R. G. Parker, and H. R. Withers. 1995. Biologically effective dose distribution based on the linear quadratic model and its clinical relevance. *Int J Radiat Oncol Biol Phys* 33:375–89.

Levegrün, S., A. Jackson, M. J. Zelefsky, et al. 2000. Analysis of biopsy outcome after three-dimensional conformal radiation therapy of prostate cancer using dose-distribution variables and tumor control probability models. *Int J Radiat Oncol Biol Phys* 47:1245–60.

Levegrün, S., A. Jackson, M. J. Zelefsky, et al. 2001. Fitting tumor control probability models to biopsy outcome after three-dimensional conformal radiation therapy of prostate cancer: Pitfalls in deducing radiobiologic parameters for tumors from clinical data. *Int J Radiat Oncol Biol Phys* 51:1064–80.

Li, X. A., J. Z. Wang, P. A. Jursinic, C. A. Lawton, and D. Wang. 2005. Dosimetric advantages of IMRT simultaneous integrated boost for high-risk prostate cancer. *Int J Radiat Oncol Biol Phys* 61:1251–7.

Li, X. A., J. Z. Wang, R. D. Stewart, S. J. Dibiase, D. Wang, and C. A. Lawton. 2008. Designing equivalent treatment regimens for prostate radiotherapy based on equivalent uniform dose. *Br J Radiol* 81:59–68.

Lind, B. K., P. Mavroidis, S. Hyödynmaa, and C. Kappas. 1999. Optimization of the dose level for a given treatment plan to maximize the complication-free tumor cure. *Acta Oncol* 38:787–98.

Ling, C. C., J. Humm, S. Larson, et al. 2000. Towards multidimensional radiotherapy (MD-CRT): Biological imaging and biological conformality. *Int J Radiat Oncol Biol Phys* 47:551–60.

Liu, D., M. Ajlouni, J. Y. Jin, et al. 2009. Analysis of outcomes in radiation oncology: An integrated computational platform. *Med Phys* 36:1680–9.

Llacer, J., J. O. Deasy, T. R. Portfeld, T. D. Solberg, and C. Promberger. 2003. Absence of multiple local minima effects in intensity modulated optimization with dose-volume constraints. *Phys Med Biol* 48:183–210.

Luxton, G., S. L. Hancock, and A. L. Boyer. 2004. Dosimetry and radiobiologic model comparison of IMRT and 3D conformal radiotherapy in treatment of carcinoma of the prostate. *Int J Radiat Oncol Biol Phys* 59:267–84.

Luxton, G., P. J. Keall, and C. R. King. 2008. A new formula for normal tissue complication probability (NTCP) as a function of equivalent uniform dose (EUD). *Phys Med Biol* 53:23–36.

Lyman, J. T. 1985. Complication probability as assessed from dose-volume histograms. *Radiat Res Suppl* 8:S13–9.

MacKay, R. I., and J. H. Hendry. 1999. The modelled benefits of individualizing radiotherapy patients' dose using cellular radiosensitivity assays with inherent variability. *Radiother Oncol* 50:67–75.

MacKay, R. I., A. Niemierko, M. Goitein, and J. H. Hendry. 1998. Potential clinical impact of normal-tissue intrinsic radiosensitivity testing. *Radiother Oncol* 46:215–6.

Marks, L. B. 1995. Extrapolating hypofractionated radiation schemes from radiosurgery data: Regarding Hall et al. (IJROBP 1999; 21:819–24) and Hall and Brenner (IJROBP 1993; 25:381–5). *Int J Radiat Oncol Biol Phys* 32:274–5.

Marks, L. B., R. K. Ten Haken, and M. K. Martel. 2010. Guest editor's introduction to QUANTEC: A users guide. *Int J Radiat Oncol Biol Phys Suppl* 76:S1–2.

Mavroidis, P., B. C. Ferreira, C. Shi, B. K. Lind, N. Papanikolaou. 2007. Treatment plan comparison between helical tomotherapy and MLC-based IMRT using radiobiological measures. *Phys Med Biol* 52:3817–36.

McGinn, C. J., R. K. Ten Haken, W. D. Ensminger, S. Walker, S. Wang, and T. S. Lawrence. 1998. Treatment of intrahepatic cancers with radiation doses based on a normal tissue complication probability model. *J Clin Oncol* 16:2246–52.

McKenna, F. W., and S. Ahmad. 2009. Fitting techniques of cell survival curves in high-dose region for use in stereotactic body radiation therapy. *Phys Med Biol* 54:1593–608.

Mehta, M., R. Scrimger, R. Mackie, B. Paliwal, R. Chappell, and J. Fowler. 2001. A new approach to dose escalation in non-small-cell lung cancer. *Int J Radiat Oncol Biol Phys* 49:23–33.

Mohan, R., G. S. Mageras, B. Baldwin, et al. 1992. Clinically relevant optimization of 3-D conformal treatments. *Med Phys* 19:933–44.

Mohan, R., and X. H. Wang. 1996. Response to Bortfeld et al. Re physical vs. biological objectives for treatment plan optimization. *Radiother Oncol* 40:186–7.

Moiseenko, V., J. Battista, and J. Van Dyk. 2000. Normal tissue complication probabilities: Dependence on choice of biological model and dose-volume histogram reduction scheme. *Int J Radiat Oncol Biol Phys* 46:983–93.

Muren, L. P., N. Jebsen, A. Gustafsson, and O. Dahl. 2001. Can dose-response models predict reliable normal tissue complication probabilities in radical radiotherapy of urinary bladder cancer? The impact of alternative radiation tolerance models and parameters. *Int J Radiat Oncol Biol Phys* 50:627–37.

Nahum, A., and G. Kutcher. 2007. Biological evaluation of treatment plans. In *Handbook of Radiotherapy Physics: Theory and Practice,* ed. P. Mayles, A. Nahum, and J. -C. Rosenwald, 731–71. Boca Raton, FL: Taylor & Francis.

Nahum, A. E., B. Movsas, E. M. Horwitz, C. C. Stobbe, and J. D. Chapman. 2003. Incorporating clinical measurements of hypoxia into tumor local control modeling of prostate cancer: Implications for the α/β ratio. *Int J Radiat Oncol Biol Phys* 57:391–401.

Nahum, A. E., and B. Sanchez-Nieto. 2001. Tumour control probability modeling: Basic principles and applications in treatment planning. *Phys Med* 17:13–23.

Newcomb, C. H., J. Van Dyk, and R. P. Hill. 1993. Evaluation of isoeffect formulae for predicting radiation-induced lung damage. *Radiother Oncol* 26:51–63.

Niemierko, A. 1997. Reporting and analyzing dose distributions: A concept of equivalent uniform dose. *Med Phys* 24:103–10.

Niemierko, A. 1999. A generalized concept of equivalent uniform dose (EUD) (abstract). *Med Phys* 26: 1101.

Niemierko, A., M. Urie, and M. Goitein. 1992. Optimization of 3D radiation therapy with both physical and biological end points and constraints. *Int J Radiat Oncol Biol Phys* 23:99–108.

Oetzel, D., P. Schraube, F. Hensley, G. Sroka-Pérez, M. Menke, and M. Flentje. 1995. Estimation of pneumonitis risk in three-dimensional treatment planning using dose-volume histogram analysis. *Int J Radiat Oncol Biol Phys* 33:455–60.

Okunieff, P., D. Morgan, A. Niemierko, and H. D. Suit. 1995. Radiation dose-response of human tumors. *Int J Radiat Oncol Biol Phys* 32:1227–37.

Ólafsson, A., R. Jeraj, and S. J. Wright. 2005. Optimization of intensity-modulated radiation therapy with biological objectives. *Phys Med Biol* 50:5357–79.

Orton, C. G., T. R. Bortfeld, A. Niemierko, and J. Unkelbach. 2008. The role of medical physicists and the AAPM in the development of treatment planning and optimization. *Med Phys* 35:4911–23.

Orton, C. G., and F. Ellis. 1973. A simplification in the use of the NSD concept in practical radiotherapy. *Br J Radiol* 46:529–37.

Park, C. S., Y. Kim, N. Lee, et al. 2005. Method to account for dose fractionation in analysis of IMRT plans: Modified equivalent uniform dose. *Int J Radiat Oncol Biol Phys* 62:925–32.

Park, C., L. Papiez, S. Zhang, M. Story, and R. D. Timmerman. 2008. Universal survival curve and single fraction equivalent dose: Useful tools in understanding potency of ablative radiotherapy. *Int J Radiat Oncol Biol Phys* 70:847–52.

Pieters, B. R., J. B. van de Kamer, Y. R. van Herten, et al. 2008. Comparison of biologically equivalent dose-volume parameters for the treatment of prostate cancer with concomitant boost IMRT versus IMRT combined with brachytherapy. *Radiother Oncol* 88:46–52.

Plataniotis, G. A., and R. G. Dale. 2008. Use of concept of chemotherapy-equivalent biologically effective dose to provide quantitative evaluation of contribution of chemotherapy to local tumor control in chemoradiotherapy cervical cancer trials. *Int J Radiat Oncol Biol Phys* 72:1538–43.

Plataniotis, G. A., and R. G. Dale. 2009. Use of the concept of equivalent biologically effective dose (BED) to quantify the contribution of hyperthermia to local tumor control in radiohyperthermia cervical cancer trials, and comparison with radiochemotherapy results. *Int J Radiat Oncol Biol Phys* 73:1538–44.

Qi, X. S., C. J. Schultz, and X. A. Li. 2006. An estimation of radiobiologic parameters from clinical outcomes for radiation treatment planning of brain tumor. *Int J Radiat Oncol Biol Phys* 64:1570–80.

Qi, X. S., C. J. Schultz, and X. A. Li. 2007. Possible fractionated regimens for image-guided intensity-modulated radiation therapy of large arteriovenous malformations. *Phys Med Biol* 52:5667–82.

Qi, X. S., V. A. Semenenko, and X. A. Li. 2009. Improved critical structure sparing with biologically based IMRT optimization. *Med Phys* 36:1790–9.

Rancati, T., C. Fiorino, G. Gagliardi, et al. 2004. Fitting late rectal bleeding data using different NTCP models: Results from an Italian multi-centric study (AIROPROS0101). *Radiother Oncol* 73:21–32.

Roberts, S. A., and J. H. Hendry. 1993. The delay before onset of accelerated tumour cell repopulation during radiotherapy: A direct maximum-likelihood analysis of a collection of worldwide tumour-control data. *Radiother Oncol* 29:69–74.

Roberts, S. A., and J. H. Hendry. 1999. Time factors in larynx tumor radiotherapy: Lag times and intertumor heterogeneity in clinical datasets from four centers. *Int J Radiat Oncol Biol Phys* 45:1247–57.

Rosenzweig, K. E., J. L. Fox, E. Yorke, et al. 2005. Results of a Phase I dose-escalation study using three-dimensional conformal radiotherapy in the treatment of inoperable nonsmall cell lung carcinoma. *Cancer* 103:2118–27.

Rowbottom, C. G., S. Webb, and M. Oldham. 1999. Is it possible to optimize a radiotherapy treatment plan? *Int J Radiat Oncol Biol Phys* 43:698–9.

Sachs, R. K., and D. J. Brenner. 1998. The mechanistic basis of the linear-quadratic formalism. *Med Phys* 25:2071–3.

Sanchez-Nieto, B., and A. E. Nahum. 2000. BIOPLAN: Software for the biological evaluation of radiotherapy treatment plans. *Med Dosim* 25:71–6.

Sanchez-Nieto, B., A. E. Nahum, and D. P. Dearnaley. 2001. Individualization of dose prescription based on normal-tissue dose-volume and radiosensitivity data. *Int J Radiat Oncol Biol Phys* 49:487–99.

Semenenko, V. A., and X. A. Li. 2008. Lyman-Kutcher-Burman NTCP model parameters for radiation pneumonitis and xerostomia based on combined analysis of published clinical data. *Phys Med Biol* 53:737–55.

Semenenko, V. A., B. Reitz, E. Day, X. S. Qi, M. Miften, and X. A. Li. 2008. Evaluation of a commercial biologically based IMRT treatment planning system. *Med Phys* 35:5851–60.

Seppenwoolde, Y., J. V. Lebesque, K. de Jaeger, et al. 2003. Comparing different NTCP models that predict the incidence of radiation pneumonitis. *Int J Radiat Oncol Biol Phys* 55:724–35.

Söderström, S., and A. Brahme. 1993. Optimization of the dose delivery in a few field techniques using radiobiological objective functions. *Med Phys* 20:1201–10.

Song, D. Y., S. H. Benedict, R. M. Cardinale, T. D. Chung, M. G. Chang, and R. K. Schmidt-Ullrich. 2005. Stereotactic body radiation therapy of lung tumors: Preliminary experience using normal tissue complication probability-based dose limits. *Am J Clin Oncol* 28:591–6.

South, C. P., M. Partridge, and P. M. Evans. 2008. A theoretical framework for prescribing radiotherapy dose distributions using patient-specific biological information. *Med Phys* 35:4599–611.

Søvik, A., E. Malinen, and D. R. Olsen. 2010. Adapting biological feedback in radiotherapy. *Semin Radiat Oncol* 20:138–46.

Søvik, A., E. Malinen, H. K. Skogmo, S. M. Bentzen, O. S. Bruland, and D. R. Olsen. 2007. Radiotherapy adapted to spatial and temporal variability in tumor hypoxia. *Int J Radiat Oncol Biol Phys* 68:1496–504.

Spalding, A. C., K. W. Jee, K. Vineberg, et al. 2007. Potential for dose-escalation and reduction of risk in pancreatic cancer using IMRT optimization with lexicographic ordering and gEUD-based cost functions. *Med Phys* 34:521–9.

START Trialists' Group, S. M. Bentzen, R. K. Agrawal, E. G. Aird, et al. 2008. The UK Standardisation of Breast Radiotherapy (START) Trial A of radiotherapy hypofractionation for treatment of early breast cancer: A randomised trial. *Lancet Oncol* 9:331–41.

Stavrev, P., D. Hristov, B. Warkentin, E. Sham, N. Stavreva, and B. G. Fallone. 2003. Inverse treatment planning by physically constrained minimization of a biological objective function. *Med Phys* 30:2948–58.

Stavreva, N., A. Niemierko, P. Stavrev, and M. Goitein. 2001. Modelling the dose-volume response of the spinal cord, based on the idea of damage to contiguous functional subunits. *Int J Radiat Biol* 77:695–702.

Steel, G. G., T. J. McMillan, and J. H. Peacock. 1989. The 5Rs of radiobiology. *Int J Radiat Biol* 56:1045–8.

Stewart, R. D., and X. A. Li. 2007. BGRT: Biologically guided radiation therapy—the future is fast approaching! *Med Phys* 34:3739–51.

Strandqvist, M. 1944. Studien uber die kumulative Wirkung der Rontgenstrahlen bei Fraktionierung. *Acta Radiol Suppl* 55:l–300.

Suwinski, R., I. Wzietek, R. Tarnawski, et al. 2007. Moderately low alpha/beta ratio for rectal cancer may best explain the outcome of three fractionation schedules of preoperative radiotherapy. *Int J Radiat Oncol Biol Phys* 69:793–9.

Tai, A., B. Erickson, K. A. Khater, and X. A. Li. 2008. Estimate of radiobiologic parameters from clinical data for biologically based treatment planning for liver irradiation. *Int J Radiat Oncol Biol Phys* 70:900–7.

Ten Haken, R. K., M. K. Martel, M. L. Kessler, et al. 1993. Use of V_{eff} and iso-NTCP in the implementation of dose escalation protocols. *Int J Radiat Oncol Biol Phys* 27:689–95.

Ten Haken, R. K., T. S. Lawrence, and L. A. Dawson. 2006. Prediction of radiation-induced liver disease by Lyman normal-tissue complication probability model in three-dimensional conformal radiation therapy for primary liver carcinoma: In regards to Xu et al. (*Int J Radiat Oncol Biol Phys* 2006;65:189–195). *Int J Radiat Oncol Biol Phys* 66:1272; author reply 1272–3.

Thames, H. D., M. Zhang, S. L. Tucker, H. H. Liu, L. Dong, and R. Mohan. 2004. Cluster models of dose-volume effects. *Int J Radiat Oncol Biol Phys* 59:1491–504.

Thieke, C., T. Bortfeld, A. Niemierko, and S. Nill. 2003. From physical dose constraints to equivalent uniform dose constraints in inverse radiotherapy planning. *Med Phys* 30:2332–9.

Thomas, E., O. Chapet, M. L. Kessler, T. S. Lawrence, and R. K. Ten Haken. 2005. Benefit of using biological parameters (EUD and NTCP) in IMRT optimization for the treatment of intrahepatic tumors. *Int J Radiat Oncol Biol Phys* 62:571–8.

Timmerman, R., J. Galvin, J. Michalski, et al. 2006. Accreditation and quality assurance for Radiation Therapy Oncology Group: Multicenter clinical trials using stereotactic body radiation therapy in lung cancer. *Acta Oncol* 45:779–86.

Tomé, W. A. 2008. Universal survival curve and single fraction equivalent dose: Useful tools in understanding potency of ablative radiotherapy: In regard to Parks et al. (*Int J Radiat Oncol Biol Phys* 2008;70:847–852). *Int J Radiat Oncol Biol Phys* 72: 1620; author reply 1620–1.

Tsougos, I., I. Grout, K. Theodorou, and C. Kappas. 2009. A free software for the evaluation and comparison of dose response models in clinical radiotherapy (DORES). *Int J Radiat Biol* 85:227–37.

Tucker, S. L. 1999. Pitfalls in estimating the influence of overall treatment time on local tumor control. *Acta Oncol* 38:171–8.

Tucker, S. L., and E. L. Travis. 1990. Comments on a time-dependent version of the linear-quadratic model. *Radiother Oncol* 18:155–63.

Tucker, S. L., F. B. Geara, L. J. Peters, and W. A. Brock. 1996. How much could the radiotherapy dose be altered for individual patients based on a predictive assay of normal-tissue radio-sensitivity? *Radiother Oncol* 38:103–13.

Tucker, S. L., M. Zhang, L. Dong, R. Mohan, D. Kuban, and H. D. Thames. 2006. Cluster model analysis of late rectal bleeding after IMRT of prostate cancer: A case-control study. *Int J Radiat Oncol Biol Phys* 64:1255–64.

van Baardwijk, A., S. Wanders, L. Boersma, et al. 2010. Mature results of an individualized radiation dose prescription study based on normal tissue constraints in stages I to III non-small-cell lung cancer. *J Clin Oncol* 28:1380–6.

Van Dyk, J., C. H. Newcomb, K. Mah, and T. J. Keane. 1990. Further comments on dose-time-fractionation considerations for lung damage. *Radiother Oncol* 18:183–4.

Wang, J. Z., and X. A. Li. 2003. Evaluation of external beam radiotherapy and brachytherapy for localized prostate cancer using equivalent uniform dose. *Med Phys* 30:34–40.

Wang, J. Z., M. Guerrero, and X. A. Li. 2003. How low is the α/β ratio for prostate cancer? *Int J Radiat Oncol Biol Phys* 55:194–203.

Wang, X. H., R. Mohan, A. Jackson, S. A. Leibel, Z. Fuks, and C. C. Ling. 1995. Optimization of intensity-modulated 3D conformal treatment plans based on biological indices. *Radiother Oncol* 37:140–52.

Warkentin, B., P. Stavrev, N. Stavreva, C. Field, and B. G. Fallone. 2004. A TCP-NTCP estimation module using DVHs and known radiobiological models and parameter sets. *J Appl Clin Med Phys* 5:50–63.

Webb, S., and A. E. Nahum. 1993. A model for calculating tumour control probability in radiotherapy including the effects of inhomogeneous distributions of dose and clonogenic cell density. *Phys Med Biol* 38:653–66.

Whelan, T. J., D. H. Kim, and J. Sussman. 2008. Clinical experience using hypofractionated radiation schedules in breast cancer. *Semin Radiat Oncol* 18:257–64.

Wheldon, T. E., C. Deehan, E. G. Wheldon, and A. Barrett. 1998. The linear-quadratic transformation of dose-volume histograms in fractionated radiotherapy. *Radiother Oncol* 46:285–95.

Widesott, L., L. Strigari, M. C. Pressello, M. Benassi, and V. Landoni. 2008. Role of the parameters involved in the plan optimization based on the generalized equivalent uniform dose and radiobiological implications. *Phys Med Biol* 53:1665–75.

Williams, S. G., J. M. Taylor, N. Liu, et al. 2007. Use of individual fraction size data from 3756 patients to directly determine the α/β ratio of prostate cancer. *Int J Radiat Oncol Biol Phys* 68:24–33.

Willner, J., A. Jost, K. Baier, and M. Flentje. 2003. A little to a lot or a lot to a little? An analysis of pneumonitis risk from dose-volume histogram parameters of the lung in patients with lung cancer treated with 3-D conformal radiotherapy. *Strahlenther Onkol* 179:548–56.

Withers, H. R. 1975. The four R's of radiotherapy. In *Advances in Radiation Biology,* ed. J. T. Lett and H. Alder, Vol. 5, 241–71. New York: Academic Press.

Withers, H. R., J. M. Taylor, and B. Maciejewski. 1988. Treatment volume and tissue tolerance. *Int J Radiat Oncol Biol Phys* 14:751–9.

Withers, H. R., H. D. Thames Jr., and L. J. Peters. 1983. A new isoeffect curve for change in dose per fraction. *Radiother Oncol* 1:187–91.

Wouters, B. G., and J. M. Brown. 1997. Cells at intermediate oxygen levels can be more important than the "hypoxic fraction" in determining tumor response to fractionated radiotherapy. *Radiat Res* 147:541–50.

Wu, P. M., D. T. Chua, J. S. Sham, et al. 1997. Tumor control probability of nasopharyngeal carcinoma: A comparison of different mathematical models. *Int J Radiat Oncol Biol Phys* 37:913–20.

Wu, Q., D. Djajaputra, Y. Wu, J. Zhou, H. H. Liu, and R. Mohan. 2003. Intensity-modulated radiotherapy optimization with gEUD-guided dose-volume objectives. *Phys Med Biol* 48:279–91.

Wu, K. L., G. L. Jiang, Y. Liao, et al. 2003. Three-dimensional conformal radiation therapy for non-small-cell lung cancer: A phase I/II dose escalation clinical trial. *Int J Radiat Oncol Biol Phys* 57:1336–44.

Wu, Q., and R. Mohan. 2002. Multiple local minima in IMRT optimization based on dose-volume criteria. *Med Phys* 29:1514–27.

Wu, Q., R. Mohan, A. Niemierko, and R. Schmidt-Ullrich. 2002. Optimization of intensity-modulated radiotherapy plans based on the equivalent uniform dose. *Int J Radiat Oncol Biol Phys* 52:224–35.

Yang, Y., and L. Xing. 2004. Clinical knowledge-based inverse treatment planning. *Phys Med Biol* 49:5101–17.

Yang, Y., and L. Xing. 2005. Towards biologically conformal radiation therapy (BCRT): Selective IMRT dose escalation under the guidance of spatial biology distribution. *Med Phys* 32:1473–84.

Yorke, E. D. 2001. Modeling the effects of inhomogeneous dose distributions in normal tissues. *Semin Radiat Oncol* 11:197–209.

Yorke, E. D., G. J. Kutcher, A. Jackson, and C. C. Ling. 1993. Probability of radiation-induced complications in normal tissues with parallel architecture under conditions of uniform whole or partial organ irradiation. *Radiother Oncol* 26:226–37.

Zaider, M., and H. I. Amols. 1999. Practical considerations in using calculated healthy-tissue complication probabilities for treatment-plan optimization. *Int J Radiat Oncol Biol Phys* 44:439–47.

Zaider, M., C. S. Wuu, and G. N. Minerbo. 1996. The combined effects of sublethal damage repair, cellular repopulation and redistribution in the mitotic cycle: I. Survival probabilities after exposure to radiation. *Radiat Res* 145:457–66.

7

Isoeffect Calculations in Adaptive Radiation Therapy and Treatment Individualization

7.1 Introduction

Adaptive radiation therapy (ART) and treatment individualization requires the identification of suitable biological objectives to facilitate the design, comparison, and ranking of alternate and refined treatments. The main biological objective of curative radiation therapy is to maximize the probability of local tumor control, while simultaneously minimizing normal tissue damage. Because parameters in dose-response models are often inaccurate or highly uncertain and because biological mechanisms are not fully understood, isoeffect calculations based on concepts such as biologically effective dose (BED; Withers, Thames Jr., and Peters 1983) or equivalent uniform dose (EUD; Niemierko 1997) are often more useful for comparing and ranking alternate and refined treatments, rather than comparing clinical outcomes that are estimated using tumor control probability (TCP) and normal tissue complication probability (NTCP) models. In this chapter, we review the key concepts motivating such biological indices and provide examples of how these concepts might be used to address clinically relevant issues and, ultimately, aid in the clinical implementation of ART and treatment individualization.

7.2 Biological Mechanisms and Tumor Control

7.2.1 DNA Damage and Cell Death

In addition to the reproductive death or outright killing of cells by radiation, intercellular communication, cell-matrix interactions, and other factors influence the collective response of malignant and normal tissues to radiation (Travis 2001; Bentzen 2006; Allan and Travis 2005; Howe et al. 2009). Nevertheless, the reproductive death or outright killing of cells, or a critical subpopulation of cells, by radiation is closely related to treatment efficacy. Reproductive death is defined as some form of molecular or cellular damage that results in permanent inhibition of cell division, whereas cell death encompasses apoptosis, necrosis, mitotic catastrophe, and other more immediate modes of cell dissolution. Reproductive death suffices to achieve local tumor control, although apoptosis and other cell-dissolution modes are certainly an effective way to permanently inhibit cell division. Decades of research on the mechanisms of formation and consequences of chromosome aberrations provide a conceptually appealing framework to link double-strand break (DSB) induction and processing to reproductive cell death (Chadwick and Leenhouts 1973; Savage 1998; Hlatky et al. 2002; Bryant 2004; Natarajan and Palitti 2008; Bedford 1991). A DSB is formed when the passage of radiation through a cell creates a cluster of DNA lesions that effectively cuts the DNA double helix into two smaller pieces. Experiments have shown that the yields of many types of clustered DNA lesions, including DSBs, are proportional to the absorbed dose up to several hundred Gy for low and high linear energy transfer (LET) radiation (Frankenberg-Schwager 1990; Rothkamm and Lobrich 2003; Frankenberg et al. 1999; Sutherland et al. 2001). Such experiments imply that single-hit mechanisms are responsible for the induction of DSBs and all other types of clustered DNA damage.

Once formed, the break-ends associated with one DSB may be rejoined to their correct partner (*correct rejoining*) or to a break-end associated with a different DSB (*incorrect rejoining*). A third alternative is that one or both break-ends may never be rejoined to another break-end (*incomplete rejoining*). The breakage and incorrect reunion process, which has also been referred to as *binary misrepair* (Curtis 1986; Tobias 1985; Carlson et al. 2008), gives rises to intra- and interchromosomal and chromatid aberrations, some of which render cells unable to produce viable progeny (Bedford 1991). The two major evolutionarily conserved pathways available to rejoin DSBs in eukaryotes are homologous recombination (HR) and nonhomologous end joining (NHEJ; Jackson 2002; Jeggo 2002). In the HR pathway, a damaged chromosome is aligned with an undamaged chromosome that has sequence homology. The genetic information stored in the undamaged chromosome is then used as a template for the repair of the damaged chromosome. Because the HR pathway has an undamaged template to guide the repair process, a DSB may be rejoined without any loss of genetic information, that is, *correct repair*. In contrast, the NHEJ pathway does not require a template with sequence homology. Instead, a series of proteins bind to the break-ends of the DSB, some of which have exo- and endonuclease activity that is responsible for the removal of damaged nucleotides, and directly rejoin (ligate) the break.

In CHO-K1 cells, the rejoining of DSBs formed by the I-SceI endonuclease produces deletions of 1–20 base pair (bp; Liang et al. 1998). Rare deletions up to 299 bp as well as the aberrant insertion of 45–205 bp also occur (Liang et al. 1998). Ionizing radiation can create DSBs composed of more than 10 or even 20 lesions (Semenenko and Stewart 2004), and the NHEJ pathway is unlikely to correctly repair complex DSBs such as those formed by ionizing radiation (Jeggo 2002). That is, correct rejoining of the DSB formed by ionizing radiation is unlikely to be synonymous with correct repair (restoration of the original base pair sequence). Instead, correct DSB rejoining most likely produces small-scale mutations (deletions, insertions, or other sequence alterations). A subset of these small-scale mutations may be lethal. Also, unrepairable damage or incomplete rejoining of a pair of DSB may contribute directly or indirectly to cell killing.

Although DSBs are widely viewed as the most critical form of DNA damage produced by radiation, other types of clustered damage are still potentially significant. For example, several experiments have shown that complex single-strand breaks or other forms of non-DSB clustered damage can sometimes be converted into a DSB as a result of unsuccessful excision repair (Semenenko, Stewart, and Ackerman 2005; Semenenko and Stewart 2005). The formation of additional DSBs through the unsuccessful excision repair of non-DSB clustered damage may result in additional binary misrepair and, ultimately, increase the yield of lethal exchanges, or at the very least, increase the yield of small-scale mutations and unrepairable or incompletely repaired DSBs. Such observations suggest that complex clustered damage and repair mechanisms other than NHEJ and HR (e.g., excision repair) impact a cell's ability to tolerate radiation damage (Wallace 1998).

Despite the potential appeal of attributing cell death exclusively to the induction and processing of DNA damage, other biological processes are known to modulate the apparent survival response of cells exposed to ionizing radiation. For example, low dose (Nagasawa and Little 1992), microbeam (Hei et al. 1997; Nelson et al. 1996; Prise et al. 1998), and medium transfer experiments (Mothersill and Seymour 1997) provide compelling evidence that cells damaged by radiation communicate with undamaged (bystander) cells through factors transmitted through the extracellular matrix (Mothersill and Seymour 1998) and through gap junctions (Azzam, De Toledo, and Little 2001, 2003). Intercellular signals communicated to bystander cells initiate many of the same effects as direct cellular irradiation, including signal-mediated death (Mothersill and Seymour 2001; Hall 2003). It is likely that radiation-damaged cells also respond to intercellular signals, although their response to these signals may or may not be the same as the ones exhibited by bystander cells.

7.2.2 Linear-Quadratic Cell Survival Model

In the linear-quadratic (LQ) model, the fraction of cells that survive absorbed dose D (Gy) is written as

$$S(D) = \exp(-\alpha D - \beta G D^2) \qquad (7.1)$$

Here, the Lea-Catcheside dose protraction factor G is given by (Sachs, Hahnfeld, and Brenner 1997)

$$G = \frac{2}{D^2} \int_{-\infty}^{\infty} dt\, \dot{D}(t) \int_{-\infty}^{t} dt'\, \dot{D}(t') e^{-\lambda(t-t')} \qquad (7.2)$$

where $\dot{D}(t)$ is the instantaneous absorbed dose rate (Gy h⁻¹) at time t(h). This dose rate function captures the temporal pattern of radiation delivery in its entirety, including split-dose and multifraction irradiation schemes. The rate of sublethal and potentially lethal damage repair is characterized by λ or, alternatively, the effective DSB repair half-time $\tau \equiv \ln 2/\lambda$. In the LQ model, cellular radiation sensitivity is characterized by three biological parameters: α(Gy⁻¹), α/β(Gy), and λ(h⁻¹) or τ(h). For the special case when (1) the time to deliver each fraction is short compared to τ and (2) the time interval between fractions is large compared to τ, the protraction factor for n daily fractions is $G \approx 1/n$ (Thames 1985). For situations in which the fraction delivery time is not short compared to the DSB repair half-time, Equation 7.3 is more appropriate (Sachs, Hahnfeld, and Brenner 1997).

$$G = \frac{g_1}{n} = \frac{2}{n(\lambda t_1)^2}(e^{-\lambda t_1} + \lambda t_1 - 1) \qquad (7.3)$$

Here, t_1 is the fraction delivery time. Even seemingly small corrections for dose rate effects, such as those expected in high dose rate (HDR) experiments, can have a significant impact on estimates of α and α/β derived from cell-survival data (Carlson et al. 2004). The time to deliver a fraction in intensity-modulated radiotherapy (IMRT) and hypofractionated treatments, such as in stereotactic body radiotherapy (SBRT), has the potential to significantly reduce treatment effectiveness when α/β is small and the half-time for repair is short (Wang et al. 2003). For permanent or temporary brachytherapy implants composed of a radioisotope with decay constant μ, the appropriate form of the dose protraction factor is (Dale 1985)

$$G = G_\infty \left\{ \frac{(1+x)}{(1-x)} - \frac{yx^2}{(1-x)^2}\left[1 - e^{-(\lambda-\mu)T}\right] \right\} \qquad (7.4)$$

Here, T is the amount of time the seeds are implanted in the patient, $G_\infty \equiv \mu/(\mu+\lambda)$, $x \equiv \exp(-\mu T)$, and $y \equiv 2\mu/(\lambda-\mu)$. As $T \to \infty$, $G \to G_\infty$.

In the limit of small doses and dose rates, the LQ can be derived from the lethal and potentially lethal (LPL) model (Curtis 1986), the repair–misrepair (RMR) model (Tobias 1985), and the repair–misrepair-fixation (RMF) model (Carlson et al. 2008) using perturbation theory and other methods. The LPL, RMR, and RMF, and hence the LQ, models are broadly consistent with the breakage and reunion theory of chromosome aberration formation (Savage 1998; Hlatky et al. 2002). When interpreted in terms of the induction and processing of DSBs (Carlson et al. 2008), the term αD represents the expected number of lethal damages arising from DSBs formed by the *same track* (hit), including possibly unrepairable DSBs, small-scale mutations (correct, but error-prone DSB rejoining), and complete or incomplete intratrack binary misrepair. This latter process occurs when multiple DSBs are formed by the same track. The term βGD^2 represents the formation of lethal exchanges through the pair-wise interaction of break-ends associated with DSBs created by different tracks, that is, intertrack binary misrepair. The biophysical interpretation of G (refer to Equation 7.2) is that break-ends associated with a DSB created at time t', if not first rejoined to their correct partners', may interact with break-ends associated with a second DSB produced at time t. Because DSB rejoining by HR requires an intact chromosome with sequence homology to act as a template for repair, the intra- and intertrack binary misrepair reactions responsible for the formation of most exchanges are likely accomplished by NHEJ rather than HR.

The mechanistic basis for the LQ survival model has been extensively reviewed in the literature (Sachs, Hahnfeld, and Brenner 1997; Brenner et al. 1998) but remains controversial (Brenner and Herbert 1997; Zaider 1998a, Zaider 1998b; Sachs and Brenner 1998). Still, DSB induction and processing provide a plausible explanation for the linear-quadratic dependence on dose of the surviving fraction and, equally important, provides a conceptual framework to understand the interplay between damage repair and dose protraction (*dose rate*) effects. Studies showing that defects in DSB repair mechanisms reduce cell survival directly implicate DSB processing in survival responses (Kuhne et al. 2004; Chavaudra, Bourhis, and Foray 2004). Also, positive correlations among chromosomal damage and clonogenic survival provide strong circumstantial evidence that DSB induction and processing is an important cell-killing mechanism. In one study, Cornforth and Bedford (1987) demonstrated that in plateau-phase cultures of AG1522 normal human fibroblasts, asymmetric exchanges exhibited a near one-to-one correlation with the logarithm of the surviving fraction, as would be predicted by Equation 7.1, when $\alpha D + \beta GD^2$ is interpreted as the expected number of lethal aberrations per cell. Studies such as this one suggest that the conversion of DSBs into lethal exchanges through intra- and intertrack binary misrepair are an especially important cell-killing mechanism and provide some justification for considering the LQ as more than a purely phenomenological model.

The mathematical simplicity of the LQ survival model is both its strength and weakness. Notwithstanding low-dose hyperradiosensitivity and adaptive responses (Joiner 1994; Joiner, Lambin, and Marples 1999; Marples et al. 2004), the strength of the LQ is that trends in clonogenic survival can often be adequately predicted in vitro and in vivo using a minimum number of adjustable parameters (α, α/β, and λ or τ). However, when contrasted to the underlying biology, the LQ model can only be viewed as a highly idealized representation of a very complex, time-dependent sequence of biophysical events and processes. That is, the aggregate effects of many time- and cell-cycle dependent processes are represented by a single set of time- and cell-cycle independent radiosensitivity parameters, α, α/β, and λ (or τ). This oversimplification of the underlying biology implies that the collective radiation response of a population of cells, even if this cell population has uniform and stable genetic traits, is determined by a distribution of values for α, α/β, and λ rather than one true or exact set of values. This inherent variability in cellular radiosensitivity is further amplified by a host of other patient- or tumor-specific factors (Bentzen, Thames, and Overgaard 1990; Bentzen 1992; Suit et al. 1992).

While several investigators have questioned the applicability of the LQ model for large fraction sizes (Guerrero and Li 2004; Park et al. 2008), the LQ model is appropriate for doses up to about 15–18 Gy, if used correctly and with caution (Brenner 2008). To reduce the possibility that treatment plans are inappropriately overoptimized, distributions of values for α, α/β, and λ should be used to estimate biological indices (e.g., EUD or BED) and outcomes rather than point estimates (Dasu, Toma-Dasu, and Fowler 2003; Warkentin et al. 2005; Moiseenko 2004; Lian and Xing 2004). The use of parameter distributions can help quantify uncertainties in outcomes arising from the inherent variability in biological parameters, as well as from the use of an oversimplified or incomplete model, such as the LQ, to predict cell killing.

7.2.3 Poisson Tumor Control Probability Model

The most commonly used TCP model is based on the Poisson distribution (Webb and Nahum 1993; Tome and Fowler 2003). With the Poisson model, the probability that all tumor cells in a region in the tissue are killed in the reproductive sense is given by

$$\text{TCP} = \prod_{i=1}^{N} \text{TCP}_i = \exp\left[-\sum_{i=1}^{N} v_i \rho_i S(D_i) e^{\gamma_i(T-T_k)}\right] \quad (7.5)$$

Here, ρ_i denotes the expected number of tumor clonogens per cubic centimeter in the ith tissue region, v_i is the volume of the ith tissue region (cubic centimeters), $S(D_i)$ is the fraction of the tumor cells in the ith tissue region that survive total treatment dose D_i, T is the overall time to complete the treatment, and T_k is the lag-time for the onset of cellular proliferation. The effects of subclinical disease on tumor control can be incorporated into Equation 7.5 by decreasing the clonogen density (ρ parameter) in regions outside the gross tumor volume. For the special case when $\rho_i = 0$ (i.e., the tissue region does not contain any tumor cells), TCP_i equals unity regardless of dose. The rate of accelerated tumor growth is characterized by a rate constant, γ, which is often conveniently expressed as $\gamma \equiv \ln(2)/T_d$. Here, T_d is the effective time for the number of tumor cells to double. Regional differences in cell proliferation within the tumor can be accommodated by adjusting the parameter γ, that is, $\gamma \rightarrow \gamma_i$. Other effects associated with differences in tumor radiosensitivity parameters among patients (interpatient heterogeneity) and within the tumor (intratumor heterogeneity) are implicit in Equation 7.5, that is, within the $S(D_i)$ term.

The estimation of radiobiological parameters from a patient population with heterogeneous radiation response parameters is notoriously challenging (Levegrun et al. 2001). Also, the Poisson TCP model does not provide a fully adequate representation of the statistical aspects of tumor growth kinetics during a course of radiation therapy (Tucker, Thames, and Taylor 1990; Tucker and Taylor 1996). The logistic or Gompertz growth kinetic models may be more appropriate than the simple exponential growth kinetics used in many tumor control models, including Equation 7.5 (Bajzer 1999; Calderon and Kwembe 1991; McAneney and O'Rourke 2007).

Despite the limitations and approximations, the Poisson TCP model provides some potentially useful constraints to guide the selection of dose distributions needed to achieve tumor control. For example, in treatment planning, we are seeking a distribution of D_i values such that the overall TCP $\cong 1$. Because the overall TCP is the product of the TCP values for all N tissue regions,

the Poisson TCP model predicts that treatment failure will occur if the TCP for any tissue region is zero. These so-called biological cold spots may arise because some regions of the tissue are underdosed or because some tumor cells are particularly resistant to radiation or are rapidly dividing. Equation 7.5 implies that the equality constraint $\text{TCP}_i = \text{TCP}^{1/N}$ may be imposed without decreasing the overall chance of achieving local tumor control. For the special case when (1) $\rho_i = \rho$, $v_i = v$, and $\gamma_i = \gamma$ for $i = 1$ to N and (2) all tumor cells are equally sensitive to radiation, the Poisson TCP model predicts that the optimal dose distribution is a uniform dose distribution, that is, $D_i = D$ for $i = 1-N$ (Webb et al. 1994). The observation that the overall TCP is a product of TCP values for a series of subvolumes also provides some justification for the use of the EUD concept (Niemierko 1997) as a surrogate dosimetric quantity for the assessment and intercomparison of treatment plans (Ebert 2000). That is, uniform and nonuniform dose distributions are biologically equivalent as long as they produce the same overall TCP.

7.3 Equivalent Prescription and Tolerance Dose

Suppose a currently accepted (*reference*) treatment involves delivery of total dose D_R over n_R fractions (fraction size $d_r = D_R/n_R$). For an alternate treatment modality or fractionation schedule, we seek the total dose D_A that will produce the same TCP in n_A fractions, that is, find D_A such that $\text{TCP}(D_A) = \text{TCP}(D_R)$. Substituting Equation 7.5 into $\text{TCP}(D_A) = \text{TCP}(D_R)$ with $N = 1$ gives $v_A\rho_A S(D_A)\exp[\gamma(T_A - T_k)] = v_R\rho_R S(D_R)\exp[\gamma(T_R - T_k)]$. Here, the tumor growth rate is assumed to be the same for the reference and alternate treatments. Regardless of the treatment modality, the initial number of cells in a tissue region of interest is the same (i.e., $v_A\rho_A = v_R\rho_R \rightarrow v\rho$). For two treatments to be equally effective at eradicating all cells, the equality constraint $S(D_A) = S(D_R)\exp[\gamma(T_R - T_A)]$ thus suffices. After substituting Equation 7.1 into the left-hand side (LHS) and right-hand side (RHS) of $S(D_A) = S(D_R)\exp[\gamma(T_R - T_A)]$ and taking the logarithm, the formula governing the equivalence of treatment A and R is $\alpha D_A + \beta G_A D_A^2 = \alpha D_R + \beta G_R D_R^2 + \gamma(T_R - T_A)$. Now, solve for D_A to obtain

$$D_A = \frac{\alpha/\beta}{2G_A}\left\{-1 + \sqrt{1 + \frac{4G_A D_R}{\alpha/\beta}\left[1 + \frac{G_R D_R}{\alpha/\beta} - \frac{\gamma(T_R - T_A)}{\alpha D_R}\right]}\right\} \quad (7.6)$$

Although Equation 7.6 was derived from the Poisson TCP model, Equation 7.6 may also be useful for estimating equivalent doses for other biological end points, such as the incidence or severity of early and late reactions in normal tissues. As long as the predominate cellular mechanism underpinning the maintenance or recovery of a tissue relates to the reproductive inactivation of a functional or critical subpopulation of cells, such as stem cells, biologically equivalent doses can plausibly be derived from Equation 7.6. So, for example, Equation 7.6 might be used to adjust published tolerance doses (Emami et al. 1991) to better account for the dose delivery characteristics of newer treatment modalities, such as SBRT, IMRT, and HDR brachytherapy procedures.

7.3.1 Equivalent Conventional and Hypofractionated Treatments

Equation 7.6 is a general formula to determine the total dose required for the alternate treatment to produce the same TCP, as well as the same surviving fraction, in a tissue region for the reference and alternate treatments. By selecting an appropriate form of the dose protraction factor, G_R and G_A, a multitude of clinically relevant questions can be addressed. To determine equivalent fraction sizes for a hypofractionated or hyperfractionated treatment, substitute $D_R = n_R d_R$, $D_A = n_A d_A$, $G_R \cong 1/n_R$, $G_A \cong 1/n_A$, and $(T_R - T_A) \cong (n_R - n_A)$ into Equation 7.6 and simplify as follows:

$$d_A = \frac{\alpha/\beta}{2}\left\{-1+\sqrt{1+\frac{4d_R}{\alpha/\beta}\cdot\frac{n_R}{n_A}\left[1+\frac{d_R}{\alpha/\beta}-\frac{\gamma/\alpha}{d_R}\left(1-\frac{n_A}{n_R}\right)\right]}\right\} \quad (7.7)$$

When $|n_R - n_A|$ is small compared to the tumor-doubling time $(= 0.693/\gamma)$, such as for slow-growing tumors and for small changes in treatment duration, Equation 7.7 reduces to

$$d_A = \frac{\alpha/\beta}{2}\left\{-1+\sqrt{1+\frac{4d_R}{\alpha/\beta}\cdot\frac{n_R}{n_A}\left(1+\frac{d_R}{\alpha/\beta}\right)}\right\} \quad (7.8)$$

An attractive aspect of Equation 7.8 is that the determination of equivalent fractionation sizes only requires the specification of one biological parameter, α/β. To correct for repopulation effects, an additional ratio of biological parameters, the ratio γ/α, must be known. The ratio γ/α determines the loss in treatment effectiveness per unit time (e.g., Gy per day) because of differences in the amount of tumor repopulation between

the reference and alternate treatments. To include the effects of fraction delivery time, Equation 7.3 should be used to compute G instead of $G \cong 1/n$ (Wang et al. 2003). The use of Equation 7.3 to compute G requires the introduction of a third biological parameter (τ or λ) into the modeling process.

Figure 7.1 shows trends in biologically equivalent fraction sizes computed using Equation 7.8. To estimate uncertainties in fraction size associated with intratumor and interpatient heterogeneity, equivalent fraction sizes are estimated 10,000 times using estimates of α/β sampled from a uniform distribution (range 1–10 Gy). Clinical judgment is incorporated into the determination of equivalent fraction sizes by selecting an appropriate reference treatment. In the left panel of Figure 7.1, a conventional 44×1.8 Gy treatment is used as the reference treatment, that is, $n_R = 44$, $d_R = 1.8$ Gy, and n_A ranges from 5 to 50 (x axis of Figure 7.1). In the right panel of Figure 7.1, a treatment consisting of 20×3 Gy fractions is used as the reference treatment ($n_R = 20$, $d_R = 3$ Gy). Combinations of fraction size and number of fractions above isoeffect lines are likely to produce increased local tumor control, whereas points below isoeffect lines suggest decreased local tumor control. Isoeffect lines for tolerance doses and normal tissue structures also delineate combinations of fraction size and number of fractions expected to produce equal levels of damage. Isoeffect calculations can thus help identify doses likely to increase or decrease the incidence and severity of treatment complications. One of the most appealing aspects of using isoeffect calculations to guide the treatment planning process is that clinical judgment, in the form of a reference treatment, is a *very effective* way to minimize the impact of uncertain biological parameters. That is, small changes away from an accepted fractionation schedule (e.g., $n_R = 44$ and $n_A = 37$) result in small uncertainties in the determination of biologically equivalent fraction sizes even when the uncertainties associated with α/β are quite large.

Figure 7.2 shows the increase in fraction size required to correct for the effects of fraction delivery time (intrafraction

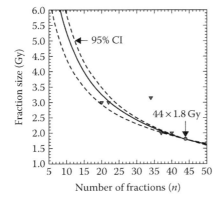
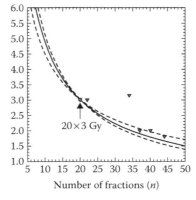

FIGURE 7.1 Effects of intratumor and interpatient heterogeneity on the determination of biologically equivalent fraction size. Equivalent fraction sizes and 95% confidence intervals based on uniform sampling of α/β 10,000 times from 1 to 10 Gy. Left panel: 44 fractions of 1.8 Gy (79.2 Gy total) used as reference treatment (filled circle) in Equation 7.8. Right panel: 20 fractions of 3 Gy (60 Gy total) used as reference treatment (filled circle) in Equation 7.8. Filled triangles indicate fraction sizes that have been tried for the treatment of prostate cancer. For fast-growing tumors, corrections for repopulation effects may be important. To correct for repopulation effects, use Equation 7.7 instead of Equation 7.8.

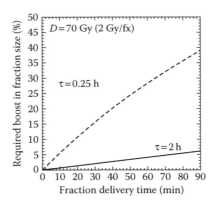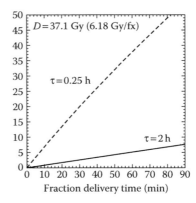

FIGURE 7.2 Effects of intrafraction double-strand break repair in a conventional and hypofractionated prostate cancer treatment ($\alpha/\beta = 1.5$ Gy). Left panel: 35 fractions of 2 Gy. Right panel: 6 fractions of 6.18 Gy. The time used to compute the dose protraction factor in Equation 7.3 is the total time to deliver one fraction rather than the beam-on time or the total time to deliver the entire course of radiation.

DSB repair) in a conventional and hypofractionated treatment. In general, corrections for fraction delivery time are most important when the fraction size is large compared to α/β. For 35 fractions of 2 Gy (70 Gy total), fraction delivery times less than 5 minutes are unlikely to require a boost in fraction size greater than 1%–3% (Figure 7.2, left panel). For a fraction delivery time of 20 minutes, the boost required to correct for intrafraction DSB repair may be as large as 10.6% for a repair half-time (τ) of 0.25 hours. As the fraction size increases, the required boost to correct for intrafraction repair increases with increasing fraction delivery time. For six fractions of 6.18 Gy (37.1 Gy total), a 20-minute fraction delivery time corresponds to a 13.4% increase in fraction size with $\tau = 0.25$ hours (Figure 7.2, right panel). For a 60-minute fraction delivery time ($\tau = 0.25$ hours), the correction for intrafraction repair increases to 37.9% (8.52 Gy per fraction instead of 6.18 Gy per fraction). Corrections for intrafraction repair may be especially important for newer treatment modalities, such as SBRT, in which very large doses are delivered over time intervals that may be as long as 60–90 minutes.

7.3.2 Equivalent Fractionated and Brachytherapy Treatments

To estimate the total dose required for a brachytherapy treatment to produce the same level of tumor control as a fractionated treatment, use Equation 7.4 to compute G_A and set $G_R \cong 1/n_R$. The initial dose rate, $\dot{D}_0$, (dose rate at the time seeds are implanted into the patient) for the brachytherapy seeds is related to the total brachytherapy dose by $\dot{D}_0 = \mu D_B/(1 - e^{-\mu T_B})$, where T_B is the length of time the seeds are implanted into the patient; the same value for T_B should be used in Equation 7.6 to correct for repopulation effects. For the special case when the brachytherapy seeds are permanently implanted into the patient, T_B should be set equal to the effective treatment time, T_E, rather than $T_B = \infty$. Otherwise, in Equation 7.6 the term that corrects for repopulation effects will inappropriately require an infinite brachytherapy dose; the unrealistic behavior of Equation 7.6 for permanent implants

arises because tumor growth is modeled using a simple exponential rather than a more appropriate Logistic or Gompertz growth kinetic model (Bajzer 1999; Calderon and Kwembe 1991; McAneney and O'Rourke 2007). Equation 7.6 can also be used to determine the total dose for a fractionated treatment equivalent to the total dose for a brachytherapy procedure, that is, use Equation 7.4 to compute G_R and set $G_A \cong 1/n_A$ (D_R would represent the total dose for the brachytherapy procedure, and D_A would denote the total dose for the fractionated treatment).

An effective treatment for permanent brachytherapy implants can be determined by finding the absorbed dose such that cell birth due to repopulation is equal to cell death from the brachytherapy seeds, that is, $S(d_E)\exp(\gamma\Delta t) = 1$. The absorbed dose delivered over time interval T_E to $T_E + \Delta t$ required to counter cell proliferation is

$$d_E = \frac{\alpha/\beta}{2}\left\{-1 + \sqrt{1 + \frac{4\gamma\Delta t}{\alpha(\alpha/\beta)}}\right\} \qquad (7.9)$$

For an exponentially decreasing dose rate characterized by decay constant μ, the effective treatment time is $T_E = -\ln[\dot{D}(T_E)/\dot{D}_0]/\mu$. Set $\dot{D}(T_E) = d_E/\Delta t$ and the effective treatment time becomes

$$T_E = -\frac{1}{\mu}\ln\left\{\frac{\alpha/\beta}{2\Delta t\dot{D}_0}\left\{-1 + \sqrt{1 + \frac{4\gamma\Delta t}{\alpha(\alpha/\beta)}}\right\}\right\} \qquad (7.10)$$

To determine the effective treatment time, the dose rate upon seed implantation must be known. Alternatively, Equations 7.6 and 7.10 can be used together in an iterative fashion to determine an appropriate initial dose rate and effective treatment time. First, calculate the total brachytherapy dose using Equation 7.6 with $T_A = T_E = 0$. Next, recalculate the dose rate using $\dot{D}_0 = \mu D_B/[1 - \exp(-\mu T_E)]$. Substitute the refined estimate for $\dot{D}_0$ back into Equation 7.10 to obtain a refined estimate of T_E. Substitute the new estimate of T_E back into Equation 7.6 for T_A and recalculate the total brachytherapy dose, and so on. Estimates of T_E and D obtained from the iterative procedure

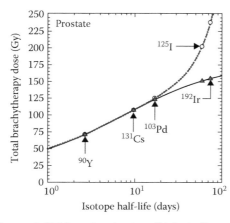 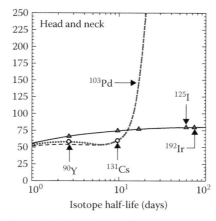

FIGURE 7.3 Effect of isotope half-life on the selection of biologically equivalent brachytherapy doses for prostate cancer (left panel) and head and neck cancer (right panel). Prostate cancer: $\alpha = 0.15$ Gy^{-1}, $\alpha/\beta = 3.1$ Gy, $T_d = 43$ days, $T_k = 0$ days (dashed line) or $T_k = 60$ days (dotted line); 44×1.8 Gy used as reference treatment in Equation 7.6. Head and neck cancer: $\alpha = 0.25$ Gy^{-1}, $\alpha/\beta = 10$ Gy, $T_d = 5$ days, $T_k = 0$ days (dashed line) or $T_k = 21$ days (dotted line) 30×2.2 Gy used as reference treatment in Equation 7.6. Solid black lines: biological equivalent doses in absence of repopulation effects ($T_d = \infty$).

typically converge after a few iterations because estimates of the initial dose rate and total dose are not overly sensitive to T_E for realistic treatment parameters.

Figure 7.3 shows biologically equivalent total brachytherapy doses with corrections (dotted and dashed lines) and without corrections (solid lines) for repopulation effects. Representative calculations are shown for a slow-growing ($T_d = 43$ days) prostate cancer (Figure 7.3, left panel) and a fast-growing ($T_d = 5$ days) head and neck cancer (Figure 7.3, right panel). Representative lag times of 60 days (prostate cancer) and 21 days (head and neck cancer) have a nominal impact on the total brachytherapy dose required for biological equivalence (compare the dotted and dashed lines). For prostate cancer, total brachytherapy doses with and without corrections for repopulation effects are within 5% for isotopes with a half-life shorter than 26 days, which includes ^{90}Y (half-life of 2.67 days), ^{131}Cs (half-life of 9.7 days), and ^{103}Pd (half-life of 17 days). Even for relatively slow-growing tumors, repopulation effects can have a substantial impact on the effectiveness of treatments using ^{125}I (half-life of 59.4 days) and ^{192}Ir (half-life of 73.8 days). For ^{125}I, the equivalent doses are 151 Gy without repopulation effects compared to 202 Gy with repopulation effects ($T_k = 60$ days, $T_E = 268.5$ days). For ^{192}Ir, the equivalent doses are 155 Gy without repopulation effects and 238 Gy with repopulation effects ($T_k = 60$ days, $T_E = 328.2$ days).

For rapidly growing tumors, such as many head and neck cancers, biologically equivalent doses of ^{131}Cs (60 Gy) and ^{103}Pd (151 Gy) may differ by a factor as high as 1.9 (Figure 7.3, right panel). For isotopes with half-lives shorter than about 12 days, equivalent doses for head and neck cancers are smaller when repopulation effects (dotted and dashed line) are included than when they are not (solid black line). This counterintuitive trend arises because equivalent brachytherapy doses are computed using a reference treatment composed of 30 fractions of 2.2 Gy. The solid black line indicates the total brachytherapy dose

equivalent to a fractionated treatment (30×2.2 Gy) without repopulation effects in both treatments (i.e., $T_d = \infty$). For isotopes with short half-lives, the brachytherapy doses with repopulation effects are smaller than the ones without repopulation effects because of the reduced chance for accelerated repopulation during the treatment. That is, the effective treatment time for a brachytherapy treatment with ^{131}Cs (30.2 days) is shorter than the effective treatment time for the fractionated treatment (39 days for a patient treated only on weekdays).

Figure 7.4 shows the initial dose rate required to deliver the total brachytherapy doses shown in Figure 7.3. For slow-growing tumors, the initial dose rate, and hence the total activity, required to deliver biologically equivalent treatments increases with decreasing half-life. For prostate cancer, initial dose rates are 0.773 Gy/h, 0.324 Gy/h, and 0.217 Gy/h for ^{90}Y, ^{131}Cs, and ^{103}Pd, respectively. For head and neck cancers, initial dose rates are slightly lower and range from 0.65 Gy/h for ^{90}Y to 0.20 Gy/h and 0.27 Gy/h for ^{131}Cs and ^{103}Pd, respectively. The differences in initial doses are due in part to differences in biological parameters (e.g., $\alpha/\beta = 1.49$ Gy vs. 10 Gy) and in part to differences in the reference treatments used for prostate (44×1.8 Gy) and head and neck (30×2.2 Gy) cancers. The choice of the most appropriate reference treatments reflects the clinical judgment of radiation oncologists and typically varies from institution to institution. For longer-lived radioisotopes, such as ^{192}Ir and ^{125}I, initial dose rates may need to be substantially increased to compensate for repopulation effects (Figure 7.4, right panel). Even for slow-growing tumors (Figure 7.4, left panel), corrections for repopulation effects may be substantial enough to warrant increasing the initial dose rate (activity). For prostate cancer treatments with ^{125}I, the correction for repopulation effects may require increasing the initial dose rate (activity) by as much as a factor of 60, that is, 2 Gy/h with repopulation effects compared to 0.032 Gy/h without repopulation effects.

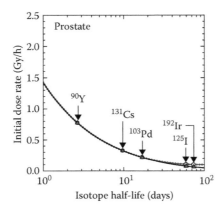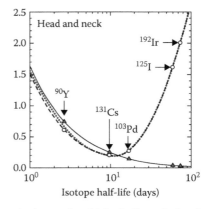

FIGURE 7.4   Effect of isotope half-life on the inital dose rate ($\propto$ activity) required to produce biologically equivalent brachytherapy doses for prostate cancer (left panel) and head and neck cancer (right panel). Prostate cancer: $\alpha = 0.15$ Gy^{-1}, $\alpha/\beta = 3.1$ Gy, $T_d = 43$ days, $T_k = 0$ days (dashed line) or $T_k = 60$ days (dotted line); 44×1.8 Gy used as reference treatment in Equation 7.6. Head and neck cancer: $\alpha = 0.25$ Gy^{-1}, $\alpha/\beta = 10$ Gy, $T_d = 5$ days, $T_k = 0$ days (dashed line) or $T_k = 21$ days (dotted line); 30×2.2 Gy used as reference treatment in Equation 7.6. Solid black lines: initial dose rate required for biological equivalence in absence of repopulation effects ($T_d = \infty$).

7.4 Effect of Fraction Size on Biologically Equivalent Dose Distributions

The selection of biologically equivalent prescription and tolerance doses can be easily determined using treatment-specific dose protraction factors and Equation 7.6. Although radiation therapy treatments are usually designed to deliver a prescribed or tolerated dose in one or more regions of the tissue, all existing treatment technologies inevitably produce a heterogeneous dose distribution across tumor targets and normal tissue structures. To design biologically equivalent dose distributions, as opposed to biologically equivalent prescription and tolerance doses, corrections for fraction size and the total number of fractions should be applied on a voxel-by-voxel basis with Equation 7.6 instead of scaling by prescription or tolerance dose.

To illustrate the potential significance of guiding the design of equivalent dose distributions using a single prescription or tolerance dose instead of applying voxel-by-voxel corrections, consider the following hypothetical dose distribution and treatment scenario. The voxel at the center of panel (a) of Figure 7.5 receives 44 fractions of 1.8 Gy. The dose to the surrounding tissue regions decrease monotonically toward 0 Gy (dark blue regions) with radial distance from the center voxel. In panel (b) of Figure 7.5, Equation 7.8 is used to determine the biologically equivalent dose in the center of the panel for 20 fractions (i.e., 2.942 Gy). Doses to the surrounding tissue regions are then determined by multiplying the dose from original treatment (panel (a)) by a constant factor equal to 1.634 = 2.942 Gy/1.8 Gy. This dose scaling scheme corresponds, for example, to increasing the number of monitor units (MU) from 180 MU to 294 MU without altering the MLC settings, beam number, and field sizes for a step-and-shoot IMRT treatment.

For comparison, panel (c) in Figure 7.5 shows the biologically equivalent dose distribution obtained when Equation 7.8 is applied on a voxel-by-voxel basis to the dose distribution from

panel (a) in Figure 7.5. The comparison of dose distributions in panels (b) and (c) in Figure 7.5 suggests that linear scaling of a dose distribution by a constant factor related to the number of MU is unlikely to produce the same overall tissue response as the original treatment even though the doses delivered at one point of interest are biologically equivalent. Therefore, scaling a dose distribution by a ratio of prescription doses, a ratio of normal tissue tolerance doses, or other physical quantities, such as a ratio of fluences or the number of MU, is *unlikely* to produce the same clinical outcome. The lack of a one-to-one correspondence between the biological effects of a dose distribution and scaling the original dose distribution by a ratio of prescription or tolerance doses arises because of nonlinear dose-response characteristics of cells and tissues.

7.5 Intratumor Heterogeneity and the Equivalent Uniform Dose Concept

Visual comparison and ranking of two- and three-dimensional dose distributions, or dose–volume histograms (DVH), often requires clinical judgment that may vary from individual to individual. The consolidation of a dose distribution into a single representative number indicative of biological response is a potentially useful supplement to visual comparisons of dose distributions. Also, treatment plan optimization using one or a few biological indicators of treatment outcome may be easier to implement and computationally more efficient than using a large array of dose constraints. The EUD concept introduced by Niemierko (1997) is one such biological indicator of treatment effectiveness. The EUD is defined as the absorbed dose, which if uniformly delivered to a tissue region of interest, will produce the same biological effect as delivery of a nonuniform dose distribution to the same tissue region of interest. The EUD concept is not specific to tumor control and may ultimately be useful for a multitude of other biological end points, including

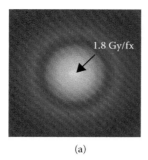 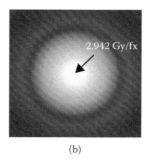 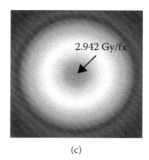

(a) (b) (c)

FIGURE 7.5 (See color insert following page 204.) Determination of biologically equivalent dose distributions for conventional and hypofractionated radiation therapy treatments. Panel (a): dose distribution for the reference treatment (44 fractions of 1.8 Gy at center; decreases monotonically to 0 Gy with radial distance away from center). Panel (b): 20 fractions of 2.942 Gy at center. Dose to other tissue regions determined by multiplying the dose from panel (a) by 1.634 (= 2.942 Gy/1.8 Gy). Panel (c): dose distribution obtained by applying Equation 7.8 on a voxel-by-voxel basis to the dose distribution in panel (a). ($n_A = 20$, $n_R = 44$, $d_R = 1.8$ Gy, $\alpha/\beta = 1.5$ Gy).

normal-tissue complications (Stewart and Li 2007), second cancers arising from therapy (Fontenot, Lee, and Newhauser 2009; Newhauser et al. 2009), and noncancer diseases such as cardiovascular, respiratory, and digestive diseases (Little 2009; Zhang, Muirhead, and Hunter 2005).

For the end point of local tumor control, or the continued reproductive viability of critical cells within a region of normal tissue, an estimate of the EUD can be derived by equating the TCP for a uniform dose distribution in the target region of interest to a non-uniform dose distribution in the same region of interest, that is,

$$TCP(EUD) = \prod_{i=1}^{N} TCP_i$$

$$\exp\left[-\sum_{i=1}^{N} v_i \rho_i S(EUD)\right] = \exp\left[-\sum_{i=1}^{N} v_i \rho_i S(D_i)\right] \quad (7.11)$$

Here, repopulation effects are neglected, and the summation is over all N voxels in the target region of interest. The target region of interest may be the entire gross tumor volume, the planning treatment volume, or a smaller region of the diseased or normal tissue. After taking the logarithm on both sides of Equation 7.11 and using the LQ model (i.e., Equation 7.1) with $G = 1$ (no dose protraction effects) to estimate S, Equation 7.11 becomes

$$\sum_{i=1}^{N} v_i \rho_i \left[\exp\left(-\alpha_i EUD - \beta_i EUD^2\right)\right]$$

$$= \sum_{i=1}^{N} v_i \rho_i \exp\left(-\alpha_i D_i - \beta_i D_i^2\right) \quad (7.12)$$

Conceptually, Equation 7.12 is applicable to any dose distribution *and* any distribution of biological parameters within the tissue region of interest. On the LHS of Equation 7.12, assume for the moment that all cells within the tissue region of interest have the same (*average*) sensitivity to radiation (i.e., $\alpha_i \to \alpha$ and $\beta_i \to \beta$) so that Equation 7.12 simplifies to

$$\exp(-\alpha EUD - \beta EUD^2) = \frac{1}{\rho V} \sum_{i=1}^{N} v_i \rho_i \exp\left(-\alpha_i D_i - \beta_i D_i^2\right) \quad (7.13)$$

In Equation 7.13, V is the total volume of the region of interest and ρ is the cell density averaged over V. Now take the logarithm of both sides of Equation 7.13 and solve for the EUD to obtain

$$EUD = \frac{1}{2} \alpha/\beta \left(-1 + \sqrt{1 - \frac{4 \ln \overline{S}}{\alpha(\alpha/\beta)}}\right) \quad (7.14)$$

Here, the average number of surviving cells in the volume of interest, $\overline{S}$, is

$$\overline{S} \equiv \frac{1}{\rho V} \sum_{i=1}^{N} v_i \rho_i \exp\left(-\alpha_i D_i - \beta_i D_i^2\right) \quad (7.15)$$

Because the derivation of Equation 7.14 neglects repopulation effects and dose protraction effects, Equations 7.14 and 7.15 are only applicable to dose distributions with the same overall treatment duration and fraction delivery time. To compare dose distributions for alternate fractionation schedules or fraction delivery times, use Equation 7.6 to apply voxel-by-voxel corrections for cell proliferation (the term proportional to γ/α in Equation 7.6) and damage repair (G factor in Equation 7.6) prior to converting the dose distribution into an EUD.

Equations 7.14 and 7.15 are premised on the pragmatic assumption that the collective response of a group of cells with different radiation sensitivities will respond to a nonuniform dose distribution (RHS of Equation 7.13) the same way that a group of cells with homogeneous (*average*) radiation sensitivity will respond to a uniform dose (LHS of Equation 7.13). Another challenging aspect of the EUD concept is that the true distribution of biological parameters within any region of normal or diseased region of tissue is unknown or, at best, poorly characterized. Factors such as oxygen concentration, cell-to-cell communication, the interactions of cells with the extracellular matrix, and even the relative numbers of diseased and normal cells in a region of tissue may all influence the ultimate ability of cells to respond to and recover from radiation damage. Larger structural and physiological differences among diseased and, seemingly, normal regions of tissue (i.e., regions with preclinical disease) may also influence a cells' ability to proliferate after sustaining radiation damage. These issues raise concerns about the practicality and usefulness of the EUD concept for comparing and ranking the overall biological effectiveness of alternate dose distributions. However, the EUD concept may still be useful for the comparison and ranking of dose distributions as long as uncertainties in EUD estimates arising from the use of approximate radiation-response models and unknown or poorly known biological parameters are small compared to the accuracy requirements for good clinical radiation dosimetry (i.e., delivered and planned doses are within about 3%–5% of each other).

To assess the nature and probable range of the uncertainties that might arise when computing an EUD for doses typical in radiation therapy, consider first the special case of a small region of tissue (~ 0.001 mm³) given a uniform absorbed dose of radiation. Also, imagine that this small region of tissue is further subdivided into 1000 voxels that are barely large enough to contain a single cell (about 10^3 μm³ per voxel). Each of the 1000 cells has unique radiation response parameters, α_i and β_i. Because the region of tissue is uniformly irradiated, $D_i = D$, Equation 7.15 reduces to

$$\overline{S} \equiv \frac{1}{\rho V} \sum_{i=1}^{N} v_i \rho_i \exp\left(-\alpha_i D - \beta_i G_i D^2\right) \qquad (7.16)$$

Here, $\rho V = 1000$ cells and $\rho_i v_i = 1$ cell. Although the actual distribution of values for α_i and β_i for a specific tumor and normal tissue may be unknown, the analysis of clinical outcomes and in vitro data does place some loose constraints on the range of probable values for α and β (Carlson et al. 2004). After computing $\overline{S}$ using Equation 7.16, the EUD is computed using Equation 7.14. In Equation 7.14, population-averaged values for α and α/β are used to represent the overall radiation sensitivity of cells within the tissue region of interest.

Figure 7.6 shows the distribution of EUD values, expressed as the ratio of EUD/D, that arise when a heterogeneous population of cells are exposed to uniform doses of radiation from 0 to 10 Gy. Because the small region of tissue in this example is, by design, uniformly irradiated, a ratio of EUD to dose that is larger or smaller than unity indicates that Equations 7.14 and 7.15 are approximate when applied to a heterogeneous population of cells. Filled triangles in Figure 7.6 represent a separate Monte Carlo simulation in which α and β (or α/β) are resampled from a uniform distribution for all 1000 cells (α is sampled from the range 0.05 Gy⁻¹–0.5 Gy⁻¹ and α/β is sampled from the range 1–10 Gy). Filled circles in Figure 7.6 represent separate Monte Carlo simulations with α uniformly sampled from the range 0.1 Gy⁻¹– 0.2 Gy⁻¹ and α/β uniformly sampled from the range 2–4 Gy. For small doses, the ratio of the EUD to dose first increases with increasing dose, then plateaus, and then starts to decrease with increasing dose. The overall trend is the same for cell populations with modest variability in radiation sensitivity (filled circles) and wide-ranging sensitivities to radiation (filled triangles).

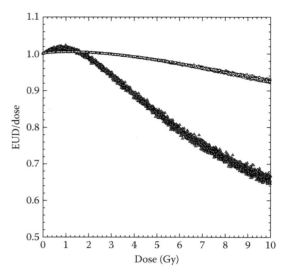

FIGURE 7.6 Effects of intratumor heterogeneity on the ratio of equivalent uniform dose (EUD) to delivered dose. Individual filled symbols denote Monte Carlo sampling of the radiation response characteristics of 1000 cells given a uniform dose of radiation (*x* axis). The EUD is computed using Equations 7.14 and 7.16. Filled triangles: α_i sampled from 0.05 Gy⁻¹ to 0.5 Gy⁻¹ and $(\alpha/\beta)_i$ sampled from 1 to 10 Gy (population-average: $\alpha = 0.275$ Gy⁻¹, $\alpha/\beta = 5.5$ Gy). Open circles: α_i sampled from 0.1 Gy⁻¹ to 0.2 Gy⁻¹; $(\alpha/\beta)_i$ sampled 2–4 Gy (population-average: $\alpha = 0.15$ Gy⁻¹, $\alpha/\beta = 3$ Gy).

The systematic trends in the EUD-to-dose ratio shown in Figure 7.6 arise because population-averaged radiosensitivity parameters do not represent adequately the overall survival response of a heterogeneous cell population. In general, αD (single-track) cell killing dominates at lower doses and βD^2 (intertrack) killing dominates at higher doses. As dose increases, some cells that are resistant to radiation at lower doses may become sensitive for large doses of radiation because of changes in the relative importance of the αD and βD^2 cell-killing mechanisms. Although the ratio of EUD to dose exhibits a systematic trend with the dose, the variability (uncertainty) in the ratio for any given dose is small (less than a few percentage) and also decreases as the size of the cell population increases.

In the limit as the cell population responds to radiation in a more and more homogeneous fashion, the ratio of EUD to dose approaches unity (compare filled triangles and filled circles in Figure 7.6). When all the cells respond the same way to radiation, $\alpha_i = \alpha$ and $(\alpha/\beta)_i = \alpha/\beta$ for all i and the ratio of EUD to dose is exactly equal to unity, regardless of the dose. This implies that Equations 7.14 and 7.15 provide accurate estimates of the EUD for uniform and nonuniform irradiation as long as the cell population has homogeneous radiation response characteristics. As the cell population becomes more and more heterogeneous, the accuracy of the EUD estimates decreases for uniform and nonuniform dose distributions. For modest levels of heterogeneity (α and α/β vary by a factor of 2 within the tissue region of interest), the results shown in Figure 7.6 (filled circles) suggest that estimates of the EUD are accurate to within about 3%–8%

for doses in the range of 0–10 Gy. The EUD concept, as represented in Equations 7.14 and 7.15, is most accurate and useful for clinical applications when the cell population within the target region of interest has uniform or modest levels of heterogeneity in their radiation response characteristics.

In addition to Equations 7.14 and 7.15, other biologically motivated algorithms to reduce a dose distribution or a distribution of EUD values into a single biological indicator of treatment effectiveness are sometimes useful for plan comparison and ranking. One algorithm that is widely used to model normal tissue complications is the generalized EUD (gEUD; Stewart and Li 2007):

$$ \text{gEUD} = \left(\sum_i f_i \text{EUD}_i^{1/a} \right)^a \tag{7.17} $$

Here, f_i is the fraction of the total volume over which EUD_i is computed, and a is an ad hoc (empirical) parameter that reflects the underlying organizational and functional structure of an at-risk organ or tissue. When a is in the range (0,1), gEUD increases monotonically with increasing values of a. When a is exactly equal to unity, the gEUD equals the average EUD. For $a > 1$, the gEUD approaches the maximum EUD anywhere within the target region of interest.

7.6 Corrections for Accidental, Random, or Systematic Deviations from a Planned Treatment

Random and systematic deviations from a planned treatment occur on a daily basis for a variety of reasons, including patient setup errors (Leong 1987) and anatomical changes in the position, volume, and shape of tumor targets between fractions (Barker et al. 2004). Such deviations can result in overdosing of normal tissues and underdosing of small or large regions of the tumor (or vice versa). To compensate for random patient setup errors, the margins of the clinical target volume (CTV) can be increased by a factor of 0.7 times the standard deviation associated with the patient setup errors (Bel, Van Herk, and Lebesque 1996; Bel et al. 1996). This method provides adequate treatment margins in IMRT for head and neck cancers when random translational and rotational setup errors are distributed according to a normal distribution (Astreinidou et al. 2005). However, a priori estimates of the standard deviation may not always be available. Increasing the CTV margins to correct for random deviations from a planned treatment tends to increase the volume of irradiated normal tissues and ultimately reduce the dose that can be delivered to tumor targets.

The advent of image-guided radiation therapy enables daily assessments of the dose delivered to tumor and normal tissue targets, for example, by using measurements of the radiation transmitted through the patient to aid in the accurate reconstruction of dose distributions on a daily basis (Verellen et al. 2008). When compared to conventional radiopaque markers,

the use of serial computed tomography (CT) bone imaging to accurately define target volume location improves the dose coverage of tumors and reduces the dose in normal tissues structures (O'Daniel et al. 2007). In addition to enhancing the dose coverage of tumor targets and reducing damage to normal tissue structures, daily imaging also provides opportunities for adaptive fractionation (Lu et al. 2008; Chen et al. 2008). Appropriate formulas to determine the size of the dose on the nth treatment day needed to correct for accidental, systematic, and random variations in a planned dose distribution on the preceding $(n-1)$ treatment days may be conveniently cast as an isoeffect calculation. Such formulas can be used to guide corrections for missed or interrupted treatments, organ motion, accidental overdosing or underdosing of a patient, and efforts to exploit adaptive fractionation.

The dose that needs to be delivered on the nth treatment day to correct for deviations from the planned dose on day 1 through day $n-1$ can be determined from the isosurvival criterion:

$$ \exp\left(-\alpha n d_p - \beta g_p n d_p^2\right) = \exp\left(-\sum_{i=1}^{n-1} \left[\alpha d_i + \beta g_i d_i^2 \right] \right) \times \exp\left(-\alpha d_n - \beta g_n d_n^2\right) \tag{7.18} $$

Here, n is the planned number of fractions, d_p is the planned fraction size, d_i is the actual dose delivered to a voxel on the ith treatment day. Alternatively, when a target region receives a uniform dose of radiation, d_i can be taken to represent the dose delivered to the entire target region on the ith treatment day. The protraction factor for delivery of the planned fraction is g_p, and g_i is the protraction factor for the delivery of the ith fraction. Equation 7.18 can be rearranged to obtain the fraction size on the nth treatment day required to produce the same surviving fraction as the planned treatment on days 1 through n, that is,

$$ d_n = \frac{\alpha/\beta}{2g_n} \left\{ -1 + \sqrt{1 + \frac{4g_n}{\alpha/\beta} \left[\left(n d_p + \frac{g_p n d_p^2}{\alpha/\beta} \right) \right.} \right. \\ \left. \left. \sqrt{-\sum_{i=1}^{n-1} \left(d_i + \frac{g_i d_i^2}{\alpha/\beta} \right)} \right] \right\}, \quad i=1,\cdots,n-1 \tag{7.19} $$

When the accumulated dose on days 1 through $n-1$ is greater than the planned dose, no dose is necessary on day n to achieve the same (or more) cell killing than the planned treatment on days 1 through n. When a patient is substantially overdosed, a treatment day may be skipped or the implementation of a dose correction should be delayed or spread over multiple treatment days. Alternate dose correction formulas for such scenarios can be easily derived from a suitably modified isosurvival criterion.

For the important special case when d_p and d_i are much smaller than α/β (or when the fraction delivery time is large compared to the half-time for sublethal damage repair so that

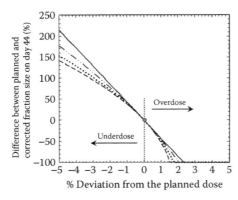
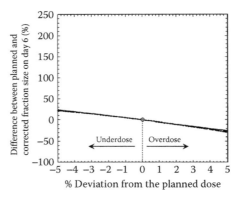

FIGURE 7.7 Effects of systematic dose delivery errors in conventional and hypofractionated prostate cancer treatments. Left panel: corrected dose on day 44 of a conventional treatment (d_p = 1.8 Gy). Right panel: corrected dose on day 6 of a hypofractionated treatment (d_p = 6.62 Gy). Solid line: corrected dose computed using Equation 7.21. Dashed line: corrected dose computed using Equation 7.19 with α/β = 1.49 Gy. Dotted line: corrected dose computed using Equation 7.19 with α/β = 3.1 Gy. Double dot-dash line: corrected dose computed using Equation 7.19 with α/β = 10 Gy.

g_p and $g_i \rightarrow 0$), the LQ model for the surviving fraction reduces to $S = \exp(-\alpha D)$ and the isosurvival criterion becomes

$$\exp\left(-\alpha n d_p\right) = \exp\left(-\alpha \sum_{i=1}^{n-1} d_i\right)\exp\left(-\alpha d_n\right) \quad (7.20)$$

From Equation 7.20, the dose that needs to be delivered on the nth treatment day to correct for deviations from the planned treatment on days 1 through $n - 1$ is

$$d_n = n d_p - \sum_{i=1}^{n-1} d_i \quad (7.21)$$

Figure 7.7 shows the fraction size needed to correct for systematic overdosing (percentage difference > 0%) or underdosing (percentage difference < 0%) on the preceding $n - 1$ treatment days. For convenience, the percentage difference between the planned and delivered fraction sizes (y axis) is normalized by dividing by the planned dose. The left panel shows the size of the correction needed on the last day of a conventional treatment ($n = 44$ fractions, d_p = 1.8 Gy), and the right panel shows the size of the correction needed on the last day of a hypofractionated treatment ($n = 6$, d_p = 6.62 Gy). As Figure 7.7 illustrates, the size of the correction on the last treatment day computed using Equation 7.21 is always greater than or equal to the size of the correction computed using Equation 7.19, regardless of whether a region of tissue is systematically overdosed or underdosed on the first $n - 1$ treatment days. For the conventional treatment (Figure 7.7, left panel), systematic underdosing and overdosing by even 1% on days 1 through 43 results in a substantial change (>38%) in the size of the correction on the last treatment day. For systematic underdosing by 5%, Equation 7.21 indicates that a fraction size of 5.67 Gy (215% increase) is needed on the last treatment day, whereas Equation 7.19 suggests that as little as 4.34 Gy (141% increase) is needed.

As α/β becomes large compared to the planned fraction size, the size of the correction predicted by Equation 7.19, which is derived using the survival model $S = \exp(-\alpha D - \beta G D^2)$, approaches the size of the correction predicted by Equation 7.21, which is derived using $S \cong \exp(-\alpha D)$. Any differences between the size of dose corrections must arise because of the term $\beta G D^2$ in the LQ survival model. In contrast to the conventional treatment (Figure 7.7, left panel), Equations 7.19 and 7.21 give nearly identical dose corrections for the hypofractionated treatment on the last treatment day for systematic underdosing or overdosing of a patient by as much as 5% (Figure 7.7, right panel). The results shown in Figure 7.7 also suggest that, for the same level of systematic errors, much larger corrections are required for conventional treatments than for hypofractionated treatments. For example, systematic underdosing or overdosing of a region of tissue by just 1% requires up to 38% change in the planned fraction size for the conventional treatment ($n = 44$, d_p = 1.8 Gy), but no more than 4.9% change in fraction size for the hypofractionated treatment ($n = 6$, d_p = 6.62 Gy).

Figure 7.8 shows the size of the average correction required to correct for random deviations from planned conventional (left panel) and hypofractionated (right panel) treatments. The percentage deviations from the planned treatments were determined by Monte Carlo sampling of the delivered dose on treatment days 1 to $n - 1$ from a normal distribution with a mean equal to zero and a specified standard deviation (x axis of Figure 7.8). Results in the left panel of Figure 7.8 are based on 25,000 Monte Carlo–simulated treatments, and results in the right panel are based on 100,000 Monte Carlo–simulated treatments. As with systematic deviations from a planned treatment, the corrected fraction size on the last treatment day predicted using Equation 7.21 is always greater than or equal to the corrected fraction size predicted using Equation 7.19. In the conventional treatment (Figure 7.4, left panel), corrections determined using Equation 7.21 are negligible (<.07%) when the standard deviation associated with the delivered dose is less than 5% per day. For large

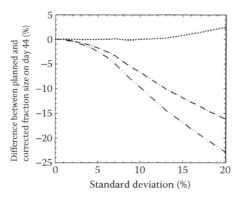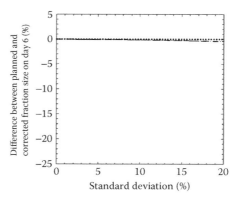

FIGURE 7.8 Effects of random dose delivery errors in conventional and hypofractionated prostate cancer treatments. Left panel: corrected fraction size on day 44 of a conventional treatment (d_p = 1.8 Gy). Right panel: corrected dose on day 6 of a hypofractionated treatment (d_p = 6.62 Gy). Dotted line: corrected dose computed using Equation 7.21. Double dot-dash line: corrected dose computed using Equation 7.19 with α/β = 1.49 Gy. Dot-dash line: corrected dose computed using Equation 7.19 with α/β = 3.1 Gy.

daily deviations from the planned treatment (>15% per day), which are unlikely on a routine basis, the average size of the correction becomes larger than 0% because a small subset of the treated patients require a systematic boost to correct for underdosing that arises due to random dose delivery errors. When Equation 7.19 is used (the biologically more plausible equation) to correct for random dose delivery errors in a conventional treatment (Figure 7.8, left panel), the corrected fraction size may be substantially smaller than the size of the correction required to deliver the same total dose as the original treatment. This result implies that a series of randomly fluctuating doses with a mean equal to the planned fraction size are not biologically equivalent to perfect delivery of the planned treatment. In the hypofractionated treatment plan (right panel of Figure 7.8), even very large (up to 20%) random deviations tend to cancel in terms of dose and biological effect.

7.7 Oxygen Effect and the Hypoxia Reduction Factor

Approximately 90% of all solid tumors have median oxygen concentrations less than the typical 40–60 mm of Hg (5%–8% O_2 concentration) found in normal tissues (Brown 1999; Vaupel and Hockel 1995). The lower levels of oxygen in many human tumors are due to structural and functional disturbances of the vasculature, which inhibit the normal delivery of blood and oxygen (Brown 1979, 2000). High levels of pretreatment tumor hypoxia have been implicated as a significant factor contributing to the failure of radiation therapy (Brizel et al. 1997; Fyles et al. 1998; Nordsmark, Overgaard, and Overgaard 1996; Sundfor et al. 2000; Movsas et al. 2002). While hypoxia has been shown to be associated with increased metastasis (Hockel et al. 1996), treatment failure in tumors with high levels of hypoxia is primarily due to the decreased sensitivity of hypoxic cells to ionizing radiation (Rofstad et al. 2000). The chemical restoration of DNA damage under reduced oxygen conditions is believed to be the main

mechanism underpinning the decreased sensitivity of hypoxic cells to ionizing radiation (Prise, Davies, and Michael 1993).

Many strategies have been proposed to overcome, or even exploit, tumor hypoxia (Brown 2000; Brown and Wilson 2004), including hypoxia-activated prodrugs, gene therapy, recombinant obligate anaerobic bacteria, and HIF-1 targeting. Ling et al. (2000) also proposed boosting the dose to a biological tumor volume to overcome radiation resistance associated with tumor hypoxia. However, many challenges hamper the clinical implementation of this strategy, especially recent evidence suggesting that the oxygen levels within the tumors exhibit substantial fluctuations over a course of radiation therapy (Lin et al. 2008). Because boosting the dose to one tumor region inevitably results in additional dose to normal tissue structures or reductions in dose to other tumor regions, dose boosts based on pretreatment determinations of tumor hypoxia are as likely to decrease overall treatment effectiveness as they are to increase it. Practical strategies to overcome tumor hypoxia through dose boosting will require noninvasive techniques to image hypoxia on a periodic basis coupled with replanning. The clinical implementation of such a strategy is likely to be impractical because of cost and labor considerations for a conventional fractionation but may be possible for SBRT or other treatments that deliver a large dose of radiation in a few fractions.

Several research groups have shown that radiation sensitivity parameters for well-oxygenated cells are related to radiation sensitivity parameters for other oxygen concentrations by $\alpha(p) = \alpha_A/\mathrm{HRF}_\alpha(p)$ and $\beta(p) = \beta_A/\mathrm{HRF}_\beta(p)^2$ (Carlson, Stewart, and Semenenko 2006; Dasu and Denekamp 1998; Nahum et al. 2003). Here, the hypoxia reduction factors, HRF_α and HRF_β, are dimensionless factors that quantify changes in radiation sensitivity as a function of partial oxygen pressure p. Although HRF_α and HRF_β may generally be independent factors, Carlson, Stewart, and Semenenko (2006) have shown that setting $\mathrm{HRF}_\alpha(p) = \mathrm{HRF}_\beta(p) = \mathrm{HRF}(p)$ provides as good a statistical fit to several in vitro data sets as independent HRF_α and HRF_β factors.

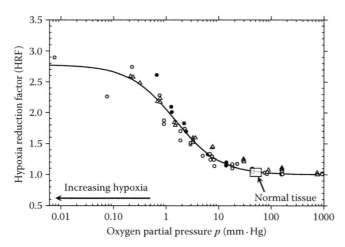

FIGURE 7.9 Estimates of the hypoxia reduction factor derived from published cell survival data as a function of oxygen partial pressure. Solid line shows a fit to the values using Equation 7.22 with $m = 2.8$ and $K = 1.5$. (Values denoted by white triangles, black circles, white circles, and gray circles derived from data reported by, respectively, Ling, C. C., et al. 1980. *Int J Radiat Oncol Biol Phys* 6:583–9; Koch, C. J., et al. 1984. *Radiat Res* 98:141–53; Whillans, D. W., and J. W. Hunt. 1982. *Radiat Res* 90:126–41; Carlson, D. J., et al. 2006. *Med Phys* 33:3105–15.)

This approximation is especially convenient when attempting to estimate radiosensitivity parameters and HRF values from sparse clinical data. As illustrated in Figure 7.9, a function of the form

$$\text{HRF}(p) = \frac{mK + p}{K + p} \tag{7.22}$$

captures trends in the HRF suggested from an analysis of published cell survival data (Ling et al. 1980; Koch, Stobbe, and Bump 1984; Whillans and Hunt 1982; Carlson, Stewart, and Semenenko 2006). Here, m is the maximum HRF and K is the partial pressure at which the HRF is equal to half its maximum value. The solid black line shown in Figure 7.9 corresponds to $m = 2.8$ and $K = 1.5$.

Because the HRF tends to increase with decreasing partial oxygen pressure, $\alpha(p)$ and $\beta(p)$ decrease with decreasing p. As $\alpha(p)$ and $\beta(p)$ decrease, larger and larger doses of radiation are needed to obtain the same surviving fraction. As with the derivation of Equation 7.6, an estimate of the dose required to achieve the same surviving fraction can be determined from the isoeffect constraint $S(D_H) = S(D_A)\exp[(\gamma_A - \gamma_H)T]$. Here, we assume that the overall duration of the treatment is the same for the hypoxic and aerobic cells, denoted by subscripts H and A, respectively. Cells in a low-oxygen environment tend to proliferate less than well-oxygenated cells, $\gamma_H \leq \gamma_A$ (for nondividing hypoxic cells, $\gamma_H = 0$). After substitution of Equation 7.1 into the LHS and RHS of the isoeffect constraint, take the logarithm and solve for D_H to obtain

$$D_H = \frac{(\alpha/\beta)_H}{2G_H}$$

$$\times \left\{ -1 + \sqrt{1 + \frac{4G_H}{(\alpha/\beta)_H} \frac{\alpha_A D_A}{\alpha_H} \left[1 + \frac{G_A D_A}{(\alpha/\beta)_A} + \frac{(\gamma_A - \gamma_H)T}{\alpha_A D_A} \right]} \right\} \tag{7.23}$$

Equation 7.23 is a general solution to determine biologically equivalent doses for hypoxic and well-oxygenated cells. For cells that become hypoxic shortly before the start of irradiation (seconds or minutes) and return to a well-oxygenated environment within an hour or two, the half-time for sublethal damage repair is about the same for aerobic and hypoxic cells (Carlson, Stewart, and Semenenko 2006); therefore, $G_H = G_A = G$. However, in general, cells maintained in a hypoxic environment for prolonged periods may not be able to repair damage as quickly as well-oxygenated cells, which would imply $G_H \leq G_A$.

Consider the special case of a single dose of radiation delivered at a high dose rate. For this case, repopulation effects are negligible, and it is also reasonable to assume that $G_H \cong G_A$. Equation 7.23 thus reduces to

$$D_H = \frac{(\alpha/\beta)_H}{2G_A} \left\{ -1 + \sqrt{1 + \frac{4G_A}{(\alpha/\beta)_H} \frac{\alpha_A D_A}{\alpha_H} \left[1 + \frac{G_A D_A}{(\alpha/\beta)_A} \right]} \right\} \tag{7.24}$$

Substituting $\alpha_H = \alpha_A/\text{HRF}(p)$ and $(\alpha/\beta)_A = (\alpha/\beta)_H/\text{HRF}(p)$ into Equation 7.24, the biologically equivalent dose for cells irradiated under partial oxygen pressure p is

$$D(p) = \frac{(\alpha/\beta)_A \text{HRF}(p)}{2G_A} \left\{ -1 + \sqrt{1 + \frac{4G_A D_A}{(\alpha/\beta)_A} \left[1 + \frac{G_A D_A}{(\alpha/\beta)_A} \right]} \right\}$$
$$= D_A \text{HRF}(p) \tag{7.25}$$

An important implication of Equation 7.25 is that the doses required to achieve the same surviving fraction, and the same level of local tumor control, are independent of α and α/β as long as repopulation effects are negligible and $G_H \cong G_A$. The maximum dose that is required to correct for the effects of hypoxia is $D(0 \text{ mm·Hg}) \cong 2.8 D_A$. Figure 7.10 shows the necessary escalation of a 2 Gy fraction to achieve equivalent tumor cell kill for a range of partial oxygen pressures.

An advantage of conventional fractionation schedules is that cells that are hypoxic during one part of the treatment may become oxygenated during other parts of the treatment because of fluctuating tumor blood flow (Brown 1979, 2000). This reoxygenation effect tends to increase overall treatment effectiveness as the number of fractions increases (Kallman 1972). While emerging technologies, such as SBRT, provide valuable physical advantages over conventional radiation therapy for patients with solitary tumors (Timmerman et al. 2003; Yamada et al. 2008), the reduced potential for reoxygenation between fractions may require additional dose to overcome the increased resistance of

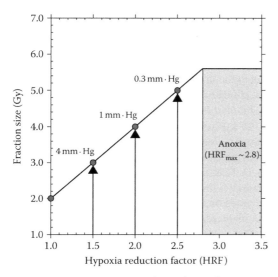

FIGURE 7.10 Fraction size required to achieve the same surviving fraction for a range of oxygen concentrations. Fraction sizes are calculated relative to a conventional 2 Gy fraction using Equations 7.22 and 7.25.

hypoxic cells to ionizing radiation. On the other hand, reductions in treatment effectiveness associated with reoxygenation may be offset by reducing the overall time available for tumor repopulation. To properly evaluate the magnitude of the competing effects of reoxygenation and repopulation in the design of alternate fractionation schedules, additional research is necessary to develop realistic models for competing effects associated with fluctuating oxygen levels during a course of radiation therapy.

7.8 Treatment Individualization and Adaptive Radiation Therapy

Because of the approximations inherent in the LQ and Poisson TCP concepts and uncertainties associated with the often unknown or poorly known radiosensitivity parameters, the attempt to individualize treatments through the direct application of TCP and NTCP modeling is ill-advised and may result in the selection of inappropriate or suboptimal treatment parameters and modalities. Because of the highly nonlinear nature of the LQ, TCP, and NTCP models, the search for biologically optimal treatments is also computationally challenging. Isoeffect calculations are potentially a very effective alternative to the direct application of TCP and NTCP models for treatment individualization. ART also provides opportunities to correct for accidental, random, or systematic deviations from a planned treatment (Section 7.6). Because the overall biological effects of a heterogeneous dose distribution across tumor targets and normal tissue structures are unlikely to scale in direct proportion to prescription or tolerance dose (e.g., see Figure 7.5 and related discussion), the design, comparison, and ranking of

treatments using metrics such as EUD may be more appropriate than comparing and ranking plans on the basis of equivalent prescription or tolerance dose.

The primary advantage of an isoeffect approach to ART and treatment individualization is that potential differences in treatment outcome are assessed in terms of easy-to-compare-and-rank quantities with the same units (i.e., the Gy) and conceptual meaning as absorbed dose. That is, a 10% or 20% increase or decrease in the EUD for a diseased and normal tissue region has about the same clinical significance as a 10% or 20% change in absorbed dose to the same region of tissue. Differences in EUD among competing treatments less than 3%–5% are unlikely to be clinically significant when compared to dose delivery errors associated with patient setup, changes in anatomical position due to intrafraction motion, or to changes in the volume or shape of a tumor target over a typical 4–6-week treatment period. For similar reasons, differences in the EUD for normal tissues less than 3%–5% are also unlikely to be clinically significant. Corrections for intra- and interfraction DSB repair (Figures 7.1 and 7.2) and repopulation effects (Figures 7.3 and 7.4) are easily incorporated into isoeffect calculations, and assessments of the effects of tumor hypoxia on alternate fractionation schedules are also possible (Section 7.7). Monte Carlo sampling of biological parameters are a useful way to assess uncertainties in the EUD and related concepts associated with intratumor and interpatient heterogeneity (Figure 7.6). An assessment of the uncertainties associated with biological indicators of treatment effectiveness can help clinicians avoid inappropriate overoptimization or individualization of treatments.

In contrast to the direct application of TCP and NTCP modeling, the initial goal of the isoeffect approach to treatment individualization and ART is to identify biologically equivalent prescription and tolerance doses, expressed perhaps in terms of patient-specific and dose-distribution specific EUDs. To improve overall treatment effectiveness, alternate fraction sizes or modalities (e.g., brachytherapy, SBRT, or IMRT) may be examined to see if they are likely to either (1) produce the same chance for local tumor control (iso-TCP) and reduce damage to normal tissue structures or (2) produce the same amount of damage in normal tissue structures (iso-NTCP) and increase the chance for local tumor control. The same level of tumor control is expected in tumor targets with the same EUD. To the extent that the reproductive inactivation of a critical subpopulation of normal cells is predictive of damage to normal tissue structures, the severity or incidence of complications may also be about the same for treatments that deliver the same EUD to the same overall volume of tissue.

Because more is known about the mechanisms and models underpinning the desired treatment outcome for diseased regions of tissue (i.e., eradication of all tumor cells) than for disease-free regions of tissue (i.e., avoidance or reduction in the severity of complications in at-risk normal tissues), strategy 1 may be more clinically prudent than strategy 2. That is, the favored plan or treatment modality is the one expected to produce about the same local tumor control (same EUD), while at the same time producing a reasonable expectation of reducing damage to normal tissue structures (smaller EUD). Then, to enhance local tumor control, while minimizing the chance for unexpected increases in normal tissue damage, the overall total dose or fraction size can be gradually increased toward the maximum tolerated or allowed dose (or EUD) for the dose-limiting normal tissue structures. Although decreases in the EUD for a normal tissue structure might plausibly indicate some reduction in the incidence or severity of normal tissue complications, the assessment of damage to normal tissue structures need not be made solely in terms of EUD. Isodose contours and the shapes of DVHs should also be compared and used to rank competing treatments. Ultimately, biological indicators of treatment effectiveness are best used as a guide, or a complementary tool, rather than as a replacement for clinical judgment and traditional physical (dosimetric) indicators of treatment effectiveness.

References

Allan, J. M., and L. B. Travis. 2005. Mechanisms of therapy-related carcinogenesis. *Nat Rev Cancer* 5:943–55.

Astreinidou, E., A. Bel, C. P. Raaijmakers, C. H. Terhaard, and J. J. Lagendijk. 2005. Adequate margins for random setup uncertainties in head-and-neck IMRT. *Int J Radiat Oncol Biol Phys* 61:938–44.

Azzam, E. I., S. M. De Toledo, and J. B. Little. 2001. Direct evidence for the participation of gap junction-mediated intercellular communication in the transmission of damage signals from alpha-particle irradiated to nonirradiated cells. *Proc Natl Acad Sci U S A* 98:473–8.

Azzam, E. I., S. M. De Toledo, and J. B. Little. 2003. Oxidative metabolism, gap junctions and the ionizing radiation-induced bystander effect. *Oncogene* 22:7050–7.

Bajzer, Z., 1999. Gompertzian growth as a self-similar and allometric process. *Growth Dev Aging* 63:3–11.

Barker, Jr., J. L., A. S. Garden, K. K. Ang, et al. 2004. Quantification of volumetric and geometric changes occurring during fractionated radiotherapy for head-and-neck cancer using an integrated CT/linear accelerator system. *Int J Radiat Oncol Biol Phys* 59:960–70.

Bedford, J. S. 1991. Sublethal damage, potentially lethal damage, and chromosomal aberrations in mammalian cells exposed to ionizing radiations. *Int J Radiat Oncol Biol Phys* 21:1457–69.

Bel, A., M. Van Herk, and J. V. Lebesque. 1996. Target margins for random geometrical treatment uncertainties in conformal radiotherapy. *Med Phys* 23:1537–45.

Bel, A., P. H. Vos, P. T. Rodrigus, et al. 1996. High-precision prostate cancer irradiation by clinical application of an offline patient setup verification procedure, using portal imaging. *Int J Radiat Oncol Biol Phys* 35:321–32.

Bentzen, S. M. 1992. Steepness of the clinical dose-control curve and variation in the in vitro radiosensitivity of head and neck squamous cell carcinoma. *Int J Radiat Biol* 61:417–23.

Bentzen, S. M. 2006. Preventing or reducing late side effects of radiation therapy: Radiobiology meets molecular pathology. *Nat Rev Cancer* 6:702–13.

Bentzen, S. M., H. D. Thames, and J. Overgaard. 1990. Does variation in the in vitro cellular radiosensitivity explain the shallow clinical dose-control curve for malignant melanoma? *Int J Radiat Biol* 57:117–26.

Brenner, D. J. 2008. The linear-quadratic model is an appropriate methodology for determining isoeffective doses at large doses per fraction. *Semin Radiat Oncol* 18:234–9.

Brenner, D. J., and D. E. Herbert. 1997. The use of the linear-quadratic model in clinical radiation oncology can be defended on the basis of empirical evidence and theoretical argument. *Med Phys* 24:1245–8.

Brenner, D. J., L. R. Hlatky, P. J. Hahnfeldt, Y. Huang, and R. K. Sachs. 1998. The linear-quadratic model and most other common radiobiological models result in similar predictions of time-dose relationships. *Radiat Res* 150:83–91.

Brizel, D. M., G. S. Sibley, L. R. Prosnitz, R. L. Scher, and M. W. Dewhirst. 1997. Tumor hypoxia adversely affects the prognosis of carcinoma of the head and neck. *Int J Radiat Oncol Biol Phys* 38:285–9.

Brown, J. M. 1979. Evidence for acutely hypoxic cells in mouse tumours, and a possible mechanism of reoxygenation. *Br J Radiol* 52:650–6.

Brown, J. M. 1999. The hypoxic cell: A target for selective cancer therapy—eighteenth Bruce F. Cain Memorial Award lecture. *Cancer Res* 59:5863–70.

Brown, J. M. 2000. Exploiting the hypoxic cancer cell: Mechanisms and therapeutic strategies. *Mol Med Today* 6:157–62.

Brown, J. M., and W. R. Wilson. 2004. Exploiting tumour hypoxia in cancer treatment. *Nat Rev Cancer* 4:437–47.

Bryant, P. E. 2004. Repair and chromosomal damage. *Radiother Oncol* 72:251–6.

Calderon, C. P., and T. A. Kwembe. 1991. Modeling tumor growth. *Math Biosci* 103:97–114.

Carlson, D. J., R. D. Stewart, X. A. Li, et al. 2004. Comparison of in vitro and in vivo alpha/beta ratios for prostate cancer. *Phys Med Biol* 49:4477–91.

Carlson, D. J., R. D. Stewart, and V. A. Semenenko. 2006. Effects of oxygen on intrinsic radiation sensitivity: A test of the relationship between aerobic and hypoxic linear-quadratic (LQ) model parameters. *Med Phys* 33:3105–15.

Carlson, D. J., R. D. Stewart, V. A. Semenenko, and G. A. Sandison. 2008. Combined use of Monte Carlo DNA damage simulations and deterministic repair models to examine putative mechanisms of cell killing. *Radiat Res* 169:447–59.

Chadwick, K. H., and H. P. Leenhouts. 1973. A molecular theory of cell survival. *Phys Med Biol* 18:78–87.

Chavaudra, N., J. Bourhis, and N. Foray. 2004. Quantified relationship between cellular radiosensitivity, DNA repair defects and chromatin relaxation: A study of 19 human tumour cell lines from different origin. *Radiother Oncol* 73:373–82.

Chen, M., W. Lu, Q. Chen, K. Ruchala, and G. Olivera. 2008. Adaptive fractionation therapy: II. Biological effective dose. *Phys Med Biol* 53:5513–25.

Cornforth, M. N., and J. S. Bedford. 1987. A quantitative comparison of potentially lethal damage repair and the rejoining of interphase chromosome breaks in low passage normal human fibroblasts. *Radiat Res* 111:385–405.

Curtis, S. B. 1986. Lethal and potentially lethal lesions induced by radiation—a unified repair model. *Radiat Res* 106:252–70.

Dale, R. G. 1985. The application of the linear-quadratic dose-effect equation to fractionated and protracted radiotherapy. *Br J Radiol* 58:515–28.

Dasu, A., and J. Denekamp. 1998. New insights into factors influencing the clinically relevant oxygen enhancement ratio. *Radiother Oncol* 46:269–77.

Dasu, A., I. Toma-Dasu, and J. F. Fowler. 2003. Should single or distributed parameters be used to explain the steepness of tumour control probability curves? *Phys Med Biol* 48:387–97.

Ebert, M. A. 2000. Viability of the EUD and TCP concepts as reliable dose indicators. *Phys Med Biol* 45:441–57.

Emami, B., J. Lyman, A. Brown, et al. 1991. Tolerance of normal tissue to therapeutic irradiation. *Int J Radiat Oncol Biol Phys* 21:109–22.

Fontenot, J. D., A. K. Lee, and W. D. Newhauser. 2009. Risk of secondary malignant neoplasms from proton therapy and intensity-modulated x-ray therapy for early-stage prostate cancer. *Int J Radiat Oncol Biol Phys* 74:616–22.

Frankenberg, D., H. J. Brede, U. J. Schrewe, et al. 1999. Induction of DNA double-strand breaks by 1H and 4He Ions in primary human skin fibroblasts in the LET range of 8 to 124 keV/microm. *Radiat Res* 151:540–9.

Frankenberg-Schwager, M. 1990. Induction, repair and biological relevance of radiation-induced DNA lesions in eukaryotic cells. *Radiat Environ Biophys* 29:273–92.

Fyles, A. W., M. Milosevic, R. Wong, et al. 1998. Oxygenation predicts radiation response and survival in patients with cervix cancer. *Radiother Oncol* 48:149–56.

Guerrero, M., and X. A. Li. 2004. Extending the linear-quadratic model for large fraction doses pertinent to stereotactic radiotherapy. *Phys Med Biol* 49:4825–35.

Hall, E. J. 2003. The bystander effect. *Health Phys* 85:31–5.

Hei, T. K., L. J. Wu, S. X. Liu, et al. 1997. Mutagenic effects of a single and an exact number of alpha particles in mammalian cells. *Proc Natl Acad Sci U S A* 94:3765–70.

Hlatky, L., R. K. Sachs, M. Vazquez, and M. N. Cornforth. 2002. Radiation-induced chromosome aberrations: Insights gained from biophysical modeling. *Bioessays* 24:714–23.

Hockel, M., K. Schlenger, B. Aral, et al. 1996. Association between tumor hypoxia and malignant progression in advanced cancer of the uterine cervix. *Cancer Res* 56:4509–15.

Howe, O., J. O'Sullivan, B. Nolan, et al. 2009. Do radiation-induced bystander effects correlate to the intrinsic radiosensitivity of individuals and have clinical significance? *Radiat Res* 171:521–9.

Jackson, S. P. 2002. Sensing and repairing DNA double-strand breaks. *Carcinogenesis* 23:687–96.

Jeggo, P. A. 2002. The fidelity of repair of radiation damage. *Radiat Prot Dosimetry* 99:117–22.

Joiner, M. C. 1994. Induced radioresistance: An overview and historical perspective. *Int J Radiat Biol* 65:79–84.

Joiner, M. C., P. Lambin, and B. Marples. 1999. Adaptive response and induced resistance. *C R Acad Sci III* 322:167–75.

Kallman, R. F. 1972. The phenomenon of reoxygenation and its implications for fractionated radiotherapy. *Radiology* 105:135–42.

Koch, C. J., C. C. Stobbe, and E. A. Bump. 1984. The effect on the Km for radiosensitization at 0 degree C of thiol depletion by diethylmaleate pretreatment: Quantitative differences found using the radiation sensitizing agent misonidazole or oxygen. *Radiat Res* 98:141–53.

Kuhne, M., E. Riballo, N. Rief, et al. 2004. A double-strand break repair defect in ATM-deficient cells contributes to radiosensitivity. *Cancer Res* 64:500–8.

Leong, J., 1987. Implementation of random positioning error in computerised radiation treatment planning systems as a result of fractionation. *Phys Med Biol* 32:327–34.

Levegrun, S., A. Jackson, M. J. Zelefsky, et al. 2001. Fitting tumor control probability models to biopsy outcome after three-dimensional conformal radiation therapy of prostate cancer: Pitfalls in deducing radiobiologic parameters for tumors from clinical data. *Int J Radiat Oncol Biol Phys* 51:1064–80.

Lian, J., and L. Xing. 2004. Incorporating model parameter uncertainty into inverse treatment planning. *Med Phys* 31:2711–20.

Liang, F., M. Han, P. J. Romanienko, and M. Jasin. 1998. Homology-directed repair is a major double-strand break repair pathway in mammalian cells. *Proc Natl Acad Sci U S A* 95:5172–7.

Lin, Z., J. Mechalakos, S. Nehmeh, et al. 2008. The influence of changes in tumor hypoxia on dose-painting treatment plans based on 18F-FMISO positron emission tomography. *Int J Radiat Oncol Biol Phys* 70:1219–28.

Ling, C. C., J. Humm, S. Larson, et al. 2000. Towards multidimensional radiotherapy (MD-CRT): Biological imaging and biological conformality. *Int J Radiat Oncol Biol Phys* 47:551–60.

Ling, C. C., H. B. Michaels, E. R. Epp, and E. C. Peterson. 1980. Interaction of misonidazole and oxygen in the radiosensitization of mammalian cells. *Int J Radiat Oncol Biol Phys* 6:583–9.

Little, M. P. 2009. Cancer and non-cancer effects in Japanese atomic bomb survivors. *J Radiol Prot* 29:A43–59.

Lu, W., M. Chen, Q. Chen, K. Ruchala, and G. Olivera. 2008. Adaptive fractionation therapy: I. Basic concept and strategy. *Phys Med Biol* 53:5495–511.

Marples, B., B. G. Wouters, S. J. Collis, A. J. Chalmers, and M. C. Joiner. 2004. Low-dose hyper-radiosensitivity: A consequence of ineffective cell cycle arrest of radiation-damaged G2-phase cells. *Radiat Res* 161:247–55.

Mcaneney, H., and S. F. O'rourke. 2007. Investigation of various growth mechanisms of solid tumour growth within the linear-quadratic model for radiotherapy. *Phys Med Biol* 52:1039–54.

Moiseenko, V., 2004. Effect of heterogeneity in radiosensitivity on LQ based isoeffect formalism for low alpha/beta cancers. *Acta Oncol* 43:499–502.

Mothersill, C., and C. Seymour. 1997. Medium from irradiated human epithelial cells but not human fibroblasts reduces the clonogenic survival of unirradiated cells. *Int J Radiat Biol* 71:421–7.

Mothersill, C., and C. Seymour. 2001. Radiation-induced bystander effects: Past history and future directions. *Radiat Res* 155:759–67.

Mothersill, C., and C. B. Seymour. 1998. Cell-cell contact during gamma irradiation is not required to induce a bystander effect in normal human keratinocytes: Evidence for release during irradiation of a signal controlling survival into the medium. *Radiat Res* 149:256–62.

Movsas, B., J. D. Chapman, A. L. Hanlon, et al. 2002. Hypoxic prostate/muscle pO2 ratio predicts for biochemical failure in patients with prostate cancer: Preliminary findings. *Urology* 60:634–9.

Nagasawa, H., and J. B. Little. 1992. Induction of sister chromatid exchanges by extremely low doses of alpha-particles. *Cancer Res* 52:6394–6.

Nahum, A. E., B. Movsas, E. M. Horwitz, C. C. Stobbe, and J. D. Chapman. 2003. Incorporating clinical measurements of hypoxia into tumor local control modeling of prostate cancer: Implications for the alpha/beta ratio. *Int J Radiat Oncol Biol Phys* 57:391–401.

Natarajan, A. T., and F. Palitti. 2008. DNA repair and chromosomal alterations. *Mutat Res* 657:3–7.

Nelson, J. M., A. L. Brooks, N. F. Metting, et al. 1996. Clastogenic effects of defined numbers of 3.2 MeV alpha particles on individual CHO-K1 cells. *Radiat Res* 145:568–74.

Newhauser, W. D., J. D. Fontenot, A. Mahajan, et al. 2009. The risk of developing a second cancer after receiving craniospinal proton irradiation. *Phys Med Biol* 54:2277–91.

Niemierko, A., 1997. Reporting and analyzing dose distributions: A concept of equivalent uniform dose. *Med Phys* 24:103–10.

Nordsmark, M., M. Overgaard, and J. Overgaard. 1996. Pretreatment oxygenation predicts radiation response in advanced squamous cell carcinoma of the head and neck. *Radiother Oncol* 41:31–9.

O'daniel, J. C., A. S. Garden, D. L. Schwartz, et al. 2007. Parotid gland dose in intensity-modulated radiotherapy for head and neck cancer: Is what you plan what you get? *Int J Radiat Oncol Biol Phys* 69:1290–6.

Park, C., L. Papiez, S. Zhang, M. Story, and R. D. Timmerman. 2008. Universal survival curve and single fraction equivalent dose: Useful tools in understanding potency of ablative radiotherapy. *Int J Radiat Oncol Biol Phys* 70:847–52.

Prise, K. M., O. V. Belyakov, M. Folkard, and B. D. Michael. 1998. Studies of bystander effects in human fibroblasts using a charged particle microbeam. *Int J Radiat Biol* 74:793–8.

Prise, K. M., S. Davies, and B. D. Michael. 1993. Evidence for induction of DNA double-strand breaks at paired radical sites. *Radiat Res* 134:102–6.

Rofstad, E. K., K. Sundfor, H. Lyng, and C. G. Trope. 2000. Hypoxia-induced treatment failure in advanced squamous cell carcinoma of the uterine cervix is primarily due to hypoxia-induced radiation resistance rather than hypoxia-induced metastasis. *Br J Cancer* 83:354–9.

Rothkamm, K., and M. Lobrich. 2003. Evidence for a lack of DNA double-strand break repair in human cells exposed to very low x-ray doses. *Proc Natl Acad Sci U S A* 100:5057–62.

Sachs, R. K., and D. J. Brenner. 1998. The mechanistic basis of the linear-quadratic formalism. *Med Phys* 25:2071–3.

Sachs, R. K., P. Hahnfeld, and D. J. Brenner. 1997. The link between low-LET dose-response relations and the underlying kinetics of damage production/repair/misrepair. *Int J Radiat Biol* 72:351–74.

Savage, J. R. 1998. A brief survey of aberration origin theories. *Mutat Res* 404:139–47.

Semenenko, V. A., and R. D. Stewart. 2004. A fast Monte Carlo algorithm to simulate the spectrum of DNA damages formed by ionizing radiation. *Radiat Res* 161:451–7.

Semenenko, V. A., and R. D. Stewart. 2005. Monte Carlo simulation of base and nucleotide excision repair of clustered DNA damage sites: II. Comparisons of model predictions to measured data. *Radiat Res* 164:194–201.

Semenenko, V. A., R. D. Stewart, and E. J. Ackerman. 2005. Monte Carlo simulation of base and nucleotide excision repair of clustered DNA damage sites: I. Model properties and predicted trends. *Radiat Res* 164:180–93.

Stewart, R. D., and X. A. Li. 2007. BGRT: Biologically guided radiation therapy-the future is fast approaching! *Med Phys* 34:3739–51.

Suit, H., S. Skates, A. Taghian, P. Okunieff, and J. T. Efird. 1992. Clinical implications of heterogeneity of tumor response to radiation therapy. *Radiother Oncol* 25:251–60.

Sundfor, K., H. Lyng, C. G. Trope, and E. K. Rofstad. 2000. Treatment outcome in advanced squamous cell carcinoma of the uterine cervix: Relationships to pretreatment tumor oxygenation and vascularization. *Radiother Oncol* 54:101–7.

Sutherland, B. M., P. V. Bennett, H. Schenk, et al. 2001. Clustered DNA damages induced by high and low LET radiation, including heavy ions. *Phys Med* 17(Suppl 1):202–4.

Thames, H. D. 1985. An 'incomplete-repair' model for survival after fractionated and continuous irradiations. *Int J Radiat Biol Relat Stud Phys Chem Med* 47:319–39.

Timmerman, R., L. Papiez, R. Mcgarry, et al. 2003. Extracranial stereotactic radioablation: Results of a phase I study in medically inoperable stage I non-small cell lung cancer. *Chest* 124:1946–55.

Tobias, C. A. 1985. The repair-misrepair model in radiobiology: Comparison to other models. *Radiat Res Suppl* 8:S77–95.

Tome, W. A., and J. F. Fowler. 2003. On the inclusion of proliferation in tumour control probability calculations for inhomogeneously irradiated tumours. *Phys Med Biol* 48:N261–8.

Travis, E. L. 2001. Organizational response of normal tissues to irradiation. *Semin Radiat Oncol* 11:184–96.

Tucker, S. L., and J. M. Taylor. 1996. Improved models of tumour cure. *Int J Radiat Biol* 70:539–53.

Tucker, S. L., H. D. Thames, and J. M. Taylor. 1990. How well is the probability of tumor cure after fractionated irradiation described by Poisson statistics? *Radiat Res* 124:273–82.

Vaupel, P., and M. Hockel. 1995. Oxygenation status of human tumors: A reappraisal of using computerized pO2 histography. In *Tumor Oxygenation,* ed. P. Vaupel, D. Kelleher and M. Gunderoth. Stuttgart, Germany: Gustav Fischer Verlag. 219–32.

Verellen, D., M. De Ridder, K. Tournel, et al. 2008. An overview of volumetric imaging technologies and their quality assurance for IGRT. *Acta Oncol* 47:1271–8.

Wallace, S. S. 1998. Enzymatic processing of radiation-induced free radical damage in DNA. *Radiat Res* 150:S60–79.

Wang, J. Z., X. A. Li, W. D. D'souza, and R. D. Stewart. 2003. Impact of prolonged fraction delivery times on tumor control: A note of caution for intensity-modulated radiation therapy (IMRT). *Int J Radiat Oncol Biol Phys* 57:543–52.

Warkentin, B., P. Stavrev, N. A. Stavreva, and B. G. Fallone. 2005. Limitations of a TCP model incorporating population heterogeneity. *Phys Med Biol* 50:3571–88.

Webb, S., P. M. Evans, W. Swindell, and J. O. Deasy. 1994. A proof that uniform dose gives the greatest TCP for fixed integral dose in the planning target volume. *Phys Med Biol* 39:2091–8.

Webb, S., and A. E. Nahum. 1993. A model for calculating tumour control probability in radiotherapy including the effects of inhomogeneous distributions of dose and clonogenic cell density. *Phys Med Biol* 38:653–66.

Whillans, D. W., and J. W. Hunt. 1982. A rapid-mixing comparison of the mechanisms of radiosensitization by oxygen and misonidazole in CHO cells. *Radiat Res* 90:126–41.

Withers, H. R., H. D. Thames Jr., and L. J. Peters. 1983. A new isoeffect curve for change in dose per fraction. *Radiother Oncol* 1:187–91.

Yamada, Y., M. H. Bilsky, D. M. Lovelock, et al. 2008. High-dose, single-fraction image-guided intensity-modulated radiotherapy for metastatic spinal lesions. *Int J Radiat Oncol Biol Phys* 71:484–90.

Zaider, M., 1998a. Sequel to the discussion concerning the mechanistic basis of the linear quadratic formalism. *Med Phys* 25:2074–5.

Zaider, M., 1998b. There is no mechanistic basis for the use of the linear-quadratic expression in cellular survival analysis. *Med Phys* 25:791–2.

Zhang, W., C. R. Muirhead, and N. Hunter. 2005. Age-at-exposure effects on risk estimates for noncancer mortality in the Japanese atomic bomb survivors. *J Radiol Prot* 25:393–404.

II

Treatment Delivery

Delivery of Intensity-Modulated Radiation Therapy

Cedric X. Yu
University of Maryland
School of Medicine

8.1 Overview

Adaptive radiation therapy is a technology that adjusts treatments according to geometric and biological variations during the course of treatment. As described in Chapters 2 and 3, such adaptation may be needed due to target movements during treatment delivery or anatomic variations between planning and delivery and among daily treatments as revealed by real-time measurements or daily imaging. The planning for the adjustments or adaptations was described in Section I of this book. This chapter summarizes different treatment delivery techniques with a focus on the delivery of intensity-modulated radiation therapy (IMRT).

IMRT is one of the most significant advances in treatment delivery in the last decade. Its ability to shape high-dose volumes to conform to the shape of the target tissues allows a decrease in irradiation-related sequelae by limiting the dose delivered to the surrounding normal tissues. The same dose-shaping capability also allows physicians to escalate doses to certain tumors to enhance local control. Both dosimetric and clinical advantages have been demonstrated for almost all common anatomical sites (Mell, Mehrotra, and Mundt 2005; Veldeman et al. 2008), though the largest number of applications has been for prostate cancer and head and neck cancers (Mell, Mehrotra, and Mundt 2005). IMRT has allowed physicians to escalate the dose to the prostate while reducing the toxicities to the rectum and bladder, resulting in improved local control and reduced complications compared with conventional three-dimensional conformal therapy (Burman et al. 1997; Junius et al. 2007; Kuban et al. 2008; Ling et al. 1996; Nutting et al. 2000; Pollack et al. 2003; Portlance et al. 2001; Zelesfsky

et al. 2002). IMRT has also shown greater capability in sparing salivary functions in patients receiving radiation therapy for head and neck cancers (Chao et al. 2001; Claus et al. 2001; Eisbruch et al. 1999; Fu et al. 2000; Sultanem et al. 2000; Veldeman et al. 2008). Rather than delivering the prescribed dose to the entire pelvis, IMRT is able to spare the small bowel, the bladder, and the rectum, resulting in significantly lower gastrointestinal toxicities (Brixey et al. 2002; Duthoy et al. 2003; Landry et al. 2002; Mundt et al. 2002; Nutting et al. 2000; Portelance et al. 2001; Roeske et al. 2000; Wong et al. 2005). The use of IMRT instead of wedge pairs for tangential whole-breast irradiation has resulted in improved dose uniformity (Harsolia et al. 2007; Hong et al. 1999; Kestin et al. 2000; Lo et al. 2000; van Asselen et al. 2001), which in turn resulted in significantly reduced acute and chronic toxicities (Harsolia et al. 2007). IMRT has also been combined with stereotactic localization for delivering radiosurgery treatments to intracranial and extracranial sites using linear accelerators (Cardinale et al. 1998; Iuchi et al. 2006; Woo et al. 1996). Other sites of application include lung (Derycke et al. 1998; Sura et al. 2008; van Sornsen et al. 2001), gastrointestinal tumors (Landry et al. 2002; Nutting et al. 2000; Sura et al. 2008), pediatric tumors (Huang et al. 2002; Kuppersmith et al. 2000; Paulino and Skwarchuk 2002), and soft-tissue sarcoma (Hong et al. 2004). Clinical use of IMRT has also brought many changes in clinical practice. Due to its ability to reduce dose to surrounding structures, IMRT has allowed physicians to escalate dose to the prostate gland to improve local control (Kuban et al. 2008; Pollack et al. 2003; Zelefsky, Fuks, and Leibel 2002). The reduced toxicity also encourages the use of hypofractionation schemes (Freedman et al. 2007; Fu et al. 2000; Iuchi et al. 2006; Junius et al.

2007). The ability of IMRT to enable more complex dose distributions has also allowed the increased use of concomitant boost and the treatment of target within targets (Freedman et al. 2007; Fu et al. 2000; Iuchi et al. 2006).

Most radiation treatments are delivered using linear accelerators, although special treatment machines have also emerged to enhance our ability to shape the dose distributions (Galvin et al. 2004). Most IMRT treatment deliveries using a linear accelerator require the use of the multileaf collimator (MLC), which was originally developed for shaping radiation fields. The ability of MLC to easily change and dynamically vary the field shapes was quickly explored for IMRT delivery. Over the years, linear accelerator vendors have not only improved the reliability of MLCs but also made the leaf width smaller. Smaller leaf widths allow the treatment planning system to use finer beamlet size during optimization for achieving better treatment plan quality (Zhang et al. 2005).

Both rotational and gantry-fixed IMRT techniques have been implemented clinically using dynamic multileaf collimation (DMLC) (Spirou et al. 1994, Boyer and Yu 1999; Brahme 1998; Burman et al. 1997; Carol 1994; Mackie et al. 1993; Yu 1995; Yu et al. 1995; Yu and Wong 1994). In gantry-fixed IMRT, multiple coplanar and noncoplanar beams at different orientations, each with spatially modulated beam intensities, are used (Bortfeld 2010; Boyer and Yu 1999; Brixey et al. 2002; Yu et al. 1995). Rotational IMRT employs either temporally modulated fan beams (Carol 1994; Mackie et al. 1993), commonly known as tomotherapy, or one or more cone-beam arcs (Yu 1995), referred to as intensity-modulated arc therapy (IMAT). This chapter describes these delivery methods and summarize the issues associated with IMRT delivery.

8.2 Fundamental Principles of Intensity-Modulated Radiation Therapy

IMRT spares critical structures by redistributing the normal tissue dose to less critical regions and reducing the high-dose volume to cover only the target. For a given case, the geometric arrangement of the target and its surrounding avoidance structures dictate the preferred beam angles and along each beam angle the preferred locations at which the radiation is delivered to the tumor. The essential role of computer optimization in IMRT planning is to derive a treatment plan that reflects these preferences in order to take advantage of the unique geometric arrangement presented by each clinical case.

Different methods employed to take advantage of the unique geometric arrangement of the target and avoidance structures presented by each clinical case lead to different forms of treatment delivery. Over the last decade, many new methods for treatment delivery using multiple fixed beams or rotating beams have been developed (Galvin et al. 2004). No matter the form of delivery that is used, the radiation dose is delivered to the target through multiple radiation fields of varying sizes and shapes. To take full advantage of the geometric arrangements of the clinical case, a large number of fields, also referred to as segments or apertures, are needed. The number of field segments is case

dependent, and in general, the more complex the geometric arrangement, the larger number of field segments is required to achieve the optimal plan (Bortfeld 2006). Where these apertures are placed—in a limited number of fields or spaced around the patient as one or more arcs—is less important.

IMRT is not the only way to take advantage of the geometric arrangements of the target and the surrounding avoidance structures. The CyberKnife system (Hara, Soltys, and Gibbs 2007; Accuray, Sunnyvale, CA), which uses small circular X-ray beams generated by a X-band linear accelerator mounted on a robot to deliver the radiation to the target, does not explicitly modulate the intensity of a beam but is able to deliver highly conformal treatments.

8.3 Intensity-Modulated Radiation Therapy Delivery Methods

8.3.1 Control Points for Dynamic or Static Delivery

To understand the delivery of different forms of IMRT, we must first understand how planning systems handle dynamic delivery and communicate with the linear accelerator (linac). In a planning system, a dynamic delivery sequence is approximated with multiple "segments," or "subfields," each defined as the delivery of a fixed number of monitor units (MUs) with a fixed aperture shape at fixed gantry and collimator angles. The use of these static subfields is also what the user observes on the planning system. These segments of subfields can be delivered either dynamically, transitioning from one segment to the next while the radiation is on, or statically, delivering each segment MUs with no mechanical movement and transitioning to the next segment after the MUs for the present segment are delivered with radiation turned off. When the former delivery scheme is used for delivering fixed intensity-modulated beams, it is often called the "sliding-window" technique because the apertures move across the field continuously during treatment delivery. The latter delivery scheme is often called the "step-and-shoot" delivery because mechanical motion and irradiation are alternated in sequence. In communicating with the linac for delivery, such segments are translated into a set of control points. How these planned segments are translated into control points determines the mode of delivery.

In the simplest translation, the first control point always has a cumulative MU of zero, with all other variables, such as field shape, gantry, and collimator angles, set to that of the first planned segment. For dynamic delivery, the second control point has the cumulative MU of the first planned segment but with the aperture shape and collimator and gantry angles of the second planned segment. In essence, the first planned static segment is converted into the dynamic transition from the first and the second control points. The transition from the second to the third control point then delivers the MUs of the second planned segment, and so on. Every pair of neighboring control points defines a dynamic delivery interval, and the corresponding planned segment is just a static sample of this delivery interval. The number of control points in a dynamic delivery is always equal to $(N + 1)$, where N is the number of planned segments.

For static or step-n-shoot delivery, the second control point repeats the aperture shape and other mechanical settings of the first control point, but with the cumulative MU equal to the MUs of the first segment. The transition from the first to the second control point represents only irradiation for delivering the MUs of the first planned segment. The third control point takes the aperture shape and other mechanical settings of the second segment, but with the same cumulative MU as the second control point. Therefore, the transition between the second and the third control point is purely mechanical movements with no change in the cumulative MU. The fourth control point then delivers the MUs assigned to the second segment. The number of control points in a static delivery is $2N$, with N being the total number of planned segments. Note that small variations to these segments to control point conversion scheme exist among different planning systems. For example, the planned segment shape and angles can be placed in the middle of two control points such that the planned segment samples at the midpoint of a delivery interval.

In all dynamic IMRT delivery, including IMAT, the cumulative delivered MUs serve as the independent variable. In sliding-window IMRT delivery, aperture shape is the only dependent variable that varies with the delivery of MUs. All the other mechanical variables, such as the collimator angle, the couch angle, and the gantry angle, are set as constants for each radiation beam. In IMAT delivery, the gantry angle also varies from one control point to the next, so that the field shape formed by the MLC varies during simultaneous gantry rotation and irradiation.

Existing linacs from different vendors have different control mechanisms in coordinating the delivery. For some machines, the linac and the MLC are controlled by separate computers that interact with each other; the control points designed for treatment delivery are divided into two groups of control parameters: (1) the MLC positions, as a function of delivered MUs, are sent to the MLC controller; (2) the gantry angle, as a function of cumulative MU, is controlled by the linac control system. If dose-rate control is required, different control mechanisms are employed by different linacs. For linacs that use gridded gun, electron pulses can be held. In addition, the corresponding pulse height and duration can also be controlled. For machines employing a nongridded gun, the dose-rate control relies mostly on controlling the electron gun current so that more or less electrons are generated and injected into the accelerating guide. Understanding the differences in how the machines execute the control points helps us design quality assurance measures for safe and accurate treatment delivery.

8.3.2 Intensity-Modulated Radiation Therapy Delivery Using Fixed Beams

8.3.2.1 Static Delivery

As described in Chapter 5 on IMRT planning, there are two ways to plan static treatments. One method is a two-step process in which the first step optimizes the intensity distributions of all beams and the second step converts the intensity distributions into deliverable overlapping field segments (Chui et al. 2001). Another method is direct aperture optimization (DAO), in which the aperture shapes and weights of the overlapping field segments are directly optimized (Shepard et al. 2002). The result of both of these methods is a set of beam apertures each with its optimized MUs. These segments are converted into control points for step-n-shoot delivery, and the linac executes these control points sequentially by alternating irradiation and segment aperture variation. As the result, different areas of the field are irradiated for different durations. For static delivery, the aperture shapes with each field are not required to be geometrically connected. This freedom allows the use of a relatively small number of segments. However, because the radiation must be turned off during the transition from one aperture shape to the next, the treatment delivery time is longer than the actual beam-on time.

8.3.2.2 Dynamic Delivery

All dynamic deliveries are planned using a two-step planning process. The feasibility of creating both positive and negative gradients by movement of two opposing collimators in a sliding-window fashion was first introduced by Convery and Rosenbloom (1992) and extended by a number of other investigators (Bortfeld, Boyer, et al. 1994; Bortfeld, Kahler, et al. 1994; Solberg 2001; Stein et al. 1994; Stein et al. 1997; Yu and Wong 1994; Yu et al. 1995). The basic principle of the sliding-window delivery method is to create the desired beam profile, $\Phi(x)$, that is aligned with a particular pair of MLC leaves, as the difference between two profiles, $\Phi_T(x)$ and $\Phi_L(x)$, created by the trailing leaf and the leading leaf respectively, that is,

$$\Phi(x) = \Phi_T(x) - \Phi_L(x)$$

An example is shown in Figure 8.1. The desired profile in Figure 8.1a can be separated into leading and trailing profiles as shown in Figure 8.1b. The vertical axis indicates "intensity," which is proportional to MUs. The horizontal axis is the position along the direction of leaf travel. The leading and trailing profiles define the positions of the leading and trailing leaves as the MUs are delivered. It is important to understand that what are actually used to control the MLC are not the profiles but the sampled points on these profiles. These sampled points form a sequence of MLC field segments, whose shapes are subject to the delivered MUs.

The component profiles shown in Figure 8.1b are not suitable for dynamic delivery because each of them contains sections that are parallel to the horizontal axis, requiring infinite leaf velocity. If we raise the component profiles by the same amount to create a slope, while maintaining the difference between the two profiles as shown in Figure 8.1c, the resultant leaf sequence will be suitable for dynamic sliding-window delivery. However, the total MU required for delivering the leaf sequence derived from Figure 8.1c is higher than that derived from Figure 8.1b. The sliding-window approach was first implemented in clinical practice

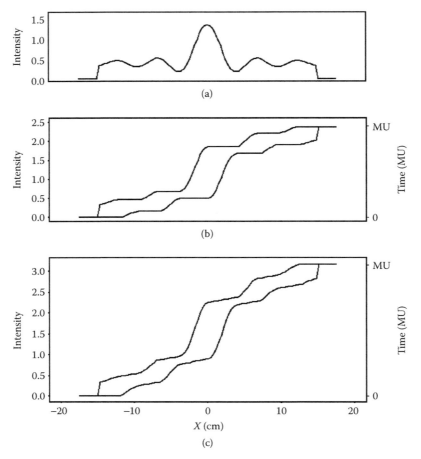

FIGURE 8.1 Converting intensity distribution into leaf trajectories.

at the Memorial Sloan-Kettering Cancer Center in New York (Ling et al. 1996).

The basic method for converting the optimized intensity distribution into sliding-window delivery assumes ideal flat beam with no head scatter and ideal MLC with no transmission or leakage. In order to deliver a predictable dose distribution, a number of other refinements are needed in an accurate dynamic MLC sequence to account for effects such as field flatness, head scatter, penumbra, leaf leakage, rounded leaf ends, and backscatter into the transmission ion chamber (AAPM 1998). Because of the tongue-and-groove feature on the sides of the MLC leaves for preventing large transmission through the leaf junctions, the actual aperture formed by the MLC is slightly smaller than what is intended whenever the side of a leaf forms part of the aperture boundary. This underdose effect is often called the *tongue-and-groove effect*. By synchronizing the leaf travel among the leaves, the underdosing effects of the tongue-and-groove design of the MLC can also be eliminated or minimized (Van Santvoort and Heijmen 1996; Webb et al. 1997). Practical leaf-sequencing algorithms must also consider physical constraints imposed by the MLC and the linear accelerator. These physical constraints include the maximum speed of leaf travel, the leaf travel limits, opposing leaf interdigitation where a leaf can cross the end of the neighboring leaf in the opposing leaf bank to

form disjointed openings, the controllability of complementing collimator jaws, the ability to close the opposing leaves, and so on.

8.3.2.3 Intensity Modulation Using Physical Modulators

Several research groups have reported on the use of a physical modulator to deliver IMRT (Hass et al. 1997; Jiang and Ayyangar 1998; Stein et al. 1997). The desired intensity distributions as produced through inverse planning can be converted into a thickness pattern and used to construct a physical compensator. The physical modulator IMRT method suffers from a relatively cumbersome and time-consuming manufacturing process and from the need to enter the treatment room to change the filter per gantry orientation, increasing the treatment time.

8.3.3 Intensity-Modulated Radiation Therapy Delivery with Rotational Beams

The dosimetric advantages of rotational treatments are illustrated by Shepard et al., who summarized the results from an optimization series performed for a C-shaped target with a sensitive structure in the concavity of C. For these simulations, all planning parameters, such as percent dose constraints, were held constant

TABLE 8.1   Impact of Number of Beam Angles on Plan Quality

# Angles	Obj. Funct. Value	Standard Deviation in Target Dose	d_{95}	Mean Dose to RAR	Total Integral Dose
3	0.665	0.124	0.747	0.488	2732.5
5	0.318	0.090	0.814	0.215	2563.3
7	0.242	0.064	0.867	0.206	2596.8
9	0.222	0.064	0.855	0.192	2598.3
11	0.202	0.058	0.879	0.186	2570.2
15	0.187	0.053	0.908	0.180	2542.9
21	0.176	0.049	0.912	0.171	2545.1
33	0.151	0.038	0.933	0.155	2543.5

Source: Reproduced from Shepard, D. M., et al. 1999. *Med Phys* 26(7):1212–21. With permission.

except for the number of beam angles (Shepard et al. 1999). The results summarized in Table 8.1 showed that each increase in the number of beam angles led to a more homogeneous dose to the tumor and a lower dose to the sensitive structure. Significant dosimetric improvements continued well beyond the number of beam angles typically used for fixed-field IMRT. It is also noteworthy that the total integral dose is nearly independent of the number of beam angles. Because of these advantages of rotational delivery, there have been continued efforts in the development of rotational techniques.

8.3.3.1 Tomotherapy

Mackie et al. (1993) proposed a form of IMRT using rotational fan beams called *tomotherapy*. This idea was quickly commercialized by NOMOS Corporation with the trade name Peacock (Carol 1994; Carol 1995). The Peacock system introduced the first commercial IMRT system using the multivane intensity-modulating collimator (MIMic), which is an add-on binary collimator that opens and closes under computer control. As the fan beam continuously rotates around the patient, the exposure time of a small width of the fan beam, or a beamlet, can be adjusted with the opening and closing of the binary collimator, allowing the radiation to be delivered to the tumor through the most preferable directions and locations of the patient. The initial commercial system by NOMOS Corporation delivered radiation treatment two slices at a time, where the treatment table had to be precisely indexed from one rotation to the next. Because of the sequential couch indexing feature, the Peacock system by NOMOS Corporation is also referred to as serial tomotherapy.

Helical tomotherapy was then developed by Tomotherapy, Inc. (Madison, Wisconsin) as a dedicated rotational IMRT system with a slip-ring rotating gantry and was made commercially available in 2002. More efficient delivery was achieved by combining continuous rotation of the intensity-modulated fan beams and couch translation resulting in a helical path relative to the patient. It is therefore often called helical tomotherapy (HT). HT offers a "turn-key" approach to IMRT implementation with the planning system specifically designed for the delivery unit. The HT utilizes all coplanar beam angles, and the intensity variations of the beamlets are not constrained by the mechanical limits of the binary MLC. Therefore,

theoretically speaking, HT provides a planning system with the freedom to use highly modulated beams to create more conformal treatment plans than MLC delivery with fixed cone beams.

In the HT system, the treatment plan is delivered through 6 MV X-ray fan beams in continuous rotation while the patient is slowly translating through the gantry aperture at a constant speed, which is predetermined by the pitch factor and the gantry period. Each gantry rotation consists of 51 equally spaced beam projections and 64 binary MLC leaves in each projection. Each MLC is individually controlled and allows full intensity modulation at each projection during dynamic gantry rotation, during which the dose rate stays constant at 880 MU per minute with small fluctuation. The gantry rotation period ranges from 15 to 60 seconds, depending on the fractionation size, modulation factor, and pitch. Thus, the degree of modulation by the binary MLC is determined by the gantry period, the modulation factor, and the inherent transit time of leaf motion (< 40 milliseconds). Different jaw widths ranging from 1 to 5 cm can be selected as a trade-off between the resolution of the intensity pattern and the delivery efficiency.

8.3.3.2 Intensity-Modulated Arc Therapy

First proposed in 1995 as an alternative to tomotherapy by Yu (1995), IMAT is also a rotational IMRT technique. Radiation is delivered with multiple superimposing coplanar or noncoplanar arcs, which can be easily implemented on a conventional linac using dynamic MLC (Yu et al. 2002). Earlier studies have shown that the dose distributions achieved by multiarc IMAT may be superior to conventional IMRT especially in stereotactic radiosurgery cases (Solberg 2001) and comparable to tomotherapy for most of the clinical cases (Cao et al. 2007). IMAT delivers optimized intensity distributions for large number of beams spaced every 5–10 degrees around the patient. Optimized intensity distributions are translated into a stack of superimposed irregular fields of uniform beam intensities and are delivered by overlapping arcs with synchronized gantry rotation and field shape variations. As the gantry is rotating around the patient and the radiation beam is on, it is important to note that the subfields of adjacent beam angles do not require the MLC leaves to travel very long distances. Ensuring such connectedness of adjacent subfields for smooth leaf motion is of great concern in the leaf-sequencing algorithm for IMAT (Earl et al. 2003; Gladwish et al. 2007; Luan et al. 2008; Wang et al. 2008). Effective planning tools for IMAT have only been developed recently (Cameron 2005; Cao et al. 2007; Cotrutz, Kappas, and Webb. 2000; Earl et al. 2003; Luan et al. 2008; Otto 2008; Shepard et al. 2007; Ulrich, Nill, and Oelfke 2007; Wang et al. 2008; Wong,Chen, and Greenland 2002).

Through the initial proof-of-principle study, IMAT could be a valid alternative to tomotherapy in terms of treatment delivery (Yu 1995). However, unlike tomotherapy, IMAT must account for restrictions on MLC movement as the gantry moves from one beam angle to the next. Because deliverability must take priority, an optimal field shape may have to be altered in order to produce smooth delivery. As a result, plan quality should be adversely affected. This restriction does not apply to tomotherapy, due to the use of binary MLC. Compared with tomotherapy, IMAT also has some advantages: (1) IMAT does not need to move the patient during treatment and avoids abutment issues as seen with serial tomotherapy; (2) IMAT retains the capability of using noncoplanar beams and arcs, which has great value for brain and head and neck tumors; and (3) IMAT uses a conventional linac, thus complex rotational IMRT treatments and simple palliative treatments can be delivered with the same treatment unit.

In proposing IMAT, Yu predicted that with the increase in the number of gantry angles, the number of intensity levels at each gantry angle can be reduced without degrading the plan quality (Yu 1995). He argued that the plan quality is a function of the total number of strata, which is defined as the product of the number of beam angles and the number of intensity levels. In other words, it is the total number of aperture shape variations that determine the plan quality. Assuming this is true, a single arc with a sufficient number of aperture shape variations would be able to create optimal treatment plans. Many subsequent works have attempted to use a single arc for IMAT. Because linac at the time cannot vary the dose rate dynamically during gantry rotation, most previous works on single-arc IMAT were done under the assumption that the machine dose rate should be constant during arc rotation. As illustrated by Jiang et al. (2005), a single arc with 36 beam aperture variations under a constant dose rate (CDR) cannot realize the optimal plan quality. To achieve the desired plan quality, we must either increase the number of field segments or apertures or allow the dose rate to vary during gantry rotation, or both.

MacKenzie and Robinson (2002) proposed a technique whereby 24 equally spaced beam orientations are optimized for sliding-window IMRT and arc delivery is performed by allowing the gantry of the linear accelerator to rotate to static gantry orientations and deliver the optimized sliding-window IMRT deliveries. Crooks et al. developed an IMAT planning algorithm that is based on the observation that the dose error for beam apertures being off by a small angular deviation is very small (Crooks et al. 2003). In their algorithm, IMRT fields were created approximately 30° apart with 56–74 segments per beam direction. The segments were spread out with the previous observation regarding angular deviations, and the plan was simulated and delivered in a single arc with deviations from the original IMRT plan in the order of 5%–10%.

Cameron et al. (2005) demonstrated the feasibility of delivering IMRT with a single arc. The technique was termed *sweeping-window arc therapy* (SWAT). Shapes of the MLC apertures prior to optimization are initialized so that the MLC leaf positions sweep across the planning target volume (PTV) with rotation. Optimization of MLC leaf positions is then performed by simulated annealing and arc weight optimization, which can be performed for a constant or variable angular dose rate.

Tang et al. (2007) showed that a multiarc IMAT could be converted into a single arc by spreading the stacked apertures to neighboring angles with little effect on the plan quality. Figure 8.2 shows the method and part of their results. A five-arc IMAT plan was created by optimizing the apertures' shapes and weights on 36 beam angles. The resulting plan had five apertures stacked at each of the beams spaced every 10 degrees. A new plan was then created by simply rearranging the stacked apertures into neighboring angles by minimizing the movement of the geometric center of the apertures as schematically shown in Figures 8.2a and b. Dose calculations for the original plan with stacked apertures and for the new plan with spaced apertures showed almost identical results for different plans. Two examples, one for a prostate case and one for a head and neck case, are shown in Figures 8.2c and d. This simple exercise elucidated that given the same number of aperture shape variations, single-arc IMAT and multiarc IMAT theoretically have the same degree of freedom for optimizing the dose distributions, if the apertures in the single-arc arrangement could be geometrically connected. It also demonstrated that in rotational delivery, the dose distribution is insensitive to small random errors in the gantry angle.

Ulrich, Nill, and Oelfke (2007) developed an optimization technique whereby arc therapy plans are optimized for a single-arc delivery. In their algorithm, arc shapes are optimized by a taboo search optimization algorithm, and arc weights are optimized by a gradient search. The algorithm demonstrates better treatment plans than an in-house IMRT optimization technique and requires a variable dose rate (VDR) delivery with gantry rotation.

By assuming that the machine dose rate can vary as needed, Otto (2008) developed a single-arc IMAT algorithm that he referred to as volumetric-modulated arc therapy (VMAT). In addition to allowing dose-rate variation, the VMAT algorithm uses progressive beam angle sampling to optimize a large number (> 100) of apertures using DAO. The apertures' shapes and weights are optimized initially for a number of coarsely spaced gantry angles with little consideration of aperture connectivity. Once the solution converges, additional gantry angles are inserted. As the angular spacing becomes smaller, the optimizer considers aperture shape connectivity both in the initialization of aperture shapes, as well as during the optimization. The shapes of the newly inserted apertures are linearly interpolated from their angular neighbors. Such coarse-to-fine sampling is called *progressive sampling*, and allows the optimization to converge faster. Because the aperture shape connectivity is ignored initially, the optimizer is given the freedom to aim for optimal dose distribution. Because the final plan ensures aperture connectivity, the optimized single arc can be delivered within 2 minutes.

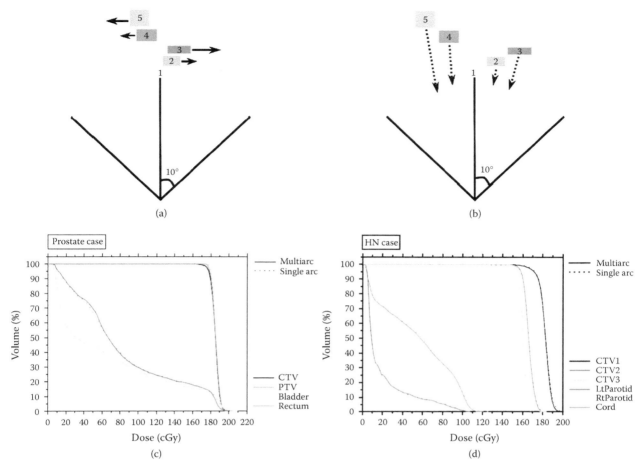

FIGURE 8.2 **(See color insert following page 204.)** (a) At every beam angle of a five-arc plan generated with 36 beams, there are five stacked apertures labeled 1 to 5. (b) The stacked apertures are spread out within the 10 degree interval with the order arranged according to the apertures center location. By spreading the stacked apertures in this manner for all 36 beams, a five arc plan is converted to a 180-field single arc plan. (c) and (d) show the DVH comparison between the original 5-arc plan and the re-configured single arc plan for a prostate cancer case and a head-neck case respectively. The dose distributions are nearly identical.

Luan et al. (2008) developed an arc-sequencing algorithm for converting continuous intensity maps into multiple arcs, using the k-link shortest path algorithm. The algorithm was tested for prostate, breast, head and neck, and lung, and it was demonstrated that the plans rivaled helical tomotherapy plans. Based on the method developed by Luan et al., Wang et al. sequenced the intensity patterns optimized for 36 beams into a single-arc delivery. These many contributions and their demonstration of superior delivery efficiency led to the present different commercial offerings of single-arc IMAT (Wang et al. 2008).

The ability of a single arc to produce dose distributions comparable to tomotherapy is commonly attributed to the freedom provided by dose-rate variation. Palma et al. found that VDR-optimized single-arc plans produced superior dose distributions to those optimized with CDR (Palma et al. 2008). In the study, the treatment plans were generated using a series of evenly spaced static beams, which is the general approach for IMAT planning. VDR delivery not only complicates delivery and quality

assurance (QA) but also limits clinical adoption because most of the existing linacs are not equipped with VDR capability.

Based on the observation that the dosimetric error introduced by displacing the beam apertures from the planned angle to a slightly different angular position is minimal (Crooks et al. 2003; Tang et al. 2007; Wang et al. 2008), Tang, Earl, and Yu (2009) hypothesized that varying the angular spacing of the apertures using CDR delivery is equivalent to varying the weights of evenly spaced beams. They draw the similarity from radio broadcasting, in which VDR delivery of evenly spaced beams resembles amplitude modulation while CDR delivery of unevenly spaced beams resembles frequency modulation. They proved such equivalence by converting RapidArc plans, which require VDR delivery, into CDR plans by assigning larger angular intervals to segments with larger MUs, and vice versa. The conversion scheme is schematically illustrated in Figure 8.3, where the vertical axis represents the dose rate in MU/degree and the horizontal axis depicts the beam angles. Each vertical bar represents a planned segment.

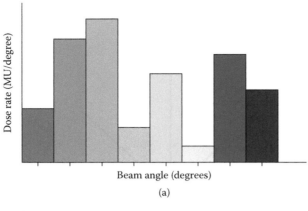

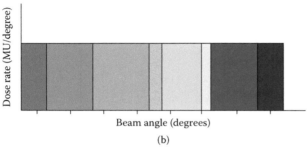

FIGURE 8.3 Variable dose rate plan with even angular spacing (a) is converted to constant dose rate plan with variable angular spacing (b) by conserving segment monitor units. (Reprinted from Tang, G., et al. 2009. *Phys Med Biol* 54:6439–56. With permission.)

VDR and CDR counterparts of the same segment have the same area, indicating the same segment MUs. Completed CDR plans were delivered and dosimetrically verified using a conventional linac without the capability of dose-rate variation. They found that the plan qualities and the delivery times of the CDR and VDR plans were comparable, which proves that single-arc IMAT can be delivered using either VDR with even angular spacing or CDR with variable angular spacing.

8.4 Considerations in Intensity-Modulated Radiation Therapy Delivery

8.4.1 Dosimetric Uncertainties of Intensity-Modulated Radiation Therapy Delivery

The dynamic delivery of IMRT introduces new issues that could affect treatment accuracy. These include the disconnect between static dose calculation in planning and dynamic delivery, the interplay between organ motion and the movement of the beam, and the trade-offs among different IMRT methods.

8.4.1.1 Relationship between Geometric and Dosimetric Uncertainties

It is important to understand that geometric errors can lead to large dosimetric errors in IMRT delivery because intensity

modulation is achieved with the superposition of many field segments. Systematic errors in the aperture shapes result in a surplus or deficit in the radiation flux to the target causing overdose or underdose. As a rule of thumb, the dosimetric uncertainty is the quotient of the leaf setting uncertainty to the MU-weighted mean width of the field segments. For example, when the sliding-window technique is used for IMRT delivery, a 1-mm leaf opening error resulting from the combined position errors of the two opposing leaves can cause a 10% error if the mean "window" width is 1 cm. Mu, Ludlum, and Xia (2008) studied the impact of MLC leaf positioning errors on IMRT plans for head and neck cancers and found that for a 1-mm systematic error, the average change in D (95%) was 4% in simple plans vs. 8% in complex plans and the average change in the dose to the parotid glands was 9% for simple plans vs. 13% for the complex plans. Rangel and Dunscombe (2009) tried to set an MLC accuracy standard based on acceptable equivalent uniform dose (EUD) variations. They found that if a 2% change in EUD of the target and 2 Gy for the organs at risk (OARs) were adopted as acceptable levels of deviation in dose due to MLC effects alone, then systematic errors in leaf position will need to be limited to 0.3 mm.

8.4.1.2 Disconnection between Planning and Delivery

In all dynamic delivery including sliding-window IMAT and tomotherapy, the treatment plans are developed by approximating the continuous dynamic delivery with static beams. Therefore, there is an intrinsic disconnection between treatment planning and treatment delivery. For example, in IMAT delivery, not only the beam aperture shape but also the machine dose rate varies dynamically. Because the field shapes are changing dynamically, the optimized aperture shape only happens at an instant during delivery. The beam aperture takes interpolated shapes the majority of the time. This also means that the MUs optimized for a fixed aperture shape at a fixed beam angle are actually delivered with different shapes at different angles.

The effects of such disconnection between planning and delivery have not been thoroughly studied. In a recent work by Tang et al. the coarsely defined static beams were interpolated into 720 beams with a fine angular spacing of 0.5 degrees, to more accurately approximate a continuous rotation (Tang et al. 2008). A Monte–Carlo based kernel superposition algorithm was used to compute the radiation doses of both the originally planned beams and 720 beams used as the surrogate of continuous delivery. It was found that for most of the treatment plans, the difference between the plans created with static beams and continuous delivery is minimal. However, for some plans, large differences were noted. Figures 8.4a and b show the discrepancies between the planned DVHs and the delivered DVHs for two plans generated by two different methods for the same case. Plan 1 (Figure 8.4a) shows large differences between the calculated dose with 36 fixed fields and the delivered dose with continuous arc delivery, simulated with interpolated field shapes and MUs to 720 beam angles. However, plan 2 (Figure 8.4b) shows little difference between the calculated and the delivered doses. Further studies of both the leaf travel histogram (Figure 8.4c) and the

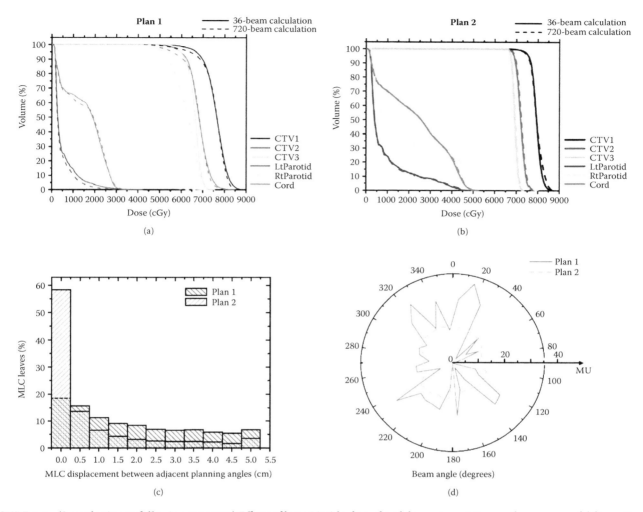

FIGURE 8.4 **(See color insert following page 204.)** Effects of large MLC leaf travel and dose-rate variation on the accuracy of delivery. For the same head and neck case, large discrepancies between the planned and delivered doses exhibited in (a) plan 1, but not in (b) plan 2. Comparison of leaf travel histogram in (c) and dose rate distribution in (d) reveals that plan 1 has longer leaf travel and larger dose rate fluctuations.

MU distribution (Figure 8.4d) show that the plan exhibiting large differences between planned and delivered doses is the one with large leaf travels and large dose-rate fluctuations.

This is essentially the same as the digital sampling problem. Large variations in MLC aperture shapes and large dose-rate variations both represent high spatial and temporal frequency. To accurately calculate such treatment plans, tighter samples, that is, more beam angles, are required in dose calculation to faithfully represent the treatment delivery.

8.4.2 Effect of Motion on Intensity-Modulated Radiation Therapy Delivery

For thoracic tumors treated with conformal therapy, breathing motion can cause the target to move outside the beam aperture, resulting in geometric misses. In breast treatment, breathing motion broadens the beam penumbra at the posterior field boundary, resulting in lower doses to the breast near the posterior field

border and higher doses to the underlying lung. To compensate for these effects, large margins are normally assigned around the target volume to ensure proper target coverage. However, with dynamic intensity modulation, the problem is more complicated due to the interplay of patient motion and the motion of the radiation fields. Yang et al. (1997) studied the problem of intrafraction organ motion for tomotherapy delivery, in which the patient is treated by a rotating slit fan beam either in a slice-by-slice or a spiral fashion. They found that ripple-like dose variations occur at the slice boundaries when the target moves in the direction of couch translation. The effect of intrafraction organ motion on the delivery of intensity modulation with dynamic multileaf collimation in a single radiation field was first studied by Yu, Jaffray, and Wong (1998). They simulated the delivery of a sliding-window IMRT to a target under cyclic motion with clinically relevant frequencies and amplitudes. The simulation revealed that the fundamental mechanism for creating such dosimetric variations in the target is the interplay between the movements of the beam aperture and the target during irradiation. The results showed that, for clinically realistic parameters, the magnitude of intensity variations in the target can be greater than 100% of the desired beam intensity. The magnitude of the photon intensity variations is strongly dependent on the speed of the beam aperture relative to the speed of the target motion and the width of the scanning beam relative to the amplitude of the target motion. In general, the dosimetric effects of target motion increase with narrowing beam aperture. With multiple beams and multiple fractions, the dosimetric errors are averaged and lower. However, the intensity patterns will also be smoothed. Such smoothed beam intensities occurring in all radiation fields will also cause large dose errors in the target.

8.4.3 Comparisons among Different Treatment Schemes

Comparisons between IMRT plans for different delivery methods have been conducted. In a comparison of tomotherapy and MLC delivery, Mavroidis et al. (2007) found that linear accelerator delivery with an MLC has a slight advantage over tomotherapy for most sites other than head and neck. Similar results have been found by Muzik, Soukup, and Alber (2008). Cao et al. (2007) compared the treatment plan quality of IMAT plans and tomotherapy plans for 10 cases including head and neck, lung, brain, and prostate. They found that these two kinds of rotational delivery methods are also equivalent for most cases. For cases where noncoplanar beams are desirable, such as for intracranial tumors and some head and neck cases, the use of partial noncoplanar arcs in IMAT was found to be more advantageous (Cao et al. 2007). Shepard et al. (2007) compared IMAT plans with IMRT and found that the employment of rotational IMRT is advantageous for most of the cases. Since single-arc IMAT has become commercially available, there have been many published comparisons among different methods of IMRT delivery (Cozzi et al. 2008; Palma et al. 2008; Tang et al. 2009). Overall, all methods are able to create clinically acceptable and highly conformal treatment plans. The highest plan quality is created with tomotherapy and IMAT with two or more arcs. Single-arc IMAT performs at least as well as IMRT using multiple fixed fields and in most of the cases rivals the plan quality of tomotherapy and IMAT with multiple arcs.

It is important to note that many other issues besides the plan quality are associated with different delivery techniques: the efficiency of planning, delivery, and quality assurance; the complexity and reliability of delivery; and the total MUs required to deliver the prescribed dose. Past attempts to improve treatment efficiency include the use of DAO (Shepard et al. 2002) and the development of single-arc IMAT (Cameron 2005; Cao et al. 2007; Cotrutz, Kappas, and Webb 2000; Crooks et al. 2003; Luan et al. 2008; MacKenzie and Robinson 2002; Otto 2008; Shepard et al. 2007; Ulrich, Nill, and Oelfke 2007; Wang et al. 2008). By optimizing the aperture shape and aperture weights directly, DAO achieves IMRT plans similar to other planning schemes in quality but with a much smaller number of segments, resulting in highly efficient delivery (Shepard et al. 2002; Ludlum and

Xia 2008). DAO also provides an efficient method for planning IMAT treatments (Earl et al. 2003; Otto et al. 2008). Among the different treatment delivery methods, single-arc IMAT is the most efficient in both MU usage and delivery time, with a typical delivery time of about 2 minutes, whereas tomotherapy is the least efficient.

8.5 Conclusion

This chapter provides an overview of the different methods used for delivering IMRT treatments. It is evident that highly conformal treatments can be achieved with different delivery schemes. This has allowed us to understand the fundamental principles of IMRT in the use of angular and location preferences to take advantage of the geometric arrangements of the target and surrounding structures. These principles of IMRT are also applicable to other forms of radiation such as electrons and protons. With the significant efficiency improvements brought by DAO and single-arc IMAT, IMRT will continue to play a key role in the era of adaptive radiation therapy.

References

AAPM Radiation Therapy Committee Task Group #53. 1998. Quality assurance for clinical radiotherapy treatment planning. *Med Phys* 25(10):1773–829.

Bortfeld, T. 2006. IMRT: A review and preview. *Phys Med Biol* 51(13):R363–79.

Bortfeld, T. 2010. The number of beams in IMRT: Theoretical investigations and implications for single-arc IMRT. *Phys Med Biol* 55(1):83–97.

Bortfeld, T., A. L. Boyer, W. Schlegel, D. L. Kahler, and T. J. Waldron. 1994. Realization and verification of three-dimensional conformal radiotherapy with modulated fields. *Int J Radiat Oncol Biol Phys* 30:899–908.

Bortfeld, T., D. Kahler, T. Waldron, and A. Boyer. 1994. X-ray field compensation with multileaf collimators. *Int J Rad Oncol Biol Phys* 28(3):723–30.

Boyer, A., and C. Yu 1999. Intensity-modulated radiation therapy with dynamic multileaf collimators. *Semin Radiat Oncol* 9(1):48–59.

Brahme, A. 1988. Optimization of stationary and moving beam radiation therapy techniques. *Radiother Oncol* 12:129–40.

Brixey, C., J. C. Roeske, A. E. Lujan, et al. 2002. Impact of intensity modulated whole pelvic radiation therapy on acute hematologic toxicity in patients with gynecologic malignancies. *Int J Radiat Oncol Biol Phys* 54:1388–96.

Burman, C., C. Chui, G. Kutcher, et al. 1997. Planning, delivery, and quality assurance of intensity-modulated radiotherapy using dynamic multileaf collimator: a strategy for large-scale implementation for the treatment of carcinoma of the prostate. *Int J Radiat Oncol Biol Phys* 39(4):863–73.

Cameron, C. 2005. Sweeping-window arc therapy: An implementation of rotational IMRT with automatic beam-weight calculation. *Phys Med Biol* 50:4317–36.

Cao, D., T. W. Holmes, M. K. Afghan, and D. M. Shepard. 2007. Comparison of plan quality provided by intensity-modulated arc therapy and helical tomotherapy. *Int J Radiat Oncol Biol Phys* 69:240–50.

Cardinale, R. M., S. H. Benedict, Q. Wu, et al. 1998. A comparison of three stereotactic radiotherapy techniques; ARCS vs. noncoplanar fixed fields vs. intensity modulation. *Int J Radiat Oncol Biol Phys* 42:431–436.

Carol, M. P. 1994. Integrated 3-D conformal multivane intensity modulation delivery system for radiotherapy. In *Proceedings of the 11th International Conference on the Use of Computers in Radiation Therapy*, ed. A. R. Hounsell, J. M. Wilkinson, and P. C. Williams, 172–73. Manchester: North Western Medical Physics Department, Christie Hospital.

Carol, M. P. 1995. Integrated 3D conformal planning/multivane intensity modulating delivery system for radiotherapy. In *3D Radiation Treatment Planning and Conformal Therapy*, ed. J. A. Purdy and B. Emami, 435–45. Madison, WI: Medical Physics Publishing.

Chao, K. S. C., J. O. Deasy, J. Markman, et al. 2001. A prospective study of salivary function sparing in patients with head-and-neck cancers receiving intensity-modulated or three-dimensional radiation therapy: Initial results. *Int J Radiat Oncol Biol Phys* 49(4):907–16.

Chui, C. S., M. F. Chan, E. Yorke, S. Spirou, and C. C. Ling. 2001. Delivery of intensity-modulated radiation therapy with a conventional multileaf collimator: Comparison of dynamic and segmental methods. *Med Phys* 28:2441–9.

Claus, F., W. De Gersem, I. Vanhoutte, et al. 2001. Evaluation of a leaf position optimization tool for intensity modulated radiation therapy of head and neck cancer. *Radiother Oncol* 61:281–6.

Convery, D. J., and M. E. Rosenbloom. 1992. The generation of intensity-modulated fields for conformal radiotherapy by dynamic collimation. *Phys Med Biol* 37(6):1359–74.

Cotrutz, C., C. Kappas, and S. Webb. 2000. Intensity modulated arc therapy (IMAT) with centrally blocked rotational fields. *Phys Med Biol* 45:2185–206.

Cozzi, L., K. A. Dinshaw, S. K. Shrivastava, et al. 2008. A treatment planning study comparing volumetric arc modulation with RapidArc and fixed field IMRT for cervix uteri radiotherapy. *Radiother Oncol* 89:180–91.

Crooks, S. M., X. Wu, C. Takita, M. Watzich, and L. Xing. 2003. Aperture modulated arc therapy. *Phys Med Biol* 48:1033–44.

Derycke, S., W. R. De Gersem, B. B. Van Duyse, and W. C. De Neve. 1998. Conformal radiotherapy of stage III non-small cell lung cancer: A class solution involving non-coplanar intensity-modulated beams. *Int J Radiat Oncol Biol Phys* 41:771–7.

Duthoy, W., W. De Gersem, K. Vergote, et al. 2003. Whole abdominopelvic radiotherapy (WAPRT) using intensity-modulated arc therapy (IMAT): First clinical experience. *Int J Radiat Oncol Biol Phys* 57:1019–32.

Earl, M. A., D. M. Shepard, S. A. Naqvi, and C. X. Yu. 2003. Inverse planning for intensity-modulated arc therapy using direct aperture optimization. *Phys Med Biol* 48:1075–89.

Eisbruch, A., R. K. Ten Haken, H. M. Kim, et al. 1999. Dose, volume, and function relationships in parotid salivary glands following conformal and intensity modulated irradiation of head and neck cancer. *Int J Radiat Oncol Biol Phys* 45:577–87.

Freedman, G. M., N. J. Meropol, E. R. Sigurdson, et al. 2007. Phase I trial of preoperative hypofractionated intensity-modulated radiotherapy with incorporated boost and oral capecitabine in locally advanced rectal cancer. *Int J Radiat Oncol Biol Phys* 67(5):1389–93.

Fu, K. K., T. F. Pajak, A. Trotti, et al. 2000. A Radiation Therapy Oncology Group (RTOG) Phase III randomized study to compare hyperfractionation and two variants of accelerated fractionation to standard fractionation radiotherapy for head and neck squamous cell carcinomas: First report of RTOG 9003. *Int J Radiat Oncol Biol Phys* 48:7–16.

Galvin, J. M., G. Ezzell, A. Eisbrauch. et al. 2004. Implementing IMRT in clinical practice: A joint document of the American Society for Therapeutic Radiology and Oncology and the American Association of Physicists in Medicine. *Int J Radiat Oncol Biol Phys* 58(5):1616–34.

Gladwish, A., M. Oliver, J. Craig, et al. 2007. Segmentation and leaf sequencing for intensity modulated arc therapy. *Med Phys* 34:1779–88.

Hara, W., S. G. Soltys, and I. C. Gibbs. 2007. CyberKnife robotic radiosurgery system for tumor treatment. *Expert Rev Anticancer Ther* 7(11):1507–15.

Harsolia, A., L. Kestin, I. Grills, et al. 2007. Intensity-modulated radiotherapy results in significant decrease in clinical toxicities compared with conventional wedge-based breast radiotherapy. *Int J Radiat Oncol Biol Phys* 68(5):1375–80.

Hass, O. C. L., J. A. Mills, K. J. Burnham, et al. 1997. Achieving conformal dose distribution via patient specific compensators (abstract). In *XII International Conference on the Use of Computers in Radiation Therapy*, Salt Lake City, UT, ed. D. D. Leavitt and G. Starkschall, 483. Madison, WI: Medical Physics Publishing.

Hong, L., K. M. Alektiar, M. Hunt, E. Venkatraman, and S. A. Leibel. 2004. Intensity-modulated radiotherapy for soft tissue sarcoma of the thigh. *Int J Radiat Oncol Biol Phys* 59(3):752–9.

Hong, L., M. Hunt, C. Chui, et al. 1999. Intensity-modulated tangential beam irradiation of the intact breast. *Int J Radiat Oncol Biol Phys* 44:1155–64.

Huang, E., B. S. Teh, D. R. Strother, et al. 2002. Intensity-modulated radiation therapy for pediatric medulloblastoma: Early report on the reduction of ototoxicity. *Int J Radiat Oncol Biol Phys* 52:599–605.

Iuchi, T., K. Hatano, Y. Narita, T. Kodama, T. Yamaki, K. Osato, 2006. Hypofractionated high-dose irradiation for the treatment of malignant astrocytomas using simultaneous integrated boost technique by IMRT. *Int J Radiat Oncol Biol Phys* 64(5):1317–24.

Jiang, S. B., and K. M. Ayyangar. 1998. On compensator design for photon beam intensity-modulated conformal therapy. *Med Phys* 25:668–75.

Jiang, Z., D. M. Shepard, M. A. Earl, G. W. Zhang, C. X. Yu, 2005. A examination of the number of required apertures for step-and-shoot IMRT. *Phys Med Biol* 50(23):5653–63.

Junius, S., K. Haustermans, B. Bussels, et al. 2007. Hypofractionated intensity modulated irradiation for localized prostate cancer, results from a phase I/II feasibility study. *Radiat Oncol* 2:29.

Kestin, L. L., M. B. Sharpe, R. C. Frazier, et al. 2000. Intensity modulation to improve dose uniformity with tangential breast radiotherapy: Initial clinical experience. *Int J Radiat Oncol Biol Phys* 48:1559–68.

Kuban, D. A., S. L. Tucker, L. Dong, et al. 2008. Long-term results of the M. D. Anderson randomized dose-escalation trial for prostate cancer. *Int J Radiat Oncol Biol Phys* 70(1):67–74.

Kuppersmith, R. B., B. S. Teh, D. T. Donovan, et al. 2000. The use of intensity modulated radiotherapy for the treatment of extensive and recurrent juvenile angiofibroma. *Int J Pediatr Otorhinolaryngol* 52:261–8.

Landry, J. C., G. Y. Yang., J. Y. Ting, et al. 2002. Treatment of pancreatic cancer tumors with intensity-modulated radiation therapy (IMRT) using the volume at risk approach (VARA): Employing dose-volume histogram (DVH) and normal tissue complication probability (NTCP) to evaluate small bowel toxicity. *Med Dosim* 27:121–9.

Ling, C. C., C. Burman, C. S. Chui, et al. 1996. Conformal radiation treatment of prostate cancer using inversely-planned intensity-modulated photon beams produced with dynamic multileaf collimation. *Int J Radiat Oncol Biol Phys* 35:721–30.

Lo, Y. C., G. Yasuda, T. J. Fitzgerald, et al. 2000. Intensity modulation for breast treatment using static multileaf collimators. *Int J Radiat Oncol Biol Phys* 46:187–94.

Luan, S., C. Wang, D. Cao, D. Z. Chen, D. M. Shepard, and C. X. Yu. 2008. Leaf-sequencing for intensity-modulated arc therapy using graph algorithms. *Med Phys* 35(1):61–9.

Ludlum, E., and P. Xia. 2008. Comparison of IMRT planning with two-step and one-step optimization: A way to simplify IMRT. *Phys Med Biol* 53:807–21.

MacKenzie, M. A., and D. M. Robinson. 2002. Intensity modulated arc deliveries approximated by a large number of fixed gantry position sliding window dynamic multileaf collimator fields. *Med Phys* 29:2359–65.

Mackie, T. R., T. Holmes, S. Swerdloff, et al. 1993. Tomotherapy: A new concept for the delivery of conformal radiotherapy. *Med Phys* 20:1709–19.

Mavroidis, P., B. C. Ferreira, C. Shi, B. K. Lind, and N. Papanikolaou. 2007. Treatment plan comparison between helical tomotherapy and MLC-based IMRT using radiobiological measures. *Phys Med Biol* 52(13):3817–36.

Mell, L. K., A. K. Mehrotra, and A. J. Mundt. 2005. Intensity-modulated radiation therapy use in the U.S. 2004. *Cancer* 104(6):1296–303.

Mu, G., E. Ludlum, and P. Xia. 2008. Impact of MLC leaf position errors on simple and complex IMRT plans for head and neck cancer. *Phys Med Biol* 53(1):77–88.

Mundt, A. J., A. E. Lujan, J. Rotmensch et al. 2002. Intensity-modulated whole pelvic radiotherapy in women with gynecologic malignancies. *Int J Radiat Oncol Biol Phys* 52:1330–7.

Muzik, J., M. Soukup, and M. Alber. 2008. Comparison of fixed beam IMRT, helical tomotherapy, and IMPT for selected cases. *Med Phys* 35(4):1580–92.

Nutting, C. M., D. J. Convery, V. P. Cosgrove, et al. 2000. Reduction of small and large bowel irradiation using an optimized intensity-modulated pelvic radiotherapy technique in patients with prostate cancer. *Int J Radiat Oncol Biol Phy* 48:649–56.

Nutting, C. M., J. L. Bedford, V. P. Cosgrove, et al. 2001. A comparison of conformal and intensity-modulated techniques for esophageal radiotherapy. *Radiother Oncol* 61:157–63.

Otto K. 2008. Volumetric modulated arc therapy: IMRT in a single gantry arc. *Med Phys* 35(1):310–7.

Palma, D., E. Vollans, K. James, et al. 2008. Volumetric modulated arc therapy for delivery of prostate radiotherapy: Comparison with intensity-modulated radiotherapy and three-dimensional conformal radiotherapy. *Int J Radiat Oncol Biol Phys* 72:996–1001.

Paulino, A. C., and M. Skwarchuk. 2002. Intensity-modulated radiation therapy in the treatment of children. *Med Dosim* 27:115–20.

Pollack, A., A. Hanlon, E. M. Horwitz, S. Feigenberg, R. G. Uzzo, and R. A. Price. 2003. Radiation therapy dose escalation for prostate cancer: A rationale for IMRT. *World J Urol* (4):200–8.

Portelance, L., K. S. Chao, P. W. Grigsby, et al. 2001. Intensity-modulated radiation therapy (IMRT) reduces small bowel, rectum, and bladder doses in patients with cervical cancer receiving pelvic and para-aortic irradiation. *Int J Radiat Oncol Biol Phys* 51:261–6.

Rangel, A., and P. Dunscombe. 2009. Tolerances on MLC leaf position accuracy for IMRT delivery with a dynamic MLC. *Med Phys* 36(7):3304–9.

Roeske, J. C., A. Lujan, J. Rotmensch, S. E. Waggoner, D. Yamada, and A. J. Mundt. 2000. Intensity-modulated whole pelvic radiation therapy in patients with gynecologic malignancies. *Int J Radiat Oncol Biol Phys* 48(5):1613–21.

Shepard, D. M., D. Cao, M. K. Afghan, and M. A. Earl. 2007. An arc-sequencing algorithm for intensity modulated arc therapy. *Med Phys* 34:464–70.

Shepard, D. M., G. Olivera, L. Angelos, O. Sauer, P. Reckwerdt, and T. R. Mackie. 1999. A simple model for examining issues in radiotherapy optimization. *Med Phys* 26(7):1212–21.

Shepard, D. M., M. A. Earl, X. A. Li, S. Naqvi, and C. X. Yu. 2002. Direct aperture optimization: A turnkey solution for step-and-shoot IMRT. *Med Phys* 29(6):1007–18.

Solberg, T. D. 2001. Dynamic arc radiosurgery field shaping: a comparison with static field conformal and noncoplanar circular arcs. *Int J Radiat Oncol Biol Phys* 49:1481–91.

Spirou, S. V., and C. S. Chui. 1994. Generation of arbitrary fluence profiles by dynamic jaws or multileaf collimators. *Med Phys* 21:1031–41.

Stein, J., K. Hartwig, S. Levegrün, et al. 1997. Intensity-modulated treatments: compensators vs. multileaf modulation. In *XII International Conference on the Use of Computers in Radiation Therapy*, Salt Lake City, UT, ed. D. D. Leavitt and G. Starkschall, 338–41. Madison, WI: Medical Physics Publishing.

Stein, J., T. Bortfeld, B. Dorschel, and W. Schlegel. 1994. Dynamic X-ray compensation for conformal radiotherapy by means of multileaf collimation. *Radiother Oncol* 32:163–73.

Sultanem, K., H. K. Shu, P. Xia, et al. 2000. Three-dimensional intensity-modulated radiotherapy in the treatment of nasopharyngeal carcinoma: The University of California-San Francisco experience. *Int J Radiat Oncol Biol Phys* 48(3):711–22.

Sura, S., V. Gupta, E. Yorke, A Jackson, H. Amols, and K. E. Rosenzweig. 2008. Intensity-modulated radiation therapy (IMRT) for inoperable non-small cell lung cancer: The Memorial Sloan-Kettering Cancer Center (MSKCC) experience. *Radiother Oncol* 87(1):17–23.

Tang, G., M. A. Earl, S. Luan, C. Wang, M. Mohiuddin, and Yu CX. 2010. Comparing radiation treatments using intensity-modulated beams, multiple arcs and single arc. *Int J Radiat Oncol Biol Phys* 76:1554–62.

Tang, G., M. A. Earl, and C. X. Yu. 2009. Variable dose rate single-arc IMAT delivered with constant dose rate and variable angular spacing. *Phys Med Biol* 54:6439–56.

Tang, G., M. A. Earl, S. Luan, et al. 2008. Stochastic versus deterministic kernel-based superposition approaches for dose calculation of intensity-modulated arcs. *Phys Med Biol* 53:4733–4746.

Tang, G., M. Earl, S. Luan, S. Naqvi, and C. X. Yu. 2007. Converting multiple-arc intensity-modulated arc therapy into a single arc for efficient delivery. *Int J Rad Oncol Biol Phys* 69(3):S673.

Ulrich, S., S. Nill, and U. Oelfke. 2007. Development of an optimization concept for arc-modulated cone beam therapy. *Phys Med Biol* 52(14):4099–119.

van Asselen, B., C. P. Raaijmakers, P. Hofman, et al. 2001. An improved breast irradiation technique using three-dimensional geometrical information and intensity modulation. *Radiother Oncol* 58:341–47.

van Santvoort, J. P. C., and B. J. M. Heijmen. 1996. Dynamic multileaf collimation without "tongue-and-groove" underdosage effects. *Phys Med Biol* 41:2091–105.

Veldeman, L., I. Madani, F. Hulstaert, G. De Meerleer, M. Mareel, and W. Neve De. 2008. Evidence behind use of intensity-modulated radiotherapy: A systematic review of comparative clinical studies. *Lancet Oncol* 9(4):367–75.

Wang, C., S. Luan, G. Tang, D. Z. Chen, M. Earl, and C. X. Yu. 2008. Arc-modulated radiation therapy (AMRT): A single-arc form of intensity-modulated arc therapy. *Phys Med Biol* 53(22):6291–303.

Webb, S., T. Bortfeld, J. Stein, et al. 1997. The effect of stair-step leaf transmission on the "tongue-and-groove" problem in dynamic radiotherapy with a multileaf collimator. *Phys Med Biol* 42:595–602.

Wong, E., D. P. D'Souza, J. Z. Chen, et al. 2005. Intensity-modulated arc therapy for treatment of high-risk endometrial malignancies. *Int J Radiat Oncol Biol Phys* 61:830–41.

Wong, E., J. Z. Chen, and J. Greenland. 2002. Intensity-modulated arc therapy simplified. *Int J Radiat Oncol Biol Phys* 53:222–235.

Woo, S. Y., W. H. Grant III, D. Bellezza, et al. 1996. A comparison of intensity modulated conformal therapy with a conventional external beam stereotactic radiosurgery system for the treatment of single and multiple intracranial lesions. *Int J Radiat Oncol Biol Phys* 35:593–7.

Yang, J. N., T. R. Mackie, P. Reckwerdt, J. O. Deasy, and B. R. Thomadsen. 1997. An investigation of tomotherapy beam delivery. *Med Phys* 24(3):425–36.

Yu, C. X, D. A. Jaffray, and J. W. Wong. 1998. Effects of intra-fraction organ motion on the delivery of dynamic intensity modulation. *Phys Med Biol* 43:91–104.

Yu, C. X. 1995. Intensity-modulated arc therapy with dynamic multileaf collimation: An alternative to tomotherapy. *Phys Med Biol* 40:1435–49.

Yu, C. X., and J. W. Wong, 1994. Dynamic photon intensity modulation. In *11th International Conference on the Use of Computers in Radiation Therapy*, ed. A. R Hounsell, J. M. Wilkinson, and P. C. Williams, 82–183. Manchester: North Western Medical Physics Department, Christie Hospital.

Yu, C. X., M. J. Symons, M. N. Du, et al. 1995. A method for implementing dynamic photon beam intensity modulation using independent jaws and multileaf collimator. *Phys Med Biol* 40:769–87.

Yu, C. X., X. A. Li, L. Ma, et al. 2002. Clinical implementation of intensity-modulated arc therapy. *Int J Radiat Oncol Biol Phys* 53(2):453–63.

Zelefsky, M. J., Z. Fuks, S. A. Leibel 2002. Intensity-modulated radiation therapy for prostate cancer. *Semin Radiat Oncol* 3:229–37.

Zhang, G., Z. Jiang, D. Shepard, M. Earl, and C. Yu. 2005. Effect of beamlet step-size on IMRT plan quality. *Med Phys* 32(11):3448–54.

Computed Tomography–Based Image-Guided Radiation Therapy Technology

Uwe Oelfke
DKFZ

Simeon Nill
DKFZ

9.1 Introduction

This chapter reviews the approaches for volumetric imaging of a patient in a treatment position, based on computed tomography (CT) technology. First, we briefly discuss the important role of volumetric imaging in the context of accurate patient setup and interfractional organ movements or deformations. Next, we explain the general differences of the major classes of imaging solutions—fan or cone beams available at either megavoltage (MV) or kilovoltage (kV) X-ray energies—before providing a detailed description of the most common technical solutions. The focus is clearly set on technical aspects while clinical applications are briefly mentioned only. Workflow issues, quality assurance procedures, and advanced imaging methods—like 4D cone-beam CT imaging—are mentioned while discussing the different modalities. Moreover, a report on some basic relationships of achievable image quality and required imaging dose for the considered technical systems is provided.

9.2 Role of Volumetric Imaging in IGRT and ART

Current modern radiotherapy approaches, like intensity-modulated radiation therapy (IMRT) or even therapy employing proton and heavy ion beams, aim to widen the therapeutic window by enhancing the dose conformity to predefined three-dimensional (3D) tumor geometries. Designing steep dose gradients between radiation targets and organs at risk compromises one of the major tasks of modern treatment planning, where an "ideal" dose delivery on an "ideal and static" patient geometry is assumed. However, prior to the advent of image-guided radiation therapy (IGRT), it was difficult to assess the quantitative accuracy between these planned ideal dose patterns and the ones that were actually delivered to the specific patient.

A whole set of geometric uncertainties in the chain of radiation therapy could be estimated only poorly, due to lack of appropriate geometrical data, documenting the patient anatomy in treatment position or even during treatment. First, and likely of primary importance for many clinical indications, is the patient- or, even more precise, the target-setup uncertainty prior to the irradiation. Assuming that outlined anatomical structures can be treated as rigid bodies, the daily location of these structures and their relative distances, for example, between planning target volumes (PTV) and organs at risk, are of utmost importance for designing a treatment plan. This knowledge about the actual irradiation geometry provides the required quantitative foundation for the design of clinical target volume to PTV margins (ICRU-50 1993; ICRU-62 1999). Second, the likelihood and impact of internal organ motion during treatment and their role in planning and treatment evaluation also need to be addressed. Third, elastic changes in anatomical structures, which can be caused by organ deformations due to variable mechanical tensions between tissues, reactions of tissues to the treatment, or weight changes of the patient during the course of treatment, have to be monitored and if necessary be corrected for by adaptive treatment strategies.

The most widespread technology to assess all these geometrical uncertainties quantitatively employs various approaches for acquiring CT data sets of the patient's anatomy in treatment position directly prior to treatment. Currently, four techniques are clinically available and used as the basis for a whole range of IGRT treatment scenarios: (1) linac-integrated kV cone-beam CTs (kV CBCTs), (2) linac-integrated MV cone-beam CTs (MV CBCTs), (3) spiral MV-fan-beam CTs, and (4) kV in-room CTs. We will review the main generic and technical features of these approaches and briefly discuss the issues of achievable image quality and required quality assurance procedures. Moreover, we will address the topic of time-resolved CT imaging. Many more details on these technologies can be found in the excellent review articles by Korreman et al. (2010), Dawson and Jaffray (2007), Jaffray (2005, 2007), Verellen et al. (2008), and Verellen et al. (2007).

9.3 General Issues in 3D-IGRT Imaging with X-Rays

There are two basic physical features that differentiate the available 3D CT imaging techniques for IGRT. These are the energy spectrum and the collimation of the X-ray beams employed for the acquisition of the CT projections. Before we describe in detail the commercially available technical solutions of volumetric CT imaging for IGRT, we will briefly discuss some of their general differences directly related to these features.

9.3.1 MV Imaging vs. kV Imaging

As imaging X-rays for IGRT one can use either the MV treatment beam or the kV spectrum of a linac-mounted diagnostic X-ray tube. The integration of the kV X-ray source and its oppositely mounted flat-panel detector requires additional QA procedures in comparison to the MV imaging approach. The most obvious is the calibration of the imaging geometry in relation to the treatment beam geometry (Bissonnette, Moseley, and Jaffray 2008).

Although the optimal MV imaging beams available refer to a bremsstrahlung spectrum of an electron beam of approximately 3.5 MeV, the well-known physical disadvantages of high-energy photon beams compared with standard kV-imaging spectra do occur. MV spectra are known to yield a reduced soft-tissue contrast resolution compared to kV spectra, if image projections are taken at comparable imaging doses (Groh et al. 2002). Moreover, an increased scatter component usually occurs due to the enhanced fraction of Compton interactions at MV energies. This effect, however, might be compensated for by the reduced signal, originating from primary photons in the kV case, especially observed for clinical indications where X-rays have to penetrate substantial layers of the tissue. The spectral response and quantum detection efficiency of the used detector systems are further factors that play a role in determining the achievable image quality per patient dose.

For MV-beam CTs, a superior image quality is naturally expected when metal implants of the patient are geometrically close to regions of interest, such that respective metal artifacts crucially deteriorate the clinically desired image quality (Hansen et al. 2006). It is well-known that more metal-penetrating MV rays lead to far less artifacts on CT images than comparable kV X-rays. This also facilitates the calibration of CT data to electron densities as required for treatment planning for MV-CTs.

9.3.2 Fan-Beam vs. Cone-Beam Imaging

The collimation of the X-ray imaging beam affects two major aspects of CT imaging. First, the occurrence of Compton scattered photons and the potential to reduce or remove the unwanted signals from individual CT projections of these systems are different. Fan-beam collimated imaging beams naturally lead to a reduced fraction of scattered photons compared to projections that were acquired with volumetric cone-beam projections. Moreover, the transmission signal of fan beams can be measured with line detectors. These can be far better shielded against unwanted signals from scattered photons than the flat-panel detectors required for cone-beam imaging. Consequently, cone-beam CT (CBCT) approaches mostly have to rely on scatter reduction algorithms in order to derive correct Hounsfield units for examining the patient's anatomy.

Another disadvantage inherent in the cone-beam approach is that the image reconstruction can only be achieved approximately mostly by various implementations of the Feldkamp–Davis–Kress (FDK) algorithm (Feldkamp, Davis, and Kress 1984). In comparison, fan-beam reconstructions of spiral projection geometries are known to be more accurate.

9.4 Technical Solutions: MV Imaging

This section discusses the technical details of the two available MV imaging systems. We focus on the technical features of the radiation source, the radiation detector, and necessary calibration issues.

9.4.1 MV-CT (Tomotherapy)

The tomotherapy unit is based on an idea from Rock Mackie, that is, combining a CT scanner with a linear accelerator (linac) on the same ring gantry. In contrast to the basic concept described in study by Mackie et al. (1993), the commercial product called HiArt II is not equipped with a kV X-ray tube but uses the treatment beam itself together with a xenon gas detector to acquire CT images of the patient in the treatment position.

Figure 9.1 shows a schematic of the tomotherapy gantry and Figure 9.2 shows the actual commercially available treatment unit. The radiation source positioned at 850 mm from the isocenter is a linear accelerator, which normally operates at 6 MV for the delivery of IMRT treatment plans. For the acquisition of a CT data set, the acceleration energy of the linac is reduced to generate a photon beam with reduced peak energy of 3.5 MV

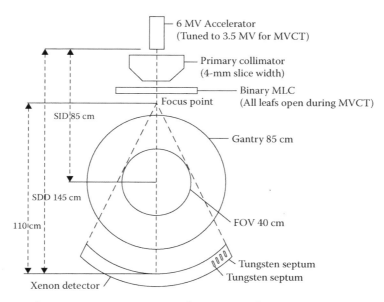

FIGURE 9.1 Schematic of the tomotherapy gantry. MVCT = megavoltage computed tomography; MLC = multileaf collimator; FOV = field of view; SID = source to isocenter distance; SDD = source to detector distance. (Courtesy of Tomotherapy Inc.)

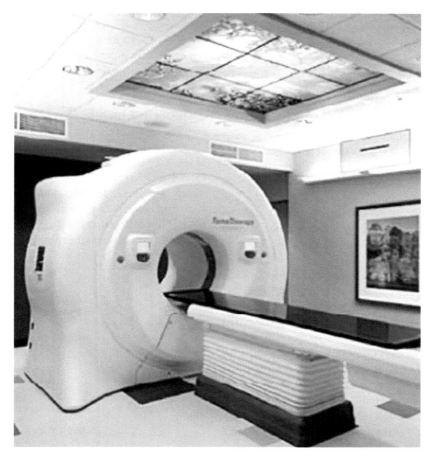

FIGURE 9.2 Tomotherapy unit: HiArt II. (Courtesy of Tomotherapy Inc.)

leading to a softer X-ray spectrum. The first beam modifier, the primary collimator, defines a 4-mm-thick slice at the isocenter plane. This collimator setting is predefined by the manufacturer and cannot be changed by the users. The binary multileaf collimator normally used to deliver the IMRT beams is fully opened for imaging to obtain a maximum field of view (FOV) of 400 mm. The transmitted radiation is detected by a modified arc-shaped CT xenon gas detector consisting of 1476 ionization cavities, each measuring 0.32 mm wide and 25.4 mm long and separated by tungsten septa of 0.32 mm wide.

To reduce the noise level of the readout signal, the charge produced within two adjacent cavities are combined to one channel leading to an effective detector size of 1.28 mm (Meeks et al. 2005). Due to the finite MV source size of a few millimeters, this approach does not lead to a reduced spatial resolution capability of the imaging system. It is important to note that the curvature of the gas detector corresponds to a circle with a diameter of 1100 mm, while the source-to-detector distance is 1420 mm. Primary photons are impinging the detector surface under a nonorthogonal angle and therefore interact not only with xenon gas but also with the tungsten septa. Most of the secondary electrons produced inside a cavity originate from photon interaction within the septa and are not created by xenon gas. The detector quantum efficiency using the modified energy spectrum compared to the treatment beam for these xenon gas detectors is approximately 20%.

The treatment beam and the imaging beam are produced by the same device, and therefore, no cross calibration between the imaging isocenter and the treatment isocenter is required in contrast to imaging systems using an additional kV source. Another advantage of the tomotherapy design is the rigid geometric relation between the imaging source and the imaging detector. Both systems are mounted on the ring gantry, and no additional correction for gantry or panel sag due to gravitational forces needs to be performed. The basic alignment of the treatment beam with the radiation isocenter is already checked during the standard QA procedures of the machine. A good overview of standard QA procedures can be found in a study by Fenwick et al. (2004).

To acquire a 3D CT, the system is continuously rotating at a fixed speed of 10 revolutions per minute and a user selectable table speed. Due to the continuous table feed, a very large volume can be scanned compared to other available linac-based imaging systems. The thin slice width of the system is very similar to a standard spiral fan-beam CT scanner. Consequently, all available CT reconstruction algorithms for such devices could in principle be applied for image reconstruction. The matrix size of the reconstructed 3D data sets is 512×512 with a voxel size of 0.78 mm and a slice thickness of 4 mm.

The image acquisition software allows the users to select between three table pitch settings of 1, 2, and 3 with a fixed slice width of 4 mm.

$$\text{pitch} = \frac{\text{table travel per rotation (mm)}}{\text{slice width (mm)}}$$

Typical patients' exposure values for three pitch settings are in the range of 10–20 mGy (Shah et al. 2008).

The tomotherapy concept has found a widespread clinical application now for many years. In recent years, a number of groups have reported about a wide range of clinical applications of IGRT (e.g., Broggi et al. 2009; Kapatoes et al. 2001; Kupelian et al. 2008; Sterzing, Herfarth, and Debus 2007; Engels et al. 2009). Also, first investigations to assess intrafraction organ motions were reported (Ngwa et al. 2009; Lu et al. 2009; Langen et al. 2008).

9.4.2 MV Cone-Beam CT

Most currently installed linear accelerators are equipped with a portal imaging device to perform 2D position verification of the patient's anatomy prior to the treatment. With the availability of amorphous flat-panel devices, commercial IGRT solutions are available that employ the treatment beam in combination with an amorphous flat-panel imager to acquire 3D MV cone-beam images (Pouliot et al. 2005; Sillanpaa et al. 2005).

The basic setup to perform MV CBCT is an amorphous silicon flat-panel detector attached to the linac on the opposite side of the treatment head at a distance of 1450 mm, for example, from the treatment source in a Siemens linac. The flat panel itself has a sensitive area of 409.6×409.6 mm and an imaging matrix of 1024×1024 with a resolution of 0.4×0.4 mm². For an imaging beam with a peak energy of 6 MV, Gd_2O_2S:Tb can be used as a scintillator material in combination with a 1-mm-buildup copper plate to improve the production of secondary electrons (Groh et al. 2002). The typical detection quantum efficiency (DQE) for such imaging systems is in the range of 2%–3%.

If the center of the detector is aligned with the central ray of the treatment beam (see Figure 9.3 left, linac gantry positioned at 180°) the maximum FOV that can be reconstructed using the FKD algorithm (Feldkamp, Davis, and Kress 1984) is approximately 270 mm in the lateral and longitudinal directions. This setup limits the FOV to an area close to the treatment isocenter, but it is impossible to reconstruct the complete treatment area of 400×400 mm² or even the complete outline of a patient's pelvis.

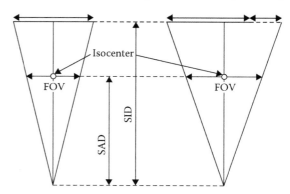

FIGURE 9.3 Extended FOV: the flat panel is shifted laterally to increase the FOV for CT image reconstruction. SAD = source to axis distance. (Reprinted from Tücking, T. 2007. *Development and Realization of the IGRT Inline Concept.* PhD thesis. Heidelberg: Universität Heidelberg. With permission.)

One possible improvement for this problem is to shift the panel laterally by at least 100 mm. With the detector in the offset position, the acquired projections are then preprocessed with a special imaging filter (Cho, Wu, and Hilal 1994) before the standard FDK reconstruction is executed. With this setup (called extended FOV), it is possible to increase the FOV to 400 mm in the lateral direction, but the it remains unchanged in the longitudinal direction.

Another technical challenge is the required synchronized readout of the panel and the delivery of the beam pulses (Pouliot et al. 2005). If this synchronization is lost and a beam pulse is delivered during the readout phase of the panel, interference patterns will appear on the 2D projections and the 3D reconstructed data set cannot be clinically used. Another issue is the reliable delivery of the beam with a very low monitor unit per beam pulse in the imaging mode compared to the treatment beam (Pouliot et al. 2005).

To acquire a MV CBCT, the linac needs to be rotated around the patient and a predefined number of projections is acquired. The range of gantry rotation angles necessary to reconstruct a 3D volume depends on the application mode. With the detector in the central position at least an interval of 180° plus the fan-beam angle (approximately 20°) must be scanned while the extended FOV acquisition always requires a complete gantry rotation. Typically the number of projections ranges between 200 and 360 per scan. The matrix size of the reconstructed 3D data sets can vary between 128×128 and 512×512 with isotropic voxel sizes between 2 and 0.5 mm. As an example of a commercially available solution, the Siemens MVision on the ARTISTE platform is shown in Figure 9.4.

During the gantry rotation around the patient, the position of the flat-panel imager position changes with respect to the central ray of the treatment beam due to the impact of gravitation and this issue can lead to significant artifacts in the reconstructed images. If the position variation of the flat panel is reproducible for every cone-beam scan, this effect can be compensated for in the CT reconstruction algorithm. For each gantry angle, the physical parameters characterizing the detector position need to be determined. One simple solution is the acquisition of 2D projections of a special calibration phantom for each gantry angle (Figure 9.5). Based on these images projection, matrices can be derived and directly be combined in the FKD backprojection algorithm (Ebert 2001; Pouliot et al. 2005; Yang et al. 2006; Robert et al. 2009). The validity of the projection matrices needs to be checked on a regular basis.

The typical dose for a MV cone-beam scan ranges from 20 mGy to 120 mGy depending on the anatomy the physician wants to see on the CB scan. For information about the position of bony structures, less dose is needed when compared to the dose required for soft-tissue visualization. Because the treatment beam is used during the cone-beam scan, it is possible to calculate the exact dose using the regular treatment planning system or take the delivered dose already into account during the optimization of the treatment plan (Morin et al. 2007).

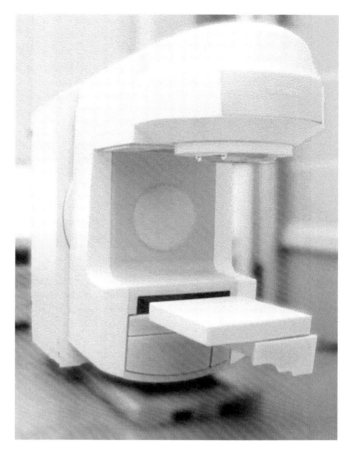

FIGURE 9.4 MV-CBCT solution from Siemens OCS: MVision. (Courtesy of Siemens Medical Solutions Oncology Care Systems.)

One potential argument against a MV cone-beam solution is also the additional dose delivered to the patient. This is a very complex issue, and more information on this topic can be found in the AAPM task group report 75 (Murphy et al. 2007). To reduce the dose delivered to the patient during a MV CBCT scan, Faddegon et al. (2008) proposed the replacement of the conventional bremsstrahlung target made of tungsten with a carbon graphite target. This measure is combined with a reduction of the peak energy of the imaging beam down to approximately to 4 MV. Further, the conventional beam flattening filter of the linac will be removed. These changes lead to a significantly softer X-ray spectrum in comparison to the 6 MV treatment beam (Figure 9.6). Using this setup, Beltran et al. (2009) reported that at the same dose level an increased contrast-to-noise ratio (CNR) of a factor of 2–3 compared to the standard treatment beam is observed. Flynn et al. (2009) showed that it is still possible to model this new imaging beam with a standard treatment planning system (TPS). Using an aluminum target instead of carbon, Robar et al. (2009) showed similar results.

Improving the photon spectrum of the imaging beam is only one option to enhance the image quality of MV cone beams. Some research groups are currently evaluating different detectors and scintillator materials to improve poor DQE of 2%–3% of

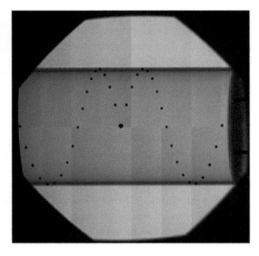

FIGURE 9.5 Left: Special calibration phantom to calculate the projection matrix for each gantry angle needed to correct for the mechanical changes during a gantry rotation. Right: 2D projection from one gantry angle. (Reprinted from Ebert, M. 2001. *Non-ideal Projection Data in X-Ray Computed Tomography*. PhD thesis. Mannheim: Universität Mannheim; Tücking, T. 2007. *Development and Realization of the IGRT Inline Concept*. PhD thesis. Heidelberg: Universität Heidelberg. With permission.)

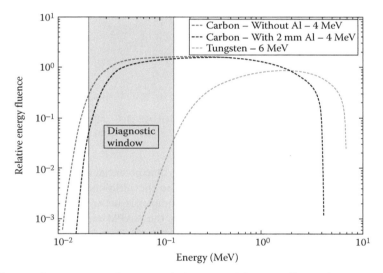

FIGURE 9.6 Relative energy fluence of 6 MeV TBL with tungsten (W) target and flattening filter and 4 MeV IBL with carbon target (C) without and with a 2-mm Al filter. Original data taken from the work by Faddegon et al. (2008) attenuated for an additional 2-mm layer of aluminum. (Reprinted from Steinke, M. 2010. *Performance Characteristics of a Novel Megavoltage Cone Beam Computed Tomography Device*. Master thesis. Heidelberg: University Heidelberg. With permission.)

currently commercially installed flat-panel devices. Using thick-segmented scintillator crystals made of BGO and CsI: Tl (Wang et al. 2009) or $CdWO_4$ (Kirvan et al. 2010) an increase in the DQE in the MV energy range of up to 25 times were observed. Similar results were also reported by Maltz and Hartmann et al. (2009) replacing the scintillator screen on an amorphous silicon panel with ultrafast ceramics detectors.

As for the tomotherapy unit, the MV cone beam is now used for widespread clinical applications. Most reports are dealing with either the evaluation of interfraction motion based on CT

image fusion (Morin et al. 2006; Pouliot 2007; Gayou and Miften 2007; Bylund et al. 2008; Koch et al. 2008) or the possibility to calculate the delivered dose by replacing the treatment planning CT with the current MV CBCT (Chen et al. 2006; Thomas et al. 2009; van Elmpt et al. 2010). As input for the recalculation of the delivered dose either the actual treatment plan or the measured portal images (pretreatment or during treatment) are used (van Elmpt et al. 2008; van Elmpt et al. 2010). Other authors are also evaluating the use of megavoltage computed tomography (MVCT) to monitor the tumor motion prior (Chang et al. 2006) or during the application of respiratory-based gated treatments (Tai et al. 2010) but also to help derive individualized treatment margins (Reitz et al. 2008).

Another new technical approach applied for kV-based (Dobbins and Godfrey 2003; Godfrey et al. 2006) and MV-based imaging (Pang and Rowlands 2005) is digital tomosynthesis. The difference compared to the standard cone beam approach is that only projections from a limited gantry angle in the range of 20–90° are acquired (Pang et al. 2008). For reconstruction, a modified FDK algorithm can be applied. Due to the restricted range of gantry angles, the acquisition takes less time. Applying this technique with the MV beam, an improved 2D beam's eye view with better spatial and contrast resolution can be achieved compared to a standard 2D MV port image. Maltz et al. (2009) used 52 nanocarbon tubes mounted below the treatment head in combination with the flat panel normally used to acquire the MV images to obtain tomosynthesis images simultaneously with the treatment. Real-time monitoring of tumor motion is a potential application of such imaging systems.

9.5 Technical Solutions: kV Imaging

A major advantage of volumetric imaging with kV-photons, enhanced soft-tissue contrast at small imaging doses has been realized with two technical solutions that have now been available for a number of years. While early approaches based on in-room CT solutions were developed already in the mid-1990s (Uematsu et al. 1996), the respective cone-beam technologies were established somewhat later by Jaffray and his coworkers (Jaffray et al. 2002; Jaffray et al. 1999). In Sections 9.5.1 and 9.5.2 we briefly review the basic features of both technologies.

9.5.1 In-Room CT

The idea of the in-room CT solution for volumetric imaging is simple: First, the required imaging device—a complete CT scanner designed for diagnostic imaging purposes—will be placed directly into the treatment room of the linac. Second, the CT scanner and the treatment machine are sharing one treatment couch carrying the patient in treatment position with all required positioning and immobilization devices. In order to minimize the required movement between the patient's position for imaging and treatment, the CT scanner is mounted on rails, such that it can slide over the patient for acquisition of CT projections.

Two different geometrical setups of linac and CT scanners are currently realized. For the original design, the y-axes of the CT-on-rails and the linac, according to International Electrotechnical Commission (IEC) convention, form an angle of 180°. Consequently, the patient is shifted from treatment position to imaging position and vice versa by a nonisocentric rotation of the treatment table by 180°. This specific geometry minimizes the required patient movement between treatment and imaging positions; however, it demands a larger-than-usual size of the linac bunker. The second setup geometry of CT scanner and linac allows for a reduced size

of the bunker by employing 90° between the two gantry rotation axes of the two devices. However, this technical solution naturally requires an additional 90° isocentric table rotation to move the patient from treatment to imaging position.

CT-on-rails technology was first developed in Japan (Uematsu et al. 1996; Uematsu et al. 1999) and was commercially available from three different vendors (Siemens OCS, Erlangen, Germany; VARIAN, Palo Alto, California; and Mitsubishi, Japan). Currently, the most widespread and commercially broadly offered solution is the CTVision solution from Siemens, which combines a diagnostic CT-scanner-on-rails (Siemens Somatom Emotion) with any of their standard linacs.

An installation of this device for a 90° geometry of linac and CT scanner is shown in Figure 9.7.

Many centers so far have gathered lots of experience with the in-room CT technique. Prominent examples include the programs at the MD Anderson Cancer Center in Houston, Texas (Court et al. 2003) and the Fox Chase Cancer Center in Philadelphia, Pennsylvania (Ma and Paskalev 2006). The main application is to identify and compensate for interfraction anatomical changes, mostly for prostate (Frank et al. 2010; Stillie et al. 2009; Knight et al. 2009; Wong et al. 2008) and head and neck cases (Barker et al. 2004; Zhang et al. 2006; Ahunbay et al. 2009; Cohen et al. 2010). Furthermore, applications for treating lung cancer were also reported (Wang et al. 2009). Without any doubt, the in-room CT solution for volumetric imaging is providing the best image quality achievable compared to all other CT-based IGRT techniques. This technology is clinically robust and utilizes more than 20 years of outstanding research and development work for diagnostic CT scanners. The calibration between Hounsfield units and relative electron densities as required for dose calculations in adaptive replanning adaptive radiation therapy (ART) strategies is well-established. Its fast image acquisition times reduce imaging artifacts due to organ motion. This IGRT modality therefore is an excellent tool for the management of intrafraction organ motion

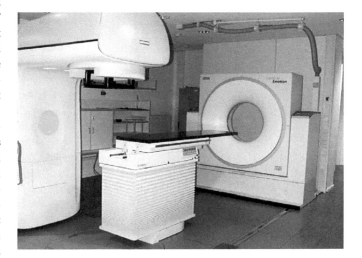

FIGURE 9.7 Example for CT-on-rails: SIEMENS CTVision with a PRIMUS accelerator (90° geometry).

and patient setup uncertainties. Moreover, it can also provide the best 4D CT images for assessing organ motion on a daily basis.

However, besides the additional bunker space, there are other practical concerns that prevent the widespread clinical use of in-room CT technology. The workflow for patient setup and eventual verification of a performed table shift requires additional movements of the patient couch. This increases the time required to perform the patient setup compared to linac-integrated CT solutions, leading to increased patient setup times. Furthermore, the registration of the in-room CT acquired images with the planning CT requires the use of additional external landmarks, visible in both images. This technology does not allow monitoring of the patient during the treatment.

9.5.2 kV Cone-Beam CT

Linac-integrated kV CBCT imaging is currently the most widespread and clinically used IGRT imaging modality. The integration of additional hardware components, a kV X-ray source together with an oppositely mounted flat-panel detector, into the treatment machine allows the acquisition of CT images in treatment position and even offers opportunities for monitoring the patient's anatomy during treatment. Performing the acquisition of CT projections with a linac gantry, whose rotation angle and rotational speed are restricted to 360° and one full rotation per minute respectively, leads to the selection of an electronic portal imaging device (EPID) as the detection system, because image acquisition of projections for a whole volume needs to be completed within minutes.

While the first attempts to develop such linac-integrated CT images derived from cone-beam projections focused on the utilization of the MV treatment beam (Hesse, Spies, and Groh 1998), the David Jaffray group pioneered the design and construction of kV CBCTs (Jaffray et al. 1999, 2002). Motivated by the advantageous imaging properties of kV X-rays compared to the MV spectrum of the treatment beam, the Jaffray group was the first to mount a kV X-ray source to a linac, solving many of the problems inherent in this technical approach to IGRT imaging. One of the early installations at William Beaumont hospital, a prototype of kV CBCT on an Elekta linac in 1999, is shown in Figure 9.8. In the remainder of this section, we will briefly discuss some technical features of the linac-integrated kV-CBCT approach, which are directly related to the achievable image quality with these devices.

A prerequisite for achieving artifact-free CT images is an accurate and stable imaging geometry, that is, the radiation source and the radiation detector have to be aligned perfectly for each rotation angle of the gantry. Unfortunately, a linac gantry cannot provide the same stability of the imaging geometry as a conventional gantry employed in diagnostic CT imaging. The gravitational sag of the linac gantry at rotation angles of 90° and 270° is especially noticeable. However, as long as the correlated deviations from the ideal imaging geometry are stable and reproducible in time, they can be accounted for by a specific calibration procedure (Ebert 2001; Cho et al. 2005; Pouliot et al. 2005; Yang et al. 2006; Robert et al. 2009).

The calibration procedure employs a specific phantom, for example, the cylindrical phantom shown in Figure 9.5. The surface of this phantom contains a set of radiopaque markers whose geometric position is exactly known. Comparing these marker positions with the reconstructed ones allows the determination of the difference between the assumed ideal imaging geometry and the actual one. This information is then used during the image reconstruction to avoid image artifacts due to a nonideal imaging geometry. As an example in Figure 9.9 we show images of a phantom with and without the proper geometric calibration.

The most common image acquisition modes are the short scan, where between 220 and 440 CT projections are acquired

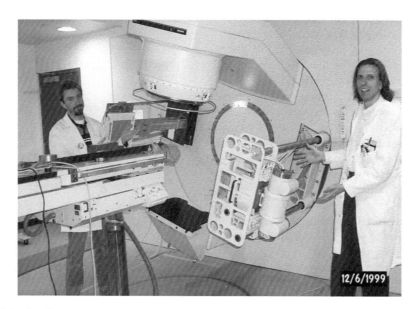

FIGURE 9.8 Prototype of the first linac-integrated kV cone-beam CT scanner at William Beamont hospital in 1999. (Courtesy of D. Jaffray (right) and B. Groh (left).)

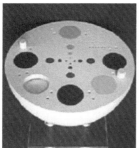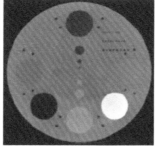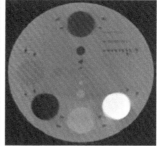

FIGURE 9.9 Central slice of a kV cone-beam CT taken for the contrast phantom on the left. The picture in the middle refers to the case where the projections were corrected for geometrical errors of the imaging geometry. The image on the right-hand side was reconstructed without geometrical calibration. The respective blurring artifacts are clearly visible.

for every degree or half-degree of gantry rotation. This protocol is often employed for head and neck cases. A full 360° scan with a number of projections ranging from 360 to 720 is usually performed for scans of the larger volumes of the abdomen. The CBCT image geometry restricts the lateral FOV to 25–27 cm, such that only the respective inner part of the patient anatomy can be reconstructed. However, this deficiency can be overcome by an extended FOV technique, in which the flat panel imager (FPI) is laterally shifted such that the FOV can be extended into the range between 40 and 50 cm. The respective data for the individual vendors can be found in a study by Korreman et al. (2010).

One specific problem of the kV-CBCT imaging technique is the introduction of a substantial amount of scattered photons (Endo et al. 2001). This effect is naturally less pronounced for smaller imaging volumes, for example for head and neck cases, but it is a severe effect for CT scans of larger volumes, for example, for CBCTs of prostate patients. The presence of scatter reduces image contrast, enhances quantum noise, and causes streaking and cupping artifacts in CBCT images. Obviously, the flat-panel detector of a CBCT cannot so easily be protected from scatter as line detectors in fan beam geometry. However, accurately aligned scatter grids in front of the detector are reported to reduce the scatter effectively (Endo et al. 2001; Letourneau et al. 2005). Moreover, the effect of missing tissue compensating bow-tie filters has been shown to improve the quality of CBCT images considerably (Graham et al. 2007; Ding et al. 2007a). The second widely used method to correct CBCT images for unwanted scatter is the application of scatter correction algorithms. A whole range of respective methods either phenomenological, or scatter kernel-based or Monte–Carlo-based algorithms have been devised (Spies et al. 2001; Siewerdsen et al. 2006; Jarry et al. 2006; Li et al. 2008; Maltz et al. 2008; Poludniowski et al. 2009; Reitz et al. 2009).

Besides the general improvement of visible image quality, these efforts naturally aim to obtain correct Hounsfield values in CBCT images, which is a prerequisite for all replanning strategies in ART. Respective calibration procedures have been investigated within recent years, and its impact on respective dose calculations has been studied (Yoo et al. 2006; Ding et al. 2007b; Yang et al. 2007; Petit et al. 2008; Guan et al. 2009).

Time-resolved volumetric imaging for linac-integrated cone-beam imaging has also been investigated by a number of research groups. Currently, no commercial product of this application is available yet. 4D CBCTs for the assessment of the actual breathing motion faces an additional challenge that the insufficient angular sampling of projections per breathing phase leads to additional streak artifacts and deteriorated the image quality. An example of this is shown in Figure 9.10. The development of 4D CBCTs started with breathing motion–correlated CBCTs (Sonke et al. 2005; Dietrich et al. 2006; Li et al. 2006; Purdie et al. 2006; Zijp, Sonke, and van Herk 2004), where a simultaneous surrogate signal or directly the diaphragm position in the projection images are employed to correlate the acquired projections with the correct breathing phase.

Image artifacts can be overcome by (1) decreasing the rotational speed of the gantry, for example, Lu et al. (2007) performed short scans of a 200°-gantry rotation in 3.3–6 minutes; or (2) by performing a motion-compensated 4D CBCT. This approach requires an estimation of the motion pattern during CBCT acquisition in the first step; this information is used within the reconstruction of the CT images in the second step. However, this method can be challenging in terms of image processing times (Li et al. 2007), which might be overcome by sophisticated a priori modeling of respiratory motion as reported in a study by Rit et al. (2009). Clinical applications of 4D CBCTs are rare but were reported as early as 2006 (Purdie et al. 2006).

All three major vendors of conventional linac-based high-precision radiotherapy have offered a kV cone-beam-CT–based IGRT solution. While VARIAN and ELEKTA have chosen to mount the imaging beam at an angle of 90° with respect to the therapy beam (as shown in Figure 9.11), SIEMENS has proposed an inline approach, in which the imaging and treatment beams are aligned along a common axis. This difference in geometrical setup is irrelevant for volumetric imaging prior to treatment; however, for further applications of monitoring intrafraction organ motion, the inline approach seems to offer certain advantages (Nill et al. 2005; Oelfke et al. 2006). However, one has to note that so far the inline approach introduced by SIEMENS only exists as a research version and is not commercially available yet.

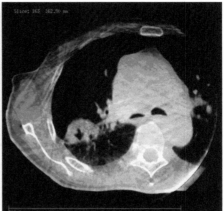
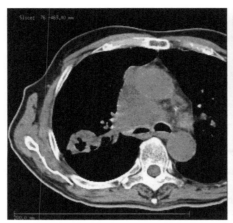

FIGURE 9.10   4D diagnostic CT in comparison to a 4D CBCT.

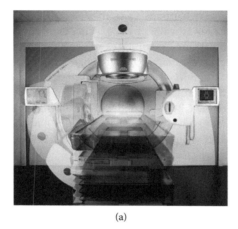
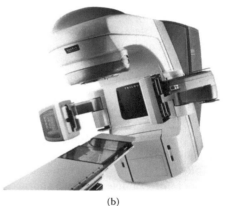

(a)   (b)

FIGURE 9.11   (a) Synergy system of ELEKTA (Courtesy of Elekta). (b) Trilogy system of VARIAN (Courtesy of Varian).

A detailed list of technical features provided by different vendors for their IGRT imaging solutions and some data characterizing the workflow for patient positioning in prostate and head and neck cases is provided in a study by Korreman et al. (2010).

9.6 Image Quality vs. Imaging Dose

As already pointed out in Section 9.3, the additional dose delivered to the patient, especially to the normal tissue, during the acquisition of the 3D data set is an important issue. This problem seems to be naturally more pronounced for MV-based imaging approaches. However, the problem of achievable image quality and required dose for a specific IGRT procedure is complex and can only be discussed here briefly. Interested readers are therefore referred to the AAPM Task Group Report 75 (Murphy et al. 2007). For any individual device, the statement, "The higher the dose the better the image quality" is generally true due to the statistical properties of photons interacting with matter. The signal-to-noise ratio (SNR) increases with the square root of the number of photons, whereas the dose delivered to the patient increases linearly with the number of primary photons.

An objective comparison of different imaging modalities based on patient data is almost impossible because the same patient cannot be scanned with all available imaging devices. Another important factor that impacts the image quality is the applied CT reconstruction algorithm, including reconstruction kernels and multiple preimage and postimage processing steps. To separate the impact of different elements in the image reconstruction chain, Stützel, Oelfke, and Nill (2008) performed a phantom study comparing four different imaging systems (standard CT scanner, kV cone beam, tomotherapy, and MV cone beam) using standardized image acquisition and reconstruction settings. For the same imaging dose, the image quality in terms of CNR, SNR, and spatial resolution was investigated. Naturally, the diagnostic CT scanner showed the best performance for the quality indicators. For phantoms with the maximum size of a head and neck patient, the kV cone-beam system performed the second best followed by the tomotherapy unit and the MV cone beam. For body-size phantoms, the difference in the image quality between the kV cone beam scanner and the tomotherapy is less pronounced. Figure 9.12 shows a transversal slice of the CatPhan 600 phantom for different imaging modalities.

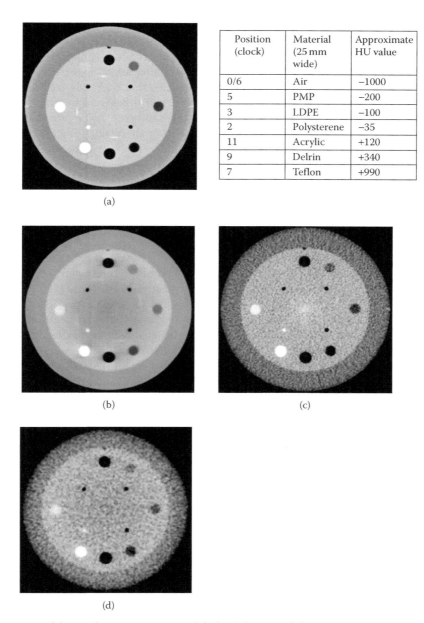

Position (clock)	Material (25 mm wide)	Approximate HU value
0/6	Air	−1000
5	PMP	−200
3	LDPE	−100
2	Polysterene	−35
11	Acrylic	+120
9	Delrin	+340
7	Teflon	+990

(a)

(b)

(c)

(d)

FIGURE 9.12 Visual impression of the Catphan 600 contrast module for different modalities. All scans are shown at identical dose and display settings. C = max, W = 800, $CTDI_w$ = 33 mGy, slice thickness = 4 mm, 3 mm for the Primatom scan. (a) Siemens Primatom scanner with H40s reconstruction kernel; (b) kV cone beam scan with 360 projections; (c) tomotherapy scan with a pitch of 1; (d) MV cone beam (360 projections).

Steinke (2010) extended this study by including a new MV imaging beam line with the softer photon spectrum. Using new MV imaging capabilities, a significant increase in image quality can be achieved compared to the standard MV setup. This finding is in good agreement with the data presented by other groups (Faddegon et al. 2008; Beltran et al. 2009).

Based on the ESTRO report (Korreman 2010), the typical dose delivered during the cone-beam image acquisition ranges from less than 1 cGy for a kV CBCT of the head and neck region up to 5 cGy for MV CBCT of pelvis. These values are based on measurements in water phantoms to derive the appropriate Computed Tomography Dose Index (CTDI) values. CTDI value is a valid indicator of the overall dose delivered to the patient but does not give any information about the spatial distribution of the imaging dose distribution (Downes et al. 2009). If multiple image acquisitions are performed during the course of treatment, the total dose delivered by the imaging part of the treatment is important to know especially for the normal tissue. Wen et al. (2007) therefore performed measurements with a thermoluminescence dosimeter measurements on the skin surface to determine the total dose delivered due to daily image guided patient setup. If the treatment beam is used to acquire a 3D cone beam, the already commissioned beam model in the TPS can be used to calculate 3D dose distribution. Generating a modified machine model within the

TPS, Flynn et al. (2009) showed that it is also possible to calculate the dose for the modified MV beam with the carbon target.

To calculate the dose for a kV cone-beam system, most publications are using some Monte Carlo framework (Downes et al. 2009; Spezi et al. 2009; Ding, Duggan, and Coffey 2008; Ding and Coffey 2009) to exactly predict the dose distribution for organs at risk, but Alaei, Ding, and Guan (2010) have presented a method in which they successfully modeled a kV source within a commercially available treatment planning system. The agreement between the calculated and measured distributions was between 0% and 19% with larger variations observed near or within bony structures. Better agreement between the calculated and the measured doses is achieved using the Monte Carlo tool boxes (Downes et al. 2009).

Ding and Coffey (2009) concluded based on the Monte Carlo dose calculation that if daily kV imaging scans are acquired the total dose to radiosensitive organs can be around 300 cGy and that the calculated 3D dose distribution provides crucial information for clinicians to make an informed decision concerning the additional imaging doses. Special attention should be paid to the dose to bones, which is two to four times higher than the dose to soft tissues (Ding, Duggan, and Coffey 2008; Ding and Coffey 2009).

9.7 Summary

Volumetric imaging of patients in a treatment position based on CT technology is one of the cornerstones of modern IGRT and ART. Uncertainties in patient positioning, the location of the internal anatomy of the patient at the time of the treatment, and organ movements are known to be crucial factors to ensuring an accurate dose delivery as specified by an optimized treatment plan. Volumetric imaging immediately prior to or even during treatment will become a prerequisite for any high-precision radiation therapy if one aims for increasingly more potent tumor dose, accomplished either by dose escalation, newly hyperfractionated treatment courses or the application of proton- and heavy-ion radiation therapy.

A wide range of volumetric CT technologies, currently clinically available in the treatment room, has been developed successfully in the last decade. Each of these devices, from the simple idea of locating a diagnostic CT scanner next to the linac so that scanner and treatment machine share a common treatment couch, to the utilization of the treatment beam and an EPID as an imaging system, to the development of the tomotherapy system or linac-integrated kV CBCT technology, provides us now with new challenges and opportunities commonly labeled as ART.

In Sections 9.4 through 9.6, we briefly discussed and introduced these main integrated imaging or therapy systems. Their mutual differences and individual strengths were described, and we provided a wide range of related literature for interested reader. Technical development, with respect to monitoring and compensating for either setup errors or interfraction anatomical changes of the patient with X-rays, has reached a mature state

and seems to be complete. Dynamic monitoring and compensation for intrafraction organ motion based on 3D imaging is still a very challenging task. For X-ray imaging, a first step in this direction might be the development of the tomosynthesis approaches that we briefly mentioned (Pang et al. 2008; Maltz et al. 2009). Another upcoming modality, capable of providing these images, may be provided by the various concepts for integrating MR imaging and a treatment machine (Raaymakers et al. 2009; Fallone et al. 2009).

References

Ahunbay, E. E., C. Peng, A. Godley, C. Schultz, and X. A. Li. 2009. An on-line replanning method for head and neck adaptive radiotherapy. *Med Phys* 36(10):4776–90.

Alaei, P., G. Ding, and H. Guan. 2010. Inclusion of the dose from kilovoltage cone beam CT in the radiation therapy treatment plans. *Med Phys* 37(1):244–8.

Barker Jr., J. L., A. S. Garden, K. K. Ang, et al. 2004. Quantification of volumetric and geometric changes occurring during fractionated radiotherapy for head-and-neck cancer using an integrated CT/linear accelerator system. *Int J Radiat Oncol Biol Phys* 59(4):960–70.

Beltran, C., R. Lukose, B. Gangadharan, A. Bani-Hashemi, and B. A. Faddegon. 2009. Image quality & dosimetric property of an investigational imaging beam line MV-CBCT. *J Appl Clin Med Phys* 10(3):3023.

Bissonnette, J. P., D. J. Moseley, and D. A. Jaffray. 2008. A quality assurance program for image quality of cone-beam CT guidance in radiation therapy. *Med Phys* 35(5):1807–15.

Broggi, S., C. Cozzarini, C. Fiorino, et al. 2009. Modeling set-up error by daily MVCT for prostate adjuvant treatment delivered in 20 fractions: Implications for the assessment of the optimal correction strategies. *Radiother Oncol* 93(2):246–52.

Bylund, K. C., J. E. Bayouth, M. C. Smith, A. C. Hass, S. K. Bhatia, and J. M. Buatti. 2008. Analysis of interfraction prostate motion using megavoltage cone beam computed tomography. *Int J Radiat Oncol Biol Phys* 72(3):949–56.

Chang, J., J. Sillanpaa, C. C. Ling, et al. 2006. Integrating respiratory gating into a megavoltage cone-beam CT system. *Med Phys* 33(7):2354–61.

Chen, J., O. Morin, M. Aubin, M. K. Bucci, C. F. Chuang, and J. Pouliot. 2006. Dose-guided radiation therapy with megavoltage cone-beam CT. *Br J Radiol* 79 Spec No 1:S87–98.

Cho, Y., D. J. Moseley, J. H. Siewerdsen, D. A. Jaffray. 2005. Accurate technique for complete geometric calibration of cone-beam computed tomography systems. *Med Phys* 32(4):968–83.

Cho, Z. H., E. X. Wu, and S. K. Hilal. 1994. Weighted backprojection approach to cone beam 3D projection reconstruction for truncated spherical detection geometry. *IEEE Trans Med Imaging* 13(1):110–21.

Cohen, R. J., K. Paskalev, S. Litwin, R. A. Price Jr., S. J. Feigenberg, and A. A. Konski. 2010. Esophageal motion during radiotherapy: quantification and margin implications. *Dis Esophagus* 23(6):473–9.

Court, L., I. Rosen, R. Mohan, and L. Dong. 2003. Evaluation of mechanical precision and alignment uncertainties for an integrated CT/LINAC system. *Med Phys* 30(6):1198–210.

Dawson, L. A., and D. A. Jaffray. 2007. Advances in image-guided radiation therapy. *J Clin Oncol* 25(8):938–46.

Dietrich, L., S. Jetter, T. Tücking, S. Nill, and U. Oelfke. 2006. Linac integrated 4D cone beam CT: First experimental results. *Phys Med Biol* 51:2939–52.

Ding, G. X., and C. W. Coffey. 2009. Radiation dose from kilovoltage cone beam computed tomography in an image-guided radiotherapy procedure. *Int J Radiat Oncol Biol Phys* 73(2):610–7.

Ding, G. X., D. M. Duggan, and C. W. Coffey. 2008. Accurate patient dosimetry of kilovoltage cone-beam CT in radiation therapy. *Med Phys* 35(3):1135–44.

Ding, G. X., D. M. Duggan, and C. W. Coffey. 2007a. Characteristics of kilovoltage x-ray beams used for cone-beam computed tomography in radiation therapy. *Phys Med Biol* 52(6):1595–615.

Ding, G. X., D. M. Duggan, C. W. Coffey, M. Deeley, D. E. Hallahan, A. Cmelak, and A. Malcolm. 2007b. A study on adaptive IMRT treatment planning using kV cone-beam CT. *Radiother Oncol* 85(1):116–25.

Dobbins III, J. T., and D. J. Godfrey. 2003. Digital x-ray tomosynthesis: Current state of the art and clinical potential. *Phys Med Biol* 48(19):R65–106.

Downes, P., R. Jarvis, E. Radu, I. Kawrakow, and E. Spezi. 2009. Monte Carlo simulation and patient dosimetry for a kilovoltage cone-beam CT unit. *Med Phys* 36(9):4156–67.

Ebert, M. 2001. *Non-Ideal Projection Data in X-Ray Computed Tomography.* PhD thesis. Mannheim: Universität Mannheim.

Endo, M., T. Tsunoo, N. Nakamori, and K. Yoshida. 2001. Effect of scattered radiation on image noise in cone beam CT. *Med Phys* 28(4):469–74.

Engels, B., M. De Ridder, K. Tournel, et al. 2009. Preoperative helical tomotherapy and megavoltage computed tomography for rectal cancer: Impact on the irradiated volume of small bowel. *Int J Radiat Oncol Biol Phys* 74(5):1476–80.

Faddegon, B. F., V. Gangadharan, B. Wu, J. Pouliot, and A. Bani-Hashemi. 2008. Low dose megavoltage cone beam CT with an unflattened 4 MV beam from a carbon target. *Med Phys* 35(12):5777–86.

Fallone, B. G., B. Murray, S. Rathee, et al. 2009. First MR images obtained during megavoltage photon irradiation from a prototype integrated linac-MR system. *Med Phys* 36(6):2084–8.

Feldkamp, L. A., L. C. Davis, and J. W. Kress. 1984. Practical cone-beam algorithm. *J Opt Soc Am A* 6:612–9.

Fenwick, J. D., W. A. Tome, H. A. Jaradat, et al. 2004. Quality assurance of a helical tomotherapy machine. *Phys Med Biol* 49(13):2933–53.

Flynn, R. T., J. Hartmann, A. Bani-Hashemi, et al. 2009. Dosimetric characterization and application of an imaging beam line with a carbon electron target for megavoltage cone beam computed tomography. *Med Phys* 36(6):2181–92.

Frank, S. J., R. J. Kudchadker, D. A. Kuban, et al. 2010. A volumetric trend analysis of the prostate and seminal vesicles during a course of intensity-modulated radiation therapy. *Am J Clin Oncol* 33(2):173–5.

Gayou, O., and M. Miften. 2007. Commissioning and clinical implementation of a mega-voltage cone beam CT system for treatment localization. *Med Phys* 34(8):3183–92.

Godfrey, D. J., F. F. Yin, M. Oldham, S. Yoo, and C. Willett. 2006. Digital tomosynthesis with an on-board kilovoltage imaging device. *Int J Radiat Oncol Biol Phys* 65(1):8–15.

Graham, S. A., D. J. Moseley, J. H. Siewerdsen, and D. A. Jaffray. 2007. Compensators for dose and scatter management in cone-beam computed tomography. *Med Phys* 34(7):2691–703.

Groh, B. A., J. H. Siewerdsen, D. G. Drake, J. W. Wong, and D. A. Jaffray. 2002. A performance comparison of flat-panel imager-based MV and kV cone-beam CT. *Med Phys* 29(6):967–75.

Guan, H., and H. Dong. 2009. Dose calculation accuracy using cone-beam CT (CBCT) for pelvic adaptive radiotherapy. *Phys Med Biol* 54(20):6239–50.

Hansen, E. K., D. A. Larson, M. Aubin, et al. 2006. Image-guided radiotherapy using megavoltage cone-beam computed tomography for treatment of paraspinous tumors in the presence of orthopedic hardware. *Int J Radiat Oncol Biol Phys* 66(2):323–6.

Hesse, B. M., L. Spies, and B. A. Groh. 1998. Tomotherapeutic portal imaging for radiation treatment verification. *Phys Med Biol* 43(12):3607–16.

ICRU-50. 1993. International Commission on Radiation Units and Measurements. ICRU Report 50: Prescribing, recording, and reporting photon beam therapy. Bethesda, MD: International Commission on Radiation Units and Measurement 3-16.

ICRU-62. 1999. International Commission on Radiation Units and Measurements. ICRU Report 62: Prescribing, recording, and reporting photon beam therapy. Bethesda, MD: International Commission on Radiation Units and Measurement 3-20.

Jaffray, D. A. 2005. Emergent technologies for 3-dimensional image-guided radiation delivery. *Semin Radiat Oncol* 15(3): 208–16.

Jaffray, D. A. 2007. Kilovoltage volumetric imaging in the treatment room. *Front Radiat Ther Oncol* 40:116–31.

Jaffray, D. A., D. G. Drake, M. Moreau, A. A. Martinez, and J. W. Wong. 1999. A radiographic antomographic imaging system integrated into a medical linear accelerator for localization of bone and soft-tissue targets. *Int J Radiat Oncol Biol Phys* 45(3):773–89.

Jaffray, D. A., J. H. Siewerdsen, J. W. Wong, and A. A. Martinez. 2002. Flat-panel cone-beam computed tomography for image-guided radiation therapy. *Int J Radiat Oncol Biol Phys* 53(5):1337–49.

Jarry, G., S. A. Graham, D. J. Moseley, D. J. Jaffray, J. H. Siewerdsen, and F. Verhaegen. 2006. Characterization of scattered radiation in kV CBCT images using Monte Carlo simulations. *Med Phys* 33(11):4320–9.

Kapatoes, J. M., G. H. Olivera, K. J. Ruchala, J. B. Smilowitz, P. J. Reckwerdt, and T. R. Mackie. 2001. A feasible method for clinical delivery verification and dose reconstruction in tomotherapy. *Med Phys* 28(4):528–42.

Kirvan, P. F., T. T. Monajemi, B. G. Fallone, and S. Rathee. 2010. Performance characterization of a MVCT scanner using multislice thick, segmented cadmium tungstate-photodiode detectors. *Med Phys* 37(1):249–57.

Knight, K., N. Touma, L. Zhu, G. M. Duchesne, and J. Cox. 2009. Implementation of daily image-guided radiation therapy using an in-room CT scanner for prostate cancer isocentre localization. *J Med Imaging Radiat Oncol* 53(1):132–8.

Koch, M., J. S. Maltz, S. J. Belongie, et al. 2008. Automatic coregistration of volumetric images based on implanted fiducial markers. *Med Phys* 35(10):4513–23.

Korreman, S., C. Rasch, H. McNair, et al. 2010. The European Society of Therapeutic Radiology and Oncology-European Institute of Radiotherapy (ESTRO-EIR) report on 3D CT-based in-room image guidance systems: A practical and technical review and guide. *Radiother Oncol* 94(2):129–44.

Kupelian, P. A., C. Lee, K. M. Langen, et al. 2008. Evaluation of image-guidance strategies in the treatment of localized prostate cancer. *Int J Radiat Oncol Biol Phys* 70(4):1151–7.

Langen, K. M., W. Lu, W. Ngwa, et al. 2008. Correlation between dosimetric effect and intrafraction motion during prostate treatments delivered with helical tomotherapy. *Phys Med Biol* 53(24):7073–86.

Létourneau, D., J. W. Wong, M. Oldham, M. Gulam, L. Watt, D. A. Jaffray, J. H. Siewerdsen, and A. A. Martinez. 2005. Cone-beam-CT guided radiation therapy: technical implementation. *Radiother Oncol* 75(3):279–86.

Li, H., R. Mohan, and X. R. Zhu. 2008. Scatter kernel estimation with an edge-spread function method for cone-beam computed tomography imaging. *Phys Med Biol* 53(23):6729–48.

Li, T., A. Koong, and L. Xing. 2007. Enhanced 4D cone-beam CT with inter-phase motion model. *Med Phys* 34(9):3688–95.

Li, T., L. Xing, P. Munro, C. McGuinness, M. Chao, Y. Yang, B. Loo, and A. Koong. 2006. Four-dimensional cone-beam computed tomography using an on-board imager. *Med Phys* 33(10):3825–33.

Lu, J., T. M. Guerrero, P. Munro, A. Jeung, P. C. Chi, P. Balter, X. R. Zhu, R. Mohan, and T. Pan. 2007. Four-dimensional cone beam CT with adaptive gantry rotation and adaptive data sampling. *Med Phys* 34(9):3520–9.

Lu, W., M. Chen, K. J. Ruchala, et al. 2009. Real-time motion-adaptive-optimization (MAO) in TomoTherapy. *Phys Med Biol* 54(14):4373–98.

Ma, C. M., and K. Paskalev. 2006. In-room CT techniques for image-guided radiation therapy. *Med Dosim* 31(1):30–9.

Mackie, T. R., T. W. Holmes, S. Swerdloff, et al. 1993. Tomotherapy: A new concept in the delivery of dynamic conformal radiotherapy. *Med Phys* 20(6):1709–19.

Maltz, J., J. Hartmann, A. Dubouloz, et al. 2009. Thick monolithic pixelated scintillator array for megavoltage imaging. *Med Phys* 36(6):2818–9.

Maltz, J. S., B. Gangadharan, M. Vidal, A. Paidi, S. Bose, B. A. Faddegon, M. Aubin, O. Morin, J. Pouliot, Z. Zheng, M. M. Svatos, and A. R. Bani-Hashemi. 2008. Focused beam-stop array for the measurement of scatter in megavoltage portal and cone beam CT imaging. *Med Phys* 35(6):2452–62.

Maltz, J. S., F. Sprenger, J. Fuerst, A. Paidi, F. Fadler, and A. R. Bani-Hashemi. 2009. Fixed gantry tomosynthesis system for radiation therapy image guidance based on a multiple source x-ray tube with carbon nanotube cathodes. *Med Phys* 36(5):1624–36.

Meeks, S. L., J. F. Harmon Jr., K. M. Langen, T. R. Willoughby, T. H. Wagner, and P. A. Kupelian. 2005. Performance characterization of megavoltage computed tomography imaging on a helical tomotherapy unit. *Med Phys* 32(8):2673–81.

Morin, O., A. Gillis, J. Chen, et al. 2006. Megavoltage cone-beam CT: System description and clinical applications. *Med Dosim* 31(1):51–61.

Morin, O., A. Gillis, M. Descovich, et al. 2007. Patient dose considerations for routine megavoltage cone-beam CT imaging. *Med Phys* 34(5):1819–27.

Murphy, M. J., J. Balter, S. Balter, et al. 2007. The management of imaging dose during image-guided radiotherapy: Report of the AAPM Task Group 75. *Med Phys* 34(10):4041–63.

Nill, S., J. Unkelbach, L. Dietrich, and U. Oelfke. 2005. Online correction for respiratory motion: Evaluation of two different imaging geometries. *Phys Med Biol* 50(17):4087–96.

Ngwa, W., S. L. Meeks, P. A. Kupelian, E. Schnarr, and K. M. Langen. 2009. Validation of a computational method for assessing the impact of intra-fraction motion on helical tomotherapy plans. *Phys Med Biol* 54(21):6611–21.

Oelfke, U., T. Tücking, S. Nill, et al. 2006. Linac-integrated kV-cone beam CT: Technical features and first applications. *Med Dosim* 31(1):62–70.

Pang, G., A. Bani-Hashemi, P. Au, et al. 2008. Megavoltage cone beam digital tomosynthesis (MV-CBDT) for image-guided radiotherapy: A clinical investigational system. *Phys Med Biol* 53(4):999–1013.

Pang, G., and J. A. Rowlands. 2005. Just-in-time tomography (JiTT): A new concept for image-guided radiation therapy. Just-in-time tomography (JiTT): A new concept for image-guided radiation therapy. *Phys Med Biol* 50(21):N323–30.

Petit, S. F., W. J. van Elmpt, S. M. Nijsten, P. Lambin, and A. L. Dekker. 2008. Calibration of megavoltage cone-beam CT for radiotherapy dose calculations: correction of cupping artifacts and conversion of CT numbers to electron density. *Med Phys* 35(3):849–65.

Poludniowski, G., P. M. Evans, V. N. Hansen, and S. Webb. 2009. An efficient Monte Carlo-based algorithm for scatter correction in keV cone-beam CT. *Phys Med Biol* 54(12):3847–64.

Pouliot, J. 2007. Megavoltage imaging, megavoltage cone beam CT and dose-guided radiation therapy. *Front Radiat Ther Oncol* 40:132–42.

Pouliot, J., A. Bani-Hashemi, J. Chen, et al. 2005. Low-dose megavoltage cone-beam CT for radiation therapy. *Int J Radiat Oncol Biol Phys* 61(2):552–60.

Purdie, T. G., D. J. Moseley, J. P. Bissonnette, M. B. Sharpe, K. Franks, A. Bezjak, and D. A. Jaffray. 2006. Respiration correlated cone-beam computed tomography and 4DCT for evaluating target motion in Stereotactic Lung Radiation Therapy. *Acta Oncol* 45(7):915–22.

Raaymakers, B. W., J. J. Lagendijk, J. Overweg, et al. 2009. Integrating a 1.5 T MRI scanner with a 6 MV accelerator: Proof of concept. *Phys Med Biol* 54(12):N229–37.

Reitz, B., O. Gayou, D. S. Parda, and M. Miften. 2008. Monitoring tumor motion with on-line mega-voltage cone-beam computed tomography imaging in a cine mode. *Phys Med Biol* 53(4):823–36.

Reitz, I., B. M. Hesse, S. Nill, T. Tücking, and U. Oelfke. 2009. Enhancement of image quality with a fast iterative scatter and beam hardening correction method for kV CBCT. *Z Med Phys* 19(3):158–72.

Rit, S., J. W. Wolthaus, M. van Herk, and J. J. Sonke. 2009. On-the-fly motion-compensated cone-beam CT using an a priori model of the respiratory motion. *Med Phys* 36(6):2283–96.

Robar, J. L., T. Connell, W. Huang, and R. G. Kelly. 2009. Megavoltage planar and cone-beam imaging with low-Z targets: Dependence of image quality improvement on beam energy and patient separation. *Med Phys* 36(9):3955–63.

Robert, N., K. N. Watt, X. Wang, and J. G. Mainprize. 2009. The geometric calibration of cone-beam systems with arbitrary geometry. *Phys Med Biol* 54(24):7239–61.

Shah, A., K. Langen, K. Ruchala, A. Cox, P. A. Kupelian, and S. L. Meeks. 2008. Patient-specific dose from megavoltage CT imaging with a helical tomotherapy unit. *Int J Radiat Oncol Biol Phys* 70(5):1579–87.

Siewerdsen, J. H., M. J. Daly, B. Bakhtiar, D. J. Moseley, S. Richard, H. Keller, and D. A. Jaffray. 2006. A simple, direct method for x-ray scatter estimation and correction in digital radiography and cone-beam CT. *Med Phys* 33(1):187–97.

Sillanpaa, J., J. Chang, G. Mageras, et al. 2005. Developments in megavoltage cone beam CT with an amorphous silicon EPID: Reduction of exposure and synchronization with respiratory gating. *Med Phys* 32(3):819–29.

Sonke, J. J., L. Zijp, P. Remeijer, and M. van Herk. 2005. Respiratory correlated cone beam CT. *Med Phys* 32(4):1176–86.

Spezi, E., P. Downes, E. Radu, and R. Jarvis. 2009. Monte Carlo simulation of an x-ray volume imaging cone beam CT unit. *Med Phys* 36(1):127–36.

Spies, L., M. Ebert, B. A. Groh, B. M. Hesse, and T. Bortfeld. Correction of scatter in megavoltage cone-beam CT. 2001. *Phys Med Biol* 46(3):821–33.

Steinke, M. 2010. Performance characteristics of a novel megavoltage cone beam computed tomography device. Master's thesis. Heidelberg: University Heidelberg.

Sterzing, F., K. Herfarth, and J. Debus. 2007. IGRT with helical tomotherapy—effort and benefit in clinical routine. *Strahlenther Onkol* 183 Spec No 2:35–7.

Stillie, A. L., T. Kron, C. Fox, A. Herschtal, A. Haworth, A. Thompson, R. Owen, K. H. Tai, G. Duchesne, and F. Foroudi. 2009. Rectal filling at planning does not predict stability of the prostate gland during a course of radical radiotherapy if patients with large rectal filling are re-imaged. *Clin Oncol (R Coll Radiol)* 21(10):760–7.

Stützel, J., U. Oelfke, and S. Nill. 2008. A quantitative image quality comparison of four different image guided radiotherapy devices. *Radiother Oncol* 86(1):20–4.

Tai, A., J. D. Christensen, E. Gore, A. Khamene, T. Boettger, and X. A. Li. 2010. Gated treatment delivery verification with on-line megavoltage fluoroscopy. *Int J Radiat Oncol Biol Phys* 76(5):1592–8.

Thomas, T. H., D. Devakumar, S. Purnima, and B. P. Ravindran. 2009. The adaptation of megavoltage cone beam CT for use in standard radiotherapy treatment planning. *Phys Med Biol* 54(7):2067–77.

Tücking, T. 2007. Development and realization of the IGRT inline concept. PhD thesis. Heidelberg: Universität Heidelberg.

Uematsu, M., T. Fukui, A. Shioda, et al. 1996. A dual computed tomography linear accelerator unit for stereotactic radiation therapy: A new approach without cranially fixated stereotactic frames. *Int J Radiat Oncol Biol Phys* 35(3):587–92.

Uematsu, M., M. Sonderegger, A. Shioda, et al. 1999. Daily positioning accuracy of frameless stereotactic radiation therapy with a fusion of computed tomography and linear accelerator (focal) unit: Evaluation of z-axis with a z-marker. *Radiother Oncol* 50(3):337–9.

van Elmpt, W., L. McDermott, S. Nijsten, M. Wendling, P. Lambin, and B. Mijnheer. 2008. A literature review of electronic portal imaging for radiotherapy dosimetry. *Radiother Oncol* 88(3):289–309.

van Elmpt, W., S. Petit, D. De Ruysscher, P. Lambin, and A. Dekker. 2010. 3D dose delivery verification using repeated cone-beam imaging and EPID dosimetry for stereotactic body radiotherapy of non-small cell lung cancer. *Radiother Oncol* doi:10.1016/j.radonc.2009.12.024

Verellen, D., M. D. Ridder, N. Linthout, K. Tournel, G. Soete, and G. Storme. 2007. Innovations in image-guided radiotherapy. *Nat Rev Cancer* 7(12):949–60.

Verellen, D., M. De Ridder, K. Tournel, et al. 2008. An overview of volumetric imaging technologies and their quality assurance for IGRT. *Acta Oncol* 47(7):1271–8.

Wang, Y., L. E. Antonuk, Q. Zhao, Y. El-Mohri, and L. Perna. 2009. High-DQE EPIDs based on thick, segmented BGO and CsI:Tl scintillators: Performance evaluation at extremely low dose. *Med Phys* 36(12):5707–18.

Wong, J. R., Z. Gao, S. Merrick, P. Wilson, M. Uematsu, K. Woo, C. W. Cheng. 2009. Potential for higher treatment failure in obese patients: correlation of elevated body mass index and increased daily prostate deviations from the radiation beam isocenters in an analysis of 1,465 computed tomographic images. *Int J Radiat Oncol Biol Phys* 75(1):49–55.

Wen, N., H. Guan, R. Hammoud, et al. 2007. Dose delivered from Varian's CBCT to patients receiving IMRT for prostate cancer. *Phys Med Biol* 52(8):2267–76.

Yang, K., A. L. Kwan, D. F. Miller, and J. M. Boone. 2006. A geometric calibration method for cone beam CT systems. *Med Phys* 33(6):1695–706.

Yang, Y., E. Schreibmann, T. Li, C. Wang, and L. Xing. 2007. Evaluation of on-board kV cone beam CT (CBCT)-based dose calculation. *Phys Med Biol* 52(3):685–705.

Yoo, S., and F. F. Yin. 2006. Dosimetric feasibility of cone-beam CT-based treatment planning compared to CT-based treatment planning. *Int J Radiat Oncol Biol Phys* 66(5):1553–61.

Zhang, L., A. S. Garden, J. Lo, K. K. Ang, A. Ahamad, W. H. Morrison, D. I. Rosenthal, M. S. Chambers, X. R. Zhu, R. Mohan, and L. Dong. 2006. Multiple regions-of-interest analysis of setup uncertainties for head-and-neck cancer radiotherapy. *Int J Radiat Oncol Biol Phys* 64(5):1559–69.

Zijp, L., J.-J. Sonke, and M. van Herk. 2004. Extraction of the respiratory signal from sequential thorax cone-beam x-ray images. *Proceedings of the 14th ICCR*, Seoul, Korea, 507–509.

Non-Computed Tomography–Based Image-Guided Radiation Therapy Technology

David Shepard
Swedish Cancer Institute

Daliang Cao
Swedish Cancer Institute

Jianzhou Wu
Swedish Cancer Institute

Jian-Yue Jin
Henry Ford Hospital System

Chad Lee
CK Solutions

Douglas Miller
Washington University

10.1 Introduction

Image-guided radiation therapy (IGRT) is a technique for improving the accuracy and precision of radiation therapy through the frequent acquisition of images in the treatment room (Dawson and Jaffray 2007; Balter and Cao 2007; Jaffray et al. 2007; Jaffray 2007; Xing et al. 2006). The IGRT is a critical tool in that it delivers highly conformal radiation therapy treatment plans that maximize the administered dose to the tumor while minimizing the dose to surrounding sensitive structures. Image-guided technologies assist us by providing accurate patient localization, distinct tumor visualization, and target volume verification.

Cone-beam computed tomography (CT) or CBCT, described in detail in Chapter 2, has become a widely adopted IGRT technology and has proven to be a valuable clinical tool. There are, however, costs involved in the adoption of CBCT in routine clinical practice, such as the following:

- CBCT increases the total treatment time.
- It exposes the patient to additional ionizing radiation.
- It provides only a snapshot of the patient position and may not reflect positioning throughout the treatment delivery.
- It requires active physician involvement in order to derive the maximum benefit from the acquired data.

In this chapter, we describe non-CT-based IGRT technologies. These technologies can serve as an alternative or as a complement to the use of CBCT and can address some of its weaknesses. The technologies described in detail in this chapter include electronic portal imaging (Section 10.2), ultrasound (Section 10.3), electromagnetic (Section 10.4), surface mapping (Section 10.5), and implanted radioactive markers (Section 10.5).

10.2 Electronic Portal Imaging Technology

10.2.1 Introduction to Electronic Portal Imaging

Flat-screen computer monitors and televisions have seen rapid technological improvements in recent years. Researchers in digital radiography have built upon these advances in the development of flat-panel image receptors. The underlying technology of flat-panel imagers is a large-area integrated circuit called an *active matrix array*, which consists of many millions of identical semiconductor elements deposited on a substrate material. An intensifying screen or a photoconductor coupled to such an active matrix array forms the basis of flat-panel X-ray image receptors (Hendee and Ritenour 2002). The active matrix array and the associated electronic circuitry are mounted on a device that replaces the X-ray cassette in screen/film imaging. Operation of the array is controlled by a digital image processor, which also stores and displays the resultant images.

Flat-panel image receptors are key components of modern electron portal imaging devices (EPIDs). For example, the linear accelerator (linac) shown in Figure 10.1 incorporates two EPIDs. The first is inline with the treatment beam and is used for megavoltage (MV) portal imaging. The MV portal imaging can be used to verify the accuracy of patient alignment and to verify the shape of the mulitleaf collimator before the delivery of each treatment beam. The second EPID is used in conjunction with a kilovoltage (kV) X-ray source. Two-dimensional (2D) images obtained with this portal imager can be used to verify the accuracy of patient alignment. Additionally, this EPID can be used to acquire data for CBCT imaging.

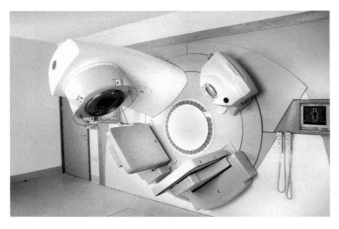

FIGURE 10.1 An Elekta Synergy linear accelerator incorporating two electronic portal imaging devices, one for the megavoltage treatment beam and the second for the kilovoltage X-ray source. (Courtesy of Elekta AB.)

Compared to film, EPIDs have numerous advantages, including fast acquisition and instantaneous display of high-quality images, reduced number of retakes due to the use of image correction techniques, ability to acquire images during treatment to record a patient's positioning throughout the treatment, and generation of digital images that are well-suited for online review and physician approval.

10.2.2 Orthogonal Imaging—Commercial Solutions

An orthogonal pair of EPIDs can be used as a tool for efficient and accurate patient alignment and makes it possible to monitor the accuracy of the alignment during each fraction of radiation (Kirby et al. 2006; Herman 2005; Cremers et al. 2004; Antonuk 2002; Herman, Kruse, and Hagness 2000). Sections 10.2.2.1 and 10.2.2.2 cover two of the most widely adopted commercial solutions using orthogonal image pairs for patient alignment. These systems are the CyberKnife (Accuray, Inc., Sunnyvale, CA) and the ExacTrac (BrainLAB AG, Feldkirchen, Germany).

10.2.2.1 CyberKnife System

The CyberKnife delivers radiosurgery and hypofractionated radiotherapy treatments by combining an X-band, compact linac; industrial robots for linac targeting and couch positioning; and a pair of orthogonal, ceiling-mounted diagnostic X-ray cameras for determining patient orientation (Figure 10.2; CyberKnife Society 2005; Heilbrun 2003; Mould 2005; Adler et al. 1997; Dieterich and Pawlicki 2008). Radiation delivery by the CyberKnife is not constrained to a fixed isocenter, but the intersection of the central axes of the diagnostic X-rays provides a reference location

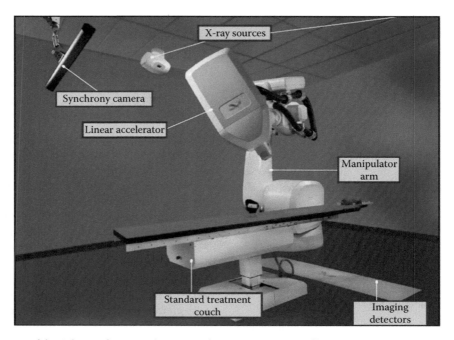

FIGURE 10.2 Components of the CyberKnife system. (Courtesy of Accuray Incorporated.)

called the *pseudoisocenter*. The CyberKnife treatment planner identifies a point in the planning CT data set that will be located at (or near to) the pseudoisocenter. Anatomical markers, such as skull or skeletal bones, or metallic fiducials near the treatment site are located during patient setup.

Orthogonal 2D images are used to reconstruct the three-dimensional (3D) orientation of a patient during treatment. This orientation is compared with the orientation during simulation, which is determined from digitally reconstructed radiographs (DRRs), and provides translational and rotational corrections. The patient orientation may be adjusted using the treatment couch. Alternatively, if the corrections are not large, the nonisocentric robot can retarget each beam to account for the calculated differences. Diagnostic images are acquired both during patient setup and periodically throughout the treatment to ensure patient orientation is current. Imaging frequency is defined by the user.

Electronic portal imaging is a fundamental component of the CyberKnife, inextricable from machine operation. This section describes the diagnostic imaging system in greater detail, as well as the targeting algorithms that are employed. Additional information regarding quality assurance and breathing corrections is also provided.

The X-ray generator geometry description is as follows: The target locating system (TLS), also called the X-ray imaging system, provides information about the location of the target throughout the treatment process. The TLS uses two X-ray sources mounted overhead on either sides of the treatment couch. The X-ray sources provide orthogonal pairs of X-ray images. These images are digitized and then compared with the reference DRR images that have been synthesized from patient CT images. The TLS includes the following components (Figure 10.3): X-ray sources

and generators, digital X-ray detectors, isopost calibration tool, and detector windows. The components are aligned within the imaging plane and are symmetric about the pseudoisocenter.

X-ray sources and generators: Two X-ray sources produce the X-ray beams for imaging the patient. They are mounted on the ceiling of the treatment room on either side of the treatment couch. The X-ray technique is adjustable by the user, up to a maximum of 125 kVp, 300 mA, and 150 milliseconds. Each X-ray beam is strongly collimated to a 20 × 20 cm field at the pseudoisocenter. Two X-ray generators supply high voltage power to each X-ray source. The X-ray generators and the X-ray generator control chassis are located either in the treatment room or in the equipment room.

Digital X-ray detectors: Two highly sensitive digital X-ray detectors are mounted within a localizing frame to ensure the accuracy of all components of the TLS. The localizing frame is mounted either flush with the floor (in-floor detectors) or slightly above the floor (on-floor detectors) in the treatment room.

Isopost calibration tool: The isopost calibration tool, the tip of which corresponds to the location of the pseudoisocenter, is mounted directly on the localizing frame. The calibration tool is used only for calibrating the CyberKnife system, and it is then removed for treatment.

Detector windows: The two detector windows are made of carbon fiber material to ensure X-ray translucency and high strength for in-floor detectors. They are designed to support up to 500 lbs (227 kg) and protect the detectors. Protective material covers the detector windows.

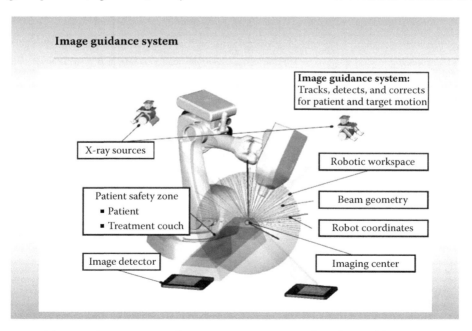

FIGURE 10.3 Components of the target locating system: the imaging center corresponds to the tip of the isopost and is also called the pseudoisocenter. (Courtesy of Accuray Incorporated.)

The CyberKnife offers a number of targeting algorithms. The six-dimensional (6D) Skull Tracking system enables the tracking of intracranial targets without the need for stereotactic frames. The 6D Skull Tracking mode involves computing the offset between live X-ray images and reference DRR images by identifying and matching skeletal features. Target tracking relies on the fixed relationship between the target volume and these skeletal features (Figure 10.4).

This rigid skeletal tracking relies strongly on information about the locations of the top of the patient's skull, the inferior cranium near the foramen magnum, and the delicate facial bones of the anterior. This translation-only correction does not provide any rotational information. Rotations are determined by generating a large number of candidate DRRs, each generated from a combination of yaw, pitch, and roll rotations of the CT data set, and comparing the live X-ray image with them to determine the closest match. Additional image processing and interpolation is used to further refine the rotational estimates.

The fiducial tracking system enables the tracking of extracranial tumors by tracking implanted fiducial markers. The fiducial tracking mode correlates the fiducial locations in reference DRR images with live X-ray images in order to extract fiducial locations. Target tracking relies on a fixed relationship between the target volume and the configuration of fiducials.

Either stainless steel screws or cylindrical gold fiducials may be used for this tracking mode. The locations of all fiducials are identified in three dimensions in the treatment planning software, which creates a 2D template of expected locations for both cameras. Conversely, the fiducial identification on both the live X-ray images may be combined into 3D positions that may be compared with the treatment planning locations in order to calculate translations and rotations. At least three fiducials are needed to determine rotations; in the case of four or more fiducials, a best fit is determined that minimizes the total deviation of all fiducials.

Rotational estimates are particularly sensitive to relative fiducial positions, primarily relative separation and near collinearity. Separations of at least 20 mm between two fiducials are recommended, as well as at least 15° between any combinations of three fiducials. It is also important to consider deformability of the tissue holding the fiducials; for example, fiducials implanted in the liver present unique tracking challenges. If any or all fiducials migrate between the time of imaging and treatment, targeting will be inaccurate. Fiducial migration is most often a concern when treating lung tumors due to the low density of lung tissue surrounding the solid tumor.

The Xsight Spine Tracking System (Accuray, Inc., Sunnyvale, California) enables tracking of skeletal structures in the spine for accurate patient positioning and treatment delivery without implanting fiducials (Antypas and Pantelis 2008; Ho et al. 2007; Muacevic et al. 2006; Thariat et al. 2009). The Xsight Spine Tracking System is capable of accurately and automatically tracking most regions of the spine, including the cervical, thoracic, and lumbar spine. The system places 81 reference locations (nodes in a 9 × 9 mesh) on the patient DRR (Figure 10.5). A band-pass filter is used to remove random noise and image artifacts from the live X-ray image in a search for anatomic spine structures that match those at the mesh nodes. This match is nonrigid and includes local displacements for each node; displacements greater than a specified value are designated "false nodes" and are eliminated in translation and rotation calculations. Although the tracking in this case is nonrigid, a large percentage of false nodes indicates appreciable deviation of the patient anatomy, despite a global match of the entire mesh.

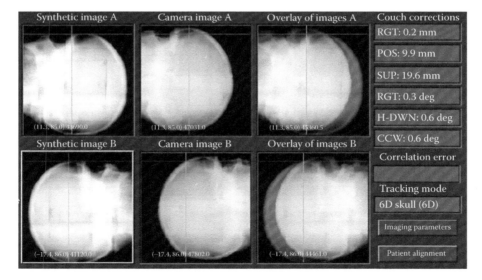

FIGURE 10.4 6D Skull Tracking for the CyberKnife system: the images on the left are digitally reconstructed radiographs; the images in the middle column are live patient images; and the images on the right are combinations of both image sets. (Courtesy of Accuray Incorporated.)

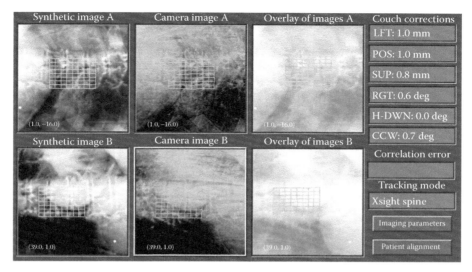

FIGURE 10.5 Xsight Spine for the CyberKnife system: corresponding anatomic points are identified on the 9 × 9 deformable mesh in order to determine translations and rotations. (Courtesy of Accuray Incorporated.)

Although the position reproducibility of the CyberKnife robot is approximately 0.1 mm, the nonisocentric nature of the CyberKnife makes the concept of "mechanical system targeting accuracy" inapplicable. Instead, the quality assurance process replicates the entire patient treatment workflow with an anthropomorphic phantom and tests the targeting accuracy of the system as a whole.

A treatment plan is created for the phantom using any of the aforementioned targeting methods, which centers the prescription dose distribution at a specified location. After placing orthogonal radiochromic films into the phantom, the treatment is delivered and the position of the dose is compared to the original plan. This "end-to-end" test is the primary targeting test for the CyberKnife system. For static targeting methods, such as 6D Skull and Xsight Spine, the guaranteed total system accuracy is less than 1 mm. For tracking methods that involve corrections for breathing, the guaranteed accuracy is less than 1.5 mm.

Failure of the CyberKnife system to achieve the guaranteed level of targeting accuracy, if reproducible, may indicate misalignment of the X-ray sources or digital detectors. The tip of the isopost is defined to be the point of intersection of the X-ray central axes. If the tip of the isopost is not located at the center pixel of the detectors, the X-ray sources require adjustment. If the tip is centered correctly, the overall system targeting error is due to a different subsystem.

For tumors that move with breathing, the CyberKnife system offers the Synchrony Respiratory Tracking System (Muacevic et al. 2007; Nioutsikou et al. 2008; Ozhasoglu et al. 2008; Seppenwoolde et al. 2007). The system is used to monitor patient breathing and adjust the position of the linac to match respiratory motion of the target. The Synchrony System utilizes a camera array mounted on the ceiling near the foot of the treatment couch to track the movement of optical markers placed on the patient's chest or abdomen. In most instances, a vest is worn to which light-emitting diodes

(LEDs) are attached. For central lung and abdominal tumor locations, gold fiducials are implanted in or close to the target. For peripheral lung locations, the Xsight Lung Tracking (XLT) System provides an alternative tracking method. (The XLT system is described in more detail later in this section.)

Once the patient's rotations are minimized using static fiducial tracking (and translations are reduced so that all fiducials are visible in the live X-ray image throughout the breathing cycle), synchrony is enabled. Subsequent X-ray exposures and 3D LED position information are matched at identical acquisition times (within less than 50 milliseconds). Typically, eight X-ray/LED pairs are obtained to form a correlation model between fiducial position and LED position (Figure 10.6). The X-ray exposures are obtained with up to 15-second separations during treatment, and the LED positions are continuously monitored. The correlation model allows the CyberKnife system to continuously predict the position of the fiducials, and this prediction is used to continuously retarget the robot to maintain submillimeter targeting accuracy.

It is important that the correlation model includes data points throughout all portions of the breathing cycle. The correlation model may be linear or curvilinear, or may consist of a loop to model hysteresis. The model is updated throughout the treatment as well, adding new points to the correlation model with every X-ray exposure. A maximum of 15 data points are used to form the model: The oldest data points are discarded as new points are added to maintain this maximum. The treatment is paused if there is any sudden change in the patient's breathing, including sneezing, coughing, speaking, or even sleep apnea. The treatment is then resumed after a new correlation model has been created.

The CyberKnife system targeting accuracy for targets that move with breathing is guaranteed to be below 1.5 mm. A fiducial-based phantom with motor control is used to perform end-to-end tests with simulated breathing patterns. Other than

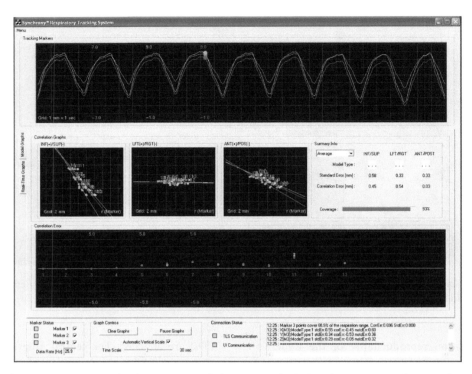

FIGURE 10.6 Example Synchrony tracking display: continuous light-emitting diode (LED) positions are shown at the top for up to three independent LEDs. The middle row shows the correlation models for each translational coordinate. The bottom row shows the correlation error for each new model point, defined as the difference between the fiducial location for that image and the predicted location of the model. (Courtesy of Accuray Incorporated.)

breathing motion, the entire quality assurance process in this case matches that for static fiducial tracking.

XLT allows accurate patient positioning and treatment without the need for implanting fiducials (Brown et al. 2009). This system is a combination of the Xsight Spine Tracking System for patient alignment and the Synchrony Respiratory Tracking System to track movement of the tumor and patient breathing patterns. This marriage of the two tracking methods maps two reference points to the pseudoisocenter, each used at different points of patient setup.

Once the patient's rotations are minimized using Xsight Spine (using the first pseudoisocenter reference), the treatment couch is automatically shifted to the second reference in order to view the target. The density difference between the solid tumor and the lung tissue creates a shadow with a unique cross section on each camera. This cross section is compared with the contoured target from the treatment plan to determine translational offsets. The Synchrony System is used to create a correlation model between the translational offsets throughout the breathing cycle and continuous 3D LED position tracking.

XLT requires a minimum target dimension of 15 mm in any direction. Smaller targets do not create a sufficiently large shadow for accurate tracking. The location of the tumor within the lung also determines whether XLT is viable. For tumors located along the same line of sight as the spine or the mediastinum on either camera, XLT may not be possible. Rarely, a tumor may be visible during one portion of the breathing cycle but may be blocked by

the spine during another portion; in this case, it may not be possible to use XLT, and fiducials must be implanted.

10.2.2.2 BrainLAB ExacTrac

The ExacTrac X-Ray 6D IGRT system in the BrainLab Novalis radiosurgical unit (BrainLAB AG, Heimstetten, Germany) is an example of a 2D X-ray IGRT system (Jin et al. 2008). The ExacTrac system integrates two main subsystems: (1) an infrared (IR)-based optical positioning subsystem and (2) a 2D radiographic kV X-ray imaging subsystem. The IR subsystem has two primary functions during patient positioning: First, it provides initial patient setup information based on the IR fiducial markers placed on the patient's surface. Second, it is used for precise adjustment of the patient's position based on the results of image registration using the X-ray subsystem.

The kV X-ray subsystem provides accurate positioning information by comparing the internal bony anatomy or the implanted fiducial markers in the setup images with that in the simulated CT images using 2D/3D imaging registration. The IR subsystem can also be used to monitor a patient's respiration and provide a signal to the linac for tracking or gating the treatment beam (Hugo, Agazaryan, and Solberg 2002).

Used in conjunction with the X-ray system, image-guided verification of the target position relative to the gating window can be performed throughout the duration of the gated delivery. The X-ray system also provides snapshots during the treatment to account for intrafraction patient motion. In addition to the IR

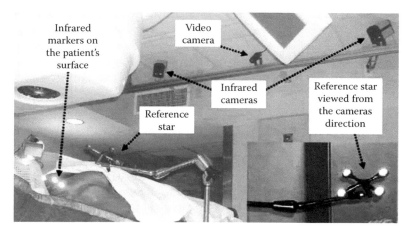

FIGURE 10.7 The infrared-camera-based ExacTrac system.

devices and the X-ray imagers, the system includes a digital video camera for monitoring each patient's position during treatment. In this section, we present a detailed description of the IR and X-ray components of the ExacTrac X-Ray 6D system, including their application in the delivery of gated radiotherapy.

The IR tracking component of the ExacTrac X-Ray 6D system includes two IR cameras, multiple passive IR-reflecting spheres placed on a patient's surface, and a reference device (aka the reference star) that contains four reflective circles (Figure 10.7). The IR cameras are rigidly mounted on a metal bar attached to the ceiling and emit an IR signal that is reflected and analyzed for positioning information.

Studies by Wang et al. (2001) have demonstrated that the position of each IR-reflecting sphere can be determined within 0.3 mm. Initial setup can be achieved by moving the couch to match the markers' positions with those recorded in a CT image. In addition, the software also provides rotational offsets along three primary axes. However, note that although the identification of the external markers by the IR system is highly accurate, patient positioning based on these markers is not reliable. The external markers have to be positioned in relatively stable locations to achieve accurate setup. The primary value of the IR system is not due to its role in initial setup, but due to its use in accurately adjusting the couch position based on the image registration. Using the reference star attached to the couch, the couch's movement can be determined precisely.

The IR system samples marker positions at a frequency of 20 Hz and, therefore, it may also be used to monitor patient motion. Because several markers (typically, five to seven markers) are used, motion signals at various body locations can be simultaneously monitored. Therefore, these markers can be used to monitor both respiratory and nonrespiratory motion (Yan et al. 2006; Jin et al. 2007).

The X-ray component consists of two floor-mounted kV X-ray tubes projecting to two corresponding flat-panel detectors mounted on the ceiling (Figure 10.8). The configuration of this X-ray system is unique compared to general diagnostic X-ray systems, because of the following factors: the X-ray

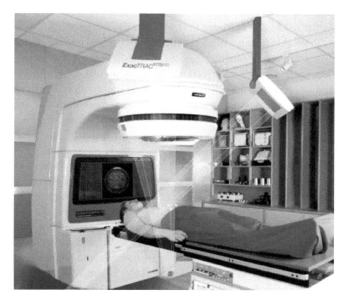

FIGURE 10.8 The X-ray subsystem showing oblique configurations of the X-ray imaging devices. (Courtesy of BrainLAB.)

tubes and corresponding detector panels are in fixed positions, the X-rays project in an oblique direction relative to the patients, and the source-to-isocenter and source-to-detector distances are relatively large (approximately 2.24 and 3.62 m, respectively).

For patient alignment, two X-ray images are obtained after a patient is initially set up with the IR subsystem. These images are then compared with the patient's 3D CT simulation images, with corresponding isocenters in terms of DRRs. This process is called image registration or image fusion. Several different image registration algorithms are available, including manual, 3D, 6D, and implanted-marker-based image fusions.

The manual match and the 3D fusion methods assume that the patient was set up with no rotational offsets. Corresponding DRRs are generated and the patient's translational shifts are computed in order to provide the best match with the X-ray

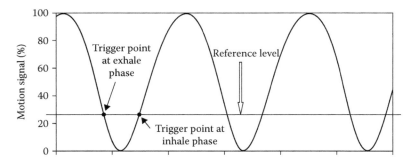

FIGURE 10.9 Specifying a reference level in a respiration trace: the level crosses the trace at two points for each respiration cycle. The point at the exhale phase is usually used for triggering the X-ray for synchronized positioning. The two points are also used to define the gating window, with the point at the exhale phase triggering on the radiation and the point at the inhale phase triggering off the radiation.

images. Therefore, the manual fusion and the 3D fusion methods use a simple 2D (X-ray) to 2D (DRR) (or 2D/2D) image fusion algorithm.

The 6D fusion software first generates a library of DRRs with positional variations in all the three translational and three rotational directions (six degrees of freedom) for the CT images. It then compares these DRRs with the corresponding X-ray images and determines the shifts that provide the greatest similarity with the corresponding X-ray images. The best match is thus determined, and the three translational and three rotational position variations are used to generate the set of DRRs and the 6D offsets. Therefore, the 6D fusion method is actually a 2D (X-ray) to 3D (CT) (or 2D/3D) image fusion algorithm. Both the 3D and 6D fusion methods also provide the option of selecting a region of interest for fusion and of excluding any structures that could potentially increase uncertainty in the fusion.

When implanted markers are used for alignment, the positional accuracy and the size of the offsets are determined by comparing the implanted marker positions in the X-ray images with those in the CT images. For all these methods, after the offsets have been determined, the patient's position can be precisely adjusted using IR guidance. A set of verification X-ray images can be obtained to check the position after readjustment.

The ExacTrac X-Ray 6D system also includes a module, the ExacTrac Adaptive Gating system, for image-guided positioning of targets in motion due to respiration and for gated treatment of such targets. It is currently designed to be used with radiopaque fiducial markers (typically, a gold coil) implanted near the target isocenter (Wurm et al. 2006; Willoughby, Forbes, et al. 2006). The coil is implanted in close proximity to the target anatomy so that it can be seen within the field of view of the X-ray localization system at the time of treatment. It is assumed that the spatial relationship between the coil and the target anatomy will remain relatively fixed (Ebe et al. 2004).

For treatment, a patient is set up in the treatment couch and IR reflective markers are attached to the abdominal region of the patient so that breathing motion can be monitored. The reference star is used not only as a reference against which the movement of the markers is measured but also to track couch location during the patient positioning process. The 3D movement of the

patient's anterior surface is tracked using the IR markers, and the anterior-posterior (AP) component of this trajectory is used as the breathing trace to monitor breathing motion. Internal target position is assumed to be correlated with this breathing motion.

Patient positioning is achieved by acquiring a set of gated X-ray localization images and comparing them with the synchronized simulation CT image. Gated-CT or four-dimensional (4D) CT has to be performed during patient simulation for gated treatments (Vedam et al. 2003; Jin and Yin 2005). A gating trigger point is set to correspond to the same phase in the breathing cycle as for the treatment planning CT. The trigger point is defined by specifying a reference level that crosses the breathing trace (Figure 10.9). However, there are usually two trigger points in a breathing cycle for a reference level. The trace direction (up or down, corresponding to inhale or exhale phases, respectively) has to be specified to define the trigger point. The trigger point is usually set at the exhale phase. Once the two localization images are acquired, the implanted-makers-based imaging registration option is used to determine the positional shifts.

A gating window is then set for the treatment and is typically centered at the end of the exhale phase. It is also defined by specifying a reference level (Figure 10.9). The radiation delivery is triggered on when the breathing trace crosses the level in the exhale phase and triggered off when the trace crosses the level in the inhale phase.

10.3 Ultrasound Technology

10.3.1 Basic Principles

Ultrasound is the term used to describe cyclic sound pressure with a frequency greater than the upper limit of human hearing. Although this limit varies from person to person, it is approximately 20 kHz (20,000 Hz) in healthy young adults. Consequently, 20 kHz serves as a useful lower frequency limit in describing ultrasound. Like audible sound waves, ultrasound propagates through a medium by compressing and decompressing the medium (Novelline and Squire 1997).

As ultrasonic waves pass from one medium to the next, the *acoustic impedance* (Z) of the two materials determines what portion is transmitted and what portion is reflected. The acoustic impedance is a product of material density and the ultrasound velocity within that material (Hendee and Ritenour 2002; Szabo 2004). At the interface between two media with different acoustic impedances, the portion of the initial wave that is reflected is called an *echo*. Analyzing the echo can reveal the depth, shape, and size of the reflecting interface.

In medical imaging, ultrasound has proven to be a very useful tool in visualizing soft tissues. The ultrasound used in imaging typically has a frequency over 1 MHz and is produced by a device called a *transducer*. The transducer converts electrical pulses into ultrasonic waves. It also measures the echoes reflected by body structures and converts echoes into electrical pulses. The converted electrical pulses are transformed by a computer to visible images.

Although ultrasound propagates well through soft tissues and fluids, it is easily disrupted by air and blocked by bone. Consequently, ultrasound imaging is limited to a subset of body parts such as the heart, pelvis, neck, and extremities. Details of the theory and application of ultrasound imaging in medicine are abundant in the literature (Hendee and Ritenour 2002; Szabo 2004).

10.3.2 Commercial Solutions

In radiation oncology, ultrasound has been used frequently as a tool to assist in accurate patient setup (Peng et al. 2008; Pinkawa et al. 2008; Bohrer et al. 2008; Lin et al. 2008; Boda-Heggemann et al. 2008; Kupelian et al. 2008; Jereczek-Fossa et al. 2007; Feigenberg et al. 2007; Peignaux et al. 2006; McNair et al. 2006; Fuller et al. 2006; Dobler et al. 2006; Paskalev et al. 2005; Jani et al. 2005; Fuss et al. 2004; Langen et al. 2003; Chinnaiyan et al. 2003; Trichter and Ennis 2003; Little et al. 2003; Chandra et al. 2003; Morr et al. 2002; Saw et al. 2002; Huang et al. 2002; Falco et al. 2002; Lattanzi et al. 2000; Lattanzi et al. 1999; Cury et al. 2006; Wu, Ling, and Ng 2000; Wein, Roper, and Navab 2005; Castro-Pareja et al. 2005). The first commercial system that used ultrasound for image-guided patient alignment was called the B-mode Acquisition and Targeting (BAT) system produced by Best Nomos (Pittsburgh, Pennsylvania). Figure 10.10 illustrates such a system. This BAT system includes a computer, an ultrasound probe, and optical cameras. The optical cameras are mounted on the ceiling or wall and are used to detect optical markers on the handle of the ultrasound probe. The ultrasound probe is mounted on an articulated arm. Any motion of the probe is detected by the cameras and sent to the computer to determine the probe position and orientation. The BAT system is calibrated such that the system knows the ultrasound image location with respect to the isocenter of the treatment machine.

When using the BAT system, the daily ultrasound images are aligned with the contours of the prostate, rectum, and bladder as outlined by the treatment planning CT image data set (see Figure 10.11). These contours along with the isocenter coordinates are exported from the treatment planning system to the BAT system before delivering the first fraction of radiation.

On each day of treatment, the patient is first aligned using skin marks before the BAT system is employed. Next, two 2D

FIGURE 10.10 The B-mode Acquisition and Targeting (BAT) system used for patient alignment with ultrasound images. (Courtesy of Best Nomos.)

ultrasound images are obtained, one in the axial plane and the other in the sagittal plane. These 2D ultrasound images are displayed on the computer monitor and, with the imported organ contours, overlaid on the images. The user then maneuvers these contours until they match the anatomy viewed in the ultrasound images. Upon completion of the manual adjustment, the BAT system calculates the 3D table shifts for fine adjustment of patient localization.

The BAT system has been widely adopted in the alignment of patients with prostate cancer (Peng et al. 2008; Pinkawa et al. 2008; Bohrer et al. 2008; Lin et al. 2008; Boda-Heggemann et al. 2008; Kupelian et al. 2008; Jereczek-Fossa et al. 2007; Feigenberg et al. 2007; McNair et al. 2006; Dobler et al. 2006; Paskalev et al. 2005; Jani et al. 2005; Fuss et al. 2004; Langen et al. 2003; Trichter and Ennis 2003; Little et al. 2003; Chandra et al. 2003; Morr et al. 2002; Saw et al. 2002; Huang et al. 2002; Falco et al. 2002; Lattanzi et al. 2000). Numerous investigators have examined its alignment accuracy (Peng et al. 2008; Pinkawa et al. 2008; Boda-Heggemann et al. 2008; Jereczek-Fossa et al. 2007; Feigenberg et al. 2007; McNair et al. 2006; Langen et al. 2003; Lattanzi et al. 1999). Studies have compared prostate alignments using the BAT system with MVCT (helical tomotherapy; Peng et al. 2008; Lin et al. 2008), kV CBCT (Boda-Heggemann et al. 2008), daily CT (Jereczek-Fossa et al. 2007; Feigenberg et al. 2007; Lattanzi et al. 1999), and portal images with implanted markers (McNair et al. 2006; Langen et al. 2003). In every study, CT images were used as the reference images. Unfortunately, there is significant

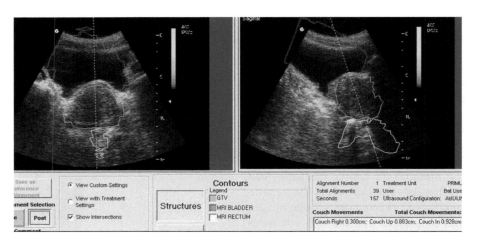

FIGURE 10.11 Axial and sagittal ultrasound images for a prostate patient aligned with the prostate, rectum, and bladder contours determined by the treatment planning CT images. (Courtesy of Best Nomos.)

variation among these studies regarding the measured accuracy of ultrasound alignment for prostate radiotherapy. For example, some studies have shown that table shifts determined by the BAT system correlate well with the shifts determined by the reference methods (Boda-Heggemann et al. 2008; Feigenberg et al. 2007; Lattanzi et al. 1999). Other studies, however, have shown a systematic disagreement between alignment using BAT and that using the reference methods (Peng et al. 2008; Lin et al. 2008; Jereczek-Fossa et al. 2007; McNair et al. 2006; Langen et al. 2003). One study concluded that the mean differences between the table shifts determined by the BAT system and the shifts determined using the reference methods were less than 2 mm in all directions (McNair et al. 2006). Other studies, however, found differences between 2 and 5 mm in all directions (Jereczek-Fossa et al. 2007; Lattanzi et al. 1999) with a maximum difference of 9 mm in the superior-inferior direction (Lattanzi et al. 1999).

Differences in the design of the tests used in these studies may account for the wide variation in the results. These differences include differences in the imaging modality used as the reference and whether or not the patients were immobilized (Feigenberg et al. 2007). The level of experience in reading ultrasound images may also have impacted the results. For example, in a study by Langen et al., eight users independently aligned the patients using the BAT system. Among the eight users, the average range of couch shifts due to contour alignment variability was 7, 7, and 5 mm in the AP, superior-inferior, and lateral directions, respectively (Langen et al. 2003). It is also known that there are daily variations in the level of bladder and rectal filling. This leads to variations in both the shape and relative locations of the bladder, rectum, and prostate, which result in alignment uncertainties. In addition, the BAT system matches 2D images to 3D structures, which naturally limits its alignment accuracy.

In addition to the BAT system, Varian Medical (Palo Alto, California) has offered the SonArray system, and Computerized Medical Systems (CMS; Maryland Heights, Missouri) has offered the I-Beam system. Both systems reconstruct 3D ultrasound

images through a series of 2D ultrasound scans, acquired by either sweeping or angling a 2D probe over the region of interest. Patient alignment parameters are calculated by matching the 3D ultrasound images with organ contours delineated on the planning CT images. Patient alignment accuracy has been assessed by comparing ultrasound-based alignment with fiducial-marker-based alignment (Fuller et al. 2006). In a study by Fuller et al. (2006), substantial differences were observed between fiducial-based and ultrasound-based alignments. The 95% limits of agreement in the x, y, and z axes were ±9.4, ±11.3, and ±13.4 mm, respectively.

With the increased availability of CBCT, both Varian and CMS have halted further development of their ultrasound-based patient alignment systems. However, Best Nomos has continued to offer the BAT system, and in 2007, they released a new BAT system called the BATCAM Multiprobe system. This new system reconstructs 3D ultrasound images for the matching of organ contours. The alignment accuracy of this new system has not yet been intensively investigated.

Another system using 3D ultrasound for image-guided patient alignment is the Clarity system, developed by Resonant Medical (Montreal, Quebec, Canada). The Clarity system differs from the previously described ultrasound alignment systems in the area of image matching. The aforementioned three systems match organ contours delineated on planning CT images to ultrasound images. The Clarity system, however, matches 3D ultrasound to 3D ultrasound for the calculation of table shifts. Typically, two Clarity systems are installed in a clinic with one in the CT simulation room and the other in the treatment room. Immediately after the CT image set is acquired for treatment planning, a 3D ultrasound is acquired with the patient in the CT-scanning position. This ultrasound image is manually segmented, and the image is exported to the Clarity system in the treatment room. Prior to each treatment, an ultrasound image is acquired with the patient in the treatment position. This daily ultrasound image is fused with the ultrasound image obtained in the CT room to reveal 3D table shifts. This workflow is designed to avoid the target definition discrepancy caused by different imaging modalities. One study shows that the alignment accuracy of the Clarity system improved when compared with the alignment using the BAT system (Cury et al. 2006).

Since ultrasound propagates well only in soft tissues and fluids, the use of ultrasound for IGRT is limited to a few treatment sites. The most common use of ultrasound in IGRT is the alignment of patients with prostate cancer. Ultrasound-based IGRT can also be applied for the alignment of patients with pelvic and breast cancers. Key concerns in the use of ultrasound for patient alignment are the user-to-user variability in reading ultrasound images and the image quality of reconstructed 3D ultrasound images. Recent research shows that the subjectivity of user alignment could be eliminated using automatic alignment tools (Wu, Ling, and Ng 2000; Wein, Roper, and Navab 2005; Castro-Pareja et al. 2005; Wang et al. 2006). Additionally, using a 3D probe instead of a 2D probe in generating 3D ultrasound images will potentially improve image quality and help improve alignment accuracy.

10.4 Electromagnetic Technology

Various researchers have examined target localization using electromagnetic signals (Houdek et al. 1992; Watanabe and Anderson 1997; Seiler et al. 2000; Balter et al. 2005). Currently, the only commercial patient alignment solution using electromagnetic localization is the Calypso 4D Localization System (Calypso Medical, Seattle, Washington). The Calypso 4D Localization System uses nonionizing alternating current (AC) electromagnetic radiation to locate and continuously track small wireless devices (called Beacon transponders) implanted in or near the tumor. Section 10.4.1 describes the basics of electromagnetic localization, and Section 10.4.2 discusses the Calypso 4D Localization System in more detail.

10.4.1 Principles of Electromagnetic Radiation

When an electric current flows through a conducting wire, a magnetic field is generated (Figure 10.12a). If the wire is wound on a coil with many turns of wire closely looped together, then a highly concentrated magnetic field can be obtained within the coil (Figure 10.12b). Because the magnetic field is proportional to the strength of the current, an AC flowing through the coil will lead to a changing magnetic field in the vicinity of the coil.

If a second coil with a closed circuit (coil B) is placed adjacent to a coil that is connected to an AC power supply (coil A), an induction current will be produced in coil B by the changing magnetic field generated from coil A, as shown in Figure 10.13. The phenomenon is also called *Faraday's law of induction*, where the induced electromotive force (EMF) within the induction is proportional to the rate of the magnetic flux varying with time in the induction coil (coil B):

$$\varepsilon = -N \frac{d\Phi_B}{dt} \tag{10.1}$$

where ε is the induced EMF, N is the number of turns in coil B, Φ_B is the magnetic flux going through a single loop of coil B, and t is time.

The AC in coil A from an AC power source results in an oscillating magnetic field within coil B. This in turn leads to an oscillating current in coil B. The amplitude of the oscillating current reaches its maximum when the oscillating frequency reaches the resonance frequency of coil B. When the AC power to coil A is switched off, the oscillating current in coil B will

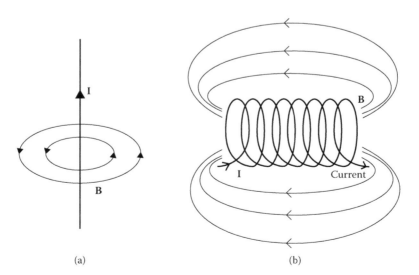

(a)   (b)

FIGURE 10.12 Illustration of Faraday's law of induction: (a) a magnetic field is produced adjacent to a conducting wire carrying a flowing current, and (b) a magnetic field is created by a coil carrying a flowing current.

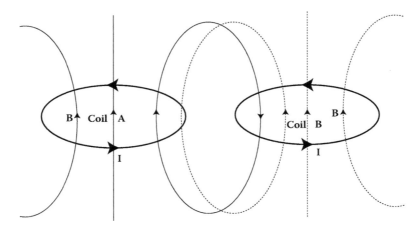

FIGURE 10.13 A plot that shows how an increasing current in coil A induces an increasing current in coil B. Here, the thick solid lines represent the coils, the thin solid lines represent the magnetic field created by coil A, and the dashed lines represent the magnetic field generated by the induction current in coil B.

continue to generate an oscillating magnetic field around it due to self-induction with a decreasing amplitude until the current dissipates over a short period of time.

10.4.2 Commercial Solution: Calypso 4D Localization System

The core Calypso system consists of three IR cameras, one mobile detecting array, and three transponders (Beacon). A localization console in the treatment room and a tracking station in the linac control area are also needed for signal processing and display. The three IR cameras are mounted on the ceiling of the treatment room to accurately locate the position of the detecting array using the multiple IR refractors on the detecting array.

The detecting array is placed over the patient; it consists of both electromagnetic power source coils and a detecting coil-array (see Figure 10.14). The field of detection of this array is 14 × 14 cm wide and extends up to 27 cm away from the detector array (Balter et al. 2005). Each implanted transponder incorporates a metal coil encased in glass and is 8 mm long and 1.85 mm in diameter (Willoughby, Kupelian, et al. 2006). Because each implanted transponder has a different number of loops in the coil, each has its own resonance frequency.

Typically, three transponders are implanted in the targeted region of the patient prior to CT simulation and radiation treatment. These transponders serve as internal surrogates for monitoring target translational and rotational motion. Clinicians commonly incorporate a period of about 14 days between the transponder implantation and the CT simulation to ensure that the transponder positions have stabilized (Willoughby, Kupelian, et al. 2006; Kupelian et al. 2007).

The localization/tracking process begins with the production of an oscillating magnetic field by the source coils, which can induce a resonance in a transponder. After the power to the source coil is switched off, the changing magnetic field produced during the relaxation in the transponder can be detected by the detecting coil-array to determine its location relative to the detecting array. Since each transponder has a different resonance frequency, the activation and detection of each transponder is performed sequentially. The actual resonance frequency for each transponder is determined by the initial oscillating magnetic field returned from the transponders before the actual target localization is performed. By repeating the aforementioned process, the 3D coordinates for each transponder can be obtained in a real-time fashion at a frequency of 10 Hz (Balter et al. 2005). As the room coordinate of the detecting array can be accurately determined by the mounted IR cameras, the position of each transponder and the target location relative to the machine isocenter can be accurately established.

Currently, the Calypso system is approved by the Food and Drug Administration (FDA) for prostate treatments only. It has been shown that this technique can provide submillimeter accuracy in detecting transponder positions with excellent system stability (Balter et al. 2005). A multi-institution trial was performed comparing the target localization results obtained using the Calypso system with those obtained using a 2D kV X-ray imaging system (BrainLAB ExacTrac; Willoughby, Kupelian, et al. 2006; Kupelian et al. 2007). The data obtained from these studies demonstrated that these two techniques provide comparable target localization results with an average of less than 2-mm deviation in 3D vector length.

One distinctive feature of the Calypso system is that it can perform real-time tracking of the target. This makes it possible to account for intrafractional target motion. Various studies have shown that the intrafraction motion for prostate can exceed 1 cm with motion primarily along the longitudinal and vertical directions (Kupelian et al. 2007). As an example, Figure 10.15 plots the maximum range of prostate motion along all three axes during radiation treatment for a selected patient over 25 fractions.

When compared with conventional image-guided techniques, a key advantage of the Calypso system is that it does not require ionizing radiation for target localization. The use

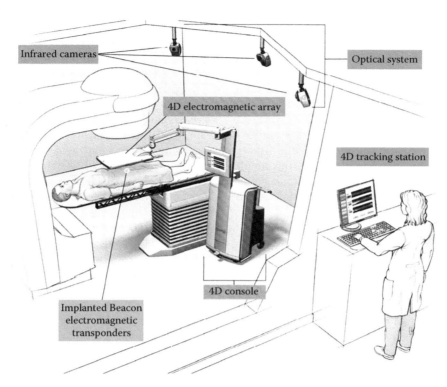

FIGURE 10.14 The Calypso 4D Electromagnetic Tracking system. (Copyright Calypso Medical Technologies Incorporated. With permission.)

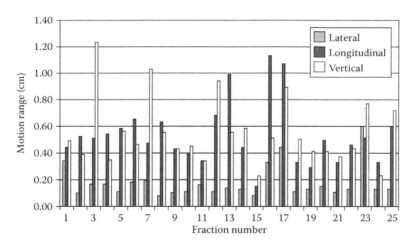

FIGURE 10.15 The maximum range of intrafractional prostate motion along all three axes for a patient over 25 fractions.

of CBCT imaging or orthogonal X-ray imaging on a daily basis can lead to a clinically significant dose over a course of fractionated radiotherapy (Wen et al. 2007; Winey, Zygmanski, and Lyatskaya 2009). The second key advantage of the Calypso system is its ability to provide real-time tracking of the target based on internal markers. One disadvantage of the Calypso system is the need to implant Beacon transponders. These transponders can migrate over the course of treatment; so, their relative positions need to be monitored.

Currently, the Calypso system can be used clinically only for prostate treatments. However, various studies have been carried out to evaluate the potential application of Calypso for other treatment sites such as lung and partial breast (Mayse et al. 2008; Sawant et al. 2009; Smith et al. 2009).

10.5 Surface Mapping Technology

10.5.1 Basic Principles

Photogrammetry, literally translated as "light-drawing-measurement," is defined as the process through which geometric properties of a particular object are determined from a photographic image. In its rudimentary form, one can calculate the distance between two points lying on a plane parallel to the plane of a photographic image using a photograph's known scale. More sophisticated techniques, or stereophotogrammetry, permit the estimation of the 3D coordinates of a point on an object through measurements made between two or more photographic images obtained from different positions. As common points of a particular object are identified on each photographic image, rays representing the camera's line of sight to a particular point are constructed. The intersection of several rays permits the triangulation of that point into three dimensions. With the assistance of robust computer algorithms, this simple process calculates the coordinates of several thousand points and serves as the foundation of optical surface imaging (OSI) as a practical technique in IGRT.

OSI systems are installed directly into the linac vault to provide initial interfraction patient localization and, if desired, to perform continuous intrafraction patient positioning during treatment (Baroni et al. 2000; Bert et al. 2005; Bert et al. 2006; Moore et al. 2003; Rogus, Stern, and Kubo 1999; Sjodahl and Synnergren 1999). Currently, there is no routine need for the incorporation of an OSI system as part of patient simulation, but new applications, including the use of the OSI system to gate therapy and determine the respiratory phase for data acquisition, are under active development.

Two camera pods are suspended from the ceiling of the treatment room (Figure 10.16). Each pod is equipped with a stereovision camera (two charge-coupled device cameras separated by a known baseline), a texture camera, a clear flash, a flash for speckle projection, and a slide projector for speckle projection. "Speckle" refers to an optically projected pseudorandom grayscale pattern to enable the 3D reconstruction of a surface. Each pod acquires 3D surface data over approximately 120° in the axial plane, from midline to posterior flank. On the basis of a proprietary calibration process, the data are merged to form a single 3D surface image of the patient. In the overlap region near the midline, the surfaces from the two pods merge smoothly with less than a 1-mm root mean square (RMS) discontinuity. A daily calibration verification procedure, easily performed during the linac warm-up period, is recommended by most vendors to check the consistency of this overlap region.

The OSI systems include proprietary software designed to facilitate patient setup, achieved by acquiring an image of the patient and aligning this new surface with a known reference topogram. Depending on clinical workflow, a topogram can be obtained prior to the first treatment session with the patient on the linac couch, at the time of patient simulation if a second imaging system is installed in this location, or by the extraction of the patient's external surface contour at the time of virtual simulation with the contour transferred via digital imaging and communications in medicine (DICOM) to the OSI system. An alignment tool allows for the comparison of the topogram with the newly acquired images. During patient alignment, the software calculates the optimal couch translation and rotation to bring a newly obtained topogram into congruence with a clinician-defined region of interest on the topogram. Any extraneous surfaces irrelevant to patient alignment (i.e., the patient's gown, intravenous [IV] lines) should be removed to ensure accuracy. The entire process takes less than 10 seconds, and can be repeated to satisfy institutional localization standards.

10.5.2 Commercial Solutions and Clinical Applications: AlignRT

The AlignRT OSI system (Vision RT, Ltd., London, United Kingdom) is a commercial solution that uses surface alignment

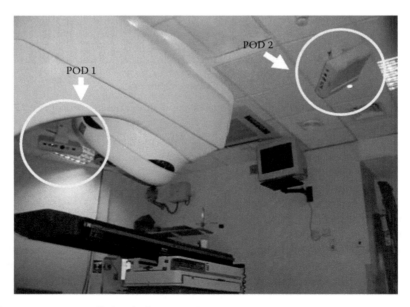

FIGURE 10.16 A two-pod camera system installed in the linear accelerator treatment vault.

technology (Hughes et al. 2009; Krengli et al. 2009; Cervino et al. 2009). A common use of AlignRT is in the localization of women receiving external beam radiation therapy as adjuvant therapy for localized breast cancer. An example of the use of AlignRT with breast localization is shown in Figures 10.17 through 10.20. This patient was treated at the Washington University in St. Louis, Missouri. In this case, a custom solid foam immobilization form (alpha cradle) was created with the patient in the treatment position. The cradle contains a strategic elevation built within to level the sternum parallel to the table. The patient was positioned supinely with both arms elevated above the head and the neck slightly extended. A left hand grip was not used in this case (Figure 10.17a). The clinician-defined region of interest (Figure 10.18) served as the reference structure for alignment and assessment of daily patient positioning. Daily patient setup was performed using laser-guided superficial skin fiducials and the AlignRT system. This was supplemented with weekly portal imaging. After the patient was placed in the alpha cradle according to the simulation specifications, the isocenter was verified using AP and lateral skin fiducials, and the source-to-skin distance (SSD) was affirmed.

The patient's topogram was recalled using the AlignRT software, and a new treatment topogram was captured using the 3D cameras, which projected a bright speckle pattern across the patient for several seconds. The acquired topogram was generated and compared with the topogram using the system's software to produce three planar translations required for precise reproduction of the patient's expected position (Figure 10.19). All orthogonal shifts (in millimeters) were applied to the patient, and a second image acquisition and analysis using AlignRT was performed for verification.

The process generally takes less than 3 minutes, and is repeated if necessary until the suggested shifts fall below 5 mm in any one plane. The AlignRT software will also calculate the volume of tissue within a specified distance from the reference topogram and localize regions of the breast that deviate beyond this threshold (Figure 10.20).

10.5.3 Surface Mapping Summary

OSI systems offer specific benefits for the development of IGRT programs including: (1) it is noninvasive; (2) data can be acquired rapidly, for either frequent verification of the patient localization or intrafraction patient monitoring; (3) it is straightforward to use for radiation therapy technicians; and (4) the ability to monitor and compare treatment progress with respect to prior sessions. Ongoing research into the validity of OSI systems as a primary method to assess respiratory motion, as well as regarding its ability to gate treatment delivery, and the validation of recorded surface motion as a surrogate for internal target

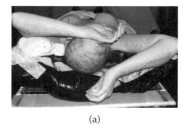
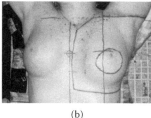
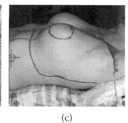

(a)   (b)   (c)

FIGURE 10.17 Cranial-caudal view of patient arm positioning: (a) hand grips are used for reproducibility according to patient comfort; (b) anterior view of patient setup for whole breast radiation therapy. Please note medial, superior, and inferior port borders. The boost volume is delineated on the left breast by the innermost contour; (c) lateral view of patient setup for whole breast radiation therapy. The lateral port edge typically follows the mid or posterior axillary line to cover the lateral breast and axillary tail.

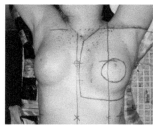
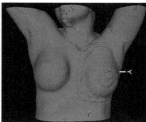
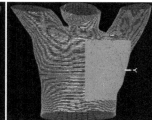

FIGURE 10.18 Region of interest defined on AlignRT: the clinician-defined whole breast volume (left panel) is transferred from the treatment planning system as an external surface contour to the AlignRT system via digital imaging and communications in medicine (center panel), where the target region of interest is defined using AlignRT for patient positioning (right panel).

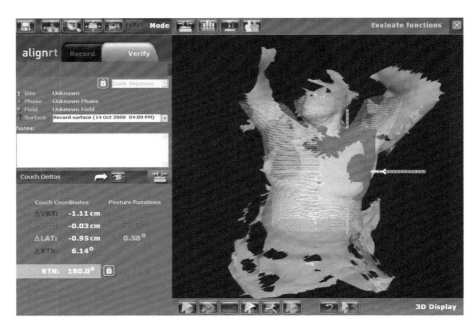

FIGURE 10.19   (See color insert following page 204.) Reference topogram compared with the treatment topogram: the reference topogram (pink) with clinician-defined region of interest is compared with the acquired treatment topogram (green). The data sets are analyzed using the AlignRT software to calculate couch shifts required to reproduce expected patient position.

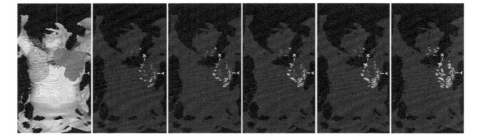

FIGURE 10.20   (See color insert following page 204.) Breast position varies between treatments. AlignRT will localize regions of the breast that deviate beyond a user-defined threshold compared to the initial reference topogram. From left to right, the location of breast position above (red), below (blue), and within (green) a specified threshold for variation between topograms of 1, 2, 3, 5, and 10 mm. For this treatment, the majority of breast tissue falls within 10 mm from the initial reference, with the greatest breast position deviation localized to the upper inner quadrant.

motion may lead to broader clinical applications of these systems (Hughes et al. 2009).

10.6 Implanted Radioactive Fiducial Marker Technology

Navotek Medical Ltd. (Yokneam, Israel) has developed a patient alignment and tracking system based on the use of a radioactive implanted fiducial. The RealEye Tracer system (Figure 10.21) is mounted on the head of the gantry and incorporates a detector system that can provide continuously updated target position information.

The Tracer trackable fiducial is a magnetic resonance imaging (MRI)-compatible implant that comes preloaded in a disposable

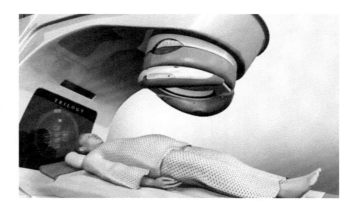

FIGURE 10.21   Diagram of the RealEye system mounted on the gantry of a Varian Trilogy linac. (Courtesy of Navotek Medical Ltd.)

implantation device featuring a 23-gauge implantation needle and a spring-loaded implantation mechanism. The Tracer (radioactive fiducial) contains 100 μCi of iridium-192 embedded within a platinum–iridium alloy. The antimigration crumpling coil design of the Tracer is intended to minimize the motion of the implanted fiducial. During treatment, the RealEye system monitors target position and provides a warning to the user if the target has moved beyond a predefined threshold.

References

Adler Jr., Jr., S. D. Chang, M. J. Murphy, J. Doty, P. Geis, and S. L. Hancock. 1997. The CyberKnife: A frameless robotic system for radiosurgery. *Stereotact Funct Neurosurg* 69(1–4 Pt 2): 124–8.

Antonuk, L. E. 2002. Electronic portal imaging devices: A review and historical perspective of contemporary technologies and research. *Phys Med Biol* 47(6):R31–65.

Antypas, C., and E. Pantelis. 2008. Performance evaluation of a CyberKnife G4 image-guided robotic stereotactic radiosurgery system. *Phys Med Biol* 53(17):4697–718.

Balter, J. M., and Y. Cao. 2007. Advanced technologies in image-guided radiation therapy. *Semin Radiat Oncol* 17(4):293–7.

Balter, J. M., J. N. Wright, L. J. Newell, et al. 2005. Accuracy of a wireless localization system for radiotherapy. *Int J Radiat Oncol Biol Phys* 61(3):933–7.

Baroni, G., G. Ferrigno, R. Orecchia, and A. Pedotti. 2000. Real-time opto-electronic verification of patient position in breast cancer radiotherapy. *Comput Aided Surg* 5(4):296–306.

Bert, C., K. G. Metheany, K. Doppke, and G. T. Chen. 2005. A phantom evaluation of a stereo-vision surface imaging system for radiotherapy patient setup. *Med Phys* 32(9):2753–62.

Bert, C., K. G. Metheany, K. P. Doppke, A. G. Taghian, S. N. Powell, and G. T. Chen. 2006. Clinical experience with a 3D surface patient setup system for alignment of partial-breast irradiation patients. *Int J Radiat Oncol Biol Phys* 64(4):1265–74.

Boda-Heggemann, J., F. M. Kohler, B. Kupper, et al. 2008. Accuracy of ultrasound-based (BAT) prostate-repositioning: A three-dimensional online fiducial-based assessment with cone-beam computed tomography. *Int J Radiat Oncol Biol Phys* 70(4):1247–55.

Bohrer, M., P. Schroder, G. Welzel, et al. 2008. Reduced rectal toxicity with ultrasound-based image guided radiotherapy using BAT (B-mode acquisition and targeting system) for prostate cancer. *Strahlenther Onkol* 184(12):674–8.

Brown, W. T., X. Wu, F. Fayad, et al. 2009. Application of robotic stereotactic radiotherapy to peripheral stage I non-small cell lung cancer with curative intent. *Clin Oncol (R Coll Radiol)* 21(8):623–31.

Castro-Pareja, C., V. Zagrodsky, L. Bouchet, and R. Shekhar. 2005. Automatic prostate localization in external-beam radiotherapy using mutual information registration of treatment planning CT and daily 3D ultrasound images. *Comput Assist Radiol Surg* 1281:435–40.

Cervino, L. I., S. Gupta, M. A. Rose, C. Yashar, and S. B. Jiang. 2009. Using surface imaging and visual coaching to improve the reproducibility and stability of deep-inspiration breath hold for left-breast-cancer radiotherapy. *Phys Med Biol* 54(22):6853–65.

Chandra, A., L. Dong, E. Huang, et al. 2003. Experience of ultrasound-based daily prostate localization. *Int J Radiat Oncol Biol Phys* 56(2):436–47.

Chinnaiyan, P., W. Tomee, R. Patel, R. Chappel, and M. Ritter. 2003. 3D-ultrasound guided radiation therapy in the post-prostatectomy setting. *Technol Cancer Res Treat* 2(5):455–8.

Cremers, F., T. Frenzel, C. Kausch, D. Albers, T. Schonborn, and R. Schmidt. 2004. Performance of electronic portal imaging devices (EPIDs) used in radiotherapy: Image quality and dose measurements. *Med Phys* 31(5):985–96.

Cury, F. L., G. Shenouda, L. Souhami, et al. 2006. Ultrasound-based image guided radiotherapy for prostate cancer: Comparison of cross-modality and intramodality methods for daily localization during external beam radiotherapy. *Int J Radiat Oncol Biol Phys* 66(5):1562–7.

CyberKnife Society. 2005. *CyberKnife Radiosurgery Practical Guide ii.* 1st ed. Sunnyvale, CA: CyberKnife Society.

Dawson, L. A., and D. A. Jaffray. 2007. Advances in image-guided radiation therapy. *J Clin Oncol* 25(8):938–46.

Dieterich, S., and T. Pawlicki. 2008. CyberKnife image-guided delivery and quality assurance. *Int J Radiat Oncol Biol Phys* 71(1 Suppl):S126–30.

Dobler, B., S. Mai, C. Ross, et al. 2006. Evaluation of possible prostate displacement induced by pressure applied during transabdominal ultrasound image acquisition. *Strahlenther Onkol* 182(4):240–6.

Ebe, K., H. Shirato, A. Hiyama, et al. 2004. Integration of fluoroscopic real-time tumor tracking system and tomographic scanner on the rail in the treatment room. *Int J Radiat Oncol Biol Phys* 60:S604.

Falco, T., G. Shenouda, C. Kaufmann, et al. 2002. Ultrasound imaging for external-beam prostate treatment setup and dosimetric verification. *Med Dosim* 27(4):271–3.

Feigenberg, S. J., K. Paskalev, S. McNeeley, et al. 2007. Comparing computed tomography localization with daily ultrasound during image-guided radiation therapy for the treatment of prostate cancer: A prospective evaluation. *J Appl Clin Med Phys* 8(3):2268.

Fuller, C. D., C. R. Thomas, S. Schwartz, et al. 2006. Method comparison of ultrasound and kilovoltage X-ray fiducial marker imaging for prostate radiotherapy targeting. *Phys Med Biol* 51(19):4981–93.

Fuss, M., B. Salter, S. Cavanaugh, et al. 2004. Daily ultrasound-based image-guided targeting for radiotherapy of upper abdominal malignancies. *Int J Radiat Oncol Biol Phys* 59(4):1245–56.

Heilbrun, M. P. 2003. *CyberKnife Radiosurgery: A Practical Guide,* 94. Sunnyvale, CA: CyberKnife Society.

Hendee, W. R., and E. R. Ritenour. 2002. *Medical Imaging Physics.* 4th ed. New York: Wiley-Liss.

Herman, M. G. 2005. Clinical use of electronic portal imaging. *Semin Radiat Oncol* 15(3):157–67.

Herman, M. G., J. J. Kruse, and C. R. Hagness. 2000. Guide to clinical use of electronic portal imaging. *J Appl Clin Med Phys* 1(2):38–57.

Ho, A. K., D. Fu, C. Cotrutz, et al. 2007. A study of the accuracy of CyberKnife spinal radiosurgery using skeletal structure tracking. *Neurosurgery* 60(2 Suppl 1):ONS147–56; discussion ONS156.

Houdek, P. V., J. G. Schwade, C. F. Serage, et al. 1992. Computer controlled stereotaxic radiotherapy system. *Int J Radiat Oncol Biol Phys* 22(1):175–80.

Huang, E., L. Dong, A. Chandra, et al. 2002. Intrafraction prostate motion during IMRT for prostate cancer. *Int J Radiat Oncol Biol Phys* 53(2):261–8.

Hughes, S., J. McClelland, S. Tarte, et al. 2009. Assessment of two novel ventilatory surrogates for use in the delivery of gated/tracked radiotherapy for non-small cell lung cancer. *Radiother Oncol* 91(3):336–41.

Hugo, G. D., N. Agazaryan, and T. D. Solberg. 2002. An evaluation of gating window size, delivery method, and composite field dosimetry of respiratory-gated IMRT. *Med Phys* 29(11):2517–25.

Jaffray, D. A. 2007. Image-guided radiation therapy: From concept to practice. *Semin Radiat Oncol* 17(4):243–4.

Jaffray, D., P. Kupelian, T. Djemil, and R. M. Mecklis. 2007. Review of image-guided radiation therapy. *Expert Rev Anticancer Ther* 7(1):89–103.

Jani, A. B., J. Gratzle, E. Muresan, and M. K. Martel. 2005. Impact on late toxicity of using transabdominal ultrasound for prostate cancer patients treated with intensity modulated radiotherapy. *Technol Cancer Res Treat* 4(1):115–20.

Jereczek-Fossa, B. A., F. Cattani, C. Garibaldi, et al. 2007. Transabdominal ultrasonography, computed tomography and electronic portal imaging for 3-dimensional conformal radiotherapy for prostate cancer. *Strahlenther Onkol* 183(11):610–6.

Jin, J. Y., M. Ajlouni, S. Ryu, Q. Chen, S. Li, and B. Movsas. 2007. A technique of quantitatively monitoring both respiratory and nonrespiratory motion in patients using external body markers. *Med Phys* 34(7):2875–81.

Jin, J. Y., and F. F. Yin. 2005. Time delay measurement for linac based treatment delivery in synchronized respiratory gating radiotherapy. *Med Phys* 32(5):1293–6.

Jin, J. Y., F. F. Yin, S. E. Tenn, P. M. Medin, and T. D. Solberg. 2008. Use of the BrainLAB ExacTrac X-Ray 6D system in image-guided radiotherapy. *Med Dosim* 33(2):124–34.

Kirby, M. C., and A. G. Glendinning. 2006. Developments in electronic portal imaging systems. *Br J Radiol* 79 Spec No 1:S50–65.

Krengli, M., S. Gaiano, E. Mones, et al. 2009. Reproducibility of patient setup by surface image registration system in conformal radiotherapy of prostate cancer. *Radiat Oncol* 1–10.

Kupelian, P., T. Willoughby, A. Mahadevan, et al. 2007. Multi-institutional clinical experience with the Calypso System

in localization and continuous, real-time monitoring of the prostate gland during external radiotherapy. *Int J Radiat Oncol Biol Phys* 67(4):1088–98.

Kupelian, P. A., T. R. Willoughby, C. A. Reddy, E. A. Klein, and A. Mahadevan. 2008. Impact of image guidance on outcomes after external beam radiotherapy for localized prostate cancer. *Int J Radiat Oncol Biol Phys* 70(4):1146–50.

Langen, K. M., J. Pouliot, C. Anezinos, et al. 2003. Evaluation of ultrasound-based prostate localization for image-guided radiotherapy. *Int J Radiat Oncol Biol Phys* 57(3):635–44.

Lattanzi, J., S. McNeeley, S. Donnelly, et al. 2000. Ultrasound-based stereotactic guidance in prostate cancer—quantification of organ motion and set-up errors in external beam radiation therapy. *Comput Aided Surg* 5(4):289–95.

Lattanzi, J., S. Mc Neeley, W. Pinover, et al. 1999. A comparison of daily CT localization to a daily ultrasound-based system in prostate cancer. Int J Radiat Oncol Biol Phys 43(4): 719–25.

Lin, S. H., E. Sugar, T. Teslow, T. McNutt, H. Saleh, and D. Y. Song. 2008. Comparison of daily couch shifts using MVCT (TomoTherapy) and B-mode ultrasound (BAT System) during prostate radiotherapy. *Technol Cancer Res Treat* 7(4):279–85.

Little, D. J., L. Dong, L. B. Levy, A. Chandra, and D. A. Kuban. 2003. Use of portal images and BAT ultrasonography to measure setup error and organ motion for prostate IMRT: Implications for treatment margins. *Int J Radiat Oncol Biol Phys* 56(5):1218–24.

Mayse, M. L., P. J. Parikh, K. M. Lechleiter, et al. 2008. Bronchoscopic implantation of a novel wireless electromagnetic transponder in the canine lung: a feasibility study. *Int J Radiat Oncol Biol Phys* 72(1):93–8.

McNair, H. A., S. A. Mangar, J. Coffey, et al. 2006. A comparison of CT- and ultrasound-based imaging to localize the prostate for external beam radiotherapy. *Int J Radiat Oncol Biol Phys* 65(3):678–87.

Moore, C., F. Lilley, V. Sauret, M. Lalor, and D. Burton. 2003. Opto-electronic sensing of body surface topology changes during radiotherapy for rectal cancer. *Int J Radiat Oncol Biol Phys* 56(1):248–58.

Morr, J., T. Dipetrillo, J. S. Tsai, M. Engler, and D. E. Wazer. 2002. Implementation and utility of a daily ultrasound-based localization system with intensity-modulated radiotherapy for prostate cancer. *Int J Radiat Oncol Biol Phys* 53(5):1124–9.

Mould, R. F. 2005. *Robotic Radiosurgery.* vol. 1. Sunnyvale, CA: CyberKnife Society Press.

Muacevic, A., C. Drexler, A. Schweikard, et al. 2007. Technical description, phantom accuracy, and clinical feasibility for single-session lung radiosurgery using robotic image-guided real-time respiratory tumor tracking. *Technol Cancer Res Treat* 6(4):321–8.

Muacevic, A., M. Staehler, C. Drexler, B. Wowra, M. Reiser, and J. C. Tonn. 2006. Technical description, phantom accuracy, and clinical feasibility for fiducial-free frameless real-time image-guided spinal radiosurgery. *J Neurosurg Spine* 5(4):303–12.

Nioutsikou, E., Y. Seppenwoolde, J. R. Symonds-Tayler, B. Heijmen, P. Evans, and S. Webb. 2008. Dosimetric investigation of lung tumor motion compensation with a robotic respiratory tracking system: An experimental study. *Med Phys* 35(4):1232–40.

Novelline, R. A., and L. F. Squire. 1997. *Squire's Fundamentals of Radiology.* 5th ed. Cambridge, MA: Harvard University Press.

Ozhasoglu, C., C. B. Saw, H. Chen, et al. 2008. Synchrony—CyberKnife respiratory compensation technology. *Med Dosim* 33(2):117–23.

Paskalev, K., S. Feigenberg, R. Jacob, et al. 2005. Target localization for post-prostatectomy patients using CT and ultrasound image guidance. *J Appl Clin Med Phys* 6(4):40–9.

Peignaux, K., G. Truc, I. Barillot, et al. 2006. Clinical assessment of the use of the Sonarray system for daily prostate localization. *Radiother Oncol* 81(2):176–8.

Peng, C., K. Kainz, C. Lawton, and X. A. Li. 2008. A comparison of daily megavoltage CT and ultrasound image guided radiation therapy for prostate cancer. *Med Phys* 35(12):5619–28.

Pinkawa, M., M. Pursch-Lee, B. Asadpour, et al. 2008. Image-guided radiotherapy for prostate cancer. Implementation of ultrasound-based prostate localization for the analysis of inter- and intrafraction organ motion. *Strahlenther Onkol* 184(12):679–85.

Rogus, R. D., R. L. Stern, and H. D. Kubo. 1999. Accuracy of a photogrammetry-based patient positioning and monitoring system for radiation therapy. *Med Phys* 26(5):721–8.

Saw, C. B., K. M. Ayyangar, W. Zhen, M. Yoe-Sein, S. Pillai, and C. A. Enke. 2002. Clinical implementation of intensity-modulated radiation therapy. *Med Dosim* 27(2):161–9.

Sawant, A., R. L. Smith, R. B. Venkat, et al. 2009. Toward submillimeter accuracy in the management of intrafraction motion: the integration of real-time internal position monitoring and multileaf collimator target tracking. *Int J Radiat Oncol Biol Phys* 74(2):575–82.

Seiler, P. G., H. Blattmann, S. Kirsch, R. K. Muench, and C. Schilling. 2000. A novel tracking technique for the continuous precise measurement of tumour positions in conformal radiotherapy. *Phys Med Biol* 45(9):N103–10.

Seppenwoolde, Y., R. I. Berbeco, S. Nishioka, H. Shirato, and B. Heijmen. 2007. Accuracy of tumor motion compensation algorithm from a robotic respiratory tracking system: A simulation study. *Med Phys* 34(7):2774–84.

Sjodahl, M., and P. Synnergren. 1999. Measurement of shape by using projected random patterns and temporal digital speckle photography. *Appl Opt* 38(10):1990–7.

Smith, R. L., K. Lechleiter, K. Malinowski, et al. 2009. Evaluation of linear accelerator gating with real-time electromagnetic tracking. *Int J Radiat Oncol Biol Phys* 74(3):920–7.

Szabo, T. L. 2004. *Diagnostic Ultrasound Imaging: Inside Out.* Amsterdam: Elsevier Academic Press.

Thariat, J., J. Castelli, S. Chanalet, S. Marcie, H. Mammar, and P. Y. Bondiau. 2009. CyberKnife stereotactic radiotherapy for spinal tumors: value of computed tomographic myelography in spinal cord delineation. *Neurosurgery* 64(2 Suppl):A60–6.

Trichter, F., and R. D. Ennis. 2003. Prostate localization using transabdominal ultrasound imaging. *Int J Radiat Oncol Biol Phys* 56(5):1225–33.

Vedam, S. S., V. R. Kini, P. J. Keall, V. Ramakrishnan, H. Mostafavi, and R. Mohan. 2003. Quantifying the predictability of diaphragm motion during respiration with a noninvasive external marker. *Med Phys* 30(4):505–13.

Wang, M., R. Rohling, N. Archip, and B. Clark. 2006. 3D ultrasound-based patient positioning for radiotherapy. *Proc SPIE* 6141:61411k.

Wang, L. T., T. D. Solberg, P. M. Medin, and R. Boone. 2001. Infrared patient positioning for stereotactic radiosurgery of extracranial tumors. *Comput Biol Med* 31(2):101–11.

Watanabe, Y., and L. L. Anderson. 1997. A system for nonradiographic source localization and real-time planning of intraoperative high dose rate brachytherapy. *Med Phys* 24(12):2014–23.

Wein, W., B. Roper, and N. Navab. 2005. Automatic registration and fusion of ultrasound with CT for radiotherapy. *Med Image Comput Comput Assist Interv* 8(Pt 2):303–11.

Wen, N., H. Guan, R. Hammoud, et al. 2007. Dose delivered from Varian's CBCT to patients receiving IMRT for prostate cancer. *Phys Med Biol* 52(8):2267–76.

Willoughby, T. R., A. Forbes, D. Buchholz, et al. 2006. Evaluation of an infrared camera and X-ray system using implanted fiducials in patients with lung tumors for gated radiation therapy. *Int J Radiat Oncol Biol Phys* 66(2):568–75.

Willoughby, T. R., P. A. Kupelian, J. Pouliot, et al. 2006. Target localization and real-time tracking using the Calypso 4D localization system in patients with localized prostate cancer. *Int J Radiat Oncol Biol Phys* 65(2):528–34.

Winey, B., P. Zygmanski, and Y. Lyatskaya. 2009. Evaluation of radiation dose delivered by cone beam CT and tomosynthesis employed for setup of external breast irradiation. *Med Phys* 36(1):164–73.

Wu, R. Y., K. V. Ling, and W. S. Ng. 2000. Automatic prostate boundary recognition in sonographic images using feature model and genetic algorithm. *J Ultrasound Med* 19(11):771–82.

Wurm, R. E., F. Gum, S. Erbel, et al. 2006. Image guided respiratory gated hypofractionated Stereotactic Body Radiation Therapy (H-SBRT) for liver and lung tumors: Initial experience. *Acta Oncol* 45(7):881–9.

Xing, L., B. Thorndyke, E. Schreibmann, et al. 2006. Overview of image-guided radiation therapy. *Med Dosim* 31(2):91–112.

Yan, H., F. F. Yin, G. P. Zhu, M. Ajlouni, and J. H. Kim. 2006. The correlation evaluation of a tumor tracking system using multiple external markers. *Med Phys* 33(11):4073–84.

11

Offline Plan Adaptation Strategies

James Balter
University of Michigan

Dan McShan
University of Michigan

11.1 Introduction

Throughout this book, various concepts of plan adaptation are reviewed, introduced, and discussed. Parallel to the need, potential, and mechanisms for such modifications lie the strategies that make such adaptation practical in a clinical setting. This chapter discusses some of the common issues related to one of the temporal scales of plan modification in which decisions and changes are not made while the patient is waiting in the treatment room. Such "offline" strategies have evolved over nearly two decades of research and clinical trials, and as such a number of the practical issues have reached a reasonable level of maturity in evaluation and implementation. Nonetheless, offline adaptation can hardly be considered common.

11.2 Strategies for Management of Changes

Figure 11.1 shows the overall flowchart for plan adaptation. This concept is not new, having been discussed by several investigators (Yan et al. 1998; de Boer and Heijmen 2001). In its basic form, the concept is simple. An initial treatment plan is developed for treating the patient. This plan is based on historical knowledge of previous patients with similar anatomy, positioning, immobilization, and disease. As such, assumptions are made, particularly with respect to the geometric variations that the patient is expected to undergo as treatment progresses. The potential exists, however, to "learn" about specific patient issues, including, but not limited to, geometric variations, and to apply this knowledge to modify the treatment plan (as well as strategies such as the frequency and tolerance of setup verification) to the benefit of the individual patient. The trick to doing so is to gain and act on this knowledge in a sufficiently timely fashion, so that

the patient can benefit from the modification of the remaining treatment fractions.

At present, one can imagine multiple forms of knowledge that can be gained to individualize a treatment. The three most obvious attributes are geometric, dosimetric, and biological. Understanding how patients deviate from their initial planning assumptions in any of these areas may encourage plan modification with potential benefit to the patients. While biological adaptation is an area of active investigation and the subject of Chapters 1, 6, and 7 in this book, in its current state it is not in routine clinical use, and thus this chapter will focus more on geometric and dosimetric indications of patient variation.

11.2.1 Early Geometric Strategies for Offline Management of Changes

Geometric offline strategies were initially proposed with respect to patient positioning rules in the early 1990s (Bel, van Herk, and Lebesque 1996). The motivating forces for offline protocols based on measurements of geometric variations were related to the speed with which measurements and decisions could be enacted at the treatment unit. At that time, portal imaging systems were the primary imaging tools for patient positioning. These systems were slow, often required high patient doses to aid imaging, suffered from poor image quality, and had limited software tools for online image analysis.

While early efforts at developing image analysis tools understood the value and need for rapid, online measurement of position, the practical clinical utilization of this emerging technology demanded exploration of methods to retrospectively evaluate patient position. Critical to the development of such offline strategies was the realization of the stochastic nature of patient positioning, specifically as related to the radiotherapy environment,

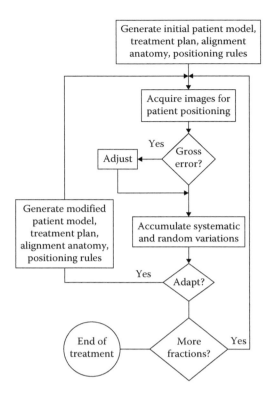

FIGURE 11.1 Schematic of workflow for offline adaptive radiotherapy.

in which the reference patient state is defined not from an average over many treatments, but rather a single sampling of the patient geometry established at simulation. As the images from the simulated patient are used for treatment planning, they are the reference state that subsequent positioning attempts to reproduce, even though the patient shape, movement, and configuration are most likely to differ from all subsequent samples at this state (Herman et al. 2001).

Offline positioning strategies developed out of this environment. Analysis of early portal images as well as digitized portal films showed distributions of patient position around the ideal location (Hunt et al. 1995; Lebesque et al. 1992). When plotted, it was typically seen that these distributions formed a cloud of positions, with a center that was typically not at the isocenter. By convention, the center of this distribution came to be called the *systematic variation* of the patient and analyses of the spread in different directions characterized components of *random variation*. Offline strategies emerged to reduce or remove systematic errors, providing significant benefit for not only reducing the overall workload in analysis of patient position but more importantly increasing the potential to perform such analysis away from the busy treatment unit.

A number of offline positioning strategies have been implemented. A driving force in the design of these strategies is the estimation of the minimal number of measurements needed to accurately estimate the systematic patient offset, as well as the frequency of monitoring of patient position after correcting for this "average" position. A number of strategies have been applied,

the most common being the No Action Level (NAL) protocol (de Boer and Heijmen 2001). In this protocol, three to five measurements of patient position are taken, the average is adjusted, and no further position monitoring is performed. A small variation of this protocol follows the systematic error correction with weekly portal images, but with a large tolerance for position variation prior to further action. This is justified by the expectation that on any given day, the patient's position contains an arbitrary offset due to the random components of variation. Both empirical investigations as well as analytical simulations have been performed to optimize the frequency and timing of measurements for geometric adaptation. Bayesian analyses have shown that the systematic error can be reasonably estimated in a very small number of measurements (two to five), while the random variation may require significantly more samples (Lam et al. 2005). Given the limited dependence of the accuracy of estimation of random error on margins, such studies support strategies such as the NAL protocol.

11.2.2 Advanced and Dosimetric Strategies for Offline Management of Changes

As geometric rules for adaptation were being introduced, research groups began investigating the dosimetric impact of geometric variations. In one of the earliest studies, Bel et al. (1996a,b) reported that, while systematic geometric variations had significant impact on the deviation between planned and delivered doses to targets, the influence of random variations (for prostate cancer treated with open fields in this case) was far lesser. This was primarily due to two factors: specifically, the nature of dose falloff and the dosimetric *coverage* definition permitting underdose or undercoverage of a small fraction of the population. While relatively straightforward, this premise and the underlying physics and statistics are still fundamental to any modern decisions to modify plans to enhance the dosimetric coverage of tumors and/or sparing of normal tissues.

A number of early investigators understood the complex nature of geometrical variations and their dosimetric impact. The next major development in adaptive therapy involved the customization of margins as well as measurement frequencies for safe and effective treatment plan implementation. By understanding the space in which the target was likely to vary in position, new margins could be made that could provide appropriate coverage in the face of random errors, while minimizing workload at the treatment unit. Although a modified plan needed to be implemented after five or so fractions, this methodology had numerous advantages in allowing for efficient and robust treatment delivery.

As in-room CT techniques and other complex localization systems have evolved, however, our knowledge of the complexities of changes that occur over a treatment course has now advanced to the point where such simple solutions are seen as insufficient in understanding the impact of true patient variations. We can now view larger rotational variations in position than previously expected from radiographic evaluation of setup.

We can also see variations in spinal articulation. In addition, we can observe changes in tumor size, ventilatory configuration of the patient, and other random and systematic changes related to the patients physiology and health state.

The primary sites that appear to be candidates for offline adaptation are the thorax and neck. Specifically, oropharyngeal tumors and certain types of lung tumors have been known to shrink considerably over a course of fractionated treatment. These changes have been well-documented, and fueled speculation that the change in patient configuration is an opportunity for plan modification to achieve increased tumor dose, reduced treatment toxicity, or both. It should be noted that very few experiments to date have shown significant benefit to these approaches, although this area is still a focus of active research and development. More importantly, the same methods necessary for plan modification to take advantage of subtle geometric changes can also be applied for more unexpected and gross variations in patient anatomy.

11.3 Components of an Offline System

Although the current intent of offline adaptation is more advanced than early methods to remove the bias of positioning from simulation, the paradigm and workflow are basically the same. Essentially, one needs to acquire the data from which a statistical assessment of trends in the baseline state can be made. Based on these data and assessments, an action plan and future set of tolerances for further refinement need to be established.

What has changed, however, is the infrastructure that is needed to support these decisions. Previously, offline measurements were typically two-dimensional image registrations of portal images and reference radiographs (simulation films or digitally reconstructed radiographs). While these images and analyses are useful for image guidance and generic margin modifications, they are insufficient for analyzing complex trends in soft-tissue location from variations in patient configuration as well as from therapeutic response. Ultimately, some form of volumetric imaging will drive offline adaptation decisions. Such images are most likely morphologic (e.g., in-room CT data from cone-beam CT [CBCT] or other systems), although functional and molecular image data from MRI, PET, and other imaging modalities may ultimately have a far greater influence on the actual implementation of meaningful plan changes.

Given that the evolving patient model through time consists of multiple volumetric images, some form of high-dimensional alignment is needed to relate these sampled patient states. Global rigid alignment, consisting of translations about three axes and rotations about three axes, will not completely resolve the differences in patient position in most of the body parts (the brain is the most likely case where rigid alignment is sufficient). Two leading options to handle variations in patient configuration are local rigid alignment and deformable alignment or modeling.

Local rigid alignment is most amenable to early adoption in a clinical setting. By selecting a point or a local region of anatomy that needs to be maintained in position or orientation, simple translation and potential rotation can be applied to maintain position. Tools for selecting such local regions include visual alignment, use of "clip boxes" that remove distal anatomy from calculation of cost metrics in automated alignment, and masking of anatomy for inclusion or exclusion, with similar effect. Figure 11.2 shows examples of local alignment to optimize cord sparing.

The consequence of local alignment is that the remaining patient anatomy distal from this region of interest may vary significantly from that initially planned. In certain instances, such a trade-off of simplicity vs. greater uncertainty at a distance may be completely justified and be of potential benefit as well. Consider stereotactic body radiotherapy treatments, in which the dose decreases significantly over relatively short distances. The more distal tissue sees very low doses, and thus large position variations in such regions will not likely increase the risk of toxicity. Instead, random variation in the paths of beams entering and exiting distal from the target will further spread out lower doses. There is a concern about dosimetric changes due to varying beam path lengths with rotations and shape changes. While these are typically second-order effects, and diminish in importance as the number of beams is increased, they should still be taken into account when considering the extent of local anatomy that must be aligned.

Deformable alignment has been discussed in Chapters 2 and 5 in this book. There are many methods currently under investigation for measuring the appropriate deformable transformation to relate sets of volumetric images of the patient acquired at different states or times. These methods are potentially very powerful and generate very high-order deformation maps in an automated fashion. Regardless of the method used, extensive visual validation of deformable alignment results will be critical to using this class of alignment tools for adaptive radiotherapy. A number of recent studies have attempted to characterize the accuracy of deformable alignment methods used in radiotherapy.

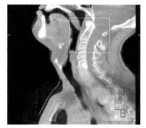

FIGURE 11.2 Local rigid alignment. Spinal positioning to provide tumor coverage and avoid cord overdose for a head and neck patient. A "clipbox" (white square) describes the region in which rigid alignment is performed to optimize the match between subregions of the treatment planning and cone-beam CT scans. Matched intensities are visible as grayscale values from mixing the overlaid colors. More distant anatomy is less accurately aligned, as can be seen by the color separation in the lower region of the spine. (Courtesy of Jan-Jakob Sonke, Netherlands Cancer Institute.)

Two of these investigations were multi-institutional and reached similar conclusions (Kashani et al. 2008; Brock 2010). Figure 11.3 shows a typical result. While the "average" accuracy of deformable alignment is around 2–3 mm, some locations have far larger errors, and these locations may vary by the type of alignment method used.

Following alignment, anatomic structures, which are defined prior to the start of treatment, need to be tracked through the series of new volumetric data acquired during treatment. This involves, at a minimum, application of the measured transformations to the structures. Automated and deformation-driven contouring has been the subject of extensive research and development in recent years (Wu, Spencer, et al. 2009; Godley et al. 2009, Xie et al. 2008; Ahunbay et al. 2008; Chao, Xie, and Xing 2008; Massoptier and Casciaro 2007; Lu et al. 2006a,b; Nanayakkara, Samarabandu, and Fenster 2006, Lu et al. 2006a,b). With deformation, an active decision about the validity of mapping all or part of an organ between images needs to be made, especially if the volume of that structure changes. Figure 11.4 shows an example of automated contour propagation from deformable alignment of a treatment planning CT scan to a series of CBCT image data sets acquired over a course of treatment.

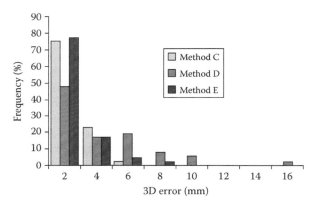

FIGURE 11.3 Histogram distributions of alignment accuracy for three different alignment methods, all employing b-splines for deformation. (Reprinted from Kashani, R., et al. 2008. *Med Phys* 35:5944–53. With permission.)

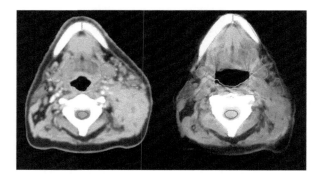

FIGURE 11.4 Treatment planning CT (left), with contours propagated via automated deformable alignment to a cone-beam CT scan from the last fraction (right) of treatment.

Given that the anatomy is defined over the course of multiple volumetric images, the next step in the adaptive therapy involves estimation of the accumulated dose delivered to tumors and normal tissues. A few methods exist and are discussed in more detail in Chapter 2. Essentially, the options are (1) map the dose from the initial treatment plan onto the newly configured anatomy; (2) recalculate the dose by applying the beam fluence from the treatment plan to the current volumetrically derived electron density map, and accumulate the dose from the mapped structures with their individual dose calculations; and (3) deform the treatment planning CT model into the coordinates of the on-site volumetric image, and use this for dose recalculation followed by deformation-driven structural dose accumulation. Option 1 is arguably the least accurate, and option 2 has the greatest potential for accuracy, given sufficiency of the in-room imaging to provide an accurate map of electron density. In the absence of this condition (e.g., incomplete sampling of CBCT due to field-of-view limitation), option 3 may provide a reasonable approximation. Figure 11.5 shows dose–volume histograms accumulated by each of the three methods described above.

In all of the options, the dose needs to be reported with respect to a frame of reference. The most logical frame is on an organ-by-organ basis. Here, a few issues arise with respect to deformation and biological models. If a "serial" organ or structure (i.e., one in which the risk of a negative event is believed to be related to highest or lowest dose to a small subvolume of the encompassed tissue) is located in or adjacent to a steep dose gradient, then local uncertainties in mapping tissue locations may erroneously overestimate or underestimate the cumulative dose, and thus lead to either needless or even potentially dangerous corrections to a treatment plan. If a more "parallel" tissue type undergoes a significant volume change, then the process of dose accumulation is ill-defined. Caution needs to be applied in analyzing dose distributions accumulated on these organs to evaluate changes in risk estimates, especially if biological corrections are to be applied for dose-per-fraction effects.

11.4 Technical Challenges

Along with the potential value of dose accumulation and plan adaptation come a number of technical and infrastructure challenges. One major issue that arises is the proper handling of data. Not only do multiple volumetric image data sets need to be stored, their geometric and temporal relationships need to be established and maintained. As the transformations relating these data are typically deformable, existing DICOM-RT support is insufficient to handle the complete information set. As the frequency of imaging and plan modification may be variable, a robust infrastructure for dose summation and reporting (e.g., the schematic of the "dose-to-date" reporting infrastructure from the University of Michigan, Figure 11.6) is necessary.

Furthermore, there may be an additional transformation applied to each image set acquired during treatment, as the patient is likely positioned via some form of rigid transformation

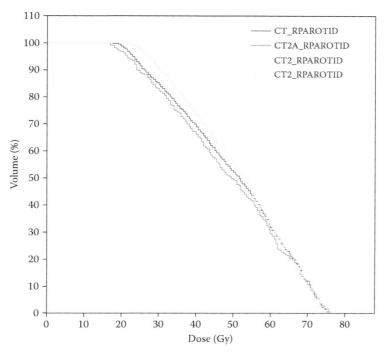

FIGURE 11.5 (See color insert following page 204.) Dose–volume histograms for a head and neck intensity-modulated radiation therapy plan evaluated on a cone beam CT image volume acquired during treatment. The original treatment plan and dose calculation is in red, and doses evaluated on the CBCT data are processed by calculation using the CBCT for density correction (green), calculation on the deformed planning CT scan (white), and deformation of the original dose calculation (brown).

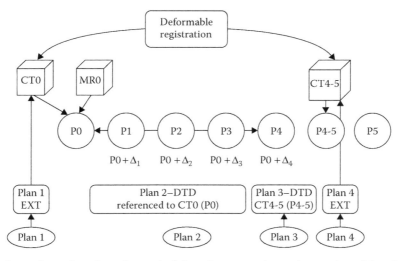

FIGURE 11.6 "Dose-to-date" paradigm, describing the methodology for summation and reporting of dose based on periodic volumetric imaging to update dose estimations and occasional plan modification.

used for alignment prior to treatment delivery. This transformation is separate from that directly aligning the treatment planning CT and in-room image data. At the least, even with the same transformation, there may have been measurement error manifest in the on-treatment positioning. At worst, higher-order transformations should be combined with the alignment performed in-room, in order that any dose accumulation is done in the frame of reference of the treated, not the imaged, patient. Thus, two transformation sets need to be related to typical image data sets, further stretching the limits of current image and data storage formats.

The workflow of offline adaptive therapy needs to be considered. It is very likely that a physician will need to approve adapted plans. The decision process may further require a step of cumulative dose reporting, followed by a decision as to whether or not to adapt, approval of (automatically generated) contours, and finally plan approval and implementation. Unless the majority of these steps can be either automated or transferred in whole or part by empowerment to other skilled individuals, the potential cost of routine plan adaptation may be prohibitive. Until such a workflow model is optimized, it is more likely that plan adaptation will occur either sparsely or even on demand in reaction to significant geometric variations observed during treatment.

11.5 Issues That Need to Be Considered for Efficacy of Offline Adaptation

One can certainly imagine a time when offline (and potentially online or live) plan adaptation becomes a part of routine practice. However, there are some fundamental issues that need to be addressed regardless of the speed, accuracy, and efficiency of the adaptive infrastructure.

One area of current debate involves the management of tumors that routinely vary in size in response to therapy (Wu et al. 2009; Woodford et al. 2007, Kupelian et al. 2005, Seibert et al. 2007). For certain tumors (e.g., nonsmall-cell lung cancer), it is generally believed that clonogenic tumor cells exist beyond the dense lesion borders on the planning CT. While generally this uncertainty is handled through the application of a clinical target volume expansion, there is no information to date describing the nature of potential changes to the spatial distribution of viable tumor cells in the periphery of a shrinking tumor. In such circumstances, assuming that the Clinical Target Volume (CTV) can be safely reduced in size in some way proportional to the reduction in gross tumor volume can be a dangerous hypothesis. Furthermore, complex interactions between tumor boundaries and other dense structures may confound the accuracy of the deformable models used to guide target shape modification (Guckenberger et al. 2009). As we consider the integration of molecular imaging to aid in initial as well as adapted plans (Feng et al. 2009), these issues will become even more prevalent.

Other issues include both the empowered decision process as well as quality assurance for routinely modified treatment plans. While this is obviously critical for online modifications of plans, the workflow of offline plan adaptation still demands very rapid processing and safe decisions to adapt plans as well as sufficient certainty that a modified plan, especially intensity-modulated radiation therapy (IMRT), is still safe to implement. Efficient user interfaces are necessary to show the impact of geometric changes on the cumulative dose to the patient, as well as estimates of the impact of modified plans on the total dose delivered. One issue that is not currently being addressed is the reporting of cumulative doses from the original and modified plans, including the uncertainty of reporting these doses. Such uncertainties include errors in image alignment, assumptions used in voxel-based dose accumulation, and assumptions about the impact of random variations in yet-undelivered fractions on dose reporting. These areas are being actively investigated, and it is hoped that user interfaces for adaptive therapy include such uncertainties in the workflow of estimating the need for, as well as implementing, offline plan modifications.

11.6 Summary

Offline adaptive processes have proven very successful as adjuncts to optimize the use of portal imaging for patient positioning. As we move into an era of routine volumetric imaging, faster computation, and access to advanced alignment tools, the technical challenges supporting routine, dose-based plan adaptation are certainly not insurmountable. These tools and the resulting workflow will present a significant paradigm shift for fractionated treatments, and it is important that appropriate levels of automation, data reporting, and quality assurance be applied as adaptive processes are implemented clinically.

References

Ahunbay, E. E., C. Peng, G. P. Chen, et al. 2008. An online replanning scheme for interfractional variations. *Med Phys* 35:3607–15.

Bel, A., M. van Herk, and J. V. Lebesque. 1996. Target margins for random geometrical treatment uncertainties in conformal radiotherapy. *Med Phys* 23:1537–45.

Bel, A., P. H. Vos, P. T. Rodrigus, et al. 1996. High-precision prostate cancer irradiation by clinical application of an offline patient setup verification procedure, using portal imaging. *Int J Radiat Oncol Biol Phys* 35:321–32.

Brock, K. K. 2010. Deformable Registration Accuracy Consortium. Results of a multi-institution deformable registration accuracy study (MIDRAS). *Int J Radiat Oncol Biol Phys* 76:583–96.

Chao, M., Y. Xie, and L. Xing. 2008. Auto-propagation of contours for adaptive prostate radiation therapy. *Phys Med Biol* 53:4533–42.

de Boer, H. C., and B. J. Heijmen. 2001. A protocol for the reduction of systematic patient setup errors with minimal portal imaging workload. *Int J Radiat Oncol Biol Phys* 50:1350–65.

Feng, M., F. M. Kong, M. Gross, et al. 2009. Using fluorodeoxy-glucose positron emission tomography to assess tumor volume during radiotherapy for non-small-cell lung cancer and its potential impact on adaptive dose escalation and normal tissue sparing. *Int J Radiat Oncol Biol Phys* 73:1228–34.

Godley, A., E. Ahunbay, C. Peng, et al. 2009. Automated registration of large deformations for adaptive radiation therapy of prostate cancer. *Med Phys* 36:1433–41.

Guckenberger, M., K. Baier, A. Richter, et al. 2009. Evolution of surface-based deformable image registration for adaptive radiotherapy of non-small cell lung cancer (NSCLC). *Radiat Oncol* 4:68.

Herman, M. G., J. M. Balter, D. A. Jaffray, et al. 2001. Clinical use of electronic portal imaging: Report of AAPM Radiation Therapy Committee Task Group 58. *Med Phys* 28:712–37.

Hunt, M. A., T. E. Schultheiss, G. E. Desobry, et al. 1995. An evaluation of setup uncertainties for patients treated to pelvic sites. *Int J Radiat Oncol Biol Phys* 32:227–33.

Kashani, R., M. Hub, J. M. Balter, et al. 2008. Objective assessment of deformable image registration in radiotherapy: A multi-institution study. *Med Phys* 35:5944–53.

Kupelian, P. A., C. Ramsey, S. L. Meeks, et al. 2005. Serial megavoltage CT imaging during external beam radiotherapy for non-small-cell lung cancer: Observations on tumor regression during treatment. *Int J Radiat Oncol Biol Phys* 63:1024–8.

Lam, K. L., R. K. Ten Haken, D. Litzenberg, et al. 2005. An application of Bayesian statistical methods to adaptive radiotherapy. *Phys Med Biol* 50:3849–58.

Lebesque, J. V., A. Bel, J. Bijhold, et al. 1992. Detection of systematic patient setup errors by portal film analysis. *Radiother Oncol* 23:198.

Lu, W., G. H. Olivera, Q. Chen, et al. 2006a. Automatic re-contouring in 4D radiotherapy. *Phys Med Biol* 51:1077–99.

Lu, W., G. H. Olivera, Q. Chen, et al. 2006b. Deformable registration of the planning image (kVCT) and the daily images (MVCT) for adaptive radiation therapy. *Phys Med Biol* 51:4357–74.

Massoptier, L., and S. Casciaro. 2007. Fully automatic liver segmentation through graph-cut technique. *Conf Proc IEEE Eng Med Biol Soc* 5243–6.

Nanayakkara, N. D., J. Samarabandu, and A. Fenster. 2006. Prostate segmentation by feature enhancement using domain knowledge and adaptive region based operations. *Phys Med Biol* 51:1831–48.

Seibert, R. M., C. R. Ramsey, J. W. Hines, et al. 2007. A model for predicting lung cancer response to therapy. *Int J Radiat Oncol Biol Phys* 67:601–9.

Woodford, C., S. Yartsev, A. R. Dar, et al. 2007. Adaptive radiotherapy planning on decreasing gross tumor volumes as seen on megavoltage computed tomography images. *Int J Radiat Oncol Biol Phys* 69:1316–22.

Wu, Q., Y. Chi, P. Y. Chen, et al. 2009. Adaptive replanning strategies accounting for shrinkage in head and neck IMRT. *Int J Radiat Oncol Biol Phys* 75:924–32.

Wu, X., S. A. Spencer, S. Shen, et al. 2009. Development of an accelerated GVF semi-automatic contouring algorithm for radiotherapy treatment planning. *Comput Biol Med* 39:650–6.

Xie, Y., M. Chao, P. Lee, et al. 2008. Feature-based rectal contour propagation from planning CT to cone beam CT. *Med Phys* 35:4450–9.

Yan, D., E. Ziaja, D. Jaffray, et al. 1998. The use of adaptive radiation therapy to reduce setup error: A prospective clinical study. *Int J Radiat Oncol Biol Phys* 41:715–20.

Online Adaptive Correction Strategies for Interfraction Variations

Ergun Ahunbay
Medical College of Wisconsin

X. Allen Li
Medical College of Wisconsin

12.1 Introduction

The daily (interfractional) variations in patient setup and anatomy have always been a concern for fractionated radiotherapy. These variations, if not accounted for, could result in suboptimal dose distributions and significant deviations from the original plan (Roeske et al. 1995; Deurloo et al. 2005; Stroom et al. 1999; O'Daniel et al. 2007; Barker et al. 2004). Image-guided radiation therapy (IGRT) has been used widely to correct (i.e., eliminate or reduce) the deteriorating effects of the interfractional variations (Jaffray et al. 2002; Jaffray 2005; Ghilezan et al. 2004; Letourneau et al. 2005; McBain et al. 2006; Yan et al. 2005; Paskalev et al. 2004; Smitsmans et al. 2005; Langen et al. 2005; Pouliot et al. 2005; Kupelian et al. 2005). A wide range of correction strategies has been developed based on the available IGRT technologies (Mohan et al. 2005; Ludlum et al. 2007; Létourneau et al. 2007; Rijkhorst et al. 2007). These correction methods can be generally classified as "online" (Court et al. 2005; Ludlum et al. 2007; Mohan et al. 2005) and "offline" approaches (Yan et al. 2000). In an online correction method, corrections to patients' treatment parameters are performed right after the daily patient information is acquired and before the daily treatment dose is delivered. This is in contrast to an offline correction, where the corrective action is taken after the daily treatment has been delivered, affecting the treatments on subsequent days. Therefore, when an online correction is applied, the delivered daily dose will be the corrected one using the very latest patient setup and anatomic information.

When the patient is set up for each fraction, the anatomy may be different from that used for the initial treatment planning. Typically, the deviations that are most harmful are the so-called systematic deviations, which are also relatively easier to account for by an either offline or online correction strategy. The random components of the deviations, although less harmful than the systematic deviations, are generally difficult to be fully accounted for and require an online correction strategy. One of the advantages of online correction strategies over offline methods is that online strategies can correct for both systematic and random variations. In addition, offline corrections may not be applicable to a course of therapy with a small number of treatment fractions, such as hypofractionated or stereotactic treatment regimens.

The chief challenge for an online correction is that it needs to be performed within an acceptable time frame while the patient is lying on the treatment table in the treatment position. This requirement limits a variety of corrective actions that are used as online strategies in today's technology. Consequently, the online correction of interfractional variations by repositioning the patient based on images acquired immediately before the treatment delivery is the current standard practice for IGRT. The online repositioning strategies practiced in most clinics are limited to correct for translational shifts only, failing to account for rotational errors, organ deformations, and independent motions between different organs. This is in spite of the fact that the current technology provides enough information to perform much more detailed modifications to the daily treatment than simple translational shifts. In principle, the data required to generate a new treatment plan for the day is available in today's CT-based IGRT practice, but by using the data only for shifting the patient, the full potential of the IGRT is not exploited.

There has been ongoing research to extend the online corrections from the table shifting to plan modifications that can correct organ rotations and deformations. Advances in computer technology allow for computationally intensive operations to be performed within a reasonable time frame. Therefore, online correction schemes that are more comprehensive than online repositioning are beginning to move into the clinic.

Corrections for rotations, in addition to translations, have been implemented, at least partially, in several technologies (e.g., tomotherapy). Ideas for rotational correction by gantry, collimator, and possibly, the table rotations are desirable because such methods do not require additional hardware and do not cause additional burden to the treatment personnel (Rijkhorst et al. 2007; Yue et al. 2005). Dedicated rotatable couch mechanisms (e.g., robotic table) have also been devised and proposed for clinical uses (Guckenberger et al. 2007).

In addition to the rotational and translational deviations, organ deformations are a major concern that can be handled by online replanning methods, including rapid online plan modifications and full-blown plan reoptimization based on the anatomy of the day (Ahunbay et al. 2008; Mestrovic et al. 2007). Quickly adjusting beam aperture shapes and weights based on the CT of the day (the anatomy of the day) is an example of an online plan modification method (Court et al. 2005; Mohan et al. 2005; Feng et al. 2006; Ahunbay et al. 2008).

In this chapter, we review the available online correction methodologies (both proposed and implemented) ranging from translational or rotational repositioning based on rigid-body registration to replanning based on aperture morphing or on full-scale reoptimization. The advantages, disadvantages, and challenging issues of these strategies will be discussed.

12.2 Online Repositioning

12.2.1 Translational Repositioning

12.2.1.1 Significance of Interfractional Patient and Organ Shifts

Lerma et al. (2009) reported that translational shifts of the patient anatomy are the major factor contributing to the interfractional variations for many anatomic sites. For example, in prostate radiotherapy, factors such as organ deformation and rotation have been reported to be relatively small or less frequent compared to organ motion (Deurloo 2005). Consequently, the most commonly used strategy for online correction of interfractional variation in today's IGRT practice is repositioning the patient by shifting the treatment couch in three translational directions based on rigid-body image registration of the planning image and the image taken on the day immediately before treatment delivery. This method intends to reposition the patient at the same position that the treatment planning image was acquired.

This strategy has been performed using either two-dimensional (2D) or three-dimensional (3D) images. 3D images with good soft-tissue contrast offer better accuracy and confidence, resulting in increased delivery accuracy, which in turn leads to reductions in the planning target volume–clinical target volume (PTV-CTV) margin (Beltran, Herman, and Davis 2008). The online repositioning method requires considerable time that is demonstrated to be manageable in routine clinical flow and tolerable for the patient. Although online translational repositioning can handle the most of the interfractional variations for many anatomic sites, it fails to compensate for organ deformation and rotation and independent motion of structures (Mohan et al. 2005; Court et al. 2005).

12.2.1.2 Procedure for Translational Shifts

The procedure for online translational repositioning commonly practiced in clinics include (1) using a dedicated software to compute the translational shifts by matching the daily images with the reference images (e.g., the planning images used for the original planning) with either bony anatomy or soft-tissue-based rigid-body registration and (2) adjusting the patient's position by moving the treatment couch according to these translational shifts prior to the treatment delivery. Image registration in online repositioning may be accomplished by either manual alignment or image intensity–based autoregistration. Autoregistration has the advantage over manual registration methods of improving consistency and reducing user-dependence. Users may manually manipulate the shifts, if necessary, even using autoregistration methods. One method of such manual manipulation is to align the anatomy from the daily CT to the volume delineated based on the planning images. Usually, good image quality of the daily images would help such manual manipulation. Generally, there is always interobserver variability (Court et al. 2004) and subjectivity associated with a manual registration, which is affected by the image quality. Compared to using the structure shape, a better approach would be to use the dose cloud or isodose lines from the original plan to align the anatomy based on the daily images. However, the dose cloud or the isodose lines do not represent the dose that the patient actually received. With fast dose-computation software and hardware becoming available, it could be possible in the near future to update the dose cloud or the isodose lines based on the daily images so that the patient could be more appropriately repositioned.

12.2.1.3 Technical Approaches

A variety of IGRT systems are available for online repositioning. Comparisons between different systems in terms of their accuracy can be found in Chapter 10 of this book or in the literature (Peng et al. 2008; Kitamura, Court, and Dong 2003). Volumetric imaging is generally believed to be more accurate than 2D planar image-based methods, such as electronic portal imaging (Li et al. 2008) and ultrasound imaging (Peng et al. 2008). For example, for pelvic regions such as the prostate and cervix, image modalities with low soft-tissue contrast are generally inaccurate. Implanting radiopaque markers can improve the visualization of the target location even with 2D planar radiographic images (van den Heuvel et al. 2006; Vigneault et al. 1997).

The most widely used IGRT imaging modalities in clinics are in-room CT and on-board CT, including (1) megavoltage (MV) fan-beam CT (Langen et al. 2005), (2) kV fan-beam CT (e.g., CT-on-rails; Court et al. 2003), (3) MV cone-beam CT (Pouliot et al. 2005), and (4) kV cone-beam CT (Jaffray et al. 2002). Generally, the kV fan-beam CT provides the best image quality, and the cone-beam CT offers the best integration and best spatial resolution in the superior-inferior (SI) direction (not dependent on the slice thickness as for fan-beam CT). The MV cone-beam CT suffers from poor soft-tissue contrast and higher imaging doses. The image acquisition and reconstruction times for the cone-beam CT are generally longer than those for the fan-beam CT. Tomosynthesis (Godfrey et al. 2006) can have a faster acquisition than the cone-beam CT by reducing the number of projections, that is, the gantry rotation range. Implanting transponder markers as an IGRT method (Balter et al. 2005; Kupelian et al. 2007) can offer advantages such as providing real-time localization information without reconstruction or reprocessing.

12.2.2 Rotational Repositioning

12.2.2.1 Significance of Interfractional Patient and Organ Rotation

In addition to the translational shifts, interfractional patient and organ rotations also exist (Rijkhorst et al. 2007; Zhang et al. 2006). These rotations have been largely ignored in the current online IGRT methods. Figure 12.1 depicts three possible rotations: yaw, pitch, and roll. *Yaw* is the rotation along the anterior–posterior (AP) axis, *pitch* is the rotation in the left–right (LR) axis, which is sometimes referred to as "tilt," and *roll* is the rotation in the SI axis. Along with the three translational directions, the repositioning consideration in both translation and rotation is often referred to as the correction for 6-degree freedom (6DoF). Hardware and software is being introduced to correct for one or more of the rotational variations. The tomotherapy system can correct for the roll error by adjusting the gantry angles of the beams (Boswell et al. 2005), which theoretically can also be applied for linear accelerator (linac)-based approaches. Devices to correct for the tilt and roll that can be mounted on top of the

existing couch have been developed (Hornick, Litzenberg, and Lam 1998). Robotic couches that can perform all six degrees of motion were proposed as early as 1965 (Stewart 1965). These couches have been implemented in clinics in conjunction with CT guidance (Guckenberger et al. 2007).

Redpath, Wright, and Muren (2008) studied the benefits of rotational corrections for bladder and prostate volumes after translational repositioning. They reported that translational repositioning accounted for the most interfractional errors, while rotational corrections had an impact in a small fraction of treatments. For prostate, the dominant anatomic rotation is around the LR axis, that is, the pitch rotation (van Herk et al. 1995; Rijkhorst et al. 2007). The daily variation in the rectum and bladder filling can cause a rotation of approximately 4° standard deviation, and days with more than 10° rotations are not uncommon. The rotation in the head and neck tumor region is mainly due to the difficulty in reproducing the head tilt, and therefore also dominated by the pitch (Zhang et al. 2006), although the magnitudes tend to be smaller. For the prostate, the concern for rotation increases when the seminal vesicles are included in the target volume.

12.2.2.2 Technical Approaches

It is generally not easy to perform rotational corrections in all three directions, that is, yaw, pitch, and roll, with a standard IGRT system. For treatments using coplanar beams, the corrective action for the roll is the easiest one, which can be done by adjusting the gantry angle of each beam according to the roll error. This approach has been implemented in the tomotherapy system (Boswell et al. 2005). Similarly, the correction for yaw can be done by adjusting couch angle. However, this adjustment is not as easy as rotating gantry with a conventional treatment couch. For the tomotherapy system, this couch rotation is not possible. Alternatively, because only a small portion of the patient's body is irradiated at any instant during the tomotherapy treatment, the translational shifts in the LR and AP-PA directions as the couch moves in the SI direction could approximately account for the rotation in yaw and pitch directions, respectively. This idea has been described by Boswell et al. (2005), but has not been implemented in the clinic.

Correction for both yaw and pitch can also be partially accomplished by rotating the collimator (Rijkhorst et al. 2007). The compensation would be complete only when the gantry angle coincides with the axis of rotation (for either pitch or yaw). The axis of yaw coincides with the collimator rotation axis for only AP or PA beams and the axis of pitch only for the lateral beams. Rijkhorst et al. (2007) mentioned that in a beam with a gantry angle γ, the best compensation for the pitch (β) would be achieved with a collimator rotation of $\alpha = \beta \cdot \sin\gamma$ and the best compensation for yaw (ρ) would be possible by $\alpha = \rho \cdot \cos\gamma$. This would not be a complete compensation as the gantry angle becomes more perpendicular to the axis of yaw or pitch, and the corrective collimator rotation reduces to 0 when the beam is completely perpendicular to the axis of patient rotation. Another effect is that the yaw or pitch rotations would cause the tumor projections in the beam's

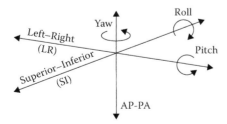

FIGURE 12.1 Explanation of three rotational directions. Pitch, roll, and yaw are around the left–right, superior–inferior, and anterior–posterior axes, respectively.

eye view (BEV) to stretch in size in the SI direction as the gantry angle becomes more perpendicular to the rotational axes of pitch or yaw, complete correction of which would necessitate scaling of the beam portals. Rijkhorst applied this formula for the pitch correction of prostate and seminal vesicles and found a method that permitted a 2 mm reduction in the PTV-CTV margin.

Wu et al. (2006) proposed a simpler correction by adjusting the collimator for only the pitch rotation of the lateral beams in addition to three translational corrections. Court et al. (2005) proposed a solution based on multileaf collimator (MLC) adjustments for the same problem, which will be described in Section 12.3.

Yue et al. (2005) proposed full 6DoF rigid-body corrections. They developed a comprehensive formula to completely reestablish the beam's original orientation relative to the patient for any beam by turning the gantry, collimator, and table angles using a conventional treatment table (with no table tilt or roll). This would be a complete correction for six degrees of motion. The shortcoming of this method is the necessity of adjusting the couch angle differently for each beam (gantry angle), which increases personal effort and generates a safety concern if performed with a remotely rotatable couch. A new effort under development aims to perform a six-degree correction without rotating the couch differently for each beam angle (Bose, Shukla, and Maltz in press). This methodology aims to find a single couch angle and translation applied for all beams that would generate the biggest correction along with adjustments to the collimator and gantry angles of each beam separately. This would be an approximation to the complete six-degree correction; however, the biggest advantage is avoiding the couch angle change for each beam. Dosimetric investigation of this approximation was performed for head and neck and prostate cases (Prah et al. 2009) and partial breast irradiation cases (Morrow et al. 2009).

Tables that can tilt and roll have also been designed. These can theoretically adjust the rotation of the table in the pitch and roll directions; however, the magnitude of rotation is limited to tilt of a couple of degrees (Guckenberger et al. 2007), which would be sufficient to correct for rotational errors in certain sites including the head and neck; however larger rotations might be encountered in the prostate region (van Herk et al. 1995). Guckenberger reported that with patient fixation by thermoplastic masks, the accuracy could be increased and maintained throughout the treatment duration for the 6DoF couches. However, without patient fixation, the drifting of the patient anatomy due to the tilt of the couch increased residual translational errors (Guckenberger et al. 2007). Another system that can offer complete rotational correction is the couch in CyberKnife (Accuray, Sunnyvale, CA) in which the beam orientation can be arbitrarily adjusted (Suzuki et al. 2007).

12.3 Online Replanning

Rigid-body repositioning (translation and rotation) can correct for a major portion of interfractional variations, but fails to fully account for other variations such as organ deformations and independent organ motion. More comprehensive methods would be needed, such as plan alteration. An ideal method would be to generate and deliver a new plan optimized based on the anatomy of the day. Theoretically, the new plan generation should be possible because the required information (volumetric imaging data) is available at the time of the treatment if CT-based IGRT is performed. However, the new plan generation process needs to be fast so that it can be completed while the patient is lying on the table (often termed *online replanning*), that is, within 5–8 minutes in today's workflow. Considering that it generally requires hours to days to generate a new plan with today's treatment planning technology, there has been considerable research on methodologies for quick plan generation or plan adaptation based on the daily anatomy. Although it is difficult to generate a plan (particularly an intensity-modulated radiation therapy [IMRT] plan) within a couple of minutes, generating a new plan for the daily anatomy does not need to start completely from scratch. The original plan that has been generated based on the same patient's anatomy at a different day is readily available, and a large portion of the information with the original plan is still applicable and therefore does not need to be regenerated.

The computing speed continuously increases, and more parallel architectures are becoming available, which would help speed up several time-consuming operations, such as dose computation, delineation of structures, plan optimization, and data transferring, in the new plan generation. There have been several applications using parallel computer architectures, such as those using computer graphics processing units to accelerate dose calculation and autosegmentation. Such innovations have the potential for up to a tenfold increase in speed (Sharp et al. 2007; Gu et al. 2009). However, some of the essential operations such as delineation of important structures are not yet completely automated, therefore, they cannot benefit fully from the speed offered by the current computer technology. Full automation is important for the online replanning process because the manual processes are not only slow but also require the dedication of human (expert) time and are prone to errors under time pressure. Full automation requires algorithms to be consistently reliable and accurate.

Létourneau et al. (2007) showed that it is possible to generate a simple palliative plan completely from scratch, that is, with no previous plan available, within 30 minutes. The authors achieved the speedy planning by integrating all processes and streamlining the operations, such as transferring information between image acquisition, planning, and delivery. The plan regeneration to account for interfractional variations (replanning), when utilizing the existence of a reference plan and the recent improvements in the computer hardware, would require even less time and therefore may become a possible online strategy. In Section 12.3.1, general methods and some representative approaches for online replanning are discussed.

12.3.1 General Methodologies

There is a trade-off between the plan quality and planning time for an online replanning strategy. In this section, we classify online replanning approaches into three groups according to their planning speed and the simplicity and complexity of the process: (1) the approaches that do not require the delineation of volumes of interest (VOI; either manually or based on deformable image registration [DIR]); (2) the strategies that require delineation of VOI but do not employ a plan optimization; and (3) the methods that require both the delineation of VOIs and plan optimization. Both VOI delineation and plan optimization can improve the quality of the correction or adaptation for the interfraction changes; however, they are time-consuming due to the difficulty of being fully automated and the requirement of a substantial amount of human intervention.

12.3.1.1 Strategies Not Requiring Volumes of Interest Delineation

VOI delineation is a time-consuming process and difficult to fully automate and thus presents a major challenge for online replanning. Manual VOI delineation, or even manual contour validation, is the main bottleneck for the speed of an online replanning process. The methods that do not require new VOI delineation are the simplest and fastest of the proposed online replanning methods so far. These methods can simply be referred to MLC-shifting methods (Court et al. 2005; Ludlum et al. 2007; Court et al. 2006). By shifting the MLC positions, the dosimetry effects of the rigid-body translation, rotation, and even deformation can be approximately compensated for. These methods acquire organ dislocation information from multiple locations in the patient anatomy and apply this dislocation information to shift the MLC positions using rigid-body registration methods, which are generally more reliable than DIR at the current time. Because of the use of the dislocation information from multiple locations and the rigid-body registration, these methods can correct for the organ deformation up to a limit.

Court et al. (2005) designed their method specifically to handle the rotation and deformation of prostate and seminal vesicle as a combined target structure. The daily changes in the prostate and seminal vesicle include the rotation around the LR axis, mainly due to different rectal and bladder fillings. This rotation results in the translations in the AP direction (Δy) as a function of the location in the SI direction (z). The $\Delta y(z)$ information is acquired by 2D slice-by-slice image registrations, after a 3D image registration is performed to determine the global shifts. The method takes advantage of the fact that for a coplanar IMRT plan, each MLC pair projects onto a specific anatomic axial slice. Allowing for arbitrary variation of local shifts $\Delta y(z)$, deformations in the AP and LR directions can be approximately considered. However, the deformation in the SI direction is ignored, which is usually less important than that in the AP direction. The method does not need VOIs or DIR to calculate the necessary shifts and only requires the 3D and 2D rigid-body registrations. A similar method was developed by Ludlum et al. (2007) to handle the simultaneous treatment of prostate and the pelvic lymph nodes (PLN).

Prostate and PLN irradiation pose a technical challenge because the prostate and the PLN can move independently of each other. The PLN is relatively fixed in proximity to vascular structures, which are presumably fixed with respect to the bony anatomy (Shih et al. 2005; Hsu et al. 2007). However, it is also well-known that the prostate motion relative to bony anatomy can be large, exceeding 1 cm (Beard et al. 1996; Roeske et al. 1995; Ten Haken et al. 1991). Therefore, any shift based on rigid-body registration that perfectly aligns either the prostate or the PLN could result in suboptimal dosimetry for the other. The method proposed by Ludlum et al. (2007) only shifts the leaf pairs that irradiate the prostate, while the leaves irradiating the PLN region remain unchanged relative to the bony anatomy. The shift information for the prostate can be generated by an IGRT system, which does not have to be CT based. Their algorithm shifts the opposing MLC leaf pairs, while the distance between each MLC leaf pair is kept unchanged. The MLC shifts are calculated from the magnitude of the prostate motion. Similar to the method proposed by Court et al. (2005), this method effectively handles the variations in the AP and LR directions, whereas variations in the SI direction are ignored.

Court et al. (2006) reported another method that can handle complicated shape changes without requiring VOI generation or DIR, similar to their earlier method. This method acquires the information of translation and rotation based on local rigid-body registration by individually registering (rigidly) each slice of the daily CT to the corresponding slice of the planning CT. The variation in the translation information among individual slices (z direction) can account for the pitch (for motion in AP) and yaw (for motion in LR) rotation (linear change in SI) and deformation (arbitrary change in the SI). However, this method fails to address the deformation in roll as well as the variations in yaw and pitch along the AP and LR directions.

These methods requiring no VOI delineation can be fully automated and can offer dosimetric improvement over the repositioning method based on rigid-body registration with a reasonably time frame and additional effort. These methods, however, are approximate approaches and were developed specifically for certain situations and sites, thus, may not be suitable as a general methodology for other anatomic sites.

12.3.1.2 Online Replanning Based on Aperture Morphing

The approaches that require daily delineation of VOIs (or anatomical deformation information) but do not employ a plan optimization are referred to as *aperture morphing methods* in this chapter (Mohan et al. 2005; Feng et al. 2006; Ahunbay et al. 2008). These methods perform corrections by adjusting the segment shapes (therefore, the intensity profiles of individual beams) using the anatomical deformation information between the daily image and the reference (e.g., the planning) image via either automated or manual delineation of VOIs. Because the anatomic deformation is a 3D entity, it needs to be projected to the BEV of each beam separately, so the beam intensity can be morphed accordingly. Morphing the aperture shapes remedies a considerable amount of dosimetric loss due to deformation, and can be relatively fast compared to performing a plan optimization. The general approach of the aperture morphing was proposed and developed by Mohan et al. (2005). Other alternative approaches are also reported in the literature (Feng et al. 2006; Ahunbay et al. 2008).

The general methodology proposed by Mohan et al. (2005) is explained in Figure 12.2. The method requires the daily VOI to be generated based on daily CT, followed by the generation of BEV projections of both the daily and original VOIs. The projections of the target volume and several critical structures are calculated. The BEV plane is segmented based on the target projection, and its overlap with critical structure projections and each of the segmented regions are assigned separate integers. A deformable registration method based on Thirion's "demons" algorithm (Thirion 1996, 1998) is applied in two dimensions to calculate the displacement vectors from each point on the reference segmented BEV to the matching point in the daily segmented BEV. These vectors are then applied to the original fluence map of the beam. The algorithm works such that, for example, if a portion of the beam fluence map that irradiates the PTV also irradiates the rectum, the intensity distribution within this PTV-rectum overlap region in the new plan beam is kept as similar as possible to the original intensity distribution in the corresponding overlap region. Therefore, the irradiation profile of the PTV and each critical structure from the individual beams are kept similar in the daily and the original plans. Finally, an MLC-sequencing procedure is applied to the deformed fluence map to generate the deliverable MLC segments for daily treatment.

Feng et al. (2006) used a similar strategy to modify the intensity distributions of beams. The method, referred to as *direct aperture deformation*, starts with a 3D DIR between the daily image and the reference image. A 3D transformation matrix is generated as a result of the deformable registration and is then "collapsed" into a planar 2D vector array in the BEV plane of each beam. This 2D vector array is used to morph the apertures at the beam directions. Different from the method proposed by Mohan et al. (2005), the beam apertures are directly modified instead of the fluence map of the beam, and the MLC-sequencing process is not needed.

Another aperture morphing approach that requires the delineation of VOIs on the daily image was proposed by Ahunbay et al. (2008). These VOIs are projected to BEV planes, and the target projections are utilized by a segment aperture morphing (SAM) algorithm that performs a linear scaling to determine the new MLC positions. Similar to the method of Feng et al. (2006), the beam apertures (instead of the fluence map) are modified directly. In addition to the aperture morphing, the method proposed by Ahunbay et al. (2008) involves a second step of segment weight optimization (SWO) that optimizes the number of monitor units (MUs) of each segment without changing its shape. The SWO is much faster than regular IMRT optimization, because it involves a much lower number of variables (number of segments vs. number of beamlets in a

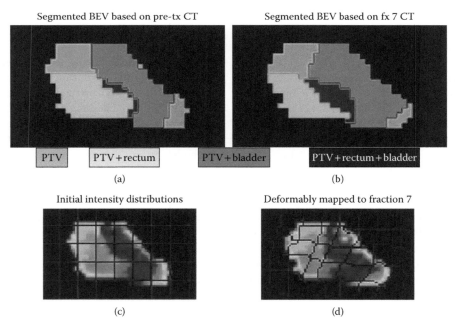

(a) Segmented BEV based on pre-tx CT — PTV | PTV+rectum | PTV+bladder

(b) Segmented BEV based on fx 7 CT — PTV+rectum+bladder

(c) Initial intensity distributions

(d) Deformably mapped to fraction 7

FIGURE 12.2 (See color insert following page 204.) Illustration of the aperture morphing method employed by Mohan, R., et al. (2005). The beam's-eye-view (BEV) apertures for each beam are segmented into regions of overlap of the planning target volume (PTV) with normal critical structures. Then, intensity distributions within each segment from the original pretreatment intensity-modulated radiotherapy plan are mapped onto the corresponding segment within the current treatment's BEV. (Reprinted from Mohan, R., et al. 2005. *Int J Radiat Oncol Biol Phys* 61:1258–66. With permission.)

plan). Ahunbay et al. (2009, 2010) reported that this approach can effectively address positioning and anatomic changes including organ deformation and can be completed within a few minutes, and thus, it can be implemented for online adaptive replanning. As an example, Figure 12.3 shows the use of this approach for a pancreas case. The daily and the original volumes of pancreas, liver, stomach and left kidney are overlaid with the planning CT. We see that the replanning method with the SAM and SWO, as proposed by Ahunbay et al. (2008), can result in significant dosimetric improvement compared to the repositioning and are practically equivalent to the replanning with full-blown reoptimization. Recently, this approach has been implemented in a commercial treatment planning system (RealART, Prowess Inc.).

The use of aperture morphing can considerably improve the dosimetry over the repositioning, the current standard of practice in IGRT (Mohan et al. 2005; Feng et al. 2006; Ahunbay et al. 2008). However, these methods are still approximate compared to a full-scale optimization, and therefore have certain limitations. The aperture morphing algorithms consider the deformation of a single target structure, which is intuitive as the intensity distributions of the beams in an IMRT plan have nonzero intensities only toward the target structure or its close vicinity (scatter range). The method proposed by Mohan et al. (2005) also considers the parts of the normal structures that overlap with the target when projected to the BEV. Although more structures and larger deformation can be considered in the aperture morphing process, the full anatomical dislocation information cannot be completely utilized by an aperture morphing process because the 3D deformation information will eventually need to

be collapsed into the 2D BEV plane by averaging. Theoretically, situations with large organ deformations and/or conflict motion of multiple structures could be better handled by a full-scale plan optimization.

12.3.1.3 Online Replanning Based on Reoptimization

One reason aperture-morphing-based replanning is not as good as full-scope reoptimization is that the aperture morphing methods tend to correct for the intensity of each beam separately, one beam at a time, while the optimization (e.g., the inverse planning of IMRT) is performed for all beams at once. In a regular IMRT plan, the dose distribution and the intensity map of each beam are not only determined by the anatomical information of the target and critical structures but also by the dose distributions from other beams. Therefore, an optimization that can consider fluence maps of all beams at the same time would be more powerful in determining the optimal dose distribution, especially for the situations with conflict motion of multiple targets and organs at risk (OARs). With optimization, we can ascertain that not only the negative impacts of the anatomical deviations can be remedied but also the positive changes that are encountered in the daily anatomy can be exploited. Furthermore, the plan optimization can consider the previously delivered doses when designing the treatment plan for the day.

Adaptive plan reoptimization may require a longer time than the aperture-morphing-based replanning, however, it can be faster than the generation of a plan from scratch. For online reoptimization, the previous optimization on the same patient can be used as a reference, and/or as a starting point. Although

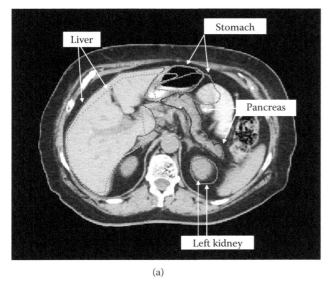

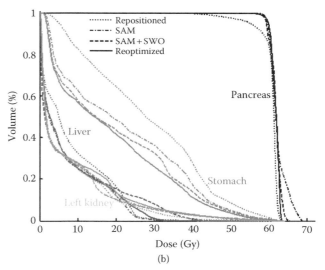

(a)   (b)

FIGURE 12.3 (a) Contours of the pancreas (target), stomach, kidney, and liver, obtained based on both the planning computed tomography (CT) and the CT of a treatment day overlaid in an axial slice of the planning CT; (b) the dose–volume histograms for the pancreas, stomach, liver, and left kidney, obtained based on the treatment CT for various scenarios. (Reprinted from Ahunbay, E. E., et al. 2008. *Med Phys* 35:3607–15. With permission.)

the patient anatomy between the planning day and the treatment day is different enough to cause dosimetric deviations, it is still quite similar as compared with a new patient. A large majority of the information (e.g. dosimetric goals and objective function) for the reference plan (e.g., original) would still be applicable for the daily plan reoptimization, eliminating the time required to generate this information.

The use of the existing original plan as a starting plan for the daily reoptimization has been proposed by Wu et al. (2004) for tomotherapy and by Mestrovic et al. (2007) for linac-based treatments. Mestrovic et al. (2007) showed that even for a case with a large deformation, starting with an existing plan would reach a clinically acceptable plan faster than starting from scratch, although the existing plan is more different from an acceptable plan than a 3D conformal plan. For a beamlet-based approach, the unsegmented fluence map immediately before the segmentation would be a better starting point for the reoptimization than the segmented fluence map. Mestrovic et al. further used the assumption that the plan derived originally based on the planning CT should be rather similar to the optimum plan based on the daily CT to accelerate the reoptimization. They proposed and tested many modifications to the regular optimization process (using the simulated annealing–based optimization [Kirkpatrick, Gellat, and Vecchi 1983]) to decrease the reoptimization time. For example, they limited the extent that the MLC positions can be modified by the optimization algorithm from their original values. Therefore, the optimization algorithm did not need to search for a wider range of possibilities, and consequently, a solution can be reached faster. Further improvements include removing or reducing the components of the "simulated annealing" optimization algorithm that aim at reducing the likelihood of being trapped in a local minimum. The assumption that the starting plan is very close to the globally optimum plan based on the daily anatomy would make the likelihood of a local minimum point very low between the two plans. Therefore, Mestrovic et al. proposed to remove the time-consuming stochastic components of the simulated annealing optimization algorithm to increase the convergence speed during the reoptimization. For example, instead of randomly determining the next leaf or beam to be modified, the leaves are tried in series. Only the modifications that yielded a better solution were kept, and all worse solutions were rejected.

Mestrovic et al. also proposed using direct aperture optimization (DAO; Shepard et al. 2002), a type of IMRT for daily replanning. DAO is an IMRT optimization technique in which the leaf positions are directly modified in the optimization instead of the beam fluences as done in the alternative method of fluence-based optimization. The leaf-sequencing process of regular beamlet-based IMRT, which may cause dosimetric degradation, is not needed for DAO. Also, DAO has been shown to result in a significant reduction in both the number of segments and the number of MUs compared to traditional IMRT. (Shepard et al. 2002; Cotrutz and Xing 2003; Bedford and Webb 2006). Therefore, DAO can be advantageous for an online replanning.

A similar study on the online adaptive reoptimization has been reported by Wu et al. (2004) for using tomotherapy. They applied an overrelaxed Cimino (ORC) algorithm (Censor et al. 1988) in the reoptimization process to reduce the number of iterations needed for a quick plan modification. ORC resulted in a faster convergence of the optimization. Wu et al. (2004) also analyzed several issues related to the computational efficiency and the convergence speed of the reoptimization. Similar to the method of Mestrovic et al. (2007), Wu et al. also proposed using the original plan as the starting plan for reoptimization.

One of the time-consuming steps during an IMRT optimization is the process of determining the right objective function that would lead to the optimal plan. Typically, the adjustment of objection function during optimization is performed via an iterative process (Deasy, Alaly, and Zakaryan 2007; Bortfeld 2006). This process is time-consuming and requires operator intervention. It is beneficial for online reoptimization that the objective function does not need to be generated from scratch. Although theoretically there should be a unique objective function that leads to the optimal plan for the anatomy of the day, using the same objective function for the online reoptimization would probably be capable of generating acceptable daily plans for most days especially if the organ deformations are not extremely large.

Wu et al. (2008) proposed a method to avoid the trial-and-error tweaking of the objective function. They hypothesized that the objective function based on dose–volume histogram (DVH) constraints is the main reason requiring the planner to iteratively adjust the objective function. They pointed out that when organ volumes change dramatically, the difficulty in meeting the dose–volume goals would also change proportionally, which would necessitate the adjustment of the objective function. Therefore, they proposed an alternative objective function formulation—that is, the weighted sum of deviations of all voxels from their original values in a dose distribution goal:

$$\sum_{i \in T} W_{T,i}(d_i^+ + d_i^-) + \sum_{i \in NT} W_{NT,i}(d_i^+)$$

where the d_i^+ and d_i^- are the positive and negative deviations from the distribution goal, respectively, for the voxel i, and $w_{T,i}$ is the weight of target voxels, and $w_{NT,i}$ is the weight of nontarget voxels. The amount of overdosage and underdosage is penalized for a target voxel, whereas only overdosage is penalized for a nontarget voxel. The dose distribution goal is the originally planned dose distribution in the planning CT that is deformed to the CT of the day. The reoptimization process is formulated as a linear goal programming (LGP) model (Chankong and Haimes 1983) to minimize the mentioned objective function. According to Wu et al. (2008), LGP is especially suitable for online plan reoptimization, mainly because a "globally optimal" solution is always guaranteed once the objective function is properly defined.

Defining the goals in terms of an already-accepted dose distribution would allow the planner to better define the desired outcome to the optimizer (e.g., hot spots at unexpected places could be better controlled); however, the problem of trial-and-error adjustment of the objective function could still be needed in a case with a large anatomy change. The DIR would stretch the "goal" dose distributions proportionally to the volume changes in structures, requiring the optimizer to proportionally generate higher or lower dose gradients. As long as the objective function is in the form of a weighted sum of individual goals, the disproportionate increase or decrease in the difficulty of achieving a goal would make the optimization to generate unexpected, therefore, potentially clinically unacceptable solutions. As a weighted sum formula, the optimum objective function (optimum set of relative weights to achieve the desired plan) is not only dependent on the relative importance but also on the relative difficulty of achieving individual goals. For treatment days with large deformations, the balance between individual goals would be lost, which might cause the objective function to be adjusted. To overcome this problem, the objective function must update its priorities based on the achievements of all goals, similar to the "prioritized prescription" method as suggested by Deasy, Alaly, and Zakaryan (2007). Explicitly stating different prioritizations for all possible combinations of individual goal achievements could be overwhelming. However, this should be less challenging for online reoptimization, because the daily possible solution region is not expected to be very large as it only covers the vicinity of the optimum solution of the original plan. A similar objective function formula for the SWO process of the online replanning for head and neck cases was proposed by Ahunbay et al. (2009). The weights of the individual objectives are updated based on the achievements of other objectives. Mainly, the weighting of any objective is reduced if that objective's achievement exceeds the achievement level of other objectives, and vice versa. The critical point is that this update should depend not only on the individual objective's achievement level but also on the other objectives' achievement levels. This prevents the break of balance between individual objectives when daily anatomy happens to be disproportionately favorable for a subset of the objectives and always ascertains that the optimized plan preserves the achievement balance of objectives as in the original (clinically optimum) plan (Ahunbay et al. 2009).

Theoretically, the adjustments to the objective function can be performed during optimization, that is, it is not necessary to wait for the optimization to complete in order to adjust the objective function and rerun the optimization. Obviously, the user interface must be able to accommodate the real-time adjustment and also (preferably) to allow for real-time visualization and evaluation of the doses while the optimization is running. As long as it is done quickly during optimization, adjusting the objective function would not pose a huge time problem. An important requirement is that the dose calculations used for the optimization (pencil beams) must be accurate enough so that a dramatic degradation in the plan quality does not happen when the dose calculation algorithm is switched to the (accurate) convolution at the end of the optimization (Deasy, Alaly, and Zakaryan 2007). Otherwise, it would be necessary to rerun the optimization after adjusting the objective function to compensate for the degradation at the end. Because it is not possible to know exactly how much the plan degradation will be at time of optimization, several trial-and-error runs could be needed. The time cost of this problem would be very large because both optimization and dose calculation would have to be performed each time. The main reason for dosimetric degradation is the reduced length of the radiation scatter tails of the pencil beams (Figure 12.4) used for IMRT optimization (Deasy, Alaly, and Zakaryan 2007). This would be especially bad for head and neck plans because the low-density air pockets would cause the loss of electronic equilibrium. Therefore, it is advisable to use pencil beams with increased scatter tails as also proposed by Mestrovic et al. (2007). SWO is immune to this problem because it does not require dose calculations for pencil beams (Ahunbay et al. 2008). Plan quality can also degrade at the end of the optimization due to MLC-sequencing process. DAO-IMRT is advantageous in this perspective, because it does not involve a leaf-sequencing process (Shepard et al. 2002).

In summary, online reoptimization is a process that would improve the daily plan quality considerably. Starting the reoptimization from an existing original plan and an existing objective function would be prudent for the speed of the process and would minimize the possibility of being stuck to a local minimum. For a planning system, it is desirable to allow for adjustment of the objective function if needed while the optimization is running, and for dose calculations for the pencil beam to be

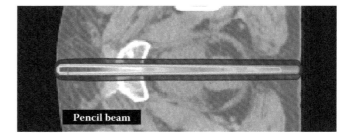

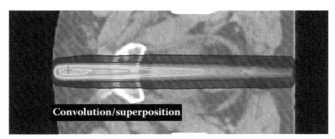

FIGURE 12.4 The difference in the dose distribution by the pencil beam and convolution/superposition algorithms. The pencil beam calculation is employed during the IMRT optimization to increase the speed of optimization. This causes deterioration of the plan quality when the dose calculation switches from pencil beam to convolution/superposition at the end of the optimization process. This deterioration may require the plan to be reoptimized, therefore increasing the planning time.

accurate with wide scatter tails. Methods also exist to overcome the problem of trial-and-error adjustment of the objective function and therefore minimize the human involvement in the online reoptimization process (Deasy, Alaly, and Zakaryan 2007; Ahunbay et al. 2009).

12.3.2 Online Dose Calculation

Being able to calculate the dose on the daily image before the plan delivery to the patient is desirable for the purposes of verification, comparison, or plan reoptimization. Although replanning without calculating daily doses is possible, visualization of the dose distribution can improve the safety and accuracy of the plan delivery. Although daily doses can be calculated on any volumetric CT, there are slight variations in dose calculation accuracy among different CT acquisition methods. MV fan-beam CT has the highest accuracy of dose calculation because the intensity of the images would be the most representative of the actual attenuation properties, and also certain artifacts such as the metallic artifact of kV CT are not present on the MVCT. Fan-beam CTs are more desirable compared to cone-beam CTs for dose calculation, because the latter has additional sources of inaccuracies such as cupping artifacts, ring artifacts, increased patient motion, and increased scatters (Yoo and Yin 2006).

The time required for calculating the dose on a new CT set can be substantial. It is possible to speed the calculation using a parallel computing architecture (Gu et al. 2009). Dose calculation is a highly parallel process in which a task can be divided into multiple independent components that can run concurrently, increasing the computing speed by many folds. Multicore systems such as the ones used for graphics acceleration have been adapted for dose calculation as well as for DIR and image reconstruction. (Sharp et al. 2007). More than 10 times acceleration in dose calculation with the use of multicore systems has been reported (de Greef et al. 2009; Sharp et al. 2007; Gu et al. 2009).

Dose calculation algorithms and hardware are constantly improving, which permits more computationally intensive operations to be applicable online. It is anticipated that the online reoptimization will become a reality in the near future, partially due to advances in computer technology.

12.3.3 Delineation of Target and Organs of Risk for Online Strategies

Having the VOIs available on the daily images would be useful for evaluation, comparison, or corrective purposes. Although VOI contours are not necessary in the current IGRT repositioning practice, a VOI-based patient alignment could improve accuracy and consistency. If the doses are also calculated on the daily image sets, the DVHs could be generated and used for quantitative evaluations before delivering the plan. The daily VOIs are also needed for some comprehensive online plan adaptations discussed in the Sections 12.3.1.2 and 12.3.1.3. However,

unless completely reliable autodelineation tools are available, VOI generation remains a major challenge and a bottleneck of the speed for an online replanning procedure.

The conventional method of VOI generation, that is, manual delineation, is not suitable for online applications because it is very time-and-effort intensive and it is prone to errors if rushed. A fully automated delineation method is desirable for online applications. Although several autodelineation methods are available, it is still not conclusive that these methods are reliable. Unless autodelineation process becomes reliable, human intervention will still be necessary, at least for validating the generated VOIs. The human intervention would inevitably compromise the speed of the autodelineation process.

Several autodelineation methods are reported in the literature, and commercial products are also available (MimVista, ABAS, Smart Segmentation, Velocity). These methods use one of two major approaches: (1) DIR (Christensen, Rabbitt, and Miller 1996; Thirion 1998) and (2) autosegmentation (generally known as deformable models; Pekar, McNutt, and Kaus 2004). In the former, a point-by-point matching between two image sets is sought, whereas the latter starts from scratch and determines the edges of a VOI based on organ specific intensity and/or shape characteristics. Both methods can utilize anatomic knowledge instead of solely relying on the image data.

Of these two approaches, DIR may be a better fit to the online VOI generation, because VOIs are already present on the reference image set that is quite similar to the one where new VOIs need to be generated (Rohlfing et al. 2005; Commowick, Gregoire, and Malandain 2008). However, there are certain problematic issues with the DIR-based segmentation (Wang et al. 2008). When relying only on the intensity matching between images, noise or low image contrast can easily cause errors in the registration. Because DIR seeks matching points with similar intensities between two image sets when there are regions that exist in one image set but do not exist in the other image set (e.g., organ shrinkage, different contents of rectum and bladder), erroneous results can be generated. Several strategies to solve this problem have also been proposed (Godley et al. 2009; Gao et al. 2006; Davis et al. 2004). The problem that DIR is trying to solve is ill-posed. Unlike rigid registration, the similarity of the registered (deformed) image to the target image by itself cannot be used as the only criteria for success, because there could be an infinite number of deformation fields that would accomplish that. The correct field is supposed to also meet some elasticity constraints (so-called regularization or smoothness criteria), which are assumed to govern the deformation of anatomical structures and are independent of the imaging data. There are some basic criteria about how the topology of the images must be preserved (e.g., there should not be tissue folding or tearing; Christensen 2001); however, the large variety of regularization approaches in the literature suggests that the issue is still inconclusive.

The second type of autodelineation, that is, autosegmentation-based methods, offer robustness due to their more global nature compared to DIR. Instead of point-to-point matching with a reference image, new volumes are generated directly on the new image by finding the high-intensity gradient regions (Cootes et al. 1992; McInerney and Terzopoulos 1996; Kass, Witkin, and Terzopoulos 1988; Pekar, McNutt, and Kaus 2004; Burnett et al. 2004). These methods (generally referred to as *deformable models*) are especially suitable for situations where the borders of the VOI are along well-defined intensity gradient lines, such as rectum and bladder (Pekar, McNutt, and Kaus 2004) or spinal cord (Burnett et al. 2004). On the other hand, the target regions are drawn by physicians based not only on the image intensity but also on other considerations, such as knowledge about the spread of disease. Most tumor volumes do not have boundaries that coincide with the intensity gradients. Even for the prostate CTV, volumes are not exactly drawn on the intensity gradients and in the SI directions where there is not enough image contrast for gradient definition. Therefore, the applicability of autosegmentation methods for autodelineation of target regions is questionable.

Although DIR seems to be the best option for online VOI generation, there are several problematic issues with the reliability of DIR. Several studies attempted to validate the accuracy of the DIR methods (Wang et al. 2008; Klein et al. 2009; Janssens et al. 2009). Wang et al. (2008) reported a >97% overlap for head and neck and lung structures. Han et al. (2008) reported a worse agreement of about 70%–80% between the manual and autogenerated VOIs. However, they investigated the DIR between different subjects, where a lower accuracy is expected. Image artifacts and inconsistencies in the original delineations also deteriorate the accuracy of results.

Unfortunately, most reported studies so far are specific to particular in-house or commercial products that were tested and have not been verified independently by other investigators. Another problem is the difficulty in finding the ground truth to test against the autocontouring methods. There always exists a substantial variability in the VOIs even when they are delineated by the same individual on multiple image sets. This was evident in an experiment performed by Wang et al. (2008). Autodelineation was performed by a DIR algorithm and the physicians were asked to modify the VOIs until they were acceptable. The final physician-modified and nonmodified auto-VOIs agreed excellently (>97% volume overlap). When the physicians drew from scratch, the agreement decreased to 85%. This shows that the human delineation inconsistency was substantially larger than the DIR errors.

The performance of the DIR methods depends on the image quality, mainly the image contrast in terms of both modality and subject contrast. The regions of head and neck and lung seem to be easier for the DIR algorithms while the prostate region is more problematic due to the low image contrast in that region. The prostate gland is not easy to delineate on a CT image because of the poor inherent contrast between prostate and other soft tissues surrounding it (Rasch et al. 1999). This is especially troublesome in the apex and base of the prostate where large uncertainties can occur (Rasch, Steenbakkers, and van Herk 2005). Diagnostic fan-beam kVCT is the best quality CT image that is available for in-room imaging (Court et al. 2003); however, even that modality can hardly provide adequate contrast for certain regions such as pelvis. The image contrast is significantly worse with MVCT imaging; therefore, accurate delineation on those image sets is even more problematic. The image contrast of the cone-beam systems, even with kV energy, are worse than the fan-beam systems due to the larger scatter contamination to the flat-panel imager (Letourneau et al. 2005). These imaging modalities would not only make the autodelineation more difficult but also make the manual drawing less accurate.

12.3.4 Dose Accumulation

DIR has been proposed as a useful tool for online applications (Zhang et al. 2007; Feng et al. 2006; Wu et al. 2006) not only for autodelineation but also for deforming the dose distribution (Wu et al. 2006) and for aperture morphing (Feng et al. 2006; Mohan et al. 2005). The end product of DIR is a 3D matrix of vectors from a reference image set to the target image set. This matrix of vectors is referred to as the *deformation field*. The deformation field needs to be obtained once, and then it can be applied to transfer any information on the reference image set (planning CT) to the target image set (daily CT). Mainly, DIR can be used as an autodelineation tool if the deformation field is applied on the already delineated VOIs of the reference image set to generate the corresponding VOIs in the daily image set. Likewise, the deformation field can be applied to the dose distribution associated with the original image set to map it onto the daily image set. This way, one can accumulate the doses delivered at different fractions and generate a combined dose distribution (Yan, Jaffray, and Wong 1999; Schaly et al. 2004). If the dose accumulation process can be performed online, the new plan optimization can be run while the accumulated doses are present; therefore, any previous suboptimalities in the dose distribution could be corrected for. The radiobiological effects of dose addition could also be incorporated into the calculations. This idea, so far, is not realized as a clinical application.

12.3.5 Various Issues for Clinical Implementation of Online Adaptive Replanning

Several other issues regarding the clinical implementation of an online replanning process are worth discussion. In the current practice of radiotherapy, there are several verification or approval procedures that are always performed by physicians before a treatment plan (either initial or modified) is delivered to the patient. Performing all these verification steps for an online plan adaptation could be impractically time-consuming. Improvements in the accuracy and reliability of the VOI generation, replanning algorithm, and dose calculation would help in relaxing some of the demands for plan verification and validation prior to delivery. However, no matter how accurate and reliable the plan generation becomes, it will still be up to an individual clinic to decide what verification process needs to be performed.

In current practice, if a plan is modified, the new plan needs physician approval before delivery. The same rule applied to an online replanning application would be prohibitively impractical. Plan approval necessity has been suggested only when the adaptive plan is significantly inferior to the original plan (Mohan et al. 2005). The other problem is the impracticality of performing a patient-specific IMRT QA after each plan modification. With prior exhaustive testing and verification, the requirement for IMRT QA can be minimized. Other possibilities are software-based QA (e.g., RadCalc [LifeLine Software, Tyler, TX]), exit-dose-based QA during delivery, or postdelivery QA. If the modification in the MLC leaf positions is small, there would be less rationale for an IMRT QA. From this perspective, a method that does not resegment the fluence maps could be preferable (Peng et al. 2009). It should also be considered that the IMRT QA performed in clinical practice is done for the whole treatment course and typically allows for 3%–5% error in dose agreement over the complete prescribed dose. Because the online generated new plan is only for a single fraction, a much larger percentage of disagreement in daily dose would need to occur to make it fail, which would be proportionally less likely to happen. From that perspective, a simpler QA method to check a gross deviation in dose and proper transfer of plan parameters instead of one that checks planar dose agreement would make more sense (Peng et al. 2009). Visual verification of segment shapes before and during the delivery is critical.

The online replanning methods can guarantee the adequate coverage of target daily; therefore, the setup portion of the PTV margin for interfractional errors can be eliminated. However, a smaller PTV-CTV margin is still necessary for the uncertainties due to intrafractional motion, delineation inconsistency, mechanical inaccuracies, and so on. This smaller PTV must be used for the online replanning and evaluations of the daily plan delivery. The CTV volumes drawn on the daily images should be expanded by this smaller PTV margin, with that volume needs to be used as the target structure. The small PTV volume (not the CTV) is the one that needs to receive the full dose on the daily dose distributions to make sure that the actual daily CTV receives full dose in reality.

12.3.6 Clinical Applications of Online Adaptive Replanning

The online replanning based on the aperture morphing algorithm as developed by Ahunbay et al. (2008) was implemented into a commercial planning system (RealART, Prowess), which was subsequently approved by the Food and Drug Administration. This new technology has been tested extensively with selected daily CTs acquired using a CT-on-rails (CTVision, Siemens) during prostate cancer treatments (Peng et al. 2009; Ahunbay et al. 2010; Holmes et al. 2010). It showed

that, compared with the repositioning plans (the plan if the standard IGRT repositioning, i.e., realigning the patient based on rigid-body registration, is applied), the online adaptive replanning with RealART improves target coverage (e.g., the minimum target doses are increased by 13% for the patients studied), while reducing the dose to critical structures (e.g., V70, the rectal volumes covered by 70Gy are reduced by 28% for the patients studied; Ahunbay et al. 2010; Peng et al. 2009; Holmes et al. 2010). The dosimetric gains with the online replanning were found to be practically equivalent to that with full-blown reoptimization (Figures 12.5 and 12.6). The concern for increased prostate motion during the implementation of the online adaptive replanning was explored (Ahunbay et al. 2010). It was found that the dosimetric effect due to the increased intrafraction motion was small as compared to the benefit from the online replanning.

Since July 2009, the online replanning process with RealART has been used clinically for selected patients and fractions at our institution, the Medical College of Wisconsin. In this implementation, a commercial DIR-based autodelineation tool (ABAS, CMS, St Louis, Missouri) is used along with manual modification to generate VOIs. A commercial MU calculation software (Radcalc, LifeLine Software, Tyler, Texas) is used to independently verify MU numbers for all segments generated by the replanning prior to the delivery. The DVHs and dose distribution for the online adaptive plan are compared with those for the repositioning plan. The adaptive plan, if it is superior to the repositioning plan, is delivered after physicians' approval. As of January 2010, the online adaptive replanning procedure has been performed for five prostate cancer patients. The time required for the entire replanning process has been reduced from 20 minutes to 10 minutes as the treatment team becomes more experienced. More system integration between imaging, planning, and delivery systems would further reduce this time required and will improve the workflow.

12.4 Future Directions

The online corrective strategies for repositioning the patient based on rigid-body registration as commonly practiced in the clinic can substantially correct for dosimetric effects of interfractional setup errors and organ motions; however, they fail to account for interfractional organ deformations and shrinkages. More comprehensive approaches, such as online replanning methods, based on either aperture morphing or full-blown reoptimization, offer advantages to fully account for the interfractional variations, eliminating the need for the PTV margin for these variations.

Online adaptive replanning enables image-plan-treat, a new paradigm that will not only allow more accurate targeting by further shrinkage of PTV margins but also encourage radiation oncologists to revise the current dose fractionation schemes to achieve greater therapeutic gain and socioeconomically better patient care. Online adaptive replanning will be particularly important for increasingly used hypofractionations (e.g., SBRT).

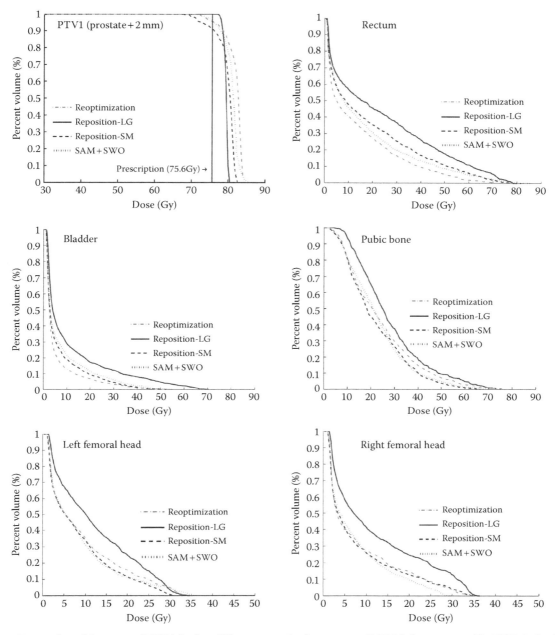

FIGURE 12.5 Dose–volume histograms (DVHs) for four different scenarios for a prostate IMRT daily treatment. The PTV1 is the daily prostate volume (clinical target volume [CTV]) plus a margin (2 mm), which needs to be applied to account for intrafractional, delineation, mechanical uncertainties, and so on. The reposition-LG is the original plan with a 5 mm margin as applied to the daily image set with rigid repositioning. The reposition-SM is the original plan with a 2 mm margin as applied to the daily image set with rigid repositioning. SAM + SWO is the fast replanning strategy with aperture morphing and segment-weight optimization. The reoptimization is the full-scale IMRT reoptimization. LG = Large; SM = Small. (Reprinted from Ahunbay, E. E., et al. 2010. *Int J Radiat Oncol Biol Phys.* With permission.)

The previously delivered doses could be included in the optimization to include the radiobiological information for subsequent treatments. Online autodelineation of VOIs is one of the major challenges. More accurate and reliable autodelineation algorithms and tools need to be developed. Improvements in the integration of imaging, planning and treatment delivery, as well as in computer technology, will be significant to the use of online adaptive replanning. An online replanning process based on an aperture morphing algorithm has been implemented in the clinic and is being used for patients. The clinical impact of using the online replanning will be revealed as more patient data are collected.

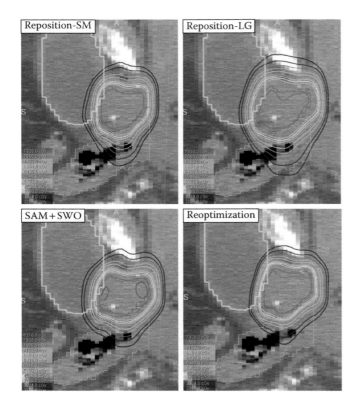

FIGURE 12.6 (See color insert following page 204.) Sagittal plane images of the daily anatomy with dose distributions from four different scenarios. The reposition-LG is the original plan with a 5-mm margin as applied to the daily image set with rigid repositioning. The reposition-SM is the original plan with a 2-mm margin as applied to the daily image set with rigid repositioning. SAM + SWO is the fast replanning strategy with aperture morphing and segment-weight optimization. The reoptimization is the full-scale IMRT reoptimization. (Reprinted from Ahunbay, E. E., et al. 2010. *Int J Radiat Oncol Biol Phys.* With permission.)

References

Ahunbay E. E., C. Peng, G. P. Chen, C. Yu, C. Lawton, and X. A. Li. 2008. An online replanning scheme for interfractional variations. *Med Phys* 35:3607–15.

Ahunbay E. E., C. Peng, A. Godley, C. Schultz, and X. A. Li. 2009. An online replanning method for head and neck adaptive radiotherapy. *Med Phys* 36:4776–90.

Ahunbay E. E., C. Peng, S. Holmes, A. Godley, C. Lawton, and X. A. Li. 2010. An online adaptive replanning method for prostate radiotherapy. *Int J Radiat Oncol Biol Phys* 77:1561–72.

Balter J. M., J. N. Wright, L. J. Newell, et al. 2005. Accuracy of a wireless localization system for radiotherapy. *Int J Radiat Oncol Biol Phys* 61:933–7.

Barker Jr, J. L., A. S. Garden, K. K. Ang, et al. 2004. Quantification of volumetric and geometric changes occurring during fractionated radiotherapy for head-and-neck cancer using an integrated CT/linear accelerator system. *Int J Radiat Oncol Biol Phys* 59:960–70.

Beard C. J., P. Kijewski, M. Bussiere, et al, 1996. Analysis of prostate and seminal vesicle motion: Implications for treatment planning. *Int J Radiat Oncol Biol Phys* 34:451–8.

Bedford J. L., and S. Webb. 2006. Constrained segment shapes in direct aperture optimization for step-and-shoot IMRT. *Med Phys* 33:944–58.

Beltran, C., M. G. Herman, and B. J. Davis. 2008. Planning target margin calculations for prostate radiotherapy based on intrafraction and interfraction motion using four localization methods. *Int J Radiat Oncol Biol Phys* 70:289–95.

Bortfeld, T. 2006. IMRT: A review and preview. *Phys Med Biol* 51:R363–79.

Bose, S., H. Shukla, and J. Maltz. 2010. Beam-centric algorithm for pretreatment patient position correction in external beam radiation therapy. *Med Phys* 37:2004–16.

Boswell S. A., R. Jeraj, K. J. Ruchala, et al. 2005. A novel method to correct for pitch and yaw patient setup errors in helical tomotherapy. *Med Phys* 32:1630–9.

Burnett S. S. C., G. Starkschall, C. W. Stevens, et al. 2004. A deformable-model approach to semi-automatic segmentation of CT images demonstrated by application to the spinal canal. *Med Phys* 31:251–63.

Censor, Y., M. D. Altschuler, W. D. Powlis, et al. 1988. On the use of Cimmino's simultaneous projections method for computing a solution of the inverse problem in radiation therapy treatment planning. *Inverse Probl* 4:607–23.

Chankong, V., and Y. Haimes. 1983. *Multiobjective Decision Making: Theory and Methodology.* Amsterdam, North-Holland: Dover Publications.

Christensen, G. E., and H. J. Johnson. 2001. Consistent image registration. *IEEE Trans Med Imag* 20:568–82.

Christensen, G. E., R. D. Rabbitt, and M. I. Miller. 1996. Deformable templates using large deformation kinematics. *IEEE Trans Image Process* 5:1435–47.

Commowick, O., V. Gregoire, and G. Malandain. 2008. Atlas-based delineation of lymph node levels in head and neck computed tomography images. *Radiother Oncol* 87:281–9.

Cootes T. F., A. Hill, C. J. Taylor, et al. 1992. The use of active shape models for locating structures in medical images. *Image Vis Comput* 12:355–66.

Cotrutz, C., and L. Xing. 2003. Segment-based dose optimization using a genetic algorithm. *Phys Med Biol* 48:2987–98.

Court, L. E., L. Dong, A. K. Lee, et al. 2005. An automatic CT-guided adaptive radiation therapy technique by online modification of multileaf collimator leaf positions for prostate cancer. *Int J Radiat Oncol Biol Phys* 62:154–63.

Court, L. E., Dong, L., Taylor, N., et al. 2004. Evaluation of a contour-alignment technique for CT-guided prostate radiotherapy: an intra- and interobserver study. *Int J Radiat Oncol Biol Phys* 59:412–8.

Court, L., I. Rosen, R. Mohan, et al. 2003. Evaluation of mechanical precision and alignment uncertainties for an integrated CT/LINAC system. *Med Phys* 30:1198–210.

Court L. E., R. B. Tishler, J. Petit, et al. 2006. Automatic online adaptive radiation therapy techniques for targets with significant shape change: A feasibility study. *Phy Med Biol* 51:2493–501.

Davis, B., D. Prigent, J. Bechtel, et al. 2004. Accommodating bowel gas in large deformation image registration for adaptive radiation therapy of the prostate. *Med Phys* 31:1780–1780.

Deasy J. O., J. R. Alaly, and K. Zakaryan. 2007. Obstacles and advances in intensity-modulated radiation therapy treatment planning. *Front Radiat Ther Oncol* 40:42–58.

de Greef, M., J. Crezee, J. C. van Eijk, et al. 2009. Accelerated ray tracing for radiotherapy dose calculations on a GPU. *Med Phys* 36:4095–102.

Deurloo K. E., R. J. Steenbakkers, L. J. Zijp, et al. 2005. Quantification of shape variation of prostate and seminal vesicles during external beam radiotherapy. *Int J Radiat Oncol Biol Phys* 61:228–38.

Feng, Y., C. Castro-Pareja, R. Shekhar, et al. 2006. Direct aperture deformation: An interfraction image guidance strategy. *Med Phys* 33:4490–98.

Gao, S., L. Zhang, H. Wang, et al. 2006. A deformable image registration method to handle distended rectums in prostate cancer radiotherapy. *Med Phys* 33:3304–12.

Ghilezan, M., D. Yan, J. Liang, et al. 2004. Online image-guided intensity-modulated radiotherapy for prostate cancer: How much improvement can we expect? A theoretical assessment of clinical benefits and potential dose escalation by improving precision and accuracy of radiation delivery. *Int J Radiat Oncol Biol Phys* 60:1602–10.

Godfrey D. J., F. F. Yin, M. Oldham, et al. 2006. Digital tomosynthesis with an on-board kilovoltage imaging device. *Int J Radiat Oncol Biol Phys* 65:8–15.

Godley, A., E. E. Ahunbay, C. Peng, and X. A. Li. 2009. Automated registration of large deformations for adaptive radiation therapy of prostate cancer. *Med Phys* 36:1433–41.

Gu, X., D. Choi, C. Men, et al. 2009. GPU-based ultra-fast dose calculation using a finite size pencil beam model. *Phys Med Biol* 54:6287–97.

Guckenberger, M., J. Meyer, J. Wilbert, et al. 2007. Precision of image-duided radiotherapy (IGRT) in six degrees of freedom and limitations in clinical practice. *Strahlenther Onkol* 183:307–13.

Han, X., M. S. Hoogeman, P. C. Levendag, et al. 2008. Atlas-based auto-segmentation of head and neck CT images. *Lect Notes Comput Sci* 5242:434–41.

Holmes, S., C. Peng, E. Ahunbay, C. Lawton, D. Wang, and X. A. Li. 2010. Dosimetric advantages of online adaptive replanning for prostate radiotherapy including seminal vesicles. *Med Phys* (Abstract) 37:3190.

Hornick D. C., D. W. Litzenberg, and K. L. Lam. 1998. A tilt and roll device for automated correction of rotational setup errors. *Med Phys* 25:1739–40.

Hsu, A., T. Pawlicki, G. Luxton, et al. 2007. A study of image-guided intensity-modulated radiotherapy with fiducials for localized prostate cancer including pelvic lymph nodes. *Int J Radiat Oncol Biol Phys* 68:898–902.

Jaffray D. A. 2005. Emergent technologies for 3-dimensional image-guided radiation delivery. *Semin Radiat Oncol* 15:208–16.

Jaffray D. A., J. H. Siewerdsen, J. W. Wong, et al. 2002. Flat-panel cone-beam computed tomography for image-guided radiation therapy. *Int J Radiat Oncol Biol Phys* 53:1337–49.

Janssens, G., J. O. de Xivry, S. Fekkes, et al. 2009. Evaluation of nonrigid registration models for interfraction dose accumulation in radiotherapy. *Med Phys* 36:4268–76.

Kass, M., A. Witkin, and D. Terzopoulos. 1988. Snakes: Active contour models. *Int J Comp Vis* 1:321–31.

Kirkpatrick, S., C. Gellat, and M. Vecchi. 1983. Optimization by simulated annealing. *Science* 220:671–80.

Kitamura, K., L. E. Court, and L. Dong. 2003. Comparison of imaging modalities for image-guided radiation therapy (IGRT). *Nippon Igaku Hoshasen Gakkai Zasshi* 63:574–8.

Klein, A., J. Andersson, B. A. Ardekani, et al. 2009. Evaluation of 14 nonlinear deformation algorithms applied to human brain MRI registration. *Neuroimage* 46:786–802.

Kupelian P. A., C. Ramsey, S. L. Meeks, et al. 2005. Serial megavoltage CT imaging during external beam radiotherapy for non-small-cell lung cancer: Observations on tumor regression during treatment. *Int J Radiat Oncol Biol Phys* 63:1024–8.

Kupelian, P., T. R. Willoughby, A. Mahadevan, et al. 2007. Multi-institutional clinical experience with the Calypso System in localization and continuous, real-time monitoring of the prostate gland during external radiotherapy. *Int J Radiat Oncol Biol Phys* 67:1088–98.

Langen K. M., Y. Zhang, R. D. Andrews, et al. 2005. Initial experience with megavoltage (MV) CT guidance for daily prostate alignments. *Int J Radiat Oncol Biol Phys* 62:1517–24.

Lerma F. A., B. Liu, Z. Wang, et al. 2009. Role of image-guided patient repositioning and online planning in localized prostate cancer IMRT. *Radiother Oncol* 93:18–24.

Létourneau, D., R. Wong, D. Moseley, et al. 2007. Online planning and delivery technique for radiotherapy of spinal metastases using cone-beam CT: Image quality and system performance. *Int J Radiat Oncol Biol Phys* 67:1229–37.

Létourneau, D., J. W. Wong, M. Oldham, et al. 2005. Cone-beam-CT guided radiation therapy: Technical implementation. *Radiother Oncol* 75:279–86.

Li, H., R. Zhu, L. Zhang, et al. 2008. Comparison of 2D radiographic images and 3D cone beam computed tomography for positioning head-and-neck radiotherapy patients. *Int J Radiat Oncol Biol Phys* 71:916–25.

Ludlum, E., G. Mu, V. Weinberg, et al. 2007. An algorithm for shifting MLC shapes to adjust for daily prostate movement during concurrent treatment with pelvic lymph nodes. *Med Phys* 34:4750–6.

McBain C. A., A. M. Henry, J. Sykes, et al. 2006. X-ray volumetric imaging in image-guided radiotherapy: The new standard in on-treatment imaging. *Int J Radiat Oncol Biol Phys* 64:625–34.

McInerney, T., and D. Terzopoulos. 1996. Deformable models in medical image analysis: A survey. *Med Image Anal* 1:91–108.

Mestrovic, A., M. Milette, A. Nichol, et al. 2007. Direct aperture optimization for online adaptive radiation therapy. *Med Phys* 34:1631–46.

Mohan, R., X. Zhang, H. Wang, et al. 2005. Use of deformed intensity distributions for online modification of image-guided IMRT to account for interfractional anatomic changes. *Int J Radiat Oncol Biol Phys* 61:1258–66.

Morrow N. V., D. E. Prah, J. R. White, H. Shukla, S. Bose, and X. A. Li. 2009. A six-degree online correction technique to account for interfraction variations in breast irradiation. *Int J Radiat Oncol Biol Phys* (Abstract) 75:S144–5.

O'Daniel, J. C., A. S. Garden, D. L. Schwartz, et al. 2007. Parotid gland dose in intensity-modulated radiotherapy for head and neck cancer: Is what you plan what you get? *Int J Radiat Oncol Biol Phys* 69:1290–6.

Paskalev, K., M. Ma, R. Jacob, et al. 2004. Daily target localization for prostate patients based on 3D image correlation. *Phys Med Biol* 49:931–9.

Pekar, V., T. R. McNutt, and M. R. Kaus. 2004. Automated model-based organ delineation for radiotherapy planning in prostatic region. *Int J Radiat Oncol Biol Phys* 60:973–80.

Peng, C., E. Ahunbay, G. Chen, C. Lawton, and X. A. Li. 2009. Dosimetric QA tests of an online replanning technique for prostate adaptive radiotherapy. *Med Phys* (Abstract) 36:2537.

Peng, C., K. Kainz, C. Lawton, and X. A. Li. 2008. A comparison of daily megavoltage CT and ultrasound image guided radiation therapy for prostate cancer. *Med Phys* 35:5619–28.

Pouliot, J., A. Bani-Hashemi, J. Chen, et al. 2005. Low-dose megavoltage cone-beam CT for radiation therapy. *Int J Radiat Oncol Biol Phys* 61:552–60.

Prah D. E., C. Peng, E. E. Ahunbay, H. Shukla, S. Bose, and X. A. Li. 2009. An online correction strategy for interfraction variations utilizing couch translation and couch, gantry, and collimator rotation. *Int J Radiat Oncol Biol Phys* (Abstract) 75:S590.

Rasch, C., I. Barillot, P. Remeijer, et al. 1999. Definition of the prostate in CT and MRI: A multi-observer study. *Int J Radiat Oncol Biol Phys* 43:57–66.

Rasch, C., R. Steenbakkers, and M. van Herk. 2005. Target definition in prostate, head, and neck. *Semin Radiat Oncol* 15:136–45.

Redpath A. T., P. Wright, and L. P. Muren. 2008. The contribution of online correction for rotational organ motion in image-guided radiotherapy of the bladder and prostate. *Acta Oncol* 47:1367–72.

Rijkhorst E. J., M. van Herk, J. V. Lebesque, et al. 2007. Strategy for online correction of rotational organ motion for intensity-modulated radiotherapy of prostate cancer. *Int J Radiat Oncol Biol Phys* 69:1608–17.

Roeske J. C., J. D. Forman, C. F. Mesina, et al. 1995. Evaluation of changes in the size and location of the prostate, seminal vesicles, bladder, and rectum during a course of external beam radiation therapy. *Int J Radiat Oncol Biol Phys* 33:1321–9.

Rohlfing, T., R. Brandt, R. Menzel, et al. 2005. Quo vadis, atlas-based segmentation? In *The Handbook of Medical Image Analysis – Volume III: Registration Models*. ed. J. Suri, D. L. Wilson and S. Laxminarayan, 435–86. New York: Kluwer Academic/Plenum Publishers.

Schaly, B., J. A. Kempe, G. S. Bauman, et al. 2004. Tracking the dose distribution in radiation therapy by accounting for variable anatomy. *Phys Med Biol* 49:791–805.

Sharp G. C., N. Kandasamy, H. Singh, et al. 2007. GPU-based streaming architectures for fast cone-beam CT image reconstruction and demons deformable registration. *Phys Med Biol* 52:5771–87.

Shepard D. M., M. A. Earl, X. A. Li, et al. 2002. Direct aperture optimization: A turnkey solution for step-and-shoot IMRT. *Med Phys* 29:1007–18.

Shih H. A., M. Harisinghani, A. L. Zietman, et al. 2005. Mapping of nodal disease in locally advanced prostate cancer: Rethinking the clinical target volume for pelvic nodal irradiation based on vascular rather than bony anatomy. *Int J Radiat Oncol Biol Phys* 63:1262–9.

Smitsmans M. H., J. de Bois, J. J. Sonke, et al. 2005. Automatic prostate localization on cone-beam CT scans for high precision image-guided radiotherapy. *Int J Radiat Oncol Biol Phys* 63:975–84.

Stewart, D. 1965. A platform with six degrees of freedom. *UK Inst Mech Eng Proc* 180:371–85.

Stroom J. C., P. C. Koper, G. A. Korevaar, et al. 1999. Internal organ motion in prostate cancer patients treated in prone and supine treatment position. *Radiother Oncol* 51:237–48.

Suzuki, O., H. Shiomi, S. Nakamura, et al. 2007. Novel correction methods as alternatives for six-dimensional correction in CyberKnife treatment. *Radiat Med* 25:31–7.

Ten Haken, R. K., J. D. Forman, D. K. Heimburger, et al. 1991. Treatment planning issues related to prostate movement in response to differential filling of the rectum and bladder. *Int J Radiat Oncol Biol Phys* 20:1317–24.

Thirion J. P. 1996. Non-rigid matching using demons. In *Proceedings of IEEE Computer Society Conference on Computer Vision and Pattern Recognition*, 245–51. Published by IEEE.

Thirion J. P. 1998. Image matching as a diffusion process: An analogy with Maxwell's demons. *Med Image Anal* 2:243–60.

van den Heuvel, F., J. Fugazzi, E. Seppi, et al. 2006. Clinical application of a repositioning scheme, using gold markers and electronic portal imaging. *Radiother Oncol* 79:94–100.

van Herk, M., A. Bruce, A. P. Guus Kroes, et al. 1995. Quantification of organ motion during conformal radiotherapy of the prostate by three dimensional image registration. *Int J Radiat Oncol Biol Phys* 33:1311–20.

Vigneault, E., J. Pouliot, J. Laverdière, et al. 1997. Electronic portal imaging device detection of radioopaque markers for the evaluation of prostate position during megavoltage irradiation: A clinical study. *Int J Radiat Oncol Biol Phys* 37:205–12.

Wang, H., A. S. Garden, L. Zhang, et al. 2008. Performance evaluation of automatic anatomy segmentation algorithm on repeat or four-dimensional computed tomography images using deformable image registration method. *Int J Radiat Oncol Biol Phys* 72:210–9.

Wu, Q., G. Thongphiew, J. Liang, et al. 2006. Geometric and dosimetric evaluations of an online image-guidance strategy for 3D-CRT of prostate cancer. *Int J Radiat Oncol Biol Phys* 64:1596–609.

Wu, C., R. Jeraj, W. Lu, et al. 2004. Fast treatment plan modification with an over-relaxed Cimmino algorithm. *Med Phys* 31:191–200.

Wu Q. J., D. Ivaldi, Z. Wang, et al. 2008. Online reoptimization of prostate IMRT plans for adaptive radiation therapy. *Phys Med Biol* 53:673–91.

Yan, D., D. A. Jaffray, and J. W. Wong. 1999. A model to accumulate fractionated dose in a deforming organ. *Int J Radiat Oncol Biol Phys* 44:665–75.

Yan, D., D. Lockman, D. Brabbins, et al. 2000. An offline strategy for constructing a patient-specific planning target volume in

adaptive treatment process for prostate cancer. *Int J Radiat Oncol Biol Phys* 48:289–302.

Yan, D., D. Lockman, A. Martinez, et al. 2005. Computed tomography guided management of interfractional patient variation. *Semin Radiat Oncol* 15:168–79.

Yoo, S., and F. Yin. 2006. Dosimetric feasibility of cone-beam CT-based treatment planning compared to CT-based treatment planning. *Int J Radiat Oncol Biol Phys* 66:1553–61.

Yue N. J., J. P. S. Knisely, H. Song, et al. 2005. A method to implement full six-degree target shift corrections for rigid body in image-guided radiotherapy. *Med Phys* 33:21–31.

Zhang, T., Y. Chi, E. Meldolesi, et al. 2007. Automatic delineation of online head-and-neck computed tomography images: Toward online adaptive radiotherapy. *Int J Radiat Oncol Biol Phys* 68:522–30.

Zhang, L., A. S. Gargen, J. Lo, et al. 2006. Multiple regions-of-interest analysis of setup uncertainties for head-and-neck cancer radiotherapy. *Int J Radiat Oncol Biol Phys* 64:1559–69.

Intrafraction Variations and Management Technologies

Laura I. Cerviño
University of California San Diego

Steve B. Jiang
University of California San Diego

13.1 Introduction

Intrafraction motion refers to organ motion during a single fraction within the course of the radiation therapy treatment. Improvements in staging, imaging, and precision radiotherapy delivery by means of three-dimensional (3D) conformal or intensity-modulated radiation therapy (IMRT) permit a higher dose to the target, and/or smaller dose to the organs at risk by conforming the dose tightly to the target. This may result in better therapeutic gain (Fang et al. 2006). There is, however, a concern related to the increased risk of geometrical misses due to intrafraction motion. Target motion can occur due to bony and internal organ movement. Prostate motion, for example, occurs due to rectal and bladder filling as well as to leg motion and clenching of pelvic floor muscles (Boda-Heggemann et al. 2008). However, as treatment fractions get shorter with the use of new technologies, the most relevant intrafraction motion is that due to respiration. Tumors in the thorax and abdomen move with breathing. It has been shown that the magnitude of the motion can be clinically significant (approximately 2–3 cm), depending on tumor location and individual patients (Keall et al. 2006). Even the motion of pelvic tumors due to respiration has been observed (Malone et al. 2000; Kitamura et al. 2002; Weiss et al. 2003). Tumor motion affects imaging, treatment planning, and treatment delivery. Motion

poses a number of special problems, including geometry uncertainty (a moving target may appear with distorted shapes and in the wrong locations during treatment planning scans; Chen, Kung, and Beaudette 2004; Yamamoto et al. 2008) and increased irradiation of normal tissues (large margins are often used to ensure that the tumor is not missed; ICRU 1999; Ramsey, Cordrey, and Oliver 1999; Langen, and Jones 2001). These sources of uncertainty may compromise the effectiveness of conformal radiotherapy for the management of such lesions. Although treatment effectiveness might depend on the technique and number of treatment fractions, it is of major concern when the treatment is done in a hypofractionated (few treatments) or single-fraction manner (Bortfeld et al. 2002). One consequence of respiration is that treatment volume has to be increased, thus unnecessarily irradiating a larger normal tissue volume with higher doses. Many breathing compensation strategies have been investigated, and continue to be under investigation. Recommendations for the use of different imaging and treatment techniques for motion management have been provided in a report by the American Association of Physicists in Medicine (AAPM) Task Group 76 (Keall et al. 2006).

Image-guided radiation therapy (IGRT) has made it possible to reduce treatment field margins in order to reduce overdose of healthy tissue and reduce the risk of normal tissue complication probability (NTCP), while at the same time preventing underdosage of the target and/or facilitating dose escalation. Image guidance, the base of several motion management techniques, provides useful information about tumor motion due to respiration. Its introduction in radiation therapy is drastically changing imaging, treatment planning, and treatment delivery with new technologies being developed and investigated (Jiang 2006a). Respiratory motion changes from day to day. Therefore, it is important to assess the tumor motion characteristics of the day by some imaging technique such as cone-beam computed tomography (CBCT) or fluoroscopy, and then set up the patient accordingly.

In this chapter, we focus on the techniques and technologies available for managing tumor motion during treatment. First, we present a review of disease sites affected by intrafraction motion. Then, we give an overview of motion management technologies currently in use or under investigation.

13.2 Intrafraction Organ Motion

Respiratory motion is the most relevant intrafraction motion, especially because treatment fraction durations decrease with the use of high dose rates and new, fast delivery therapies such as volumetric-modulated arc therapy (VMAT). The lung is, due to its tumor prevalence, the most relevant organ affected by respiratory motion, but it is not the only site or the most affected site. Other thoracic and abdominal anatomic sites, such as breast, liver, pancreas, kidneys, and even prostate, are also affected by respiratory motion. Motion can be in any direction, although, in most cases, craniocaudal (CC) direction is the most prevalent direction of motion. Tumor motion reported in the bibliography has been assessed with several imaging modalities, such as ultrasound, fluoroscopy, four-dimensional (4D)-CT, and 4D magnetic resonance imaging (4D-MRI). Sections 13.2.1 through 13.2.9 discuss thoracic and abdominal organs affected by intrafraction motion.

13.2.1 Lung

The lung is the most prevalent tumor site affected by respiration. For this reason, most studies of tumor motion with respiration focus on the lungs, and most efforts in developing new techniques for motion management in radiation therapy focus on lung treatments. Lung tumor motion is both patient- and location-specific. Tumors can move up to 2–3 cm during normal respiration and even more during deep inspirations. Due to the complicated motion patterns of lung tumors, only CC motion, which is in most cases the larger motion, is presented in the results here.

The most comprehensive study to date on tumor motion is the one performed by Seppenwoolde et al. (2002), who measured tumor motion in three directions (CC, anteroposterior [AP], and lateral) for different locations in the lung (upper lobe, lower lobe, and middle lobe) and made the additional distinction of whether a tumor was or was not attached to any anatomical structure. This study used two fluoroscopy units and evaluated lung tumor motion in 20 patients. The observed motion during normal breathing was 2 ± 2 mm for tumors in the upper lobe and 12 ± 6 mm for tumors in the lower lobe.

Ekberg et al. performed a study in 20 lung cancer patients using fluoroscopy and observed a CC tumor motion of 3.9 mm (range: 0–12 mm; Ekberg et al. 1998; Holmberg et al. 1998). Plathow et al. (2004) used dynamic MRI to evaluate 3D tumor motion in 20 patients with solitary lung tumors and observed that tumor motion depends greatly on tumor location; tumor motions of 4.3 ± 2.4 mm for tumors in the upper lobe, 7.2 ± 1.8 mm in the middle lobe, and 9.5 ± 4.9 mm in the lower lobe were measured (Ley et al. 2004). They also observed motion during deep breathing, obtaining 4 ± 2 mm in the upper lobe, 17 ± 12 mm in the middle lobe, and 24 ± 17 mm in the lower lobe. IGRT techniques are used for either tracking or treating a tumor only when the tumor is in a given position (in the beam path).

13.2.2 Diaphragm

Diaphragm motion has been widely studied, and it has been observed that it is of a large amplitude in both regular and deep breathing. An early study by Wade (1954) using fluoroscopy as the imaging method on 10 patients reported a motion of 17 ± 3 mm during quiet breathing and 99 ± 16 mm during deep inspirations. Another study with fluoroscopy by Weiss, Baker, and Potchen (1972) on 25 patients in the supine position showed an average diaphragm motion of 13 ± 5 mm during quiet respiration. They also observed that this motion is larger in the supine position than in the erect position.

Using ultrasound, Davies et al. (1994) found in nine patients a motion of 12 ± 7 mm (range: 7–28 mm) during quiet respiration and 43 ± 10 mm (range: 25–57 mm) during deep inspiration. Using cine MRI, Korin et al. (1992) observed that the average diaphragm motion in 15 patients was 13 mm during normal breathing and 39 mm during deep breathing. A more recent CT study by Hanley et al. (1999) shows that the average motion of the diaphragm in five patients during normal breathing is 26.4 mm (range: 18.8–38.2 mm).

13.2.3 Liver

Radiotherapy of the liver has traditionally had a limited role in the treatment of liver cancer, due mainly to the poor whole-liver radiation tolerance and tumor motion (Ingold et al. 1965). However, with the advancements of 3D conformal radiotherapy, IMRT and IGRT, treatment can be applied without affecting the whole liver. Respiratory motion in the liver has a larger effect than the effect observed previously in lung tumor motion. Studies on liver respiratory motion have observed different motion ranges. Weiss, Baker, and Potchen (1972) observed in 12 patients, by means of a scintillation camera, a hepatic motion of 11 ± 3 mm. Harauz and Bronskill (1979) studied liver motion in 51 randomly selected individuals by means of scintigraphy, and observed an average motion of 14 mm.

Suramo, Päivänsalo, and Myllylä (1984) observed liver respiratory motion with a linear array real-time ultrasound transducer during normal and deep breathing in 50 individuals, which was 25 mm (range: 10–40 mm) and 55 mm (range: 30–80 mm)

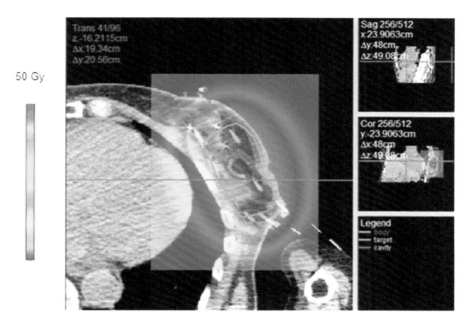

FIGURE 1.7 Example of the estimated impact of dose variations on subclinical disease control: a brachytherapy (high-dose-rate afterloading) case to treat breast carcinoma (50 Gy dose in 10 fractions). The estimated regional control rate is 93.9%, based on the assumption that 40% of patients have at least one subclinical microscopic focus (cf. Figure 1.8).

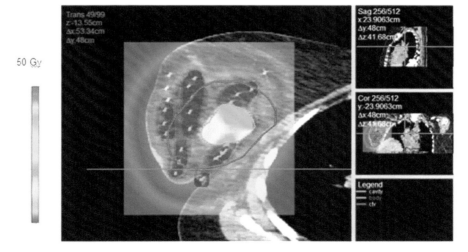

FIGURE 1.8 Example of the estimated impact of dose variations on subclinical disease control: a brachytherapy (high-dose-rate afterloading) case to treat breast carcinoma (50 Gy dose in 10 fractions). The estimated regional control rate is 84.2%, based on the assumption that 40% of patients have at least one subclinical microscopic focus (cf. Figure 1.7).

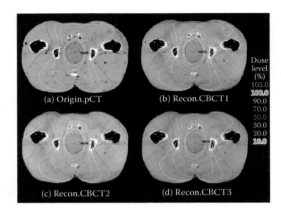

FIGURE 2.8 The VMAT dose distributions on a transverse slice from the following: (a) the original plan on the pCT and (b–d) the reconstituted plan on the CBCTs (CBCT1, CBCT2, CBCT3) of the pelvis phantom set up with three predefined errors, respectively. (Reprinted from Qian, J., et al. 2010. *Phys Med Biol* 55:3597–610. With permission.)

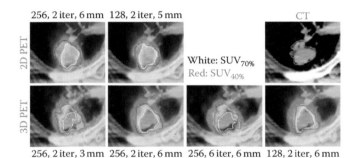

FIGURE 3.5 Impact of different scan acquisition and image reconstruction parameters on target definition: An example is shown for the same patient with 2D and 3D FLT PET scan acquisition and a different set of reconstruction parameters (labels: reconstruction matrix, number of iterations, and postfilter width). Note the high discrepancies between the SUV contours (SUV$_{40\%}$), which would result in significantly different treatment volumes (volumes from 26 to 32 cc and SUV$_{mean}$ from 3.1 to 4.1). The SUV$_{70\%}$ contours are even more different, indicating large discrepancies if one was to design a biologically conformal treatment plan (volumes from 4 to 11 cc and SUV$_{max}$ from 3.8 to 5.3).

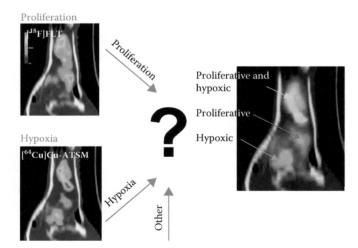

FIGURE 3.7 How to combine biological phenotypes. This figure shows how two phenotypes determine three distinctive regions—regions of high hypoxia/low proliferation, low hypoxia/low proliferation, and high hypoxia/high proliferation. Each of these regions will likely respond differently to therapy and would require a different dose painting strategy.

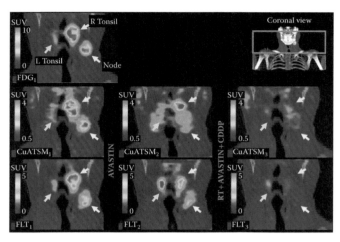

FIGURE 3.10 Complexity of the positron emission tomography response to a combined radiation therapy/molecular targeted therapy. The behavior and magnitude of the response is variable and might depend on the pretreatment status of different biological phenotypes. For example, note that hypoxia (as indicated by the Cu-diacetyl-bis(N4-methylthiosemicarbazone) uptake) is reduced after avastin therapy in the nodal volume but remains high in the primary tumor.

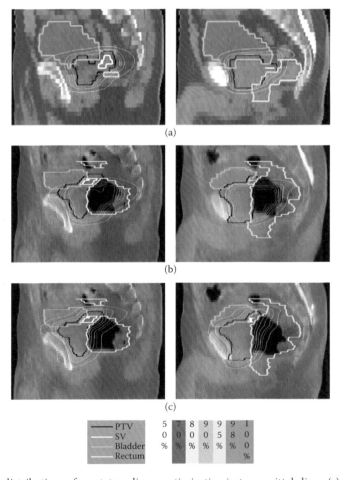

FIGURE 5.3 Comparison of dose distributions of prostate online reoptimization in two sagittal slices: (a) initial optimized plan on planning computed tomography (CT), (b) initial plan applied to treatment CT, and (c) adapted plan on treatment CT. Note the large anatomy changes between planning CT and treatment CT. PTV = planning target volume; SV = seminal vesicles. (Reprinted from Wu, Q. J., et al. 2008. *Phys Med Biol* 53(3):673–91. With permission.)

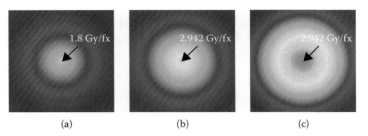

FIGURE 7.5 Determination of biologically equivalent dose distributions for conventional and hypofractionated radiation therapy treatments. Panel (a): dose distribution for the reference treatment (44 fractions of 1.8 Gy at center; decreases monotonically to 0 Gy with radial distance away from center). Panel (b): 20 fractions of 2.942 Gy at center. Dose to other tissue regions determined by multiplying the dose from panel (a) by 1.634 (= 2.942 Gy/1.8 Gy). Panel (c): dose distribution obtained by applying Equation 7.8 on a voxel-by-voxel basis to the dose distribution in panel (a) ($n_A = 20$, $n_R = 44$, $d_R = 1.8$ Gy, $\alpha/\beta = 1.5$ Gy).

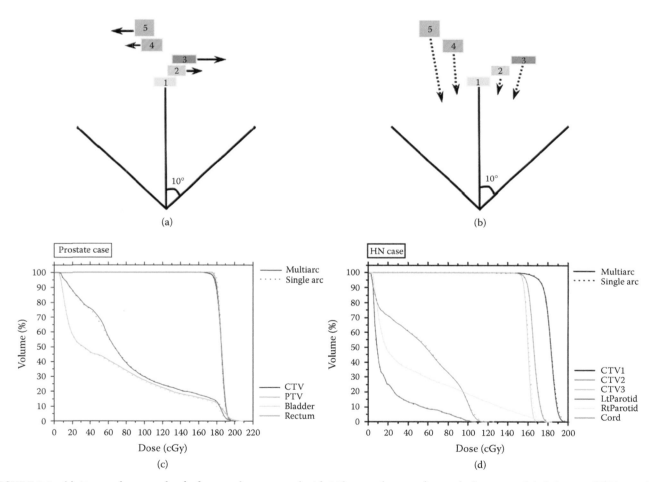

FIGURE 8.2 (a) At every beam angle of a five-arc plan generated with 36 beams, there are five stacked apertures labeled 1 to 5. (b) The stacked apertures are spread out within the 10 degree interval with the order arranged according to the apertures center location. By spreading the stacked apertures in this manner for all 36 beams, a five arc plan is converted to a 180-field single arc plan. (c) and (d) show the DVH comparison between the original 5-arc plan and the re-configured single arc plan for a prostate cancer case and a head-neck case respectively. The dose distributions are nearly identical.

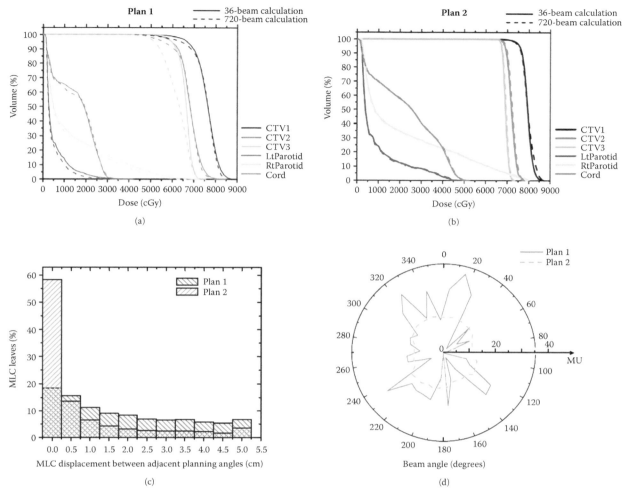

FIGURE 8.4 Effects of large MLC leaf travel and dose-rate variation on the accuracy of delivery. For the same head and neck case, large discrepancies between the planned and delivered doses exhibited in (a) plan 1, but not in (b) plan 2. Comparison of leaf travel histogram in (c) and dose rate distribution in (d) reveals that plan 1 has longer leaf travel and larger dose rate fluctuations.

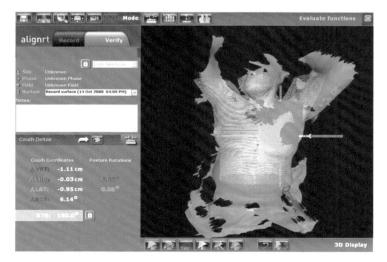

FIGURE 10.19 Reference topogram compared with the treatment topogram: the reference topogram (pink) with clinician-defined region of interest is compared with the acquired treatment topogram (green). The data sets are analyzed using the AlignRT software to calculate couch shifts required to reproduce expected patient position.

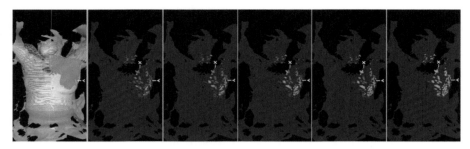

FIGURE 10.20 Breast position varies between treatments. AlignRT will localize regions of the breast that deviate beyond a user-defined threshold compared to the initial reference topogram. From left to right, the location of breast position above (red), below (blue), and within (green) a specified threshold for variation between topograms of 1, 2, 3, 5, and 10 mm. For this treatment, the majority of breast tissue falls within 10 mm from the initial reference, with the greatest breast position deviation localized to the upper inner quadrant.

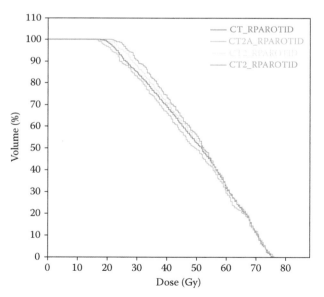

FIGURE 11.5 Dose–volume histograms for a head and neck intensity-modulated radiation therapy plan evaluated on a cone beam CT image volume acquired during treatment. The original treatment plan and dose calculation is in red, and doses evaluated on the CBCT data are processed by calculation using the CBCT for density correction (green), calculation on the deformed planning CT scan (white), and deformation of the original dose calculation (brown).

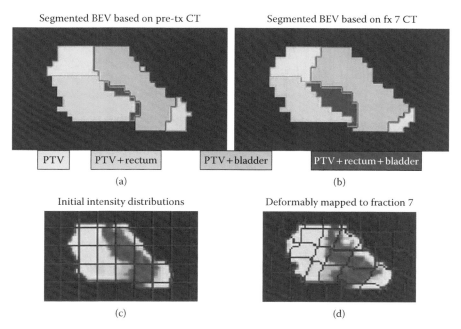

FIGURE 12.2 Illustration of the aperture morphing method employed by Mohan, R., et al. (2005). The beam's-eye-view (BEV) apertures for each beam are segmented into regions of overlap of the planning target volume (PTV) with normal critical structures. Then, intensity distributions within each segment from the original pretreatment intensity-modulated radiotherapy plan are mapped onto the corresponding segment within the current treatment's BEV. (Reprinted from Mohan, R., et al. 2005. *Int J Radiat Oncol Biol Phys* 61:1258–66. With permission.)

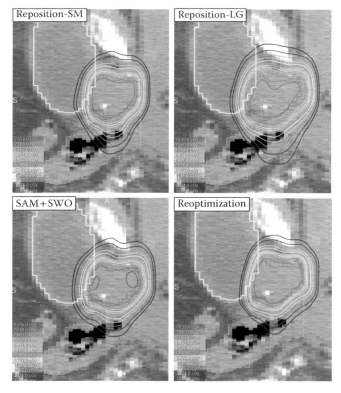

FIGURE 12.6 Sagittal plane images of the daily anatomy with dose distributions from four different scenarios. The reposition-LG is the original plan with a 5-mm margin as applied to the daily image set with rigid repositioning. The reposition-SM is the original plan with a 2-mm margin as applied to the daily image set with rigid repositioning. SAM + SWO is the fast replanning strategy with aperture morphing and segment-weight optimization. The reoptimization is the full-scale IMRT reoptimization. (Reprinted from Ahunbay, E. E., et al. 2010. *Int J Radiat Oncol Biol Phys*. With permission.)

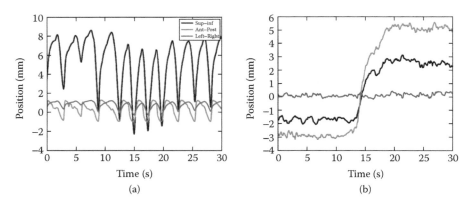

FIGURE 14.1 Examples of the complexity of tumor motion that can be encountered during radiotherapy treatments for (a) lung and (b) prostate.

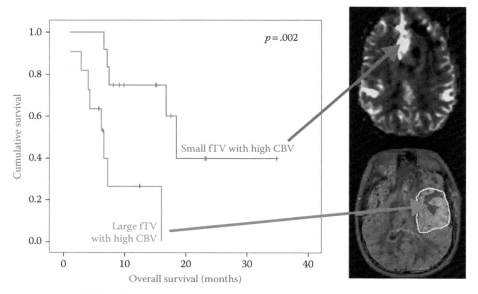

FIGURE 16.1 Survival curves stratified based on the fractional tumor volumes (fTV) with elevated cerebral blood volume (CBV) prior to radiation therapy. Patients with a small fractional tumor volume with high CBV had better chances of survival than those with a large fractional tumor volume with high CBV. Hot colors (red and yellow) indicate high values of CBV.

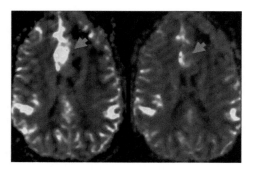

FIGURE 16.3 Cerebral blood volume (CBV) maps in a patient with high-grade glioma prior to radiation therapy (RT; left) and 3 weeks after starting fractionated RT (right). Note the reduction of high CBV in the tumor volume after 3 weeks of fractionated RT as an early indicator of response.

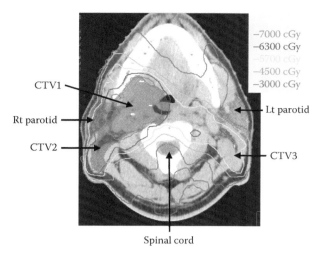

FIGURE 17.1 Intensity-modulated radiation therapy plan for head and neck cancer treatment. A typical plan consists of several target volumes to be treated at different dose levels and adjacent normal critical structures. In this case, the primary clinical target volume (CTV1) is to be treated at 7000 cGys, the intermediate-risk target (CTV2) at 6300 cGys, and the low-risk target (CTV3) at 5700 cGys, respectively. The normal structures include the left and right parotid, spinal cord, and oral cavity (not shown) in this axial view. Due to the close proximity of normal structures, treatment design is usually a compromise between target coverage and the dose tolerance of nearby normal structures. High conformality in dose distribution is desired to achieve a high therapeutic ratio.

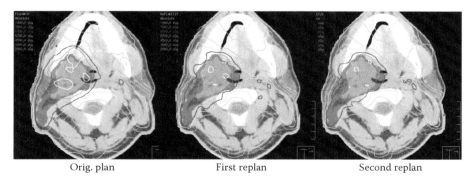

FIGURE 17.11 Typical example of dose recalculation on a daily computed tomography (CT) image acquired on the 25th treatment fraction. Left: the original plan was calculated; middle: the first replan (designed on treatment fraction number 15) was calculated; and right: the second replan was designed on this CT. We can see that the original plan has a larger treatment margin and higher doses in the target. The first replan has a smaller difference compared to the second replan; however, the parotid sparing and lower body dose can still be seen in the second replan.

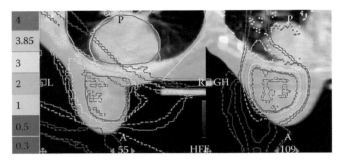

FIGURE 18.4 Isodose distributions for a TomoTherapy-treated prone-positioned partial-breast irradiation patient. The isodose distributions are calculated for a pretreatment megavoltage computed tomography (MVCT) image (dashed lines) with the original planned isodose distributions (solid lines) superimposed on the MVCT image. (Reprinted from Kainz, K., et al. 2009. *Int J Radiat Oncol Biol Phys* 74:275–82. With permission.)

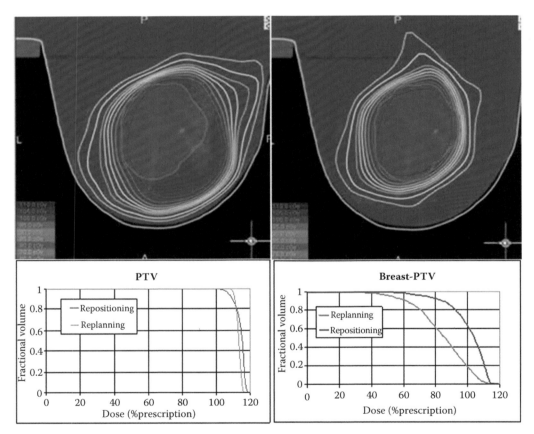

FIGURE 18.6 Image-plan-treat paradigm for image-guided radiation therapy. The upper left part of the figure shows the originally planned isodose distribution relative to the planning target volume (PTV) as apparent from the daily pretreatment computed tomography (CT), in the instance where the pretreatment CT is merely shifted to align with the plan CT, The upper right shows a more conformal isodose distribution obtained from generating a new treatment plan based on the pretreatment CT. The lower left shows PTV dose–volume histograms (DVHs) obtained using repositioning alone and using replanning, where the replanned DVH indicates more uniform PTV coverage; and the lower right shows DVHs for the uninvolved ipsilateral breast, for both the repositioning and replanning schemes.

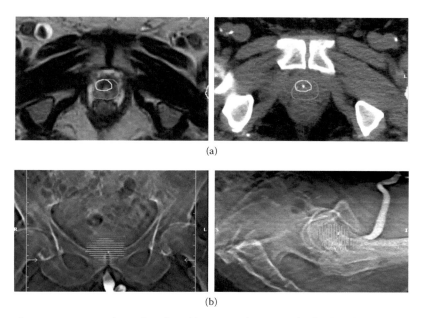

FIGURE 21.8 Comparison of prostate contours drawn based on (a) computed tomography (CT) and magnetic resonance imaging (MRI) and (b) contour projections on DRRs: yellow—contours based on CT; blue—contours based on MRI.

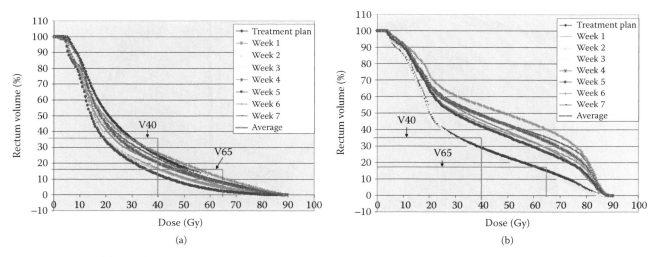

FIGURE 21.14 Rectal dose–volume histograms (DVHs) based on the recomputed dose distributions using the CT-on-rails scans immediately after an intensity-modulated radiation therapy treatment. In (a) the original treatment plan was based on an empty rectum and in (b) the original treatment plan was based on a large rectum. The "average" DVH combines the doses received in all fractions. An empty rectum represented the most difficult scenario to plan, but resulted in better rectal dose distributions when rectal volume increases in subsequent treatments.

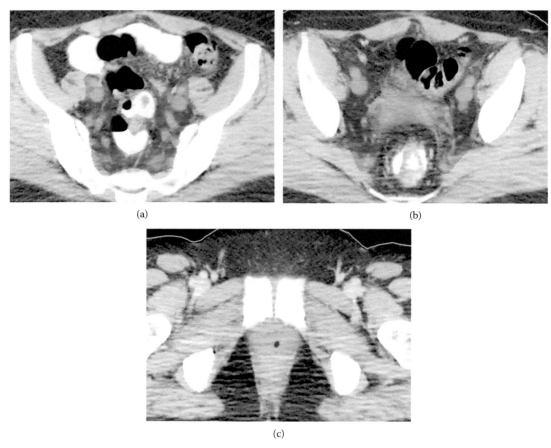

FIGURE 22.1 (a) Superior clinical target volume (CTV)—upper external and internal iliac (red) and presacral (blue). (b) Middle CTV—external and internal iliac (red) and parametrial/vaginal (green). (c) Inferior CTV—vaginal.

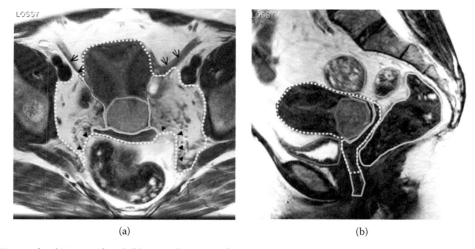

FIGURE 22.2 (a) T2-weighted MR axial and (b) sagittal images of one patient demonstrating gross tumor volume (GTV; red), cervix (pink), uterus (blue), vagina (yellow), parametrium (green), bladder (purple), rectum (light blue), sigmoid (orange). Black arrow heads refer to uterosacral ligaments and mesorectal fascia. Open arrow heads refer to the broad ligament and top of the fallopian tube. Dashed white line represents the clinical target volume (CTV). (Reprinted from Lim, K., W. Small Jr., L. Portelance, et al. 2009. *Int J Radiat Oncol Biol Phys*. With permission.)

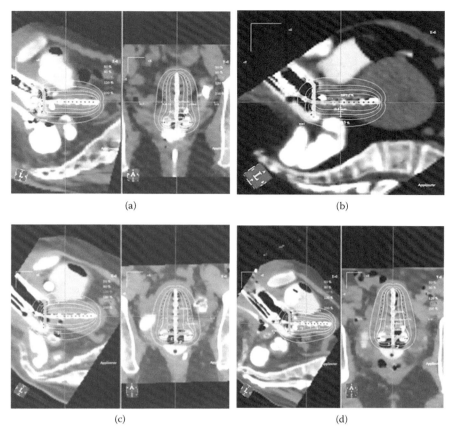

FIGURE 22.4 (a) Sagittal and coronal CT plan with pear-shaped dose distribution and dose specification points; (b) Sagittal CT image revealing sigmoid above rectal retractor; (c) and (d) Sagittal and coronal CT scans in the same patient for two different fractions revealing change in position of organs at risk (OARs) relative to dose distribution.

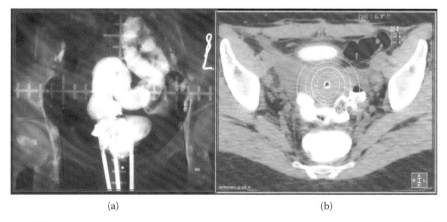

FIGURE 22.5 (a) Orthogonal film displaying a circuitous sigmoid with point doses that are difficult to evaluate. (b) Axial CT relates the dose distribution to the adjacent sigmoid loop and allowed dose adjustment to prevent potential sigmoid toxicity.

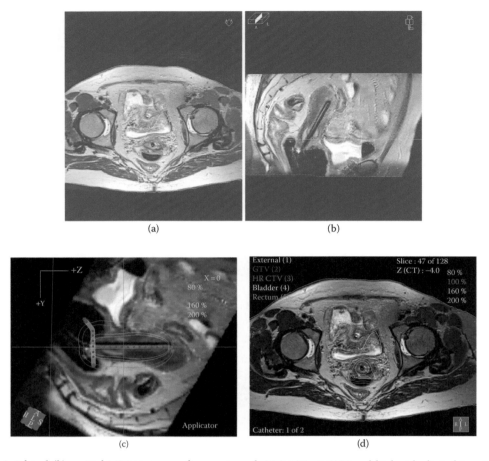

(a)   (b)

(c)   (d)

FIGURE 22.6 (a) Axial and (b) sagittal MRI scans revealing contoured GEC-ESTRO GTV and high-risk clinical target volume (HRCTV). Isodose distribution displayed on this same patient in (c) sagittal and (d) axial planes.

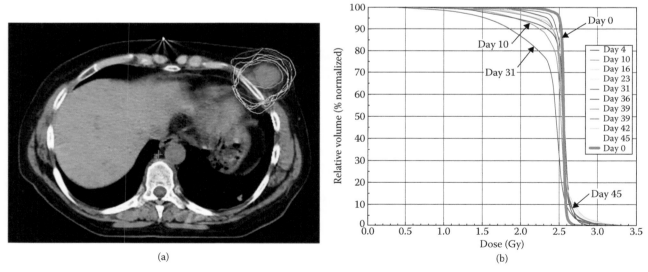

(a)   (b)

FIGURE 23.1 Interfractional variations in clinical target volume (CTV) for a chest wall sarcoma. (a) A series of daily CTVs (yellow) contoured from daily megavoltage computed tomography (MVCT) scans. The planning CTV (green) and primary target volume (PTV; red) are overlaid on the axial KVCT image. The CTV in the middle of the treatment nearly doubled in volume compared to the planning CTV (day 0). (b) A series of verification DVHs for CTVs based on daily MVCTs (using tomotherapy-planned adaptive software). The thicker red line represents the dose-volume histogram (DVH) for the original/planned CTV.

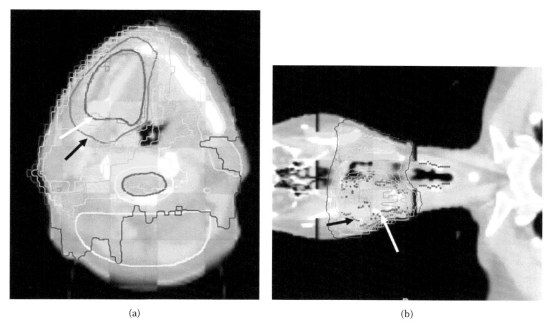

(a)   (b)

FIGURE 23.2 Primary myofibroblastic sarcoma of the right mandible. (a) Axial view and (b) coronal view of the patient depicting clinical target volume (CTV1; tumor bed), CTV2 (operative bed), and isodose lines (blue 18 Gy, cyan 30 Gy, green 42 Gy, light green 48 Gy, orange 54 Gy, dark orange 57 Gy, red 60 Gy). In both (a) and (b), the black arrow indicates CTV (purple contour) and the white arrow indicates primary target volume (PTV; red contour).

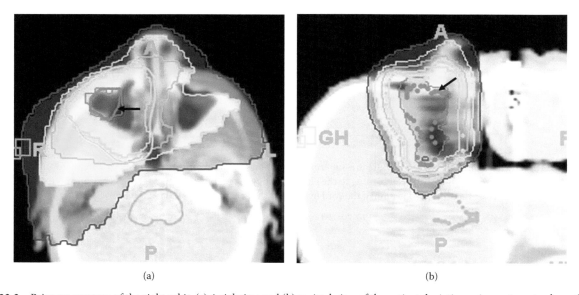

(a)   (b)

FIGURE 23.3 Primary sarcoma of the right orbit. (a) Axial view and (b) sagittal view of the patient depicting primary target volume (PTV) and isodose regions (dark blue 10 Gy, light blue 20 Gy, green 30 Gy, yellow 40 Gy, red 50 Gy). The patient was treated preoperatively with tomotherapy. In both (a) and (b), the black arrow indicates PTV (red contour).

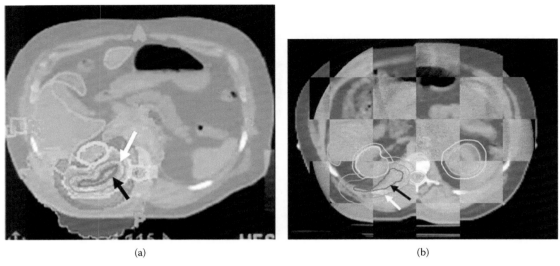

(a)   (b)

FIGURE 23.4 Recurrent retroperitoneal sarcoma (axial view). (a) Tomotherapy plan depicting gross tumor volume (GTV; indicated by black arrow), primary target volume (PTV; indicated by white arrow), and isodose regions (light blue 14 Gy, dark blue 25 Gy, cyan 35 Gy, orange 45 Gy, pink 50 Gy). Note the absence of even 14 Gy to the majority of the kidney despite its close proximity to the PTV. (b) Daily megavoltage computed tomography (MVCT) superimposed on planning CT scan. Black arrow indicates GTV (red contour) and white arrow indicates PTV (pink contour). Right kidney is contoured in cyan; left kidney, in green.

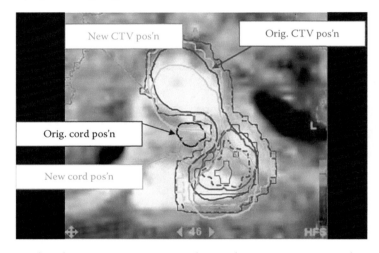

FIGURE 23.6 The target-volume and cord contours are superimposed upon the pretreatment megavoltage computed tomography (MVCT) image corresponding to their original locations (according to the planning kilovoltage computed tomography (kVCT) and to their apparent locations on the MVCT image. Note that the new location of the cord overlaps with the original position of the 50.4-Gy CTV.

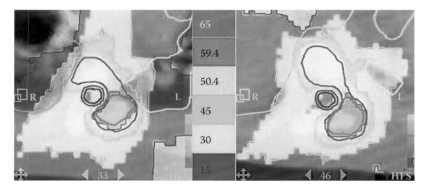

FIGURE 23.8 Isodose distributions for an adaptive tomotherapy plan using a megavoltage computed tomography (MVCT) scan acquired in the middle of the treatment course, and the modified target volume and cord contours based upon the MVCT image. Dose levels in Gy are included.

respectively. Also using ultrasound, Davies et al. (1994) measured a motion of 10 ± 8 mm (range: 5–7 mm) during normal breathing and 37 ± 8 mm (range: 21–57 mm) during deep breathing in a smaller study with eight individuals. Balter et al. (1996) performed a motion analysis based on CT with nine patients, observing a motion of 17 mm during normal breathing. The liver along with breast and lung are tumor sites that will benefit greatly from IGRT techniques for motion management.

13.2.4 Breast

The breast surface moves mainly in an AP direction, although motion during regular breathing is only 2–4 mm. Because this motion is relatively small, motion management techniques for breast treatment focus on the sparing of healthy tissue rather than immobilizing the tumor. Treatment using the deep inspiration breath hold (DIBH) technique has the advantage of increasing the distance from heart to chest wall. Breast motion during respiration is characterized by chest wall excursion, which is motion in the AP direction. Motion, however, depends on a patient's position. Motion in the supine position has been measured by several researchers.

Pedersen et al. (2004) observed in 15 patients an average chest wall excursion of 2.5 mm (range: 1–4 mm) during free breathing (FB) and 12.6 mm (range: 8–20 mm) when performing deep inspirations, by means of infrared markers on the xiphoid process. In a related study, Korreman et al. (2005) observed in 17 patients that these motions were 2.5 mm (range: 1–4.3 mm) and 12.6 mm, respectively. Stock et al. (2006) performed an analysis in 35 patients, in which they observed the motion of infrared markers on the surface of the thorax; AP motions of 1.8 ± 0.9 mm during shallow breathing and 5.8 ± 3.6 mm during deep breathing, and CC motions of 1.6 ± 0.8 mm during shallow breathing and 6.9 ± 3.3 mm for deep inspirations were observed. Cerviño et al. (2009) observed in 20 individuals by means of surface imaging that the chest wall excursion for deep inspirations was 11.3 ± 5.3 mm. Reduced tissue motion in the prone position compared to the supine position has been observed using imaging techniques such as MRI (Doyle, Howe, and Griffiths 2000), ultrasound (El-Fallah, Plantec, and Ferrara 1997), and cine CT (Becker, Patel, and Mackie 2006; Morrow et al. 2007).

13.2.5 Pancreas

The pancreas, as an upper abdominal organ, suffers from both respiratory motion and gastrointestinal motion. Respiratory motion occurs due to the contraction of the diaphragm, as in the case of other abdominal organs. It has been observed that different parts of the pancreas have different motion modes and amplitudes (Bhasin et al. 2006). As with liver, the effect of pancreas motion is larger than that observed in lung. In a study conducted on 50 patients, Suramo, Päivänsalo, and Myllylä (1984) observed with ultrasound that the pancreas moves by an average of 20 mm (range: 10–30 mm) during normal respiration and 43 mm (20–80 mm) during maximum respiration. In a different

ultrasound study by Bryan et al. (1984) with 36 patients, it was observed that the average pancreas motion during normal breathing was 18 mm. Bhasin et al. (2006) analyzed pancreas motion with respiration using fluoroscopy. They observed in 22 patients that the pancreas can move from 1 to 34 mm during deep breathing, and concluded in their study that the pancreas moves craniocaudally with respiration and the head moves medially on inspiration.

13.2.6 Kidneys

The kidneys present large motion with respiration. Kidney motion has been evaluated with radiographs, ultrasound, CT, and MRI. In the same study mentioned in Section 13.2.5, Suramo, Päivänsalo, and Myllylä (1984) observed, with ultrasound, CC kidney motion in 50 patients. Motion of both left and right kidneys during regular breathing was 19 mm (range: 1–4 mm), and motion during maximum respiration was 40 mm (range: 20–70 mm) for the right kidney and 41 mm (range: 20–70 mm) for the left kidney. Davies et al. (1994) measured kidney motion in eight patients with a linear ultrasound array, observing a CC motion of 11 ± 4 mm (range: 5–16 mm). In CT images of nine patients, Balter et al. (1996) measured a kidney motion during normal breathing of 18 mm for both left and right kidneys.

13.2.7 Prostate

Prostate intrafraction motion has been linked not only to respiratory motion but also to rectal motion, as discussed in a detailed review on the topic by Byrne (2005). Padhani et al. (1999) performed a study on intrafraction rectal and prostatic motion in 55 patients using axial cine MRI. They observed a total of 33 AP prostate motions in 16 patients, and obtained a good correlation between rectal and prostatic motions. The median AP prostate motion was anterior by 4.2 mm (range: 5–14 mm; Padhani et al. 1999). A later study by Ghilezan et al. (2005) using alternate transversal and sagittal cine MRI in six patients also emphasized the role of rectal wall movement in prostate motion. Both studies found that prostate motion due to rectal wall movement happens primarily in the AP direction.

Mah et al. (2002) performed another cine MRI study on 42 patients and observed large displacements of the prostate, up to 12 mm, which appeared to have been driven by peristalsis in the rectum. In addition to the influence of rectal motion on prostate motion, and although prostate is not treated with special respiratory motion management techniques, respiratory motion of the diaphragm has been linked to intrafraction prostate motion. Dawson et al. (2000) did a fluoroscopy study on four patients with implanted radiopaque markers, and observed a CC prostate motion of 0.9–5.1 mm and an AP motion of up to 3.5 mm during normal breathing in the prone position. During deep breathing, the CC motion was found to be 3.8–10.5 mm. Motion in the supine position was smaller.

Malone et al. (2000) used fluoroscopy and implanted gold markers in 20 patients who were immobilized prone in thermoplastic shells; the prostate was displaced 3.3 ± 1.8 mm (range: 1.0–10.2 mm) with respiration. They observed that respiratory-associated prostate movement decreased significantly when the thermoplastic shells were removed. Souchon et al. (2003) reported in a study, in which they used ultrasound, motions of the prostate due to respiration of up to 5 mm.

13.2.8 Head and Neck

Organs in the head and neck, such as the larynx, move during swallowing. Portal images have shown that although swallowing produces a large motion of up to 20 mm, it has a duration of only 0.45% of the total irradiation time (van Asselen et al. 2003). The dose using small fields is only reduced by 0.5% (Hamlet, Ezzell, and Aref 1994). Therefore, no action on motion management for swallowing has been recommended. Breathing has a smaller effect on tumor motion but is more frequent, and recommendations on adjusting the treatment margins have been made (van Asselen et al. 2003).

13.2.9 Cardiac Motion

Cardiac motion can cause nearby tumors to move. In a study of lung tumor motion using a set of two orthogonal fluoroscopic systems and implanted gold markers, Seppenwoolde et al. (2002) observed that in 7 out of 20 patients a measurable motion in the range of 1–4 mm was caused by the cardiac beat. These tumors were located near the heart or attached to the aortic arch. An earlier study using ultrafast CT by Ross et al. (1990) in 20 patients showed that 5 of 6 hilar lesions showed significant lateral motion (average 9.2 mm) with cardiac contraction. In a fluoroscopy study by Eckberg et al. (1998) in 20 lung cancer patients, it was observed that for several tumors located close to the heart, the cardiac beat made a major contribution to tumor movement. To date, there are no motion management strategies to deal with cardiac motion in radiation therapy.

13.3 Respiratory Monitoring

Most of the newest technologies for motion management discussed in Sections 13.5 through 13.8, such as breath holding, beam gating, and tumor tracking, rely on the existence and knowledge of a breathing signal. Ideally, the motion of the tumor would give this breathing signal. However, knowledge of tumor motion itself is not usually available. Therefore, surrogates have to be used to derive the breathing signal. Surrogates can be grouped into internal and external surrogates (Jiang 2006a). Internal surrogate methods include tracking implanted fiducial markers in or near the tumor to generate the breathing signal, whereas external surrogate methods include measuring motion of infrared markers on the abdomen or chest, surface motion derived from surface imaging cameras, airflow during breathing, and so on.

Currently, the only clinically available system that can obtain a breathing signal from internal surrogates is the real-time tumor-tracking radiation therapy (RTRT) system developed jointly by Mitsubishi Electric Co. (Tokyo, Japan) and Hokkaido University.

It has been used for beam gating treatments (treatments in which the radiation beam is turned on only when the tumor is in the path of the beam; Shirato et al. 1999; Shirato et al. 2000). Gold seeds are implanted in or near the tumor so that their position is automatically tracked in all three dimensions with the aid of a stereotactic kilovoltage X-ray imaging system. The imaging system acquires simultaneous orthogonal fluoroscopic images at a rate of 30 Hz.

Breathing can be externally monitored using external surrogates in different ways, which has led to the development of different systems. Early methods explored in the mid-1990s by Kubo and his colleagues for gating included the use of thermistors, thermocouples, strain gauge methods, and a pneumotachograph (Kubo, and Hill 1996). They also developed a system jointly with Varian Medical Systems Inc. (Palo Alto, California; Kubo et al. 2000) that tracks infrared reflective markers on a patient's abdomen using a video camera. This system was later commercialized by Varian and called the Real-Time Position Management (RPM) respiratory gating system. The RPM system has been extensively implemented and investigated clinically at a number of centers. Similar systems have been implemented, like ExacTrac Gating/Novalis Gating, from BrainLab (Feldkirchen, Germany), which, in addition to external markers, incorporates X-ray imaging capabilities for determining the internal anatomy position and verifying the internal anatomy position during treatment. Siemens Medical Systems (Concord, California) has developed a linear accelerator (linac) gating interface that receives the respiratory signal from a belt around the patient with a pressure cell that senses pressure changes as the patient breathes (Li, Stepaniak, and Gore 2006). More recently, 3D surface imaging video cameras have been used to obtain, track, and monitor images from the patient's surface. A commercially available 3D camera system comes from VisionRT, Ltd. in London, United Kingdom (Bert et al. 2005; Bert et al. 2007; Metheany et al. 2005).

In breath-holding techniques, the most commonly used device to monitor breath hold is a spirometer (Hanley et al. 1999; Mah et al. 2000; Rosenzweig et al. 2000). Spirometers measure the time-integrated airflow and, therefore, provide a baseline lung volume (e.g., end of exhale). Voluntary breath hold can be monitored using RPM. Mostly used for lung gating treatments, it has also been used in breath-hold treatment of the upper abdomen and breast (Berson et al. 2004; Pedersen et al. 2004). More recently, magnetic sensors and optical tracking systems have been proposed for monitoring breath hold (Remouchamps et al. 2007; Cerviño et al. 2009). As in respiratory gating, a key issue is the accuracy of externally placed breath-hold monitors in predicting internal positions of the tumor and nearby organs. The external-to-internal correlation can change, not only by alterations in breathing pattern but also by changes in the internal anatomy such as abdominal contents changes, ascites, and tumor growth or shrinkage with treatment. Thus, the external-to-internal constancy requires verification at simulation and throughout the treatment course.

Besides the aforementioned techniques used for gating and breath holding, new respiration monitoring strategies are being investigated for tumor tracking. One technique, as in gating, involves the fluoroscopic tracking of radiopaque fiducial markers implanted inside or near the tumor (Shirato et al. 2003; Tang, Sharp, and Jiang 2007). The accuracy of this technology is better than 1.5 mm for tracking moving targets, which is much higher than the accuracy of the external surrogates approach (Shirato et al. 2000). Marker tracking based on nonionizing electromagnetic fields, using small wireless transponders implanted in human tissue, has also been implemented (Balter et al. 2005). One commercial system with electromagnetic transponders is the 4D Localization System of Calypso Medical. However, no matter how marker tracking is realized, as long as percutaneous marker implantation is involved the clinical implementation of this technology in lung cancer radiotherapy is limited due to the risk of pneumothorax (Arslan et al. 2002; Geraghty et al. 2003). Tracking of lung tumors can also be achieved without implanted fiducial markers with fluoroscopic and electron portal imaging device (EPID) images. Template-matching techniques, optical flow, active shape model, and principal component analysis combined with artificial neural network techniques have achieved promising results when the tumor has reasonably high contrast and clear boundary in the images (Cui et al. 2007; Xu et al. 2007; Xu et al. 2008; Lin et al. 2009).

13.4 Breathing Coaching

A weak point in using external surrogates to derive tumor position is the assumption that a good and constant relationship between the surrogate and the tumor motion exists. This relationship may change over time, both inter- and intrafractionally (Seppenwoolde et al. 2007). At the same time, it is important to maintain a reproducible breathing pattern in order to provide adequate treatment in an efficient way. It has been shown that treatment efficiency can be improved with the cooperation of the patient and with visual and audio respiratory coaching (Mageras et al. 2001; Kini et al. 2003; George et al. 2006; Neicu et al. 2006; Baroni et al. 2007). Visual coaching consists of providing the patient with a video of the breathing signal, whereas audio coaching consists of providing the patient with audible instructions on how to breathe. Audio coaching alone improves periodicity of the patient's breathing pattern, but not amplitude or baseline drift, which can be modified using visual coaching (Kini et al. 2003; Neicu et al. 2006). Breath-holding techniques can also benefit from visual coaching in terms of reproducibility and stability of the breath hold (Cerviño et al. 2009).

13.5 Classification of Motion Management Techniques

Techniques for the management of respiratory tumor motion have been developed following two different lines of thought. The main idea of the first class of techniques is to allow the tumor to move freely relative to the treatment beam, and to try to integrate the motion effect (geometrically or dosimetrically) into the

treatment plan. The first class of techniques includes population-based internal margin (IM), patient-specific IM, the internal target volume (ITV) method, and IMRT optimization using a motion probability density function (PDF; Trofimov et al. 2005; Jiang 2006a). The main purpose of the second class of techniques is to stop the tumor motion relative to the treatment beams, which includes two categories: (1) techniques that allow free tumor motion but adjust the treatment equipment to maintain a constant target position in the beam's eye view when the beam is on, through respiratory gating, beam tracking, or couch-based motion compensation (Jiang 2006a); and (2) techniques that control tumor motion, using techniques such as breath holding, forced shallow breathing, or abdominal compression.

Motion management techniques discussed in Sections 13.5 through 13.8 refer to techniques that are applied during treatment delivery. However, some of these techniques are also used during imaging and treatment simulation. It is important to note that respiration patterns can vary from day to day, and that tumor deformation or regression might occur as well. It is therefore important to confirm on a day-by-day basis that respiratory motion is compatible with the treatment plan. This could be done, for example, by imaging the patient with fluoroscopy or CBCT and setting up the patient so that the day-specific motion range coincides with the motion range in the plan.

13.6 Motion-Encompassing Techniques

Motion-encompassing techniques consist of developing a treatment plan by adding margins to the target volume to encompass tumor motion. The techniques presented in Sections 13.6.1 through 13.6.4 go from less to more patient-specific margins, and less to more patient-specific motion information is used. A major drawback of these approaches is that tumor motion is measured only at the time of simulation; however, motion can vary greatly from simulation to treatment time, as well as between different treatment days. Therefore, it is important to use some image guidance techniques to monitor tumor motion during treatment or to ensure a regular and stable breathing pattern by using breath coaching techniques.

13.6.1 Population-Based Margins

Motion management is, in many institutions, accounted for by the use of margins that encompass tumor motion. The definition of relevant terminology is included in the International Commission on Radiation Units and Measurements (ICRU) Report 50 (ICRU 1993). Gross tumor volume (GTV) includes the demonstrated tumor; clinical target volume (CTV) is constructed by adding margins around the GTV in order to include possible neighboring tumor involvement; and finally, planning target volume (PTV) is defined as the CTV with an added margin to account for geometrical variations such as patient movement, setup errors, and organ motion.

Addition of population-based margins has been a standard clinical practice for many years. However, population-based margins used to expand GTV to PTV only work for the "average" patient. In patients whose tumor motion is larger than the population average, the target will be underdosed. On the other hand, in patients with target motion smaller than the population average, the surrounding healthy tissues will be overdosed. Because, in general, margins are larger than tumor motion, this method to encompass tumor motion often leads to unnecessary normal tissue irradiation.

13.6.2 Patient-Specific Margins

If some information regarding tumor motion in a patient is available, margins can be designed on a patient-by-patient basis. IGRT plays an important role in the safe adjustment of margins so as not to overdose healthy tissue and/or underdose a target. Tumor motion can be measured with fluoroscopy or 4D-CT scanning, so that patient-specific IMs, which are often asymmetric, can be designed according to the measurements.

13.6.3 Internal Target Volume

The ICRU Report 62 defines ITV as the volume that encompasses the CTV and an IM (ICRU 1999). The IM is added to compensate expected physiologic movements and variations in size, shape, and position of the CTV during therapy in relation to an internal reference point. The ITV is normally asymmetric around the CTV. Because the ITV concept includes motion, it cannot be detected with a single static image of the patient. Instead, it has to be determined with imaging techniques that show the entire range of tumor motion. These techniques include fluoroscopy, slow CT scans, 4D-CT or respiration-correlated CT scans, 4D-MRI, and combined inhale and exhale breath-hold CT scans (Liu et al. 2007). When CT scans are used to obtain the ITV, a maximum intensity projection (MIP) is normally used to obtain the tumor-motion-encompassing volume (Underberg et al. 2005). In addition to ITV, a setup margin (SM) needs to be added around the ITV to account for uncertainties in patient positioning and therapeutic beam alignment during treatment planning and for all treatment sessions. The ITV with the addition of the SM forms the PTV.

13.6.4 Intensity-Modulated Radiation Therapy Optimization Using Motion Probability Density Function

One way to account for real patient-specific tumor motion in IMRT is to include the motion into plan optimization. As mentioned earlier, motion of the tumor can be determined via 4D-CT. A PDF of the tumor motion can then be derived from it, by checking the positions of the tumor and the organs at risk in the different instants of the CT scan. This PDF is convolved with the pencil beam kernel during inverse optimization (Unkelbach and Oelfke 2004; Trofimov et al. 2005). Regions of the patient where both tumor and organs at risk are located with certain probability are irradiated with lower doses than the prescribed

tumor dose. Irradiating with a higher dose regions that are always occupied by tumors compensates for this lower dose.

13.7 Techniques for Control of Tumor Motion

13.7.1 Breath Hold

Breath-holding techniques consist of freezing the tumor position by a voluntary or forced breath hold while the radiation beam is on. Used mainly for lung and breast radiation treatments, this has also been used for other organs such as liver and kidneys (Suramo, Päivänsalo, and Myllylä 1984; Schwartz et al. 1994; Hanley et al. 1999; Wong et al. 1999; Mah et al. 2000; Rosenzweig et al. 2000; Stromberg et al. 2000; Barnes et al. 2001; Dawson et al. 2001; Kim et al. 2001; Sixel, Aznar, and Ung 2001; Remouchamps et al. 2003; Berson et al. 2004; Pedersen et al. 2004). Different breath-hold techniques present different advantages, with tumor immobilization being a common advantage of all. A key issue in breath-hold treatments is that the treatment plan should always be designed according to the treatment technique. Therefore, if treatment is going to be delivered at a specified point in the breathing cycle, the plan and simulation should be performed at the same point in the breathing cycle. In radiation therapy, the aim is to achieve the same breath-hold position between fields during a single treatment fraction and between fractions. Therefore, reproducibility of the breath hold needs to be guaranteed. One of the main problems of breath holding is the compliance of the patient. As the duration of a treatment beam is generally 15–30 seconds, many patients are able to have the beam dose delivered within one breath hold. However, whereas some patients, such as breast cancer patients, usually comply well with breath-holding requirements in treatment, it may be difficult for lung cancer patients to hold their breath due to their compromised pulmonary status. Patient cooperation and patient comfort need to be considered when selecting this treatment method.

Most breath-holding methods use some means of monitoring each breath hold. These methods range from spirometry to more sophisticated surface motion. In addition, verbal coaching as well as visual coaching can be provided to the patient in order to improve reproducibility of the breath hold. Sections 13.7.1.1 through 13.7.1.3 discuss DIBH, voluntary breath hold, and forced breath-holding techniques.

13.7.1.1 Deep Inspiration Breath Hold

DIBH is a technique that consists of an FB interval followed by a breath hold at approximately 100% vital capacity during a prescribed period. DIBH has been explored mainly for lung and breast cancer treatments. It has two different features: (1) deep inspiration, which increases the distance from the breast to the heart and reduces lung density, thereby increasing normal tissue sparing and reducing lung toxicity and risk of pneumonitis (Barnes et al. 2001); and (2) breath hold, which immobilizes the tumor (Hanley et al. 1999; Barnes et al. 2001).

13.7.1.2 Voluntary Breath Hold

Voluntary breath hold, or self-held breath hold, can be achieved with or without respiratory monitoring. The patient voluntarily holds the breath at a specified point in the breathing cycle, and the beam is turned on during the breath hold. The duty cycle is larger than that for FB respiratory-gated techniques because delivery of radiation is continuous during the breath hold. It has been observed that maximum reproducibility is achieved at deep inhale or deep exhale.

13.7.1.3 Forced Breath Hold

Assisting devices can be used for improving the reproducibility of breath hold. A spirometer is used for monitoring breathing. An occlusion valve can be added to the spirometer, forcing the breath hold when desired and leading to forced breath hold. This idea led to the development of what is known as *active breathing control* (ABC). ABC was first developed at the William Beaumont Hospital in Michigan, and is currently widely used in DIBH treatments, although the ABC device can suspend breathing at any predetermined position. ABC has been used in the treatment of diverse tumor sites at breath hold, such as lung, liver, and breast, and Hodgkin's disease (Wong et al. 1999; Stromberg et al. 2000; Dawson et al. 2001; Sixel et al. 2001; Remouchamps, Letts, Vicini, et al. 2003; Remouchamps, Letts, Yan, et al. 2003). The patient is verbally coached to achieve steady breathing. Treatments at moderate DIBH (mDIBH), which consists of holding the breath at approximately 75% of the vital capacity, have been shown to achieve good reproducibility of internal organ position when using ABC, while being comfortable to the patient (Remouchamps, Letts, Yan, et al. 2003; Remouchamps, Letts, Vicini, et al. 2003).

13.7.2 Shallow Breathing and Immobilization

A simple method to minimize organ breathing motion in the upper abdomen and thoracic area is to use abdominal compression, which is used in extracranial stereotactic radiosurgery or *stereotactic body radiotherapy* (SBRT) when tumor motion exceeds clinical goals. SBRT is a radiotherapy treatment method to deliver a high dose of radiation to the target with a high degree of precision within the body (Potters et al. 2004; Murray, Forster, and Timmerman 2007). SBRT with immobilization was first applied to treatments of the lung, liver, and some extrahepatic lesions (Lax et al. 1994; Blomgren et al. 1995). The standard technique makes use of a stereotactic body frame with a vacuum pillow for body immobilization, with an attached plate that applies compression to the abdomen, thereby reducing diaphragmatic excursions while still permitting limited normal respiration. There are a few commercial systems, such as Elekta Stereotactic Body Frame, BodyFIX frame, and Precision Therapy.

Some institutions have designed their own systems (Lax et al. 1994; Lee et al. 2003). The body frame is used for defining and fixing a coordinate system for stereotactic treatment, while abdominal compression is used to manage tumor motion.

The abdominal plate is attached to the body frame by a rigid arc, and a scaled screw is used to achieve reproducible pressure. Usually, the maximum pressure that the patient can tolerate for the duration of the treatment is used. During CT scan, the range of motion of the tumor by the pulmonary and cardiac motions is determined fluoroscopically. Negoro et al. (2001) observed tumor motion in 18 lung cancer patients (20 lesions) and applied abdominal compression (or diaphragm control) to 10 who showed a tumor motion range above 5 mm. Tumor motion was reduced from a mean of 12.3 mm during free respiration to 7 mm with compression. There was one patient who incremented tumor motion when compression was applied, probably due to an effort to overcome compression. Motion after abdominal compression is applied needs to be fluoroscopically determined, in order to adjust pressure.

Because of high doses delivered in just a few fractions, treatment delivery precision is very important. Approaches that help in the accurate delivery of SBRT include whole-body immobilization and image-based assessment of patient and target positions immediately prior to the delivery of radiation. Two major factors have to be considered for the image guidance of SBRT. First, setups of patients for treatments of the body stem are rarely as reliable and accurate as setups for intracranial stereotactic radiation therapy. Second, target motion (either intrafraction or interfraction) frequently occurs. In order to assess the accuracy of patient position, verification portal films are usually taken before each treatment (Wulf et al. 2000; Lee et al. 2003). However, other approaches are available and under investigation, such as kilovoltage CT, megavoltage CT, onboard kilovoltage CBCT, and ultrasound guidance (Fuss et al. 2007). Measurements of diaphragm motion under fluoroscopy on different days can be made to verify reproducibility (Keall et al. 2006).

13.8 Techniques for Synchronization of Beam with Tumor Motion

13.8.1 Beam Gating

Respiratory gating is a motion management technique that consists of delivering radiation only during the portion of the breathing cycle when the tumor is in the path of the beam (Ohara et al. 1989; Inada et al. 1992; Kubo et al. 2000; Kubo and Wang 2000; Shirato et al. 2000; Jiang 2006a). Because tumors move and treatment is limited to a breathing cycle portion, the treatment plan needs to be adjusted to the treatment geometry. In addition, respiratory gating relies on the existence and knowledge of a breathing signal and the feasibility to trigger the beam during the desired portion (phase or amplitude) of the given breathing signal. Commonly used in the lungs, it is also being used for treatment of other organs such as the liver (Wagman et al. 2003). The performance of linacs with gating capabilities has been extensively evaluated for gated photon beams (Kubo et al. 1996; Ramsey et al. 1999; Kubo et al. 2000; Hugo, Agazaryan, and Solberg 2002; Duan et al. 2003; Kriminski, Li, and Solberg 2006).

Gating systems can be categorized depending on the surrogates they use into internal and external gating (Jiang 2006a). Internal gating consists of using internal surrogates such as implanted fiducial markers in the tumor or nearby regions to generate the gating signal, whereas external gating uses surrogates such as infrared markers on the abdomen or chest, surface motion derived from surface imaging cameras, airflow during breathing, and so on. As mentioned in Section 13.3, the only internal gating system currently clinically available is the RTRT system developed jointly by Mitsubishi Electric Co. and Hokkaido University (Shirato et al. 1999; Shirato et al. 2000). When fiducials are within a small range of their target position (from simulation), the treatment beam is turned on.

In external gating with surface motion as a surrogate, both chest motion and abdominal motion have been used as surrogates, the latter one being the preferred method due to its higher amplitude. Examples of commonly used external gating methods include the RPM system by Varian and the pressure belt system by Siemens. These systems are described in Sections 13.8.1.3 and 13.8.1.4.

13.8.1.1 Gated Computed Tomography and Four-Dimensional Computed Tomography

A CT scan is normally used for treatment planning. Regular FB CT scans do not properly reflect tumor motion. Depending on the tumor and the scanning speeds, a different variety of artifact may appear in the FB CT image (Chen, Kung, and Beaudette 2004; Lewis and Jiang 2009). One way to mitigate this motion artifact is to use gated CT, which consists of imaging only at a specified breathing phase (Shen et al. 2003; D'Souza et al. 2005). Treatment should be delivered according to the phase selected for imaging. A more powerful technique is 4D imaging, which is currently the most widely used method for treatment planning in gated treatments. Longer scans at each couch position are acquired, and the acquisition time is synchronized with a breathing signal. A 4D-CT scan is acquired, showing a different CT image for different phases in the breathing cycle. Normally, 10–20 phases are used. A reduced number of phases are selected from the scan for gated treatment, and usually MIP is used for lung cancer to combine the phases and define the gated ITV. Just a few breathing phases are considered and, therefore, the gated ITV is smaller than the one derived from motion-encompassing methods. 4D-CT is based on the hypothesis that breathing produces a regular and reproducible signal. Because breathing is not always regular, and because of the mechanical limitations of CT scanners, artifacts might appear in the image. Breathing coaching techniques have been developed to aid patients in achieving a regular and reproducible breathing signal in order to have an artifact-free image. Visual and audio coaching techniques have been studied (Mageras et al. 2001; Kini et al. 2003; Nelson et al. 2005; George et al. 2006; Neicu et al. 2006; Baroni et al. 2007; Cerviño et al. 2009).

13.8.1.2 Gating Window

The *gating window* is defined as the range in the surrogate signal (such as the 3D marker position in the case of internal

gating and the AP marker motion in the case of external gating) when the beam is turned on during treatment. When the surrogate signal falls within the gating window, the gating signal is 1, and otherwise it is 0. Therefore, the gating window converts the respiratory signal into a gating signal, and this gating signal controls the linac. The surrogate signal can be multidimensional; however, the gating signal is always one dimensional. For each dimension of the surrogate signal, there are two values that define the boundary of the gating window in that dimension. For internal gating, the gating window is often a small rectangular solid corresponding with the 3D position of the implanted fiducial marker. For external gating, the gating window can be defined by two AP positions or two phase values of the surface marker, which correspond to two types of external gating: (1) amplitude gating (when using the position) and (2) phase gating (when using the phase). Due to the irregularity of patient breathing, the phase value and the surface marker position often do not have a perfect one-to-one correspondence and, therefore, amplitude gating and phase gating might generate different gating signals.

Two important factors to be taken into account when specifying the gating window are the linac duty cycle and the residual motion of the tumor in the gating window. The duty cycle is the ratio of the beam-on time to the total treatment time and, therefore, it is 100% in a nongated treatment. The objective of gating is to maximize the duty cycle and to minimize the residual motion of the tumor in the gating window. These goals are achieved better at end-of-inhale (EOI) and end-of-exhale (EOE). It has been shown that gating at EOE and EOI are equivalent in terms of lung dose and toxicity, and with proper coaching, in terms of duty cycle (Berbeco et al. 2006; Wu et al. 2008). However, gated treatments are more frequently performed at EOE. The PTV–CTV margins have to be added in accordance with the residual motion during the beam-on time. A typical duty cycle value in gated treatments is 30%–50% (Jiang 2006a,b).

13.8.1.3 Varian Real-Time Position Management Optical System

As discussed in Section 13.3, the RPM system was developed in the 1990s and was commercialized by Varian Medical Systems Inc. It belongs to the external gating systems category and has been widely implemented clinically. It is used for linac gating and as a patient respiration monitoring and coaching tool. The system consists of a plastic block with two or six passive infrared reflective markers that is placed on the patient's anterior abdominal surface and is monitored by a charge-coupled-device video camera mounted on the treatment room wall. The surrogate signal is the abdominal surface motion. Whereas the two-reflector block provides only AP motion information, the six-reflector block provides 3D information of its motion. Although phase gating is used for 4D-CT acquisition, treatment is performed with amplitude gating. Once the patient has been set up, a periodicity filter checks the regularity of the breathing waveform. When breathing is regular, a gating window is selected according to treatment planning (e.g., if the treatment planning has been done at EOE, the gating window is selected also at EOE). Then, the gated treatment starts with visual or audio coaching if available. When the breathing signal lies outside the gating window, the beam is stopped, and it is resumed once the signal again enters the gating window. Therapists are instructed to closely monitor the RPM signal. If the patient has problems following the coaching signal, the treatment may be interrupted and resumed after the problem is solved.

13.8.1.4 Siemens Anzai Pressure Belt System

The Anzai system utilizes a pressure sensor (load cell; dimensions: 30 mm in diameter and 9.5 mm in thickness) to detect the external respiratory motion by detecting pressure changes in real time (Li, Stepaniak, and Gore 2006). The sensor is attached to a belt that is positioned on the abdomen of the patient. The respiratory abdominal motion signal (amplitude and phase) detected by the pressure sensor is recorded by the sensor port and transferred to the wave deck. This motion signal is then transferred to the control computer for processing and display. The predetermined gating parameters from 4D-CT are also displayed in the computer. As with the RPM system, the display can be shown to the patient via video goggles. The Anzai gating system is interfaced with the Siemens linac via an open gating portal and it can output the gating signal that triggers the beam on and off. The gating software has a predictive filter to calculate the patient's average breathing pattern in order to account for any abnormalities that may occur during gating (coughing, baseline shift, change in breathing amplitude in amplitude gating, etc.). When an abnormality is detected, the predictive filter will send a signal that automatically holds the beam in the "off" position until normal breathing resumes. The beam will be off for at least two consecutive breathing cycles.

13.8.2 Couch-Motion Compensation

Couch-motion compensation is an idea that has not yet been clinically applied. It consists of adjustments of the treatment couch opposite to tumor motion during treatment delivery (D'Souza, Naqvi, and Yu 2005). It involves determining accurately the position of the tumor in real time and a robotically controlled couch coupled to the target tracking system. The motion is opposite to tumor motion as the patient breathes, thereby keeping the target at a fixed position in space. Although not commercially available, D'Souza et al. (D'Souza, Naqvi, and Yu 2005; D'Souza and McAvoy 2006) have proved the feasibility of this technique. There are still some concerns about this technique: couch motion will compensate for intrafraction motion, however, it will not compensate for tumor deformation. In addition, the patient might react to couch motion and might try to compensate it. To minimize this, immobilization devices and patient coaching could be used. But the main difficulty comes in implementing fast couch motion and dealing with the associated position verification uncertainties. Mean tumor position, however, can be compensated for due to its slow-motion nature (Trofimov et al. 2008; Wilbert et al. 2008).

13.8.3 Tumor Tracking

The beam tracking technique involves following the target dynamically with the radiation beam (Murphy 2004). Although the RTRT system from Mitsubishi and Hokkaido University can track the gold markers implanted in the tumor in real time, it is used only for gating and no tumor tracking is performed.

Beam tracking was first implemented in a robotic radiosurgery system (Adler et al. 1999; Ozhasoglu et al. 2000; Schweikard et al. 2000; Murphy 2002; Murphy et al. 2003). The Synchrony Respiratory Tracking System integrated with the CyberKnife robotic linac (Accuray, Inc., Sunnyvale, California) is the only robotic linac used currently for radiation therapy. Before treatment, the 3D position of the tumor is determined by automatically detecting implanted gold markers in orthogonal X-ray images at the same time as the signal from an external surrogate on the patient's surface is recorded. The X-ray images and the recorded signal are used to build a mathematical model that relates the external and internal motions.

During treatment, the tumor position is continuously predicted from the external surrogate and the model built. The model is updated for every radiation beam by acquiring new X-ray images.

In linac-based radiotherapy, tumor motion can be compensated using a dynamic multileaf collimator (MLC; Keall et al. 2001; Neicu et al. 2003; Papiez 2003; Suh et al. 2004; Keall et al. 2005; Papiez and Rangaraj 2005; Rangaraj and Papiez 2005; Webb 2005a,b; Wijesooriya et al. 2005; Neicu et al. 2006; Sawant, Venkat, and Srivastava 2008). The dynamic MLC continuously aligns or reshapes the treatment aperture so as to compensate for tumor motion. Theoretical and empirical investigations have shown the great potential of this technique; however, no clinical implementation exists to date. Other ideas have been proposed for heavy ion therapy, such as magnetic deflection of the pencil beam for lateral compensation and energy modulation for longitudinal compensation (Grozinger et al. 2006; Bert et al. 2007).

13.9 Future Directions

Respiratory motion poses a challenge for the radiation therapy of some organs, such as the lungs, liver, and pancreas. Although motion management techniques have already been applied to some of these treatments, there is still much work to do in order to improve the treatment of mobile tumors. Some future directions for tumor motion management radiation therapy include new techniques in imaging and in treatment delivery.

In the area of imaging, one technique currently under development is the integration of online MRI with a linac (hybrid MRI–linac; Lagendijk et al. 2008), which will be able to provide real-time imaging during radiation therapy. Advantages of MRI include radiation-free imaging and good visualization of soft tissues. We believe that another technique improvement will be the use of real-time volumetric imaging during treatment delivery (current tumor imaging methods used in the treatment room provide only planar images).

In the area of treatment delivery, future developments will be based on tumor location rather than on surrogates; but there is also room for the development of good models for tumor motion based on surrogates, which can be used for treatment, such as biomechanical models. Delivery methods such as "snapshot" radiotherapy (using very high dose rates for treatment) would clearly reduce the effect of tumor motion. In addition, and although beam tracking and dynamic compensation have been extensively researched, no optimal solution for the clinical implementation of tumor tracking has been found.

References

Adler, J. Jr., M. J. Murphy, S. D. Chang, and S. L. Hancock. 1999. Image-guided robotic radiosurgery. *Neurosurgery* 44(6):1299–307.

Arslan, S., A. Yilmaz, B. Bayramgurler, et al. 2002. CT-guided transthoracic fine needle aspiration of pulmonary lesions: Accuracy and complications in 294 patients. *Med Sci Monit* 8(7):493–7.

Balter, J. M., R. K. Ten Haken, T. S. Lawrence, K. L. Lam and J. M. Robertson. 1996. Uncertainties in CT-based radiation therapy treatment planning associated with patient breathing. *Int J Radiat Oncol Biol Phys* 36(1):167–74.

Balter, J. M., J. N. Wright, L. J. Newell, et al. 2005. Accuracy of a wireless localization system for radiotherapy. *Int J Radiat Oncol Biol Phys* 61(3):933–7.

Barnes, E. A., B. R. Murray, D. M. Robinson, et al. 2001. Dosimetric evaluation of lung tumor immobilization using breath hold at deep inspiration. *Int J Radiat Oncol Biol Phys* 50(4):1091–8.

Baroni, G., M. Riboldi, M. F. Spadea, et al. 2007. Integration of enhanced optical tracking techniques and imaging in IGRT. *J Radiat Res* 48:A61–74.

Becker, S. J., R. R. Patel, and T. R. Mackie. 2006. Accelerated partial breast irradiation with helical tomotherapy: Prone or supine setup? *Int J Radiat Oncol Biol Phys* 66(3, Supp 1): S230.

Berbeco, R. I., S. Nishioka, H. Shirato, and S. B. Jiang. 2006. Residual motion of lung tumors in end-of-inhale respiratory gated radiotherapy based on external surrogates. *Med Phys* 33(11):4149–56.

Berson, A. M., R. Emery, L. Rodriguez, et al. 2004. Clinical experience using respiratory gated radiation therapy: Comparison of free-breathing and breath-hold techniques. *Int J Radiat Oncol Biol Phys* 60(2):419–26.

Bert, C., K. G. Metheany, K. Doppke, and G. T. Chen. 2005. A phantom evaluation of a stereo-vision surface imaging system for radiotherapy patient setup. *Med Phys* 32(9):2753–62.

Bert, C., N. Saito, A. Schmidt, et al. 2007. Target motion tracking with a scanned particle beam. *Med Phys* 34(12):4768–71.

Bhasin, D. K., S. S. Rana, S. Jahagirdar, and B. Nagi. 2006. Does the pancreas move with respiration? *J Gastroenterol Hepatol* 21:1424–7.

Blomgren, H., I. Lax, I. Naslund, and R. Svanstrom. 1995. Stereotactic high dose fraction radiation therapy of extracranial tumors using an accelerator. Clinical experience of the first thirty-one patients. *Acta Oncol* 34(6):861–70.

Boda-Heggemann, J., F. Kohler, H. Wertz, et al. 2008. Intrafraction motion of the prostate during an IMRT session: A fiducial-based 3D measurement with cone-beam CT. *Radiat Oncol* 3(1):37.

Bortfeld, T., K. Jokivarsi, M. Goitein, J. Kung, and S. B. Jiang. 2002. Effects of intra-fraction motion on IMRT dose delivery: Statistical analysis and simulation. *Phys Med Biol* 47(13):2203–20.

Bryan, P. J., S. Custar, J. R. Haaga, and V. Balsara. 1984. Respiratory movement of the pancreas: An ultrasonic study. *J Ultrasound Med* 3(7):317–20.

Byrne, T. E. 2005. A review of prostate motion with considerations for the treatment of prostate cancer. *Med Dosim* 30(3):155–61.

Cerviño, L. I., S. Gupta, M. A. Rose, C. Yashar, and S. B. Jiang. 2009. Using surface imaging and visual coaching to improve reproducibility and stability of deep-inspiration breath hold for left breast cancer radiotherapy. *Phys Med Biol* 54(22):6853–65.

Chen, G. T., J. H. Kung, and K. P. Beaudette. 2004. Artifacts in computed tomography scanning of moving objects. *Semin Radiat Oncol* 14(1):19–26.

Cui, Y., J. G. Dy, G. C. Sharp, B. Alexander, and and S. B. Jiang. 2007. Multiple template-based fluoroscopic tracking of lung tumor mass without implanted fiducial markers. *Phys Med Biol* 52(20):6229–42.

Davies, S. C., A. L. Hill, R. B. Holmes, M. Halliwell, and P. C. Jacson. 1994. Ultrasound quantitation of respiratory organ motion in the upper abdomen. *Br J Radiol* 67:1096–102.

Dawson, L. A., K. K. Brock, S. Kazanjian, et al. 2001. The reproducibility of organ position using active breathing control (ABC) during liver radiotherapy. *Int J Radiat Oncol Biol Phys* 51(5):1410–21.

Dawson, L. A., D. W. Litzenberg, K. K. Brock, et al. 2000. A comparison of ventilatory prostate movement in four treatment positions. *Int J Radiat Oncol Biol Phys* 48(2):319–23.

Doyle, V. L., F. A. Howe, and J. R. Griffiths. 2000. The effect of respiratory motion on CSI localized MRS+. *Phys Med Biol* 45(8):2093–104.

D'Souza, W. D., Y. Kwok, C. Deyoung, et al. 2005. Gated CT imaging using a free-breathing respiration signal from flow-volume spirometry. *Med Phys* 32(12):3641–9.

D'Souza, W. D., and T. J. McAvoy. 2006. An analysis of the treatment couch and control system dynamics for respiration-induced motion compensation. *Med Phys* 33(12):4701–9.

D'Souza, W. D., S. A. Naqvi, and C. X. Yu. 2005. Real-time intra-fraction-motion tracking using the treatment couch: A feasibility study. *Phys Med Biol* 50(17):4021–33.

Duan, J., S. Shen, J. B. Fiveash, et al. 2003. Dosimetric effect of respiration-gated beam on IMRT delivery. *Med Phys* 30(8):2241–52.

Ekberg, L., O. Holmberg, L. Wittgren, G. Bjelkengren, and T. Landberg. 1998. What margins should be added to the clinical target volume in radiotherapy treatment planning for lung cancer? *Radiother Oncol* 48(1):71–7.

El-Fallah, A. I., M. B. Plantec, and K. W. Ferrara. 1997. Ultrasonic measurement of breast tissue motion and the implications for velocity estimation. *Ultrasound Med Biol* 23(7):1047–57.

Fang, L. C., R. Komaki, P. Allen, et al. 2006. Comparison of outcomes for patients with medically inoperable Stage I non-small-cell lung cancer treated with two-dimensional vs. three-dimensional radiotherapy. *Int J Radiat Oncol Biol Phys* 66(1):101–16.

Fuss, M., J. Boda-Heggemann, N. Papanikolau, and B. J. Salter. 2007. Image-guidance for stereotactic body radiation therapy. *Med Dosim* 32(2):102–10.

George, R., T. D. Chung, S. S. Vedam, et al. 2006. Audio-visual biofeedback for respiratory-gated radiotherapy: Impact of audio instruction and audio-visual biofeedback on respiratory-gated radiotherapy. *Int J Radiat Oncol Biol Phys* 65(3):924–33.

Geraghty, P. R., S. T. Kee, G. McFarlane, et al. 2003. CT-guided transthoracic needle aspiration biopsy of pulmonary nodules: Needle size and pneumothorax rate. *Radiology* 229(2):475–81.

Ghilezan, M. J., D. A. Jaffray, J. H. Siewerdsen, et al. 2005. Prostate gland motion assessed with cine-magnetic resonance imaging (cine-MRI). *Int J Radiat Oncol Biol Phys* 62(2):406–17.

Grozinger, S. O., E. Rietzel, Q. Li, et al. 2006. Simulations to design an online motion compensation system for scanned particle beams. *Phys Med Biol* 51(14):3517–31.

Hamlet, S., G. Ezzell, and A. Aref. 1994. Larynx motion associated with swallowing during radiation therapy. *Int J Radiat Oncol Biol Phys* 28(2):467–70.

Hanley, J., M. M. Debois, D. Mah, et al. 1999. Deep inspiration breath-hold technique for lung tumors: The potential value of target immobilization and reduced lung density in dose escalation. *Int J Radiat Oncol Biol Phys* 45(3):603–11.

Harauz, G., and M. J. Bronskill. 1979. Comparison of the liver's respiratory motion in the supine and upright positions: Concise communication. *J Nucl Med* 20(7):733–5.

Hugo, G. D., N. Agazaryan, and T. D. Solberg. 2002. An evaluation of gating window size, delivery method, and composite field dosimetry of respiratory-gated IMRT. *Med Phys* 29(11):2517–25.

ICRU. 1993. Prescribing, recording, and reporting photon beam therapy. ICRU Report 50, International Commission on Radiation Units and Measurements, Bethesda, MD.

ICRU. 1999. Prescribing, recording, and reporting photon beam therapy (supplement to ICRU Report 50). ICRU Report 62, International Commission on Radiation Units and Measurements, Bethesda, MD.

Inada, T., H. Tsuji, Y. Hayakawa, A. Maruhasi, and H. Tsujii. 1992. Proton irradiation synchronized with respiratory cycle. *Nippon Igaku Hoshasen Gakkai Zasshi* 52(8):1161–7 (in Japanese).

Ingold, J., G. Reed, H. Kaplan, and M. Bagshaw. 1965. Radiation hepatitis. *Am J Roentgenol* 93:200–8.

Jiang, S. B. 2006a. Radiotherapy of mobile tumors. *Semin Radiat Oncol* 16(4):239–486.

Jiang, S. B. 2006b. Technical aspects of image-guided respiration-gated radiation therapy. *Med Dosim* 31(2):141–51.

Keall, P. J., S. Joshi, S. S. Vedam, et al. 2005. Four-dimensional radiotherapy planning for DMLC-based respiratory motion tracking. *Med Phys* 32(4):942–51.

Keall, P. J., V. R. Kini, S. S. Vedam, and R. Mohan. 2001. Motion adaptive x-ray therapy: A feasibility study. *Phys Med Biol* 46(1):1–10.

Keall, P. J., G. S. Mageras, J. M. Balter, et al. 2006. The management of respiratory motion in radiation oncology report of AAPM Task Group 76. *Med Phys* 33(10):3874–900.

Kim, D. J., B. R. Murray, R. Halperin, and W. H. Roa. 2001. Held-breath self-gating technique for radiotherapy of non-small-cell lung cancer: A feasibility study. *Int J Radiat Oncol Biol Phys* 49(1):43–9.

Kini, V. R., S. S. Vedam, P. J. Keall, et al. 2003. Patient training in respiratory-gated radiotherapy. *Med Dosim* 28(1):7–11.

Kitamura, K., H. Shirato, Y. Seppenwoolde, et al. 2002. Three-dimensional intrafractional movement of prostate measured during real-time tumor-tracking radiotherapy in supine and prone treatment positions. *Int J Radiat Oncol Biol Phys* 53(5):1117–23.

Korin, H. W., R. L. Ehman, S. J. Riederer, J. P. Felmlee, and R. C. Grimm. 1992. Respiratory kinematics of the upper abdominal organs: A quantitative study. *Magn Reson Med* 23(1):172–8.

Korreman, S. S., A. N. Pedersen, T. J. Nøttrup, L. Specht, and H. Nyström. 2005. Breathing adapted radiotherapy for breast cancer: Comparison of free breathing gating with the breath-hold technique. *Radiother Oncol* 76(3):311–8.

Kriminski, S., A. N. Li, and T. D. Solberg. 2006. Dosimetric characteristics of a new linear accelerator under gated operation. *J Appl Clin Med Phys* 7(1):65–76.

Kubo, H. D., and B. C. Hill. 1996. Respiration gated radiotherapy treatment: A technical study. *Phys Med Biol* 41(1):83–91.

Kubo, H. D., P. M. Len, S. Minohara, H. Mostafavi. 2000. Breathing-synchronized radiotherapy program at the University of California Davis Cancer Center. *Med Phys* 27(2):346–53.

Kubo, H. D., and L. Wang. 2000. Compatibility of Varian 2100C gated operations with enhanced dynamic wedge and IMRT dose delivery. *Med Phys* 27(8):1732–8.

Lagendijk, J. J. W., B. W. Raaymakers, A. J. E. Raaijmakers, et al. 2008. MRI/linac integration. *Radiother Oncol* 86(1):25–9.

Langen, K. M., and D. T. L. Jones. 2001. Organ motion and its management. *Int J Radiat Oncol Biol Phys* 50(1):265–78.

Lax, I., H. Blomgren, I. Naslund, and R. Svanstrom. 1994. Stereotactic radiotherapy of malignancies in the abdomen. Methodological aspects. *Acta Oncol* 33(6):677–83.

Lee, S. W., E. K. Choi, and H. J. Park, et al. 2003. Stereotactic body frame based fractionated radiosurgery on consecutive days for primary or metastatic tumors in the lung. *Lung Cancer* 40(3):309–15.

Lewis, J. H., and S. B. Jiang. 2009. A theoretical model for respiratory motion artifacts in free-breathing CT scans. *Phys Med Biol* 54(3):745–55.

Li, X. A., C. Stepaniak, and E. Gore. 2006. Technical and dosimetric aspects of respiratory gating using a pressure-sensor motion monitoring system. *Med Phys* 33(1):145–54.

Lin, T., L. I. Cerviño, X. Tang, N. Vasconcelos, and S. B. Jiang. 2009. Fluoroscopic tumor tracking for image-guided lung cancer radiotherapy. *Phys Med Biol* 54(4):981–92.

Liu, H. H., P. Balter, T. Tutt, et al. 2007. Assessing respiration-induced tumor motion and internal target volume using four-dimensional computed tomography for radiotherapy of lung cancer. *Int J Radiat Oncol Biol Phys* 68(2):531–40.

Mageras, G. S., E. Yorke, K. Rosenzweig, et al. 2001. Fluoroscopic evaluation of diaphragmatic motion reduction with a respiratory gated radiotherapy system. *J Appl Clin Med Phys* 2(4):191–200.

Mah, D., G. Freedman, B. Milestone, et al. 2002. Measurement of intrafractional prostate motion using magnetic resonance imaging. *Int J Radiat Oncol Biol Phys* 54(2):568–575.

Mah, D., J. Hanley, K. E. Rosenzweig, et al. 2000. Technical aspects of the deep inspiration breath-hold technique in the treatment of thoracic cancer. *Int J Radiat Oncol Biol Phys* 48(4):1175–85.

Malone, S., J. M. Crook, W. S. Kendal, and J. Szanto. 2000. Respiratory-induced prostate motion: Quantification and characterization. *Int J Radiat Oncol Biol Phys* 48(1):105–9.

Morrow, N. V., C. Stepaniak, J. White, J. F. Wilson, X. A. Li. 2007. Intra- and interfractional variations for prone breast irradiation: An indication for image-guided radiotherapy. *Int J Radiat Oncol Biol Phys* 69(3):910–7.

Murphy, M. J. 2002. Fiducial-based targeting accuracy for external-beam radiotherapy. *Med Phys* 29(3):334–44.

Murphy, M. J. 2004. Tracking moving organs in real time. *Semin Radiat Oncol* 14(1):91–100.

Murphy, M. J., S. D. Chang, I. C. Gibbs, et al. 2003. Patterns of patient movement during frameless image-guided radiosurgery. *Int J Radiat Oncol Biol Phys* 55(5):1400–8.

Murray, B., K. Forster, and R. Timmerman. 2007. Frame-based immobilization and targeting for stereotactic body radiation therapy. *Med Dosim* 32(2):86–91.

Negoro, Y., Y. Nagata, T. Aoki, et al. 2001. The effectiveness of an immobilization device in conformal radiotherapy for lung tumor: Reduction of respiratory tumor movement and evaluation of the daily setup accuracy. *Int J Radiat Oncol Biol Phys* 50(4):889–98.

Neicu, T, R. Berbeco, J. Wolfgang, and S. B. Jiang. 2006. Synchronized moving aperture radiation therapy (SMART): Improvement of breathing pattern reproducibility using respiratory coaching. *Phys Med Biol* 51(3):617–36.

Neicu, T., H. Shirato, Y. Seppenwoolde, S. B. Jiang. 2003. Synchronized moving aperture radiation therapy (SMART): Average tumour trajectory for lung patients. *Phys Med Biol* 48(5):587–98.

Nelson, C, G. Starkschall, P. Balter, et al. 2005. Respiration-correlated treatment delivery using feedback-guided breath hold: A technical study. *Med Phys* 32(1):175–81.

Ohara, K., T. Okumura, M. Akisada, et al. 1989. Irradiation synchronized with respiration gate. *Int J Radiat Oncol Biol Phys* 17(4):853–7.

Ozhasoglu, C., M. J. Murphy, G. Glosser, et al. 2000. Real-time tracking of the tumor volume in precision radiotherapy and body radiosurgery: A novel approach to compensate for respiratory motion. *Proc 14th Int Conf on Computer Assisted Radiology and Surgery* (2000), San Francisco, CA.

Padhani, A. R., V. S. Khoo, J. Suckling, et al. 1999. Evaluating the effect of rectal distension and rectal movement on prostate gland position using cine MRI. *Int J Radiat Oncol Biol Phys* 44(3):525–33.

Papiez, L. 2003. The leaf sweep algorithm for an immobile and moving target as an optimal control problem in radiotherapy delivery. *Math Comput Model* 37(7–8):735–45.

Papiez, L., and D. Rangaraj. 2005. DMLC leaf-pair optimal control for mobile, deforming target. *Med Phys* 32(1):275–85.

Pedersen, A. N., S. Korreman, H. Nystrom, and L. Specht. 2004. Breathing adapted radiotherapy of breast cancer: Reduction of cardiac and pulmonary doses using voluntary inspiration breath-hold. *Radiother Oncol* 72(1):53–60.

Plathow, C., S. Ley, C. Fink, et al. 2004. Analysis of intrathoracic tumor mobility during whole breathing cycle by dynamic MRI. *Int J Radiat Oncol Biol Phys* 59(4):952–9.

Potters, L., M. Steinberg, C. Rose, et al. 2004. American Society for Therapeutic Radiology and Oncology and American College of Radiology Practice Guideline for the performance of stereotactic body radiation therapy. *Int J Radiat Oncol Biol Phys* 60(4):1026–32.

Ramsey, C. R., I. L. Cordrey, and A. L. Oliver. 1999. A comparison of beam characteristics for gated and nongated clinical x-ray beams. *Med Phys* 26(10):2086–91.

Rangaraj, D., and L. Papiez. 2005. Synchronized delivery of DMLC intensity modulated radiation therapy for stationary and moving targets. *Med Phys* 32(6):1802–17.

Ramsey, C. R., D. Scaperoth, D. Arwood, and A. L. Oliver. 1999. Clinical efficacy of respiratory gated conformal radiation therapy. *Med Dosim* 24(2):115–9.

Remouchamps, V. M., D. P. Huyskens, I. Mertens, et al. 2007. The use of magnetic sensors to monitor moderate deep inspiration breath hold during breast irradiation with dynamic MLC compensators. *Radiother Oncol* 82(3):341–8.

Remouchamps, V. M., N. Letts, F. A. Vicini, et al. 2003. Initial clinical experience with moderate deep-inspiration breath hold using an active breathing control device in the treatment of patients with left-sided breast cancer using external beam radiation therapy. *Int J Radiat Oncol Biol Phys* 56(3):704–15.

Remouchamps, V. M., N. Letts, D. Yan, et al. 2003. Three-dimensional evaluation of intra- and interfraction immobilization of lung and chest wall using active breathing control: A reproducibility study with breast cancer patients. *Int J Radiat Oncol Biol Phys* 57(4):968–78.

Remouchamps, V. M., F. A. Vicini, M. B. Sharpe, et al. 2003. Significant reductions in heart and lung doses using deep inspiration breath hold with active breathing control and intensity-modulated radiation therapy for patients treated with locoregional breast irradiation. *Int J Radiat Oncol Biol Phys* 55(2):392–406.

Rosenzweig, K. E., J. Hanley, D. Mah, et al. 2000. The deep inspiration breath-hold technique in the treatment of inoperable non-small-cell lung cancer. *Int J Radiat Oncol Biol Phys* 48(1):81–7.

Ross, C. S., D. H. Hussey, E. C. Pennington, W. Stanford, and J. F. Doornbos. 1990. Analysis of movement of intrathoracic neoplasms using ultrafast computerized tomography. *Int J Radiat Oncol Biol Phys* 18:671–7.

Sawant, A., R. Venkat, and V. Srivastava. 2008. Management of three-dimensional intrafraction motion through real-time DMLC tracking. *Med Phys* 35(5):2050–61.

Schwartz, L. H., J. Richaud, L. Buffat, E. Touboul, and M. Schlienger. 1994. Kidney mobility during respiration. *Radiother Oncol* 32(1):84–6.

Schweikard, A., G. Glosser, M. Bodduluri, M. J. Murphy, and J. R. Adler. 2000. Robotic motion compensation for respiratory movement during radiosurgery. *Comput Aided Surg* 5(4):263–77.

Seppenwoolde, Y., R. I. Berbeco, S. Nishioka, H. Shirato, and B. Heijmen. 2007. Accuracy of tumor motion compensation algorithm from a robotic respiratory tracking system: A simulation study. *Med Phys* 34(7):2774–84.

Seppenwoolde, Y., H. Shirato, K. Kitamura, et al. 2002. Precise and real-time measurement of 3D tumor motion in lung due to breathing and heartbeat, measured during radiotherapy. *Int J Radiat Oncol Biol Phys* 53(4):822–34.

Shen, S., J. Duan, J. B. Fiveash, et al. 2003. Validation of target volume and position in respiratory gated CT planning and treatment. *Med Phys* 30(12):3196–205.

Shirato, H., T. Harada, T. Harabayashi, et al. 2003. Feasibility of insertion/implantation of 2.0-mm-diameter gold internal fiducial markers for precise setup and real-time tumor tracking in radiotherapy. *Int J Radiat Oncol Biol Phys* 56(1):240–7.

Shirato, H., S. Shimizu, T. Kunieda, et al. 2000. Physical aspects of a real-time tumor-tracking system for gated radiotherapy. *Int J Radiat Oncol Biol Phys* 48(4):1187–95.

Shirato, H., S. Shimizu, T. Shimizu, T. Nishioka, and K. Miyasaka. 1999. Real-time tumour-tracking radiotherapy. *Lancet* 353(9161):1331–2.

Sixel, K. E., M. C. Aznar, and Y. C. Ung. 2001. Deep inspiration breath hold to reduce irradiated heart volume in breast cancer patients. *Int J Radiat Oncol Biol Phys* 49(1):199–204.

Souchon, R., O. Rouvière, A. Gelet, et al. 2003. Visualisation of HIFU lesions using elastography of the human prostate in vivo: Preliminary results. *Ultrasound Med Biol* 29(7):1007–15.

Stock, M., K. Kontrisova, K. Dieckmann, et al. 2006. Development and application of a real-time monitoring and feedback system for deep inspiration breath hold based on external marker tracking. *Med Phys* 33(8):2868–77.

Stromberg, J. S., M. B. Sharpe, L. H. Kim, et al. 2000. Active breathing control (ABC) for Hodgkin's disease: Reduction in normal tissue irradiation with deep inspiration and implications for treatment. *Int J Radiat Oncol Biol Phys* 48(3):797–806.

Suh, Y., B. Yi, S. Ahn, et al. 2004. Aperture maneuver with compelled breath (AMC) for moving tumors: A feasibility study with a moving phantom. *Med Phys* 31(4):760–6.

Suramo, I., M. Päivänsalo, and V. Myllylä. 1984. Cranio-caudal movements of the liver, pancreas and kidneys in respiration. *Acta Radiol Diagn* 25(2):129–31.

Tang, X., G. C. Sharp, and S. B. Jiang. 2007. Fluoroscopic tracking of multiple implanted fiducial markers using multiple object tracking. *Phys Med Biol* 52(14):4081–98.

Trofimov, A., E. Rietzel, H. M. Lu, et al. 2005. Temporo-spatial IMRT optimization: Concepts, implementation and initial results. *Phys Med Biol* 50(12):2779–98.

Trofimov, A., C. Vrancic, T. C. Y. Chan, G. C. Sharp, and T. Bortfeld. 2008. Tumor trailing strategy for intensity-modulated radiation therapy of moving targets. *Med Phys* 35(5):1718–33.

Underberg, R. W., F. J. Lagerwaard, B. J. Slotman, J. P. Cuijpers, and S. Senan. 2005. Use of maximum intensity projections (MIP) for target volume generation in 4DCT scans for lung cancer. *Int J Radiat Oncol Biol Phys* 63(1):253–60.

Unkelbach, J., and U. Oelfke. 2004. Inclusion of organ movements in IMRT treatment planning via inverse planning based on probability distributions. *Phys Med Biol* 49(17):4005–29.

van Asselen, B., C. P. J. Raaijmakers, J. J. W. Lagendijk, and C. H. J. Terhaard. 2003. Intrafraction motions of the larynx during radiotherapy. *Int J Radiat Oncol Biol Phys* 56(2):384–90.

Wade, O. L. 1954. Movements of the thoracic cage and diaphragm in respiration. *J Physiol* 124(2):193–212.

Wagman, R., E. Yorke, E. Ford, et al. 2003. Respiratory gating for liver tumors: Use in dose escalation. *Int J Radiat Oncol Biol Phys* 55(3):659–68.

Webb, S. 2005a. The effect on IMRT conformality of elastic tissue movement and a practical suggestion for movement compensation via the modified dynamic multileaf collimator (DMLC) technique. *Phys Med Biol* 50(6):1163–90.

Webb, S. 2005b. Limitations of a simple technique for movement compensation via movement-modified fluence profiles. *Phys Med Biol* 50(14):N155–61.

Weiss, E. H. Vorwerk, S. Richter, and C. F. Hess. 2003. Interfractional and intrafractional accuracy during radiotherapy of gynecologic carcinomas: A comprehensive evaluation using the ExacTrac system. *Int J Radiat Oncol Biol Phys* 56(1):69–79.

Weiss, P. H., J. M. Baker, and E. J. Potchen. 1972. Assessment of hepatic respiratory excursion. *J Nucl Med* 13(10):758–9.

Wijesooriya, K., C. Bartee, J. V. Siebers, S. S. Vedam, and P. J. Keall. 2005. Determination of maximum leaf velocity and acceleration of a dynamic multileaf collimator: Implications for 4D radiotherapy. *Med Phys* 32(4):932–41.

Wilbert, J., J. Meyer, K. Baier, et al. 2008. Tumor tracking and motion compensation with an adaptive tumor tracking system (ATTS): System description and prototype testing. *Med Phys* 35(9):3911–21.

Wong, J. W., M. B. Sharpe, D. A. Jaffray, et al. 1999. The use of active breathing control (ABC) to reduce margin for breathing motion. *Int J Radiat Oncol Biol Phys* 44(4):911–9.

Wu, J., H. Li, R. Shekhar, M. Suntharalingam, and W. D'Souza. 2008. An evaluation of planning techniques for stereotactic body radiation therapy in lung tumors. *Radiother Oncol* 87(1):35–43.

Wulf, J., U. Hädinger, U. Oppitz, B. Olshausen, and M. Flentje. 2000. Stereotactic radiotherapy of extracranial targets: CT-simulation and accuracy of treatment in the stereotactic body frame. *Radiother Oncol* 57(2):225–36.

Xu, Q., R. J. Hamilton, R. A. Schowengerdt, B. Alexander, and S. B. Jiang. 2008. Lung tumor tracking in fluoroscopic video based on optical flow. *Med Phys* 35(12):5351–9.

Xu, Q., R. J. Hamilton, R. A. Schowengerdt, and S. B. Jiang. 2007. A deformable lung tumor tracking method in fluoroscopic video using active shape models: A feasibility study. *Phys Med Biol* 52(17):5277–93.

Yamamoto, T., U. Langner, B. W. Loo Jr., J. Shen, P. J. Keall. 2008. Retrospective analysis of artifacts in four-dimensional CT images of 50 abdominal and thoracic radiotherapy patients. *Int J Radiat Oncol Biol Phys* 72(4):1250–8.

Methods for Real-Time Tumor Position Monitoring and Radiotherapy Tracking

Paul Keall
Stanford University

14.1 Rationale for Real-Time Tumor Position Monitoring and Radiotherapy Tracking

The musculoskeletal, respiratory, cardiac, gastrointestinal, and genitourinary systems cause human bodies to move voluntarily and involuntarily over a variety of time scales. These time scales span subseconds to minutes and are relevant to radiation therapy treatment. The body naturally will move the tumor with respect to the treatment beam if the motion is not monitored and corrected for. Examples, albeit for the upper end of motion, of patient-derived lung (Suh et al. 2008) and prostate tumors (Langen et al. 2008) are shown in Figure 14.1. In both cases, the motion is complex and multidimensional. The average position of the tumor changes with time.

If this motion is not accounted for during imaging, planning, and delivery of radiotherapy, geometric errors will occur in the beam-target alignment, leading to dosimetric errors. Dosimetric errors are a double-edged sword—missing the tumor means hitting normal tissue with higher-than-planned doses. A lower dose to the tumor means a lesser chance of local tumor control and potentially metastatic control, whereas a higher dose to normal tissue means a higher chance of acute and late treatment-related toxicity.

Because of the concerns and consequences of dosimetric errors, accounting for intrafraction tumor motion is the subject of much basic research, clinical investigations, and commercial development. There are many methods to manage motion during treatment: respiratory motion examples include breath hold, abdominal compression, and respiratory gating; prostate immobilization examples include rectal balloons. However, the motion management method described in this chapter is a real-time tumor position monitoring and radiotherapy tracking. Throughout this chapter the term *monitoring* refers to the act of measuring or inferring the tumor position and the term *tracking* refers to the act of aligning the radiation beam and the tumor. Figure 14.2 shows the general components of a real-time tumor position monitoring and radiotherapy tracking system. The position of a moving tumor is measured or inferred by a monitoring device. Where applicable, for example, for thoracic targets with quasiperiodic motion, the reaction time, or system latency, is partially corrected for by estimating the future motion of the target. The estimated position is then used to determine the adjustments of, for example, the multileaf collimator (MLC) leaves, couch, or steering magnets required to align the beam and the tumor.

In this chapter there are two main sections: Real-Time Tumor Position Monitoring Methods and Radiotherapy Tracking Methods. As expanded upon in the integration section, the

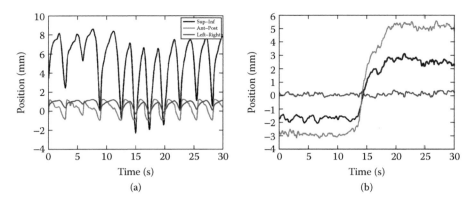

FIGURE 14.1 **(See color insert following page 204.)** Examples of the complexity of tumor motion that can be encountered during radiotherapy treatments for (a) lung and (b) prostate.

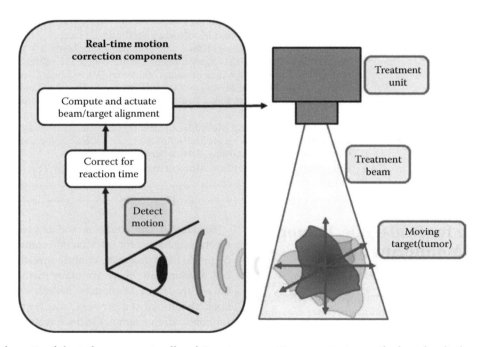

FIGURE 14.2 A schematic of the tasks common to all real-time tumor positions monitoring methods and radiotherapy tracking systems. (Reprinted from Dieterich, S., et al. 2008. *Med Phys* 35:5684–94. With permission.)

monitoring and tracking components can be considered separately, and in principle, any of the monitoring methods can be integrated with any of the radiotherapy tracking methods.

14.2 Real-Time Tumor Position Monitoring Methods

14.2.1 The Ideal Position Monitoring Method

For position monitoring methods, the characteristics of an ideal system should be considered in order to realize the approximations made in currently available systems and also to point future development directions for these systems. An ideal real-

time position monitoring method should have the following desirable characteristics:

- *Volumetric*: To obtain three-dimensional information.
- *High spatial resolution*: To improve tumor and normal tissue identification.
- *High temporal resolution*: To allow information to be acquired quickly with low latency to reduce error induced by the system reaction time.
- *High fidelity*: The image represents the object without geometric distortions or artifacts.
- *Ability to register images*: Deformable image registration algorithms can accurately and automatically register images

to allow contour and dose information to be transferred between the planning and guidance images.

- *No interference with delivery system*: The monitoring method does not interfere physically (e.g., with the hardware between the beam and patient) or otherwise (e.g., radiofrequency interference) with the delivery system.
- *Noninvasive*: No implanted markers are required.
- *No imaging dose*: The patient does not receive extra dose, for example from kilovoltage beams that are not therapeutic and often irradiate a much larger volume than the target.
- *Ability to optimize and compute dose*: Computed tomography (CT) numbers are directly measured or accurately estimated.
- *Reduces treatment time*: Guidance is integrated with delivery to allow fast setup and smooth efficient delivery.
- *Cheap*: It should have low operational costs.

The ideal system, of course, is not currently available; however, many systems are in use and being developed for real-time tumor position monitoring, examples of which are discussed in Sections 14.2.3 through 14.2.10. A variety of clinical configurations of available real-time monitoring systems are given in Figure 14.3. These systems span no image-guided radiotherapy (IGRT) utilizing skin-marker-based localization as the most rudimentary system in clinical use through to multiple data streams including one or more X-ray imaging systems and optical monitoring.

14.2.2 Implanting Fiducial Markers

Not all of the techniques for real-time tumor position monitoring require marker implantation, but a number of them do, and hence it is worthwhile to state the advantages and disadvantages of marker implantation (Table 14.1). Markerless internal target position monitoring methods have been developed and applied to lung lesions (Park et al. 2009; Lin et al. 2009). The clinically approved Xsight Lung Tracking System from Accuray combines occasional oblique X-ray image pairs in which the tumor is segmented in each image with continuous respiratory

monitoring to determine the 3D target position in real time. This system is applicable for some lung lesions. A challenge for gantry-mounted rotating X-ray imaging systems is that the delineation of the tumor can be difficult in some views.

14.2.3 Electromagnetic Tumor Position Monitoring

Electromagnetic tumor position monitoring involves the placement of wireless or wired transponders in or near the tumor. Tumor localization is performed by relating the measured positions of one or more transponders with the relation between the transponders and the isocenter determined from treatment planning CT imaging. The in-room measurements are performed in two steps:

1. *Determining the location of the transponders with respect to an electromagnetic source or receiver coil array*: In this step, the source coils generate an oscillating field inducing a resonance in the transponders. When the applied oscillating field is switched off, the transponder continues to give a decaying resonant signal that can be measured and used to establish its position and orientation. Each transponder has a different resonant frequency allowing multiple internal positions to be measured.
2. *Determining the location of the array with respect to the linear accelerator (linac) isocenter*: The location of the array can be determined by either a known physical relation to another device, such as a couch that has its own calibration with respect to the treatment isocenter, or having a second system to determine the position of the array with respect to the treatment isocenter. The Calypso system uses an infrared optical tracking system to determine the position and orientation of the array with respect to the treatment isocenter. A combination of these two measurements allows the determination of the target's position. Established system calibration procedures are used for routine quality assurance. Figure 14.3 shows a volunteer setup with the Calypso system.

The positions of the individual transponders can be used in several ways: to determine the center of the mass of the transponders to yield translation information, estimate rotation, and estimate internal deformation. The Calypso system has an update rate of 10–25 Hz and has submillimeter accuracy (Balter et al. 2005). Many clinical investigations have found that the extent and frequency of prostate motion during radiotherapy delivery can be easily monitored and used for motion management (Kupelian et al. 2007; Willoughby et al. 2006). More recent work has investigated the preclinical integration of electromagnetic monitoring for lung applications (Mayse et al. 2008) and also the integration of the real-time position monitoring system with the dynamic MLC to correct for the measured translational offsets (Sawant et al. 2009; Smith et al. 2009). Electromagnetic monitoring is discussed in Chapter 11.

TABLE 14.1 Costs and Benefits of Implanting Fiducial Markers

Costs	Benefits
Time for fiducial placement	Improved tumor targeting
Toxicity of fiducial placement	Improved normal tissue sparing
Cost of fiducial placement	Lower integral dose from smaller clinical target volume-planning target volume (CTV-PTV) margin
Mistargeting due to migration	
Possible increased imaging dose	
Possible increased treatment time	
Possible MR artifacts	

The costs need to be outweighed by the benefits.
MR = magnetic resonance.

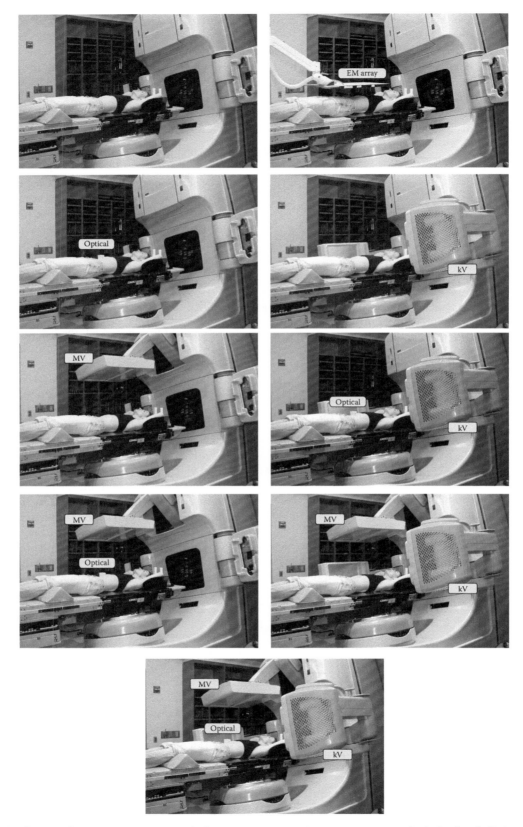

FIGURE 14.3　A subject in the treatment position with the various tumor position monitoring methods displayed. (EM = electromagnetic, kV = kilovoltage, MV = megavoltage).

14.2.4 Optical Position Monitoring

Optical position monitoring is a continuous, real-time, radiation-free method to monitor the patient's position and motion during radiotherapy imaging and treatment. Single or multiple in-room cameras use reflected light from the patient's skin or add reflective markers to determine the position of the object with respect to the camera. Calibration procedures are performed to determine the relative location of the camera coordinate system and the marker coordinate system.

The optical signal can be used for monitoring patient motion and can have automatic or manual intervention thresholds, for example, if the patient shift is detected to be more than 3 mm. The signal can also be used to trigger a gated beam on/off signal in a linac. Figure 14.3 shows a volunteer setup with the Varian real-time position management optical system in which the marker block is placed on the patient's chest or abdomen. This is one of the first systems in use and can be used to trigger a gating signal. Multiple surface marker systems and surface monitoring systems not requiring surface markers are now available.

It is hoped that optical monitoring will eventually replace traditional laser alignment as a more accurate and faster means of patient setup. The potential to provide surface information and in-room positional feedback, not just of the target region, and to ensure that the angle and positions of the neck and extremities are reproducibly aligned should increase throughput and anatomic reproducibility. For intracranial and some head and neck applications, surface imaging alone should be sufficient as the sole guidance method. For other sites, the integration with other internal imaging modalities will likely become more routine, particularly for thoracic and abdominal sites (discussed in Section 14.2.7). Where anatomic changes are such that they require adaptation, (i.e., tumor shrinkage as opposed to musculoskeletal anatomy misalignment), volumetric imaging will be required.

14.2.5 Kilovoltage Tumor Position Monitoring

The importance of kilovoltage (kV) guidance in radiotherapy has been recognized since the first development of linacs (Figure 14.4). Currently, almost all linac manufacturers offer kV imaging options, typically for pretreatment guidance, some examples of which are given in Figure 14.5. The prevalence of these systems indicates a large potential for the use of kV imaging for real-time position monitoring.

Pioneering scientific and clinical work in the application of real-time kV localization has been performed by Shirato et al. at Hokkaido University in Japan (Shirato, Shimizu, Kitamura, et al. 2000; Shirato, Shimizu, Kunieda et al. 2000; Shirato et al. 1999; Seppenwoolde et al. 2002). Their system consists of four kV systems (any two of which offer an unobstructed view of the patient at a given time), an image processor unit, a gating control unit, and an image display unit. The system recognizes the positions of implanted marker(s) in patients at 30 Hz. The marker(s) is inserted in or near the tumor using image-guided implantation.

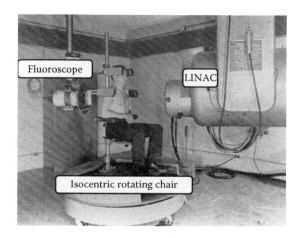

FIGURE 14.4 Example of early use of kilovoltage guidance in radiotherapy at Stanford circa 1956. The linac could be moved and replaced with a kilovoltage imaging tube to allow fluoroscopy.

The linac is gated to irradiate the tumor only when the marker is within a given tolerance from its planned coordinates relative to the isocenter. Phantom experiments have demonstrated that the geometric accuracy of the tumor-tracking system is better than 1.5 mm. This system has been used for liver, lung, and prostate patients.

The CyberKnife system offers occasional (approximately every 30 seconds) stereoscopic imaging, which allows the correction of translational and rotation tumor drifts during treatment, though it cannot as yet be used for real-time applications. An interesting preclinical approach to real-time tumor localization using a single rotating kV imager has been studied in simulation and experimental conditions by Poulsen et al. (Poulsen, Cho, and Keall 2008, 2009; Poulsen et al. 2008; Poulsen et al. 2010). In their method, a probability density function is created. In addition to the motion magnitudes in each direction of internal landmarks, such as fiducial markers, the correlation of motion in different directions is also computed. From any view, two of the dimensions can be determined in the beam view. With the collation model, the third dimension can be estimated. Submillimeter accuracy in clinical realistic experimental conditions has been reported. An issue with continuous kV imaging is additional dose to the patient, approximately 0.18 mGy per image for detecting implanted markers (Shirato, Shimizu, Kunieda, et al. 2000), and hence, the development of systems integrated with respiratory monitoring systems may be useful to maintain accuracy with lower imaging doses.

14.2.6 Megavoltage Tumor Position Monitoring

Most modern radiotherapy linacs have megavoltage (MV) imaging systems suitable for quasicontinuous radiographic monitoring while the treatment beam is on. Currently, the use of these systems for real-time monitoring has been limited though studies have been performed on preclinical and clinical applications (Keall et al. 2004; Berbeco et al. 2007; Berbeco et al.

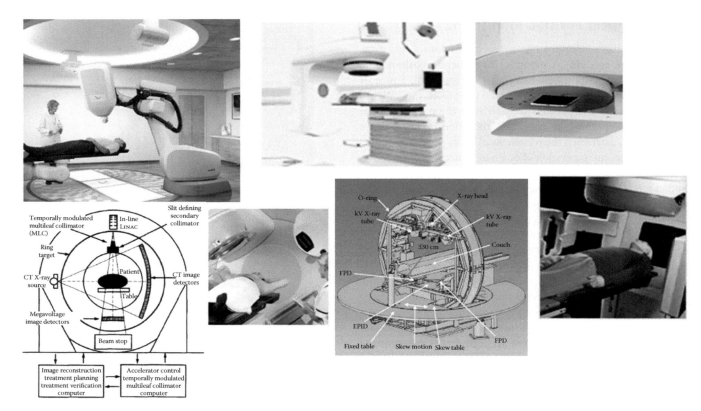

FIGURE 14.5 Examples of modern linac concepts and systems with kilovoltage image guidance.

2008; Park et al. 2009). The MV imaging has been integrated with MLC tracking in a phantom study (Poulsen et al. 2009), showing a system latency of 370 milliseconds and submillimeter accuracy.

The advantages of MV tumor position monitoring are

- No additional IGRT equipment is needed.
- No additional imaging dose to patients is required.
- The imaging system is the treatment beam; therefore, there is no need for further alignment or calibration with other coordinate systems.

Challenges for MV tumor position monitoring are

- Reduction in field of view and contrast with respect to kV images.
- Further reduction of aperture with intensity-modulated delivery (fixed field and arc).
- The target position is not known before the beam is on; therefore, methods to quickly detect and adapt treatment are needed for treatment initiation, between beams and after beam holds.

To address some of the challenges for MV imaging, one possible approach is to develop treatment plans in which one or more internal landmarks or fiducial markers are constrained to be visible during all or part of the treatment delivery (Ma et al. 2009).

14.2.7 Combined Kilovoltage, Megavoltage, and Optical Tumor Position Monitoring

As shown in Figures 14.3 and 14.5, there are a variety of different combinations of imaging systems that can be combined with optical monitoring, including a single kV system, single MV system, combined kV and MV system, and two or more kV systems. The CyberKnife Synchrony system combines occasional (30–60 seconds) stereoscopic imaging of implanted markers with quasicontinuous optical monitoring, allowing a target position estimation of the marker positions with very small systematic error and <2 mm precision (Seppenwoolde et al. 2007; Suh et al. 2008; Hoogeman et al. 2009). Similar accuracy and precision of the Synchrony system have been reported for an experimental linac system by combining the kV and MV imaging systems with an optical monitor (Cho et al. 2010). These systems build an internal or external motion correlation model that is updated during treatment as new information becomes available.

With simultaneous stereoscopic imaging, we can unambiguously identify radiopaque landmarks. However, when only a single imager is available, resolving the position along the beam direction becomes difficult. This unresolved position is a particular concern when the imager is orthogonal to the treatment beam, such as for some of the systems in Figure 14.5. To estimate the unresolved position, prior information can be used, resulting in accurate 3D targeting (Cho et al. 2008). The development and implementation of algorithms that combine kV, MV, and optical signals is an area of intense research and development.

14.2.8 Radioactive Tumor Position Monitoring

Radioactive tumor position monitoring can be performed using two emission techniques: single photon emission and positron emission tomography (PET). A positron system was conceptually studied in which planar detectors were placed around the patient and a small radioactive implanted marker provided the signal. A new company, Navotek, is developing an implanted marker-based emission system in which directional dependent detectors are placed on the treatment head of the linac, allowing fast and precise internal target positioning. This system is discussed in Chapter 11 of this book.

A company at the early phase of development, Reflexion Medical, is taking the positron emission one step further and seeking to use the PET signals from patients rather than markers to provide an image of the tumor that can be targeted in near real time using a linac. The potential for PET-image guidance during treatment is particularly exciting, especially as the radiation oncology field moves from anatomic to functional targeting. PET images of physiological targets, such as hypoxic regions, could allow selective dose boosting of particularly resistant or aggressive areas of the tumor (Ling et al. 2000).

14.2.9 Ultrasound Tumor Position Monitoring

Ultrasound position monitoring offers many of the desirable features listed for an ideal position monitoring method in Section 14.2.1 (Hsu et al. 2005; Wu et al. 2006; Sawada et al. 2004; Fuss et al. 2007). This method is relatively cheap, and the ultrasound system operation is compatible with the linac operation (Hsu et al. 2005). There have been significant developments in ultrasound technology—particularly 3D probes, resolution, image quality, and acquisition frequency. In addition, advances in robotic devices allow for remote probe placement, a requirement for operation in a linac vault. Several in-room navigation options are available to relate the coordinate system of the ultrasound images to the linac isocenter. Ultrasound monitoring is discussed in detail in Chapter 11 of this book.

14.2.10 Magnetic Resonance Imaging Tumor Position Monitoring

Magnetic resonance (MR) imaging has many of the features listed for the ideal monitoring method and is under active development in university and commercial settings (Dempsey et al. 2006; Fallone et al. 2007; Lagendijk et al. 2008). In 2009, the Alberta and Utrecht groups reported that for the first time MR images have been acquired with the radiation beam on, a major milestone toward the realization of this technology.

Major challenges to be addressed are the integration of the two systems, radiofrequency interference, shielding of both systems,

the effect of the linac on the quality of MR images, and the effect of the main magnetic field on the beam generation and radiation transport. Geometric distortions, including distortions due to chemical shift that are difficult to correct for, as well as cost are other challenges to the widespread dissemination of this technology.

14.3 Radiotherapy Tracking Methods

If real-time target position information is available, there are a number of different ways in which the target and beam can be aligned: simply by moving the patient via couch correction, or the beam via mechanical motion using the MLC, or the linac head by a robotic or gimbaled system. For particle therapy, the beam can be steered magnetically.

The various motion adaptation systems in use and under consideration involve two fundamentally different types of control loop. The CyberKnife robotic treatment system (Murphy, 2004; Schweikard, Shiomi, and Alder 2004; Ozhasoglu et al. 2000; Seppenwoolde et al. 2007), the MLC (McClelland et al. 2007; McQuaid and Webb 2006; Webb and Binnie 2006; Webb 2006; Papiez and Rangaraj 2005; Papiez, Rangaraj, and Keall 2005; Rangaraj and Papiez 2005; McMahon, Papiez, and Rangaraj 2007; Keall et al. 2006; Keall et al. 2001; Neicu et al. 2003; Sawant et al. 2008; Zimmerman et al. 2008), the gimbaled linac (Kamino et al. 2006), and proton and heavy ion treatment systems (Bert et al. 2007; Grozinger et al. 2006) all use an open-loop control design. In an open-loop system, the corrective response has no effect on the target behavior that signaled the response. In other words, there is no feedback between the corrective signal and the adaptive response. In contrast, a couch compensation system (D'Souza and McAvoy 2006; D'Souza et al. 2005; Qiu et al. 2007; Wilbert et al. 2008) has a closed-loop control system. When the tumor moves within the patient, the detection system triggers a compensating movement of the patient (and thus the tumor) via the treatment couch. Thus, there is a feedback through which the adaptive response may influence the target movement (Dieterich et al. 2008).

Closed-loop systems such as a movable couch can have higher-order dynamics that involve multiple system response time constants in addition to the dead time before the couch begins its corrective response (D'Souza and McAvoy 2006). In an open-loop system, the total reaction time must be compensated by predicting the future position of the target so that the beam reaches the future target position at the same time as the tumor. A closed control loop needs to predict the dead time before the system responds to a change in the tumor position (D'Souza and McAvoy 2006).

14.3.1 Ideal Radiotherapy Tracking Method

As with position monitoring methods, for radiotherapy tracking methods the characteristics of an ideal system should be considered in order to realize the approximations made in currently available systems and also to point future development directions for these systems. The desirable characteristics of an ideal radiotherapy tracking method include the following:

- *Integration of all sources of imaging information*: Real-time and prior CT, CBCT, surface, and kV and MV imaging information is used in an optimal way to estimate patients' poses at the time of treatment through to the end of treatment.
- *Integration of all sources of uncertainty*: The error is given on a patient-specific basis, which can be used for personalized margin adaptation.
- *Low system latency*: The system can respond quickly without a significant reaction time that may reduce the treatment accuracy.
- *Optimization in real-time using all available degrees of freedom*: The MLC, couch, gantry, collimator, energy, dose rate, etc., can all be seen as degrees of freedom to create a revised optimal plan that changes with time and includes previous dose and estimates of the patient pose and uncertainties
- *Modification of delivery appropriately*: The desired beam delivery is executed with high mechanical accuracy and stable radiation beam characteristics.

An ideal system does not exist; however, several systems are in use and being developed for real-time tumor radiotherapy tracking, examples of which are discussed below.

14.3.2 Robotic Linac

The CyberKnife robotic treatment system is the only commercially available system for radiotherapy tracking (Murphy 2004; Schweikard, Shiomi, and Alder 2004; Ozhasoglu et al. 2000; Seppenwoolde et al. 2007). The immediate difference between the CyberKnife and other linac technology is the use of a Kuka robot, with very precise six degrees of freedom control. An X-band linac is attached to the robot. Two fixed X-ray imaging systems and a real-time respiratory motion tracking camera provide the data needed to control the robot.

14.3.3 Couch

Most linacs have couches with at least four degrees of freedom, three translations and one rotation; some couches or additional couch overlays allow six degrees of freedom motion—three translations and three rotations. In principle, each of the couch motors can be controlled in real time. Therefore, the couch can be used to compensate for detected target motion by shifting the patient to continuously align the target with the radiation beam. Several projects are underway for motion compensation using the couch. Couch compensation systems are closed-loop control systems (D'Souza and McAvoy 2006; D'Souza et al. 2005; Qiu et al. 2007; Wilbert et al. 2008). When the tumor moves within the patient's body, the detection system triggers a compensating movement of the patient (and thus the tumor) via the treatment couch. Thus, there is feedback through which the adaptive

response may influence the target movement. In an open-loop system, the total reaction time must be compensated by predicting the future position of the target so that the beam reaches the future position of the target at the same time as the tumor. A closed control loop needs to predict the dead time before the system begins to respond to a change in the tumor position (D'Souza and McAvoy 2006).

14.3.4 Dynamic Multileaf Collimator

Most modern linacs have dynamic multileaf collimators (DMLCs) designed for conformal and intensity-modulated radiotherapy delivery. The same technology can be adapted to deliver radiation to moving targets by adjusting the positions of the MLC based on real-time localization information. An advantage of the DMLC over other approaches is that each individual MLC leaf can be independently controlled, and thereby can be used as a degree of freedom to account for tumor rotation and deformation. Figure 14.6 shows how the MLC beam can in principle be adapted to account for translation, rotation, and deformation of the target.

Experimentally, use of the MLC has been demonstrated on four different vendor platforms: Accuknife (Liu et al. 2009), Siemens (Tacke, Nill, and Oelfke 2007), TomoTherapy (Gustavo Olivera, Tomotherapy, July 2008, pers. comm.), and Varian (Keall et al. 2001; Keall et al. 2006; Sawant et al. 2008; Zimmerman et al. 2008) though none is clinically available yet.

14.3.5 Gimbaled Linac

Another method to align the radiation beam with the patient is to use rotational motors to steer the linac itself, and thus the entire treatment beam moves (Kamino et al. 2006). In their implementation, the linac head and the rotational motors are attached to a rotating o-ring gantry, which also contains X-ray images for target localization. The system has been tested for dynamic tracking with submillimeter accuracy observed.

14.3.6 Particle Beams

For particle therapy, magnetic steering is needed to deflect the treatment beam in order to align the beam with the detected target position. This approach has no moving parts and is potentially very fast to respond, with reported delay times of 80 milliseconds (Bert et al. 2007; Grozinger et al. 2006). The additional complexity of particle therapy is that for optimal target coverage, the beam energy may need to be quickly modified, and the magnetic deflection may need to be changed.

14.4 Integrating Tumor Position Monitoring and Radiotherapy Tracking Systems

As Figure 14.2 shows, in principle, any of the tumor position monitoring methods can be used as an input to any of the radiotherapy delivery tracking systems. For example, the CyberKnife

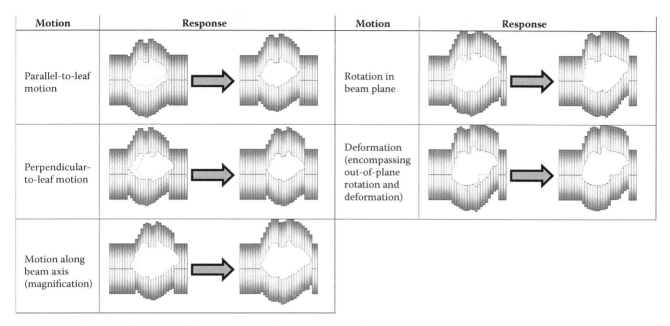

FIGURE 14.6 Schematic illustration of the various types of target motion and the desired change in multileaf collimator configuration to account for each type of motion. (Reprinted from Sawant, A., et al. 2008. *Med Phys* 35:2050–61. With permission.)

system currently allows four different motion inputs: (1) X-ray based on skeletal anatomy, (2) X-ray based on fiducials, (3) combined respiratory signal and X-ray imaging based on implanted fiducials, and (4) combined respiratory signal and X-ray imaging based on tumor imaging directly. A research version of DMLC tracking has been experimentally investigated for eight different inputs, including electromagnetic, kV and MV X-ray alone, kV and MV combined, and kV and MV combined with respiratory monitoring. Thus, the total number of possibilities for clinical implementation of real-time tumor targeting is quite high.

The heart of a real-time tumor tracking system is the control loop that connects the tumor detection system to the targeting alignment system. The control loop must receive target coordinate measurements, test that the target identification is authentic (so as not to go off chasing false motion), filter the target data to avoid disruptive transients, and then send the delivery system a prediction of where the tumor will be when the targeting adjustment has been completed. Tumor tracking requires motors that can guide the beam or couch faster than most respiratory motion and a fast secondary feedback mechanism for each motor used to control the beam (or the patient) placement (Dieterich et al. 2008).

The future displacement of a breathing motion can be predicted in several ways: (1) with a mathematical model of the breathing signal (Seppenwoolde et al. 2002); (2) with a biomechanical model of the breathing process (Low et al. 2005); or (3) with heuristic learning algorithms that mimic the breathing process as it is observed (Murphy, Jalden, and Isaksson 2002; Vedam et al. 2004). The last category includes neural

network prediction algorithms, which have been studied by several groups (Isaksson, Jalden, and Murphy 2005; Murphy, Jalden, and Isaksson 2002; Sharp et al. 2004; Vedam et al. 2004). Neural networks can adapt to changes in the breathing pattern as they occur, making them an attractive way to deal with the nonstationary character of real breathing (Dieterich et al. 2008).

Real-time systems must react to change at least as fast as it occurs. A system's reaction time depends on a number of factors. Time is required to detect a change in the target position, process and filter the position data, communicate with the beam alignment system, and complete the repositioning of either the beam or the patient. If a single motion detection method (e.g., fluoroscopic imaging) is used in a first-order open-loop system, then all these time delays combine linearly in series to give the total reaction time. On the other hand, if two different motion detection systems are used in parallel to provide partially redundant target position data, then only the faster system's delay adds into the reaction time of an open-loop system. For example, the CyberKnife uses an optical tracking system to provide fast, continuous estimates of a tumor's respiratory-induced motion, with the estimate updated by periodic X-ray imaging (Schweikard, Shiomi, and Alder 2004). This is an example of a hybrid detection method. The total system reaction time includes the optical processing time while the time spent processing the X-ray images can be done outside the duty cycle of the control loop. This allows the targeting system to invest more time in sophisticated image processing without a corresponding degradation of response time (Dieterich et al. 2008).

14.5 Summary

The clinical practice of real-time tumor position monitoring and radiotherapy tracking is evolving; however, there is significant user knowledge, predominantly with the CyberKnife system. The experience through the use of this and other systems, in terms of patients' and treatment teams' acceptance of the technology, development of quality assurance procedures, and understanding of using real-time automated systems for treating the patients—always under the watchful care of experienced human operators—provides the field a firm foundation from which to build as real-time systems become more mainstream and are used to improve care of an increasing fraction of the patient population.

Real-time systems are very complex; however, with appropriate attention to workflow, real-time systems can equal or even improve upon the throughput of patients as the setup and delivery time can be reduced though real-time knowledge of target location, reducing trial-and-error image and shift approaches.

Some of the real-time position monitoring and targeting methods require increased cost with additional equipment; however, with others, the better use of existing information, addition of software, and control of delivery devices using currently available conventional systems can facilitate real-time guidance. If improved workflow can be achieved without additional hardware, there is a compelling economic as well as clinical case to use these systems. It is important for the treatment team to develop, learn, and incorporate knowledge of the safe and appropriate use of real-time systems into education programs. Much of this knowledge has yet to be fully developed, given the infancy of the field. We anticipate that real-time tumor position monitoring and targeting will play a central role as a mainstream tool in the future of radiation oncology.

References

Balter, J. M., J. N. Wright, L. J. Newell, et al. 2005. Accuracy of a wireless localization system for radiotherapy. *Int J Radiat Oncol Biol Phys* 61:933–7.

Berbeco, R. I., F. Hacker, D. Ionascu, and H. J. Mamon. 2007. Clinical feasibility of using an EPID in CINE mode for image-guided verification of stereotactic body radiotherapy. *Int J Radiat Oncol Biol Phys* 69:258–66.

Berbeco, R. I., F. Hacker, C. Zatwarnicki, et al. 2008. A novel method for estimating SBRT delivered dose with beam's-eye-view images. *Med Phys* 35:3225–31.

Bert, C., N. Saito, A. Schmidt, N. Chaudhri, D. Schardt, and E. Rietzel. 2007. Target motion tracking with a scanned particle beam. *Med Phys* 34:4768–71.

Cho, B. C., P. R. Poulsen, A. Sawant, D. Ruan, and P. Keall. 2010. Real-time target position estimation using stereoscopic kV/MV imaging and external respiratory monitoring for dynamic MLC tracking. *Int J Radiat Oncol Biol Phys*.

Cho, B., Y. Suh, S. Dieterich, and P. J. Keall. 2008. A monoscopic method for real-time tumour tracking using combined occasional X-ray imaging and continuous respiratory monitoring. *Phys Med Biol* 53:2837–55.

D'Souza, W. D., and T. J. McAvoy. 2006. An analysis of the treatment couch and control system dynamics for respiration-induced motion compensation. *Med Phys* 33:4701–9.

D'Souza, W. D., S. A. Naqvi, and C. X. Yu. 2005. Real-time intra-fraction-motion tracking using the treatment couch: A feasibility study. *Phys Med Biol* 50:4021–33.

Dempsey, J., B. Dionne, J. Fitzsimmons, et al. 2006. A real-time MRI guided external beam radiotherapy delivery system. *Med Phys* 33:2254.

Dieterich, S., K. Cleary, W. D'Souza, M. Murphy, K. H. Wong, and P. Keall. 2008. Locating and targeting moving tumors with radiation beams. *Med Phys* 35:5684–94.

Fallone, G., M. Carlone, B. Murray, S. Rathee, and S. Steciw. 2007. Investigations in the design of a novel linac-MRI system. *Int J Radiat Oncol Biol Phys* 69:S19.

Fuss, M., J. Boda-Heggemann, N. Papanikolau, and B. J. Salter. 2007. Image-guidance for stereotactic body radiation therapy. *Med Dosim* 32:102–10.

Grozinger, S. O., E. Rietzel, Q. Li, C. Bert, T. Haberer, and G. Kraft. 2006. Simulations to design an online motion compensation system for scanned particle beams. *Phys Med Biol* 51:3517–31.

Hoogeman, M., J. B. Prevost, J. Nuyttens, J. Poll, P. Levendag, and B. Heijmen. 2009. Clinical accuracy of the respiratory tumor tracking system of the CyberKnife: Assessment by analysis of log files. *Int J Radiat Oncol Biol Phys* 74:297–303.

Hsu, A., N. R. Miller, P. M. Evans, J. C. Bamber, and S. Webb. 2005. Feasibility of using ultrasound for real-time tracking during radiotherapy. *Med Phys* 32:1500–12.

Isaksson, M., J. Jalden, and M. J. Murphy. 2005. On using an adaptive neural network to predict lung tumor motion during respiration for radiotherapy applications. *Med Phys* 32:3801–9.

Kamino, Y., K. Takayama, M. Kokubo, et al. 2006. Development of a four-dimensional image-guided radiotherapy system with a gimbaled X-ray head. *Int J Radiat Oncol Biol Phys* 66:271–8.

Keall, P. J., H. Cattell, D. Pokhrel, et al. 2006. Geometric accuracy of a real-time target tracking system with dynamic multi-leaf collimator tracking system. *Int J Radiat Oncol Biol Phys* 65:1579–84.

Keall, P. J., V. R. Kini, S. S. Vedam, and Mohan R. 2001. Motion adaptive X-ray therapy: A feasibility study. *Phys Med Biol* 46:1–10.

Keall, P. J., A. D. Todor, S. S. Vedam, et al. 2004. On the use of EPID-based implanted marker tracking for 4D radiotherapy. *Med Phys* 31:3492–9.

Kupelian, P., T. Willoughby, A. Mahadevan, et al. 2007. Multi-institutional clinical experience with the Calypso system in localization and continuous, real-time monitoring of the prostate gland during external radiotherapy. *Int J Radiat Oncol Biol Phys* 67:1088–98.

Lagendijk, J. J. W., B. W. Raaymakers, A. J. E. Raaijmakers, et al. 2008. MRI/linac integration. *Radiother Oncol* 86:25–29.

Langen, K. M., T. Willoughby, S. L. Meeks, et al. 2008. Observations on real-time prostate gland motion using electromagnetic tracking. *Int J Radiat Oncol Biol Phys* 71:1084–90.

Lin, T., R. Li, X. Tang, J. G. Dy, and S. B. Jiang. 2009. Markerless gating for lung cancer radiotherapy based on machine learning techniques. *Phys Med Biol* 54:1555–63.

Ling, C. C., J. Humm, S. Larson, et al. 2000. Towards multidimensional radiotherapy (MD-CRT): Biological imaging and biological conformality. *Int J Radiat Oncol Biol Phys* 47:551–60.

Liu, Y., C. Shi, B. Lin, C. S. Ha, and N. Papanikolaou. 2009. Delivery of four-dimensional radiotherapy with TrackBeam for moving target using an AccuKnife dual-layer MLC: Dynamic phantoms study. *J Appl Clin Med Phys* 10:2926.

Low, D. A., Parikh P. J., W. Lu, et al. 2005. Novel breathing motion model for radiotherapy. *Int J Radiat Oncol Biol Phys* 63:921–9.

Ma, Y., L. Lee, O. Keshet, P. Keall, and L. Xing. 2009. Four-dimensional inverse treatment planning with inclusion of implanted fiducials in IMRT segmented fields. *Med Phys* 36:2215–21.

Mayse, M. L., P. J. Parikh, K. M. Lechleiter, et al. 2008. Bronchoscopic implantation of a novel wireless electromagnetic transponder in the canine lung: A feasibility study. *Int J Radiat Oncol Biol Phys* 72:93–8.

McClelland, J. R., S. Webb, D. McQuaid, D. M. Binnie, and D. J. Hawkes. 2007. Tracking "differential organ motion" with a "breathing" multileaf collimator: Magnitude of problem assessed using 4D CT data and a motion-compensation strategy. *Phys Med Biol* 52:4805–26.

McMahon, R., L. Papiez, and D. Rangaraj. 2007. Dynamic-MLC leaf control utilizing on-flight intensity calculations: A robust method for real-time IMRT delivery over moving rigid targets. *Med Phys* 34:3211–23.

McQuaid, D., and S. Webb. 2006. IMRT delivery to a moving target by dynamic MLC tracking: Delivery for targets moving in two dimensions in the beam's eye view. *Phys Med Biol* 51:4819–39.

Murphy, M. J. 2004. Tracking moving organs in real time. *Semin Radiat Oncol* 14:91–100.

Murphy, M. J., J. Jalden, and M. Isaksson. 2002. Adaptive filtering to predict lung tumor breathing motion during image-guided radiation therapy. *Proceedings of the 16th International Congress on Computer-Assisted Radiology and Surgery*, Paris, 539–44.

Neicu, T., H. Shirato, Y. Seppenwoolde, and S. B. Jiang. 2003. Synchronized moving aperture radiation therapy (SMART): Average tumour trajectory for lung patients. *Phys Med Biol* 48:587–98.

Ozhasoglu C., M. J. Murphy, G. Glosser, et al. 2000. Real-time tracking of the tumor volume in precision radiotherapy and body radiosurgery: A novel approach to compensate for respiratory motion. *Proceedings of the 14th International Congress on Computer-Assisted Radiology and Surgery*, San Francisco, CA, 691–6.

Papiez, L., and D. Rangaraj. 2005. DMLC leaf-pair optimal control for mobile, deforming target. *Med Phys* 32:275–85.

Papiez, L., D. Rangaraj, and P. Keall. 2005. Real-time DMLC IMRT delivery for mobile and deforming targets. *Med Phys* 32:3037–48.

Park, S. J., D. Ionascu, F. Hacker, H. Mamon, and R. Berbeco. 2009. Automatic marker detection and 3D position reconstruction using cine EPID images for SBRT verification. *Med Phys* 36:4536–46.

Poulsen, P. R., B. Cho, and P. J. Keall. 2008. A method to estimate mean position, motion magnitude, motion correlation, and trajectory of a tumor from cone-beam CT projections for image-guided radiotherapy. *Int J Radiat Oncol Biol Phys* 72:1587–96.

Poulsen, P. R., B. Cho, and P. J. Keall. 2009. Real-time prostate trajectory estimation with a single imager in arc radiotherapy: A simulation study. *Phys Med Biol* 54:4019–35.

Poulsen, P. R., B. Cho, K. Langen, P. Kupelian, and P. J. Keall. 2008. Three-dimensional prostate position estimation with a single X-ray imager utilizing the spatial probability density. *Phys Med Biol* 53:4331–53.

Poulsen, P. R., B. C. Cho, D. Ruan, A. Sawant, P. J. Keall. 2010. Dynamic multileaf collimator tracking of respiratory target motion based on a single kilovoltage imager during arc radiotherapy. *Intl J Radiant Oncol Biol Phys* 77:600–7.

Poulsen, P. R., B. Cho, A. Sawant, and P. Keall. 2009. Time analysis of image-based dynamic MLC tracking. *Med Phys* 36:2764.

Qiu, P., W. D. D'Souza, T. J. McAvoy and K. J. Ray Liu. 2007. Inferential modeling and predictive feedback control in real-time motion compensation using the treatment couch during radiotherapy. *Phys Med Biol* 52:5831–54.

Rangaraj, D., and L. Papiez. 2005. Synchronized delivery of DMLC intensity modulated radiation therapy for stationary and moving targets. *Med Phys* 32:1802–17.

Sawada, A., K. Yoda, M. Kokubo, T. Kunieda, Y. Nagata, and M. Hiraoka. 2004. A technique for noninvasive respiratory gated radiation treatment system based on a real time 3D ultrasound image correlation: A phantom study. *Med Phys* 31:245–50.

Sawant, A., R. L. Smith, R. B. Venkat, et al. 2009. Toward submillimeter accuracy in the management of intrafraction motion: the integration of real-time internal position monitoring and multileaf collimator target tracking. *Int J Radiat Oncol Biol Phys* 74:575–82.

Sawant, A., R. Venkat, V. Srivastava, et al. 2008. Management of three-dimensional intrafraction motion through real-time DMLC tracking. *Med Phys* 35:2050–61.

Schweikard, A., H. Shiomi, and J. Adler. 2004. Respiration tracking in radiosurgery. *Med Phys* 31:2738–41.

Seppenwoolde, Y., R. I. Berbeco, S. Nishioka, H. Shirato, and B. Heijmen. 2007. Accuracy of tumor motion compensation algorithm from a robotic respiratory tracking system: A simulation study. *Med Phys* 34:2774–84.

Seppenwoolde, Y., H. Shirato, K. Kitamura, et al. 2002. Precise and real-time measurement of 3D tumor motion in lung due to breathing and heartbeat, measured during radiotherapy. *Int J Radiat Oncol Biol Phys* 53:822–34.

Sharp, G. C., S. B. Jiang, S. Shimizu, and H. Shirato. 2004. Prediction of respiratory tumour motion for real-time image-guided radiotherapy. *Phys Med Biol* 49:425–40.

Shirato, H., S. Shimizu, K. Kitamura, et al. 2000. Four-dimensional treatment planning and fluoroscopic real-time tumor tracking radiotherapy for moving tumor. *Int J Radiat Oncol Biol Phys* 48:435–42.

Shirato, H., S. Shimizu, T. Kunieda, et al. 2000. Physical aspects of a real-time tumor-tracking system for gated radiotherapy. *Int J Radiat Oncol Biol Phys* 48:1187–95.

Shirato, H., S. Shimizu, T. Shimizu, T. Nishioka, and K. Miyasaka. 1999. Real-time tumour-tracking radiotherapy. *Lancet* 353:1331–2.

Smith, R. L., A. Sawant, L. Santanam, et al. 2009. Integration of real-time internal electromagnetic position monitoring coupled with dynamic multileaf collimator tracking: an intensity-modulated radiation therapy feasibility study. *Int J Radiat Oncol Biol Phys* 74:868–75.

Suh, Y., S. Dieterich, B. Cho, and P. J. Keall. 2008. An analysis of thoracic and abdominal tumour motion for stereotactic body radiotherapy patients. *Phys Med Biol* 53 3623–40.

Tacke, M., S. Nill, U. Oelfke. 2007. Real-time tracking of tumor motions and deformations along the leaf travel direction with the aid of a synchronized dynamic MLC leaf sequencer. *Phys Med Biol* 52:N505–12.

Vedam, S. S, P. J. Keall, A. Docef, D. A. Todor, V. R. Kini, and R. Mohan. 2004. Predicting respiratory motion for four-dimensional radiotherapy. *Med Phys* 31:2274–83.

Webb, S. 2006. Quantification of the fluence error in the motion-compensated dynamic MLC (DMLC) technique for delivering intensity-modulated radiotherapy (IMRT). *Phys Med Biol* 51:L17–21.

Webb, S, and D. M. Binnie. 2006. A strategy to minimize errors from differential intrafraction organ motion using a single configuration for a "breathing" multileaf collimator. *Phys Med Biol* 51:4517–31.

Wilbert, J., J. Meyer, K. Baier, et al. 2008. Tumor tracking and motion compensation with an adaptive tumor tracking system (ATTS): System description and prototype testing. *Med Phys* 35:3911–21.

Willoughby, T. R., P.A. Kupelian, J. Pouliot, et al. 2006. Target localization and real-time tracking using the Calypso 4D localization system in patients with localized prostate cancer. *Int J Radiat Oncol Biol Phys* 65:528–34.

Wu, J., O. Dandekar, D. Nazareth, P. Lei, W. D'Souza, and R. Shekhar. 2006. Effect of ultrasound probe on dose delivery during real-time ultrasound-guided tumor tracking. *Conf Proc IEEE Eng Med Biol Soc* 1:3799–802.

Zimmerman, J., S. Korreman, G. Persson, et al. 2008. DMLC motion tracking of moving targets for intensity modulated arc therapy treatment: A feasibility study. *Acta Oncol* 48:245–50.

15

Quality Assurance in Adaptive Radiation Therapy

Zheng Chang
Duke University Medical Center

Jennifer O'Daniel
Duke University Medical Center

Fang-Fang Yin
Duke University Medical Center

15.1 Introduction

"Adaptive radiation therapy is a closed-loop radiation treatment process where the treatment plan can be modified using a systematic feedback of measurements" (Yan et al. 1997, p. 123). Recent advances in technology enable users to systematically monitor treatment variations during the course of treatment and based on those measurements adapt the treatment plan for an individual patient (Yan et al. 1997; Wu et al. 2008). This way, radiation doses can be more precisely delivered to the tumor while sparing adjacent healthy tissue. Such adaptation can be based on biological changes in the tumor or geometric changes of the tumor and normal structures. Therefore, the effectiveness of adaptive radiation therapy relies on the accuracy of these measurements.

To ensure measurement accuracy, it is necessary to generate a comprehensive quality assurance (QA) program. "The 'quality' of radiation oncology can be defined as the totality of features or characteristics of the radiation oncology service that bear on its ability to satisfy the stated or implied goal of effective patient care" (Kutcher et al. 1994, p. 585). The comprehensive

QA program is used to maintain and monitor the performance characteristics of the treatment system, which includes, but is not limited to, the treatment machine, imaging technology, and the planning system. If necessary, action should be taken to correct any unacceptable deviations from the baseline values acquired during acceptance testing and commissioning. Deviation from the baseline values could compromise patient treatment, resulting in suboptimal treatment response and undesirable complication effects. The quality of radiation oncology is therefore directly affected by the acceptance testing and commissioning process. The significance of the acceptance testing and commissioning process is well-acknowledged, and the corresponding procedures have been published in the literature (Nath et al. 1994; Svensson et al. 1984; Das et al. 2008).

After acceptance testing and commissioning, system parameters may vary from the corresponding baseline values for many reasons that can be generally divided into two categories: (1) gradual small changes (typically due to aging of the machine), or (2) sudden large changes (due to accidents or mechanical failures). During a machine's lifetime, various components will need to be replaced, which may introduce significant deviations from the baseline parameters. A comprehensive QA program monitors the variation in treatment system performance by using established QA tests and correcting the unacceptable deviations to maintain the quality of radiation therapy (Klein et al. 2009).

In this chapter, various QA tests are presented for adaptive radiation therapy, consisting of radiation therapy machine-specific QA, intensity-modulated radiotherapy (IMRT) QA, gating system QA, X-ray-image-based treatment guidance system QA, non-X-ray-image-based treatment guidance system QA, online adaptive correction QA, and offline adaptive correction QA.

15.2 Radiation Therapy Machine-Specific Quality Assurance

Since radiation therapy came into being, various treatment modalities have been developed and employed in clinical practice. In modern radiation therapy, these machines are generally divided into two major categories: (1) radioisotope-based and (2) medical accelerator-based machines. Examples of the radioisotope-based units include cobalt-60 teletherapy units, Gamma Knife units, and high dose rate (HDR) units. Examples of medical accelerator-based units include standard medical linear accelerators, TomoTherapy (TomoTherapy, Inc., Madison, Wisconsin) units, and CyberKnife (Accuray, Inc., Sunnyvale, California) units. Figure 15.1 illustrates a standard medical linear accelerator unit referred to as the Novalis Tx system (Varian Medical Systems, Inc., Palo Alto, California; and BrainLab, Heimstetten, Germany). Our discussion will mainly focus on medical accelerator-based machines.

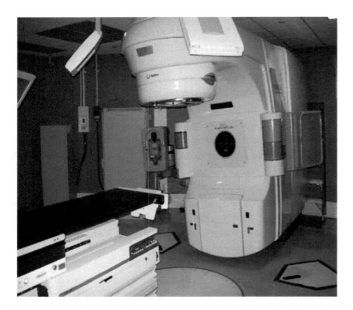

FIGURE 15.1 The Novalis Tx system equipped BrainLAB ExacTrac system, Varian high-definition multi-leaf collimator (HD120 MLC), MV electronic portal imager device (EPID), and KV on-board-imager (OBI).

15.2.1 Machine-Specific Quality Assurance for a Standard Medical Linear Accelerator Unit

To ensure the accuracy of treatment, the American Association of Physicists in Medicine (AAPM) recommends that various periodic QA tests be performed on a standard linear accelerator (Kutcher et al. 1994; Klein et al. 2009). These QA tests include comprehensive dosimetry, safety, and mechanical checks, which are summarized in Table 15.1 with the corresponding criteria of such tests. Due to the limited scope of this chapter, only major linear accelerator QA tests are selected and discussed here. Multileaf collimator (MLC) QA will be discussed in Section 15.3.

15.2.1.1 Mechanical Check: Machine Mechanical/Radiation Isocenter

The mechanical and radiation isocenter of the linear accelerator should be verified annually using multiple "star-shot" films. To test the couch/collimator rotations, the film is laid flat on the couch, the collimator is closed to a symmetrical slit (2–5 mm, machine-dependent), and the film is exposed at a range of couch/collimator angles. A pinprick at the center of the collimator crosshair marks the nominal machine isocenter. The same setup is used to test the gantry rotation except that the film is placed upright in the plane of rotation. The resultant beam intersections on the film should occur within a diameter of 2 mm (task group, TG no. 143) or within 1 mm of baseline (TG no. 142). Due to the tight margins of stereotactic radiosurgery (SRS)/stereotactic radiotherapy (SRT) treatments, the machine's radiation/mechanical isocenter should be tested daily, typically using a Winston–Lutz test (Lutz, Winston, and Maleki 1988; Winston and Lutz 1988; Schell et al. 1995). The setup for a Winston–Lutz test is illustrated in Figure 15.2, where the simulated target ball is placed at the isocenter and the gantry/couch is rotated in different combinations as the ball is imaged. Accuracy should be kept to within 1 mm (Lutz, Winston, and Maleki 1988; Winston and Lutz 1988; Schell et al. 1995). Collimator accuracy should similarly be within 1 mm (SRS/SRT, daily test) or 2 mm (standard, monthly test), and can be checked by rotating the collimator through its range of motion and tracking the crosshair walkout (Kutcher et al. 1994; Klein et al. 2009).

15.2.1.2 Mechanical Check: Laser Alignment

The accuracy of the laser alignment to isocenter should be checked daily (±2 mm standard, ±1 mm SRS/SRT). The coincidence of the lasers with the collimator crosshair and with each other within 30 cm of isocenter should be ≤2 mm (Kutcher et al. 1994; Klein et al. 2009).

15.2.1.3 Mechanical Check: Accuracy of Gantry, Collimator, and Couch Angle/Position Indicators

The gantry angle indicator can be checked with a calibrated level by setting the gantry at vertical and horizontal positions (±1°). The collimator angle indicator can be checked with a level by

TABLE 15.1 List of QA Tests of Medical Linear Accelerator According to AAPM TG No. 40 and AAPM TG No. 142

	Test	Tolerance
	Standard Linear Accelerator	
	Mechanic	
M1	Mechanical isocenter of gantry rotation	1-mm radius
M2	Coincidence of mech. and rad. gantry iso	1 mm
M3	Laser alignment	2 mm (1 mm for IMRT and <1 mm for SRS/SBRT TG no. 142)
M4	Optical distance indicator	2 mm
M5	Gantry angle indicator	1°
M6	Collimator angle indicator	1°
M7	Crosshair centricity	1-mm radius
M8	Coincidence of mech. and rad. crosshair	1 mm
M9	Patient support assembly (PSA) movement	2 mm
M9	PSA travel maximum range movement in all directions	2 mm
M10	PSA rotation angle	1°
M11	PSA isocenter	1-mm radius
M12	Coincidence of mech. and rad. PSA iso	1 mm
M13	PSA level and sag	0.2°/0.3°
M14	Field size defined by light field	2 mm or 1%
M15	Light and radiation field coincidence	2 mm
M16	Graticule	2 mm
M17	Accessory functional check	Functional
M18	Safety interlock	Functional
	MLC	
L1	Field size defined by light field: MLC	2 mm or 1%
L2	Light and radiation field coincidence: MLC	2 mm
L3	Static MLC precision test	1 mm
L4	dMLC Test	Specs
L5	Leaf transmission	0.5%
L6	MLC transmission (average of leaf and interleaf transmission)	±0.5% from baseline
L7	MLC spoke shot	≤1.0-mm radius
L8	Segmental IMRT (step and shoot) test	<0.35-cm maximum error root mean square (RMS), 95% of error counts <0.35 cm
L9	Moving window IMRT (4 cardinal gantry angles)	<0.35-cm maximum error RMS, 95% of error counts <0.35 cm
R1	RapidArc MLC precision test	2 mm
R2	Modulation of dose rate, gantry speed, and MLC speed during RapidArc delivery	2%
	Dosimetric	Tolerance/With Commissioning
D1	Photon/electron beam output factors (TG no. 51)	2% (1% TG no. 142)
D2	Photon beam wedge factors	2%
D3	Electron beam applicator factors	2%
D4	Output vs. gantry angle	2% (1% TG no. 142)
D5	Transmission of tray, holders, and other accessories	1%
D6	MU linearity	1%
D7	Virtual source distance	2 cm
D8	X-ray output constancy vs. dose rate	±2% from baseline
D9	Constancy of daily output, flatness, and symmetry, percentage depth dose (PDD)	3%
D10	Monthly output constancy	2%
D11	Arc mode uniformity	0.2 MU/0.5° (0.2 MU/1.0°/2% whichever is greater TG no. 142)
D12	Percentage depth doses and profiles	2%/3%/2 mm (1% TG no. 142)

Source: Kutcher, G. J., L. Coia, M. Gillen, et al. 1994. *Med Phys* 21:581–618 (for TG no. 40); Klein, E. E., J. Hanley, J. Bayouth, et al. 2009. *Med Phys* 36:4197–212 (for TG no. 142).

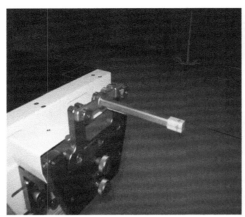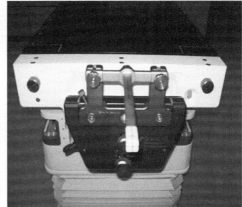

FIGURE 15.2   The setup from two different views for a Winston-Lutz test with the simulated target ball at the isocenter.

setting the gantry at the horizontal position and rotating the collimator to neutral and perpendicular positions (±1°). The couch angle indicators can be checked by rotating the couch to align with the projection of the crosshair along the longitudinal and lateral directions (±1° standard, 0.5° SRS/SRT). Couch movement accuracy can be checked with a ruler by moving the couch to lateral, vertical, and longitudinal directions by a known distance (2 mm standard, 1 mm SRS/SRT). All position/angle indicators should be tested on a monthly basis (Kutcher et al. 1994; Klein et al. 2009).

15.2.1.4 Mechanical Check: Light Field Size/ Radiation Field Size Accuracy

Light field size accuracy can be checked by using graph paper set at the isocenter level. The light field formed by jaws and by MLCs should be checked on a monthly basis (±2 mm or 1%, whichever is greater; Kutcher et al. 1994). The radiation field size accuracy can be tested by using ready-pack film with the light field edge marked before the exposure. Alternatively, the test can be performed using an online portal imager. The coincidence of the radiation fields formed by jaws and by MLCs with the light field should be checked on a monthly basis (±2 mm or 1% on a side, whichever is greater; Kutcher et al. 1994; Klein et al. 2009).

15.2.1.5 Dosimetric Check: Linear Accelerator Output

The output of a linear accelerator should be checked daily, monthly, and annually. Daily checks may be done by the machine operators using a simple daily QA device, notifying the physicist to investigate if changes are ≥3%, and not using the machine if changes are ≥5%. On a monthly basis, the physicist should verify the output within ±2% using a calibrated ionization chamber and solid water or water tank arrangement (Kutcher et al. 1994). Annually, the output should be checked and recalibrated as necessary using a standard protocol, such as AAPM task group (TG) no. 51 (Almond et al. 1999).

15.2.1.6 Dosimetric Check: Beam Energy

The beam energy can be checked on a monthly basis by using an ion chamber to measure the output at two different depths, such as 5 and 10 cm, and compare the ratio of these two readings to its baseline value. A more thorough energy check, to be done annually, involves scanning the percentage depth dose with beam scanning equipment. According to AAPM TG no. 40, results should be within 2% of the baseline value (Kutcher et al. 1994), although the recently published AAPM TG no. 142 recommends a tighter tolerance, 1% of the baseline value (Klein et al. 2009).

15.2.1.7 Dosimetric Check: Flatness and Symmetry

The flatness and symmetry can be checked daily or monthly with a multiple channel QA device or with an imaging device (film, electronic portal dosimeter). Annually, a more thorough test should be done by scanning the profiles in water. AAPM TG no. 40 recommends tolerances of 2% for photon flatness, 3% for electron flatness, and 3% for symmetry of both electron and photon (Kutcher et al. 1994). AAPM TG no. 142 recommends the discrepancies be within 1% of the baseline value (Klein et al. 2009).

15.2.2 Machine-Specific Quality Assurance for a TomoTherapy Unit

A second kind of medical accelerator-based machine is the TomoTherapy unit (TomoTherapy Inc.). Compared with the standard medical accelerator-based machines, TomoTherapy units are designed to deliver radiation treatment through a helical rotation mode. More specifically, a linear accelerator is mounted on a ring gantry in a TomoTherapy unit, analogous to the X-ray tube mounted on the computed tomography (CT) ring gantry. During treatment delivery, the ring gantry rotates at a constant speed while the patient on a flattop couch moves translationally along the gantry rotation axis. In a TomoTherapy unit, the radiation beam is collimated into a fan beam. The fan beam is further modulated into 64 narrow beamlets with 64 binary leaf collimators, which either close or open during treatment delivery. Due to its unique design, the TomoTherapy unit demands various special QA tests to monitor the characteristics of the system, which are summarized in Table 15.2 with the corresponding criteria of the tests (Broggi et al. 2008). Due to the limited scope of this chapter, major unique QA tests are selected and discussed in Sections 15.2.2.1 through 15.2.2.4.

15.2.2.1 Mechanical Check: Accuracy of Virtual Isocenter Alignment

Due to the design of a TomoTherapy unit, a patient is not initially aligned with the actual radiation isocenter. Instead, the patient is aligned with a location referred to as the "virtual isocenter" and then automatically moved to the true radiation isocenter, which is typically 70 cm away from the virtual isocenter along the gantry rotation axis. To ensure the accuracy of the treatment, it is therefore crucial to verify the correlation between the virtual isocenter and the radiation isocenter. A phantom with fiducial markers can be used to perform the test.

15.2.2.2 Dosimetric Check: Output and Energy Constancy

Similar to a standard linear accelerator, output and energy constancy are measured for a static open field size. For a TomoTherapy unit, the maximum open field size of 40 × 5 cm is commonly used to perform this test. During the test, a square solid water phantom is placed at source-to-surface distance (SSD) of 85 cm on the flattop couch. A cylindrical ionization chamber is used to measure radiation doses at various depths, usually including the depth of maximum dose (1.5 cm for the 6-MV beam), 10 cm, and 20 cm from the surface of the solid water phantom. Energy constancy can be calculated as the ratio between the dose measurements at 10-cm and 20-cm depths. The measurements are then compared with the baseline values determined during commissioning.

15.2.2.3 Dosimetric Check: Rotational Output Reproducibility for a Simple Intensity-Modulated Radiotherapy Plan

Due to the dose delivery design of a TomoTherapy unit, the static output consistency may not represent the consistency of radiation delivery for a clinical case. To verify rotational delivery consistency, a cylindrical ionization chamber is placed at the center of a cylindrical solid water phantom to measure the dose during the delivery of a simple IMRT test case. The measurements are compared with the commissioning baseline values.

15.2.2.4 Dosimetric Check: Completion Procedure Check

In treatment using a TomoTherapy unit, radiation is delivered in helical mode, with the gantry rotating as the couch moves translationally. When the treatment is interrupted, it is essential

TABLE 15.2 List of Recommended QA Tests for a TomoTherapy Unit

	Test	Tolerance
	Mechanical Accuracy	
M1	Virtual isocenter alignment	1 mm
M2	Field divergence vs. gantry plane	0.5 mm
M3	Jaw twist	1°
M4	Field center constancy with jaw size	1 mm
M5	Gantry position accuracy	1°
M6	Star shot of radiation isocentricity	1 mm
M7	Field size accuracy: Set field vs. irradiated field for different gantry angle	1 mm
M8	Couch movement accuracy	1 mm
M9	Gantry–couch synchronization	1 mm
M10	Gantry (static)–couch synchronization	1 mm
M11	Gantry–MLC synchronization	1°
M12	Couch speed uniformity	<2%
	Dosimetric Accuracy	
D1	Static output	1.5%
D2	Rotational output	1.5%
D3	Transversal/longitudinal profiles checks	1%
D4	Energy check: PDD curves	1%
D5	Rotational output reproducibility	1%
D6	Rotational output linearity vs. irradiation time	1%
D7	Rotational output reproducibility vs. gantry rotation period	1%
D8	Output reproducibility in dynamic condition (vs. couch speed)	1%
D9	Output reproducibility for a simple IMRT plan	1%
D10	Completion procedure check	1.5%

Source: Broggi, S., G. M. Cattaneo, S. Molinelli, et al. 2008. *Radiother Oncol* 86:231–41.

to generate a completion procedure in order to complete the radiation delivery. The generation of the completion procedure shall be verified by delivering the same IMRT plan with and without interruption. The difference in dose between these two scenarios shall be evaluated to be within an acceptable tolerance level.

15.2.3 Machine-Specific Quality Assurance for a CyberKnife Unit

Another kind of medical accelerator-based machine is the CyberKnife unit, which is the only commercially-available robotic radiosurgery unit. A 6-MV linear accelerator mounted on a robotic arm uses real-time image guidance to adapt for inter- and intrafractional patient motion. Specific technology has been developed to enable the tracking of multiple treatment sites including brain, spine, prostate, and lung tumors. All these features require specific QA protocols and the ongoing development of appropriate techniques. Recommended readings include Dieterich and Pawlicki (2008) and the soon-to-be-published AAPM TG no. 135.

The basic QA goals for the CyberKnife are similar to those for a standard linear accelerator as discussed in Section 15.2.1: mechanical accuracy, dosimetric accuracy, and general safety. A daily check of the temperature/pressure output correction is recommended. Dosimetric verifications include the use of small-field dosimetry techniques as with linear accelerators that utilize field diameters below 4 cm (Pappas et al. 2008).

Two unique aspects of CyberKnife treatment delivery are the robot-mounted linear accelerator and the active automatic image-guided tracking used during treatment. There is currently no recommended QA for the robot itself, and it will require many years of usage before the first manufacturer-recommended QA time point is specified (Dieterich and Pawlicki 2008). The location of the robot "perch" should be verified vs. the lasers daily. The automatic image alignment and robotic targeting accuracy combination should be tested both daily (using the manufacturer-provided automatic QA tool) and monthly (using the isocenter-path/search2 check and the ball-cube end-to-end targeting test procedure). As additional automatic alignment protocols are developed, additional specific QA tests should also be designed.

15.3 Intensity-Modulated Radiotherapy Quality Assurance

QA tests for IMRT encompass the commissioning tests of IMRT capability for linear accelerators, periodic QA tests to ensure continuing functionality, and patient-specific QA measurements done for each IMRT treatment plan. Over the past few years, various published studies and reports have addressed this important issue (Klein et al. 2009; Ezzell et al. 2003; Boyer, Biggs, et al. 2001; Boyer, Butler, et al. 2001). In this section, we briefly describe the recommended QA protocols for IMRT and provide references for further in-depth study.

IMRT delivery techniques can be divided into four main categories: (1) segmental IMRT, (2) dynamic IMRT, (3) rotational IMRT, and (4) physical-attenuator-based IMRT (Ezzell et al. 2003). Segmental IMRT ("step-and-shoot" treatment) alternates between delivering radiation and moving the MLC leaves to create new segments. Dynamic IMRT ("sliding window") allows the MLC leaves to travel unidirectionally across the field while radiation is being delivered at a static gantry angle. Rotational IMRT varies combinations of the gantry angle, MLC leaf position, MLC leaf speed, and/or dose rate while radiation is being delivered, a technique that encompasses both fan-beam delivery (Best Nomos' Peacock and TomoTherapy) and cone-beam delivery (Elekta's volumetric modulated arc therapy [VMAT] and Varian's RapidArc). Finally, physical attenuator-based IMRT relies on physical modulators attached to the gantry head to create the desired fluence. However, since this last technique is becoming rare due to its time-consuming nature, we focus on the QA for the first three IMRT techniques.

15.3.1 Machine-Specific Quality Assurance

15.3.1.1 Quality Assurance for Segmental Intensity-Modulated Radiotherapy

Segmental IMRT requires the measurement of MLC leaf position accuracy, leaf gap, leaf penumbra, leaf transmission, the monitor unit (MU)-dose linearity, and field flatness/symmetry for small MU deliveries.

15.3.1.1.1 Multileaf Collimator Leaf Position Accuracy

Determining the leaf position accuracy can be complicated, because often there is an offset between the light field and the radiation field edge due to rounded leaf ends or to the travel motion of double-focused leaves. This offset should be measured and then film can be irradiated by matching multiple strips of radiation at the 50% dose position for a variety of gantry and collimator angles. Positional displacements of 0.2 mm can be detected (Chui, Spirou, and LoSasso 1996; Low et al. 2001). Alternatively, a "picket fence" film may be used, as shown in Figure 15.3a, with multiple irradiations of 1-mm-wide strips allowing the detection of positioning errors of 0.5 mm or greater (Chui, Spirou, and LoSasso 1996). The AAPM recommends static MLC positioning errors of 1 mm or less, as illustrated in Table 15.1 (Klein et al. 2009).

15.3.1.1.2 Multileaf Collimator Leaf Gap

The leaf gap is the minimum distance between a pair of closed leaves. Using film, the gap should be measured for all leaf pairs and the transmission quantified. A transmission of up to 25% is typical for certain linear accelerators (Boyer et al. 2001). The penumbra in the direction of leaf travel should be measured with either film or diodes and the information entered into the treatment planning system. The transmission both through an individual leaf and between leaves should be measured with film or a large-volume ionization chamber at multiple gantry/collimator

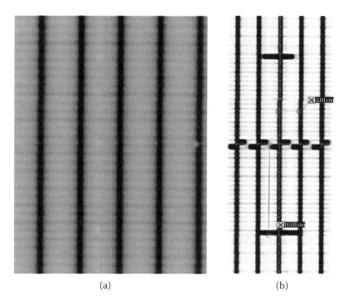

FIGURE 15.3 (a) Picket fence test pattern and (b) monthly MLC leaf position accuracy test pattern by digital measurements.

angles and at different extensions across the treatment field. The average transmission should be <2% (Boyer 2001) and should vary by ±0.5% from baseline annually (Klein et al. 2009).

Annual and monthly/quarterly QA tests are recommended by the AAPM (Klein et al. 2009; Ezzell et al. 2003; Boyer et al. 2001). It is important that the leaf setting vs. light field vs. radiation field (±1 mm) is verified monthly or quarterly with at least two different MLC field sizes. An additional pattern provides a quick check of the positioning accuracy of all leaves (Figure 15.3b). Sites with a high volume of IMRT treatments will often choose to do these tests on a monthly basis. The leaf travel speed should also be measured via MLC log file evaluation, and it should remain consistent within 0.5 cm per second. Annually, the leaf setting vs. light field vs. radiation congruency should be checked for multiple gantry/collimator angles. The penumbra should be verified with either film or water tank measurements. The inter-leaf leakage (<3%), average transmission (<2%), and abutted leaf transmission (<25%) should also be remeasured on an annual basis. Finally, the MLC log file error histogram report of leaf position deviations should be evaluated (95% of deviations must be <0.35 cm; Klein et al. 2009).

15.3.1.2 Quality Assurance for Dynamic Intensity-Modulated Radiotherapy

Dynamic IMRT QA encompasses all the tests for segmental IMRT QA with a few additional procedures (Ezzell et al. 2003; LoSasso, Chui, and Ling 2001). Because the leaves move during dose delivery, the accuracy of leaf speed and dose rate should be verified. Films can be irradiated to a given dose with different combinations of leaf speeds and dose rates across the field. The dose deposited on the film should be uniform when normalized to an open field. The minimum leaf gap during leaf motion may be different from the gap for static leaves and should be measured

with transmission quantified. Dynamic IMRT treatments are more sensitive to leaf gap offsets than segmental IMRT, and it is recommended that the leaf gap be accurate to within 0.2 mm (LoSasso, Chui, and Ling 2001).

Monthly QA should include segmental IMRT QA plus a consistency check of a selected dynamic IMRT field using film, ion chamber, or the clinic-chosen IMRT QA device as described later in Section 15.3.2. Additional annual QA tests involve the evaluation of output vs. gantry angle with varying leaf speed/dose rate and verification of the moving leaf gap for each leaf pair with varying gantry angle (Ezzell et al. 2003; LoSasso, Chui, and Ling 2001).

15.3.1.3 Quality Assurance for Rotational Intensity-Modulated Radiotherapy

Rotational IMRT can be delivered either in a fan-beam mode or in a cone-beam mode. Rotational IMRT with fan-beam delivery chiefly consists of helical tomotherapy (TomoTherapy Inc.). The patient couch is moved through the bore of the CT-scanner-like machine as the internal gantry rotates and delivers the radiation. Couch motion, gantry rotation, the on/off switching of radiation, and MLC leaf positions must all work in concert. Commissioning of such a machine is complex (Balog et al. 2003; Balog, Olivera, and Kapatoes 2003). QA includes testing of MLC leaf leakage (≤1%), leaf latency (±2%), MLC alignment (±1%), and MLC "twist" (visual assessment; Fenwick et al. 2004). The synchronization of leaf position with gantry rotation (±1°) should be verified on a monthly basis. A report from AAPM TG no. 148 on QA for helical tomotherapy will be available in the near future.

Cone-beam rotational IMRT can be delivered by either Elekta's VMAT, or Varian's RapidArc. Recent publications have described commissioning and QA for this technology (Ling et al. 2008; Bedford and Warrington 2009). To deliver cone-beam rotational IMRT, the linear accelerator mechanical QA requires tighter calibration of gantry angle (±0.5°) and isocenter (±1 mm). The output of a static field vs. output of a rotating gantry (±1%) as well as the output from a rotating gantry and dynamic MLC (dMLC) field (±2°) should also be measured. Similar to dynamic IMRT, the accuracy of leaf positioning at various gantry positions as well as with a rotating gantry should be verified with film (±1 mm). The output should remain consistent (±2%) as dose rate, gantry speed, and leaf speed are varied during arc delivery. The linear accelerator commissioning tests should be repeated on an annual basis. The dMLC tests should be repeated on a monthly basis, at least until more experience has been gained with this new technology.

15.3.2 Patient-Specific Intensity-Modulated Radiotherapy Quality Assurance

Due to the complexity of treatment delivery, each IMRT plan undergoes QA testing before it is used on a patient. This testing includes verification of the absolute dose and the dose distribution. The MU calculations, transfer of information from the treatment planning system to the record-and-verify system to the machine for delivery, and the delivery itself are verified. However, any mistakes in designing the IMRT plan (e.g., overdosing a critical organ or forgetting to override the density of contrast agents) will not be detected.

Ionization chambers, detector arrays, and film are used as the standard tools for patient-specific IMRT QA. Measurements can be done in tandem, with both the ionization chamber and the film contained in an appropriate phantom (e.g., CIRS Model 002H5 IMRT Phantom for Film and Ion Chamber [CIRS, Inc., Norfolk, VA]). The AAPM recommends an action level of 3%–4% for ionization chamber measurements in high-dose low-gradient regions. Placement of the chamber near a sharp dose gradient may be unavoidable in small fields, in which case the use of a compact ionization chamber (e.g., volume $\leq.05$ cm^3) to minimize the volume averaging effect is recommended. Two-dimensional (2D) planar dose measurement is typically evaluated using the gamma index (Low et al. 1998). Common gamma parameters are 3% dose distance and 3-mm distance to agreement with a threshold dose of 10% or lower; acceptance criteria is generally 90%–95% of pixels passing the gamma criteria (Nelms and Simon 2007).

Although film provides accurate QA results with excellent spatial resolution (Childress et al. 2005), the method can be both time consuming and difficult. Many new products have entered the market recently to replace film in day-to-day QA use. The electronic portal imaging device (EPID) used to take megavoltage (MV) images for patient alignment may also be used to verify IMRT treatments (Renner et al. 2003; Chen et al. 2007). The MapCheck device (Sun Nuclear Corp., Melbourne, Florida) uses a 2D plane of diodes, whereas the MatriXX (IBA Dosimetry, Bartlett, Tennessee) uses a similar arrangement of ionization chambers (Buonamici et al. 2007; Herzen et al. 2007). The Delta4 device (ScandiDos, Inc., Ashland, Virginia) measures two perpendicular 2D planes of dose using diodes and interpolates between them to approximate the three-dimensional (3D) dose distribution (Sadagopan et al. 2009). Full 3D dose measurements may be done using gel dosimeters, such as Presage (Heuris Pharma LLC, Skillman, New Jersey) and Bang (MGS Research, Inc., Madison, Connecticut), though the time involved and the cost are currently prohibitive for making these devices part of daily clinical QA (Sakhalkar et al. 2009; Lopatiuk-Tirpak et al. 2008). Some clinics employ independent MU calculation software to verify the absolute dose (Boyer et al. 2001).

The rotational delivery of IMRT adds the requirement that the dosimeter have minimal angular dependency. A film and ionization chamber combination is appropriate for commissioning QA, but it may be overly time consuming for patient-specific QA. The Delta4 dosimeter response varies by ±0.5% with gantry angle, but it has larger disagreements with ion chamber readings (±2.5%), possible due to volume averaging effects (Bedford et al. 2009). A new cylindrical dosimeter with 124 diodes arranged in rings (ArcCheck; Sun Nuclear Corp., Melbourne, Florida) has also been introduced for IMRT QA (Letourneau et al. 2009). Gel dosimetry is another viable option (Thomas et al. 2009).

15.4 Respiratory Gating System Quality Assurance

It is well-known that respiratory motion has considerable impact on radiation therapy of patients with thoracic, abdominal, and pelvic tumors because they move when the patient breathes. In order to minimize the effect of respiratory motion, various respiratory management techniques have been proposed and investigated. In respiratory management, QA plays a critical role in ensuring the accuracy of radiation treatment, as discussed in the report of AAPM TG no. 76 (Keall et al. 2006). The report describes various techniques used in the management of respiratory motion, and proposes general recommendations to different methods of respiratory motion management (Keall et al. 2006). In addition to management techniques, AAPM TG no. 76 also contains QA recommendations for these techniques, summarized in Table 15.3.

Although there are different methods for implementing respiratory motion management, all the techniques share a common principle that the radiation beam shall be synchronized with the patient's respiration, which is generally referred to as "respiratory gating." According to AAPM TG no. 142, dynamic phantoms that simulate human organ motions associated with respiration are recommended to test target localization and treatment delivery.

Currently, there are several gating systems commercially available for clinical use. Examples include an infrared (IR)-camera-based respiratory gating system (real-time position management [RPM]; Varian Medical Systems Inc.), a pressure-sensor-based motion-monitoring system (AZ-773V; Anzai Medical, Inc., Tokyo, Japan), an active breath control (ABC)-based motion-monitoring system (ABC; Elekta Oncology Systems, Inc., Crawley, United Kingdom), and an IR tracking and stereo X-ray camera system (BGI, Inc., Colorado, and Accuray Inc.).

As illustrated in Figure 15.4, the Varian RPM respiratory gating system consists of a marker block, an IR light ring that emits IR light, a charge-coupled device (CCD) tracking camera used to visualize the relative position of the block, and a workstation that displays and records the motion data. The marker block consists of two reflective fiducials that are placed 3 cm apart. The marker block is often placed on the patient's chest or abdomen. The

TABLE 15.3 List of QA Tests for a Respiratory Gating System According to AAPM TG No. 142

	Test	Tolerance
	Respiratory Gating	
RG1	Beam output constancy	2%
RG2	Phase/amplitude beam control	Functional
RG3	In-room respiratory monitoring system	Functional
RG4	Temporal accuracy of phase/amplitude gate-on	100 m of expected value
RG5	Calibration of surrogate for respiratory phase/amplitude	100 m of expected value
RG6	Interlock testing	Functional

Source: Klein, E. E., J. Hanley, J. Bayouth, et al. 2009. *Med Phys* 36:4197–212.

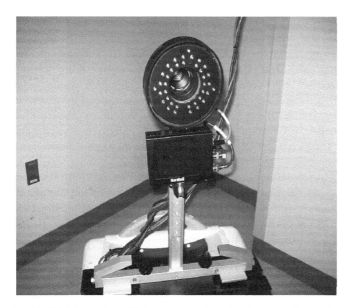

FIGURE 15.4 Varian RPM system consists of an in-room view finder on the bottom and a tracking camera on the top, which includes an IR source and a CCD detector (Cardenas, A., J. Fontenot, K. M. Forster, et al. 2004. Quality assurance evaluation of delivery of respiratory-gated treatments. *J Appl Clin Med Phys* 5:55–61.).

reflective fiducial markers are tracked using the IR light source and CCD detector. In this way, the motion of the block is considered as a surrogate for respiration-induced tumor motion.

The pressure sensor-based motion-monitoring system uses a pressure sensor to monitor respiratory motion, similar to the reflective fiducials in the RPM system. The pressure sensor-based system generally consists of a belt, a pressure sensor inserted in the belt, and a gating control computer. During the gated treatment, a belt that is fixed on the patient with the sensor inserted within can be used to detect respiratory motion, similar to the principle of the RPM system.

In the ABC-based motion-monitoring system, a patient's breathing is monitored using ABC devices. The respiratory motion is minimized by actively controlling the airflow of the patient.

The IR tracking and stereo X-ray camera system combines the techniques of IR tracking and X-ray image guidance. In this system, the motion of the fiducial marker implanted near or within the tumor is used as a surrogate of the tumor. An external respiratory IR-based monitoring system is used to monitor the patient's breathing in real time and to correlate external respiratory motion with the motion of the implanted marker. In this way, the uncertainties between respiratory motion and tumor motion can be effectively minimized. To ensure the accuracy of the gated treatment, the respiratory gating system shall be installed on both the CT simulator and the radiotherapy unit.

Although there are different systems that implement respiratory motion management, all the techniques share the same principle of QA, which is illustrated in AAPM TG no. 142. Due to the limited scope of this chapter, only the major QA tests are selected and discussed in Sections 15.4.1 through 15.4.3.

15.4.1 Output Constancy of Gated Beams

As illustrated in Table 15.3, beam output constancy shall be verified periodically. The test can be performed using an ion chamber at a specific depth (e.g., 10 cm) measuring doses with and without respiratory gating. The discrepancies shall be within ±2% for gating windows ≥500 ms (Klein et al. 2009; Keall et al. 2006).

15.4.2 Functionality and Temporal Accuracy of Phase/Amplitude Beam Control

In a gating system, the beam shall be controlled in order to precisely and effectively deliver radiation dose to the target while minimizing the uncertainties caused by respiratory motion. For example, in the RPM system, the beam is controlled using a beam switch box, which is often placed near the treatment console. The beam switch box can automatically turn on or off the radiation beam in either the phase-gating or amplitude-gating modes. As discussed in AAPM TG no. 142, the temporal accuracy of phase/amplitude gating can be tested using a radiopaque target attached to motion phantom, where the geometric center of a radiopaque target was known at each phase/amplitude relative to the beam central axis. These images can be acquired during gated beam-on time with a radiographic film or an EPID. The spatial discrepancies between measurements and expectations can be used to calculate the temporal accuracy. The temporal accuracy shall be within 100 m if the moving object travels at speeds no greater than 20 mm/s. With such speeds, the 100-m tolerance for temporal accuracy would result in 2 mm of positional uncertainty (Keall et al. 2006).

15.4.3 Calibration of Surrogate for Respiratory Phase/Amplitude

As discussed in AAPM TG no. 142 and AAPM TG no. 76, there are different types of surrogates of respiratory pattern used clinically, including optical IR reflective fiducial markers, strain-gauge belts with pressure sensors, and spirometry. The phase and amplitude indicated by the surrogate shall not change significantly over time. Testing can be performed using a motion phantom. The monitored data of the surrogate shall be compared with the expected motion data. The temporal accuracy shall be within 100 ms if the moving object travels at speeds no greater than 20 mm/s (Klein et al. 2009; Keall et al. 2006).

15.5 Quality Assurance for X-Ray-Image-Based Treatment Guidance Systems

In modern radiation therapy, image guidance has been commonly used to locate a tumor target with increased accuracy so that radiation doses can be delivered more precisely to the tumor while minimizing doses to critical structures. Imaging systems can be used to monitor variations during the course of treatment.

The treatment can then be adapted on a daily basis to the measured deviations and/or reoptimized based on the additional information from the imaging systems. Various imaging devices are employed clinically: MV 2D-EPIDs, MV 3D fan-beam CT, kilovoltage (kV) onboard imaging (OBI) for 2D imaging and 3D cone-beam CT (CBCT), in-room 3D-CT scanners such as the CT-on-rails, the BrainLAB 6D ExacTrac system (BrainLAB), and the CyberKnife kV alignment system (Accuray Inc.). In this chapter, our discussion focuses on three representative imaging modalities: (1) planar imaging (MV EPID, kV 2D-OBI), (2) CBCT (MV and kV), and (3) the BrainLAB 6D ExacTrac system. As shown in Figure 15.1, devices corresponding to all of these imaging devices are available on a Novalis Tx system unit. The QA tests and corresponding criteria are summarized in Table 15.4.

15.5.1 Planar Imaging

One common option for taking 2D planar images is to use the MV EPID (Yin et al. 2006). In particular, the EPID used in the Novalis Tx system is a commercially available amorphous silicon imaging device (aS1000; Varian Medical Systems Inc.) mounted on the linear accelerator using robotic (Exact) arms.

TABLE 15.4 List of X-Ray-Image-Based Treatment Guidance QA Tests According to AAPM TG No. 142

	Test	Tolerance
	kV and MV (EPID) Imaging	
I1	Collision interlocks	Functional
I2	Positioning/repositioning	≤2 mm
I3	Imaging and treatment coordinate coincidence (single gantry angle)	≤2 mm
I4	Scaling	≤2 mm (≤1 mm for SRS/SBRT TG no. 142)
I5	Spatial resolution	Baseline
I6	Contrast	Baseline
I7	Uniformity and noise	Baseline
	CBCT (kV)	
I4	Collision interlocks	Functional
I5	Imaging and treatment coordinate coincidence	≤2 mm
I6	Positioning/repositioning	≤2 mm
I7	Geometric distortion	≤1 mm
I8	Spatial resolution	Baseline
I9	Contrast	Baseline
I10	HU constancy	Baseline
I11	Uniformity and noise	Baseline
	ExacTrac System	
E1	Isocenter calibration	Functional
E2	X-ray calibration	Functional
E3	ExacTrac calibration verification	≤2 mm

Source: Klein, E. E., J. Hanley, J. Bayouth, et al. 2009. *Med Phys* 36:4197–212.

The EPID system consists of (1) an image detection unit including the detector and its accessory electronics, (2) an image acquisition unit including acquisition electronics and interface hardware, and (3) a workstation shared with OBI applications. The MV EPID can be used for 2D radiographic acquisition or fluoroscopic image acquisition and is positioned by extending and retracting the robotic imager arm. After comparing the real-time EPID images with the planned digitally reconstructed radiographs (DRRs), the patient's position can be appropriately adjusted for radiotherapy via a couch shift.

A second common planar imaging option is kV OBI (Yin et al. 2006). As shown in Figure 15.1, the OBI system consists of a kV X-ray source (KVS) and a kV amorphous silicon detector (KVD) that are both mounted on the linear accelerator using robotic (Exact) arms. The OBI system provides three imaging modes: (1) 2D radiographic acquisition, (2) 2D fluoroscopic image acquisition, and (3) 3D-CBCT acquisition. In the applications of 2D OBI, various X-ray techniques with different milliampere seconds and peak voltage (kVp) can be selected to reach an optimal image contrast. Similar to the MV EPID image analysis, shifts in patient position are determined by comparing the planned DRRs with the kV images, and the couch is moved as needed. Another option is to mount an oblique pair of kV sources/detectors on the wall, ceiling, or floor of the treatment room. These QA techniques would apply to the 2D imaging devices as well.

15.5.1.1 Imaging Isocenter Accuracy

The isocenter of the imaging system should match the isocenter of the linear accelerator within 1 mm (SRS/SRT) or 2 mm (standard; Klein et al. 2009). It can be checked by setting a small radiopaque target at the isocenter and imaging with two orthogonal angles (e.g., anterior–posterior, and right–lateral) with the target at the center of the digital crosshair. In general, when the concept of isocenter is not applicable, it is critical to correlate the coordinate systems between the imaging system and the delivery system.

15.5.1.2 Imager Positioning Accuracy

The positioning accuracy of an image detector (and kV source for OBI systems) should be checked on a monthly basis by measuring the isocenter-to-detector/source distance with a ruler (±2 mm standard, ±1 mm SRS/SRT). The offset of the image detector/source center from the projection of the isocenter should also be measured on a monthly basis (±2 mm standard, ±1 mm SRS/SRT; Klein et al. 2009; Yoo et al. 2006).

15.5.1.3 Repositioning Accuracy

A phantom can be imaged and aligned by a known amount with image matching. The couch is shifted based on this alignment. The accuracy of the couch movement should be within 2 mm (standard) or 1 mm (SRS/SRT; Klein et al. 2009).

15.5.1.4 Image Quality

On a monthly basis, the spatial resolution, contrast, uniformity, and noise of the planar imaging system should be evaluated by imaging the QA phantom that was used for the acceptance tests. The image quality should remain within the baseline specifications provided by the manufacturer.

15.5.2 Three-Dimensional Imaging

Three-dimensional images may be generated either as traditional CT images from an in-room CT unit or as CBCT images from hundreds of X-ray projections acquired over a range of gantry rotations by either the MV EPID or the kV OBI system (Court et al. 2003; Sonke, Remeijer, and van Herk 2006). After the CT reconstruction is completed, the corresponding planning CT and daily CT images are aligned. Once the desired match is made, the couch shift parameters are transferred to the linear accelerator and the couch is moved remotely to position the patient for treatment.

The mechanical tests for planar imaging discussed in Section 15.5.1 also apply to CT imaging. The 3D image quality should be tested on a monthly basis with a QA CT phantom (Figure 15.5) for Hounsfield unit (HU) constancy, spatial resolution, contrast, uniformity, and noise (baseline manufacturer specifications). The geometric distortion of the CT images should also be checked (≤2 mm standard, ≤1 mm SRS/SRT; Klein et al. 2009).

15.5.3 BrainLAB ExacTrac System

As shown in Figure 15.1, the BrainLAB ExacTrac system includes IR and video detectors as well as dual diagnostic kV X-ray tubes with amorphous silicon digital detectors. The IR cameras are

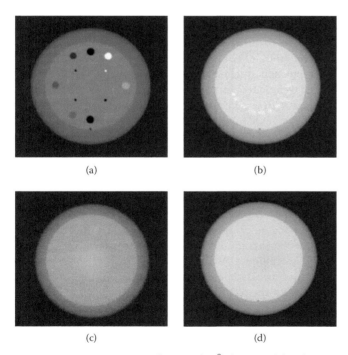

(a)   (b)

(c)   (d)

FIGURE 15.5 CBCT QA with a Catphan® phantom (The Phantom Laboratory, Inc., Salem, NY). (a) Hounsfield unit consistency. (b) High contrast resolution. (c) Low contrast resolution. (d) Noise and uniformity.

used to monitor IR reflective body markers placed on the patient and an IR reflective reference star, which is attached to the treatment couch. The radiographic kV devices consist of two ceiling-mounted X-ray imagers. Matching tools in three dimensions (translation only) and six dimensions (translation and rotation) are available to align the kV images with the corresponding DRRs. The robotic couch can correct the patient position in six dimensions, including pitch, roll, and yaw, with the guidance of the IR monitoring system. To verify the patient position, kV imaging is repeated after shifting the couch.

The BrainLab ExacTrac system should be calibrated and verified daily including isocenter and X-ray calibrations. The isocenter calibration is performed with a $10 \times 10 \times 2$ cm³ solid block with 5 IR markers on its anterior surface, as shown in Figure 15.6(a). Next, the X-ray calibration is performed with a BrainLAB X-ray calibration phantom, as shown in Figure 15.6(b). The phantom is positioned to the previously calibrated isocenter with the guidance of its IR markers. Two oblique X-rays are acquired to visualize the radiopaque disks in the phantom. Their locations are used to determine mapping projection parameters, which translate a 3D object into a 2D projection. At this point, the calibration of ExacTrac system is complete. Calibration on a daily basis is recommended to minimize the sensitivity of the system to small shifts in position of the cameras, sources, and detectors. As a final check, the isocenter phantom should be aligned and reimaged; the measured shift should be within 1 mm.

15.6 Quality Assurance for Non-X-Ray-Image-Based Treatment Guidance Systems

In addition to the X-ray-image-based guidance systems discussed in Section 15.5, a few commercially available systems rely on other sources of information to position patients for treatment. Photogrammetry uses 3D surface imaging (either the patient surface alone or with additional markers placed on the skin) for daily alignment, particularly for patients undergoing accelerated partial breast irradiation (Bert et al. 2006). Another localization device employs implanted electromagnetic transponders to track the position of the prostate (Willoughby et al. 2006; Santanam et al. 2008). In Sections 15.6.1 and 15.6.2, we

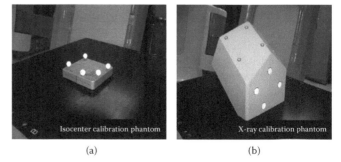

(a)   (b)

FIGURE 15.6 BrainLAB calibration phantoms: (a) isocenter calibration phantom; (b) x-ray calibration phantom.

focus on QA for one commercially available system for each technique: AlignRT (Vision RT, Inc., Boston, Massachusetts) for photogrammetry and the Calypso 4D Localization System (Calypso Medical Technologies, Inc., Seattle, Washington) for electromagnetic tracking. The QA procedures described in these sections may be generalized to similar alignment systems. Additionally, a future publication from AAPM TG no. 147 will address QA for nonradiographic radiotherapy localization and positioning systems.

15.6.1 Quality Assurance for the AlignRT System

The AlignRT device uses multiple cameras to produce a 3D surface image of the patient, which may be used both for patient setup and for monitoring the patient position during treatment. Bert et al. (2005) performed a series of commissioning tests on the AlignRT system, investigating its stability, alignment accuracy, surface modeling accuracy, and gating ability. They reported that the positional measurements made by AlignRT differed from the known locations by 0.95 ± 0.58 mm. Additional QA protocols recommended by the vendor consist of both daily and monthly QA. Monthly QA protocols can be separated into two parts: (1) camera calibration to ensure the cameras are working properly and that their field of view is not obstructed and (2) isocenter calibration to check the consistency of the system's localization. On a daily basis, a QA phantom is aligned to the room's lasers and imaged by the AlignRT system to check that the match line of visual data from the cameras is ≤1 mm.

15.6.2 Quality Assurance for the Calypso 4D Localization System

The Calypso 4D Localization System consists of three electromagnetic transponders, implanted in the prostate or prostate bed, which can be continuously monitored by an electromagnetic array placed over the patient during radiotherapy. A positional accuracy and precision of ≤0.5 mm at a range of distances from the array has been reported (Balter et al. 2005). A series of acceptance tests for this system have also been proposed (Santanam et al. 2009).

First, the integrity of information transfer between the treatment planning system (TPS) and the Calypso workstation was established by taking a CT scan of a test phantom implanted with three transponders, contouring the CT image in our TPS, and exporting the information to the Calypso workstation where it was independently verified. Second, the localization and tracking accuracies were measured by displacing the phantom by a known distance (expected accuracy ≤0.5 mm) for a range of distances (up to 25 cm) beneath the array. Third, the accuracy of motion measurements was tested using a motion platform (expected accuracy ≤1mm) with varying speeds. Fourth, the operational range of detection was verified (nominally 14 cm laterally, 14 cm longitudinally, and 27 cm vertically). Fifth, operational test cases were created to check that the appropriate

software warnings (e.g., transposed transponders) and abilities (e.g., exclusion of a transponder) functioned. Sixth, MV, kV, and CBCT images were taken with the array in place to determine the magnitude of the introduced image artifact.

Monthly and daily QA protocols are based on the recommendations from the Calypso user's manual (Calypso 2008). Each month, a camera calibration (testing camera functionality and range of vision) and a system calibration (reestablishing the correlation between the Calypso coordinate reference frame and the machine isocenter) followed by a calibration verification (comparing the array-measured isocenter to the machine isocenter, accepted deviation of ±1 mm) must be done. Calypso calibration relies on lasers to represent the machine isocenter; so, laser accuracy should be validated each month prior to performing the Calypso QA tests. Additionally, the acceptance tests of measuring a known displacement (±2 mm) and of software functionality are repeated. On a daily basis, a small phantom with implanted fiducials is used to verify the machine isocenter is within 2 mm of the Calypso isocenter.

15.7 Quality Assurance for Online Adaptive Correction

Tumor shrinkage and patient weight loss may cause significant anatomic changes in certain patients during radiotherapy. Additionally, systematic positional displacements may be introduced by the planning CT image if the patient's anatomy was not in a "typical" arrangement. When such changes are of significant magnitude relative to the planning margins, potentially serious dosimetric consequences will result. As discussed in Sections 15.5 and 15.6, the development of in-room imaging systems has improved the ability to adapt radiation treatment to anatomical variations and systematic displacements. The adaptive image-guided techniques aim to achieve better planning target volume coverage with less irradiated normal tissues. A complete adaptive image-guided radiotherapy strategy involves daily imaging on the treatment couch, segmentation of structures, reoptimization, and treatment evaluation. In this section and Sections 15.8 and 15.9, QA for adaptive correction will be discussed, first focusing on QA for online adaptive correction and then addressing QA for offline adaptive correction.

In online adaptive correction, time is of critical concern. The simplest and quickest online correction is to reduce any patient alignment errors immediately prior to treatment using various imaging techniques. Patient alignment errors can be determined by registering 2D radiologic images with planning DRRs, or CT images with the planning CT images. Patient setup errors can be corrected by 3D translational shifts or six-dimensional (6D) shifts, including pitch, roll, and yaw. The accuracy of the online adaptive correction carried out by treatment guidance systems is ensured by QA. These treatment guidance systems include, but are not limited to, onboard kV imaging, the BrainLab ExacTrac system, and the Calypso 4D Localization System. The basic QA protocols for these positional imaging systems are discussed in Sections 15.5 and 15.6. When these technologies are used on a daily basis for online adaptive corrections, assuring their accuracy becomes even more crucial. It is prudent to utilize an intercomparison of two alignment technologies to validate the operation of both. In this section, we present an example of a CBCT and BrainLAB ExacTrac intercomparison.

The BrainLAB ExacTrac system can be verified against a CBCT system with a radiosurgery anthropomorphic head phantom (Computerized Imaging Reference Systems, Inc., Norfolk, Virginia), shown in Figure 15.7. The phantom has bone, brain, soft tissue, and air volumes with shapes and attenuation characteristics simulating a human head. The phantom was scanned on a General Electric Lightspeed multislice spiral CT scanner (General Electric Medical Systems, Waukesha, Wisconsin) with a slice thickness of 1.25 mm. Two independent setup plans were generated in BrainLAB and Eclipse (for CBCT), and share the same isocenter and CT images. The phantom was initially set up with the laser marks on the surface, and then repositioned with the ExacTrac system by a 6D shift. After the shift was applied, another two X-ray images were taken for positioning verification immediately followed by a CBCT scan using the high-quality head mode at the slice thickness of 1 mm. The postshift X-ray images were automatically registered with the corresponding DRRs to get the residual errors using the ExacTrac 6D fusion software. The CBCT images were compared with the planning CT images to measure residual errors using the 6D fusion feature in the Varian ARIA Offline Review software (Varian Medical Systems Inc.). The discrepancies between ExacTrac and CBCT should be within 1 mm and 0.5°. Although currently the online adaptive is focused on the geometric correction, it would be desirable in the future to incorporate dosimetric correction by replanning treatment based on the anatomy of the day. The QA needed for this approach will be described in Section 15.8.

FIGURE 15.7 ExacTrac system is verified against CBCT system with a CIRS radiosurgery anthropomorphic head phantom.

15.8 Quality Assurance for Offline Adaptive Correction

Adaptive correction takes information regarding daily patient alignment and uses it to adapt the radiotherapy plan. Online adaptive correction must occur in a short period of time in order to facilitate a reasonable time for patient treatments. Offline adaptive corrections, occurring when the patient is not waiting for treatment, may be more involved. Currently, online adaptive corrections are generally limited to translations and rotations of patients to maintain their anatomical position relative to the radiation delivered. Offline adaptive corrections may utilize the daily positional information to adapt the radiation treatment by either replanning on the original planning CT image with patient-specific margins or potentially replanning on a treatment CT image (from TomoTherapy, CT-on-rails, MV CBCT, or kV CBCT). The daily and planning images may be linked via deformable registration. This section reviews the recommended QA procedures for deformable registration and for dose calculation on a CBCT. The localization accuracy and quality of the daily images are discussed in Section 15.5. The use of CBCT for planning purpose needs to be extremely cautious due to uncertainties in CBCT dosimetry information. Generally, it is not recommended unless strict guidelines are available.

Dose calculation uncertainties in CBCT information are caused by patient setup errors and organ deformation, both discussed in Section 15.7, and by relative electronic density variation. The calibration of CT's HUs to electronic density is accomplished with an electronic density CT phantom. After the initial calibration, AAPM TG no. 66 recommends a monthly constancy check (Mutic et al. 2003). An annual recalibration is the conservative approach (Langen et al. 2005). Because CBCT images are more affected by scatter than fan-beam CT images, it is important that the calibration phantom be constructed of tissue-equivalent material (except for the varying electronic density portions) and that the appropriate diameter is used for the body segment being imaged (Hatten, McCurdy, and Greer 2009). Typically, HU consistency tolerance limits are given as the baseline established during commissioning. Alternatively, energy- and modality-specific tolerances are proposed that will provide ≤2%, 2 mm dose variation (Kilby, Sage, and Rabett 2002).

Deformable image registration may be used to automatically create contours on the daily pretreatment images and to track the dose delivered to the patient. The accuracy of deformable registration can significantly change based on the algorithm and the parameters chosen. The accuracies of eight deformable registration algorithms in use were evaluated at different institutions (Kashani et al. 2008). The average error ranged from 1.5 to 3.9 mm, whereas the maximum error was as large as 15.4 mm. These uncertainties could be clinically significant, particularly when evaluating the dose delivered in a high-gradient region. Currently, there are no standard QA recommendations for deformable image registration. At minimum, an institution should validate its algorithm utilizing a phantom with known deformations (Wang et al. 2005), an image with known simulated deformations (Wang et al. 2005), a comparison of anatomical landmarks (Coselmon et al. 2004; Rietzel and Chen 2006; Voroney et al. 2006), or a comparison of physician-drawn contours with deformed contours (Brock et al. 2005; Wang et al. 2008). The deformable registration algorithm should be revalidated with any change in technology (new CT scanner, change in CT scan parameters, etc.). The AAPM recently formed TG no. 132 to provide additional guidance on this subject.

To ensure the accuracy of the entire adaptive therapy workflow, it is recommended that an end-to-end test be performed periodically. The end-to-end test can be performed with a phantom containing ion chambers, thermoluminescent dosimeters (TLDs), and/or films. The phantom is used to simulate a patient, and is scanned with CT or CBCT. An adaptive treatment plan is generated to deliver an adequate dose to a predefined target. The phantom is then positioned and localized on the treatment system. The adaptive treatment plan is delivered to the phantom after the position of the phantom is verified. The ion chambers and TLDs in the phantom can be used to record the delivered dose, and films in the phantom can be used to verify the localization accuracy.

15.9 Quality Assurance for Radiotherapy Software

In addition to performing QA for the hardware components involved in patient alignment and radiation treatment delivery, it is important to perform QA on the software that operates these systems. Depending on the system involved, this can be quite a complicated task. As a first step, QA protocols may be developed based on the acceptance testing and commissioning of the software. AAPM TG no. 53 enumerates the crucial tasks that must be verified including 2D/3D imaging, 3D beam physical definitions, 3D dose calculations, and functions of all plan evaluation tools (Fraass et al. 1998). For example, methods for verifying the accuracy of deformable registration were discussed in Section 15.8.

15.10 Conclusions

A successful QA program for adaptive radiotherapy shall be carried out by a group of well-trained professionals, including radiation oncologists, radiation oncology physicists, medical dosimetrists, and therapists. Responsibilities and roles in any QA program have been comprehensively discussed in the literature (Kutcher et al. 1994). Furthermore, adequate documentation shall also be conducted as a part of the QA program.

In this chapter, various QA tests were discussed for adaptive radiation therapy, covering machine-specific QA, IMRT QA, gating system QA, X-ray-image-based treatment guidance system QA, non-X-ray-image-based treatment guidance system QA, online adaptive correction QA, and offline adaptive correction QA. Although the implementation of QA tests could vary depending on individual institutions, it is important for

any institute to establish an adequate QA program for adaptive radiotherapy to monitor imaging, dosimetry, and the mechanical accuracy of the system.

References

Almond, P. R., P. J. Biggs, B. M. Coursey, et al. 1999. AAPM's TG-51 protocol for clinical reference dosimetry of high-energy photon and electron beams. *Med Phys* 26:1847–70.

Balog, J., T. R. Mackie, D. Pearson, et al. 2003. Benchmarking beam alignment for a clinical helical tomotherapy device. *Med Phys* 30:1118–27.

Balog, J., G. Olivera, and J. Kapatoes. 2003. Clinical helical tomotherapy commissioning dosimetry. *Med Phys* 30:3097–106.

Balter, J. M., J. N. Wright, L. J. Newell, et al. 2005. Accuracy of a wireless localization system for radiotherapy. *Int J Radiat Oncol Biol Phys* 61:933–7.

Bedford, J. L., Y. K. Lee, P. Wai, et al. 2009. Evaluation of the Delta4 phantom for IMRT and VMAT verification. *Phys Med Biol* 54:N167–76.

Bedford, J. L., and A. P. Warrington. 2009. Commissioning of volumetric modulated arc therapy (VMAT). *Int J Radiat Oncol Biol Phys* 73:537–45.

Bert, C., K. G. Metheany, K. Doppke, et al. 2005. A phantom evaluation of a stereo-vision surface imaging system for radiotherapy patient setup. *Med Phys* 32:2753–62.

Bert, C., K. G. Metheany, K. P. Doppke, et al. 2006. Clinical experience with a 3D surface patient setup system for alignment of partial-breast irradiation patients. *Int J Radiat Oncol Biol Phys* 64:1265–74.

Boyer, A., P. Biggs, J. Galvin, et al. 2001. *Basic Applications of Multileaf Collimators: Report of Task Group No. 50 Radiation Therapy Committee.* Madison, WI: Medical Physics Publishing.

Boyer, A., E. B. Butler, T. A. DiPetrillo, et al. 2001. Intensity-modulated radiotherapy: Current status and issues of interest. *Int J Radiat Oncol Biol Phys* 51:880–914.

Brock, K. K., M. B. Sharp, L. A. Dawson, et al. 2005. Accuracy of finite element model-based multi-organ deformable image registration. *Med Phys* 32:1647–59.

Broggi, S., G. M. Cattaneo, S. Molinelli, et al. 2008. Results of a two-year quality control program for a helical tomotherapy unit. *Radiother Oncol* 86:231–41.

Buonamici, F. B., A. Compagnucci, L. Marrazzo, et al. 2007. An intercomparison between film dosimetry and diode matrix for IMRT quality assurance. *Med Phys* 34:1372–9.

Calypso Medical Technologies, Inc. 2008. *Calypso® 4D Localization System™ User's Manual.* Hannnover, Germany: MDSS GmbH.

Cardenas, A., J. Fontenot, K. M. Forster, et al. 2004. Quality assurance evaluation of delivery of respiratory-gated treatments. *J Appl Clin Med Phys* 5:55–61.

Chen, Y., J. M. Moran, D. A. Roberts, et al. 2007. Performance of a direct-detection activation matrix flat panel dosimeter (AMFPD) for IMRT measurements. *Med Phys* 34:4911–22.

Childress, N. L., R. A. White, C. Bloch, et al. 2005. Retrospective analysis of 2D patient-specific IMRT verifications. *Med Phys* 32:838–50.

Chui, C. S., S. Spirou, and T. LoSasso. 1996. Testing of dynamic multileaf collimation. *Med Phys* 23:635–41.

Coselmon, M. M., J. M. Balter, D. L. McShan, et al. 2004. Mutual information based CT registration of the lung at exhale and inhale breathing states using thin-plate splines. *Med Phys* 31:2942–8.

Court, L., I. I. Rosen, R. Mohan, and L. Dong. 2003. Evaluation of the mechanical precision and alignment uncertainties for an integrated CT/LINAC system. *Med Phys* 30:1198–210.

Das, I., J. C. Cheng, R. J. Watts, et al. 2008. Accelerator beam data commissioning equipment and procedures: Report of the TG-106 of the Therapy Physics Committee of the AAPM. *Med Phys* 35:4186–215.

Dieterich, S., and T. Pawlicki. 2008. Cyberknife image-guided delivery and quality assurance. *Int J Radiat Oncol Biol Phys* 71:S126–30.

Ezzell, G. A., J. M. Galvin, D. Low, et al. 2003. Guidance document on delivery, treatment planning, and clinical implementation of IMRT: Report of the IMRT Subcommittee of the AAPM Radiation Therapy Committee. *Med Phys* 30:2089–115.

Fenwick, J. D., W. A. Tome, H. A. Jaradat, et al. 2004. Quality assurance of a helical tomotherapy machine. *Phys Med Biol* 49:2933–53.

Fraass, B., K. Doppke, M. Hunt, et al. 1998. American Association of Physicists in Medicine Radiation Therapy Committee Task Group 53: Quality assurance for clinical radiotherapy treatment planning. *Med Phys* 25:1773–829.

Hatten, J., B. McCurdy, and P. B. Greer. 2009. Cone beam computerized tomography: The effect of calibration of Hounsfield unit number to electron density on dose calculation accuracy for adaptive radiation therapy. *Phys Med Biol* 54:N329–46.

Herzen, J., M. Todorovic, F. Cremers, et al. 2007. Dosimetric evaluation of a 2D pixel ionization chamber for implementation in clinical routine. *Phys Med Biol* 52:1197–208.

Kashani, R., M. Hub, J. M. Balter, et al. 2008. Objective assessment of deformable image registration in radiotherapy: A multi-institution study. *Med Phys* 35:5944–53.

Keall, P. J., G. S. Mageras, J. M. Balter, et al. 2006. The management of respiratory motion in radiation oncology: Report of AAPM Task Group 76. *Med Phys* 33:3874–900.

Kilby, W., J. Sage, and V. Rabett. 2002. Tolerance levels for quality assurance of electron density values generated from CT in radiotherapy treatment planning. *Phys Med Biol* 47:1485–92.

Klein, E. E., J. Hanley, J. Bayouth, et al. 2009. Task Group 142 report: Quality assurance of medical accelerators. *Med Phys* 36:4197–212.

Kutcher, G. J., L. Coia, M. Gillen, et al. 1994. Comprehensive QA for radiation oncology: Report of AAPM Radiation Therapy Committee Task Group 40. *Med Phys* 21:581–618.

Langen, K. M., S. L. Meeks, D. O. Poole, et al. 2005. The use of megavoltage CT (MVCT) images for dose recomputations. *Phys Med Biol* 50:4259–76.

Letourneau, D., J. Publicover, J. Kozelka, et al. 2009. Novel dosimetric phantom for quality assurance of volumetric modulated arc therapy. *Med Phys* 36:1813–21.

Ling, C. C., P. Zhang, Y. Archambault, et al. 2008. Commissioning and quality assurance of RapidArc radiotherapy delivery system. *Int J Radiat Oncol Biol Phys* 72:575–81.

Lopatiuk-Tirpak, O., K. M. Langen, S. L. Meeks, et al. 2008. Performance evaluation of an improved optical computed tomography polymer gel dosimeter system for 3D dose verification of static and dynamic phantom deliveries. *Med Phys* 35:3847–59.

LoSasso, T., C. S. Chui, and C. C. Ling. 2001. Comprehensive quality assurance for the delivery of intensity modulated radiotherapy with a multileaf collimator used in the dynamic mode. *Med Phys* 28:2209–19.

Low, D. A., W. B. Harms, S. Mutic, et al. 1998. A technique for the quantitative evaluation of dose distributions. *Med Phys* 25:656–61.

Low, D. A., J. W. Sohn, E. E. Klein, et al. 2001. Characterization of a commercial multileaf collimator used for intensity modulated radiation therapy. *Med Phys* 28:752–6.

Lutz, W., K. R. Winston, and N. Maleki. 1988. A system for stereotactic radiosurgery with a linear accelerator. *Int J Radiat Oncol Biol Phys* 14:373–81.

Mutic, S., J. R. Palta, E. K. Butler, et al. 2003. Quality assurance for computed-tomography simulators and the computed-tomography-simulation process: Report of the AAPM Radiation Therapy Committee Task Group No. 66. *Med Phys* 30:2762–92.

Nath, R., P. J. Biggs, F. J. Bova, et al. 1994. AAPM code of practice for radiotherapy accelerators: Report of AAPM Radiation Therapy Task Group No. 45. *Med Phys* 21:1093–121.

Nelms, B. E., and J. A. Simon. 2007. A survey on planar IMRT QA analysis. *J Appl Clin Med Phys* 8:76–90.

Pappas, E., T. G. Maris, F. Zacharopoulou, et al. 2008. Small SRS photon field profile dosimetry performed using a PinPoint air ion chamber, a diamond detector, a novel silicon-diode array (DOSI), and polymer gel dosimetry: Analysis and intercomparison. *Med Phys* 35:4640–8.

Renner, W. D., M. Sarfaraz, M. A. Earl, et al. 2003. A dose delivery verification method for conventional and intensity-modulated radiation therapy using measured field fluence distributions. *Med Phys* 30:2996–3005.

Rietzel, E., and G. T. Y. Chen. 2006. Deformable registration of 4D computed tomography data. *Med Phys* 33:4423–30.

Sadagopan, R., J. A. Bencomo, R. L. Martin, et al. 2009. Characterization and clinical evaluation of a novel IMRT quality assurance system. *J Appl Clin Med Phys* 10(2):104–19.

Sakhalkar, H. S., J. Adamovics, G. Ibbott, et al. 2009. A comprehensive evaluation of the PRESAGE/optical-CT 3D dosimetry system. *Med Phys* 36:71–82.

Santanam, L., K. Malinowski, J. Hubenshmidt, et al. 2008. Fiducial-based translational localization accuracy of electromagnetic tracking system and on-board kilovoltage imaging system. *Int J Radiat Oncol Biol Phys* 70:892–99.

Santanam, L., C. Noel, T. R. Willoughby, et al. 2009. Quality assurance for clinical implementation of an electromagnetic tracking system. *Med Phys* 36:3477–86.

Schell, M., F. J. Bova, D. Larson, et al. 1995. *Stereotactic Radiosurgery: Report of Task Group 42 Radiation Therapy Committee.* Woodbury, New York: American Institute of Physics.

Sonke, J. J., P. Remeijer, and M. van Herk. 2006. In-room cone beam computed tomography. In *Integrating New Technologies into the Clinic: Monte Carlo and Image-Guided Radiation Therapy,* ed. B. H. Curran, J. M. Balter and I. J. Chetty, 543–64. Madison, Wisconsin: Medical Physics Publishing.

Svensson, G. K., N. A. Baily, R. Loevinger, et al. 1984. *Physical Aspects of Quality Assurance in Radiation Therapy, Task Group Report 13.* Woodbury, New York: American Institute of Physics.

Thomas, A. S., C. G. Clift, J. C. O'Daniel, et al. 2009. A 3D solution for advanced photon arc therapy quality assurance. *Med Phys* 36:2748.

Voroney, J. P., K. K. Brock, C. Eccles, et al. 2006. Prospective comparison of computed tomography and magnetic resonance imaging for liver cancer delineation using deformable image registration. *Int J Radiat Oncol Biol Phys* 66:780–91.

Wang, H., L. Dong, J. C. O'Daniel, et al. 2005. Validation of an accelerated "demons" algorithm for deformable image registration in radiation therapy. *Phys Med Biol* 50:2887–905.

Wang, H., A. S. Garden, L. Zhang, et al. 2008. Performance evaluation of automatic anatomy segmentation algorithm on repeat or four-dimensional computed tomography images using deformable image registration method. *Int J Radiat Oncol Biol Phys* 72:210–9.

Willoughby, T. R., P. A. Kupelian, J. Pouliot, et al. 2006. Target localization and real time tracking using the Calypso 4D Localization System in patients with localized prostate cancer. *Int J Radiat Oncol Biol Phys* 65:528–34.

Winston, K. R., and W. Lutz. 1988. Linear accelerator as a neurosurgical tool for stereotactic radiosurgery. *Neurosurgery* 22:454–64.

Wu, Q. J., D. Thongphiew, Z. Wang, et al. 2008. On-line re-optimization of prostate IMRT plans for adaptive radiation therapy. *Phys Med Biol* 53:673–91.

Yan, D., F. Vicini, J. Wong, et al. 1997. Adaptive radiation therapy. *Phys Med Biol* 42:123–32.

Yin, F. F., Z. Wang, S. Yoo, et al. 2006. In-room radiographic imaging for localization. In *Integrating New Technologies into the Clinic: Monte Carlo and Image-Guided Radiation Therapy,* ed. B. H. Curran, J. M. Balter and I. J. Chetty, 491–500. Madison, Wisconsin: Medical Physics Publishing.

Yoo, S., G. Y. Kim, R. Hammoud, et al. 2006. A quality assurance program for the on-board imagers. *Med Phys* 33:4431–47.

III

Clinical Applications

16

Promises of Functional Imaging for Adaptive Radiation Therapy of Central Nervous System Tumors

Yue Cao
University of Michigan

16.1 Introduction

Cancer in the central nervous system consists of primary tumors (e.g., gliomas, meningiomas) and metastatic tumors from various sources throughout the body. Such metastases may manifest in the brain or invade the spinal canal. Treatment may be curative or palliative, with local control of metastases having the potential to significantly impact quality of life for certain patients. Gliomas have proven especially challenging, with very low long-term local control and survival rates from conventional therapies.

Intracranial tumors may be treated with fractionation regimens ranging from 1 (e.g., stereotactic radiosurgery) through 30 or more treatment fractions. Whereas some courses of radiation therapy (RT; e.g., full cranial irradiation) are not geometrically conformal, others involve extensive planning to deliver high tumor doses while attempting to spare either individual structures within the brain and head or to reduce the overall volume of the brain exposed to high doses. Intensity-modulated RT (IMRT) has been applied to the treatment of multiple intracranial metastases, large paraspinal tumors, and gliomas, with some investigations of the latter using radiation intensity variations intended to deliver a higher dose per fraction to the primary gross target volume (GTV) while delivering a lower but still substantial dose to a large surrounding clinical target volume (CTV) in hopes of improving treatment effectiveness by irradiating a larger region without dramatically increasing the risk of treatment-related toxicity (Wang et al. 2007; Clark et al. 2010; Hsu et al. 2009; Thilmann et al. 2001).

The geometric localization of tumors in the brain for treatment in general is very precise. Whereas immobilization for typical treatments may reasonably reproduce the patients' position as well as restrict their movement, the advent of precise and image-guided

processes for positioning patients has provided the additional needed infrastructure for high accuracy. Stereotactic radiosurgery has the potential to provide accuracies of 2 mm or better, and more conventional image-guided localization approaches even this benchmark, with cone-beam computed tomography (CBCT) having improved the ability to resolve position variations due to residual rotations of patients inside of masks.

In general, and specifically for more-hypofractionated treatments, little, if any, deformation is expected in the patient geometry on the brain over the course of treatment. The goal of adaptive therapy for these cases is to consider dose arrangements prior to treatment, or modifications of total dose and distribution during treatment, that are based on the variations of tissue (normal and tumor) properties as assessed by molecular and physiological imaging processes. This chapter explores some of the indications for therapy modification, as well as some of the methods under investigation to provide guidance for treatment customization.

For spinal and paraspinal treatments, there is some limited potential for geometric modification in addition to individualization based on biological and response metrics. This is due to the less-rigid regional geometry involving some of the larger tumors. This chapter does not discuss such geometric variations, but the issues are the same as those discussed in Chapters 12 and 13.

16.2 Indications for Therapy Modification

Molecular, functional, and metabolic imaging has been tested intensively for biomarkers of therapy responses and outcomes. The general investigational hypothesis is that biological imaging has superior sensitivity and specificity over biological processes induced by therapy in indicating early response to treatment, normal tissue injury, or both. These indications can be prognostic or predictive. Prognostic indicators can be utilized to stratify patients for different treatment strategies, whereas predictive indicators can be used for therapy modification, for example, treatment intensification for nonresponders. Similarly, imaging has been used to assess individual variations in sensitivity to radiation-induced normal tissue toxicity, and thereby to adjust radiation doses to organs at risk.

Several imaging techniques have demonstrated prognostic or predictive value for response of brain tumors to therapy and as such may serve as indicators for therapy modification. The most promising predictive imaging techniques for brain tumors include proton magnetic resonance (MR) spectroscopy imaging (MRSI), MR-based perfusion and vascular permeability imaging, ^{11}C-methionine (^{11}C-MET) positron emission tomography (PET), and ^{123}I-methyl-tyrosine single-photon CT (SPECT). The detailed methodologies for image acquisition using these techniques have been reviewed in the literature (Tofts et al. 1999; Ostergaard et al. 1999; Nelson et al. 2002; Cao, Sundgren, et al. 2006; Jacobs, Dittmar, et al. 2002), and thus the focus of this chapter is on the value of these imaging techniques and their derived quantitative metrics for adaptive therapy of brain tumors.

16.2.1 Prognostic Indicators from Functional Imaging

16.2.1.1 Tumor Vasculature by Dynamic Contrast-Enhanced Magnetic Resonance Imaging

One of the important approaches for prognosis arises from characterizing brain tumor vascularization and perfusion.

Normal cerebral vasculature provides oxygen and nutrients to and removes metabolic by-products from tissue in order to maintain brain functions. Similarly, tumor cell growth and proliferation depend on adequate tumor vasculature. The rapid growth of the malignant tumor drives the genesis of blood vessels, resulting in immature, leaky, and compromised vascular networks in the tumor. Thus, the neovascularity of glioma may represent one aspect of tumor aggressiveness, and could be a prognostic factor for outcome prior to or early in the course of a therapeutic regimen.

Perfusion and vascular permeability of brain tumors have been assessed by dynamic contrast-enhanced (DCE) or dynamic susceptibility contrast (DSC) MR imaging (MRI). Quantitative metrics, such as perfusion, vascular permeability, and blood volume, can be derived from the DCE/DSC MRI and pharmacokinetic models.

Neovascularization exhibits significant heterogeneity in a typical glioma. A rapidly growing periphery forms a rim, manifesting high vascular volume, high blood flow, and highly leaky vasculature, whereas the central portion of the tumor consists of a core with regressed vessels exhibiting both low perfusion and low vascular volume. Despite the challenges encountered in the analysis of heterogeneous vascular properties of high-grade gliomas, several studies have shown that vascular volume and permeability of the gliomas prior to RT predict outcome. A prospective MRI perfusion study of 23 patients with high-grade gliomas showed that the fractional tumor volume with high vascular density prior to RT predicts survival (Cao, Tsien, et al. 2006; Figure 16.1). The extent of leaky vasculature in the high-grade gliomas before chemoradiation treatment has been shown to predict survival better than the tumor volumes defined in anatomic MRI (Cao, Negash, et al. 2006). A recent retrospective

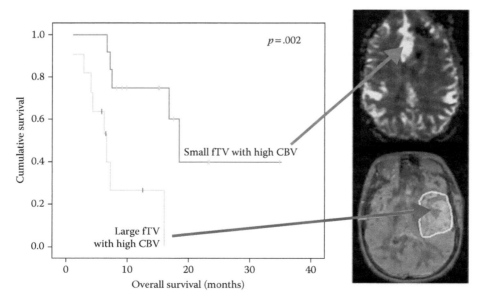

FIGURE 16.1   (See color insert following page 204.) Survival curves stratified based on the fractional tumor volumes (fTV) with elevated cerebral blood volume (CBV) prior to radiation therapy. Patients with a small fractional tumor volume with high CBV had better chances of survival than those with a large fractional tumor volume with high CBV. Hot colors (red and yellow) indicate high values of CBV.

analysis of 189 patients with high- or low-grade gliomas shows that the mean relative blood volume of the tumor prior to therapy is a prognostic factor for time to tumor progression, independent of the histological grade (Law et al. 2008). These positive clinical findings provide enticing support for the use of the vascular abnormality derived from DCE or DSC MRI in defining a target volume for radiation dose boosting, although there is no thorough analysis to date of the association of patterns of failure with the vascular abnormality prior to treatment.

16.2.1.2 Choline Metabolism by Magnetic Resonance Proton Spectroscopy Imaging

Abnormal chemical compounds and metabolites in brain tumors, particularly those containing choline, have been investigated by two-dimensional (2D) or three-dimensional (3D) MRSI (Wald et al. 1997; Preul et al. 1996; Negendank et al. 1996). It has been found that choline signals increase and N-acetylaspartate (NAA) decreases in malignant gliomas. Further, these elevated choline signals are often found beyond the extent of the abnormality defined on post-Gd T1-weighted images or even T2 or fluid attenuation inversion recovery (FLAIR) images (Li et al. 2002; Vigneron et al. 2001; Pirzkall et al. 2004; Figure 16.2). Histological studies confirm that the choline peak signals are normalized to contralateral creatine and choline or ipsilateral NAA are correlated with tumor cell densities (Croteau et al. 2001; McKnight et al. 2002).

The prognostic values of choline signals in malignant gliomas measured by MRSI as well as in radiation treatment planning have been evaluated by several studies (Oh et al. 2004; Graves et al. 2000; Chan et al. 2004; Park et al. 2007). A study evaluated the spatial overlap between the target volume defined for gamma knife radiosurgery and the metabolic abnormalities defined by a choline and NAA index (CNI) in 26 patients with recurrent glioblastoma multiforme (GBM) (Chan et al. 2004). This investigation found that a 50% or greater volumetric overlap of radiosurgical targets with metabolic lesions resulted in better survival and longer time to progression. In another study of 28 patients with newly diagnosed GBM treated by fractionated RT and chemotherapy, the metabolic abnormality volume defined by CNI > 2.5 within the T2 hyperintense abnormality volume pre-RT predicts survival (Oh et al. 2004). In another study from the same group, the spatial association between the postoperative but pre-RT metabolic abnormality defined by CNI > 2 and the new contrast-enhancing lesions developed after RT was evaluated (Pirzkall et al. 2004). Out of 10 patients who did not have contrast-enhanced lesions after resection and prior to RT, 8 developed new contrast-enhancing lesions after RT. The area of the new contrast-enhancing lesions was clearly located in the pre-RT CNI abnormality in four patients, appeared in close proximity in three, and distant in one. The volume of the pre-RT CNI abnormality appeared to be associated with the time to presentation of new enhancing lesions. However, for the other 17 patients who had contrast-enhancing lesions prior to RT, there was no report on the relationship between pattern of failure and pre-RT CNI abnormality. In another study of nine patients who had GBM and were treated with RT, at the

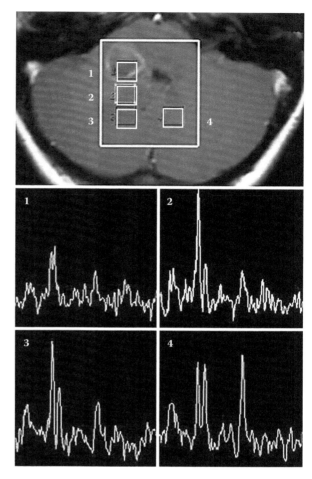

FIGURE 16.2 Proton spectra of a patient with newly diagnosed glioblastoma multiform. Compared to the spectrum of normal tissue (voxel 4), both choline and N-acetylaspartate (NAA) signals were decreased in voxel 1 located within the contrast-enhanced tumor, possibly due to the mixing of tumor cells and necrosis. For the spectra in voxels 2 and 3, which extended posteriorly beyond the contrast-enhanced tumor, choline signals were elevated but NAA signals were decreased. (Courtesy of Pia Sundgren.)

time of relapse 82% of voxels with metabolic abnormality defined by a ratio of choline to NAA (CNR) > 2 pre-RT corresponded with either persistent or new contrast-enhancing lesions, and 71% of the voxels with CNR > 2 pre-RT were still in the area of the CNR abnormality (Laprie et al. 2008).

Although choline abnormalities in malignant gliomas have been reported for over a decade, there is no prospective study to date that evaluates the clinical value of MRSI for target definition and/or boost volume definition of radiation treatment. However, some of the challenges encountered in the use of MRSI for radiation treatment planning have been discussed in the literature (Pirzkall et al. 2004; Pirzkall et al. 2001). One concern is that, although in approximately 60% of patients with malignant gliomas CNI2 (CNI at greater than two standard deviations) abnormality exists beyond the T2 hyperintense abnormality, the volume of CNI2 is rather small (7–8 cm³ in average), preferentially followed by white matter (WM) fibers ipsilaterally, and in some cases contralaterally (Pirzkall et al. 2004). A uniform expansion of the margin can result in both overdose in the region of normal tissue and underdose in the tumor. Another concern is which level of CNI or other metabolic indices should be outlined (Pirzkall et al. 2001). The third concern pointed out by Pirzkall et al. (2004) is that a 2.5-mm margin needs to be added in order to compensate for the uncertainty arising from the limited spatial resolution of MRSI, which is usually 1 cm^3 per voxel. Finally, increasing the robustness of 3D MRSI techniques is still a challenge that must be overcome in order to propagate this technique in typical clinical environments.

16.2.1.3 Metabolic Positron Emission Tomography and Single-Photon Computed Tomography

Several metabolic PET or SPECT tracers, including 18F-fluorodeoxyglucose (FDG; Holzer et al. 1993; Kaschten et al. 1998; Herholz et al. 1993; Holthoff et al. 1993), ^{11}C-MET (Herholz, Kracht, and Heiss 2003), ^{123}I-L-alpha-methyltyrosine (IMT), and 3′-deoxy-3′-(18)F-fluorothymidine (FLT; Jacobs et al. 2005) have been investigated in patients with gliomas (Jacobs, Dittmar, et al. 2002; Jacobs, Winkler, et al. 2002).

FDG is a Food and Drug Administration (FDA)-approved commercially available PET tracer. It has been studied intensively for gliomas since a decade ago (Holzer et al. 1993; Herholz et al. 1993; Holthoff et al. 1993). Its applications in the treatment of gliomas are hindered by the poor ratio of tumor to background brain tissue due to the high level of uptake of FDG in a normal brain cortex.

MET is a natural amino acid avidly taken up by glioma cells, but with very little uptake in normal brain tissue. MET uptake is probably facilitated by the increased activation of the L-mediated and A-mediated amino acid transport located in the endothelial cell membrane (Kracht et al. 2003). A study of MET PET with stereotactic biopsy in gliomas has shown that not only does increased MET uptake occur in solid tumors but also MET PET is able to detect infiltrating tumor tissues with a sensitivity of 87% and a specificity of 89% (Kracht et al. 2004). Also, MET uptake has been correlated with microvascular density in gliomas (Kracht et al. 2003).

Several studies have compared the spatial extent of MET uptake with the MRI abnormality in gliomas (Voges et al. 1997; Grosu, Weber, Riedel, et al. 2005). In 63%–74% of the patients studied, increased MET uptake was found beyond the contrast-enhancing lesions and at distances up to 45 mm from the contrast-enhanced periphery. In 50% of these patients, MET uptake was seen even beyond the T2 hyperintense abnormality (Grosu et al. 2005). Conversely, MRI enhancing lesions were also found beyond the space defined by increased MET uptake, consistent with observed discrepancies between tumor volumes defined by MRI and MET PET in gliomas.

Preoperative evaluation of 54 gliomas by PET with FDG and ^{11}C-MET shows that for both tracers the ratios of uptakes of the tumor to normal cortical/WM region are correlated with survival (Kaschten et al. 1998). In this study, MET had stronger predictive power, possibly due to better differentiation of the tumor from the background tissue uptake (Kaschten et al. 1998). The prognostic value of SPECT imaging using the amino acid analog IMT is more apparent in patients who had resection of gliomas 4–6 weeks before SPECT, as the tumor-to-tissue uptake ratio predicts survival, but not in the patients who had biopsy only (Weber et al. 2001). The next question is whether the high amino acid tracer uptake region in the glioma prior to therapy presents a higher risk for failure. A prospective study from the University of Michigan evaluated this possibility (Lee et al. 2009). This investigation defined MET GTV as a region in which the ratio of tumor to normal tissue was 1.5 or greater. Such MET GTVs, when incompletely encompassed by high radiation doses that were planned based on MRI and CT, resulted in higher risk for noncentral failure than those fully covered by high doses, suggesting a treatment planning strategy that incorporates MET PET in GTV delineation. Finally, a study developed and tested a treatment strategy for recurrent high-grade gliomas using stereotactic hypofractionated reirradiation (5 × 6 Gy) based on MET PET/IMT SPECT for the GTV definition (Grosu, Weber, Franz, et al. 2005). In 84% of the 44 patients who had recurrent gliomas, MET PET or IMT SPECT was available for treatment planning and was registered with CT and MRI. The GTVs included the increased amino acid tracer uptake, but no expansion to create the CTVs. The planning target volume encompassed the GTV plus a 3-mm margin. The investigators found that treatment planning based on the combination of MET PET (IMT SPECT), MRI, and CT was associated with improved survival in comparison to treatment planning using CT and MRI alone. However, further multivariate analysis showed that time of the interval to retreatment and temozolomide remained significant prognostic factors for survival (Grosu, Weber, Franz, et al. 2005).

A clinical question is whether we can use MET PET for adaptive RT. Although there is no published data to date on clinical trials of plan modification for RT, MET PET has been used to monitor the metabolic response of chemotherapy for malignant gliomas (Herholz, Kracht, and Heiss 2003; Galldiks et al. 2006). In a study of the effect of temozolomide treatment, the patients with a decline in MET uptake during therapy (after three cycles of temozolomide treatment) had either stable disease or longer time to progression than those with an increase in MET uptake (Galldiks et al. 2006). Although not definitive, such a finding provides compelling evidence for the potential role of MET PET in target definition for RT boosting following chemotherapy, as well as in intratherapeutic modification during a course of RT.

16.2.2 Adaptive Therapy Based on Functional Imaging

The potential for effective modification of radiation patterns or total dose based on events that occur after the beginning of treatment is unclear. One thing that is certain is that if adaptive therapy of central nervous system (CNS) tumors is to be effective, it will be guided by nonmorphological response measures.

Tumor volume changes in gliomas during the course of RT are minimal, and are therefore not a sensitive indicator for response. Molecular, functional, and metabolic imaging has greater potential for the early identification of tumors that respond to RT.

As discussed in Section 16.2.1.1, vascular permeability, blood volume, and blood flow in gliomas assessed by DCE and DSC MRI are prognostic factors for response. Vasculature in high-grade gliomas exhibits significant heterogeneity with blood volume and flow high in the peripheral region, which can be possibly correlated with rapid tumor growth, and low in the region likely consistent with vasogenic edema or necrotic tissue. In a prospective study of 23 patients with high-grade gliomas receiving fractionated RT, investigators at the University of Michigan found that elevation in the low-blood-volume region at week 1 vs. pre-RT and reduction in the high-blood-volume region at week 3 vs. week 1 were associated with better survival (Cao, Tsien, et al. 2006; Figure 16.3). This study suggests that the changes in tumor blood volume during the course of RT have the potential to indicate which tumor is responsive to RT. Further, subvolumes of tumors with high cerebral blood volume (CBV) might be resistant to radiation and might benefit from intensified RT. Another MRI technique, diffusion MRI, may also be sensitive in predicting response and outcome of high-grade gliomas. The fractional tumor volume exhibiting increased apparent diffusion coefficients at 3 weeks was the strongest predictor of patient survival at 1 year (Moffat et al. 2005; Hamstra et al. 2008). Diffusion imaging may be valuable for early indication of tumor response vs. resistance to RT.

There are a number of difficulties associated with using the aforementioned methodologies for intracourse adjustment of radiation patterns or total dose for CNS tumors. The diffusion-based response map, which presents a patched or Swiss-cheese-like pattern, may not be sufficiently specific for geometric definition of the subvolume with high tumor cellular densities and, therefore, for delineation of a radiation boost volume. Perfusion-based CBV measures are strongly correlated with pretherapeutic measures and those occurring early in the

course of treatment, particularly for nonresponsive tumors. The aims of therapy modification in the middle of the treatment course are two-fold, specifically (1) increasing the intensity of treatment for nonresponsive tumors and (2) decreasing the potential toxicity for treatment of responsive tumors without compromising tumor control. Functional imaging acquired in the early course of treatment can aid in identifying responsive vs. nonresponsive tumors as well as the related responding or nonresponding subtumor volumes. There are increasing data to support the role of functional imaging in this area. However, the difference between the radiation boost volumes defined on functional images acquired pre-RT and during the course of RT for the nonresponsive tumor is usually not large enough to proceed with the modification of the therapy. For the responsive tumor, adapting treatment to focus on the redefined radiation boost volume in the middle of the treatment course has the potential to reduce doses in normal tissue. However, risks and benefits of modification have to be worked out given the uncertainties in the acquisition of functional imaging parameters extracted by image processing, treatment planning, patient setup, and treatment delivery.

16.2.3 Image-Based Toxicity Assessment

Normal tissue toxicity is a major concern in brain irradiation. Clinical reactions have been divided into acute (days to weeks after irradiation), early delayed (2–6 months after irradiation), and late (6 months to years after the completion of radiation treatment; Schultheiss et al. 1995; Sheline, Wara, and Smith 1980; Hopewell and Wright 1970). Acute reactions are usually transient and often controlled or relieved by corticosteroids. The neurological symptoms occurring in the period of 2–6 months after irradiation are believed to be primarily due to intralesional reactions, probably indicative of tumor response, and/or perilesional reactions, such as edema or demyelination in the WM. However, WM lesions in the periventricular region also start to appear on conventional MRI or CT during this interval (Constine et al. 1988; Tsuruda et al. 1987). Necrosis, particularly in the WM, near or within the high-dose region, begins to develop in this time interval, following high doses and large brain volume radiation exposure (Van Tassel et al. 1995; Kumar et al. 2000). The late effect includes focal or diffuse necrosis, which becomes apparent a year after the completion of RT. Similarly, WM lesions are found beyond the high-dose region 12 months after RT. Complications include worsening neurological signs or symptoms, seizures, and increased intracranial pressure. Recently, attention has been given to neurocognitive dysfunction in memory and executive function as a late radiation effect (Klein et al. 2002; Brown et al. 2003). Following whole brain irradiation, substantial decline of memory function occurs 4–6 months after the completion of RT (Chang et al. 2008). It is not clear whether the brain tissue necrosis and the decline in cognitive functions are reversible. There is no effective treatment yet for brain tissue necrosis following irradiation; therefore, it is important to understand the progressive time course of these late

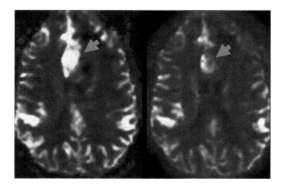

FIGURE 16.3 (See color insert following page 204.) Cerebral blood volume (CBV) maps in a patient with high-grade glioma prior to radiation therapy (RT; left) and 3 weeks after starting fractionated RT (right). Note the reduction of high CBV in the tumor volume after 3 weeks of fractionated RT as an early indicator of response.

effects as well as their dose and dose-volume relationships, and thereby adapt the treatment to attempt to avoid radiation injury in critical structures and connecting networks. Molecular, functional, and metabolic imaging can be effective biomarkers for predicting late effects (Cao et al. 2009; Sundgren and Cao 2009; Sundgren et al. 2009).

Diffusion tensor imaging (DTI) is the most sensitive technique that assesses WM integrity and histopathologic changes before structural changes are visible on any other imaging modalities. DTI is able to assess water diffusion as well as anisotropic diffusion in tissue structures (Basser and Pierpaoli 1996; Basser and Pierpaoli 1998; Song et al. 2002; Khong et al. 2005). Diffusion tensor imaging has been used to assess WM injury in pediatric and adult patients treated with brain radiation. In children, craniospinal dose as well as the age at treatment is definitely found to affect the integrity of WM fibers in the temporal lobe, hippocampus, and thalamus, which is assessed by fractional anisotropy (FA) of DTI (Khong et al. 2005; Dellani et al. 2008). It seems that the decreased integrity of WM in pediatric patients treated with craniospinal radiation for cancers compared to normal age-matched children has an effect on intelligence quotient (IQ) scores, after effects of age at treatment, craniospinal dose, and time interval since treatment are accounted for (Khong et al. 2006). Currently, these findings are from the cited cross-sectional studies. Prospective studies with long-term follow-ups are required to test the hypotheses generated by these preliminary investigations.

In adult patients who undergo partial or whole brain RT, similar to children, radiation effects on normal-appearing WM have been detected by DTI during RT and after the completion of RT (Nagesh et al. 2008; Welzel et al. 2008). In a longitudinal study of 25 patients who were treated for brain tumors with focal brain RT, progressive decreases in FA from the start of RT to 45 weeks after the start of therapy were observed in large WM fibers of the genu and splenium of the corpus callosum (Nagesh et al. 2008). Most importantly, the decrease in FA was dependent on dose. Further analysis showed progressive increases in diffusivity perpendicular to the long axis of axons but little increase in parallel diffusivity, suggesting demyelination predominant after WM irradiation. In another study of 26 patients who underwent prophylactic cranial irradiation, decreases in FA of several WM anatomic sites, including frontal WM, corona radiate, and cerebellum, were observed at the end of RT and 6 weeks after RT, the extent of which appears to depend on risk factors of vascular diseases (Welzel et al. 2008). However, the question of how these observed WM changes are associated with neurocognitive function changes remains to be answered.

Cerebral microvasculature is sensitive to radiation (Cao et al. 2009; Taki et al. 2002; Hahn et al. 2009). Histopathologic studies reveal that lifting of endothelia from the basement membrane, dilation and thickening of blood vessels, endothelial cell nuclear enlargement, and hypertrophy of perivascular astrocytes are among the first effects of irradiation (Calvo et al. 1988; Reinhold et al. 1990; Okeda et al. 2003). The initial injury of vessels is followed by the formation of platelet matrix and thrombi, which eventually results in occlusion and thrombosis in microvessels within weeks to months (Verheij et al. 1994; Belka et al. 2001). Cerebral microvascular injury has been long recognized to occur acutely, and precedes subacute demyelination and reactive astrocytic and microglial responses (Price et al. 2001; Ljubimova et al. 1991; Pena, Fuks, and Kolesnick 2000; Li et al. 2003). Therefore, studying vascular injury such as dose and dose-volume effects, time course, and neurocognitive consequence and its relationship to WM injury could be fruitful for understanding radiation-induced neurotoxicity (Cao et al. 2009; Hahn et al. 2009). In a prospective study of DCE MRI for the prediction of radiation-induced neurocognitive dysfunction (Cao et al. 2009), vascular volumes and blood-brain barrier (BBB) permeability were increased significantly in the high-dose regions during RT, followed by a decrease after RT. Changes in both vascular volume and BBB permeability correlated with the doses accumulated at the time of scans at weeks 3 and 6 during RT and 1 month after RT. The effect of the dose volume on the vascular volume was also observed as a large effect of high doses with a large brain volume that is receiving the high dose. These early vascular effects of RT are also correlated significantly with verbal learning scores 6 months after RT. These data suggest that the early changes in cerebral vasculature may predict delayed alterations in verbal learning and total recall, which are important components of neurocognitive functions. In another study using FDG and ^{15}O-PET, decreased FDG uptake and increased perfusion in the brain region receiving a dose greater than 40 Gy were observed 3 weeks after RT. The Wisconsin card sorting test showed a significant correlation of decreased FDG uptake with increased errors and perseveration in test performance (Hahn et al. 2009).

These data demonstrate that dose and dose volume have effects on cerebral tissue and neurocognitive functions, and such effects can be detected early and monitored longitudinally by various imaging techniques. However, more work needs to be done to determine what structures and interconnected WM networks are sensitive to various radiation doses, thereby providing guidance for radiation treatment planning and delivery.

16.3 Summary

There have been a large variety of intriguing studies on the use of various forms of functional imaging to modify radiation delivery to CNS tumors, most notably for gliomas. The findings to date strongly support the customization of treatment prior to the start of therapy to optimize tumor doses and limit toxicity risk and provide support for the possibility of images acquired early in the course of treatment to further reduce individual patient risk of toxicity. The translation of these early findings to large-scale clinical trials and eventually to standards of practice is still in a very early stage, and significant changes will be needed in the imaging and image analysis tools used for RT for these findings to be of practical use.

References

Basser, P. J., and C. Pierpaoli. 1996. Microstructural and physiological features of tissues elucidated by quantitative-diffusion-tensor MRI. *J Magn Reson B* 111:209–19.

Basser, P. J., and C. Pierpaoli. 1998. A simplified method to measure the diffusion tensor from seven MR images. *Magn Reson Med* 39:928–34.

Belka, C., W. Budach, R. D. Kortmann, and M. Bamberg. 2001. Radiation induced CNS toxicity: Molecular and cellular mechanisms. *Br J Cancer* 85:1233–9.

Brown, P. D., J. C. Buckner, and J. R. O'Fallon, et al. 2003. Effects of RT on cognitive function in patients with low-grade glioma measured by the folstein mini-mental state examination. *J Clin Onco* 21:2519–24.

Calvo, W., J. W. Hopewell, H. S. Reinhold, and T. K. Yeung. 1988. Time- and dose-related changes in the white matter of the rat brain after single doses of X-rays. *Br J Radiol* 61:1043–52.

Cao, Y., V. Nagesh, and D. Hamstra, et al. 2006. The extent and severity of vascular leakage as evidence of tumor aggressiveness in high-grade gliomas. *Cancer Res* 66:8912–7.

Cao, Y., P. C. Sundgren, C. I. Tsien, T. T. Chenevert, and L. Junck. 2006. Physiologic and metabolic magnetic resonance imaging in gliomas. *J Clin Oncol* 24:1228–35.

Cao, Y., C. I. Tsien, and V. Nagesh, et al. 2006. Clinical investigation survival prediction in high-grade gliomas by MRI perfusion before and during early stage of RT. *Int J Radiat Oncol Biol Phys* 64:876–85.

Cao, Y., C. I. Tsien, and P. Sundgren, et al. 2009. DCE MRI as a biomarker for prediction of radiation-induced neurocognitive dysfunction. *Clinical Cancer Research* 15:1747–54.

Chan, A. A., A. Lau, and A. Pirzkall, et al. 2004. Proton magnetic resonance spectroscopy imaging in the evaluation of patients undergoing gamma knife surgery for Grade IV glioma. *J Neurosurg* 101:467–75.

Chang, E. L., J. S. Wefel, K. R. Hess, et al. 2009. Neurocognition in patients with brain metastases treated with radiosurgery or radiosurgery plus whole-brain irradiation: a randomised controlled trial. *Lancet Oncol* 10:1037–44.

Clark, G. M., R. A. Popple, P. E. Young, and J. B. Fiveash. 2010. Feasibility of single-isocenter volumetric modulated arc radiosurgery for treatment of multiple brain metastases. *Int J Radiat Oncol Biol Phys* 76:296–302.

Constine, L. S., A. Konski, S. Ekholm, S. McDonald, and P. Rubin. 1988. Adverse effects of brain irradiation correlated with MR and CT imaging. *Int J Radiat Oncol Biol Phys* 15:319–30.

Croteau, D., L. Scarpace, and D. Hearshen, et al. 2001. Correlation between magnetic resonance spectroscopy imaging and image-guided biopsies: Semiquantitative and qualitative histopathological analyses of patients with untreated glioma. *Neurosurgery* 49:823–9.

Dellani, P. R., S. Eder, and J. Gawehn, et al. 2008. Late structural alterations of cerebral white matter in long-term survivors of childhood leukemia. *J Magn Reson Imaging* 27:1250–5.

Galldiks, N., L. W. Kracht, and L. Burghaus, et al. 2006. Use of ^{11}C-methionine PET to monitor the effects of temozolomide chemotherapy in malignant gliomas. *Eur J Nucl Med Mol Imaging* 33:516–24.

Graves, E. E., S. J. Nelson, and D. B. Vigneron, et al. 2000. A preliminary study of the prognostic value of proton magnetic resonance spectroscopic imaging in gamma knife radiosurgery of recurrent malignant gliomas. *Neurosurgery* 46:319–26; discussion 326–18.

Grosu, A. L., W. A. Weber, and M. Franz, et al. 2005. Reirradiation of recurrent high-grade gliomas using amino acid PET (SPECT)/CT/MRI image fusion to determine gross tumor volume for stereotactic fractionated RT. *Int J Radiat Oncol Biol Phys* 63:511–9.

Grosu, A. L., W. A. Weber, and E. Riedel, et al. 2005. L-(methyl-11C) methionine positron emission tomography for target delineation in resected high-grade gliomas before RT. *Int J Radiat Oncol Biol Phys* 63:64–74.

Hahn, C. A., S. M. Zhou, and R. Raynor, et al. 2009. Dose-dependent effects of radiation therapy on cerebral blood flow, metabolism, and neurocognitive dysfunction. *Int J Radiat Oncol Biol Phys* 73:1082–7.

Hamstra, D. A., K. C. Lee, B. A. Moffat, T. L. Chenevert, A. Rehemtulla, and B. D. Ross. 2008. Diffusion magnetic resonance imaging: An imaging treatment response biomarker to chemoradiotherapy in a mouse model of squamous cell cancer of the head and neck. *Transl Oncol* 1:187–94.

Herholz, K., L. W. Kracht, and W. D. Heiss. 2003. Monitoring the effect of chemotherapy in a mixed glioma by C-11-methionine PET. *J Neuroimaging* 13:269–71.

Herholz, K., U. Pietrzyk, and J. Voges, et al. 1993. Correlation of glucose consumption and tumor cell density in astrocytomas: A stereotactic PET study. *J Neurosurg* 79:853–8.

Holthoff, V. A., K. Herholz, and F. Berthold, et al. 1993. In vivo metabolism of childhood posterior fossa tumors and primitive neuroectodermal tumors before and after treatment. *Cancer* 72:1394–403.

Holzer, T., K. Herholz, J. Jeske, and W. D. Heiss. 1993. FDG-PET as a prognostic indicator in radiochemotherapy of glioblastoma. *J Comput Assist Tomogr* 17:681–7.

Hopewell, J. W., and E. A. Wright. 1970. The nature of latent cerebral irradiation damage and its modification by hypertension. *Br J Radiol* 43:161–7.

Hsu, F., H. Carolan, and A. Nichol, et al. 2010. Whole brain RT with hippocampal avoidance and simultaneous integrated boost for 1–3 brain metastases: A feasibility study using volumetric modulated arc therapy. *Int J Radiat Oncol Biol Phys* 76(5):1480–5.

Jacobs, A. H., C. Dittmar, A. Winkler, G. Garlip, and W. D. Heiss. 2002. Molecular imaging of gliomas. *Mol Imaging* 1:309–35.

Jacobs, A. H., A. Thomas, and L. W. Kracht, et al. 2005. 18F-fluoro-L-thymidine and 11C-methylmethionine as markers of increased transport and proliferation in brain tumors. *J Nucl Med* 46:1948–58.

Jacobs, A. H., A. Winkler, and C. Dittmar, et al. 2002. Molecular and functional imaging technology for the development of efficient treatment strategies for gliomas. *Technol Cancer Res Treat* 1:187–204.

Kaschten, B., A. Stevenaert, and B. Sadzot, et al. 1998. Preoperative evaluation of 54 gliomas by PET with fluorine-18-fluorodeoxyglucose and/or carbon-11-methionine. *J Nucl Med* 39:778–85.

Khong, P. L., L. H. Leung, and G. C. Chan, et al. 2005. White matter anisotropy in childhood medulloblastoma survivors: Association with neurotoxicity risk factors. *Radiology* 236:647–52.

Khong, P. L., L. H. Leung, and A. S. Fung, et al. 2006. White matter anisotropy in post-treatment childhood cancer survivors: Preliminary evidence of association with neurocognitive function. *J Clin Oncol* 24:884–90.

Klein, M., J. J. Heimans, and N. K. Aaronson, et al. 2002. Effect of RT and other treatment-related factors on mid-term to long-term cognitive sequelae in low-grade gliomas: A comparative study. *Lancet* 360:1361–8.

Kracht, L. W., M. Friese, and K. Herholz, et al. 2003. Methyl-[11C]-l-methionine uptake as measured by positron emission tomography correlates to microvessel density in patients with glioma. *Eur J Nucl Med Mol Imaging* 30:868–73.

Kracht, L. W., H. Miletic, and S. Busch, et al. 2004. Delineation of brain tumor extent with [11C]L-methionine positron emission tomography: Local comparison with stereotactic histopathology. *Clin Cancer Res* 10:7163–70.

Kumar, A. J., N. E. Leeds, and G. N. Fuller, et al. 2000. Malignant gliomas: MR imaging spectrum of radiation therapy- and chemotherapy-induced necrosis of the brain after treatment. *Radiology* 217:377–84.

Laprie, A., I. Catalaa, and E. Cassol, et al. 2008. Proton magnetic resonance spectroscopic imaging in newly diagnosed glioblastoma: Predictive value for the site of postradiotherapy relapse in a prospective longitudinal study. *Int J Radiat Oncol Biol Phys* 70:773–81.

Law, M., R. J. Young, and J. S. Babb, et al. 2008. Gliomas: Predicting time to progression or survival with cerebral blood volume measurements at dynamic susceptibility-weighted contrast-enhanced perfusion MR imaging. *Radiology* 247:490–8.

Lee, IH, M. Piert, and D. Gomez-Hassan, et al. 2009. Association of 11C-methionine PET uptake with site of failure after concurrent temozolomide and radiation for primary glioblastoma multiforme. *Int J Radiat Oncol Biol Phys* 73:479–85.

Li, Y. Q., P. Chen, A. Haimovitz-Friedman, R. M. Reilly, and C. S. Wong. 2003. Endothelial apoptosis initiates acute blood-brain barrier disruption after ionizing radiation. *Cancer Res* 63:5950–6.

Li, X., Y. Lu, A. Pirzkall, T. McKnight, and S. J. Nelson. 2002. Analysis of the spatial characteristics of metabolic abnormalities in newly diagnosed glioma patients. *J Magn Reson Imaging* 16:229–37.

Ljubimova, N. V., M. K. Levitman, E. D. Plotnikova, and L. Eidus. 1991. Endothelial cell population dynamics in rat brain after local irradiation. *Br J Radiol* 64:934–40.

McKnight, T. R., M. H. von dem Bussche, and D. B. Vigneron, et al. 2002. Histopathological validation of a three-dimensional magnetic resonance spectroscopy index as a predictor of tumor presence. *J Neurosurg* 97:794–802.

Moffat, B. A., T. L. Chenevert, and T. S. Lawrence, et al. 2005. Functional diffusion map: A noninvasive MRI biomarker for early stratification of clinical brain tumor response. *Proc Natl Acad Sci USA* 102:5524–9.

Nagesh, V., C. I. Tsien, and T. L. Chenevert, et al. 2008. Radiation-induced changes in normal-appearing white matter in patients with cerebral tumors: A diffusion tensor imaging study. *Int J Radiat Oncol Biol Phys* 70:1002–10.

Negendank, W. G., R. Sauter, and T. R. Brown, et al. 1996. Proton magnetic resonance spectroscopy in patients with glial tumors: A multicenter study. *J Neurosurg* 84:449–458.

Nelson, S. J., E. Graves, and A. Pirzkall, et al. 2002. In vivo molecular imaging for planning radiation therapy of gliomas: An application of 1H MRSI. *J Magn Reson Imaging* 16:464–76.

Oh, J., R. G. Henry, and A. Pirzkall, et al. 2004. Survival analysis in patients with glioblastoma multiforme: Predictive value of choline-to-N-acetylaspartate index, apparent diffusion coefficient, and relative cerebral blood volume. *J Magn Reson Imaging* 19:546–54.

Okeda, R., S. Okada, A. Kawano, S. Matsushita, and T. Kuroiwa. 2003. Neuropathology of delayed encephalopathy in cats induced by heavy-ion irradiation. *J Radiat Res (Tokyo)* 44:345–52.

Ostergaard, L., F. H. Hochberg, and J. D. Rabinov, et al. 1999. Early changes measured by magnetic resonance imaging in cerebral blood flow, blood volume, and blood-brain barrier permeability following dexamethasone treatment in patients with brain tumors. *J Neurosurg* 90:300–5.

Park, I., G. Tamai, and M. C. Lee, et al. 2007. Patterns of recurrence analysis in newly diagnosed glioblastoma multiforme after three-dimensional conformal radiation therapy with respect to pre-radiation therapy magnetic resonance spectroscopic findings. *Int J Radiat Oncol Biol Phys* 69:381–9.

Pena, L. A., Z. Fuks, and R. N. Kolesnick. 2000. Radiation-induced apoptosis of endothelial cells in the murine central nervous system: Protection by fibroblast growth factor and sphingomyelinase deficiency. *Cancer Res* 60:321–7.

Pirzkall, A., X. Li, and J. Oh, et al. 2004. 3D MRSI for resected high-grade gliomas before RT: Tumor extent according to metabolic activity in relation to MRI. *Int J Radiat Oncol Biol Phys* 59:126–37.

Pirzkall, A., T. R. McKnight, and E. E. Graves, et al. 2001. MR-spectroscopy guided target delineation for high-grade gliomas. *Int J Radiat Oncol Biol Phys* 50:915–28.

Preul, M. C., Z. Caramanos, and D. L. Collins, et al. 1996. Accurate, noninvasive diagnosis of human brain tumors by using proton magnetic resonance spectroscopy. *Nat Med* 2:323–5.

Price, R. E., L. A. Langford, E. F. Jackson, L. C. Stephens, P. T. Tinkey, and K. K. Ang. 2001. Radiation-induced morphologic changes in the rhesus monkey (*Macaca mulatta*) brain. *J Med Primatol* 30:81–7.

Reinhold, H. S., W. Calvo, J. W. Hopewell, and A. P. van der Berg. 1990. Development of blood vessel-related radiation damage in the fimbria of the central nervous system. *Int J Radiat Oncol Biol Phys* 18:37–42.

Schultheiss, T. E., L. E. Kun, K. K. Ang, and L. C. Stephens. 1995. Radiation response of the central nervous system. *Int J Radiat Oncol Biol Phys* 31:1093–112.

Sheline, G. E., W. M. Wara, and V. Smith. 1980. Therapeutic irradiation and brain injury. *Int J Radiat Oncol Biol Phys* 6:1215–28.

Song, S. K., S. W. Sun, M. J. Ramsbottom, C. Chang, J. Russell, and A. H. Cross. 2002. Dysmyelination revealed through MRI as increased radial (but unchanged axial) diffusion of water. *Neuroimage* 17:1429–36.

Sundgren, P. C., and Y. Cao. 2009. Brain irradiation: Effects on normal brain parenchyma and radiation injury. *Neuroimaging Clin N Am* 19:657–68.

Sundgren, P. C., V. Nagesh, and A. Elias, et al. 2009. Metabolic alterations: A biomarker for radiation-induced normal brain injury—an MR spectroscopy study. *J Magn Reson Imaging* 29:291–7.

Taki, S., K. Higashi, and M. Oguchi, et al. 2002. Changes in regional cerebral blood flow in irradiated regions and normal brain after stereotactic radiosurgery. *Ann Nucl Med* 16:273–7.

Thilmann, C., A. Zabel, K. H. Grosser, A. Hoess, M. Wannenmacher, and J. Debus. 2001. Intensity-modulated RT with an integrated boost to the macroscopic tumor volume in the treatment of high-grade gliomas. *Int J Cancer* 96:341–9.

Tofts, P. S., G. Brix, and D. L. Buckley, et al. 1999. Estimating kinetic parameters from dynamic contrast-enhanced T(1)-weighted MRI of a diffusable tracer: Standardized quantities and symbols. *J Magn Reson Imaging* 10:223–32.

Tsuruda, J. S., K. E. Kortman, W. G. Bradley, D. C. Wheeler, W. Van Dalsem, and T. P. Bradley. 1987. Radiation effects on cerebral white matter: MR evaluation. *AJR Am J Roentgenol* 149:165–71.

Van Tassel, P., J. M. Bruner, and M. H. Maor, et al. 1995. MR of toxic effects of accelerated fractionation radiation therapy and carboplatin chemotherapy for malignant gliomas. *AJNR Am J Neuroradiol* 16:715–26.

Verheij, M., L. G. Dewit, M. N. Boomgaard, H. J. Brinkman, and J. A. van Mourik. 1994. Ionizing radiation enhances platelet adhesion to the extracellular matrix of human endothelial cells by an increase in the release of von Willebrand factor. *Radiat Res* 137:202–7.

Vigneron, D., A. Bollen, and M. McDermott, et al. 2001. Three-dimensional magnetic resonance spectroscopic imaging of histologically confirmed brain tumors. *Magn Reson Imaging* 19:89–101.

Voges, J., K. Herholz, and T. Holzer, et al. 1997. 11C-methionine and 18F-2-fluorodeoxyglucose positron emission tomography: A tool for diagnosis of cerebral glioma and monitoring after brachytherapy with 125I seeds. *Stereotact Funct Neurosurg* 69:129–35.

Wald, L. L., S. J. Nelson, and M. R. Day, et al. 1997. Serial proton magnetic resonance spectroscopy imaging of glioblastoma multiforme after brachytherapy. *J Neurosurg* 87:525–34.

Wang, S. J., M. Choi, C. D. Fuller, B. J. Salter, and M. Fuss. 2007. Intensity-modulated radiosurgery for patients with brain metastases: A mature outcomes analysis. *Technol Cancer Res Treat* 6:161–8.

Weber, W. A., S. Dick, and G. Reidl, et al. 2001. Correlation between postoperative 3-[(123)I]iodo-L-alpha-methyltyrosine uptake and survival in patients with gliomas. *J Nucl Med* 42:1144–50.

Welzel, T., A. Niethammer, and U. Mende, et al. 2008. Diffusion tensor imaging screening of radiation-induced changes in the white matter after prophylactic cranial irradiation of patients with small cell lung cancer: First results of a prospective study. *AJNR Am J Neuroradiol* 29:379–83.

Adaptive Radiation Therapy for Head and Neck Cancer

Lei Dong
The University of Texas M.D. Anderson Cancer Center

David L. Schwartz
The University of Texas M.D. Anderson Cancer Center

17.1 Introduction

In this chapter, we review characteristics of patients' anatomy changes for head and neck cancers, and evaluate different adaptive radiotherapy scenarios.

17.1.1 Clinical Rationales and Typical Treatment Approaches for Head and Neck Radiotherapy

With intensity-modulated radiation therapy (IMRT; Purdy 2001; Wu et al. 2000), it is possible to treat very complicated target volumes at variable dose levels while sparing adjacent normal structures. This form of advanced radiation therapy technique has long been realized as an outstanding method for treating head and neck (H&N) cancer. It has been shown that IMRT is the most commonly used modality for H&N radiotherapy (RT; Mell, Roeske, and Mundt 2003). Implementation of IMRT requires adequate selection and delineation of target volumes based on different imaging modalities, appropriate specification and dose prescription regarding dose–volume constraints, and proper knowledge of setup uncertainties.

One key requirement for IMRT is target definition, as inverse planning algorithms used in IMRT require target and avoidance information to optimize deliverable beam parameters. The importance of target delineation in H&N has been subject to many prior studies and discussions (Gregoire, Levendag, et al. 2003; Chao et al. 2007; Gregoire, De Neve, et al. 2007; Eisbruch and Gregoire 2009; Nangia et al. 2009). The gross target volume (GTV) determines the volume containing the highest tumor cell density and receiving the highest prescribed dose. The GTV is typically contoured based on computed tomography (CT) images (contrasted or noncontrasted) or magnetic resonance (MR) images, or with the assistance of 18F-fluorodeoxyglucose (FDG)-positron emission tomography (PET) images (Ashamalla et al. 2007; Ashamalla et al. 2005). The high-risk clinical treatment volume (CTV) for gross disease is defined by a margin surrounding the GTV (Eisbruch et al. 2002). The CTV for lymph node groups are contoured based on nodal groups at risk (Eisbruch and Gregoire 2009; Gregoire, Levendag, et al. 2003; Gregoire et al. 2006).

In addition to target delineation, normal-tissue contouring is also important to minimize the dose to adjacent normal structures. It is well-known that radiation can also cause side effects, such as xerostomia, dysphagia, and aspiration. It has been demonstrated that saliva production is closely related to the irradiated volumes (Eisbruch et al. 1999; Chao, Deasy, et al. 2001). Similarly, dysphagia and aspiration are also related to irradiated volumes for certain functional structures (Eisbruch et al. 2004; Eisbruch et al. 2007; Feng et al. 2007). The mean dose and the partial volumes receiving specified doses can be determined from dose–volume histograms (DVHs). It was found that the use of IMRT to meet parotid dose constraints improves both objective and subjective measures of xerostomia (Chao, Majhail, et al. 2001; Lee et al. 2002; Eisbruch et al. 2003). It is apparent that the partial sparing of the salivary glands, made possible by IMRT, achieves substantial gains both in the retention of salivary production and in the symptoms of xerostomia. Eisbruch and colleagues (Eisbruch et al. 2004; Feng et al. 2007) also associated dose with the superior pharyngeal constrictors and supraglottic larynx with postradiation aspiration. Therefore, it is important to apply adequate dose constraints to normal structures in order to minimize treatment-related toxicities.

In most situations, IMRT for H&N cancer is designed to use the simultaneous integrated boost (SIB) technique (Wu et al. 2000), which is designed to deliver radiation to several targets at the same time with different intensities for the same fractionation schedule. This means the dose per fraction will be different for different targets. A typical IMRT plan for tonsil cancer treatment is shown in Figure 17.1. The CTV targets are shown in different color-filled contours. We can see that extremely conformal IMRT plans can be designed. Prescription isodose lines tightly wrap around each target volume. Nevertheless, nearby normal structures still received doses that were not insignificant due to their proximity to target volumes and physical limitations in dose fall-off gradient. The ability to deliver such high dose conformality throughout a treatment course is critical to maintain the theoretical advantage of IMRT for H&N treatment.

17.1.2 Margins and Treatment Designs— Balancing the Risk and Benefit

Although IMRT shows promise as a radiation procedure aimed at increasing therapeutic gain, in the H&N area it still presents a number of challenges and issues that have yet to be resolved. One such challenge is the margin design. Due to setup errors and anatomical changes, it is necessary to treat a larger volume: the planning target volume (PTV), which accounts for treatment uncertainties. Unfortunately, the use of PTV will inevitably overlap with adjacent normal structures, which creates less-than-ideal IMRT plans. Therefore, dosimetric compromises between increasing target doses and minimizing adjacent organ doses are inevitable. In order to maximize the therapeutic index, accurate knowledge of setup uncertainties and normal organ dose tolerances, and the optimal selection and delineation of treatment targets are required.

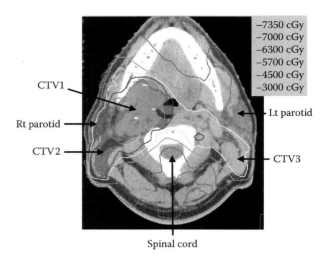

FIGURE 17.1 (See color insert following page 204.) Intensity-modulated radiation therapy plan for head and neck cancer treatment. A typical plan consists of several target volumes to be treated at different dose levels and adjacent normal critical structures. In this case, the primary clinical target volume (CTV1) is to be treated at 7000 cGys, the intermediate-risk target (CTV2) at 6300 cGys, and the low-risk target (CTV3) at 5700 cGys, respectively. The normal structures include the left and right parotid, spinal cord, and oral cavity (not shown) in this axial view. Due to the close proximity of normal structures, treatment design is usually a compromise between target coverage and the dose tolerance of nearby normal structures. High conformality in dose distribution is desired to achieve a high therapeutic ratio.

With highly conformal treatments such as IMRT, marginal tumor recurrence is the highest risk associated with either inadequate target volume delineation or overly optimistic margin selection. Therefore, it is critical to address the risk of target miss and balance it with organ-sparing goals. The questions that remain to be answered, which are the focus of this chapter, are

- How much and what are the characteristics of setup errors in the H&N region?
- What are the dosimetric consequences?
- How much and how quickly can patient anatomy (both target volumes and normal structures) change during a course of treatment in approximately 6–7 weeks?
- Is a treatment plan "What you see is what you get"?
- Will adaptive replanning and treatment improve the therapeutic index for an individual patient?
- What treatment strategies are available to improve existing treatment procedures?
- What are the clinical challenges for adaptive RT of H&N cancer patients?

17.2 Setup Uncertainties

17.2.1 Setup Uncertainties and Immobilization Techniques

Numerous publications have presented data on setup uncertainties for H&N cancers (Hong et al. 2005; Prisciandaro et al. 2004;

Gregoire, Daisne, et al. 2003; Manning et al. 2001; Hatherly et al. 2001; Bel et al. 1995; Willner et al. 1997; Karger et al. 2001), and setup correction protocols have been proposed to improve the effectiveness of online corrections (Brock, McShan, and Balter 2002; de Boer et al. 2001; Bel et al. 2000; Pisani et al. 2000; Hatherly, Smylie, and Rodger 1999). Typically, H&N cancer patients are immobilized using thermoplastic masks, which are custom made to fit an individual patient. In addition, a head rest will be used to ensure a patient's neck position is reproducible (Linthout et al. 2006; Boda-Heggemann et al. 2006). Other immobilization techniques, such as bite blocks and vacuum pillows, are occasionally used. In the cited early reports, portal images were exclusively used to measure setup accuracy. Portal imaging is a two-dimensional (2D) projection X-ray technique that uses megavoltage (MV) high-energy X-rays and usually produces poor image quality at therapeutic energies. With the recent availability of kilovoltage (kV) X-rays mounted orthogonal to the therapy beam line, it is possible to acquire high-quality setup images for daily image guidance (Li et al. 2008). One additional advantage of in-room kV imaging is that the acquired projection images are very similar to the digitally reconstructed radiographs (DRRs) calculated using the simulation CT. This potentially reduces the systematic error due to the appearance difference in the reference DRR and the MV portal images. However, being a 2D technique, the method assumes that the same object is imaged and measured using the two project images, which may not be true. At different gantry angles, the same bony landmark may not be as visible as other bony landmarks. As a result, people are generally using different bony landmarks, instead of the same one, to calculate three-dimensional (3D) shifts, which can introduce additional errors if the relative positions of the two landmarks are changed.

17.2.2 Relative Movement of Bony Anatomy

Perhaps the most significant development in recent years is the use of in-room 3D imaging devices for daily setup. In-room CT scanners, tomotherapy-based MVCT, and the latest gantry-mounted cone-beam CT have been introduced as the means for 3D imaging–based verification just prior to treatment with the patient immobilized in the treatment position (Court et al. 2003; Jaffray et al. 1999; Kuriyama et al. 2003; Mackie et al. 2003).

One unique requirement for 3D verification is the selection of a region of interest (ROI) to determine its shift relative to the reference treatment plan. However, H&N anatomy is complicated by the semi-independent movement of various bony components. The rigid skull is attached to a semirigid mandible and successive levels of the cervical spine and upper thoracic spine, which are attached by flexible ligaments (Ahn et al. 2009). Ahn et al. (2009) have estimated that there are at least 54 degrees of freedom in the movement in the H&N region. Zhang et al. (2006) were the first to use in-room CT scanner and 3D–3D image registration to analyze the relative movements of different ROIs. Three bony ROIs were defined: (1) C2 vertebral bodies, (2) C6 vertebral bodies, and (3) the palatine process of the maxilla (PPM), which is attached to the base of the skull. Although shifts were found to be highly correlated relative to the marked isocenter on the immobilization mask, noticeable differences on the order of 2–6 mm were observed between any two ROIs, indicating the flexibility, rotational effect, or both in the H&N region. Ahn et al. (2009) also observed that the largest positional variability measured by means of vector displacement was in the mandible and lower cervical spine.

The most extensive study on setup uncertainties in subregions of the H&N anatomy was performed by the Netherlands Cancer Institute (van Kranen et al. 2009). In this study, eight ROIs were defined and analyzed for 38 patients who had received IMRT treatment and had routine cone-beam CT scans during their treatments. Figure 17.2a shows the eight subregions in the H&N site in the sagittal plane, which include the occiput bone; upper, central, and lower parts of the neck; vertebra caudal of C7; jugular notch; larynx; and mandible. The large dotted box in Figure 17.2a represents the clinically used ROI for patient setup. This includes almost the entire H&N area, including the lower neck. Figure 17.2b

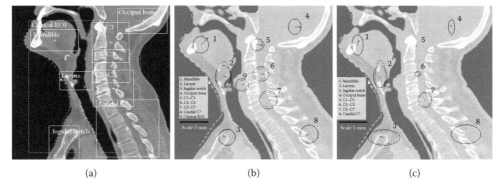

(a)	(b)	(c)

FIGURE 17.2 The eight subregions in the head and neck site chosen to study relative setup variations in 38 patients. (a) The large dotted box represents the clinically used region of interest (ROI) for patient positioning. (b) The residual setup uncertainties in local regions if the single large ROI was chosen as the reference ROI for patient setup. The systematic and random errors are plotted as dotted and solid ellipses, respectively, with the length of the major axes corresponding to one standard deviation. (c) The residual setup uncertainties if the vertebrae C1–C3 were chosen as the reference for patient setup. Note increase in uncertainty with longer distances from the reference ROI. (Reprinted from van Kranen, S., et al. 2009. *Int J Radiat Oncol Biol Phys* 73:1566–73. With permission.)

shows the residual setup uncertainties in these local regions if the single large ROI was chosen as the reference ROI for patient setup. The systematic and random errors are plotted as dotted and solid ellipses. It can be seen that different subregions move differently due to the flexing of bony structures in various regions. The local residual error can be up to 3.4 mm (one standard deviation [SD]). This translates to a population-based geometric margin of approximately 4–7 mm even under daily image guidance. The distribution of residual error is closely related to which ROI will be used as the reference ROI. It would be beneficial to use an ROI close to the primary tumor site (Zhang et al. 2006). For example, the vertebrae C1–C3 can be chosen as the reference for typical oropharyngeal cancer patients. The residual error distribution for this situation is shown in Figure 17.2c. It is noted that there are increased uncertainties with longer distances from the reference ROI. This study also agrees with other studies (Zhang et al. 2006; Ahn et al. 2009) in that the largest uncertainties appear to be in the lower neck and the mandibular regions.

17.2.3 Intrafractional Changes in Certain Head and Neck Anatomy

It is worth noting that the largest systematic setup uncertainty in H&N region occurs in the larynx, as can be seen from Figure 17.2. Typically, intrafractional motion is not an issue for bony structures in H&N patients because patients are usually immobilized using masks with the assistance of a head rest and perhaps bite blocks during each treatment session. However, internal motion due to swallowing and tongue movement can be significant. A study by Zhang et al. (2006) showed that the use of a stent (mouthpiece) had an immobilization effect on structures near the oral cavity, which resulted in the reduction of setup uncertainties (especially in the superior–inferior [SI] direction). Unfortunately, there was no effective immobilization of the larynx.

There are many early studies regarding the movement of the larynx. Because both the larynx and the base of the tongue are attached to the hyoid bone, tongue movement may also result in displacement of the larynx. Inspiration and expiration also causes displacement of the larynx, although the displacement is usually small during normal breathing. Using videofluoroscopy, the swallowing of liquid bolus showed that the larynx can move 20–25 mm in the cranial–caudal direction and 3–8 mm in the anterior–posterior (AP) direction (Cook et al. 1989; Dantas et al. 1990; Leonard et al. 2000). The duration of swallowing for normal subjects is approximately 1 second, and the frequency of swallowing in the supine position is once every 1–2 minutes (Kendall et al. 1998; Hamlet, Ezzell, and Aref 1994). The intrafractional motions of the larynx can be studied using an electronic portal imager (van Asselen et al. 2003). Images are obtained every 200 milliseconds, resulting in a movie of images for each beam. Overall, van Asselen et al. (2003) found that the incidence and total duration of swallowing was low, and will not likely cause any dosimetric deviations during treatment. Nevertheless, they found quite significant movement in this region. During 95% of the irradiation time, the tip of the epiglottis moves within a range of 7.1 mm.

Nonperiodic soft-tissue variations can also be studied using in-room 3D imaging. At the MD Anderson Cancer Center in Texas, researchers investigated internal systematic and random errors caused by swallowing. Taking advantage of a fast helical CT scanner (CT-on-rails) in the treatment room, the position of thyroid cartilage (as a landmark for larynx) relative to that of hyoid bone (landmark for lower oropharynx) was analyzed in 17 oropharyngeal patients prospectively enrolled in an institutional image-guided adaptive RT protocol. A total of 555 daily CTs were analyzed (approximately 33 CTs per patient). It was found that the hyoid bone and the thyroid cartilage moved significantly during the treatment course (up to 16 mm) relative to the vertebrae, most notably in the SI direction. The systematic errors relative to baseline position from simulation for individual patients were large (up to 12 mm), with a SD for SI error of 4.5 mm. Median absolute random shifts (SD) for the hyoid were 3.3 (2.9) mm, 1.4 (1.4) mm, and 1.3 (1.1) mm in the SI, AP, and left–right directions, respectively.

The effect of the systematic error is illustrated in Figure 17.3. Because the swallowing action is usually infrequent and is of short duration (approximately 1 second; van Asselen et al. 2003; Dantas et al. 1990; Hamlet, Ezzell, and Aref 1994), its intrafractional dosimetric effect is much less than its systematic impact: The simulation CT could be biased toward an infrequent anatomical pose as the consequence of swallowing. The larynx may receive a higher dose during treatment if the hyoid or the thyroid

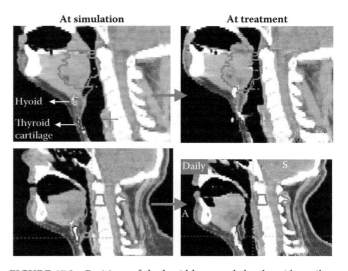

FIGURE 17.3 Positions of the hyoid bone and the thyroid cartilage at various computed tomography (CT) scanning times. The positions can change noticeably during the simulation or during the course of treatment. Because the swallowing action is usually infrequent and is of short duration, its intrafractional dosimetric effect is much less than its systematic effect. The simulation CT could be biased toward an infrequent anatomical pose as a consequence of swallowing. If the hyoid or cartilage is captured at the most inferior position (top row), the larynx may receive a higher dose during treatment. In the opposite situation, if the hyoid and cartilage are captured at the most superior position during CT simulation (bottom row), a primary target near the base of tongue could be underdosed during treatment.

cartilage is inadvertently captured at the most inferior position (top row of Figure 17.3). In the opposite situation, if the hyoid and thyroid are captured at the most superior position during CT simulation (bottom row), a primary target near the base of the tongue could be underdosed during treatment.

17.3 Anatomical Changes during Treatment Course

It has long been recognized that some patients receiving radiation therapy to the H&N region will have significant anatomical changes during their course of treatment, including shrinking primary tumors or nodal masses, postsurgical changes/edema, and changes in overall body habitus/weight loss (Suit and Walker 1980; Barkley and Fletcher 1977; Sobel et al. 1976; Trott 1983). An example of such a case is shown in Figure 17.4. Large anatomical changes will render the original IMRT plan less conformal than what was originally intended.

17.3.1 Volumetric and Positional Changes in Gross Target Volume

Barker et al. (2004) used repeat in-room CT to study GTV changes. Computed tomography scans were acquired three times a week in 14 patients. Manual contouring was used to evaluate GTV changes. It was found that GTV decreased throughout fractionated radiation therapy, at a median rate of 0.2 cm³ per treatment day (range: 0.01–1.95 cm³/day). Figure 17.5a shows the volume change for primary tumors, and Figure 17.5c shows the volume change for lymph nodes. When measured as a percentage of initial volume, GTVs decreased at a median rate of 1.8% (of initial volume) per treatment day (range: 0.2%–3.1%

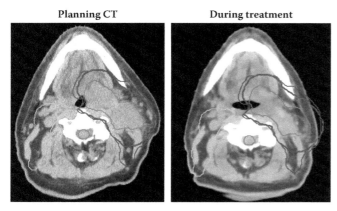

Planning CT   **During treatment**

FIGURE 17.4 Anatomical changes can be significant during the course of treatment for head and neck cancer patients. In this example, the planning computed tomography (CT) scan and clinical target volume (CTV) contours are shown on the left. The midcourse CT (3 weeks into treatment) shows significant volume reduction. The same CTV contour (the overlays after rigid registration) does not match well with the CT image on the right. A part of the CTV was in the air, which made the initial intensity-modulated radiation therapy plan suboptimal.

per day). Figure 17.5b and d shows the percentage of volume loss during the course of radiation therapy for primary tumors and treated lymph nodes. Because the graphs are normalized to the initial volume, the curves show a greater variation. It was found that both the primary tumors and the involved lymph nodes (treated as GTV) were losing volume at approximately the same rate of 1.7%–1.8% per treatment day. On the last day of radiation treatment, this corresponded to a median total relative loss of approximately 70% of the initial GTV (range: 10%–92%). It should be noted, however, that the interobserver variation in contouring the GTV could be relatively substantial for small primary tumors or nodes.

It is also interesting to observe that the rate of volume loss is highly proportional to the initial target volume. Figure 17.6 shows a linear relationship (Barker et al. 2004). It is important to point out that such a relationship could help to identify candidate patients who may benefit from the adaptive RT procedure.

17.3.2 Volumetric and Positional Changes in Normal-Tissue Structures

It is noted that parotid glands also decrease in volume during the course of radiation therapy. Barker et al. (2004) observed that the median parotid volume loss was 0.2 cm³/day or 0.6%/day of the initial volume (range: 0.04–0.84 cm³/day or 0.2%–1.8%/day). At the end of treatment, the median parotid volume loss was 28.1% (range: from 5.9% to 53.6%). The center of mass for the parotid gland also shifts medially to both sides over time. At the end of treatment, the median parotid medial shift was 3.1 mm (range: −0.3–9.9 mm for 14 cases). Figure 17.7 shows an example of parotid volume variations during a course of treatment using the daily in-room CT for each fraction.

Lee et al. (2008b) used tomotherapy MV-CT scans acquired during daily treatment to assess geometric changes in parotid glands. A deformable registration algorithm was applied to 330 daily MV-CT images (10 patients) to create deformed parotid contours. The day-to-day variations (1 SD of errors) in the center-of-mass distance and volume were 1.61 mm and 4.36%, respectively. The volumes tended to decrease with a median total loss of 21.3% (6.7%–31.5%) and a median change rate of 0.7%/day (0.4%–1.3%/day). Parotids migrated toward the patient center with a median total distance change of −5.26 mm (0.00–16.35 mm) and a median change rate of −0.22 mm/day (0.02 to −0.56 mm/day).

A repeat CT imaging study was performed by a group from the University of Florida, in which 82 H&N cancer patients underwent 4 CT scans: (1) one before RT, (2) one after 3 weeks or at the 15th fraction, (3) one upon completing RT, and (4) one at 2 months after RT (Wang et al. 2009). The average volume loss in parotid glands after 3 weeks of RT, at the end of RT, and 2 months after RT vs. volume loss before RT were 20.01%, 26.93%, and 27.21%, respectively. Parotid and submandibular glands did not continue to shrink after completion of RT. These gland volume reductions correlated significantly with the mean dose to the irradiated glands: volume loss at higher doses (> 30 Gys) to the glands was significantly larger than that at low doses (< 30 Gys; $p < .001$).

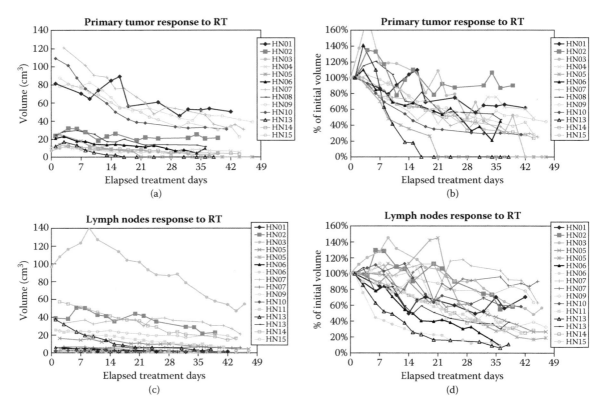

FIGURE 17.5 Gross tumor volume (GTV) changes over time among patients with head and neck cancers: Both primary tumor ((a) and (b)), and lymph nodes greater than 2 cm³ of volume ((c) and (d)) show similar trends. The GTVs decreased at a median rate of 0.2 cm³ or 1.8% of initial volume per treatment day. (Reprinted from Barker Jr., J. L., et al. 2004. *Int J Radiat Oncol Biol Phys* 59:960–70. With permission.)

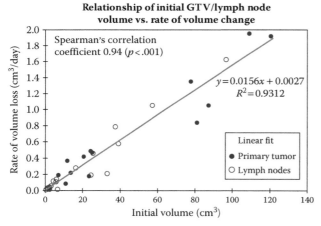

FIGURE 17.6 Relationship between initial target volume and the rate of volume loss for a group of head and neck cancer patients receiving radiation therapy. Initial volume was measured by the first computed tomography scan. The solid circles represent primary tumors, and the open circles represent lymph nodes. The rate of volume change is highly correlated with the initial volumes measured ($p < .001$). This relationship indicates that patients with large initial GTVs or lymph nodes are usually good candidates for adaptive radiotherapy because their anatomical changes tend to be more significant during treatment. (Reprinted from Barker Jr., J. L., et al. 2004. *Int J Radiat Oncol Biol Phys* 59:960–70. With permission.)

17.3.3 Weight Loss and Volume Changes

Nearly all patients lost weight throughout their course of radiation or chemoradiation. Barker et al. (2004) found that the median weight change from the start to completion of treatment was 7.1% (range: +5.2% to −13%) in their study. Reductions in external skin contours at the level of the C2 reference point and at the base of the skull were found to highly correlate with weight loss. Median weight loss correlates significantly ($p < .001$) with median parotid medial displacement over time ($p < .001$; Barker et al. 2004). This indicates that skin contours and weight loss could be used as indicators for significant anatomical changes.

17.4 Image-Guided Approaches for Head and Neck Cancer

The term IGRT (image-guided radiation therapy) is broadly interpreted as a set of imaging technologies that allow treatment decisions to be made on the basis of acquired images. Image guidance is usually defined as the use of in-room imaging to make setup corrections, in particular, positional shifts of the treatment target. Image guidance usually does not involve the modification of the original treatment plan, which means that IGRT aims at the correction of setup errors and the reduction of CTV-to-PTV margin. Although there are other approaches for

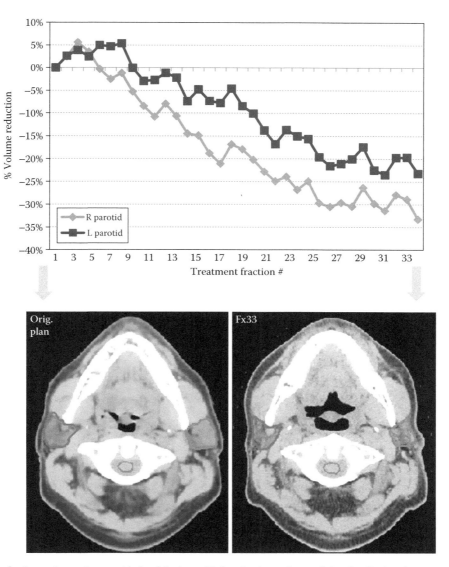

FIGURE 17.7 Example of volume change in parotid gland during a 33-fraction intensity-modulated radiation therapy treatment. The top figure shows the percent of volume change for each parotid as a function of treatment fraction. The bottom pictures show an axial computed tomography slice of the parotid before radiotherapy (RT; bottom left) and after 33 fractions of RT (bottom right).

H&N patient setup, in-room stereoscopic X-ray imaging and the recent volumetric imaging are two major techniques of modern RT of H&N patients.

17.4.1 Daily Image-Guided Setup with (2D) Radiographic X-Ray Imaging and (3D) Volumetric Imaging

Radiographic X-ray imaging uses two projection X-ray images to determine the shift of a 3D object in space. The configuration of a modern kV imaging system usually consists of a gantry-mounted X-ray source, which is orthogonal to the therapeutic X-ray beam direction (Jaffray and Siewerdsen 2000). The volumetric imaging system is based on the tomographic imaging

principle, which is to acquire many projections at different gantry angles and use an image reconstruction algorithm to restore the 3D representation of the true object. Due to the high contrast of bony structures and stationary anatomy, cone-beam CT usually gives sufficient image quality for daily image guidance or treatment planning (Yoo and Yin 2006; Seet et al. 2009; Thomas et al. 2009).

17.4.2 2D vs. 3D Setup for Head and Neck Treatment

In general, 2D imaging is much quicker than 3D imaging in acquiring projection images at two gantry angles; reviewing such 2D images is also a much more familiar process for physicians or therapists (similar to the review of portal films). As indicated in Section 17.2.1, a 2D projection method assumes that the same object can be identified in two projection angles. However, this may not be the case because the same bony landmark may not be easily identified in these two views. People have been using different bony landmarks that might be more visible for each projection X-ray image. In addition, out-of-plane rotations cannot be easily identifiable in 2D imaging. On the other hand, 3D images by CT techniques more closely depict the actual anatomy than do the 2D kV image pairs. Li at al. (2008) compared 2D alignment and 3D alignment and concluded that differences between the two alignments were mainly caused by the relative flexibility of certain H&N structures and possibly by rotation.

17.4.3 Alignment Strategies for Head and Neck Image-Guided Radiotherapy

The IGRT workflow for H&N patients is not different from that for other treatment sites, with an exception that an ROI should be explicitly identified by the treating physician as a part of a patient's treatment directive. Previous studies have provided sufficient evidence that the relative movement of different ROIs can be significant in the H&N region, as illustrated by Figure 17.2. Ignoring these differences in different regions would introduce additional setup uncertainties (van Kranen et al. 2009; Li et al. 2008). Therefore, it is important to realize this requirement. This ROI-based alignment strategy should be adopted for each H&N patient using IGRT for daily setup corrections.

The selection of an ROI depends on the treatment case, which is a balance between target coverage and sparing of normal structure. Typically, spinal cord sparing is critical for most H&N cases; therefore, vertebral body is a good choice as an alignment object, in general. Two typical examples of ROI selection are shown in Figure 17.8. The first case is an oropharyngeal patient. The appropriate ROI is the C1–C3 vertebral body, which is shown in red contours in Figure 17.8 (top row). The second example is a sinus case (bottom row), in which the PTV is perhaps a more important structure for direct alignment than

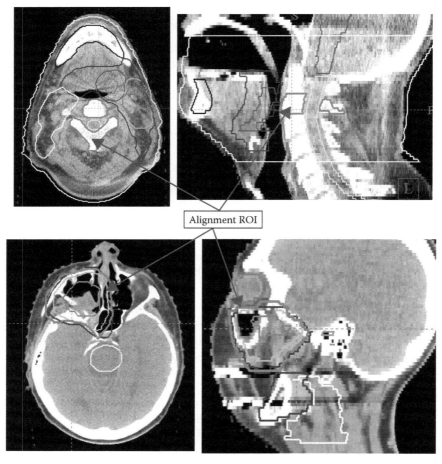

Alignment ROI

FIGURE 17.8 Region-of-interest (ROI) selection for head and neck (H&N) image-guided radiation therapy (IGRT). Due to the relative movement in different regions of H&N anatomy, it is important to identify an alignment ROI for IGRT. The selection of ROI should be based on the primary clinical goal, which can be either target coverage or sparing of normal structure. The top row shows a base-of-tongue cancer case, in which the C2 vertebra was selected as the alignment object. For a sinus cancer case (bottom row), the planning target volume is used as the alignment ROI for patient setup.

other structures far away from the PTV. In general, it would be better to use soft-tissue GTV as the alignment target for setup of H&N patients. Asymmetric tumor shrinkage can affect the relationship between the isocenter and other conformal avoidance structures such as the spinal cord. In addition, different CTVs may vary in their shapes, positions, and volumes. Without modifying the original plan, it may be more appropriate to use bony landmarks for image-guided setup.

17.5 Image-Guided Adaptive Radiotherapy for Head and Neck Cancer

Radiation therapy has gone through a series of technological revolutions in the last few decades. With IMRT, it became possible to produce highly conformal dose distributions, such as the one illustrated in Figure 17.1. These techniques utilize 3D anatomical information extracted from images of various types (CT, MR, PET, etc.) acquired a few days prior to the first treatment. However, the locations, shapes, and sizes of the tumor(s), and normal anatomy have been found to change during the course of treatment, primarily due to daily positioning uncertainties of various ROIs and physiological and/ or clinical factors. The latter include tumor shrinkage, weight loss, volumetric changes in normal organs, nonrigid variations in different bony structures, etc., which have been discussed in Sections 17.3.1 and 17.3.2. The traditional assumption that the anatomy discerned from 3D-CT images acquired for planning purposes prior to the treatment course is applicable for every fraction may not adequately account for interfractional changes and may limit the ability to fully exploit the potential of highly conformal treatment modalities such as IMRT. In fact, it has been argued that IMRT is more susceptible to anatomical changes than conventional 3D conformal therapy (Li and Xing 2000; Samuelsson, Mercke, and Johansson 2003; Hector, Webb, and Evans 2000), and there is concern that highly conformal, high-dose IMRT dose distributions designed based on a single CT data set acquired for planning purposes may lead to unforeseen complications or marginal misses of target volumes. This improved capability in dose conformality necessitates better target localization (IGRT), as well as the ability to adapt to inter- or intrafractional changes, both in the design and in the delivery of treatments.

17.5.1 Workflow for Image-Guided Adaptive Radiotherapy

In contrast to image-guided setup for repositioning treatment fields, adaptive RT is defined as changing the radiation treatment plan delivered to a patient during a course of RT to account for temporal changes in anatomy (e.g., tumor shrinkage, weight loss) or changes in function (e.g., hypoxia). Adaptive RT requires that a new radiation plan that can occur at three different timescales be made for the patient: (1) offline between fractions,

(2) online immediately prior to a fraction, and (3) in real time during a treatment fraction.

Because a new treatment plan will be designed in adaptive RT, it is essential that CT images (or equivalent volumetric images) be acquired to provide up-to-date information for replanning. It is conceivable that adaptive RT is closely linked to image guidance processes because the acquired images in the IGRT procedure could be conveniently used for monitoring changes in anatomy and designing new plans.

A possible workflow for image-guided adaptive RT is shown in Figure 17.9. In-room CT imaging is essential to acquire the patient's 3D anatomy for each treatment session. The 3D images can be used for online image-guided setup. The solid lines indicate the image guidance procedure, which will control the position of the treatment couch for realigning the patient relative to the isocenter. Additionally, the 3D images for the day of treatment can be sent to a treatment planning system where a new plan can be designed and reloaded back to the therapy machine for treatment delivery. This later procedure is shown in dotted lines in the procedure workflow. The new plan can be used immediately (online correction) or used for future treatment (offline correction).

17.5.2 Deformable Image Registration for Autosegmentation

For using IMRT in adaptive replanning, it is essential to have treatment targets and critical structures available for inverse planning. The existing contouring practice may not fit this process because delineating these structure contours from scratch is time consuming. Manual contouring is also subject to intra- or interobserver variations (Geets et al. 2005), which can affect the consistency of treatment planning. Recent developments in deformable image registration for atlas-based autosegmentation proves to be an effective method for adaptive RT (Lu et al. 2004; Wang et al. 2005; Chao et al. 2007; Castadot et al. 2008; Zhang et al. 2007; Nithiananthan et al. 2009). Deformable image registration is a geometric mapping process that creates one-to-one correspondence between two images of the same but deformed object. If the contours exist in one of the reference CT images, deformable transformation can be used to morph these reference contours on to the newly acquired CT images with minimal effort needed for manual adjusting. This process works well for adaptive RT because the original treatment plan acts as the reference image. The original contours with all necessary target volumes and critical structures on the simulation CT are previously approved.

An example of using deformable image registration for autosegmentation is shown in Figure 17.10. The process starts with a rigid alignment of bony structure (C2 vertebra) between the reference planning CT (left) and the daily in-room CT (middle and right). The necessary planning contours are overlaid on the daily CT not only to verify setup accuracy but also to evaluate if there are significant changes in the daily anatomy. If the changes are significant, as illustrated in the middle picture

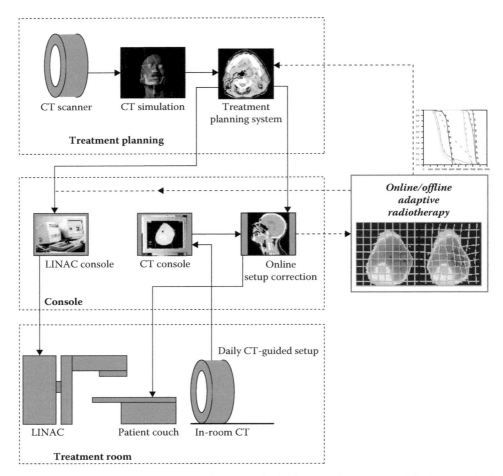

FIGURE 17.9 Workflow diagram for using in-room computed tomography (CT) or cone-beam CT-guided adaptive radiotherapy: The first level of treatment modification is a simple couch shift to correct for daily setup errors (CT-guided setup). The patient's anatomy was assumed to be unchanged. Nonrigid changes in tumor volumes and normal organs can be corrected by using an online or offline adaptive replanning process, which is shown using dotted lines.

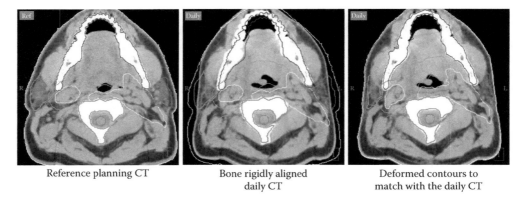

Reference planning CT   Bone rigidly aligned   Deformed contours to
                                        daily CT   match with the daily CT

FIGURE 17.10 Adaptive Radiation Therapy (ART) process for patient treatment. The process starts with a rigid alignment (using the C2 vertebra) between the reference planning computed tomography (CT) and the daily in-room CT (left and middle). The planning contours are overlaid on the daily CT not only to verify setup accuracy but also to evaluate if there are anatomical changes between the original plan and the daily anatomy. If the changes are significant, as illustrated in the middle picture, a deformable image registration will be performed to propagate the planning contours to the daily anatomy. The resultant contours are shown in the right picture. The entire transformation takes less than 30 seconds.

of Figure 17.10 (the contours do not match with the daily CT images), a deformable image registration will be performed to propagate the planning contours into the daily anatomy. The resultant contours are shown to the right of the figure. The CT to CT–based deformable image registration is fully automatic using the consistency in the CT Hounsfield unit. The entire transformation takes less than 30 seconds, which makes possible its use for online or offline IMRT replanning. Contouring from scratch could take several hours (Chao et al. 2007), which may not be practical for the timely adaptive process.

17.5.3 Dosimetric Benefit of Adaptive Radiotherapy

As described in Section 17.3, there are noticeable changes in the anatomy of H&N cancer patients who have undergone radiation therapy. A treatment plan created on the initial planning CT may no longer be optimal for changing anatomy during the treatment, and the actual radiation dose delivered to the patient may be significantly different from that specified in the original plan (O'Daniel et al. 2007).

With deformable image registration for dose accumulation, it is possible to evaluate the existing practice without replanning and compare it with IGRT and adaptive RT. O'Daniel et al. (2007) studied the differences between planned and delivered parotid gland and target doses in a group of H&N cancer patients receiving IMRT treatment. The clinical IMRT plans, designed with 3–4 mm planning margins, were recalculated on the repeat CT images. In-house deformable image registration software was used to map daily dose distributions to the original treatment plan and to calculate a cumulative delivered dose distribution for each patient. They found that without using IGRT, dose to the parotid gland can be increased above the planned dose by 5–7 Gys in 45% of the patients. The use of IGRT aligned to the cervical bone (C2 vertebra) can lead to dose reductions in 91% of patients (median: 2 Gys; range: 0.3–8.3 Gys; 15 of 22 parotids improved). However, the parotid dose from bone alignment was still greater than planned (median: 1 Gy; $p = .007$) due to volume shrinkage. The study also found that the current PTV margin of 3–4 mm provided adequate target coverage without daily IGRT.

Using daily MV imaging in 10 tomotherapy patients, Lee et al. (2008a) analyzed changes in parotid gland dose resulting from anatomical changes throughout a course of RT using a deformable image registration method. They found that the daily parotid mean dose of the 10 patients differed from the plan dose by an average of 15%. At the end of treatment, 3 of the 10 patients were estimated to have received a greater than 10% higher mean parotid dose than that in the original plan (range: 13%–42%), whereas the remaining 7 patients received doses that differed by less than 10% (range: −6%–8%). The dose difference was correlated with a migration of the parotids toward the high-dose region.

Wu et al. (2009) performed a comprehensive adaptive replanning simulation study to evaluate the differences between planned and delivered dose and to investigate different replanning strategies. In the study, 11 patients received 6 weekly helical CTs. It was found that the cumulative doses to targets were preserved even at the 0-mm PTV margin. Significant increase in parotid doses was observed without using replanning. These authors reported that 1 midcourse adaptive replanning improved parotid mean dose sparing by 3%, 2 replannings by 5%, and 6 replannings by 6%, assuming replanning occurs 1 week prior to actual treatment delivery. If 6 weekly replannings are used immediately for treatment, parotid dose sparing can be improved by 8%. However, the study assumed that each new plan must be perfectly executed without additional setup errors or nonrigid changes.

17.5.4 Clinical Experience of Adaptive Radiotherapy of Head and Neck Cancer

Before we discuss various alternative adaptive treatment strategies, it would be useful to share our clinical experience at the MD Anderson Cancer Center. We conducted a prospective clinical trial to test the feasibility and clinical benefit of adaptive RT for oropharyngeal cancer. For practical reasons, we designed our planning strategy to incorporate in-house standard IMRT planning margin into the initial (baseline) IMRT plan. Typical volumetric CTV-to-PTV expansions were 3–4 mm. Our previous clinical experience supports such a practice since we have observed that setup errors are usually large during the first few treatment fractions when the patient is not used to the immobilization mask or the treatment procedure. In addition, simple couch shifts may not be able to correct nonrigid changes, such as neck curvature or chin angle variations. Thus, we have adopted the standard PTV expansion at the start of treatment, regardless of any potential downstream corrections.

In-room CT-guided IGRT has been used for each treatment session, as illustrated in Figure 17.9. Daily in-room CT images acquired on a CT-on-rails system were transferred to an IGRT decision system, which provided data and directions for the alteration of setup or correction of daily treatment or both. Then, the couch position was corrected to align the patient relative to the isocenter prior to each treatment fraction. This rigid alignment step corrects for any setup errors based on the bony landmark of cervical vertebrae in the C1–C3 region, as illustrated in Figure 17.10.

When therapists performed this online image guidance procedure, they were asked to evaluate the overlaid skin contours delineated from the initial treatment plan. Any mismatch between the initial skin contour and the daily CT image represents significant anatomical changes, most likely due to weight loss and other treatment-related responses. In addition, other anatomical contours, such as parotid, GTV, and CTV can be evaluated in the same way to detect any visible changes between the original contours and a patient's up-to-date anatomy. Because the original contours are usually a good representation of the initial conformal IMRT plan, the side-by-side contour evaluation triggers subsequent replanning activities. A screen capture of the patient setup with the contour overlaid was saved in our electronic medical record system. Attending physicians have the chance to evaluate each daily IGRT setup.

If significant discrepancy was found between the CT image and the overlaid contours, either a therapist or a physician informed the physicists that a planning evaluation will be conducted. For such an evaluation, the CT images are transferred to a treatment planning system. In-house deformable image registration is performed to transfer the initial contours (GTV, CTV, parotid, spinal cord, brainstem, and other normal structures) to the new CT image set. The deformable image registration and its validation in H&N anatomy were reported previously (Wang et al. 2008). In addition to contours, isocenter information and the original IMRT plan are loaded into a plan based on the newly acquired CT image. This plan will serve as a baseline to evaluate any improvement that can be achieved by (adaptive) replanning. A copy of this plan is made, and new planning parameters are adjusted based on the new anatomy configuration. Usually, this is a case-by-case situation in which additional planning constraints may be added to improve the quality of the new plan.

We decided to use a PTV margin of 0 mm in all subsequent replanning. This is based on our experience that patient setup is much more stabilized in subsequent treatments, and the initial large margin provides room for any small subsequent deviations. Our experience from previous planning studies confirmed that 3–4 mm PTV expansion margins were too generous with daily image guidance (O'Daniel et al. 2007). Mainly due to shrinkage, the conventional parotid dose limits (based on Radiation Therapy Oncology Group [RTOG] guidelines) are more difficult to meet in replanning. There is no dosimetric guideline for replanning. Therefore, the planning is based on the best achievable plan, instead of on any predefined guidelines.

An example of such a dose recalculation and replanning is shown in Figure 17.11. This is a case in which a second replan is being investigated. The recalculated original plan, calculated on the daily CT acquired on the 25th treatment fraction, is shown to the left. The first replan (designed on treatment fraction number 15) was also calculated and is shown in the middle of the figure. The newly designed second replan is shown to the right. It can be seen that the original plan has a larger treatment margin and higher doses to the target. Due to shrinkage, there is less attenuation in each beam. Therefore, the original plan tends to produce more hot spots than the original design in the new anatomy. In Figure 17.11, the first replan has significantly improved dose conformality not only due to better matching with the anatomy but also due to the use of a smaller treatment margin. The first replan has a smaller difference compared to the second replan. Nevertheless, the second plan still produced better parotid sparing and lower body dose, which was the primary reason that our physicians approved the second replan.

We follow a conventional in-house quality assurance (QA) procedure for any new IMRT plan. Typically, the newly approved plan is used in the next fraction. The QA measurement is typically done on the day when the treatment plan is approved or on the same day as its first use. We have an in-house policy that IMRT QA should be performed and reviewed before the third fraction of treatment. Over 9000 IMRT QA plans have been measured in our practice, and large dosimetric discrepancies are rare (< 0.2%). Therefore, we do not require IMRT QA measurement done prior to the plan's first use.

A preliminary analysis of 724 daily CT images in 22 patients was performed. By treatment completion, mean parotid shrinkage was 26%, consistent with our original published findings (Barker et al. 2004). The CTV regressed by an average of 9%. Importantly, not every patient enjoyed tumor response to treatment. Two of 22 patients (10%) had an initial increase in tumor volume by more than 8%, which resulted in early replanning and subsequent second replanning. In one such patient, GTV increased 50% between the planning CT and the first day of treatment, resulting in almost immediate replanning (second fraction).

All patients received at least one replan Adaptive RT strategy 1 (ART1) and eight patients (36%) received two replans Adaptive RT strategy 2 (ART2) during their course of treatment. The median trigger point for the first adaptive plan was the 16th

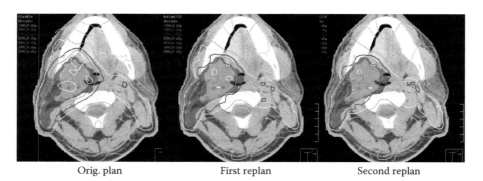

Orig. plan   First replan   Second replan

FIGURE 17.11 **(See color insert following page 204.)** Typical example of dose recalculation on a daily computed tomography (CT) image acquired on the 25th treatment fraction. Left: the original plan was calculated; middle: the first replan (designed on treatment fraction number 15) was calculated; and right: the second replan was designed on this CT. We can see that the original plan has a larger treatment margin and higher doses in the target. The first replan has a smaller difference compared to the second replan; however, the parotid sparing and lower body dose can still be seen in the second replan.

treatment fraction (range: 2–28), at which point the median bilateral parotid volumes had shrunk by an average of 16% and the combined CTVs had shrunk by 5%. For ART2 patients, the median trigger point for the first replan was the 11th fraction (range: 2–15) and that for the second replan was the 22nd fraction (range: 11–25), at which point the bilateral parotid volumes and CTVs had shrunk by 24% and 14%, respectively. The timing distribution of the first replan and the second replan is plotted in Figure 17.12.

It does not appear that a fixed replanning fraction exists; replanning fractions vary. An example of volume variations for a patient who had two replans is shown in Figure 17.13. Although there are fluctuations in daily volume variations for various structures, there is a trend that anatomical change is greater in the first half of treatment than in the second half. In this patient, the parotid change was the main cause for the first replanning and CTV change was the cause for the second replanning. The elapsed time interval from triggering in-room CT imaging to subsequent delivery of the prompted ART plan was 1.7 days (median: 2 days; range: 1–4 days), excluding weekends.

17.6 Future Development of Adaptive Strategies

We will now discuss various scenarios and rationales for adaptive RT. Adaptive RT is still an evolving technology and clinical practice. It is expected that more refinement will be possible with further advancements in imaging technologies, IMRT planning techniques, and in our clinical knowledge. The future development of adaptive RT can be categorized into various aspects, discussed in Sections 17.6.1 through 17.6.5.

17.6.1 Clinical Objectives and Targets of Interest

Most existing adaptive RT procedures focus on morphological anatomical changes and discrepancies between the original plan and the delivered plan. Adaptive RT may improve IMRT conformality and the sparing of normal tissue. Although this is a reasonable concern, specific clinical goals are still lacking, such as tumor control and normal-tissue toxicity. In addition, target delineation is still considered the bottleneck for this technology. Adaptive RT provides additional capability in delivering highly conformal treatment based on feedback or control theory. Accurate target definition is highly desirable for the success of this technology.

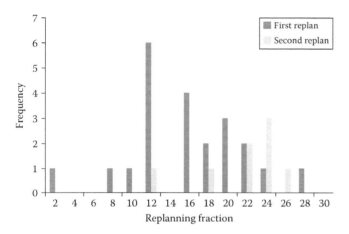

FIGURE 17.12 Distribution of the triggering fraction for replanning. The distribution is plotted for both the first replan and the second replan. The replanning time seems to be individualized with a wide spread. The median replanning fractions are the 16th and 22nd fractions for the first and second adaptive plan, respectively.

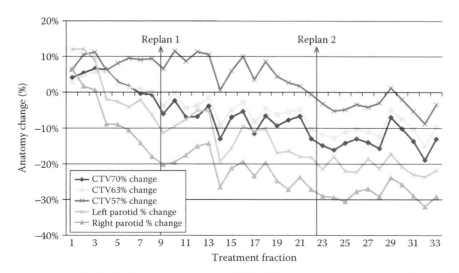

FIGURE 17.13 Volumetric changes for high-risk clinical target volume (CTV; CTV70), intermediate-risk CTV (CTV63), low-risk CTV (CTV57), and parotid gland for a patient who had two replans during the course of treatment. The first replan occurred on fraction number 9, and the second replan happened on fraction number 23.

17.6.2 Incorporating Functional and Biological Imaging in Treatment Adaptation

One way to improve the quality of target information is to use functional imaging. Many previous studies have shown that functional imaging, such as FDG-PET, provides additional information over gross anatomy (CT) imaging. An even more appealing use of FDG-PET for "dose painting" can be integrated into the adaptive RT procedure. Using the imaged residual metabolically active tumor after a part of the treatment course has been delivered was suggested by the Brussels group (Gregoire and Haustermans 2009; Gregoire, Haustermans, et al. 2007; van Baardwijk et al. 2006). The residual activity may represent parts of the tumor that are more radioresistant than the parts that ceased to be PET avid after some radiation had been delivered. If FDG-PET imaging during therapy indeed represents a resistant subvolume, the FDG-PET avid tumor can be used as target for dose escalation, which theoretically should improve the outcome. Similar to FDG-PET, tumor subvolumes defined by hypoxia imaging have been proposed as targets for integrated boosts as "hypoxia-guided radiotherapy" (Sovik et al. 2007; Sovik, Malinen, and Olsen 2009). Compared to FDG-PET, which is spatially stable within the tumor before and throughout RT, hypoxia imaging shows significant spatial variability due to reoxygenation and changes in the locations of tumoral hypoxia. In general, existing functional imaging is still at a premature stage. However, the techniques are promising and perhaps can play a more significant role in future adaptive RT.

17.6.3 Online vs. Offline Adaptive Radiotherapy

The online adaptive RT approach makes adjustments to the treatment plan based on images acquired during the current treatment session. The offline adaptive RT approach is one in which the intervention is determined from an accumulation of information that may be drawn from previous treatment sessions or other times of measurement. The online approach is generally categorized as having a greater capacity to increase precision with an associated increase in effort and treatment time. Online replanning is significantly more challenging due to time constraints, as it must be done while the patient is still immobilized on the treatment couch waiting for treatment. Fortunately for H&N treatment sites, anatomical changes are gradual. There is no great demand for real-time intervention. The only exception is the situation we reported earlier, in which a patient's anatomy can change on the first day of treatment due to the longer interval between CT simulation and the first day of treatment. Therefore, offline adaptive RT appears to be a more practical approach for treating H&N cancer.

17.6.4 Correction for Nonrigid Setup Error

In H&N RT patients, it is common that bony structures exhibit nonrigid changes that can affect the conformality of the original IMRT plan. If these nonrigid changes are systematic (i.e., they were caused by inappropriate simulation procedures), adaptive replanning can be used to correct nonrigid systematic setup errors although the tumor volume does not change. Our experience has shown that random nonrigid setup errors are difficult to correct. Resetting the patient setup position usually does not fully correct these errors. If the desired accuracy requires online correction of random (nonrigid) setup errors, online replanning may be necessary.

17.6.5 Autoreplanning

Replanning is still a time-consuming process. If adaptive RT is routinely used in the clinic, it is necessary to reduce this burden by providing more effective autoreplanning techniques. Although replanning is relatively easier than planning from scratch, trial-and-error is still necessary to fine-tune planning parameters to get a satisfactory plan. Many research groups have studied autoreplanning algorithms. In almost all such implementations, deformable image registration is a critical component. Mohan et al. (2004) used an IMRT intensity-warping technique to adapt an IMRT plan based on the changes in the anatomy in the beam's eye-view projection. They demonstrated that an autoplanning was possible as a rival to planning from scratch. However, a deformed intensity map may not always be deliverable. Ahunbay et al. (2009) proposed a two-step procedure for autoreplanning. The first step used an aperture-morphing technique to transform a multileaf collimator (MLC) leaf segment based on anatomical changes. The second step applies a segment weight optimization, which reoptimizes the entire plan. The entire process takes about 5–8 minutes. Autoreplanning is important for online adaptive RT, and it will also benefit offline replanning to reduce workload.

17.7 Summary

In summary, adaptive RT is a process closely linked to feedback-based control theory. It has three basic components: (1) detection of changes, (2) method of intervention, and (3) management of overall clinical goal. The successful implementation of each component will decide the overall success of clinical application. Existing adaptive RT is a much more labor-intensive process; the practicality of this technology will become a reality when it is deployed in routine clinical practice. Early clinical results and simulation studies demonstrate the benefit of adaptive RT for H&N cancer patients. With emerging clinical data from functional imaging and treatment response, we expect adaptive RT to evolve further and gradually increase its clinical utilization in the future.

References

Ahn, P. H., A. I. Ahn, C. J. Lee, J. Shen, E. Miller, A. Lukaj, E. Milan, R. Yaparpalvi, S. Kalnicki, and M. K. Garg. 2009. Random positional variation among the skull, mandible, and cervical spine with treatment progression during head-and-neck radiotherapy. *Int J Radiat Oncol Biol Phys* 73:626–33.

Ahunbay, E. E., C. Peng, A. Godley, C. Schultz, and X. A. Li. 2009. An on-line replanning method for head and neck adaptive radiotherapy. *Med Phys* 36:4776–90.

Ashamalla, H., A. Guirgius, E. Bieniek, S. Rafla, A. Evola, G. Goswami, R. Oldroyd, B. Mokhtar, and K. Parikh. 2007. The impact of positron emission tomography/computed tomography in edge delineation of gross tumor volume for head and neck cancers. *Int J Radiat Oncol Biol Phys* 68:388–95.

Ashamalla, H., S. Rafla, K. Parikh, B. Mokhtar, G. Goswami, S. Kambam, H. Abdel-Dayem, A. Guirguis, P. Ross, and A. Evola. 2005. The contribution of integrated PET/CT to the evolving definition of treatment volumes in radiation treatment planning in lung cancer. *Int J Radiat Oncol Biol Phys* 63:1016–23.

Barker Jr., J. L., A. S. Garden, K. K. Ang, J. C. O'Daniel, H. Wang, L. E. Court, W. H. Morrison, et al. 2004. Quantification of volumetric and geometric changes occurring during fractionated radiotherapy for head-and-neck cancer using an integrated CT/linear accelerator system. *Int J Radiat Oncol Biol Phys* 59:960–70.

Barkley, H. T., and G. H. Fletcher. 1977. The significance of residual disease after external irradiation of squamous-cell carcinoma of the oropharynx. *Radiology* 124:493–5.

Bel, A., R. Keus, R. E. Vijlbrief, and J. V. Lebesque. 1995. Setup deviations in wedged pair irradiation of parotid gland and tonsillar tumors, measured with an electronic portal imaging device. *Radiother Oncol* 37:153–9.

Bel, A., O. Petrascu, I. Van de Vondel, L. Coppens, N. Linthout, D. Verellen, and G. Storme. 2000. A computerized remote table control for fast on-line patient repositioning: Implementation and clinical feasibility. *Med Phys* 27:354–8.

Boda-Heggemann, J., C. Walter, A. Rahn, H. Wertz, I. Loeb, F. Lohr, and F. Wenz. 2006. Repositioning accuracy of two different mask systems-3D revisited: Comparison using true 3D/3D matching with cone-beam CT. *Int J Radiat Oncol Biol Phys* 66:1568–75.

Brock, K. K., D. L. McShan, and J. M. Balter. 2002. A comparison of computer-controlled versus manual on-line patient setup adjustment. *J Appl Clin Med Phys* 3:241–7.

Castadot, P., J. A. Lee, A. Parraga, X. Geets, B. Macq, and V. Grégoire. 2008. Comparison of 12 deformable registration strategies in adaptive radiation therapy for the treatment of head and neck tumors. *Radiother Oncol* 89:1–12.

Chao, K. S. C., S. Bhide, H. Chen, J. Asper, S. Bush, G. Franklin, V. Kavadi, et al. 2007. Reduce in variation and improve efficiency of target volume delineation by a computer-assisted system using a deformable image registration approach. *Int J Radiat Oncol Biol Phys* 68:1512–21.

Chao, K. S., J. O. Deasy, J. Markman, J. Haynie, C. A. Perez, J. A. Purdy, and D. A. Low. 2001. A prospective study of salivary function sparing in patients with head-and-neck cancers receiving intensity-modulated or three-dimensional radiation therapy: Initial results. *Int J Radiat Oncol Biol Phys* 49:907–16.

Chao, K. S., N. Majhail, C. J. Huang, J. R. Simpson, C. A. Perez, B. Haughey, and G. Spector. 2001. Intensity-modulated radiation therapy reduces late salivary toxicity without compromising tumor control in patients with oropharyngeal carcinoma: A comparison with conventional techniques. *Radiother Oncol* 61:275–80.

Cook, I. J., W. J. Dodds, R. O. Dantas, B. Massey, M. K. Kern, I. M. Lang, J. G. Brausseur, and W. J. Hogan. 1989. Opening mechanisms of the human upper esophageal sphincter. *Am J Physiol: Gastrointestinal and Liver Physiology* 257:G748–59.

Court, L., I. Rosen, R. Mohan, and L. Dong. 2003. Evaluation of mechanical precision and alignment uncertainties for an integrated CT/LINAC system. *Med Phys* 30:1198–210.

Dantas, R. O., M. K. Kern, B. T. Massey, W. J. Dodds, P. J. Kahrilas, J. G. Brasseur, I. J. Cook, and I. M. Lang. 1990. Effect of swallowed bolus variables on oral and pharyngeal phases of swallowing. *Am J Physiol: Gastrointestinal and Liver Physiology* 258:G675–81.

de Boer, H. C., J. R. van Sornsen de Koste, C. L. Creutzberg, A. G. Visser, P. C. Levendag, and B. J. Heijmen. 2001. Electronic portal image assisted reduction of systematic set-up errors in head and neck irradiation. *Radiother Oncol* 61:299–308.

Eisbruch, A., R. L. Foote, B. O'Sullivan, J. J. Beitler, and B. Vikram. 2002. Intensity-modulated radiation therapy for head and neck cancer: Emphasis on the selection and delineation of the targets. *Semin Radiat Oncol* 12:238–49.

Eisbruch, A., and V. Gregoire. 2009. Balancing risk and reward in target delineation for highly conformal radiotherapy in head and neck cancer. *Semin Radiat Oncol* 19:43–52.

Eisbruch, A., P. C. Levendag, F. Y. Feng, D. Teguh, T. Lyden, P. I. Schmitz, M. Haxer, I. Noever, D. B. Chepeha, and B. J. Heijmen. 2007. Can IMRT or brachytherapy reduce dysphagia associated with chemoradiotherapy of head and neck cancer? The Michigan and Rotterdam experiences. *Int J Radiat Oncol Biol Phys* 69:S40–2.

Eisbruch, A., N. Rhodus, D. Rosenthal, B. Murphy, C. Rasch, S. Sonis, C. Scarantino, and D. Brizel. 2003. The prevention and treatment of radiotherapy-induced xerostomia. *Semin Radiat Oncol* 13:302–8.

Eisbruch, A., M. Schwartz, C. Rasch, K. Vineberg, E. Damen, C. J. Van As, R. Marsh, F. A. Pameijer, and A. J. Balm. 2004. Dysphagia and aspiration after chemoradiotherapy for head-and-neck cancer: Which anatomic structures are affected and can they be spared by IMRT? *Int J Radiat Oncol Biol Phys* 60:1425–39.

Eisbruch, A., R. K. Ten Haken, H. M. Kim, L. H. Marsh, and J. A. Ship. 1999. Dose, volume, and function relationships in parotid salivary glands following conformal and intensity-modulated irradiation of head and neck cancer. *Int J Radiat Oncol Biol Phys* 45:577–87.

Feng, F. Y., H. M. Kim, T. H. Lyden, M. J. Haxer, M. Feng, F. P. Worden, D. B. Chepeha, and A. Eisbruch. 2007. Intensity-modulated radiotherapy of head and neck cancer aiming to reduce dysphagia: Early dose-effect relationships for the swallowing structures. *Int J Radiat Oncol Biol Phys* 68:1289–98.

Geets, X., J. F. Daisne, S. Arcangeli, E. Coche, M. De Poel, T. Duprez, G. Nardella, and V. Gregoire. 2005. Inter-observer variability in the delineation of pharyngo-laryngeal tumor, parotid glands and cervical spinal cord: Comparison between CT-scan and MRI. *Radiother Oncol* 77:25–31.

Gregoire, V., J. F. Daisne, X. Geets, and P. Levendag. 2003. Selection and delineation of target volumes in head and neck tumors: Beyond ICRU definition. *Rays* 28:217–24.

Gregoire, V., W. De Neve, A. Eisbruch, N. Lee, D. Van Den Weyngaert, and D. Van Gestel. 2007. Intensity-modulated radiation therapy for head and neck carcinoma. *Oncologist* 12:555–64.

Gregoire, V., A. Eisbruch, M. Hamoir, and P. Levendag. 2006. Proposal for the delineation of the nodal CTV in the node-positive and the post-operative neck. *Radiother Oncol* 79:15–20.

Gregoire, V., and K. Haustermans. 2009. Functional image-guided intensity modulated radiation therapy: Integration of the tumour microenvironment in treatment planning. *Eur J Cancer* 45:459–460.

Gregoire, V., K. Haustermans, X. Geets, S. Roels, and M. Lonneux. 2007. PET-based treatment planning in radiotherapy: A new standard? *J Nucl Med* 48(1 Suppl):68S–77S.

Gregoire, V., P. Levendag, K. K. Ang, J. Bernier, M. Braaksma, V. Budach, C. Chao, et al. 2003. CT-based delineation of lymph node levels and related CTVs in the node-negative neck: DAHANCA, EORTC, GORTEC, NCIC, RTOG consensus guidelines. *Radiother Oncol* 69:227–36.

Hamlet, S., G. Ezzell, and A. Aref. 1994. Larynx motion associated with swallowing during radiation therapy. *Int J Radiat Oncol Biol Phys* 28:467–70.

Hatherly, K., J. Smylie, and A. Rodger. 1999. A comparison of field-only electronic portal imaging hard copies with double exposure port films in radiation therapy treatment setup confirmation to determine its clinical application in a radiotherapy center. *Int J Radiat Oncol Biol Phys* 45:791–6.

Hatherly, K. E., J. C. Smylie, A. Rodger, M. J. Dally, S. R. Davis, and J. L. Millar. 2001. A double exposed portal image comparison between electronic portal imaging hard copies and port films in radiation therapy treatment setup confirmation to determine its clinical application in a radiotherapy center. *Int J Radiat Oncol Biol Phys* 49:191–8.

Hector, C. L., S. Webb, and P. M. Evans. 2000. The dosimetric consequences of inter-fractional patient movement on conventional and intensity-modulated breast radiotherapy treatments. *Radiother Oncol* 54:57–64.

Hong, T. S., W. A. Tome, R. J. Chappell, P. Chinnaiyan, M. P. Mehta, and P. M. Harari. 2005. The impact of daily setup variations on head-and-neck intensity-modulated radiation therapy. *Int J Radiat Oncol Biol Phys* 61:779–88.

Jaffray, D. A., D. G. Drake, M. Moreau, A. A. Martinez, and J. W. Wong. 1999. A radiographic and tomographic imaging system integrated into a medical linear accelerator for localization of bone and soft-tissue targets. *Int J Radiat Oncol Biol Phys* 45:773–89.

Jaffray, D. A., and J. H. Siewerdsen. 2000. Cone-beam computed tomography with a flat-panel imager: Initial performance characterization. *Med Phys* 27:1311–23.

Karger, C. P., O. Jakel, J. Debus, S. Kuhn, and G. H. Hartmann. 2001. Three-dimensional accuracy and interfractional reproducibility of patient fixation and positioning using a stereotactic head mask system. *Int J Radiat Oncol Biol Phys* 49:1493–504.

Kendall, K. A., S. W. McKenzie, R. J. Leonard, and C. Jones. 1998. Structural mobility in deglutition after single modality treatment of head and neck carcinomas with radiotherapy. *Head Neck* 20:720–5.

Kuriyama, K., H. Onishi, N. Sano, T. Komiyama, Y. Aikawa, Y. Tateda, T. Araki, and M. Uematsu. 2003. A new irradiation unit constructed of self-moving gantry-CT and linac. *Int J Radiat Oncol Biol Phys* 55:428–35.

Lee, C., K. M. Langen, W. Lu, J. Haimerl, E. Schnarr, K. J. Ruchala, G. H. Olivera, S. L. Meeks, P. A. Kupelian, T. D. Shellenberger, and R. R. Mañon. 2008a. Assessment of parotid gland dose changes during head and neck cancer radiotherapy using daily megavoltage computed tomography and deformable image registration. *Int J Radiat Oncol Biol Phys* 71:1563–71.

Lee, C., K. M. Langen, W. Lu, J. Haimerl, E. Schnarr, K. J. Ruchala, G. H. Olivera, S. L. Meeks, P. A. Kupelian, T. D. Shellenberger, and R. R. Mañon. 2008b. Evaluation of geometric changes of parotid glands during head and neck cancer radiotherapy using daily MVCT and automatic deformable registration. *Radiother Oncol* 89:81–8.

Lee, N., P. Xia, J. M. Quivey, K. Sultanem, I. Poon, C. Akazawa, P. Akazawa, V. Weinberg, and K. K. Fu. 2002. Intensity-modulated radiotherapy in the treatment of nasopharyngeal carcinoma: An update of the UCSF experience. *Int J Radiat Oncol Biol Phys* 53:12–22.

Leonard, R. J., K. A. Kendall, S. McKenzie, M. I. Gonçalves, and A. Walker. 2000. Structural displacements in normal swallowing: A videofluoroscopic study. *Dysphagia* 15:146–52.

Linthout, N., D. Verellen, K. Tournel, and G. Storme. 2006. Six dimensional analysis with daily stereoscopic X-ray imaging of intrafraction patient motion in head and neck treatments using five points fixation masks. *Med Phys* 33:504–13.

Li, J. G., and L. Xing. 2000. Inverse planning incorporating organ motion. *Med Phys* 27:1573–8.

Li, H., X. R. Zhu, L. Zhang, L. Dong, S. Tung, A. Ahamad, K. S. C. Chao, et al. 2008. Comparison of 2D radiographic images and 3D cone beam computed tomography for positioning head-and-neck radiotherapy patients. *Int J Radiat Oncol Biol Phys* 71:916–25.

Lu, W., M. L. Chen, G. H. Olivera, K. J. Ruchala, and T. R. Mackie. 2004. Fast free-form deformable registration via calculus of variations. *Phys Med Biol* 49:3067–87.

Mackie, T. R., J. Kapatoes, K. Ruchala, W. Lu, C. Wu, G. Olivera, L. Forrest, et al. 2003. Image guidance for precise conformal radiotherapy. *Int J Radiat Oncol Biol Phys* 56:89–105.

Manning, M. A., Q. Wu, R. M. Cardinale, R. Mohan, A. D. Lauve, B. D. Kavanagh, M. M. Morris, and R. K. Schmidt-Ullrich. 2001. The effect of setup uncertainty on normal tissue sparing with IMRT for head-and-neck cancer. *Int J Radiat Oncol Biol Phys* 51:1400–9.

Mell, L. K., J. C. Roeske, and A. J. Mundt. 2003. A survey of intensity-modulated radiation therapy use in the United States. *Cancer* 98:204–11.

Mohan, R., X. Zhang, C. Wang, H. Wang, H. Liu, D. Kuban, K. Ang, and L. Dong. 2004. Deforming intensity distributions to incorporate inter-fraction anatomic variations for image-guided IMRT. *Int J Radiat Oncol Biol Phys* 60:S226–7.

Nangia, S., K. S. Chufal, A. Tyagi, A. Bhatnagar, M. Mishra, and D. Ghosh. 2010. Selective nodal irradiation for head and neck cancer using intensity-modulated radiotherapy: Application of RTOG consensus guidelines in routine clinical practice. *Int J Radiat Oncol Biol Phys* 76:146–53.

Nithiananthan, S., K. K. Brock, M. J. Daly, H. Chan, J. C. Irish, and J. H. Siewerdsen. 2009. Demons deformable registration for CBCT-guided procedures in the head and neck: Convergence and accuracy. *Med Phys* 36:4755–64.

O'Daniel, J. C., A. S. Garden, D. L. Schwartz, H. Wang, K. K. Ang, A. Ahamad, D. I. Rosenthal, et al. 2007. Parotid gland dose in intensity-modulated radiotherapy for head and neck cancer: Is what you plan what you get? *Int J Radiat Oncol Biol Phys* 69:1290–6.

Pisani, L., D. Lockman, D. Jaffray, D. Yan, A. Martinez, and J. Wong. 2000. Setup error in radiotherapy: On-line correction using electronic kilovoltage and megavoltage radiographs. *Int J Radiat Oncol Biol Phys* 47:825–39.

Prisciandaro, J. I., C. M. Frechette, M. G. Herman, P. D. Brown, Y. I. Garces, and R. L. Foote. 2004. A methodology to determine margins by EPID measurements of patient setup variation and motion as applied to immobilization devices. *Med Phys* 31:2978–88.

Purdy, J. A. 2001. Intensity-modulated radiotherapy: Current status and issues of interest. *Int J Radiat Oncol Biol Phys* 51:880–914.

Samuelsson, A, C. Mercke, and K. A. Johansson. 2003. Systematic set-up errors for IMRT in the head and neck region: Effect on dose distribution. *Radiother Oncol* 66:303–11.

Seet, K. Y. T., A. Barghi, S. Yartsev, and J. Van Dyk. 2009. The effects of field-of-view and patient size on CT numbers from cone-beam computed tomography. *Phys Med Biol* 54:6251–62.

Sobel, S., P. Rubin, B. Keller, and C. Poulter. 1976. Tumor persistence as a predictor of outcome after radiation therapy of head and neck cancers. *Int J Radiat Oncol Biol Phys* 1:873–80.

Sovik, A., E. Malinen, and D. R. Olsen. 2009. Strategies for biologic image-guided dose escalation: A review. *Int J Radiat Oncol Biol Phys* 73:650–8.

Sovik, A., E. Malinen, H. K. Skogmo, S. M. Bentzen, O. S. Bruland, and D. R. Olsen. 2007. Radiotherapy adapted to spatial and temporal variability in tumor hypoxia. *Int J Radiat Oncol Biol Phys* 68:1496–504.

Suit, H. D., and A. M. Walker. 1980. Assessment of the response of tumours to radiation: Clinical and experimental studies. *Br J Cancer Supplement* 41:1–10.

Thomas, T. H. M., D. Devakumar, S. Purnima, and B. P. Ravindran. 2009. The adaptation of megavoltage cone beam CT for use in standard radiotherapy treatment planning. *Phys Med Biol* 54:2067–77.

Trott, K. R. 1983. Human tumour radiobiology: Clinical data. *Strahlentherapie* 159:393–7.

van Asselen, B., C. P. Raaijmakers, J. J. Lagendijk, and C. H. Terhaard. 2003. Intrafraction motions of the larynx during radiotherapy. *Int J Radiat Oncol Biol Phys* 56:384–90.

van Baardwijk, A., B. G. Baumert, G. Bosmans, M. van Kroonenburgh, S. Stroobants, V. Gregoire, P. Lambin, and D. De Ruysscher. 2006. The current status of FDG-PET in tumour volume definition in radiotherapy treatment planning. *Cancer Treat Rev* 32:245–60.

van Kranen, S., S. van Beek, C. Rasch, M. van Herk, and J. J. Sonke. 2009. Setup uncertainties of anatomical sub-regions in head-and-neck cancer patients after offline CBCT guidance. *Int J Radiat Oncol Biol Phys* 73:1566–73.

Wang, H., L. Dong, J. O'Daniel, R. Mohan, A. S. Garden, K. K. Ang, D. A. Kuban, M. Bonnen, J. Y. Chang, and R. Cheung. 2005. Validation of an accelerated "demons" algorithm for deformable image registration in radiation therapy. *Phys Med Biol* 50:2887–905.

Wang, H., A. S. Garden, L. Zhang, X. Wei, A. Ahamad, D. A. Kuban, R. Komaki, J. O'Daniel, Y. Zhang, R. Mohan, and L. Dong. 2008. Performance evaluation of automatic anatomy segmentation algorithm on repeat or four-dimensional computed tomography images using deformable image registration method. *Int J Radiat Oncol Biol Phys* 72:210–9.

Wang, Z. H., C. Yan, Z. Y. Zhang, C. P. Zhang, H. S. Hu, J. Kirwan, and W. M. Mendenhall. 2009. Radiation-induced volume changes in parotid and submandibular glands in patients with head and neck cancer receiving postoperative radiotherapy: A longitudinal study. *Laryngoscope* 119:1966–74.

Willner, J., U. Hadinger, M. Neumann, F. J. Schwab, K. Bratengeier, and M. Flentje. 1997. Three dimensional variability in patient positioning using bite block immobilization in 3D-conformal radiation treatment for ENT-tumors. *Radiother Oncol* 43: 315–21.

Wu, Q., Y. Chi, P. Y. Chen, D. J. Krauss, D. Yan, and A. Martinez. 2009. Adaptive replanning strategies accounting for shrinkage in head and neck IMRT. *Int J Radiat Oncol Biol Phys* 75:924–32.

Wu, Q., M. Manning, R. Schmidt-Ullrich, and R. Mohan. 2000. The potential for sparing of parotids and escalation of biologically effective dose with intensity-modulated radiation treatments of head and neck cancers: A treatment design study. *Int J Radiat Oncol Biol Phys* 46:195–205.

Yoo, S., and F. F. Yin. 2006. Dosimetric feasibility of cone-beam CT-based treatment planning compared to CT-based treatment planning. *Int J Radiat Oncol Biol Phys* 66:1553–61.

Zhang, T., Y. Chi, E. Meldolesi, and D. Yan. 2007. Automatic delineation of on-line head-and-neck computed tomography images: Toward on-line adaptive radiotherapy. *Int J Radiat Oncol Biol Phys* 68:522–30.

Zhang, L., A. S. Garden, J. Lo, K. K. Ang, A. Ahamad, W. H. Morrison, D. I. Rosenthal, M. S. Chambers, X. R. Zhu, R. Mohan, and L. Dong. 2006. Multiple regions-of-interest analysis of setup uncertainties for head-and-neck cancer radiotherapy. *Int J Radiat Oncol Biol Phys* 64:1559–69.

18

Adaptive Radiation Therapy for Breast Cancer

Kristofer Kainz
Medical College of Wisconsin

X. Allen Li
Medical College of Wisconsin

Julia White
Medical College of Wisconsin

18.1 Introduction

For most patients with early-stage breast cancer, a long-established alternative to mastectomy is breast conservation therapy, in which a segmental mastectomy (lumpectomy) is performed, followed by irradiation of the ipsilateral breast (Fisher et al. 2002; Veronesi et al. 2002; Horiguchi et al. 2002; van Dongen et al. 2000; Abe et al. 1995; Holli et al. 2001; Liljegren et al. 2000; Clark et al. 1996; Forrest et al. 1996). As with any irradiation technique, for breast radiotherapy (RT), it is desirable to administer dose uniformly throughout the planning target volume (PTV), limit high-dose spots that otherwise result in a poor cosmetic outcome (Buchholz et al. 1997; Gray et al. 1991; Moody et al. 1994; Taylor et al. 1995; Cheng, Das, and Baldassarre 1994), and to adequately limit the dose received by thoracic organs at risk (OARs) so as to minimize the risk of treatment-related complications (Fisher et al. 1989; Lingos et al. 1991; Wallgren 1992). Achieving breast dose uniformity is challenging given the irregular size and shape of the breast, and inter- and intrafractional variations during breast RT. The positioning and sparing of OARs also complicates efforts toward target dose uniformity. Another major concern with

breast RT is radiation doses to OARs. The inter- and intrafractional variations necessitate large margins to extend clinical target volume (CTV) to PTV that, in turn, increase doses to OARs and thereby radiation-related toxicities. Motivated by the need to improve target dose uniformity and to reduce dose to OARs, technology and methodology for breast RT have been advancing rapidly in recent years. In this chapter, a review is presented of these challenges as well as the techniques developed to overcome them, among these being intensity-modulated RT (IMRT; Mihai et al. 2005; Hong et al. 1999; Teh et al. 2001; van Asselen et al. 2001; Partridge et al. 2001; Hurkmans et al. 2002; Cho et al. 2002; Lo et al. 2000; Vicini et al. 2002) and image-guided RT (IGRT; Jaffray 2007; Morrow et al. 2007; Weed et al. 2004), which are the two most important enabling technologies for adaptive RT.

18.2 Challenges and Concerns in Breast Radiotherapy Delivery

The inter- and intrafractional variations are substantial during the course of breast RT delivery. These variations reduce the accuracy of treatment targeting and increase the dose to OARs.

The secondary effect of radiation to normal structures is a major concern for breast cancer patients.

18.2.1 Interfraction Variability

Similar to other sites, interfractional changes in breast RT include patient positional uncertainties and anatomic variations. To accommodate any variations in the setup of the patient prior to each fraction, as detected via the comparison of planar or volumetric treatment images with planning data, the conventional method is to move the couch to apply appropriate translational shifts along the three cardinal directions (lateral, longitudinal, and vertical). However, on a day-to-day basis there may be additional setup errors as well as changes to the patient's internal anatomy that require more corrections including rotational corrections along these same three axes (e.g., pitch, roll, and yaw). Morrow et al. (2009) investigated the dosimetric impact for breast patients treated in the prone position of allowing six degrees of freedom for patient setup correction, including rotational corrections as well as translational shifts. The patients were treated for either a PTV boost subsequent to whole-breast irradiation (WBI) or for a dedicated partial-breast irradiation (PBI) regimen. Alignment of each daily pretreatment kilovoltage computed tomography (kVCT; acquired with CT-on-rails) with the plan kVCT was done using the lumpectomy-site seroma and at least three postlumpectomy surgical clips. The required rotation angles were as great as 4° for the collimator, 5° for the couch, and 1° for the gantry. Figure 18.1 demonstrates, for a sample patient, the improved alignment of the lumpectomy site with the planned PTV contour when using rotation and translation rather than translation changes alone. Also, using rotation and translation instead of translation alone resulted in values for D_{95} and D_{98} (doses that covered 95% and 98% of the PTV, respectively) that were increased by 2%–5% and 3%–7% on average, respectively. For fractions that required larger rotations, the effect on D_{95} and D_{98} was as great as 15%–20%.

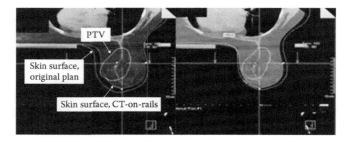

FIGURE 18.1 Improved alignment of the lumpectomy site with the planning target volume (PTV) contour when using rotation and translation rather than translation alone. For a patient being treated to a PTV surrounding the lumpectomy site (indicated by postsurgical clips), the left frame shows the pretreatment kilovoltage computed tomography images corrected for setup error using only translation, and the right frame shows the result of using both translation and rotation. Note that the PTV is better confined within the patient anatomy when rotation was included.

According to a study by Flannery et al. (2009) in which patients being administered WBI had received two sets of CT scans, the first just before treatment began and the second after 21 to 23 fractions, 86% of patients exhibited a decrease in the apparent volume of the lumpectomy cavity. Among all patients in their study, the median volume decrease was 11.2 cm³, and the median percentage change was −32%. For patients who are prescribed WBI followed by a boost to the lumpectomy cavity, the WBI plan and the boost plan are routinely performed on the same (pretreatment) CT image set. Without a dedicated resimulation for the boost plan, there is a risk that the boost fractions may irradiate more ipsilateral breast tissue than is necessary to achieve optimal local control and place the patient at higher risk of a poor cosmetic outcome.

Analyses of daily kVCT data for PBI patients by Ahunbay et al. (2009) demonstrated considerable variation in the location, shape, and volume of the lumpectomy cavity over the course of treatment. Figure 18.2 shows the overlap of lumpectomy-site contours once alignments are made based on the center of mass of the lumpectomy cavity for one patient. Significant interfractional deformation and variation in size are apparent; on average, the overlap between the lumpectomy site on the plan CT and on the daily pretreatment CT was 64%. An estimate of the required expansion margin to accommodate these variations and cover at least 95% of the lumpectomy cavity was 5.3 mm if center-of-mass–based registration was performed, and 12.2 mm if registration was done based on postsurgical clips. If reduced CTV-to-PTV margins are desired, adaptive replanning is required.

18.2.2 Intrafraction Variability

For both WBI and PBI, motion of the ipsilateral breast during treatment can be a concern, even if the breast remains encompassed by tangent fields during irradiation. This is because beam intensity maps for whole-breast treatment fields are generally made nonuniform (either using wedges, multileaf collimator [MLC]-defined subfields within the open tangent field, or IMRT) and the intensity variations are designed to account for tissue heterogeneities within the breast (Ding et al. 2007).

Observations of the effect of respiratory motion on the locations of the breast target and OARs have been made based on studies using four-dimensional CT (4D-CT). A study by

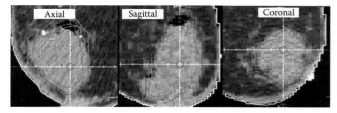

FIGURE 18.2 Superimposition of the contours of the lumpectomy site as apparent on the daily pretreatment kilovoltage computed tomography (kVCT). The figure shows the contours of the lumpectomy site superimposed on the CT image of the lumpectomy site for a partial-breast irradiation patient as apparent on the daily pretreatment kVCT. The lumpectomy site contours are aligned on their respective centers of mass.

Qi et al. (2007) indicated that for supine-positioned patients, displacement of the lumpectomy site over the respiratory cycle has been observed to be as great as 9 mm. Further, for PBI, treatment planning recalculations indicate that this displacement can lead to a decrease in the PTV coverage by as much as 7% relative to the initial treatment plan. For cases in which internal mammary chain (IMC) nodes were intended to receive 45 Gy, V_{45} (volume covered by 45 Gy) was observed to vary by as much as 28% over the respiratory cycle. This indicates that either a larger margin to encompass the IMC nodes is necessary (doing so could increase the dose received by the heart and lungs) or a means to account for respiratory motion, such as gated delivery, is required. Also, a study by Ding et al. (2007) showed that according to measurements from 4D-CT images, the displacement of the breast with radiation was as great as 2.3 cm. For WBI plans, variation of D_{95} was seen to be considerable; for one patient it decreased from 40 Gy to approximately 25 Gy (Ding et al. 2007).

With regard to dose to the heart, observations have generally shown that respiratory motion has minimal effect. Ding et al. (2007) observed that heart dose distribution did not change significantly with respiratory cycle or radiation beam-on timing. At most, there was only a variation of 1 Gy in the maximum point dose to the heart, and variations of the heart V_5 (volume covered by 5 Gy) over the respiratory cycle were within 1% (Ding et al. 2007). Note that cardiac motion occurs with a greater frequency than respiratory motion. Since the patient's heart beats more rapidly than he or she breathes, the heart will appear "blurred out" over its full range of cardiac motion. Thus, one explanation for the minimal effect of respiratory motion on heart dose distribution is that the apparent size and position of the heart in any given 4D-CT phase image is invariant, and thus the calculated dose distribution to the heart will change little with respiratory phase. An investigation by Yue et al. (2007) found that the volume and shape of the heart as well as the ipsilateral breast do not change significantly during respiration, which may explain why the heart dose volume does not significantly change either.

18.2.3 Risk of Secondary Effects of Radiation

Irradiation of various OARs is a major concern in breast RT. Contralateral breast dose avoidance is critical especially for younger breast cancer patients, since such patients have more time to develop a secondary tumor in the contralateral breast. For conventional WBI, dose to the contralateral breast may occur due to radiation scattered off of physical wedges; for this reason, the use of virtual or dynamic wedges is preferred for breast treatment planning. The contralateral breast may receive dose from treatment fields that intercept parts of it; thus, the contralateral breast is kept out of any direct beam path. Dose may also occur due to secondary neutrons that are generated in the treatment head of the linear accelerator (linac) when using photon beams with energy greater than approximately 18 MV; although higher-energy photon beams are often necessary to accommodate larger breast volumes, for IMRT where the monitor units per field can be substantial the beam energy is restricted to 6 MV.

For WBI of a patient in the supine position, dose to the ipsilateral lung is unavoidable given the concave curvature of the ipsilateral breast around the chest wall and lung. For conventional tangent-based breast planning, the volume of the ipsilateral lung receiving lower dose is only slightly greater than that receiving higher dose. However, for IMRT techniques that require a greater number of beam angles to achieve dose conformity, greater volumes of the lung are encompassed by primary beams and thus the low-dose coverage to the lung is greater. This is of concern given reports that respiratory complications may correlate with relatively large volumes of the lung receiving doses of 10 Gy (Lee et al. 2003) or 5 Gy (Wang et al. 2006).

Dose to the heart is another concern. The Early Breast Cancer Trialists' Collaborative Group demonstrated that breast cancer patients having received RT exhibited a greater risk of fatal cardiac events than those who had not (Abe et al. 2005). Dose to the left descending coronary artery is of concern, as it resides along the anterior surface of the heart and is thus nearest to the posterior edge of the tangents' field for left-breast treatments. Sparing the heart from radiation dose is further warranted as several chemotherapy agents are known to exhibit cardiac toxicity independent of RT. Considerable attention to minimizing dose to the heart is paid to patients undergoing left-breast treatments, given both the greater proximity of the heart to the left breast and the possibility of cardiac motion or respiratory motion displacing the heart into the treatment fields. In the effort to reduce radiation doses to OARs, radiation treatment planning and delivery technology is being developed to account for inter- and intrafractional variations, so that the PTV margins can be reduced/eliminated and the radiation-related toxicities minimized.

18.3 Treatment Planning

To address the effects of inter- and intrafractional variations and to improve the effectiveness of breast RT, efforts have been made to increase the accuracy of treatment targeting during treatment planning. These efforts include consistent delineation of the target and OARs, reducing respiratory motion, use of advanced multimodality imaging, stable and reproducible patient positioning, and immobilization.

18.3.1 Delineation of Targets and Organs at Risk

Accurate delineation of the volumes of the targets and OARs is a critical prerequisite for conformal RT, as all subsequent decisions regarding treatment planning and delivery are based on these volumes. A few studies have reported that there are significant variations in defining target volumes for breast RT (Li, Tai, et al. 2009; Landis et al. 2007; Petersen et al. 2007). To reduce such variations, the Radiation Therapy Oncology Group (RTOG) has provided a guideline and has established a breast cancer atlas for RT (www.rtog.org). A brief description of the RTOG guideline for delineating the targets and OARs is provided in Sections 18.3.1.1 and 18.3.1.2.

18.3.1.1 Delineating Targets

The RTOG guideline for delineating targets includes the following aspects:

- Lumpectomy cavity (i.e., gross tumor volume [GTV]) should include seroma and surgical clips when present.
- Lumpectomy CTV should include lumpectomy GTV plus 10–15 mm of surrounding breast tissue excluding pectoralis muscle, all chest wall muscles, ribs, and patient surface.
- Breast CTV should include the apparent CT glandular breast tissue, incorporate anatomic borders (Table 18.1), and include the lumpectomy CTV.
- Chest wall CTV should incorporate the anatomic borders described in Table 18.1 and include the mastectomy scar.
- Regional nodal CTV will depend on the specific clinical case and it should take into account the anatomic borders described in Table 18.2. The three levels of the axilla can overlap caudal to cranial.

18.3.1.2 Delineating Organs at Risk

The OARs should incorporate the anatomic borders described in Table 18.3.

18.3.2 Considerations for Reducing Respiratory Motion

To minimize PTV margins and/or reduce dose to OARs, it is desirable to reduce respiratory motion during breast RT. Several strategies have been applied to reduce the effects of respiratory motion discussed in Section 18.2.2 on the location of the breast target volume. One technique is to position the patient in a prone rather than a supine position; this has been demonstrated to substantially reduce the intrafraction motion of the anterior chest wall to a submillimeter level (Morrow et al. 2007), and thus reduce the uncertainties introduced by respiration.

The 4D-CT has been shown to be useful in developing gated RT plans to reduce dose to the heart for left-sided breast patients (Wennberg et al. 2005). Indeed, a 4D-CT study could be a necessary component of the simulation process to determine beforehand whether a gating technique would be beneficial in reducing cardiac toxicity.

The effects of respiratory motion may also be minimized by accommodating patient respiration during planning and treatment. Audio breathing coaching may help the patient maintain a periodic breathing pattern, although audio coaching might prompt some patients to exhibit breathing amplitudes that are larger or smaller than those of normal free breathing (Neicu et al. 2006). Similar techniques include active breathing control (ABC; Frazier et al. 2004), respiratory-gated RT, deep inspiration breath hold (Remouchamps et al. 2003), the use of external markers or image guidance to track the target motion in real time, and dynamic MLC tracking (Ding et al. 2007). For the latter, time-correlated 4D-CT images are obtained and dynamic MLC leaf sequences are generated for specific respiratory phases. For each respiratory-phase 4D-CT image, the dose is calculated from the MLC pattern, and a composite dose is calculated using deformable registration of the 4D-CT images. In practice, it has been observed that the duration of the respiratory cycle period (between 4 and 8 seconds) and the beam-on timing (the start

TABLE 18.1   Breast and Chest Wall Contour: Anatomic Boundaries

	Cranial	Caudal	Anterior	Posterior	Lateral	Medial
Breast[a]	Clinical reference + second rib insertion[b]	Clinical reference + loss of CT apparent breast	Skin	Excludes pectoralis muscles, chest wall muscles, ribs	Clinical reference + midaxillary line typically, excludes latissimus (lat.) dorsi muscle[c]	Sternal–rib junction[d]
Breast + chest wall[e]	Same	Same	Same	Includes pectoralis muscles, chest wall muscles, ribs	Same	Same
Chest wall[f]	Caudal border of the clavicle head	Clinical reference + loss of CT apparent contralateral breast	Skin	Rib–pleural interface (includes pectoralis muscles, chest wall muscles, ribs)	Clinical reference/ midaxillary line typically, excludes latissimus dorsi muscle[g]	Sternal–rib junction[h]

[a] Appropriate lumpectomy for breast only treatment.

[b] Cephalad border is highly variable depending on breast size and patient position. The lateral aspect can be more cephalad than the medial aspect depending on breast shape and patient position.

[c] Lateral border is highly variable depending on breast size and amount of ptosis.

[d] Medial border is highly variable depending on breast size and amount of ptosis. Clinical reference needs to be taken into account; should not cross midline.

[e] CTV after appropriate lumpectomy for more locally advanced cases includes those with clinical stages IIb and III who receive neoadjuvant chemotherapy and lumpectomy and who have sufficient disease risk to require postmastectomy radiation, had mastectomy been done.

[f] CTV after appropriate mastectomy.

[g] Lateral border meant to estimate the lateral border of the previous breast. Typically extends beyond the lateral edge of the pectoralis muscles but excludes the latissimus dorsi muscle; should encompass the lateral extent of mastectomy scar.

[h] Clinical reference marks need to be taken into account. The chest wall typically should not cross midline. Medial extent of mastectomy scar should typically be included.

TABLE 18.2 Regional Nodal Contours: Anatomic Boundaries

	Cranial	Caudal	Anterior	Posterior	Lateral	Medial
Supra clavicular	Caudal to the cricoid cartilage	Junction of brachioceph.–axillary vns./caudal edge clavicle head[a]	—	Anterior aspect of the scalene muscles (m.)	Cranial: lateral edge of SCM muscle; caudal: junction first rib–clavicle	Excludes the thyroid and trachea
Axilla-level I	Axillary vessels cross lateral edge of pectoralis (Pec.). minor m.	Pec. major muscle insert into ribs[b]	Plane defined by anterior surface of Pec., Maj. m., and Lat. dorsi m.	Anterior surface of subscapularis m.	Medial border of latissimus dorsi m.	Lateral border of Pec. minor m.
Axilla-level II	Axillary vessels cross medial edge of Pec. minor m.	Axillary vessels cross lateral edge of Pec. minor m.[c]	Anterior surface Pec. minor m.	Ribs and intercostal muscles	Lateral border of Pec. minor m.	Medial border of Pec. minor m.
Axilla-level III	Pec. Minor m. insert on cricoids	Axillary vessels cross medial edge of Pec. minor m.[d]	Posterior surface Pec. major m.	Ribs and intercostal muscles.	Medial border of Pec. minor m.	Ribs
Internal mammary	Superior aspect of the medial first rib	Cranial aspect of the fourth rib	[e]	[e]	[e]	[e]

[a] Supraclavicular caudal border meant to approximate the superior aspect of the breast/chest wall field border.
[b] Axillary level I caudal border is clinically at the base of the anterior axillary line.
[c] Axillary level II caudal border is the same as the cephalad border of level I.
[d] Axillary level III caudal border is the same as the cephalad border of level II.
[e] Internal mammary lymph nodes encompass the internal mammary/thoracic vessels.

TABLE 18.3 OARs (Heart and Lung): Anatomic Boundaries

	Cranial	Caudal	Anterior	Posterior	Lateral	Medial
Heart	Inferior aspect of the left pulmonary artery	Loss of CT apparent heart[a]	Pericardium	Excludes descending aorta, esophagus, and vertebral body	Pericardium	Pericardium
Lung[b]	Pleura	Pleura	Pleura	Pleura	Pleura	Pleura

[a] Heart caudal border: The heart blends with the diaphragm and the liver at its caudal end. Adjusting the window/level can assist in discerning the heart vs. these other organs.
[b] Lung volume: The lung volume within the pleura surface excluding ribs, mediastinum, and diaphragm can be autocontoured by most planning systems.

time of the radiation relative to the respiratory cycle) correlated with target dose heterogeneity but otherwise had limited impact on target dose coverage or heart dose avoidance. Target dose coverage and heterogeneity tended to be adversely affected when respiratory cycles were longer than 6 seconds or when the amplitude of the breast motion was greater than 7 mm (Ding et al. 2007). Since the volume and shape of the heart were not observed by Yue et al. (2007) to change significantly with respiration, dose deformation tools would not be necessary for the calculation of heart dose.

18.3.3 Imaging for Treatment Planning

Conventional CT, three-dimensional (3D)-CT, acquired with the patient breathing normally (nongated), is typically used as a basis for breast treatment planning. However, the 3D-CT acquisition may span over several respiratory cycles, and the motion of the breast and associated OARs may contribute to artifacts in the planning image. The 4D-CT may resolve these artifacts, and also provide information regarding the motion of the target and OARs as a function of the patient's respiratory cycle. As anatomic information must be gathered throughout the respiratory cycle, the duration of a thoracic 4D-CT scan is typically 3–5 minutes.

By convention, a respiratory cycle is divided into 10 phases, expressed as percentages of a full respiratory cycle; the data from a 4D-CT scan are reconstructed such that the resulting images correspond to one of the 10 phases of the respiratory cycle. The end-inhalation point of the respiratory cycle (with the maximum lung volume) is defined to be the 0% phase (or 100% phase). The 50% phase is often assumed to be the end-exhalation point (minimum lung volume) of the respiratory cycle. However, depending on the shape of the patient's respiratory trace (e.g., rapid inhalation and slow exhalation), the end-exhalation point may be better characterized by a different respiratory phase.

Magnetic resonance imaging (MRI) has proven useful in refining breast cancer staging; for example, MRI can confirm findings from positron emission tomography (PET) that are suspicious for distant metastasis. Al-Hallaq et al. (2008) have advocated that for patients considered for PBI, the use of MRI in the treatment planning process may help to identify secondary lesions (multifocal, multicentric, or in the contralateral breast) outside the primary tumor site that would otherwise not be identified by mammography and removed during surgery, and thus serve to rule out PBI and suggest WBI.

For intermediate-stage breast cancer patients (stage II or stage III), occult metastasis of the disease is of concern. Given that breast cancers are known to exhibit strong glucose uptake, the management of those patients may be improved by using ^{18}F-labeled 2-fluoro-2-deoxy-D-glucose (^{18}F-FDG) PET fused with CT to identify axillary lymph node involvement not apparent from CT alone, involvement of the supraclavicular and IMC lymph nodes (which are typically not dissected during surgery), or distant metastases (Groheux et al. 2008; Bral et al. 2008). Further, PET may be useful in measuring the metabolic activity of the primary tumor, and thus may help to evaluate the effectiveness of chemotherapy.

PET has been proposed as an alternative to the considerably more invasive axillary lymph node dissection as a means to identify axillary lymph node metastasis. Indeed, PET has been demonstrated to exhibit high specificity (a large number of true negatives relative to false positives) with regard to detecting axillary node involvement. However, micrometastases (lesions less than 2 mm in size) are not readily observed in PET/CT and, therefore, PET/CT is not recommended for staging purposes (as an alternative to axillary lymph node dissection or sentinel lymph node biopsy) with regard to axillary node involvement (Barranger et al. 2003; Fehr et al. 2004; Wahl et al. 2004).

CT is included and coregistered with the PET scan in order to provide additional morphological information and, in the case of PET/CT, more accurate attenuation corrections for the PET data. A typical PET or PET/CT protocol requires patients to fast at least 6 hours prior to the procedure and to exhibit a sufficiently low blood glucose level at the time of ^{18}F-FDG uptake. After a 1-hour wait following FDG infusion (typically administered in the contralateral arm), the patient would lie with both arms positioned superiorly so as to separate the breasts from the heart and to enable favorable imaging of the axillary region. Supine positioning is standard for whole-body PET, although prone positioning with the breasts suspended downward may yield more favorable image results in the breast region. The CT and PET data are obtained from the base of the skull to the middle of the thigh. Fusion of the PET image with the CT image depends on consistent patient positioning for both image acquisitions, especially if the PET is preoperative and the CT is postoperative (Bral et al. 2008). Performing the PET scan using a flat tabletop, instead of the curved tabletop used for most diagnostic PET scanners, also aids in better registration with the planning CT image. Thus, PET/CT is advantageous over stand-alone PET. Respiratory-correlated acquisition is not typically done for breast patients.

18.3.4 Use of High-Density Fiducials

Following the segmental mastectomy procedure to remove a GTV and an acceptable margin from the ipsilateral breast, fiducials in the form of high-density metallic surgical clips are left behind. Often, the presence of fiducials within the breast is incidental, although surgeons, in collaboration with radiation oncologists, can place the fiducials strategically so that they define the lumpectomy site better and are less likely to migrate.

Although fiducials have the advantage of being very easy to identify radiographically, especially with kV-imaging techniques, among their shortcomings is a tendency to migrate as well as an inherent undersampling of the lumpectomy bed; a considerable volume is represented using a relatively small number of points. Weed et al. (2004) estimated that the inherent uncertainty in delineating the lumpectomy bed when using surgical clips is approximately 3 mm.

18.3.5 Patient Positioning and Immobilization

In most cases, external-beam breast RT is administered with the patient lying on his or her back. The patient may be immobilized with both arms outstretched and positioned superior to the head. An advantage of supine positioning is that it allows for easy access of electron cones, if electrons are desired for tumor-bed boosts or IMC irradiation. For portal imaging, this allows for clear visualization of the treatment field. Also, it is not necessary for a planning system to account for the presence of the couch with regard to treatment-beam attenuation calculation.

However, there are several breast RT contingencies where supine positioning is not appropriate. It is not ideal for patients with large, pendulous breasts; for such cases, gravity tends to shift the breast tissue laterally or superiorly, making day-to-day reproducibility of ipsilateral breast positioning difficult. Also, for the tangent fields to encompass the ipsilateral breast, the volume of the ipsilateral lung and (for left-breast cases) the heart included in the direct beam may be prohibitively large. Further, the presence of skin folds can lead to acute skin reactions.

An alternative positioning technique is to have the patient lie on his or her stomach, with an aperture within the table or an immobilization device to enable the ipsilateral breast to fall away from the chest wall. The contralateral breast remains directly between the patient's anterior surface and the platform upon which the patient lies, and is pushed as laterally as possible. One way to achieve this is to place a Plexiglas plate with an aperture to accommodate the ipsilateral breast across two Styrofoam blocks. Both of the patient's arms are immobilized in an outstretched position superior to the head.

With regard to both WBI and PBI, a key advantage of prone positioning is that, for some patients, it allows better dose sparing of the chest wall, lung, and heart as the ipsilateral breast and the lumpectomy bed are further separated from these OARs. This leads to a reduction in the irradiated volume of the heart and lung. Prone positioning also reduces the respiration-correlated chest wall motion. Qi et al. (2007) observed that prone positioning reduced chest wall motion in the anterior–posterior (AP) direction at both the anterior lung surface and the posterior lung surface. With regard to PBI, this may enable a reduction of the CTV-to-PTV expansion for the tumor bed (Qi et al. 2007). As reported by others (Mahe et al. 2002; Formenti et al. 2004) as well as researchers at the Medical College of Wisconsin

(Bergom et al. 2009), when treating patients with large breasts, prone positioning has been shown to improve cosmesis and reduce late toxicity relative to supine positioning while achieving a homogeneous target dose distribution and comparable local tumor control.

A disadvantage of prone positioning is that electron beams cannot be used due to the presence of the treatment couch; further, the accessibility of other treatment beams is limited. Thus, for conventional RT, prone positioning is typically not used when treating advanced-stage disease where nodal irradiation is required. Also, the reproducibility of the ipsilateral breast contour is problematic, with interfractional variations up to 1.5 cm observed. Image guidance and adaptive planning may still be necessary in order to correct for this variation.

In some cases with the patient positioned supine, the tumor bed may not be readily accessible for a boost treatment. An example of this is when the distal extent of the tumor bed is beyond the therapeutic depth for the highest-energy electron beam available on the linac. Similar in concept to prone positioning, the patient may be immobilized on the side, shifting the tumor bed to a location closer to the skin surface while still making it accessible for an electron cone. Ludwig et al. (2010) reported that when resimulating a patient in the lateral decubitus position, the maximum depth of the boost target volume was reduced on average by 2.1 cm. Further, the mean skin dose decreased from approximately 90% on average to approximately 85% on average. With regard to tumor-bed boosts for WBI, a disadvantage of using this technique is that a second CT simulation is necessary.

An alternative technique to enable electron boost treatments for deep-seated tumor beds is to compress the ipsilateral breast so as to reduce the thickness of tissue proximal to the tumor bed. One way to do this is to use a compression paddle made from relatively low-density materials, such as a mammography compression paddle, held in place by a stand mounted on the immobilization device, and rescanning, planning, and treating the patient in this position. Schinkel et al. (2008) have reported that using the compression paddle resulted in a decrease of the maximum depth of the boost target volume by 1.1 cm on average. Another advantage of using compression is that the skin surface is made flatter, thus minimizing dose inhomogeneities that otherwise arise when electron beams are incident on uneven skin surfaces. As with the lateral decubitus technique, the compression technique requires the acquisition of a second CT image.

18.4 Irradiation Techniques

18.4.1 Whole-Breast Irradiation

Breast conservation therapy with lumpectomy (segmental mastectomy of the GTV) followed by ipsilateral breast irradiation has demonstrated equivalent survival and comparable local control relative to radical mastectomy (Fisher et al. 2002; Veronesi et al. 2002; Horiguchi et al. 2002; van Dongen et al. 2000; Abe et al. 1995). What has motivated the irradiation of the whole ipsilateral breast, rather than just the lumpectomy cavity, is the observation that occult disease may be present in regions of the ipsilateral breast remote from the primary site (Rosen et al. 1975).

Typical prescriptions for WBI include 45–50.4 Gy total dose (1.8–2 Gy per fraction). Four randomized trials have evaluated hyperfractionated radiation compared to standard 1.8–2 fractionation for WBI and have demonstrated comparable local control. The most common accelerated WBI regimen applied is the per-fraction dose of 2.66 Gy, resulting in a total dose of 42.56 Gy (Whelan et al. 2002).

Traditionally, for each tangent direction, pairs of wedged fields have been used to account for variations in tissue thickness along the AP and superior–inferior directions. Since the variability of breast-tissue thickness may be more complex than what could be accommodated by a simple wedge, an alternative is to produce for each tangent direction a customized falloff of the beam intensity using MLCs. This can be achieved during the planning process by identifying areas of high dose that arise using the two open tangent fields alone and creating subfields (the so-called field-in-field) that block out the areas of high dose. Whole-breast dose homogeneity is achieved by adjusting the weight of the subfield relative to the open field. Despite the improved dose homogeneity, however, dose calculation may be more uncertain if the subfield aperture is very small in area or if the number of monitor units of the subfield is very low relative to that for the open tangent field.

In recent years, IMRT has been demonstrated to yield homogeneous dose to the ipsilateral breast. Although 3D conformal radiation therapy (3D-CRT) techniques using cardiac blocking tend to give slightly better heart dose avoidance at lower doses, suitable reduction of high-dose coverage of the heart can be achieved using IMRT (Landau et al. 2001; Hurkmans et al. 2002; Ahunbay et al. 2007). Also, Freedman et al. (2009) have reported that acute radiation-induced dermatitis is less prevalent in patients receiving whole-breast treatment via inverse-planned intensity-modulated tangent fields than in those treated with conventional wedged tangent fields.

Inverse-planned IMRT can be performed using only the two tangent fields; alternatively, breast dose homogeneity and conformity may be improved by using additional fields, so long as those fields do not encompass substantial volumes of the contralateral breast and heart. An arc therapy or tomotherapy treatment can further improve the dose homogeneity to the ipsilateral breast; however, the plan may involve beams that are oriented such that the contralateral breast or heart is in the direct beam path, potentially leading to increased doses to these structures that may not be clinically acceptable.

In certain situations, higher doses of radiation may be given to the lumpectomy site in an attempt to control the small foci of disease that may remain following the original excision (Bartelink et al. 2001). Beyond the whole-breast dose of 45–50.4 Gy, a boost dose of 10–16 Gy to the tumor bed (plus an appropriate margin) may be administered so as to bring the cumulative tumor-bed dose to between 60 and 66.6 Gy.

For advanced-stage disease, breast RT may need to incorporate the regional lymph node groups (such as the axillary, supraclavicular, and IMC lymph nodes). The challenges introduced by this incorporation include a larger treatment area and greater difficulty in keeping the OAR dose below tolerance. For conventional 3D-CRT planning, multiple multimodality fields are often required, along with a means to account for isodose mismatches at the field junctions.

Treatment of postmastectomy chest wall using electrons has been demonstrated (Kudchadker et al. 2002), and this can be a desirable technique given that the distal falloff of dose with depth is more favorable for electron beams than for photons (advantageous for sparing of the heart and lung), and given that most linacs will accommodate multiple electron energies ranging from 4 to 21 MeV, with corresponding therapeutic depths ranging from 1.3 to 7 cm. Among the challenges associated with electron therapy is the requirement for fields that are not obliquely incident on the skin surface (a problem that can in part be resolved using custom bolus or electron arc therapy) as well as a variable distal extent of the breast over the treatment volume (which can in part be resolved using custom bolus or intensity modulation). Multiple abutted electron fields with varying energies can be used, although the dosimetry at the field junctions is complicated. Custom bolus can be used to compensate for the tissue variations, although skin sparing is sacrificed and cosmesis becomes less favorable.

18.4.2 Partial-Breast Irradiation

A review of the pattern of failure of local recurrences within the breast following breast-conservation therapy with WBI indicates that the majority of failures occur at or near the original lumpectomy site, whereas failures arising elsewhere in the breast are relatively uncommon. This suggests that the primary effect of postlumpectomy WBI is the reduction of recurrences near the lumpectomy site, rather than addressing any subclinical disease that may be present throughout the ipsilateral breast.

For PBI, the dose delivery is often accelerated to 3–6 Gy per fraction over 5–10 fractions, administered 1 or 2 times per day. If treatments are given twice daily, the fractions are given at least 6 hours apart. This has the advantage of reducing the overall duration of the RT treatment regimen from the 5–6 weeks typical for WBI. Other advantages of PBI treatment are that it should lead to reduced dose to OARs, such as the contralateral breast, heart, and lungs, and to improved cosmesis given that skin reactions may be confined to smaller areas of the breast surface.

The Accelerated Partial-Breast Irradiation Consensus Statement Task Force of the Health Services Research Committee of the American Society for Radiation Oncology has recommended patient selection criteria for PBI (Smith et al. 2009). Beyond the core recommendations for prospective PBI patients (no prior RT, no history of certain collagen vascular diseases, not pregnant, and commitment to long-term follow-up), the task force defines varying levels of suitability for PBI based on patient demographics and characteristics of the disease. For example, patients with advanced-stage disease, larger tumor size, multicentric disease, or nodal involvement may be considered unsuitable for PBI. The NSABP B-39/RTOG 0413 protocol has established criteria for target and OAR dose volumes for PBI planning. Ideally, according to RTOG 0413 (Vicini et al. 2009), the PTV should not exceed 20%–25% of the total ipsilateral breast volume.

Interstitial brachytherapy is the first method developed for the delivery of PBI, and thus it has a much longer follow-up history. For catheter-based brachytherapy techniques, in-breast failure rates have been observed to be low; RTOG 95-17 reported a recurrence rate of 4% (Arthur et al. 2008). The multicatheter brachytherapy PBI technique uses 15–20 narrow flexible catheters that are inserted around the lumpectomy cavity in two to three planes on average. During a treatment fraction, a high-activity ^{192}Ir high-dose rate (HDR) brachytherapy source is stepped internally through each catheter to form an isodose distribution encompassing the lumpectomy cavity. A typical prescription is a total dose of 34 Gy administered in 10 fractions. A disadvantage of this technique is that catheter placement, planning, and treatment delivery are technically difficult. As a result, relatively few institutions have expertise in administering PBI using this method.

A simpler PBI brachytherapy technique is MammoSite, which was approved by the U.S. Food and Drug Administration (FDA) in 2002. The ^{192}Ir HDR source is propagated within a single double-lumen catheter with a balloon at the closed end. Once the catheter is inserted, the balloon is expanded to fill the lumpectomy cavity. The ^{192}Ir source dwells within the catheter long enough to deliver a prescribed dose of radiation to 1 cm of tissue adjacent to the balloon. As with multicatheter brachytherapy, for MammoSite the typical prescription dose is 34 Gy total dose administered in 10 fractions, and the local control rates are similar. MammoSite has the advantage of having technically easier catheter placement and treatment planning, along with more reproducible placement of the balloon from fraction to fraction. Newer multichannel applicators have been developed; among these, the Contura Multi-Lumen Balloon (SenoRx, Inc., Irvine, CA) and the ClearPath-HDR system (North American Scientific, Inc., Chatsworth, CA) deserve special mention. Electron intraoperative RT has also been applied to PBI. Veronesi et al. (2001) have described a technique in which a single fraction of 17–21 Gy is administered following tumor excision using 3–9-MeV electrons. A disadvantage of brachytherapy is that additional surgical intervention is required for placing the catheters. This increases patient discomfort, introduces additional scarring, and leads to a greater probability of infection. Also, brachytherapy procedures are expensive and complex, requiring a radiation oncologist, a dosimetrist or physicist with specific training, a specialized treatment delivery system (such as a HDR remote afterloader), a modified treatment suite including the necessary barriers and interlocks to protect staff from radiation exposure, and stringent adherence to governmental regulations regarding the handling of radioactive materials.

Although external-beam PBI is relatively newer than brachytherapy, it has potentially broader applicability in that only a 3D imaging system and a linac are required, and that it is appealing to patients who wish to avoid surgical intervention. Another advantage of external-beam PBI over brachytherapy is that the former (especially IMRT PBI) can achieve a more conformal and homogeneous dose over the PTV, thus potentially reducing the probability of side effects such as fat necrosis.

The feasibility and reproducibility of external-beam PBI techniques in multiple institutions was demonstrated by RTOG 0319 (Vicini, Pass, and Wong, 2003). A typical external-beam PBI prescription is 38.5 Gy total dose to the PTV administered in 10 fractions over a time span of 5–8 days. Two fractions are given per day with a minimum 6-hour separation between fractions. Radiobiological models suggest that the biologically equivalent dose (BED) for this fractionation scheme is equivalent to 45 Gy total dose given over 25 fractions, assuming the α/β ratio (the dose at which the linear [α] and quadratic [β] components of radiation damage are equal) is equal to 10 Gy for the breast tumor.

As external-beam PBI is relatively new, there are several open areas for investigation. First, whether the dose to OARs such as the lung, heart, and contralateral breast can be kept well below their tolerances has yet to be determined. For IMRT techniques such as helical tomotherapy, minimizing the dose to the contralateral breast is a challenge (Kainz et al. 2009). Also, whether external-beam PBI can be reproducibly administered to the same volume of the breast for each fraction to an accuracy level comparable to, or better than, that achievable with brachytherapy must be established. This requirement motivates study into reproducibility in immobilization of the ipsilateral breast, accommodation for intrafraction patient motion, and image guidance to refine the position of the patient and to possibly modify the treatment beams. Most external-beam PBI is planned and delivered using 3D-CRT techniques. The beam arrangements are usually noncoplanar, and three to five beams are used. For inverse-planned external-beam PBI, these same noncoplanar beams may be intensity modulated, or a more thorough series of beams may be applied, such as the continuous coverage from axial or helical tomotherapy (Kainz et al. 2009).

As the target volume is often small in PBI, the chance that the radiation ports will miss part of the target due to positioning uncertainty is greater. Also, the consequences of geometric misses are greater if there are fewer fractions over which to administer the dose. Thus, for accelerated PBI, there is a strong motivation to use image guidance for accurate target definition and treatment planning, treatment delivery, and posttreatment follow-up.

18.5 Image Guidance for External-Beam Radiation Therapy

Given the great degree of variation that has been observed in positioning of the ipsilateral breast, pretreatment images are acquired prior to each fraction whenever image guidance is available. There is evidence of considerable systematic error in the day-to-day setup of breast patients, and the magnitude of this error can be of the order of 1–2 cm and is patient specific. Traditionally, planar portal images, acquired along either the

cardinal orthogonal directions or along the same directions as the treatment beam angles, are compared with either films from a conventional simulation (before the era of CT-based planning) or digitally reconstructed radiographs (DRRs) generated by the treatment planning system using the planning CT image set. If volumetric image guidance (e.g., CT-based guidance) is available, the 3D image is compared with the planning image in order to determine the appropriate pretreatment shift.

For the purposes of daily correction of patient positioning, an IGRT modality that enables good visualization of the lumpectomy cavity is ideal. If this is not available, good visualization of the postsurgical clips is essential, as their positions correlate well with the location of the lumpectomy cavity (Weed et al. 2004).

18.5.1 Planar Imaging

Conventionally, prior to the first fraction, portal images (film or electronic portal images) are obtained for each of the treatment fields as well as orthogonal images (AP and lateral) for comparison with DRRs generated by the treatment planning system. Subsequently, in treatment, orthogonal portal images for comparison with DRRs are acquired at least once every five fractions. Planar portal imaging is conventionally used to assess the coverage of the ipsilateral breast and to evaluate the amount of lung within the treatment field. A key limitation of planar portal imaging is that it does not provide 3D information.

For many linacs, verification images can only be acquired using the low-energy megavoltage beam (e.g., 6 MV) that is also used for treatment. Thus, spatial resolution and low-contrast resolution are limited. This presents a challenge given the relative uniformity of density of tissues within the breast; also, postsurgical clips are not readily apparent on MV portal images. For treatment machines with kV-imaging capability (e.g., Varian OBI; Varian Medical Systems, Palo Alto, California), kV-based planar imaging may allow the pretreatment refinement of patient positioning with an accuracy and reproducibility that is actually comparable to cone-beam CT (CBCT)-based patient repositioning (Fatunase et al. 2008). Although both kV and MV planar imaging are effective in distinguishing bony anatomy from soft tissue, for sites in which soft tissue dominates (e.g., breast, pelvis), kV planar imaging may be more useful. Onboard fluoroscopy, although not widely used for the treatment of breast cancer, might potentially be used for gated RT to the breast, so as to minimize heart or lung dose.

18.5.2 Fan-Beam Kilovoltage Computed Tomography (Computed Tomography-on-Rails)

Kilovoltage CT (kVCT) imaging provides favorable low-contrast resolution, which is useful in distinguishing the seroma site from the surrounding ipsilateral breast. Also readily apparent are any rotations of the patient's torso and any distortion of the ipsilateral breast.

Fan-beam kVCT IGRT systems usually take the form of CT-on-rails systems, in which separate kVCT imaging and

MV treatment systems are both installed in the same treatment room. An example is the Siemens CTVision CT-on-rails system (Siemens Medical Systems, Malvern, Pennsylvania). The patient couch is common to both systems, and the pedestal angle is rotated 180° to accommodate one unit or the other. During the kVCT scan, the couch is kept stationary and the kVCT scanner moves (on rails, hence the name) such that the patient and the couch pass through the scanner bore.

Prior to the treatment fraction, following a CT-on-rails scan acquisition, manual registration can be done based on either bony anatomy or the locations of the postsurgical clips as apparent on the CT-on-rails image and the planning CT image. Day-to-day variation in the magnitude of the shifts can be as great as approximately 2 cm; this demonstrates the utility of kV CT-on-rails for IGRT of the breast (Morrow et al. 2008).

A study by Robbins et al. (2009) at the Medical College of Wisconsin evaluated the refinement of patient positioning using daily pretreatment CT-on-rails images with the planning CT image set (a 3D alignment technique), relative to the comparison of daily pretreatment portal images with DRRs acquired from the planning CT image set (a 2D alignment technique). For the alignment of daily CT images with the planning CT image set, registration was based on alignment of the center of mass of the lumpectomy cavity apparent in both image sets. For the comparison of portal images with the plan DRRs, the daily portal images were simulated by constructing DRRs from the daily CT-on-rails images. Alignment of these simulated daily portal images with the plan DRRs was done based on the coincidence of the chest wall, surgical clips, or the patient exterior. Figure 18.3 illustrates the variation in daily setup positioning for the 2D-based techniques relative to IGRT using CT-on-rails. 2D-based alignment techniques exhibited systematic and random uncertainties ranging from 3.2 to 4.5 mm relative to the 3D-based alignments using the center of mass of the lumpectomy site.

18.5.3 Fan-Beam Megavoltage Computed Tomography (TomoTherapy)

Low-contrast resolution is not as favorable in MV imaging as it is in kV imaging. However, MV imaging may still be useful in distinguishing the breast from the chest wall (Kainz et al. 2009). High-density fiducials, such as surgical clips, may also be visible, provided the reconstructed slice thickness is not too great so that partial-volume averaging effects occlude the clips on the MVCT image. A slice thickness should be chosen accordingly, although it should not be so narrow as to result in a prohibitively long reconstruction time or produce an excessively large image file on disk.

Fan-beam MVCT is readily available on the TomoTherapy Hi-Art system (TomoTherapy, Inc., Madison, Wisconsin). The MVCT imaging system functions as a third-generation helical CT scanner, in that there is a detector array situated on the same slip ring on which the radiation source (magnetron, linac, and target) is located, and diametrically opposite the target. The

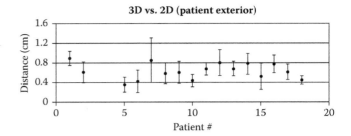

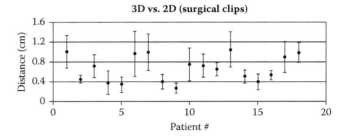

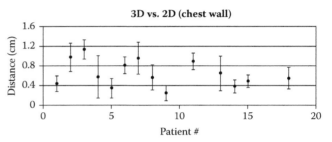

FIGURE 18.3 Plots of the mean and standard deviation of the differences in postshift isocenter position for two-dimensional-based alignment techniques. The techniques are external surface–based (top), clips-based (middle), and chest wall–based (bottom) relative to that for the three-dimensional-based (CT-on-rails image-guided radiation therapy) alignment using the center of mass of the seroma bed.

geometry of the target and detector array is such that the resulting MVCT image has a field of view of diameter 40 cm. Along the craniocaudal direction, the MVCT scan length can be made arbitrarily long. For the MVCT imaging procedure, the accelerating potential of the electron beam is reduced from approximately 6 MV (otherwise used for treatment) to approximately 3.25 MV, in an attempt to provide better low-contrast resolution. Three different reconstructed slice thicknesses are available: 2, 4, and 6 mm.

18.5.4 Megavoltage Cone-Beam Computed Tomography

CBCT is distinct from fan-beam CT in that, rather than requiring the gantry to make multiple rotations using a narrow craniocaudal beam, a wider craniocaudal beam is used (the width being established by the dimensions of the electronic portal imaging device) and only one gantry rotation is required for image reconstruction. Compared to other onboard IGRT modalities, a major advantage of MV CBCT is that since most linacs are equipped with electronic portal imaging devices and the gantry is capable of 360° rotation, MV CBCT can be readily made available on a linac by installing the appropriate image reconstruction software.

However, the MV CBCT approach has several disadvantages. One is that the scan acquisition time for MV CBCT is long due to the limited rotation speed of the several-ton gantry. This can lead to increased motion artifacts (such as those due to respiration) in the resulting image, which is an issue when scanning the region of the breast. Also, for a single scan, the craniocaudal field of view for MV CBCT is limited to the maximum field size at isocenter, which is typically 40 cm. This may or may not be sufficient to cover the breast and associated nodal regions. Further, since for MV CBCT the imaging beam is not as tightly collimated in the craniocaudal direction as it is for fan-beam CT, the MV-CBCT image includes more scatter, which increases the noise and reduces the low-contrast resolution of the image. Relative to kV CBCT, MV CBCT has the inherent disadvantage of a higher-energy imaging beam and thus reduced low-contrast resolution, although some MV CBCT–equipped linacs overcome this disadvantage by including a lower-energy MV X-ray beam dedicated solely for imaging purposes.

Another disadvantage of MV CBCT, specific to IGRT for breast patients, arises from the fact that for most patients, the treatment isocenter is displaced considerably lateral to the patient midline. During MV-CBCT imaging, in order to prevent the gantry from colliding with the patient, the patient must be shifted medially before acquiring the MV-CBCT image and reshifted laterally (including any IGRT-guided corrections to the initial patient setup) after the MV-CBCT scan and before treatment. Such couch indexing can potentially introduce setup errors. A possible remedy for this is to initially set up the patient out of the treatment position intentionally, with the linac's isocenter residing within the patient's midsagittal plane, acquire the MV-CBCT scan, and include the gross lateral table shift into the overall MV-CBCT image registration with the planning CT. A remaining drawback is that a postcorrection CBCT scan cannot be acquired once the shift is made (Kim, Wong, and Yan 2007).

18.5.5 Kilovoltage Cone-Beam Computed Tomography

Manufacturers of RT linacs are increasingly incorporating onboard kV imaging systems, with a kV-imaging axis positioned either 90° or 180° relative to the MV treatment beam. As with the MV beam, the kV beam also lends itself to CBCT. Among the kV-CBCT systems that are available and have been applied to image guided partial breast irradiation (IG-PBI) are the Varian Onboard Imaging (OBI) system (Zhang et al. 2009) and the Elekta Synergy X-Ray Volume Imaging (XVI) system (Elekta CMS Software, Maryland Heights, Missouri; White et al. 2007). Onboard kV CBCT provides volumetric information with favorable contrast resolution for bony anatomy (as is achievable with MV CBCT) as well as for soft tissue.

The kV-CBCT images can be acquired prior to treatment and compared with the planning CT to refine patient setup or to detect significant interfractional anatomic changes that may require replanning. Such comparisons can be based on the alignment of the soft tissue of the tumor bed, any available surgical clips, or the external surface of the breast.

For PBI, kV CBCT has been demonstrated to reduce both the systematic errors and the random error in patient setup along the lateral, AP, and craniocaudal directions (White et al. 2007). The necessary CTV-to-PTV expansion margin, as determined using the formula reported by van Herk (2004), was shown to be reduced from 8.8 mm for a skin-mark-based setup to 3.6 mm for CBCT. Current practice regarding PBI is such that a 1-cm expansion margin is typically used, and the expansion margin required for a skin-mark-based setup is consistent with this practice. However, kV CBCT shows promise in reducing this margin (White et al. 2007), thereby enabling a lower dose to the rest of the ipsilateral breast as well as possibly fewer fractions for PBI.

One advantage of kV CBCT over MV CBCT, and possibly over MV portal imaging, is a reduced dose to the patient per scan; this is beneficial in minimizing the dose to the contralateral breast. Kan et al. (2008) have reported phantom-dose thermoluminescence dosimeter (TLD) measurements for two different kV-CBCT techniques: (1) a "standard mode" using 125 kV, 80 mA, and 25 milliseconds; and (2) a "low-dose-rate mode" using 125 kV, 40 mA, and 10 milliseconds. For chest scans, the mean absorbed dose per scan to the breast was approximately 4.7 and 1.1 cGy, to the lung approximately 5.3 and 1.2 cGy, and to the heart approximately 6.7 and 1.5 cGy, for the standard mode and the low-dose-rate mode, respectively. Fan-beam CT dose measurements indicated the dose per scan to be approximately 3 cGy to the heart and ranging from 1.1 to 2.7 cGy for all other organs.

18.5.6 Tomosynthesis

Digital tomosynthesis (DTS) is a tomographic imaging technique in which the data used to reconstruct the image is compiled over a smaller gantry rotation angle. Whereas for CT, attenuation information is required from projections spanning the full 360° around the object, DTS images can be acquired using gantry-rotation-angle ranges of the order of 45°. For linacs with CBCT capability, the data required for a DTS image is a subset of that required for a CBCT image.

If the gantry-rotation axis is parallel to the patient's craniocaudal dimension and the gantry-angle excursion of the DTS acquisition is centered about the AP view, the data can be used to reconstruct stacks of coronal images; similarly, for an acquisition centered about the lateral view, a set of sagittal DTS images can be reconstructed. To obtain axial DTS images, one would require a scan centered about the craniocaudal view and the gantry-rotation axis to be perpendicular to the patient's craniocaudal dimension; because the patient's length would severely restrict the possible range of gantry motion, high-quality axial DTS images cannot be obtained. In practice, the DTS scan orientation and reconstruction can be chosen along an appropriate oblique direction such that the tumor bed, breast tissue, ribs, and lungs are well-separated. For refinement of patient setup, a pretreatment DTS image can be compared with a reference DTS image that is compiled using DRRs from the treatment planning system.

A general advantage of DTS is that it can yield more anatomic information than planar imaging using either kV or MV X-rays. DTS has several advantages over CBCT, stemming from the reduced scan angle of DTS. First, for DTS, the dose to the patient per scan is lower, since for DTS the range of gantry angles for the scan can be chosen such that dose to key OARs, such as the contralateral breast and contralateral lung, is minimized. Reducing the scan angle also reduces the scan acquisition time (typically, obtaining a CBCT scan requires over 1 minute), a feature that is beneficial in reducing motion artifacts. Also, a reduced scan angle for DTS results in improved geometric clearance of the gantry relative to the patient. For CBCT, since the target volume can be lateral if the couch is shifted accordingly so as to place the PTV at the isocenter for a CBCT scan, the gantry may collide with the patient over a full rotation. Further, if the couch is kept in a medial position for a CBCT scan, the contralateral breast and lung will receive the same dose from the CBCT scan as do the ipsilateral breast and lung. Zhang et al. (2009) have demonstrated that the localization accuracy of DTS is comparable to that achieved with CBCT, consistent to within approximately 1–2 mm. Among the disadvantages of DTS relative to CBCT is that whereas the resolution in a CBCT image is approximately the same in all directions, for a DTS image slice, the resolution is poorer along the off-plane direction than along the two in-plane directions. Also, there is not suitable volume information within a DTS image to calculate tumor or OAR volume. Periodic volumetric CT scans would still be required for this.

18.5.7 Ultrasound-Based Image Guidance

Three-dimensional ultrasound is a noninvasive and nonionizing imaging technique that may be used to improve the accuracy of breast irradiation. For MammoSite brachytherapy treatments, ultrasound images may be used to indicate any interfractional changes in the balloon diameter. For external-beam radiation therapy, 3D ultrasound images can be coregistered with the planning CT image, and this may help to reduce interobserver variability in identifying the postlumpectomy bed. Berrang et al. (2009) reported that the accuracy of the ultrasound-to-CT image registration is within 2 mm. Uncertainties in ultrasound-based localization, relative to CT-based localization, were found to be clinically insignificant, demonstrating the feasibility of using ultrasound-based IGRT for daily pretreatment definition of the tumor-bed location (Wong et al. 2008). Accurate registration of breast images is a challenge due to the relative mobility of the breast (e.g., respiration) as well as the absence of reliable anatomic landmarks as a reference for registration (although registration can be done with respect to treatment-room coordinates or CT-scanner coordinates).

18.6 Image Guidance for Brachytherapy

Traditionally for multicatheter brachytherapy, the 3D reconstruction of the catheters within the breast and their dose calculation is done using two planar films. A disadvantage of this technique is that a complete 3D volume and position of the target volume are not readily available; such information is critical for dose coverage and dose verification with regard to PBI brachytherapy. Although postsurgical clips may help delineate the target volume, there is still the limitation that relatively few points in space are used to characterize a large and possibly complex-shaped volume. Although ultrasound may help define the target volume, distortions of the ipsilateral breast and target volume may arise due to compression from the ultrasound probe. This motivates the use of CT for the purpose of PBI brachytherapy planning. For the planning CT scan, a plastic template (identical in shape to the eventual catheter-loading template, but less dense and more pliable) is placed on the patient's breast to identify potential catheter placement locations (Major et al. 2009). Also, pretreatment CT images may serve to rule out a patient as a candidate for PBI if the lumpectomy site is not readily discernible within the image.

18.7 Adaptive Radiation Therapy

18.7.1 Offline Verification Calculations and Adaptive Strategy

A PBI case that was planned and treated on the TomoTherapy Hi-Art system is presented here as an example to explain offline adaptive strategy. The pretreatment MVCT images acquired for setup-refinement purposes can be used for offline calculations, along with the planned helical tomotherapy beam intensity distribution and a CT number-to-electron density table appropriate for the MVCT image. Such calculations serve to verify that target coverage and OAR avoidance remain consistent with the intent of the original treatment plan and/or to identify deviations from the original plan that require a modification to the treatment plan, thereby providing a robust quality assurance procedure throughout the hypofractionated PBI treatment regimen. The modified (adaptive) plan may be generated using the MVCT or a new kVCT. Kainz et al. (2009) reported the results of MVCT–based verification calculations for four prone-positioned PBI patients treated on the TomoTherapy system in 8–10 fractions. They found the MVCT–based verification calculations to be consistent with the original kVCT–based plan calculations, as illustrated by the isodose comparisons in Figure 18.4 and the dose–volume histogram (DVH) comparison in Figure 18.5.

18.7.2 Online Adaptive Replanning

When using treatment modalities for which onboard volumetric image-guided capability is available, the same images that are used for patient setup verification can be used as a basis for dose verification calculations on the treatment planning system

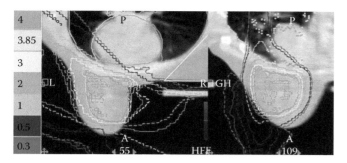

FIGURE 18.4 (See color insert following page 204.) Isodose distributions for a TomoTherapy-treated prone-positioned partial-breast irradiation patient. The isodose distributions are calculated for a pretreatment megavoltage computed tomography (MVCT) image (dashed lines) with the original planned isodose distributions (solid lines) superimposed on the MVCT image. (Reprinted from Kainz, K., et al. 2009. *Int J Radiat Oncol Biol Phys* 74:275–82. With permission.)

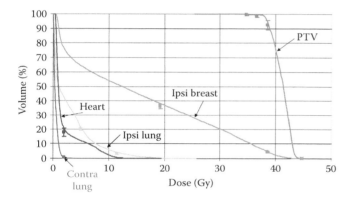

FIGURE 18.5 Plan dose-volume histograms (DVHs; solid lines) and the dose-volume points for a TomoTherapy-treated prone-positioned partial-breast irradiation patient. The figure shows the plan DVHs (solid lines) and the dose-volume points from a verification calculation based on one of the pretreatment megavoltage computed tomography scans. (Reprinted from Kainz, K., et al. 2009. *Int J Radiat Oncol Biol Phys* 74:275–82. With permission.)

(Kainz et al. 2009), and may possibly be used to create a modified treatment plan online (when the patient is lying on the table). This may be useful in instances where the daily verification image indicates dramatic interfractional changes to the geometry of the ipsilateral breast or postlumpectomy bed.

Given the aforementioned observations of interfractional changes to the relative location, shape, and volume of the lumpectomy site as apparent from daily volumetric pretreatment imaging, it is thus desirable to adaptively replan the CTV irradiation in order to preserve the sparing of OARs. Li, Ahunbay, et al. (2009) have studied the feasibility of online replanning for PBI treatments. The overall process required 7–10 minutes for each fraction, a clinically practical amount of time, and consisted of the following steps: (1) acquiring a CT scan at the start of the treatment session; (2) registering that pretreatment CT with the planning CT, and in the process using the planned target, OAR contours, and deformable registration to obtain revised contours for the daily CT image; (3) verifying the revised contours relative to the daily CT; (4) adjusting the apertures for the beam segments to accommodate the revised contours; (5) recalculating dose distributions for the revised beam apertures; (6) optimizing the relative weights for the beams and beam apertures; and (7) transferring these revised beams to the linac console for delivery.

Comparisons of the lumpectomy cavity contours among all fractions indicated interfractional displacements up to 2 cm and fraction-to-fraction CTV overlap as low as 70%. Figure 18.6 demonstrates that the generation of an adaptive plan leads to target dose uniformity and uninvolved ipsilateral breast avoidance that is superior to that achieved by merely repositioning the patient based on daily-CT-to-plan-CT comparison. This "image-plan-treat" paradigm for IGRT enables a reduction in the CTV-to-PTV expansion margin, thereby sparing the uninvolved breast and nearby OARs, as well as hypofractionation for PBI.

18.8 Imaging for Postradiotherapy Follow-Up

For most patients receiving breast conservation radiation therapy, post-RT surveillance is done using mammography, with ultrasound as warranted, to look for signs of in-breast recurrence or new contralateral disease. MRI may be able to distinguish recurrent disease from fibrosis in dense breasts on mammography. In instances where collagen vascular disease (CVD) is discovered after RT, or if RT is necessary despite the presence of CVD, MRI may be used for follow-up imaging (Seale et al. 2008). With regard to post-RT surveillance of the supraclavicular region, Hoeller et al. (2004) observed that although MRI cannot definitively distinguish radiation-induced plexopathy from fibrosis, it can be used to exclude tumor recurrence in this region.

Morphological imaging techniques such as mammography, CT, and MRI detect primary, recurrent, or metastatic disease based on observable size and density changes alone. Functional imaging techniques such as PET are sensitive to changes in metabolic activity. Neoplasms are expected to exhibit abnormal metabolic behavior before demonstrating abnormal morphological behavior; thus, PET may have an advantage in early detection of recurrent disease. Another advantage of FDG-PET imaging for post-RT surveillance is that for a single whole-body scan following a single FDG infusion, the patient can be screened for local recurrence, lymph node metastases, and distant metastases all at once with high sensitivity. The disadvantages of whole-body PET are that specificity is rather low (rarely exceeding 80%) and localization of the area of enhanced update is challenging; however, both of these difficulties can be improved using PET/CT in place of stand-alone PET. There are several circumstances that complicate PET interpretation, such as the presence of infectious disease, muscle activity, brown fat, and bowel activity. Further, the detection rate of bone metastases tends to be low for PET (Lind et al. 2004).

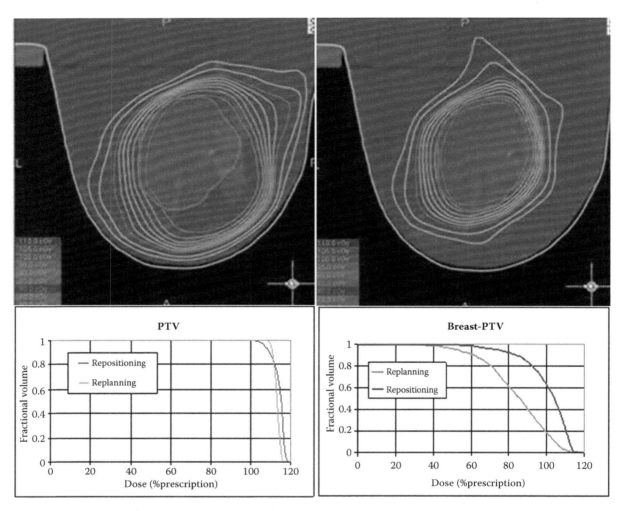

FIGURE 18.6   (See color insert following page 204.) Image-plan-treat paradigm for image-guided radiation therapy. The upper left part of the figure shows the originally planned isodose distribution relative to the planning target volume (PTV) as apparent from the daily pretreatment computed tomography (CT), in the instance where the pretreatment CT is merely shifted to align with the plan CT, The upper right shows a more conformal isodose distribution obtained from generating a new treatment plan based on the pretreatment CT. The lower left shows PTV dose–volume histograms (DVHs) obtained using repositioning alone and using replanning, where the replanned DVH indicates more uniform PTV coverage; and the lower right shows DVHs for the uninvolved ipsilateral breast, for both the repositioning and replanning schemes.

18.9 Future Developments

Numerous newer methods of breast RT that hold promise to improve the therapeutic ratio over historical treatment are under investigation. Improved targeting with image guidance and adaptive methods can help improve local control and avoidance of OAR. Investigations so far reveal that inter- and intrafractional variations during radiation therapy of breast cancer exist and they can be drastic. Image-guided treatment delivery and adaptive approaches are desirable. As of this writing, the technologies and techniques for IGRT and adaptive RT of the breast as described in Sections 18.5 through 18.7 are still under investigation. Research and clinical trials are ongoing to determine whether such novel strategies can accommodate the changes to the patient anatomy over the course of treatment, improve the conformity and uniformity of radiation dose to target volume,

reduce the dose to nearby critical structures and along with it the likelihood of adverse reactions to RT treatment, and ultimately improve treatment outcome.

References

Abe, O., R. Abe, K. Asaishi, et al. 1995. Effects of radiotherapy and surgery in early breast cancer: An overview of the randomized trials. *N Engl J Med* 333:1444–55.

Abe, O., R. Abe, K. Enomoto, et al. 2005. Effects of radiotherapy and of differences in the extent of surgery for early breast cancer on local recurrence and 15-year survival: An overview of the randomised trials. *Lancet* 366:2087–106.

Ahunbay, E., G. P. Chen, P. Jursinic, et al. 2007. Direct aperture optimization based IMRT for whole breast irradiation. *Int J Radiat Oncol Biol Phys* 67:1248–58.

Ahunbay, E. E., J. Robbins, A. Godley, J. White, and X. A. Li. 2009. Interfractional change of lumpectomy cavity during partial breast irradiation [abstract]. *Int J Radiat Oncol Biol Phys* 75:S142–3.

Al-Hallaq, H. A., L. K. Mell, J. A. Bradley, et al. 2008. Magnetic resonance imaging identifies multifocal and multicentric disease in breast cancer patients who are eligible for partial breast irradiation. *Cancer* 113:2408–14.

Arthur, D. W., K. Winter, R. R. Kuske, et al. 2008. A Phase II trial of brachytherapy alone after lumpectomy for select breast cancer: Tumor control and survival outcomes of RTOG 95-17. *Int J Radiat Oncol Biol Phys* 72:467–73.

Barranger, E., D. Grahek, M. Antoine, F. Montravers, J.-N. Talbot, and S. Uzan. 2003. Evaluation of fluorodeoxyglucose positron emission tomography in the detection of axillary lymph node metastases in patients with early-stage breast cancer. *Ann Surg Oncol* 10:622–27.

Bartelink, H., J. C. Horiot, P. Poortmans, et al. 2001. Recurrence rates after treatment of breast cancer with standard radiotherapy with or without additional radiation. *N Engl J Med* 345:1378–87.

Bergom, C., T. Kelly, J. Wagner, N. Morrow, J. F. Wilson, and J. White. 2009. Prone whole breast irradiation using 3D conformal radiotherapy in women undergoing breast conservation for early disease yields high rates of excellent-good cosmetic outcomes in large-pendulous breasts [abstract]. *Proceedings of the 91st Annual Meeting of the American Radium Society,* Vancouver, BC, Canada, April 25–29, 2009, 3.

Berrang, T. S., P. T. Truong, C. Popescu, et al. 2009. 3D ultrasound can contribute to planning CT to define the target for partial breast radiotherapy. *Int J Radiat Oncol Biol Phys* 73:375–83.

Bral, S., V. Vinh-Hung, H. Everaert, P. De Coninck, and G. Storme. 2008. The use of molecular imaging to evaluate radiation fields in the adjuvant setting of breast cancer. *Strahlenther Onkol* 184:100–4.

Buchholz, T. A., E. Gurgoze, W. S. Bice, et al. 1997. Dosimetric analysis of intact breast irradiation in off-axis planes. *Int J Radiat Oncol Biol Phys* 39:261–7.

Cheng, C. W., I. J. Das, and S. Baldassarre. 1994. The effect of the number of computed tomographic slices on dose distributions and evaluation of treatment planning systems for radiation therapy of intact breast. *Int J Radiat Oncol Biol Phys* 30:183–95.

Cho, B. C., C. W. Hurkmans, E. M. Damen, et al. 2002. Intensity modulated versus non-intensity modulated radiotherapy in the treatment of the left breast and upper internal mammary lymph node chain: A comparative planning study. *Radiother Oncol* 62:127–36.

Clark, R. M., T. Whelan, M. Levine, et al. 1996. Randomized clinical trial of breast irradiation following lumpectomy and axillary dissection for node-negative breast cancer: An update. Ontario Clinical Oncology Group. *J Natl Cancer Inst* 88:1659–64.

Ding, C., X. Li, M. S. Huq, C. B. Saw, D. E. Heron, and N. J. Yue. 2007. The effect of respiratory cycle and radiation beam-on timing on the dose distribution of free-breathing breast treatment using dynamic IMRT. *Med Phys* 34:3500–9.

Fatunase, T., Z. Wang, S. Yoo, et al. 2008. Assessment of the residual error in soft tissue setup in patients undergoing partial breast irradiation: Results of a prospective study using cone-beam computed tomography. *Int J Radiat Oncol Biol Phys* 70:1025–34.

Fehr, M. K., R. Hornung, Z. Varga, et al. 2004. Axillary staging using positron emission tomography in breast cancer patients qualifying for sentinel lymph node biopsy. *Breast J* 10:89–93.

Fisher, B., S. Anderson, J. Bryant, et al. 2002. Twenty-year follow-up of a randomized trial comparing total mastectomy, lumpectomy, and lumpectomy plus irradiation for the treatment of invasive breast cancer. *N Engl J Med* 347:1233–41.

Fisher. B., C. Redmond, R. Poisson, et al. 1989. Eight-year results of a randomized clinical trial comparing total mastectomy, and lumpectomy with or without irradiation in the treatment of breast cancer. *N Engl J Med* 320:822–8.

Flannery, T. W., E. M. Nichols, S. B. Cheston, et al. 2009. Repeat computed tomography simulation to assess lumpectomy cavity volume during whole-breast irradiation. *Int J Radiat Oncol Biol Phys* 75:751–6.

Formenti, S. C., N. T. Thuong, J. D. Goldberg, et al. 2004. Prone accelerated partial breast irradiation after breast conserving surgery: Preliminary clinical results and dose-volume histogram analysis. *Int J Radiat Oncol Biol Phys* 60:493–504.

Forrest, A. P., H. J. Stewart, D. Everington, et al. 1996. Randomized controlled trial of conservation therapy for breast cancer: 6-year analysis of the Scottish trial. *Lancet* 348:708–13.

Frazier, R. C., F. A. Vicini, M. B. Sharpe, et al. 2004. Impact of breathing motion on whole breast radiotherapy: A dosimetric analysis using active breathing control. *Int J Radiat Oncol Biol Phys* 58:1041–7.

Freedman, G. M., T. Li, N. Nicolaou, Y. Chen, C. C.-M. Ma, and P. R. Anderson. 2009. Breast intensity-modulated radiation therapy reduces time spent with acute dermatitis for women of all breast sizes during radiation. *Int J Radiat Oncol Biol Phys* 74:689–94.

Gray, J. R., B. McCormick, L. Cox, et al. 1991. Primary breast irradiation in large-breasted or heavy women: Analysis of cosmetic outcome. *Int J Radiat Oncol Biol Phys* 21:347–54.

Groheux, D., J. L. Moreti, G. Baillet, et al. 2008. Effect of ^{18}F-FDG PET/CT imaging in patients with clinical stage II and II breast cancer. *Int J Radiat Oncol Biol Phys* 71:695–704.

Hoeller, U., M. Bonacker, A. Bajrovic, W. Alberti, and G. Adam. 2004. Radiation-induced plexopathy and fibrosis: Is magnetic resonance imaging the adequate diagnostic tool? *Strahlenther Onkol* 10:650–54.

Holli, K., R. Saaristo, J. Isola, et al. 2001. Lumpectomy with or without postoperative radiotherapy for breast cancer with favourable prognostic features: Results of a randomized study. *Br J Cancer* 84:164–9.

Hong, L., M. Hunt, C. Chui, et al. 1999. Intensity-modulated tangential beam irradiation of the intact breast. *Int J Radiat Oncol Biol Phys* 44:1155–64.

Horiguchi, J., Y. Iino, Y. Koibuchi, et al. 2002. Breast-conserving therapy versus modified radical mastectomy in the treatment of early breast cancer in Japan. *Breast Cancer* 9:160–5.

Hurkmans, C. W., B. C. J. Cho, E. Damen, L. Zijp, and B. J. Mijnheer. 2002. Reduction of cardiac and lung complication probabilities after breast irradiation using conformal radiotherapy with or without intensity modulation. *Radiother Oncol* 62:163–71.

Jaffray, D. A. 2007. Image-guided radiation therapy: From concept to practice. *Semin Radiat Oncol* 17:243–4.

Kainz, K., J. White, J. Herman, and X. A. Li. 2009. Investigation of helical tomotherapy for partial-breast irradiation of prone-positioned patients. *Int J Radiat Oncol Biol Phys* 74:275–82.

Kan, M. W. K., L. H. T. Leung, W. Wong, and N. Lam. 2008. Radiation dose from cone beam computed tomography for image-guided radiation therapy. *Int J Radiat Oncol Biol Phys* 70:272–279.

Kim, L. H., J. Wong, and D. Yan. 2007. On-line localization of the lumpectomy cavity using surgical clips. *Int J Radiat Oncol Biol Phys* 69:1305–09.

Kudchadker, R. J., K. H. Hogstrom, A. S. Garden, M. D. McNeese, R. A. Boyd, and J. A. Antolak. 2002. Electron conformal radiotherapy using bolus and intensity modulation. *Int J Radiat Oncol Biol Phys* 53:1023–37.

Landau, D., E. J. Adams, S. Webb, and G. Ross. 2001. Cardiac avoidance in breast radiotherapy: A comparison of simple shielding techniques with intensity-modulated radiotherapy. *Radiother Oncol* 60:247–55.

Landis, D. M., W. Luo, J. Song, et al. 2007. Variability among breast radiation oncologists in delineation of the post-surgical lumpectomy cavity. *Int J Radiat Oncol Biol Phys* 67:1299–308.

Lee, H. K., A. A. Vaporciyan, J. D. Cox, et al. 2003. Postoperative pulmonary complications after preoperative chemoradiation for esophageal carcinoma: Correlation with pulmonary dose-volume histogram parameters. *Int J Radiat Oncol Biol Phys* 57:1317–22.

Li, X. A., E. Ahunbay, A. Godley, N. V. Morrow, J. F. Wilson, and J. White. 2009. An online replanning technique for breast adaptive radiation therapy [abstract]. *Int J Radiat Oncol Biol Phys* 75:S71.

Li, X. A., A. Tai, D. Arthur, et al. 2009. Variability of target and normal structure delineation for breast-cancer radiotherapy: A RTOG multi-institutional and multi-observer study. *Int J Radiat Oncol Biol Phys* 73:944–51.

Liljegren, G., L. Holmberg, J. Bergh, et al. 2000. 10-year results after sector resection with or without postoperative radiotherapy for stage I breast cancer: A randomized trial. *J Clin Oncol* 17:2326–33.

Lind, P., I. Igerc, T. Beyer, P. Reinprecht, and K. Hausegger. 2004. Advantages and limitations of FDG PET in the follow-up of breast cancer. *Eur J Nucl Med Mol Imaging* 31:S125–34.

Lingos, T. I., A. Recht, F. A. Vicini, et al. 1991. Radiation pneumonitis in breast cancer patients treated with conservative surgery and radiation therapy. *Int J Radiat Oncol Biol Phys* 21:355–60.

Lo, Y. C., G. Yasuda, T. J. Fitzgerald, et al. 2000. Intensity modulation for breast treatment using static multi-leaf collimators. *Int J Radiat Oncol Biol Phys* 46:187–94.

Ludwig, M. S., M. D. McNeese, T. A. Buchholz, G. H. Perkins, and E. A. Strom. 2010. The lateral decubitus breast boost: Description, rationale, and efficiency. *Int J Radiat Oncol Biol Phys* 76:100–3.

Mahe, M-A, J-M Classe, F. Dravet, et al. 2002. Preliminary results for prone position breast irradiation. *Int J Radiat Oncol Biol Phys* 52:156–60.

Major, T., G. Frohlich, K. Lovey, J. Fodor, and C. Polgar. 2009. Dosimetric experience with accelerated partial breast irradiation using image-guided interstitial brachytherapy. *Radiother Oncol* 90:48–55.

Mihai, A, E. Rakovitch, K. Sixel, et al. 2005. Inverse vs. forward breast IMRT planning. *Med Dosim* 30:149–54.

Moody, A. M., W. P. M. Mayles, J. M. Bliss, et al. 1994. The influence of breast size on late radiation effects and association with radiotherapy dose inhomogeneity. *Radiother Oncol* 33:106–12.

Morrow, N. V., D. E. Prah, J. R. White, H. Shukla, S. Bose, X. A. Li. 2009. A six-degree online correction technique to account for interfractional variations in breast irradiation [abstract]. *Int J Radiat Oncol Biol Phys* 75:S144–5.

Morrow, N. V., C. Stepaniak, J. White, J. F. Wilson, and X. A. Li. 2007. Intra- and interfractional variations for prone breast irradiation: An indication for image-guided radiotherapy. *Int J Radiat Oncol Biol Phys* 69:910–7.

Morrow, N. V., J. White, J. J. Rownd, and X. A. Li. 2008. IGRT with CT-on-rails for prone breast irradiation [abstract]. *Int J Radiat Oncol Biol Phys* 72:S522.

Neicu, T., R. Berbeco, J. Wolfgang, and S. B. Jiang 2006. Synchronized moving aperture radiation therapy (SMART): Improvement of breathing pattern reproducibility using respiratory coaching. *Phys Med Biol* 51:617–36.

Partridge, M., S. Aldridge, E. Donovan, et al. 2001. An intercomparison of IMRT delivery techniques: A case study for breast treatment. *Phys Med Biol* 46:N175–85.

Petersen, R. P., P. T. Truong, H. A. Kader, et al. 2007. Target volume delineation for partial breast radiotherapy planning: Clinical characteristics associated with low interobserver concordance. *Int J Radiat Oncol Biol Phys* 69:41–8.

Qi, X., A. Sood, J. White, A. Bauer, R. Tao, and X. Li. 2007. Dosimetric impacts of respiratory motion in breast and nodal irradiation [abstract]. *Int J Radiat Oncol Biol Phys* 69:S137–8.

Remouchamps, V. M., F. A. Vicini, M. B. Sharpe, L. L. Kestin, A. A. Martinez, and J. W. Wong. 2003. Significant reductions on heart and lung doses using deep inspiration breath hold with active breathing control and intensity-modulated radiation therapy for patients treated with locoregional breast irradiation. *Int J Radiat Oncol Biol Phys* 55:392–406.

Robbins, J. R., E. E. Ahunbay, J. White, and X. A. Li. 2009. Dosimetric advantages of lumpectomy cavity guided patient positioning in partial breast irradiation [abstract]. *Int J Radiat Oncol Biol Phys* 75:S650.

Rosen, P. P., A. A. Fracchia, J. A. Urban, et al. 1975. "Residual" mammary carcinoma following simulated partial mastectomy. *Cancer* 35:739–47.

Schinkel, C. G., E. A. Strom, J. L. Johnson, et al. 2008. Use of a compression cevice to improve radiation boost dose coverage of the tumor bed for treatment of breast cancer [abstract]. *Int J Radiat Oncol Biol Phys* 72:S187.

Seale, M., W. Koh, M. Henderson, R. Drummond, and J. Cawson. 2008. Imaging surveillance of the breast in a patient diagnosed with scleroderma after breast-conserving surgery and radiotherapy. *Breast J* 14:379–81.

Smith, B. D., D. W. Arthur, T. A. Buchholz, et al. 2009. Accelerated partial breast irradiation consensus statement from the American Society for Radiation Oncology (ASTRO). *Int J Radiat Oncol Biol Phys* 74:987–1001.

Taylor, M. E., C. A. Perez, K. J. Halverson, et al. 1995. Factors influencing cosmetic results after conservation therapy for breast cancer. *Int J Radiat Oncol Biol Phys* 31:753–764.

Teh, B. S., H. H. Lu, S. Sobremonte, et al. 2001. The potential use of intensity modulated radiotherapy (IMRT) in women with pectus excavatum desiring breast-conserving therapy. *Breast J* 7:233–9.

van Asselen, B., C. P. Raaijmakers, P. Hofman, et al. 2001. An improved breast irradiation technique using three-dimensional geometrical information and intensity modulation. *Radiother Oncol* 58:341–7.

van Dongen, J. A., A. C. Voogd, I. S. Fentiman, et al. 2000. Long-term results of a randomized trial comparing breast-conserving therapy with mastectomy: European Organization for Research and Treatment of Cancer 10801 trial. *J Natl Cancer Inst* 92:1143–50.

van Herk, M. 2004. Errors and margins in radiotherapy. *Semin Radiat Oncol* 14:52–64.

Veronesi, U., R. Orecchia, A. Luini, et al. 2001. A preliminary report of intraoperative radiotherapy (IORT) in limited-stage breast cancers that are conservatively treated. *Eur J Cancer* 37:2178–83.

Veronesi, U., N. Cascinelli, L. Mariani, et al. 2002. Twenty-year follow-up of a randomized study comparing breast-conserving surgery with radical mastectomy for early breast cancer. *N Engl J Med* 347:1227–32.

Vicini, F., J. White, D. Arthur, et al. 2009. NSABP Protocol B-39/RTOG Protocol 0413: A randomized phase III study of conventional whole breast irradiation (WBI) versus partial breast irradiation (PBI) for women with stage 0, I, or

II breast cancer, 2004. http://www.rtog.org/members/protocols/0413/0413.pdf/ (accessed August 30, 2010).

Vicini, F. A., H. Pass, and J. Wong. 2003. A phase I/II trial to evaluate three dimensional conformal radiation therapy (3D-CRT) confined to the region of the lumpectomy cavity for stage I and II breast carcinoma, 2003. http://www.rtog.org/members/protocols/0319/0319.pdf. (accessed August 30, 2010).

Vicini, F. A., M. Sharpe, L. Kestin, et al. 2002. Optimizing breast cancer treatment efficacy with intensity-modulated radiotherapy. *Int J Radiat Oncol Biol Phys* 54:1336–44.

Wahl, R. L., B. A. Siegel, R. E. Coleman, and C. G. Gatsonis. 2004. Prospective multicenter study of axillary nodal staging by positron emission tomography in breast cancer: A report of the staging breast cancer with PET study group. *J Clin Oncol* 22:277–85.

Wallgren, A. 1992. Late effects of radiotherapy in the treatment of breast cancer. *Acta Oncol* 31:237–42.

Wang, S.-L., Z. Liao, A. A. Vaporciyan, et al. 2006. Investigation of clinical and dosimetric factors associated with postoperative pulmonary complications in esophageal cancer patients treated with concurrent chemoradiotherapy followed by surgery. *Int J Radiat Oncol Biol Phys* 64:692–9.

Weed, D. W., D. Yan, A. A. Martinez, et al. 2004. The validity of surgical clips as a radiographic surrogate for the lumpectomy cavity in image-guided accelerated partial breast irradiation. *Int J Radiat Oncol Biol Phys* 60:484–92.

Wennberg, B., M. Hussain, R. Odh, et al. 2005. Heart complication following left-sided breast cancer radiotherapy: A gated CT study aiming to understand which patients might benefit from gated RT treatment [abstract]. *Radiother Oncol* 76:S115–6.

Whelan, T., R. MacKenzie, J. Julian, et al. 2002. Randomized trial of breast irradiation schedules after lumpectomy for women with lymph node-negative breast cancer. *J Natl Cancer Inst* 94:1143–50.

White, E. A., J. Cho, K. A. Vallis, et al. 2007. Cone beam computed tomography guidance for setup of patients receiving accelerated partial breast irradiation. *Int J Radiat Oncol Biol Phys* 68:547–54.

Wong, P., R. Heimann, D. Hard, J. Archambault, T. Muanza, and K. Sultanem. 2008. A multi-institutional comparison study evaluating the use of 3D-ultrasound for defining the breast tumor bed for IGRT in chemotherapy versus non-chemotherapy patients [abstract]. *Int J Radiat Oncol Biol Phys* 72:S179–80.

Yue, N. J., X. Li, S. Beriwal, D. E. Heron, M. R. Sontag, and M. S. Huq. 2007. The intrafraction motion induced dosimetric impacts in breast 3D radiation treatment: A 4DCT based study. *Med Phys* 34:2789–800.

Zhang, J., Q. J. Wu, D. J. Godfrey, T. Fatunase, L. B. Marks, and F. F. Yin. 2009. Comparing digital tomosynthesis to cone-beam CT for position verification in patients undergoing partial breast irradiation. *Int J Radiat Oncol Biol Phys* 73:952–7.

Adaptive Radiation Therapy for Lung Cancer

Jun Lian
University of North Carolina

Lawrence B. Marks
University of North Carolina

19.1 Introduction

Lung cancer is a leading cause of cancer morbidity and mortality in the United States, accounting for approximately 219,000 new diagnoses and 159,000 deaths in 2009. Lung cancer accounts for more deaths than breast, prostate, and colon cancers combined (Jemal et al. 2009). Radiation therapy remains one of the main treatment options for lung cancer. Overall, the outcomes with radiation therapy alone for unresectable lung cancer are poor, with 5-year survival rate of approximately 5%–15% (Jemal et al. 2009; Dosoretz et al. 1996; Coy and Kennelly 1980; Dosoretz et al. 1993; Armstrong and Minsky 1989; Dosoretz et al. 1992; Kaskowitz et al. 1993). Technical difficulties for thoracic irradiation and radiation-induced toxicities are among the challenges that hinder improvements in treatment outcomes. Adaptive radiation therapy (ART) is being developed to address these challenges. By adapting to the specific anatomical and biological properties of a given patient or tumor, ART has the potential to deliver individualized treatments.

Lung tumors are subject to respiratory motion, which negatively affects imaging, treatment planning, and delivery for radiation therapy. Failure to consider respiratory motion may lead to inaccurate target definition in treatment planning and underdosing of the target and/or overdosing of the normal tissues during the treatment delivery (Keall et al. 2006). The variation in tissue density in the thoracic region makes dose calculation somewhat challenging. Ignoring these heterogeneities may result in dose discrepancies of ≈5%–15% (Xiao et al. 2009; Ding et al. 2007). These issues are patient-specific and need to be specifically addressed. Accumulated clinical data indicates that the radiation-induced lung toxicities (e.g., pneumonitis) limit the delivery of therapeutic dose to the target (Ghafoori et al. 2008; Marks et al. 2003). Understanding the dose–volume responses of the lung is essential in optimizing radiation therapy for lung cancer. In this chapter, we review and discuss some of the recent developments on these challenging issues toward the ART of lung cancer.

19.2 Technical Challenges of Thoracic Radiotherapy

19.2.1 Interfraction Changes in Lung Tumor Position

This section addresses translational movements of internal targets (e.g., assuming a rigid body). Change in volume is addressed in Section 19.2.2. Juhler-Nottrup et al. (2008) acquired three gated computed tomography (CT) images of 10 patients during the treatment course of 30 fractions. The first CT was acquired before the treatment, the second at approximately fraction 15, and the third at the end of treatment. The three-dimensional (3D) mobility vector of lung tumor was 5.1 ± 2.1 mm (with images matched based on bony landmarks). Chang et al. (2007)

compared the simulation CT and weekly megavoltage (MV) cone-beam CT (CBCT) and found the average centroid displacements of tumor were 2.5 ± 2.7 mm, –2 ± 2.7 mm, and –1.5 ± 2.6 mm in the right–left (LR), anterior–posterior (AP) and superior–inferior (SI) directions, respectively. Table 19.1 summarizes the studies of interfraction lung tumor motion of several research groups. These findings are largely in concordance with each other, with 3D displacement ~ 2–7.5 mm.

19.2.2 Interfraction Change in the Tumor Volume

Several research groups have reported on the changes of lung tumor volume during radiation treatment (Fox et al. 2009; Feng et al. 2009; Bosmans et al. 2006; Chang et al. 2007; Kupelian et al. 2005; Siker, Tome, and Mehta 2006; Underberg et al. 2006). Fox et al. (2009) compared respiration correlated kilovoltage (kV) spiral CT scans taken at 30 Gy (range, 8–40 Gy) and 50 Gy (range, 42–66 Gy) to their corresponding preradiotherapy (pre-RT) scans in 22 patients with non-small-cell lung cancer (NSCLC). Compared to the pre-RT scan, the median gross tumor volume (GTV) was reduced by 25% and 44% at the 30 and 50 Gy scans, respectively (Figure 19.1). Also, using kV spiral CT

scans and comparing the target volume of the scan pre-RT and the scan at 58–60 Gy, Juhler-Nottrup et al. (2008) noted 19% and 34% volume reductions in primary lung tumors and metastatic lymph nodes, respectively. MVCT of helical tomotherapy can be used to monitor changes in tumor volume too. Kupelian et al. (2005) measured the rate of regression of 10 NSCLC patients who were all treated definitively with 2 Gy per fraction (total dose not reported). From an analysis of an average of 27 scans per patient (range, 9–35), they found that the average decrease in volume was 1.2% per day (range, 0.6%–2.3%) over approximately 60 elapsed days from the first treatment day. The lowest rate of shrinkage was observed for the smallest lesion, and the highest rate was observed for the largest lesion.

Feng et al. (2009) quantified changes in fluorodeoxyglucose-avid tumor volume on positron emission tomography (PET)/CT during the course of RT. In their study, 10 patients underwent PET/CT scan before RT and in mid-RT (after 40–50 Gy; Figure 19.2). The mean decreases in CT and PET tumor volumes were 26% (range, +15% to –75%) and 44% (range, +10% to –100%), respectively. They designed boost plans in five patients based on smaller residual tumor volumes on mid-RT PET, which they suggested an escalated total dose up to 102 Gy or a reduction in normal tissue complication

TABLE 19.1 Interfraction Motion of Non-Small Cell Lung Cancer

Author/Center	Number of Patients	Imaging Protocol	Imaging Interval (between Initial and Subsequent Evaluations; Days Unless Otherwise Stated)	Degree of Interfraction Motion, in the Three Primary Axes and the Associated 3D Vector Displacement (mm)			
				SI Direction	AP Direction	LR Direction	3D Vector
Juhler-Nottrup/The Finsen Center, Denmark (Juhler-Nottrup et al. 2008)	10	Gated-CT	14–16 and 29–30	–	–	–	5.1 (range, 1.8–7.1)
Britton/MD Anderson Cancer Center (Britton et al. 2005)	8	4D-CT	Weekly	5.4 ± 3.2	4.5 ± 2.7	3.0 ± 2.3	7.6[a]
Chang/Memorial Sloan Kettering Cancer Center (Chang et al. 2007)	8	MV CBCT	Weekly	1.5 ± 2.6	2.0 ± 2.7	2.5 ± 2.7	3.5[a]
Haasbeek/VU University Medical Center, the Netherlands (Haasbeek et al. 2007)	59	4D-CT	2–12	–	–	–	2.0 (range, 0.2–4.5)
Matsugi/Kyoto University (Matsugi et al. 2009)	4 upper lobe	4D-CT	5–16	1.9 ± 1.1	1.8 ± 0.9	1.0 ± 0.8	2.8[a]
Matsugi/Kyoto University (Matsugi et al. 2009)	4 lower lobe	4D-CT	5–16	3.4 ± 2.3	1.3 ± 1.5	0.4 ± 0.5	3.7[a]
Cheung/University of Toronto (Cheung et al. 2003)	10	Breath-hold CT	Daily in the first week	1.1 ± 3.5	1.2 ± 2.3	0.3 ± 1.8	1.7[a]
Onish/Yamanashi Medical University (Onishi et al. 2003)	20	Breath-hold CT	–	2.2	1.4	1.3	2.9[a]

Note: – Not reported.

[a] Derived from the data of motion in three axes.

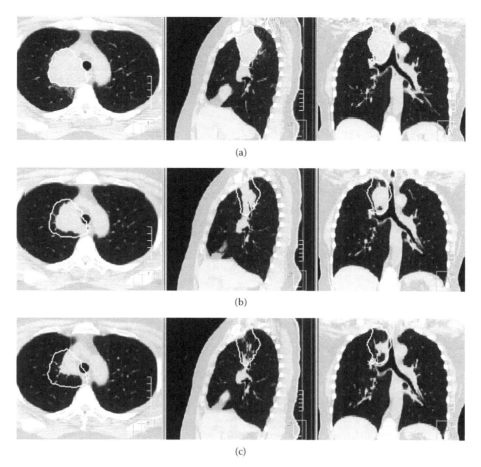

FIGURE 19.1 Reduction of lung tumor gross tumor volume (GTV). Computed tomography images were acquired (a) pre-RT with initial GTV (b) at 30 Gy, with initial contour displayed, showing 42% volume reduction, and (c) at 50 Gy, with initial contour displayed, showing 71% volume reduction. (Reprinted from Fox, J., et al. 2009. *Int J Radiat Oncol Biol Phys* 74(2):341–8. With permission.)

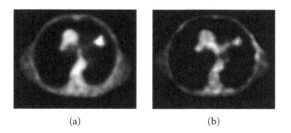

FIGURE 19.2 An example of the change in positron emission tomography tumor volume (a) between pretreatment and (b) after 40–50 Gy during the course of radiation therapy. (Reprinted from Feng, M., et al. 2009. *Int J Radiat Oncol Biol Phys* 73(4):1228–34. With permission.)

probability (NTCP) of 0.4%–3% (mean, 2%). More results are summarized in Table 19.2. It is hard to draw certain conclusions regarding the amount of volume reduction during RT from these studies, because they used different dose prescriptions, patients, and imaging techniques. We believe that the findings are approximately concordant with GTV shrinkage of 10%–45% during RT.

19.2.3 Intrafraction Motion of Lung Tumors

Breathing motion poses special challenges to targeting tumors in or around the lung (Shirato et al. 2004). The degree of intrafraction motion with free breathing has been extensively reported (Keall et al. 2006). Barnes et al. (2001) examined tumor motion with fluoroscopic imaging. They found lower lobe lesions, located closer to the diaphragm, had a greater mean SI motion when compared to upper, middle, and mediastinal lesions (18.5 mm vs. 7.5 mm). Seppenwoolde et al. (2002) measured 3D trajectories for 21 patients via dual real-time fluoroscopic imaging of implanted fiducial markers in or near the tumor (Figure 19.3). They found motion in SI direction is more dominant (range, 0–25 mm) than in the AP (range, 0–8 mm) and LR (range, 0–3 mm) directions. Table 19.3 shows the studies of intrafraction lung tumor displacement with free breathing. Thus, the degree of intrafraction motion is typically in the range of 5–20 mm. The most prominent displacements were often reported to be more in the lower lobe of the lung and to be mostly in the SI direction. The degree of motion can be greater with patients purposefully taking large breaths.

TABLE 19.2 Interfraction Volume Change of Non-Small Cell Lung Cancer

Author/Center	Number of Patients	RT Protocol: Total Dose, Dose Per Fraction (Gy)	Imaging Technique	Dose Delivered When Imaging and Time Interval between Initial and Subsequent Evaluations	Volume Reduction
Juhler-Nottrup/The Finsen Center, Denmark (Juhler-Nottrup et al. 2008)	10[a]	60 Gy, 2 Gy per fraction	Gated helical CT	≈ 60 Gy in 6 weeks	Lung tumor: 19%; mediastinal tumor: 34%
Fox/Johns Hopkins (Fox et al. 2009)	22	50–74 Gy, 2 Gy per fraction	4D-CT	30 Gy in 3 weeks[b]; 50 Gy in 5 weeks[c]	30 Gy: 25% 50 Gy: 44%
Kupelian/MD Anderson (Kupelian et al. 2005)	22	2 Gy per fraction[d]	Tomotherapy MVCT	Daily	1.2% per day
Siker/University of Wisconsin (Siker et al. 2006)	25	22–80 Gy[e]	Tomotherapy MVCT	End of tx	28% (range, 1%–79%)
Haasbeek/VU University Medical Center, the Netherlands (Haasbeek et al. 2009)	59	60 Gy, 20 Gy or 12 Gy per fraction; SBRT	4D-CT	After first fraction; mean 6.6 days (range, 2–12)	8.1%
Woodford/The University of Western Ontario (Woodford et al. 2007)	17	60–64 Gy, 2 Gy per fraction	Tomotherapy MVCT	60 Gy in 6 weeks	38% (range, 12%–87%)
Feng/University of Michigan (Feng et al. 2009)	10	90 Gy, 2 Gy per fraction	FDG-PET/CT	40–50 Gy in 4–5 weeks	CT: 26% (range, 15%–75%); PET: 44% (range 10%–100%).

[a] 8/10 with metastatic disease
[b] Range, 8–40 Gy
[c] Range, 42–66 Gy
[d] Total dose not reported
[e] Variable fraction size

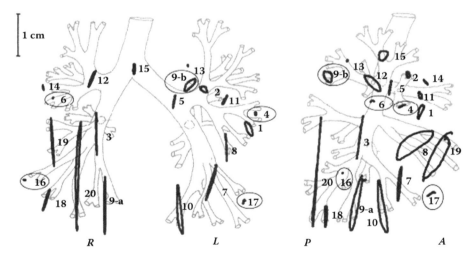

FIGURE 19.3 Orthogonal projections of the trajectories of 21 tumors on (left) the coronal plane and (right) the sagittal plane. The tumors are displayed at the approximate position, based on the localization mentioned in the treatment chart. Tumors that were attached to bony structures are circled. (Reprinted from Seppenwoolde, Y., et al. 2002. *Int J Radiat Oncol Biol Phys* 53(4):822–34. With permission.)

With breath-control techniques, the degree of intrafraction motion may be reduced. Panakis et al. (2008) used fluoroscopy to measure the intrafraction tumor movement with free-breathing and active breathing control (ABC) in 12 patients. The random contribution of periodic tumor motion was reduced 96% (from 8 to 0.3 mm) in the SI direction with the use of ABC. The magnitudes of motion in the other two directions were also

decreased from ~5 to ~0.5 mm. Barnes et al. (2001) found that the SI tumor motion of patients with free breathing and deep inspiration breath hold (DIBH) was 12.9 and 2.8 mm, respectively; this is a 70% mean reduction in motion. More results are shown in Table 19.4. In summary, the degree of intrafraction lung tumor motion with breath control is in the range of ~1–10 mm, which is likely dependent upon the control technique that is used.

TABLE 19.3 Intrafraction Lung Tumor Motion with Free Breathing

Author/Center	Number of Patients/ Tumors	Imaging Protocol	Degree of Intrafraction Motion, in the Three Primary Axes and the Associated 3D Vector Displacement (mm)			
			SI Direction	AP Direction	LR Direction	3D Vector
Seppenwoolde/Netherlands Cancer Inst. (Seppenwoolde et al. 2002)	17 (3, metastatic)	Fluoroscopy	5.8 (0–25)	1.5 (0–3)	2.5 (0–8)	6.5[a]
Hugo/William Beaumont (Hugo et al. 2009)	17	4D-CT[b]	7.0	2.0	1.0	8.0
Barnes/Cross Cancer Institute (Barnes et al. 2001)	4 Lower lobe	Fluoroscopy	18.5 (9–32)	–	–	–
Barnes/Cross Cancer Institute (Barnes et al. 2001)	4 Middle or upper lobe	Fluoroscopy	7.5 (2–11)	–	–	–
Chen/Cleveland Clinic Foundation (Chen et al. 2001)	20	Fluoroscopy	(0–50)	–	–	–
Grills/William Beaumont (Grills et al. 2003)	18	Fluoroscopy	(2–30)	(0–10)	(0–6)	–
Plathow/University of Heidelberg (Plathow et al. 2004)	9 Lower lobe	Dynamic MRI	9.5 (4.5–16.4)	6.1 (2.5–9,8)	6.0 (2.9–9.8)	12.8[a]
Plathow/University of Heidelberg (Plathow et al. 2004)	4 Middle lobe	Dynamic MRI	7.2 (4.3–10.2)	4.3 (1.9–7.5)	4.3 (1.5–7.1)	9.4[a]
Plathow/University of Heidelberg (Plathow et al. 2004)	6 Upper lobe	Dynamic MRI	4.3 (2.6–7.1)	2.8 (1.2–5.1)	3.4 (1.3–5.3)	6.2[a]
Panakis/The Royal Marsden NHS Foundation Trust (Panakis et al. 2008)	12	Fluoroscopy	8.0 (0–21)	5.1 (0–9)	4.2 (0–9)	10.4[a]
Britton/MD Anderson Cancer Center(Britton et al. 2005)	8	Weekly 4D-CT[b]	8.6 ± 1.9	3.9 ± 0.8	1.9 ± 0.5	9.6[a]
Stevens/MD Anderson Cancer Center(Stevens et al. 2001)	22	Radiographs	4.5 (0–22)[c]	–	–	–
Erridge/Western General Hospital (Erridge et al. 2003)	25	Radiographs	12.5 (5–34)	9.4 (5–21)	7.3 (3–13)	17.3[a]
Hanley/Memorial Sloan-Kettering Cancer Center (Hanley et al. 1999)	3	Breath-hold CT[d]	12 (1–20)	5 (0–13)	1 (0–1)	13.0[a]

Note: – Not reported.

[a] Derived from displacement of each axis.

[b] 4D-CT is acquired during the simulation. It is used to estimate the motion during the therapy. The reported displacements are mean values calculated from the CTs of all patients.

[c] The mean tumor motion for all 22 patients was 4.5 ± 5.0 mm (range, 0–22), but 10 patients demonstrated no SI motion. For 12 patients with motion, the mean tumor motion was 8.3 ± 3.7 mm (range, 3–22).

[d] Patients received shallow inspiration breath-hold CT and shallow expiration breath-hold CT. The displacement was measured on two CTs, which represents the maximum displacement of the tumor during free breathing.

19.2.4 Dose Calculation of Lung Radiotherapy

19.2.4.1 Dose Calculation Issues Due to the Motion

Inter- and intrafraction motions can affect the delivered radiation dose in interesting and unforeseen ways. First, consider the simplest situation of a conformal non-intensity-modulated radiation therapy (non-IMRT) field. Inter- and intrafraction motions can blur the dose distribution near the field edges. This simple phenomenon can occur anywhere in the body but is of particular concern in the thoracic region due to prevalence and magnitude of intrafraction respiration–associated motion. This is analogous to keeping the shutter of a camera open while filming a moving object. The amount of blurring depends on the amplitude and the characteristics of the motion as well as the sharpness of static dose distribution. McCarter and Beckham (2000) showed the significant difference of beam penumbra through a simulation study on a 10 × 10 cm field. Without correcting the motion, the real dose can be about 10% lower than anticipated at a point 4 cm from the central axis of the field.

TABLE 19.4 Intrafraction Lung Tumor Motion with Breathing Control

Author/Center	Number of Patients/ Tumors	Breathing Control	Imaging Protocol	Degree of Intrafraction Motion, in the Three Primary Axes and the Associated 3D Vector Displacement (mm)			
				SI Direction	AP Direction	LR Direction	3D Vector
Purdie/Princess Margaret (Purdie et al. 2007)	7 Middle or lower lobe	Full-body vacuum pillow	kV CBCT[b]	–	–	–	6.7
Purdie/Princess Margaret (Purdie et al. 2007)	21 Upper lobe	Full-body vacuum pillow	kV CBCT[b]	–	–	–	4.2
Bissonnett/Princess Margaret (Bissonnette et al. 2009)	18	Abdominal compression if the observed tumor excursion exceeded 10 mm	Respiration correlated kV CBCT[b]	–	–	–	6.2
Barnes/Cross Cancer Institute (Barnes et al. 2001)	4 lower lobe	DIBH	Fluoroscopy	2.4 (1.4–3.8)	–	–	–
Barnes/Cross Cancer Institute (Barnes et al. 2001)	4 middle or upper lobe	DIBH	Fluoroscopy	2.6 (1.0–3.3)	–	–	–
Panakis/The Royal Marsden NHS Foundation Trust (Panakis et al. 2008)	12	ABC	Fluoroscopy	0.3 (0–1)	0.6 (0–2)	0.6 (0–3)	0.9[a]

Note: – Not reported.

[a] Derived from displacement of each axis.

[b] CTs were acquired at the beginning, midpoint, and end of the treatment, and they were used to estimate intrafraction motion.

Second, modern RT delivery often involves moving modulators (dynamic wedges or multileaf collimators [MLCs]), and/or moving gantries. There may be an interplay between the respiratory (or other) intrafraction motion and the movement of these devices. This theoretically can lead to unusual and unforeseen effects such as marked overdosing or underdosing of different regions. In broad terms, because the movement of the delivery mechanisms and respirations are totally independent, there can be, by chance, synergistic effects. The more the beams are used, or the longer the treatments, the *less* likely it is that any of these synergistic effects will be dominant (i.e., the synergistic effects, if present, will occur only during a small fraction of the overall treatment).

Conversely, for rapid treatment times, or a limited number of beams, unusual random synergy can have a dominant effect (i.e., there are fewer other beams, and less treatment time, to "drown out" the effects of synergy). The interplay between respiration and a dynamic wedge was studied by Pemler et al. (2001). For a breathing cycle with 3-cm amplitude and a 6-second period, the maximum dose variation was estimated to be 2.5%, 7%, and 16% for 10°, 30°, and 60° wedges, respectively. For an MLC-based IMRT, the mechanism of the interplay effect is different when the angle between tumor motion and MLC motion is 0° or 90°. Yu, Jaffray, and Wong (1998) considered scenarios of MLC moving along the *same direction* as tumor motion. Using an example of 30 fractions and assuming that each starts with a randomly selected breathing phase, they found ≈10% dose errors. Theoretically, there will be a different interaction each day as the phases of the IMRT delivery and respiration will vary from fraction to fraction. Jiang et al. (2003) simulated MLC motion *perpendicular* to the tumor motion, a more likely situation, and noted the possibility for severe interplay for treatments with a small number of beams and fractions. The maximum dose variation was 30% for 1 field and 1 fraction, 18% for 5 fields and 1 fraction, and less than 1%–2% for 5 fields and 30 fractions.

The thorax presents an added variable that makes the dose calculation more complex—variations in tissue density. Not only are structures moving during the treatment, but the density of the tissues, and hence their associated beam attenuation, will vary. If the densities of all structures were uniform and the patient's size changes with respiration were ignored, we can estimate the target movements within a stationary dose cloud. However, the movement of interthoracic structures will deform the originally calculated dose cloud (Bortfeld, Jiang, and Rietzel 2004). Fortunately, this effect is likely to be relatively modest because the degree of attenuation variation itself is modest, and the interplay is likely to vary over time, thus reducing its overall effect during a course of radiation treatment. Indeed, Mexner et al. (2009) studied the total accumulated dosimetric effect over the respiratory cycle using deformable image registration to analyze 10 patients' 4D data. They concluded that density variations need not be considered during 4D dose calculations and that planning based on the mid-ventilation CT with an appropriate margin was a sufficient method to account for respiration-induced anatomy variations.

19.2.4.2 Dose Calculation Issues Due to the Tissue Heterogeneity

Dose calculation on a static thoracic image is challenging, largely due to uncertainties of electronic disequilibrium at the lung-air interfaces (Lax et al. 2006; Matsuo et al. 2007;

Panettieri et al. 2007; Ding et al. 2007; Xiao et al. 2009). Failure to consider tissue inhomogeneities can lead to inaccurate dose estimates and inadequate tumor doses, often related to increases in the penumbra within low-density materials. This is particularly a problem if the field margins around a target are relatively narrow.

Xiao et al. (2009) evaluated the dosimetric effects of heterogeneity correction of 59 patients accrued to RTOG 0236 (NSCLC SBRT) protocol. They found significant differences between the calculated doses submitted by multiple institutions and the actual heterogeneity-corrected doses. Figure 19.4 shows the isodose distributions with and without heterogeneity correction for one of the cases submitted. With the heterogeneity correction applied, the prescription isodose line contracted and the 50% isodose line expanded, which is also reflected in the dose–volume histogram (DVH) of the planning target volume (PTV). The percent of volume receiving 60 Gy decreased from 95% to 60% in this case. Even more modest changes in dose can theoretically alter the tumor control probability (TCP) and NTCP, if these probabilities have a steep slope (Vanderstraeten et al. 2006; Fraass, Smathers, and Deye 2003; Papanikolaou et al. 2004).

19.3 Technologies Used for Addressing Challenges of Thoracic Radiation Therapy

19.3.1 Interfraction Changes in Lung Tumor Position

19.3.1.1 Assessments

Photon-based imaging assessments include 2D planar radiographs (traditional and convenient) and volumetric CT imaging. Either can be generated via a MV beam (i.e., the treatment source) or a kV source. The latter provides somewhat better tissue contrast. 2D localization images in patients with lung cancer are typically aligned based on bony landmarks, or possibly the visible soft-tissue tumor against the low-density

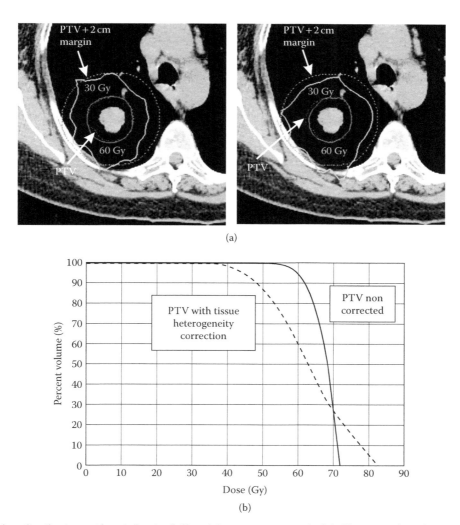

(a)

(b)

FIGURE 19.4 (a) Isodose distributions with unit density (left) and density corrections (right). (b) Dose–volume histograms for planning target volume (PTV) coverage with unit density and density correction. (Reprinted from Xiao, Y., et al. 2009. *Int J Radiat Oncol Biol Phys* 73(4):1235–42. With permission.)

lung, or at times an implanted marker. 3D imaging with kV or MVCT (e.g., CT-on-rails or CBCT) provides additional anatomical information for better localization. Both 2D and 3D imaging have limitations related to respiration—albeit of different types. 2D imaging is fast and accurately represents the anatomy at a given point in time, but is not necessarily representative of the entire respiratory cycle. Multiple 2D images (e.g., fluoroscopy) address this limitation.

On the other hand, 3D imaging is time-consuming (a full rotation for a CBCT takes approximately 30–60 seconds), and resultant images may be blurry if they are obtained during free breathing. This limitation can be addressed via breath-hold CBCT. This is possible, but requires the patient to hold their breath for a prolonged time for multiple image acquisitions to be used to generate a complete image set. Further, this is only useful if the patient is to be treated in a similar breath-hold state. Alternatively, CT images can be sorted based on the respiratory cycle, that is, 4D-CT. This approach is being widely recommended in patients with lung cancer (Hugo et al. 2006; Sonke et al. 2005). 4D helical or axial CT images can be obtained during treatment planning and 4D respiratory-correlated CBCT can be acquired during patient setup. Imaging needs to be coordinated with a respiratory signal obtained *during* image acquisition. 3D-CT images can be obtained either during only specific phases of the respiratory cycle (i.e., prospective gated acquisition) or throughout the respiratory cycle and then retrospectively sorted into different portions of the respiratory cycle. Respiratory-correlated CT reduces motion artifacts in CT images and provides respiratory motion information facilitating 4D-RT (Sonke et al. 2005).

19.3.1.2 Management

This section will discuss how to manage the interfraction motion of lung tumors. One can provide a margin around the CTV (i.e., a PTV) to compensate for the motion. The International Commission on Radiation Units and Measurements (ICRU) Report 50 (ICRU 1993) and the Supplement Report 62 (ICRU 1999) describe the current recommendation for the incorporation of tumor motion into radiation therapy planning. Briefly, the GTV is a tumor that is apparently visible on imaging studies, the clinical tumor volume (CTV) is the GTV plus a margin to account for local microscopic spread of malignant cells, and the internal target volume (ITV) is the CTV plus a margin for internal motion (e.g., respiratory; Guerrero et al. 2004; van Sornsen de Koste et al. 2001; ICRU 1993; ICRU 1999). van Herk discussed treatment margins based on physical and biological considerations (van Herk 2004; van Herk et al. 2003). They recommend 0.25 A (amplitude of respiration) caudally and 0.45 A cranially when there is respiratory motion. When patient setup is based on the tumor itself, not the surrogate, such as a bony landmark, this may facilitate a reduction in the PTV.

Pre-RT imaging can be used to assess target location with a shift of the table to bring the treatment isocenter to the planned isocenter. Multiple options for pre-RT imaging are described in Section 19.3.1.1.

Pre-RT imaging can also be used to assess target location with an adaptation of the plan by repositioning of the jaws or MLC of the initial beams into a new position to reflect the "new" treatment isocenter. Theoretically, this approach can also address changes in the tumor shape and size. This "adaptive" approach has been reported for patients with prostate cancer in retrospective simulation studies (Rijkhorst et al. 2007; Court et al. 2005; Wu et al. 2008; Fu et al. 2009). In practice, this may *not* be feasible for lung lesions, in part because the magnitude of intrafraction motion of lung tumors usually well exceeds that of interfraction motion. This adaptive approach is cumbersome as it requires resegmentation of the 3D images and recalculation of the treatment plan.

19.3.2 Interfraction Changes in the Tumor Volume

19.3.2.1 Assessment

Changes in the size and shape of tumors can be assessed via 3D imaging such as CT, magnetic resonance imaging (MRI), and PET (Biederer et al. 2009; Feng et al. 2009). Similarly, changes in normal lung anatomy can be detected via CT and MRI. Single-photon emission CT (SPECT) has also been used as a means to measure regional lung function. Changes in SPECT-defined perfusion can thus also be used as an assessment of changes in regional lung function during RT (Kong et al. 2005).

19.3.2.2 Management

In most instances, there are no changes made to plans to address changes in target volume or shape. If there are reductions in the size of the tumor, we can alter the RT plan (i.e., shrink the fields). The logic behind this approach is questionable. When a grossly involved region of the lung becomes normal in appearance during radiation treatment, due to tumor shrinkage, that region of the lung may contain a residual microscopic cancer. Thus, shrinkage of the field size may not be justifiable. This certainly is an area in need of further study.

It might be reasonable to limit "supranormal" boost doses to residual abnormalities as part of a clinical study. For example, if one were to do a study of conventional therapy with a boost for unresectable lesions, it might be reasonable to base the boost volume on the residual radiological abnormality, rather than on the pre-RT imaging. Further, in cases where the normal tissue exposure is "too high" to deliver a therapeutic dose, shrinking the field during RT, to reflect tumor shrinkage, might be a reasonable compromise to facilitate treatment if the alternative is no treatment.

The one situation where meaningful changes in the target volume can be readily justified is in patients with ambiguous target volumes at the start of therapy and where the target volume becomes more easily defined during therapy. A typical scenario is a tumor associated with collapsed or atelectatic lung. Tumor masses will often obstruct adjacent airways, leading to a collapse of the distal lung; distinguishing the tumor from the collapsed lung can be challenging. In these cases, the initial RT field may be made large to cover all of the gross abnormality. During therapy, the airway may open, and the boundaries of the target may become more apparent. Subsequent alterations in the target volume are reasonable in this setting. Further, such "reaeration" of parts of the lung can be associated with large intrathoracic anatomical shifts, and adjustments in the field size are prudent. The need to adjust the fields due to large shifts can often be readily detected on planar imaging. However, the replanning itself usually requires 3D imaging. This "reaeration" scenario is most relevant in patients with central tumors that compress the major airways.

19.3.3 Intrafraction Motion of Lung Tumors

19.3.3.1 Assessment

Intrafraction changes in lung position can be assessed by several modalities (Evans 2008) including

2D radiographs: Fluoroscopy and cine images of electronic portal imaging device. The tumor itself, or a surrogate such as the diaphragm or an implanted radiopaque marker, can be detected.

Respiratory-correlated 4D-CT images: With this approach, multiple 4D image sets are obtained during multiple free-breathing respiratory cycles (Bissonnette et al. 2009). Each of the multiple CT images, and hence the tumor position in 3D space, is sorted into different bins defined by different phases of the respiratory cycle, detected via spirometry or the location of a surrogate structure such as the chest wall. Then, during therapy, the position of the tumor can be inferred by monitoring the position of the chest wall or via spirometry. In this way, the intrafraction motion of the tumor can be estimated.

Nonionizing radiation methods: Implantable RF coils can emit a signal whose location can be determined via triangulation (Seiler et al. 2000; Balter et al. 2005).

Table 19.5 summarizes the image techniques used for lung RT. Refer to Chapters 9 and 10 for more details.

19.3.3.2 Management

Approaches to address lung intrafraction motion include (Jiang 2006a; Mageras, Yorke, and Jiang 2006)

Placement of a physical "safety" margin around the CTV to define an ITV: The size of the margin can be determined by pre-RT assessment of tumor motion (e.g., fluoroscopy or 4D-CT). The margin formula was described previously in Section 19.3.1.2. However, the patient-specific motion information measured for treatment planning may not accurately represent the situation during treatment delivery (van Herk 2004). The image acquired from the treatment can be used to reevaluate the margin.

Reducing tumor motion by controlling the patient's breathing: For example, this would include ABC (where the flow of air into the patient's airway is restricted to a

prescribed volume to obtain the desired intrathoracic anatomy) and voluntary DIBH (where the patient essentially holds his or her breath). Both require the patient's tolerance, compliance, and active participation, which may be difficult for patients with lung cancer and compromised lung function.

Allowing the tumor motion, but controlling the RT delivery pattern and/or device temporally and spatially so that the target maintains the desired relative position within the treatment beam(s): This can be achieved via gating of the beam (i.e., turning the beam on/off during specific phases of the respiratory cycle) and tracking (moving the RT beam with the respiratory cycle; Jiang 2006b).

Respiratory gating is relatively accurate, tolerable, and practical for many patients (Jiang 2006a). The first internal gating system was developed by Hokkaido University and Mitsubishi Electronics Co., Japan. Gold fiducial markers implanted to tumor are tracked by their real-time tumor tracking treatment system. The system consists of four X-ray units, and a tracking signal is used to control the beam on/off of linear accelerator (Shirato et al. 1999; Shirato et al. 2000; Shimizu et al. 2001). In the United States, an external gating system was developed by the University of California at Davis and Varian Medical Systems, Inc. in Palo Alto, California. Its real-time position management system uses a video camera to detect reflected infrared light from two markers placed on the surface of the patient. Computer software processes the video signal to control the accelerator (Kubo and Hill 1996; Kubo et al. 2000a; Kubo and Wang 2000b).

Tumor tracking directs the beam to move dynamically to follow the moving target. It was first implemented in a robotic radiosurgery system (CyberKnife, Accuray Inc., Sunnyvale, California; Adler et al. 1999; Schweikard et al. 2000; Murphy et al. 2003; Murphy 2002). In linear-accelerator-based RT, the leaf position was modified dynamically to compensate the motion of the target (Keall et al. 2006; Webb 2005; Trofimov et al. 2005). Figure 19.5 shows the integration of kV/MV imaging with dynamic MLC target tracking (Cho et al. 2009). D'Souza et al. proposed the third way in tracking to move the couch during the treatment to compensate for tumor motion (D'Souza, Naqvi, and Yu 2005; Qiu et al. 2007). The implementation of dynamic tracking in the conventional linear accelerator (linac) is technically complicated both in hardware and software. A prototype real-time beam tracking system has been evaluated in the clinic (Liu et al. 2009).

19.3.4 Dose Calculation to Address Issues of Lung Radiotherapy

19.3.4.1 Dose Calculation to Account for Motion

The dose delivered to a moving lung tumor is affected by blurring, interplay, and dose cloud deformation effects as described in Section 19.2.4.1. The interplay and dose cloud deformation have been suggested to have small effects in a

TABLE 19.5 Image Techniques That Have Been Used During Lung Radiotherapy

Modality	Technique	Time	Dose	Lung Tumor Visibility	Assesses Interfraction Motion	Assesses Intrafraction Motion	Pros	Cons
Radiograph	kV or MV	5 minutes	2–4 cGy/film	Poor; good with marker	Yes	No	Inexpensive	Lacks 3D information
Fluoroscopy	kV or MV	During treatment	kV: 2–10 cGy/min; MV: tx beam	Poor; good with marker	Yes	Yes	Real-time monitoring	Lacks 3D information; radiation dose
4D-CT	kV or MV	10–15 minutes	kV: 3–5 cGy; MV: 4–15 cGy	Good	Yes	No[a]	3D information	Time; radiation dose
4D-PET	Positron	1 hour	200–400 MBq for FDG PET	Used for localize	Yes	No	Functional information	Low resolution
4D-MRI	Magnetic field	0.5 hour	None	Good	Yes	No	High detail in the soft tissues	Distortion
Implantable coil	RF	During treatment	None	Surrogate	Yes	Yes	No dose	Lacks 3D information; invasive; possible coil migration
External surrogate indicators	Infrared/visible light and pressure sensor	During treatment	None	Surrogate	Yes	Yes	No dose	Lacks 3D information; hard to correlate with internal organ

[a] A few researchers repeated CBCTs at the beginning, midpoint, and end of the treatment when the radiation was not delivered. They assumed the tumor motion was similar with the beam on or off. They estimated the intrafraction motion from these CTs. (Bissonnette, J. P., K. N. Franks, T. G. Purdie, et al. 2009. *Int J Radiat Oncol Biol Phys* 75(3):688–95.)

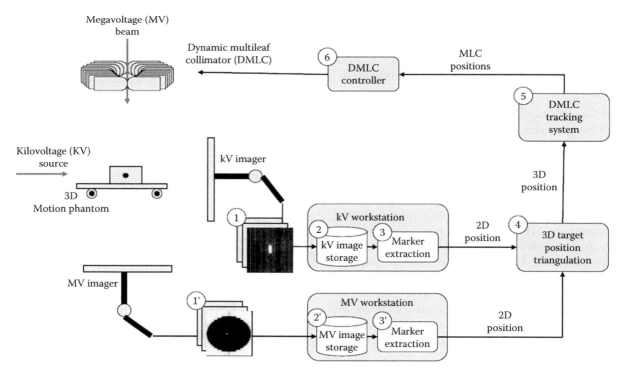

FIGURE 19.5 Flowchart of X-ray image-guided real-time dynamic multileaf collimator tracking. (Reprinted from Cho, B., et al. 2009. *Int J Radiat Oncol Biol Phys* 74(3):859–67. With permission.)

practical setting when multiple fields and fractions are used (Bortfeld, Jiang, and Rietzel 2004; Mexner et al. 2009; Li et al. 2008). The blurring of dose distribution can be theoretically alleviated by the motion management strategy described in Sections 19.3.1.2 and 19.3.3.2. The effect can also be included in the dose calculation of planning by convoluting the static dose with a characteristic motion kernel (Trofimov et al. 2005; Li and Xing 2000; Bortfeld, Jiang, and Rietzel 2004).

19.3.4.2 Dose Calculation to Account for Heterogeneity

For real-time adaptive therapy, with recalculation based on daily imaging, the efficiency of dose computation, along with the accuracy, may become an important practical issue. The accuracy of patient dose predictions based on the static image has continuously improved by moving from simple scatter and inhomogeneity corrections over pencil-beam algorithms to point kernel-based convolution superposition method (CS), and lastly the Monte Carlo simulation (MC). Experiments involving phantom measurements have already shown that MC calculations are more accurate than conventional systems, including CS algorithms, near low-density materials, and around air cavities (Krieger and Sauer 2005; Arnfield et al. 2000; Martens et al. 2002; Paelinck et al. 2005; Vanderstraeten et al. 2006). A large number of general-purpose MC algorithms have been developed for simulating the transport of electrons and photons. These codes are considered too slow for routine treatment planning purposes. Several research groups have reported on the use of parallelization of MC techniques over multiple computers or processors to provide more reasonable turn-around times for simulation in clinical research (Chetty et al. 2007; Tyagi, Bose, and Chetty 2004; Leal et al. 2004). A few vendors, including CMS, Elekta, BrainLab, Nucletron, Accuray, and Varian, have efficiently incorporated the MC method into their treatment planning systems. When tested on a 30-cm³ phantom with 5-mm-voxel size, some can accomplish MC calculation in 1 minute for a 10×10 field using a single Intel P-IV 3 GHz processor. While the MC is being improved for its use in the RT, the computationally efficient convolution and superposition has become the mainstream dose-calculation algorithm. Vanderstraeten et al. (2006) evaluated the accuracy of two commercially available convolution or superposition algorithms, Pinnacle CS and Helax-TMS collapsed cone (CC), for 10 IMRT lung patients planned with 6 MV and 18 MV. They found within target the difference between convolution/superposition and MC is all within 5%, but one patient with Helax-TMS CC. Within organs at risk (OAR), the number of patients with dose deviation above 5% was 2 for 6 MV plan and 6 for the 18 MV plan. This implies that the widely used convolution/superposition-based inhomogeneity algorithm is generally accurate for target dose calculation; however, for the lung it may underestimate the dose by about 5%. The pencil-beam algorithm should be avoided if possible because it deviates from the MC calculations by more than 10% (Krieger and Sauer 2005; Vanderstraeten et al. 2006).

19.4 Radiation Treatment Based on Patient-Specific Functional/Anatomical Factors and Patient-Specific Normal Tissue Risk Assessments

19.4.1 Treatment Planning with General Physiological Considerations

Conventional radiation-treatment planning is based on anatomical images (typically CT). The widespread use of DVHs, which discard spatial information, fails to consider possible regional differences in the function of many organs. This is routinely done for organs where the physician can readily consider spatial variations in functional importance. For example, it is clearly understood that different portions of the brain are of variable importance for different functions. It is known that a "hot spot" in the brain stem is much more of a concern than would be a similarly sized hot spot in the frontal lobe. Thus, DVHs of subregions of the brain are often considered during RT planning. However, spatial variations in function are more difficult to consider in most organs.

In a normal lung, there are known spatially based functional variations. For example, the ventilation/perfusion match is thought to be better in the lower lobes than in the upper lobes (due primarily to gravity-related factors). We and others have considered additional ways to quantify spatial differences in lung function (Marks et al. 1997; Marks et al. 1999; Marks et al. 1993; Boersma et al. 1993). SPECT perfusion scans provide a 3D map of capillary perfusion throughout the lung. Because the main function of the lung is to facilitate gas exchange between the alveoli and the capillaries, perfusion scans are often considered to be a reasonable representation of the 3D distribution of "function." This may not be ideal because gas exchange requires both ventilation and perfusion. However, the lung has intrinsic compensatory mechanisms to try to minimize ventilation or perfusion mismatch. In poorly ventilated lung regions, which therefore become somewhat hypoxic, the blood vessels constrict, thereby redirecting the blood flow to better-ventilated lung regions. This supports the contention that perfusion imaging is a reasonable surrogate for function (Osborne et al. 1982; Osborne et al. 1985). Theoretically, ventilation imaging may also be useful. However, because the bronchi cannot constrict to the degree to which the capillaries constrict, ventilated lung regions that are poorly perfused are relatively more common compared to poorly ventilated regions that are well-perfused (West 2008). Similarly, the normal physiology of the heart informs us that the left ventricle is generally more "important" than the other chambers of the heart. For RT-induced coronary artery disease, the course of the arteries in the pericardial area is known and thus represents an important "region of interest" during treatment planning. There are also known spatial variations in function of the kidney and femur (Marks 1996).

19.4.2 Treatment Planning with Patient-Specific Physiological Considerations

There may be physiological differences *between* patients that afford the opportunity to adapt therapy in a more customized manner to individual patients. We have considered interpatient differences in the spatial variation in regional lung function. As part of a prospective clinical study in patients with lung cancer, participants underwent pre-RT SPECT lung perfusion scans. Perfusion abnormalities were commonly found. Because a normal lung is well-perfused, the areas with hypoperfusion were studied in more detail. Each area of hypoperfusion was characterized as being (1) at the site of primary tumor, (2) adjacent to the primary tumor (presumably reflecting mass effect of the primary tumor on regional blood vessels, and associated obstruction of the blood flow to distal regions), or (3) apart from the primary tumor (typically representing emphysema in other regions of the lung).

In an analysis of 56 patients, decreases in perfusion were observed at the tumor, the site adjacent to the tumor, and the site apart from the tumor in 94%, 74%, and 42% of patients, respectively (Marks et al. 1995). Essentially, almost all patients with an intrathoracic mass had hypoperfusion at the site of that mass. Hypoperfusion adjacent to a tumor was seen in most patients. The size of the hypoperfused area was largest in patients with central lesions involving the hilar or mediastinal regions. Hypoperfusion apart from an intrathoracic mass was seen largely in patients with relatively low pulmonary function tests (PFTs) (Marks et al. 1995). Figure 19.6 illustrates an example of CT and SPECT images of a patient with a central lung tumor and adjacent hypoperfusion (Marks et al. 2000).

If one believes the areas that are poorly perfused pre-RT are "destined" to remain poorly perfused, it would be logical to orient the treatment beams such that they preferentially pass through the hypoperfused regions of lung. However, if these poorly perfused lung regions might become perfused (i.e., undergo reperfusion), this might *not* be a logical approach. Because the areas of hypoperfusion that are adjacent to a central lesion might be related to the obstruction of regional vessels and because these regional vessels may become unobstructed as the tumor shrinks in response to treatment, it might not be logical to purposely orient radiation beams through areas of hypoperfusion that are immediately adjacent to a central tumor. Conversely, areas of hypoperfusion apart from the gross tumor are usually associated with emphysema (an irreversible process) and are thus likely permanent. Directing a radiation beam purposely through these areas would therefore be the most logical.

Incorporating such patient-specific functional information into the treatment planning process is appealing. However, there are some logistic and practical limitations, particularly with conventional 3D planning (i.e., large fields that each encompasses the entire target volume). For example, in patients with a large central lung lesion, it is almost always most efficient to deliver the majority of the dose through large AP portals. The use of oblique beams typically increases the irradiated lung volume (either anatomical lung or functional lung). In practice, we

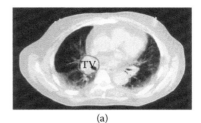

(a)

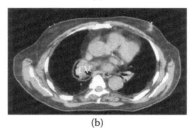

(b)

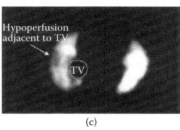

(c)

FIGURE 19.6 An example of a patient with a central lung tumor involving the right hilar area is shown in panels (a) (lung window) and (b) (soft tissue window). There is hypoperfusion seen adjacent to the tumor on the corresponding single-photon emission computed tomography (SPECT) image (panel (c)). The images are visually registered. The contour of the gross tumor volume seen on the computed tomography scan is traced onto the SPECT. (Reprinted from Marks, L. B., et al. 2000. *Cancer* 88(9):2135–41. With permission.)

have found that the application of functional lung information in the treatment planning process is most practical in patients with relatively small- to medium-sized targets (i.e., when there is flexibility to use alternative beam orientations), as well as relatively poor PFTs (i.e., when there is a greater desire or need to minimize radiation-induced lung injury). In our prior analysis, the *potential* utility of improving the treatment plan with the inclusion of functional information was much more prevalent than was the actual use of this approach (Marks et al. 2000; Marks et al. 1995; Marks et al. 1999).

With IMRT, there is some increased flexibility to consider function-based imaging during treatment planning, even for larger lesions. For IMRT, the beam orientation, shape of beamlets, and relative contribution of each beamlet can be variables for an optimizer to consider when seeking the "optimal" solution to spare the functional lung while maintaining the prescription dose on the tumor. Das et al. (2004) and Christian et al. (2005) have assessed the utility of SPECT-based IMRT optimization. In a theoretical analysis, they reported that it was possible to incorporate SPECT functional lung imaging into inverse treatment planning to limit dose to healthy "perfused" lung. McGuire et al. (2006) compared IMRT treatment plans with and without SPECT guidance for five patients with lung cancer. They found that SPECT-guided plans produced a more favorable dose functional histogram of lung than the non-SPECT-guided plan. The volumes of functional lung receiving 20 Gy (FV_{20}) and 30 Gy (FV_{30}) were reduced for all patients' SPECT-guided plans by an average of 14% and 11%, respectively. Similarly, incorporating perfusion information into IMRT planning, Shioyama et al. (2007) reported the median reductions in the mean doses to the 50- and 90-percentile hyperperfusion lung in the functional plan were 2.2 and 4.2 Gy, respectively, compared with those in the anatomical plans. Bates et al. (2009) found that there were statistically significant reductions in FV_{20}, V_{20}, and mean functional lung dose when IMRT planning was supplemented by functional information (Bates et al. 2009). Note that these studies (including our own) present almost "circular" arguments. Plans are optimized based on a functional-imaging-based parameter, and the resultant plans are deemed "superior" using that same functional-imaging-based evaluation metric.

With SPECT-guided IMRT, the anatomical lung is divided into subregions, based on their relative perfusion. The inverse planning optimizer is instructed to selectively spare the better-perfused region of lung, essentially at the expense of the lesser-perfused lung regions. It is customary in the studies reported thus far to segregate the lung into subvolumes as noted during the optimization process (Marks et al. 1999; Das et al. 2004). This approach, therefore, essentially groups the lung regions based on function. Theoretically, perfusion may be better considered a continuous variable, rather than a discrete variable. However, the available commercial treatment planning systems can more readily consider this issue with defined subregions and can handle the consideration of a continuous variable less well.

19.4.3 Risk-Adapted Therapy

A conventional treatment approach for physicians is to define the target volume and the prescription dose. Treatment planning is performed with the hope of minimizing the risk of normal tissue injury, while maintaining a certain level of desired target coverage. When the normal tissue risks are deemed excessive, physicians typically will alter the target volume and/or the prescribed dose. Therefore, risk-based patient-specific treatment planning is a long-standing tradition. The advent of 3D treatment planning tools and tumor control and NTCP modeling provides us more opportunities for a more patient-specific approach to prescribing dose or volume. For example, investigators at the University of Michigan performed a dose-escalation study for patients with lung cancer where the prescribed radiation dose was dictated by the effective volume (V_{eff}, from the Lyman–Kuture–Burman model; Kong et al. 2005a; Hayman et al. 2001; Narayan et al. 2004). For a given bin of V_{eff}, a prescription dose was chosen with a certain level of predicted NTCP (Ten Haken et al. 1993; Semenenko and Li 2008). With this approach, smaller lesions were treated to higher doses, and larger lesions were treated to lower doses, both with the expected same level

of complication probability. Their results suggest that radiation doses of 92.4 and 102.9 Gy can be delivered safely to limited lung volumes with minimal toxicity (Narayan et al. 2004). A similar approach could be used in many other disease settings.

As our understanding of the causes of radiation-associated normal tissue injury increase, and our predictive models become more accurate (Ghafoori et al. 2008; Das et al. 2003; Fan et al. 2001; Marks et al. 2003), we have a potential opportunity to refine this entire risk-based treatment planning approach. There are a variety of clinical and dosimetric parameters that have been suggested to be predictive for radiation-induced lung injury (Rodrigues et al. 2004; Das et al. 2008; Marks 2002; Hernando et al. 2001). For example, several studies suggest that lower lobe regions have a higher rate of pneumonitis than do upper lobe regions, even after correcting for dosimetric parameters such as mean lung dose (Graham et al. 1999; Hope et al. 2006). Bradley et al. (2007) formally created a predictive model based on mean lung dose and SI GTV position. This model performed well on a combined data set from the Washington University and Radiation Oncology Group (RTOG) trial 9311. Conceivably, future prescription volume or doses might be adapted to a patient's specific risk profile based on many clinical variables. In the future, there may be "genetic fingerprints" that relate to a patient's inherent radiosensitivity that may also be incorporated into these risk-based treatment planning approaches.

For the lung, additional clinical information that may alter the risk includes pre-RT lung function and the degree of tumor-induced impairment in lung function. However, sophisticated models that consider these parameters are neither available nor validated. In addition, for patients undergoing RT for lung cancer, radiation-induced lung injury is not the only concern. There are also risks to the brachial-plexus, esophagus, heart, and soft tissue. Ideally, a comprehensive risk-based model that considers each of these normal tissue organs would be desired. Investigators have made suggestions along these lines (Langer, Morrill, and Lane 1998; Jain et al. 1993; Niemierko and Goitein 1991). These approaches considered the uncomplicated TCP (UTCP), which is based on the statistical models of dose response such as TCP and NTCP. However, an additional layer of sophistication was considered as well. The clinical impact of a grade 2 esophagus injury may be different than a grade 2 lung injury. Further, the relative importance of these different normal tissue reactions may vary between patients. Miften et al. (2008) suggested assigning a "utility value" to the states of health, where this reflects the "severity" of the complication regarding quality of life. They generalized the formalism of UTCP to incorporate the total quality adjusted life years (QALY) expectance. For a particular RT plan, the total QALY represent the product of duration-weighted states of health. The $UTCP_{QALY}$ is formulated as

$$UTCP_{QALY} = TCP \prod_{i=1}^{N} [1 - Durationfactor_i \cdot$$
$$(1 - Utilityfactor_i) \cdot NTCP_i]$$

where the index i indicates one of N states of health in a complication. The duration factor is the ratio of the average duration of each state of health relative to the average patient life expectancy. The utility factor quantifies the relative quality of life for each state of health in complication.

TABLE 19.6   Summary of Issues of Lung Radiotherapy

Issues	Magnitude
Interfraction motion	2–7.5 mm
Intrafraction motion—no breathing control	5–20 mm
Intrafraction motion—with breathing control	1–10 mm
Tumor shrinkage	10%–45% during radiation therapy
Convolving motion with dose calculation	<2% for multiple fields over many fractions (e.g., 30)
Heterogeneity corrected dose calculation	5%—convolution/superposition
	10%—pencil beam

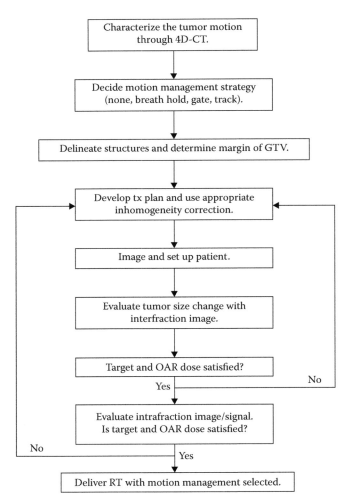

FIGURE 19.7   An idealized workflow of image-guided adaptive radiotherapy for lung cancer. GTV = gross tumor volume; OAR = organ at risk.

19.5 Idealized Workflow of Image-Guided Adaptive Radiotherapy for Lung Cancer

A few advanced adaptive RT methods for lung cancer mentioned in Sections 19.3.1.2, 19.3.2.2, and 19.3.3.2, such as MLC reshaping, replanning, respiratory-gated and tumor-tracking motion management, and functional image-based treatment planning, are technically complicated and may prolong the whole process of RT. Although intuitively they should be useful tools to accomplish physicians' intention (i.e., deliver dose to the tumor while minimizing harm on normal structures), they can take an enormous amount of time or resource and may be impractical for daily clinical use. Further, there is presently no strong scientific evidence to support that the cure rates will meaningfully increase with any of these technical advances. The magnitude of several of the issues discussed here is summarized in Table 19.6. When adaptive therapy and associated imaging techniques and motion management strategies become clinically efficient, they may be more readily studied and implemented more widely in the clinic. Figure 19.7 schematically illustrates an "idealized" workflow of image-guided adaptive therapy for lung cancer.

19.6 Conclusions

In patients with lung cancer, physicians and planners should consider the potential impact of intra- and interfraction motions on the prescribed treatment plan. In practice, this is usually accomplished with adjustments to the target volumes (i.e., ITVs and PTVs) to reflect uncertainties associated with motion. In some settings (tight fields due to poor lung function, or a high dose per fraction), it might be reasonable to more routinely and aggressively manage these factors with (for example) a breath-hold or gating technique. In patients with large or central lung lesions that shrink during therapy, and with associated anatomical shifts, it might be necessary to replan the patient, or at least assess the impact of the anatomical shifts on the delivered doses. Functional imaging can be used to differentiate normal tissues and guide the treatment planning to preferentially spare the more healthy regions of OAR. Functional imaging may afford a more meaningful way to adapt therapy to each patient based on his or her unique physiology.

Additional work is clearly needed to better understand the impact of motion on treatment delivery and to develop or assess efficient adaptive strategies to manage the associated uncertainties with our best efforts.

References

Adler Jr., J. R., M. J. Murphy, S. D. Chang, et al. 1999. Image-guided robotic radiosurgery. *Neurosurgery* 44(6):1299–307.

Armstrong, J. G., and B. D. Minsky. 1989. Radiation therapy for medically inoperable stage I and II non-small cell lung cancer. *Cancer Treat Rev* 16(4):247–55.

Arnfield, M. R., C. H. Siantar, J. Siebers, et al. 2000. The impact of electron transport on the accuracy of computed dose. *Med Phys* 27(6):1266–74.

Balter, J. M., J. N. Wright, L. J. Newell, et al. 2005. Accuracy of a wireless localization system for radiotherapy. *Int J Radiat Oncol Biol Phys* 61(3):933–7.

Barnes, E. A., B. R. Murray, D. M. Robinson, et al. 2001. Dosimetric evaluation of lung tumor immobilization using breath hold at deep inspiration. *Int J Radiat Oncol Biol Phys* 50(4):1091–8.

Bates, E. L., C. M. Bragg, J. M. Wild, et al. 2009. Functional image-based radiotherapy planning for non-small cell lung cancer: A simulation study. *Radiother Oncol* 93(1):32–6.

Biederer, J., J. Dinkel, G. Remmert, et al. 2009. 4D-Imaging of the lung: Reproducibility of lesion size and displacement on helical CT, MRI, and cone beam CT in a ventilated ex vivo system. *Int J Radiat Oncol Biol Phys* 73(3):919–26.

Bissonnette, J. P., K. N. Franks, T. G. Purdie, D. J. Moseley, J. J. Sonke, D. A. Jaffray, L. A. Dawson, and A. Bezjak. 2009. Quantifying interfraction and intrafraction tumor motion in lung stereotactic body radiotherapy using respiration-correlated cone beam computed tomography. *Int J Radiat Oncol Biol Phys* 75 (3):688–95.

Boersma, L. J., E. M. Damen, R. W. de Boer, et al. 1993. A new method to determine dose-effect relations for local lung-function changes using correlated SPECT and CT data. *Radiother Oncol* 29(2):110–6.

Bortfeld, T., S. B. Jiang, and E. Rietzel. 2004. Effects of motion on the total dose distribution. *Semin Radiat Oncol* 14(1):41–51.

Bosmans, G., A. van Baardwijk, A. Dekker, et al. 2006. Intra-patient variability of tumor volume and tumor motion during conventionally fractionated radiotherapy for locally advanced non-small-cell lung cancer: A prospective clinical study. *Int J Radiat Oncol Biol Phys* 66(3):748–53.

Bradley, J. D., A. Hope, I. El Naqa, et al. 2007. A nomogram to predict radiation pneumonitis, derived from a combined analysis of RTOG 9311 and institutional data. *Int J Radiat Oncol Biol Phys* 69(4):985–92.

Britton, K. R., Y. Takai, M. Mitsuya, et al. 2005. Evaluation of inter- and intrafraction organ motion during intensity modulated radiation therapy (IMRT) for localized prostate cancer measured by a newly developed on-board image-guided system. *Radiat Med* 23(1):14–24.

Chang, J., G. S. Mageras, E. Yorke, et al. 2007. Observation of interfractional variations in lung tumor position using respiratory gated and ungated megavoltage cone-beam computed tomography. *Int J Radiat Oncol Biol Phys* 67(5):1548–58.

Chen, Q. S., M. S. Weinhous, F. C. Deibel, et al. 2001. Fluoroscopic study of tumor motion due to breathing: Facilitating precise radiation therapy for lung cancer patients. *Med Phys* 28(9):1850–6.

Chetty, I. J., B. Curran, J. E. Cygler, et al. 2007. Report of the AAPM Task Group No. 105: Issues associated with clinical implementation of Monte Carlo-based photon and electron external beam treatment planning. *Med Phys* 34(12):4818–53.

Cheung, P. C., K. E. Sixel, R. Tirona, et al. 2003. Reproducibility of lung tumor position and reduction of lung mass within the planning target volume using active breathing control (ABC). *Int J Radiat Oncol Biol Phys* 57(5):1437–42.

Cho, B., P. R. Poulsen, A. Sloutsky, et al. 2009. First demonstration of combined kV/MV image-guided real-time dynamic multileaf-collimator target tracking. *Int J Radiat Oncol Biol Phys* 74(3):859–67.

Christian, J. A., M. Partridge, E. Nioutskiou, et al. 2005. The incorporation of SPECT functional lung imaging into inverse radiotherapy planning for non-small cell lung cancer. *Radiother Oncol* 77(3):271–7.

Court, L. E., L. Dong, A. K. Lee, et al. 2005. An automatic CT-guided adaptive radiation therapy technique by online modification of multileaf collimator leaf positions for prostate cancer. *Int J Radiat Oncol Biol Phys* 62(1):154–63.

Coy, P., and G. M. Kennelly. 1980. The role of curative radiotherapy in the treatment of lung cancer. *Cancer* 45(4):698–702.

Das, S. K., S. Chen, J. O. Deasy, et al. 2008. Combining multiple models to generate consensus: application to radiation-induced pneumonitis prediction. *Med Phys* 35(11):5098–109.

Das, S., T. Cullip, G. Tracton, et al. 2003. Beam orientation selection for intensity-modulated radiation therapy based on target equivalent uniform dose maximization. *Int J Radiat Oncol Biol Phys* 55(1):215–24.

Das, S. K., M. M. Miften, S. Zhou, et al. 2004. Feasibility of optimizing the dose distribution in lung tumors using fluorine-18-fluorodeoxyglucose positron emission tomography and single photon emission computed tomography guided dose prescriptions. *Med Phys* 31(6):1452–61.

Ding, G. X., D. M. Duggan, B. Lu, et al. 2007. Impact of inhomogeneity corrections on dose coverage in the treatment of lung cancer using stereotactic body radiation therapy. *Med Phys* 34(7):2985–94.

Dosoretz, D. E., D. Galmarini, J. H. Rubenstein, et al. 1993. Local control in medically inoperable lung cancer: An analysis of its importance in outcome and factors determining the probability of tumor eradication. *Int J Radiat Oncol Biol Phys* 27(3):507–16.

Dosoretz, D. E., M. J. Katin, P. H. Blitzer, et al. 1992. Radiation therapy in the management of medically inoperable carcinoma of the lung: Results and implications for future treatment strategies. *Int J Radiat Oncol Biol Phys* 24(1):3–9.

Dosoretz, D. E., M. J. Katin, P. H. Blitzer, et al. 1996. Medically inoperable lung carcinoma: The role of radiation therapy. *Semin Radiat Oncol* 6(2):98–104.

D'Souza, W. D., S. A. Naqvi, and C. X. Yu. 2005. Real-time intra-fraction-motion tracking using the treatment couch: A feasibility study. *Phys Med Biol* 50(17):4021–33.

Erridge, S. C., Y. Seppenwoolde, S. H. Muller, et al. 2003. Portal imaging to assess set-up errors, tumor motion and tumor shrinkage during conformal radiotherapy of non-small cell lung cancer. *Radiother Oncol* 66(1):75–85.

Evans, P. M. 2008. Anatomical imaging for radiotherapy. *Phys Med Biol* 53(12):R151–91.

Fan, M., L. B. Marks, D. Hollis, et al. 2001. Can we predict radiation-induced changes in pulmonary function based on the sum of predicted regional dysfunction? *J Clin Oncol* 19(2):543–50.

Feng, M., F. M. Kong, M. Gross, et al. 2009. Using fluorodeoxyglucose positron emission tomography to assess tumor volume during radiotherapy for non-small-cell lung cancer and its potential impact on adaptive dose escalation and normal tissue sparing. *Int J Radiat Oncol Biol Phys* 73(4):1228–34.

Fox, J., E. Ford, K. Redmond, et al. 2009. Quantification of tumor volume changes during radiotherapy for non-small-cell lung cancer. *Int J Radiat Oncol Biol Phys* 74(2):341–8.

Fraass, B. A., J. Smathers, and J. Deye. 2003. Summary and recommendations of a National Cancer Institute workshop on issues limiting the clinical use of Monte Carlo dose calculation algorithms for megavoltage external beam radiation therapy. *Med Phys* 30(12):3206–16.

Fu, W., Y. Yang, N. J. Yue, et al. 2009. A cone beam CT-guided online plan modification technique to correct interfractional anatomic changes for prostate cancer IMRT treatment. *Phys Med Biol* 54(6):1691–703.

Ghafoori, P., L. B. Marks, Z. Vujaskovic, et al. 2008. Radiation-induced lung injury. Assessment, management, and prevention. *Oncology (Williston Park)* 22(1):37–47.

Graham, M. V., J. A. Purdy, B. Emami, et al. 1999. Clinical dose-volume histogram analysis for pneumonitis after 3D treatment for non-small cell lung cancer (NSCLC). *Int J Radiat Oncol Biol Phys* 45(2):323–9.

Grills, I. S., D. Yan, A. A. Martinez, et al. 2003. Potential for reduced toxicity and dose escalation in the treatment of inoperable non-small-cell lung cancer: A comparison of intensity-modulated radiation therapy (IMRT), 3D conformal radiation, and elective nodal irradiation. *Int J Radiat Oncol Biol Phys* 57(3):875–90.

Guerrero, T., G. Zhang, T. C. Huang, et al. 2004. Intrathoracic tumour motion estimation from CT imaging using the 3D optical flow method. *Phys Med Biol* 49(17):4147–61.

Haasbeek, C. J., F. J. Lagerwaard, J. P. Cuijpers, et al. 2007. Is adaptive treatment planning required for stereotactic radiotherapy of stage I non-small-cell lung cancer? *Int J Radiat Oncol Biol Phys* 67(5):1370–4.

Haasbeek, C. J., B. J. Slotman, and S. Senan. 2009. Radiotherapy for lung cancer: Clinical impact of recent technical advances. *Lung Cancer* 64(1):1–8.

Hanley, J., M. M. Debois, D. Mah, et al. 1999. Deep inspiration breath-hold technique for lung tumors: The potential value of target immobilization and reduced lung density in dose escalation. *Int J Radiat Oncol Biol Phys* 45(3):603–11.

Hayman, J. A., M. K. Martel, R. K. Ten Haken, et al. 2001. Dose escalation in non-small-cell lung cancer using three-dimensional conformal radiation therapy: Update of a phase I trial. *J Clin Oncol* 19(1):127–36.

Hernando, M. L., L. B. Marks, G. C. Bentel, et al. 2001. Radiation-induced pulmonary toxicity: A dose-volume histogram analysis in 201 patients with lung cancer. *Int J Radiat Oncol Biol Phys* 51(3):650–9.

Hope, A. J., P. E. Lindsay, I. El Naqa, et al. 2006. Modeling radiation pneumonitis risk with clinical, dosimetric, and spatial parameters. *Int J Radiat Oncol Biol Phys* 65(1):112–24.

Hugo, G. D., J. Campbell, T. Zhang, and D. Yan. 2009. Cumulative lung dose for several motion management strategies as a function of pretreatment patient parameters. *Int J Radiat Oncol Biol Phys* 74(2):593–601.

ICRU. 1993. *Report 50: Prescribing, Recording, and Reporting Photon Beam Therapy.* Washington, DC: International Commission on Radiation Units and Measurements.

ICRU. 1999. *Report 62: Prescribing, Recording, and Reporting Photon Beam Therapy (Supplement to ICRU Report 50).* Washington, DC: International Commission on Radiation Units and Measurements.

Jain, N. L., M. G. Kahn, R. E. Drzymala, et al. 1993. Objective evaluation of 3-D radiation treatment plans: A decision-analytic tool incorporating treatment preferences of radiation oncologists. *Int J Radiat Oncol Biol Phys* 26(2):321–33.

Jemal, A., R. Siegel, E. Ward, Y. Hao, J. Xu, and M. J. Thun. 2009. Cancer statistics, 2009. *CA Cancer J Clin* 59(4):225–49.

Jiang, S. B. 2006a. Radiotherapy of mobile tumors. *Semin Radiat Oncol* 16(4):239–48.

Jiang, S. B. 2006b. Technical aspects of image-guided respiration-gated radiation therapy. *Med Dosim* 31(2):141–51.

Jiang, S. B., C. Pope, K. M. Al Jarrah, et al. 2003. An experimental investigation on intra-fractional organ motion effects in lung IMRT treatments. *Phys Med Biol* 48(12):1773–84.

Juhler-Nottrup, T. S., S. Korreman, A. N. Pedersen, et al. 2008. Interfractional changes in tumour volume and position during entire radiotherapy courses for lung cancer with respiratory gating and image guidance. *Acta Oncol* 47(7):1406–13.

Kaskowitz, L., M. V. Graham, B. Emami, K. J. Halverson, and C. Rush. 1993. Radiation therapy alone for stage I non-small cell lung cancer. *Int J Radiat Oncol Biol Phys* 27(3):517–23.

Keall, P. J., G. S. Mageras, J. M. Balter, et al. 2006. The management of respiratory motion in radiation oncology report of AAPM Task Group 76. *Med Phys* 33(10):3874–900.

Kong, F. M., M. Kessler, F. Kirk, et al. 2005. Using FDG-PET and V/Q SPECT to assess changes in tumor activity and pulmonary function during radiotherapy of non-small cell lung cancer. *Int J Radiat Oncol Biol Phys* 63(Supplement 1):S224.

Kong, F. M., R. Ten, K. Haken, M. J. Schipper, et al. 2005a. High-dose radiation improved local tumor control and overall survival in patients with inoperable/unresectable non-small-cell lung cancer: Long-term results of a radiation dose escalation study. *Int J Radiat Oncol Biol Phys* 63(2):324–33.

Krieger, T., and O. A. Sauer. 2005. Monte Carlo- versus pencil-beam-/collapsed-cone-dose calculation in a heterogeneous multi-layer phantom. *Phys Med Biol* 50(5):859–68.

Kubo, H. D., and B. C. Hill. 1996. Respiration gated radiotherapy treatment: A technical study. *Phys Med Biol* 41(1):83–91.

Kubo, H. D., P. M. Len, S. Minohara, and H. Mostafavi. 2000a. Breathing-synchronized radiotherapy program at the University of California Davis Cancer Center. *Med Phys* 27(2):346–53.

Kubo, H. D., and L. Wang. 2000b. Compatibility of Varian 2100C gated operations with enhanced dynamic wedge and IMRT dose delivery. *Med Phys* 27(8):1732–8.

Kupelian, P. A., C. Ramsey, S. L. Meeks, et al. 2005. Serial megavoltage CT imaging during external beam radiotherapy for non-small-cell lung cancer: Observations on tumor regression during treatment. *Int J Radiat Oncol Biol Phys* 63(4):1024–8.

Langer, M., S. S. Morrill, and R. Lane. 1998. A test of the claim that plan rankings are determined by relative complication and tumor-control probabilities. *Int J Radiat Oncol Biol Phys* 41(2):451–7.

Lax, I., V. Panettieri, B. Wennberg, et al. 2006. Dose distributions in SBRT of lung tumors: Comparison between two different treatment planning algorithms and Monte-Carlo simulation including breathing motions. *Acta Oncol* 45(7):978–88.

Leal, A. F., R. Sanchez-Doblado, M. Arrans, et al. 2004. Monte Carlo simulation of complex radiotherapy treatments. *Comput Sci Eng* 6:60–8.

Li, H. S., I. J. Chetty, C. A. Enke, et al. 2008. Dosimetric consequences of intrafraction prostate motion. *Int J Radiat Oncol Biol Phys* 71(3):801–12.

Li, J. G., and L. Xing. 2000. Inverse planning incorporating organ motion. *Med Phys* 27(7):1573–8.

Liu, Y., C. Shi, B. Lin, C. S. Ha, and N. Papanikolaou. 2009. Delivery of four-dimensional radiotherapy with TrackBeam for moving target using an AccuKnife dual-layer MLC: Dynamic phantoms study. *J Appl Clin Med Phys* 10(2):2926.

Mageras, G. S., E. Yorke, and S. B. Jiang. 2006. "4D" IMRT delivery. In *Image-Guided IMRT*, ed. T. Bortfeld, R. Schmidt-Ullrich, W. De Neve, and D. E. Wazer. Berlin: Springer.

Marks, L. B. 1996. The impact of organ structure on radiation response. *Int J Radiat Oncol Biol Phys* 34(5):1165–71.

Marks, L. B. 2002. Dosimetric predictors of radiation-induced lung injury. *Int J Radiat Oncol Biol Phys* 54(2):313–6.

Marks, L. B., D. Hollis, M. Munley, G. Bentel, et al. 2000. The role of lung perfusion imaging in predicting the direction of radiation-induced changes in pulmonary function tests. *Cancer* 88(9):2135–41.

Marks, L. B., M. T. Munley, D. P. Spencer, et al. 1997. Quantification of radiation-induced regional lung injury with perfusion imaging. *Int J Radiat Oncol Biol Phys* 38(2):399–409.

Marks, L. B., G. W. Sherouse, M. T. Munley, G. C. Bentel, and D. P. Spencer. 1999. Incorporation of functional status into dose-volume analysis. *Med Phys* 26(2):196–9.

Marks, L. B., D. P. Spencer, G. C. Bente, et al. 1993. The utility of SPECT lung perfusion scans in minimizing and assessing the physiologic consequences of thoracic irradiation. *Int J Radiat Oncol Biol Phys* 26(4):659–68.

Marks, L. B., D. P. Spencer, G. W. Sherouse, et al. 1995. The role of three dimensional functional lung imaging in radiation treatment planning: The functional dose-volume histogram. *Int J Radiat Oncol Biol Phys* 33(1):65–75.

Marks, L. B., X. Yu, Z. Vujaskovic, W. Small Jr., R. Folz, and M. S. Anscher. 2003. Radiation-induced lung injury. *Semin Radiat Oncol* 13(3):333–45.

Martens, C., N. Reynaert, C. De Wagter, et al. 2002. Underdosage of the upper-airway mucosa for small fields as used in intensity-modulated radiation therapy: A comparison between radiochromic film measurements, Monte Carlo simulations, and collapsed cone convolution calculations. *Med Phys* 29(7):1528–35.

Matsugi, K., Y. Narita, A. Sawada, et al. 2009. Measurement of interfraction variations in position and size of target volumes in stereotactic body radiotherapy for lung cancer. *Int J Radiat Oncol Biol Phys* 75(2):543–8.

Matsuo, Y., K. Takayama, Y. Nagata, et al. 2007. Interinstitutional variations in planning for stereotactic body radiation therapy for lung cancer. *Int J Radiat Oncol Biol Phys* 68(2):416–25.

McCarter, S. D., and W. A. Beckham. 2000. Evaluation of the validity of a convolution method for incorporating tumour movement and set-up variations into the radiotherapy treatment planning system. *Phys Med Biol* 45(4):923–31.

McGuire, S. M., S. Zhou, L. B. Marks, M. Dewhirst, F. F. Yin, and S. K. Das. 2006. A methodology for using SPECT to reduce intensity-modulated radiation therapy (IMRT) dose to functioning lung. *Int J Radiat Oncol Biol Phys* 66(5):1543–52.

Mexner, V., J. W. Wolthaus, M. van Herk, E. M. Damen, and J. J. Sonke. 2009. Effects of respiration-induced density variations on dose distributions in radiotherapy of lung cancer. *Int J Radiat Oncol Biol Phys* 74(4):1266–75.

Miften, M., O. Gayou, D. S. Parda, R. Prosnitz, and L. B. Marks. 2008. Using quality of life information to rationally incorporate normal tissue effects into treatment plan evaluation and scoring. In *Late Effects of Cancer Treatment on Normal Tissues*, ed. P. Rubin, L. S. Constine, L. B. Marks, P. Okunieff. Heidelberg, Germany: Springer.

Murphy, M. J. 2002. Fiducial-based targeting accuracy for external-beam radiotherapy. *Med Phys* 29(3):334–44.

Murphy, M. J., S. D. Chang, I. C. Gibbs, et al. 2003. Patterns of patient movement during frameless image-guided radiosurgery. *Int J Radiat Oncol Biol Phys* 55(5):1400–8.

Narayan, S., G. T. Henning, R. K. Ten Haken, M. A. Sullivan, M. K. Martel, and J. A. Hayman. 2004. Results following treatment to doses of 92.4 or 102.9 Gy on a phase I dose escalation study for non-small cell lung cancer. *Lung Cancer* 44(1):79–88.

Niemierko, A., and M. Goitein. 1991. Calculation of normal tissue complication probability and dose-volume histogram reduction schemes for tissues with a critical element architecture. *Radiother Oncol* 20(3):166–76.

Onishi, H., K. Kuriyama, T. Komiyama, et al. 2003. A new irradiation system for lung cancer combining linear accelerator, computed tomography, patient self-breath-holding, and patient-directed beam-control without respiratory monitoring devices. *Int J Radiat Oncol Biol Phys* 56(1):14–20.

Osborne, D., R. Jaszczak, R. E. Coleman, K. Greer, and M. Lischko. 1982. In vivo regional quantitation of intrathoracic Tc-99m using SPECT: Concise communication. *J Nucl Med* 23(5):446–50.

Osborne, D., R. J. Jaszczak, K. Greer, M. Lischko, and R. E. Coleman. 1985. SPECT quantification of technetium-99m microspheres within the canine lung. *J Comput Assist Tomogr* 9(1):73–7.

Paelinck, L., N. Reynaert, H. Thierens, W. De Neve, and C. De Wagter. 2005. Experimental verification of lung dose with radiochromic film: Comparison with Monte Carlo simulations and commercially available treatment planning systems. *Phys Med Biol* 50(9):2055–69.

Panakis, N., H. A. McNair, J. A. Christian, et al. 2008. Defining the margins in the radical radiotherapy of non-small cell lung cancer (NSCLC) with active breathing control (ABC) and the effect on physical lung parameters. *Radiother Oncol* 87(1):65–73.

Panettieri, V., B. Wennberg, G. Gagliardi, M. A. Duch, M. Ginjaume, and I. Lax. 2007. SBRT of lung tumours: Monte Carlo simulation with PENELOPE of dose distributions including respiratory motion and comparison with different treatment planning systems. *Phys Med Biol* 52(14):4265–81.

Papanikolaou, N., J. Battista, A. Boyer, et al. 2004. Tissue inhomogeneity corrections for megavoltage photon beams. In *AAPM Report No 85, Task Group No 65 of the Radiation Therapy Committee of the American Association of Physicists in Medicine.* Madison, WI: Medical Physics Publishing.

Pemler, P., J. Besserer, N. Lombriser, R. Pescia, and U. Schneider. 2001. Influence of respiration-induced organ motion on dose distributions in treatments using enhanced dynamic wedges. *Med Phys* 28(11):2234–40.

Plathow, C., S. Ley, C. Fink, et al. 2004. Analysis of intrathoracic tumor mobility during whole breathing cycle by dynamic MRI. *Int J Radiat Oncol Biol Phys* 59(4):952–9.

Purdie, T. G., J. P. Bissonnette, K. Franks, et al. 2007. Cone-beam computed tomography for on-line image guidance of lung stereotactic radiotherapy: Localization, verification, and intrafraction tumor position. *Int J Radiat Oncol Biol Phys* 68(1):243–52.

Qiu, P., W. D. D'Souza, T. J. McAvoy, and K. J. Ray Liu. 2007. Inferential modeling and predictive feedback control in real-time motion compensation using the treatment couch during radiotherapy. *Phys Med Biol* 52(19): 5831–54.

Rijkhorst, E. J., M. van Herk, J. V. Lebesque, and J. J. Sonke. 2007. Strategy for online correction of rotational organ motion for intensity-modulated radiotherapy of prostate cancer. *Int J Radiat Oncol Biol Phys* 69(5):1608–17.

Rodrigues, G., M. Lock, D. D'Souza, E. Yu, and J. Van Dyk. 2004. Prediction of radiation pneumonitis by dose-volume histogram parameters in lung cancer: A systematic review. *Radiother Oncol* 71(2):127–38.

Schweikard, A., G. Glosser, M. Bodduluri, M. J. Murphy, and J. R. Adler. 2000. Robotic motion compensation for respiratory movement during radiosurgery. *Comput Aided Surg* 5(4):263–77.

Seiler, P. G., H. Blattmann, S. Kirsch, R. K. Muench, and C. Schilling. 2000. A novel tracking technique for the continuous precise measurement of tumour positions in conformal radiotherapy. *Phys Med Biol* 45(9):N103–10.

Semenenko, V. A., and X. A. Li. 2008. Lyman-Kutcher-Burman NTCP model parameters for radiation pneumonitis and xerostomia based on combined analysis of published clinical data. *Phys Med Biol* 53(3):737–55.

Seppenwoolde, Y., H. Shirato, K. Kitamura, et al. 2002. Precise and real-time measurement of 3D tumor motion in lung due to breathing and heartbeat, measured during radiotherapy. *Int J Radiat Oncol Biol Phys* 53(4):822–34.

Shimizu, S., H. Shirato, S. Ogura, et al. 2001. Detection of lung tumor movement in real-time tumor-tracking radiotherapy. *Int J Radiat Oncol Biol Phys* 51(2):304–10.

Shioyama, Y., S. Y. Jang, H. H. Liu, et al. 2007. Preserving functional lung using perfusion imaging and intensity-modulated radiation therapy for advanced-stage non-small cell lung cancer. *Int J Radiat Oncol Biol Phys* 68(5):1349–58.

Shirato, H., Y. Seppenwoolde, K. Kitamura, R. Onimur, and S. Shimizu. 2004. Intrafractional tumor motion: Lung and liver. *Semin Radiat Oncol* 14(1):10–8.

Shirato, H., S. Shimizu, K. Kitamura, et al. 2000. Four-dimensional treatment planning and fluoroscopic real-time tumor tracking radiotherapy for moving tumor. *Int J Radiat Oncol Biol Phys* 48(2):435–42.

Shirato, H., S. Shimizu, T. Shimizu, T. Nishioka, and K. Miyasaka. 1999. Real-time tumour-tracking radiotherapy. *Lancet* 353(9161):1331–2.

Siker, M. L., W. A. Tome, and M. P. Mehta. 2006. Tumor volume changes on serial imaging with megavoltage CT for non-small-cell lung cancer during intensity-modulated radiotherapy: How reliable, consistent, and meaningful is the effect? *Int J Radiat Oncol Biol Phys* 66(1):135–41.

Sonke, J. J., L. Zijp, P. Remeijer, and M. van Herk. 2005. Respiratory correlated cone beam CT. *Med Phys* 32(4):1176–86.

Stevens, C. W., R. F. Munden, K. M. Forster, et al. 2001. Respiratory-driven lung tumor motion is independent of tumor size, tumor location, and pulmonary function. *Int J Radiat Oncol Biol Phys* 51(1):62–8.

Ten Haken, R. K., M. K. Martel, M. L. Kessler, et al. 1993. Use of Veff and iso-NTCP in the implementation of dose escalation protocols. *Int J Radiat Oncol Biol Phys* 27(3):689–95.

Trofimov, A., E. Rietzel, H. M. Lu, et al. 2005. Temporo-spatial IMRT optimization: Concepts, implementation and initial results. *Phys Med Biol* 50(12):2779–98.

Tyagi, N., A. Bose, and I. J. Chetty. 2004. Implementation of the DPM Monte Carlo code on a parallel architecture for treatment planning applications. *Med Phys* 31(9):2721–5.

Underberg, R. W., F. J. Lagerwaard, H. van Tinteren, J. P. Cuijpers, B. J. Slotman, and S. Senan. 2006. Time trends in target volumes for stage I non-small-cell lung cancer after stereotactic radiotherapy. *Int J Radiat Oncol Biol Phys* 64(4):1221–8.

Vanderstraeten, B., N. Reynaert, L. Paelinck, et al. 2006. Accuracy of patient dose calculation for lung IMRT: A comparison of Monte Carlo, convolution/superposition, and pencil beam computations. *Med Phys* 33(9):3149–58.

van Herk, M. 2004. Errors and margins in radiotherapy. *Semin Radiat Oncol* 14(1):52–64.

van Herk, M., M. Witte, J. van der Geer, C. Schneider, and J. V. Lebesque. 2003. Biologic and physical fractionation effects of random geometric errors. *Int J Radiat Oncol Biol Phys* 57(5):1460–71.

van Sornsen de Koste, J. R., F. J. Lagerwaard, R. H. Schuchhard-Schipper, et al. 2001. Dosimetric consequences of tumor mobility in radiotherapy of stage I non-small cell lung cancer: An analysis of data generated using 'slow' CT scans. *Radiother Oncol* 61(1):93–9.

Webb, S. 2005. The effect on IMRT conformality of elastic tissue movement and a practical suggestion for movement compensation via the modified dynamic multileaf collimator (dMLC) technique. *Phys Med Biol* 50(6):1163–90.

West, J. B. 2008. *Pulmonary Pathophysiology: The Essentials, Respiratory Physiology.* Philadelphia, PA: Wolters Kluwer Health/Lippincott Williams & Wilkins.

Woodford, C., S. Yartsev, A. R. Dar, G. Bauman, and J. Van Dyk. 2007. Adaptive radiotherapy planning on decreasing gross tumor volumes as seen on megavoltage computed tomography images. *Int J Radiat Oncol Biol Phys* 69(4):1316–22.

Wu, Q. J., D. Thongphiew, Z. Wang, et al. 2008. On-line re-optimization of prostate IMRT plans for adaptive radiation therapy. *Phys Med Biol* 53(3):673–91.

Xiao, Y., L. Papiez, R. Paulus, et al. 2009. Dosimetric evaluation of heterogeneity corrections for RTOG 0236: Stereotactic body radiotherapy of inoperable stage I-II non-small-cell lung cancer. *Int J Radiat Oncol Biol Phys* 73(4):1235–42.

Yu, C. X., D. A. Jaffray, and J. W. Wong. 1998. The effects of intrafraction organ motion on the delivery of dynamic intensity modulation. *Phys Med Biol* 43(1):91–104.

20

Adaptive Radiation Therapy for Liver Cancer

An Tai
Medical College of Wisconsin

X. Allen Li
Medical College of Wisconsin

Laura A. Dawson
University of Toronto

20.1 Introduction

Primary liver cancer is the third leading cause of cancer deaths worldwide with mortality estimated at 598,000 in 2002 (Parkin et al. 2005). Recent statistics from the American Cancer Society estimated that approximately 22,620 new cases of liver and intra-hepatic bile duct cancers will be diagnosed in the United States in 2009 (ACS 2009). The majority of primary liver cancers in North America are hepatocellular carcinoma (HCC) with many of the remaining being cholangiocarcinoma. The incidence rates of primary liver cancer have historically been lower in the United States than in Asian countries. However, the disease rates are increasing as a result of increasing incidence of the hepatitis B and C infections. HCC has tripled in the United States from 1975 to 2005 (Altekruse, McGlynn, and Reichman 2009). In addition, liver metastases from solid malignancies are a large source of morbidity. Colorectal cancer, which accounted for about 1 million new cases in 2002 (Parkin et al. 2005) and 146,970 new cases in the United States in 2009 (ACS 2009), often metastasizes to the liver, sometimes as the only metastatic disease site.

Surgical resection has been considered the most successful treatment for liver cancers with a 5-year survival rate ranging from 31% to 54% (Esnaola et al. 2003; Ercolani et al. 2003; Shah et al. 2007) for HCC patients and from 33% to 47% (Cummings, Payes, and Cooper 2007; Fong et al. 1999; Wei et al. 2006) for patients with colorectal metastasis to the liver. Unfortunately, a great majority of patients are not suitable for surgery at diagnosis due to either technical or medical considerations (Cummings, Payes, and Cooper 2007; Liu, Chen, and Asch 2004).

Historically, radiation therapy (RT) has only been used for its palliative benefit (e.g., pain relief) in the context of whole liver radiation (Leibel et al. 1987). An early clinical trial effort by the Radiation Therapy Oncology Group (RTOG 8405) to escalate whole liver dose from 27 Gy to 36 Gy using hyperfractionation of 1.5 Gy per fraction twice a day for treating liver metastases was hampered, and the study closed early because 5 of 51 patients (10% ± 7.3%) developed clinical radiation liver toxicity following 33 Gy (Russell et al. 1993). The median survival from this trial was about 4 months with no difference among the different dose groups. It was further established that there is about a 5% risk of radiation-induced liver disease (RILD) following whole liver irradiation at 28–32 Gy in conventional fractions (Emami et al. 1991; Lawrence et al. 1995), which limits the role of radiation as a curative modality when delivered to the whole liver because a much higher dose is required to control the gross disease of liver tumors.

Following the previous experience of whole liver irradiation, Mohiuddin, Chen, and Ahmad (1996) demonstrated that it was safe and beneficial to deliver a boost tumor dose (up to 60 Gy) to the partial liver after treating the whole liver to a tolerable dose (≤30 Gy) established previously. In this series of patients with metastatic liver tumors from colorectal cancer, the median survival times were 4 and 14 months for those patients with whole liver irradiation alone and those receiving additional tumor boost dose, respectively, indicating that higher doses may be delivered to partial liver for improving patient survival with limited normal issue toxicities.

In the late 1980s, Lawrence et al. (Lawrence, Tesser, and Ten Haken 1990; Lawrence et al. 1991; Lawrence et al. 1992) from the University of Michigan pioneered the work to treat liver cancer with curative intent using three-dimensional (3D) conformal RT (3D-CRT) following a dose-escalation strategy for maximizing tumor control, while keeping normal liver complications under an acceptable limit. The treatment was delivered at 1.5–1.65 Gy/fraction twice a day with concurrent chemotherapy of intra-arterial hepatic fluorodeoxyuridine (FUdR) or bromodeoxyuridine (BUdR). Initially, the prescription dose was individualized (from 48 Gy to 72.6 Gy) based on the percentage of normal liver receiving more than 50% of the isocenter dose. Promising results were reported, as the median survival was 19 months for hepatobiliary cancers (HCC and cholangiocarcinoma; Robertson et al. 1993) and 20 months for colorectal liver metastases (Robertson et al. 1995).

Radiation-induced liver disease (RILD) was historically referred to as radiation hepatitis and a veno-occlusive disease. However, RILD is not associated with inflammation and is unique from other veno-occlusive diseases. The treatment options for RILD are limited, and in severe cases, it leads to liver failure and patient death. Efforts have been made to predict probability of RILD based on models of normal tissue complication probability (NTCP; McGinn et al. 1998; Jackson et al. 1995). The NTCP models can be used to determine an individualized prescription dose based on an acceptable limit of RILD once the model parameters are extracted from clinical data of normal liver complication. The parameters of the Lyman NTCP model for a dose fractionation scheme of 1.5 Gy per fraction twice a day were derived by Dawson et al. (2002) from data of 203 patients treated at the University of Michigan. The same parameters were also obtained by Cheng et al. (2004) and by Xu et al. (2006) for a fractional dose of 2 and 4.6 Gy, respectively. The dose-escalation strategy based on the NTCP model calculation has been implemented clinically with a correction for the difference in the dose per fraction (Dawson et al. 2000; Dawson, Eccles, and Craig 2006).

So far, the best outcome of RT for liver cancer has been achieved in particle therapy using carbon-ion and proton beams, presumably due to the ability to deliver higher doses to the tumor while substantially sparing the healthy liver. Kato et al. (2004) reported a 5-year survival of 25% obtained from 24 HCC patients treated using carbon-ion beams with a dose-escalation scheme up to 79.5 GyE (Gray equivalents) in 15 fractions. Fukumitsu et al. (2009) obtained a 5-year survival of 38.7% from 51 HCC patients treated using proton beams with 66 GyE in 10 fractions. Unfortunately, proton or carbon-ion therapy is currently limited to a few centers worldwide because of its cost. However, such encouraging outcomes may convey an important message that RT to liver can achieve clinical outcomes similar to surgery if high enough doses can be delivered.

The previous experience of 3D-CRT, especially the individualized treatment planning based on NTCP calculations, leads to the rationale for the dose-escalation trials of liver cancers. Historically, a liver tumor was treated with a fractionated scheme of 1.5–2 Gy per fraction to reduce normal tissue injury. The advent of new technology such as image-guided radiotherapy (IGRT; Dawson and Jaffray 2007) and respiratory motion management (Keall et al. 2006) has made it possible to spare more healthy liver, while delivering a high dose to the tumor. As a result, hypofractionated radiation treatment and stereotactic body radiotherapy (SBRT) have become popular for treating liver tumors in select patients. Many patients with liver metastases are treated with SBRT delivering high doses to tumors in three to six fractions. Recent clinical trials with a prescription dose of 60 Gy in three to six fractions suggest that it is safe to treat liver patients with SBRT, with a median overall survival of 17.6 months (Lee, Kim, et al. 2009) and 20.5 months (Rusthoven et al. 2009) for liver metastases, and 11.7 months for HCC, and 15 months for intrahepatic cholangiocarcinoma (Tse et al. 2008). A RTOG trial treating liver metastases with hypofractionation in 10 fractions was conducted with results pending (Katz et al. 2005). With limited follow-up time at this time, the incidence of RILD and grade ≥3 toxicities are rare after SBRT. Dose-limiting toxicity was not exceeded in a trial of three-fraction SBRT at the highest prescribed dose of 60 Gy (Schefter et al. 2005) and was not observed in a trial of six-fraction SBRT for HCC or liver metastases (Tse et al. 2008; Lee, Kim, et al. 2009). However, more clinical trials and longer follow-up are necessary to evaluate the benefit and safety of SBRT in the treatment of liver cancer due to the potential for increased late toxicity.

The progress outlined above has redefined the role of external beam radiotherapy (EBRT) in the treatment of liver cancers (Hawkins and Dawson 2006). The optimal radiation treatment for the liver cancer needs to be adaptive (highly individualized), which includes (1) individualized motion management for accurate treatment delivery and (2) individualized dose and dose fractionation for treatment planning. In this chapter, we will review the development and the current status for using EBRT to treat unresectable liver malignancies. We will then address the difficulties that we are facing and discuss strategies to overcome these difficulties.

20.2 Biologically Based Treatment Planning

20.2.1 Target Definition

Ideally, breath-hold, triphasic contrast-enhanced computed tomography (CT), and/or magnetic resonance imaging (MRI) of the patient should be obtained with the patient in the treatment position so that target delineation, such as gross tumor volume (GTV), is based on the best diagnostic quality imaging at the time of radiation planning. The arterial phase of contrast enhancement is best for the delineation of HCCs, while the venous phase is best for delineations of portal vein thrombosis from HCC and for most metastases. Contrast-enhanced MRI can provide complimentary information to CT in target definition, as MRI has high soft-tissue contrast resolution.

There is little clinical data to guide what the most appropriate clinical target volume (CTV) around the GTV is to ensure the highest-risk regions containing occult microscopic foci are treated. One radiological–pathological study evaluated the degree of microscopic extension for tumor beyond the HCC capsule in patients who were treated with surgery (Wang et al. 2009). In this radiological–pathological study, 149 HCC patients with a median tumor diameter of 5.8 cm (range, 1–22 cm), low platelet count, large tumor size, increased tumor stage, and elevated tumor marker were correlated with increased risk of microscopic extension beyond the tumor capsule. Approximately 50% of the patients had no microscopic extension, 44% had <2 mm of microscopic extension, and 9% had up to 4 mm of microscopic extension. For tumors <5 cm, 96% had <2 mm of microscopic extension. Thus, a CTV margin of 5 mm appears to be adequate for most HCCs, and smaller margins can be used for tumors <5 cm in maximum diameter. A caveat to this is that the pathological fixation shrinkage was not corrected for, so the actual degree of microscopic extension could be up to 15% larger in vivo.

20.2.2 Outcome Modeling for Tumor Response to Radiation

Various dose fractionations (ranging from 1.5 to 8 Gy per fraction) were used to treat HCC using photon beams in the world with a median survival ranging from 10 to 25 months (Cheng et al. 2000; Guo and Yu 2000; Seong et al. 2003; Li et al. 2003; Wu, Liu, and Chen 2004; Liu et al. 2004; Zeng et al. 2004; Ben-Josef et al. 2005; Park et al. 2005; Liang et al. 2005; Kim et al. 2006; Mornex et al. 2006; Zhou et al. 2007). An important component in individualized treatment planning for the liver is optimizing dose fractionation schemes for the patient to maximize therapeutic gain. To do this, it is necessary to understand how the liver tumor and healthy tissue respond to the radiation dose. For the phase II trial at the University of Michigan, Ben-Josef et al. (2005) reported that the total dose was the only significant predictor of survival although a weak correlation between the delivered dose and the tumor size was observed. Similarly, in a multivariate analysis of data for HCC patients treated with 1.8 Gy per fraction, Seong et al. (2003) concluded that the dose was the most significant factor affecting the patient survival.

Outcome data from multiple institutions treating HCC with 3D-CRT are listed in Table 20.1 (some exceptions exist, see the table notes). Data from liver metastases treated with SBRT will be discussed in Section 20.4. Most of the HCC patients in Asia were treated with transarterial chemoembolization (TACE) in addition to RT (either pre-RT or post-RT), whereas patients in the University of Michigan study were concurrently treated with FUdR. Patients with a liver function score of Child-Pugh C were generally not eligible for radiation treatment, and there is little experience with treatment of patients with Child-Pugh score B.

Unlike tumor size, staging, liver function scores and so on, there is no clear evidence showing that TACE and FUdR significantly alter patients' survival (Ben-Josef and Lawrence 2005). Data from treatments with proton and carbon-ion beams are also listed in Table 20.1, although the tumor sizes of these patients are relatively smaller than those of patients

treated with photons (Kato et al. 2004; Fukumitsu et al. 2009; Bush et al. 2004; Kawashima et al. 2005; Chiba, Tokuuye, and Matsuzaki 2005). Because fractional doses are different, it is more meaningful to compare the outcome data as a function of biologically effective dose (BED; Table 20.1). BED is expressed to be $BED = \left(1 + \frac{d}{\alpha/\beta}\right)D - \gamma T/\alpha$ in the linear quadratic (LQ) model (Fowler 1989), in which d and D are the fractional dose and total dose, respectively. α/β, α, and γ are three radiobiological parameters, and γ is related to the potential tumor doubling time (T_p) by $\gamma = 0.693/T_p$. It implies that BED would be increased by increasing either the prescribed dose or delivering the prescribed dose in a shorter period and fewer fractions. The radiation treatment time (T) can be calculated with $D/d \times 1.4$ in Table 20.1 if it is not reported in the literature.

The 3-year overall survival rates in Table 20.1 are plotted as a function of BED in Figure 20.1 using the radiobiological parameters derived by Tai et al. (2008), who analyzed clinical survival data of primary liver cancer from a few institutions following a model that associates the survival rate with the radiobiological parameters in the LQ model. These parameters were estimated from this study to be $\alpha/\beta = 15 \pm 2$ Gy, $\alpha = 0.010 \pm 0.001$ Gy^{-1}, $T_d = 128 \pm 12$ day. The error bars in Figure 20.1 show one standard deviation of the statistic error. In the BED calculation, the median prescription dose and median fractional dose were used if they were available. Otherwise, the middle value of the given dose range was used.

Most of the photon beam data (except Ben-Josef et al. 2005 and Tse et al. 2008) in Table 20.1 are from retrospective studies, while most of the particle therapy data (except Chiba, Tokuuye, and Matsuzaki 2005) are from prospective studies. Because of the nature of retrospective and multiple institutional studies, the interpretation of Figure 20.1 is hindered by the heterogeneity of patients, liver function, tumor size or staging, dose prescription methods, and so on in different institutions. The data of the photon beams do not show a clear dependence on BED. However, the proton/carbon data show much higher survival rates, which may be associated with both high BEDs and small tumor sizes. The fact that BEDs for Kato et al. (2004) and Wu, Liu, and Chen (2004) are almost the same implies that patient selection, underlying liver disease etiology, and tumor size may contribute to the survival difference. In fact, the patients with tumor volume <125 cc in the study by Liang et al. (2005) and tumor size <5 cm in the study by Wu, Liu, and Chen (2004) achieved a 3-year survival rate as high as 75% and 85%, respectively. This indicates that better clinical outcomes for HCC patients may be achievable with a high BED for a select group of patients.

The tumor sizes of patients treated with the proton or carbon beams are comparable to those of the patients treated with resection (Esnaola et al. 2003; Ercolani et al. 2003), which have 3-year survival rates between 50% and 62%, indicating similar outcomes between liver resection and proton or carbon radiotherapy for HCC. A few outstanding questions may be raised in light of the previous experience of treating HCC with RT.

TABLE 20.1 Patient Survival after Radiation Therapy for Hepatocellular Carcinoma/PLC

Study	Patients (*n*) (CPS)	Median Tumor Size (cm)/ Volume (Range)	Median Dose (Range), Dose/fx (Gy)[a] Treatment Time (Days)	MS (Mon.)	OS Rate (%)
Cheng 2000	25 23 (A) 2 (B)	10.3 (3.7–18)	46.9 (36–61.1) 1.8–2/fx 35	19.2	54 (1 y) 41 (2 y)
Guo 2000	107 77 (A) 30 (B)	10.2 (5–18)	25–55 1.6–2/fx 31	18	59 (1 y) 28 (2 y) 16 (3 y)
Seong 2003	158 117 (A) 41 (B)	9 ± 3	48.2 (25.2–60) 1.8/fx 37	10	42 (1 y) 20 (2 y) 14 (3 y)
Li 2003	45 No Child C	8.5 (4.1–12.7)	50.4 1.8/fx 39	23.5	68 (1 y) 48 (2 y) 23 (3 y)
Wu 2004	94 43 (A) 51 (B)	10.7 (3–18)	56 (48–60) 4–8/fx 22	25	94 (1 y) 54 (2 y) 26 (3 y)
Liu 2004	44 32 (A) 12 (B)	Tumor size (patients) <5 (16) 5–10 (16) >10 (12)	50.40 (39.6–60) 1.8/fx 39	15.2	60 (1 y) 40 (2 y) 32 (3 y)
Zeng 2004	54 44 (A) 10 (B)	479 ± 80 cc	50 (36–60) 2/fx 35	20	72 (1 y) 42 (2 y) 24 (3 y)
Ben-Josef 2005	35 Child A	10	60.8 (40–90) 1.5/fx 28	15.2	57 (1 y)[b] 17 (3 y)[b]
Park 2005	59 38 (A) 8 (B)	9.6 (1–21.5)	45.3 (30–55) 2–3/fx 25	10	47 (1 y) 27 (2 y) 14 (3 y)
Liang 2005[c]	128 108 (A) 20 (B) No Child C	459 ± 430 cc	53.6 ± 6.6 4.9/fx 28	20	65 (1 y) 43 (2 y) 33 (3 y)
Kim 2006	70 56 (A) 14 (B)	7.5 (2–17)	54 (44–54) 2–3/fx 30	10.8	43 (1 y) 18 (2 y)
Zhou 2007	50 48 (A) 2 (B)	144 (31–792) cc	43 (30–54) 2/fx 30	17	60 (1 y) 38 (2 y) 28 (3 y)
Tse 2008	31 Child A only	173 (9–1913) cc	36 (24–54) 6/fx 14	11.7	48 (1 y)
Kato 2004	24 16 (A) 8 (B)	5 (21.–8.5)	66 (49.5–79.5) 3.3–5.3/fx 35	37	92 (1 y) 50 (3 y) 25 (5 y)
Bush 2004	34 CPS ≤10	5.6 (1.5–10)	63 4.2/fx 21	N/A	55 (2 y)
Kawashima 2005	30 20 (A) 10 (B)	4.5 (2.5–8.2)	76 (50–87.5) 3.8 (2.9–5)/fx 35	N/A	77 (1 y) 66 (2 y) 62 (3 y)
Chiba 2005	162 82 (A) 62 (B) 10 (C)	3.8 (1.5–14.5)	72 (50–88) 4.5 (2.9–6)/fx 29	32	44 (3 y) 24 (5 y)
Fukumitsu 2009	51 41 (A) 10 (B)	2.8 (0.8–9.3)	66 6.6/fx 14	34	49 (3 y) 39 (5 y)

PLC = primary liver cancer; CPS = Child-Pugh score; MS = median survival; OS = overall survival.

[a] The range will be given if the median value is not available.

[b] Includes patients with cholangiocarcinoma and liver metastases.

[c] Reported as PLC.

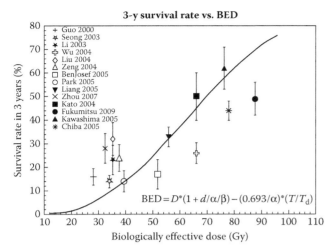

3-y survival rate vs. BED

$$BED = D*(1 + d/\alpha/\beta) - (0.693/\alpha)*(T/T_d)$$

FIGURE 20.1 Three-year overall survival rate as a function of biologically effective dose (BED) for primary liver tumors. BED and the curve were calculated using the model described by Tai, A., et al. (2008). Note that in the studies by Wu, D. H., et al. (2004), Liu et al. (2004), and Zeng, Z. C., et al. (2004) the follow-up time was recorded from the beginning of diagnosis, whereas in other studies it was recorded from the start of treatment.

Will the clinical outcome be further improved if BED can be escalated beyond those in Figure 20.1? Can the RT with photon beams with escalated dose achieve a clinical outcome similar to surgery for resectable patients? How can we determine the maximum tolerable dose for an individual patient? Future clinical trials may help to answer the first two questions. The last question is related to our understanding of the complications of organ at risk (OAR) in liver treatment, especially the normal liver tolerance to radiation.

20.2.3 Normal Tissue Toxicity and Normal Tissue Complication Probability-Based Treatment Planning

RILD is the one of the most serious potential dose-limiting complications following liver cancer irradiation. RILD is a clinical syndrome characterized by anicteric hepatomegaly, ascites, and elevated (2x) liver enzymes (particularly serum alkaline phosphatase), typically occurring 2 weeks to 4 months after completion of radiation to the liver (Lawrence et al. 1995). This end point (classic RILD) was found among patients who have otherwise fairly well-functioning livers (Child-Pugh A), as those reported by the Michigan group (Dawson and Ten Haken 2005). For those patients who have poorer liver function scores (Child-Pugh B and C) as mostly reported in Asia, another end point of elevated transaminases (5x) was also used to define the onset of RILD (Cheng et al. 2004), which has been referred to as nonclassic RILD in contrast to the definition of the Michigan group. The nonclassic RILD may also include general liver function decline as a potential toxicity in HCC and reactivation of hepatitis.

In a famous article by Emami et al. (1991), the doses that lead to 5% of RILD are 30 Gy for whole liver irradiation, 35 Gy for two-thirds liver irradiation, and 50 Gy for one-third liver irradiation. Since the article's publication in 1991, more clinical data have become available leading to revised liver tolerances for partial liver irradiation. Table 20.2 lists the RILD and grade ≥ 3 gastrointestinal (GI) toxicity data from multiple institutions.

Many dosimetric parameters have been proposed to determine the tolerable prescription dose in order to reduce the risk of RILD. The quantity $V_{50\%}$, which is the fraction of the nontumor liver treated with more than 50% of prescribed dose, was first introduced by Robertson et al. (1993) and later used by others (Park et al. 2002; Wu, Liu, and Chen 2004; Lee, Seong, et al. 2009). Lee, Seong, et al. (2009) reported that the incidence of RILD was 11.1%, 10.3%, and 18.2%, respectively, for a $V_{50\%}$ of less than 33%, 33%–66%, and more than 66%, respectively. Cheng et al. (2000) also suggested using the indocyanine green retention rate at 15 minutes (ICG-R15), which is a sensitive marker for liver reserve function as another measure in addition to the volume of nontumor liver. However, data in a study by Lee, Seong, et al. (2009) did not show a significant correlation between liver complication and ICG-R15. A few other dosimetric parameters, such as the mean dose to normal liver (MDTNL) and V_{30} (percentage volume receiving ≥ 30 Gy) of the healthy liver, have also been used to attempt to minimize liver dose and RILD risk. $V_{30} \leq 60\%$ (Kim et al. 2007) and MDTNL ≤ 30 Gy (Li et al. 2003; Zhou et al. 2007) were suggested for the conventional dose per fraction of 1.8–2 Gy. For the hypofractionated treatment at 4.6 Gy per fraction, these constraints were $V_{30} \leq 28\%$ and MDTNL ≤ 23 Gy for patients of Child-Pugh A (Liang et al. 2006).

The group (Ben-Josef et al. 2005) at the University of Michigan used the Lyman NTCP model to determine the prescription dose for each individual patient by allowing an NTCP less than 10%–15%. The Lyman model parameters that were used are $n = 0.97$, $m = 0.12$, and $TD_{50}(1) = 39.8$ Gy for the primary liver tumor and $TD_{50}(1) = 45.8$ Gy for the metastatic liver tumor, respectively (Dawson et al. 2002). Here $TD_{50}(1)$ is the tolerance dose associated with a 50% NTCP for uniform whole liver irradiation, m characterizes the steepness of the dose response at $TD_{50}(1)$ and reflects the inhomogeneity of patient population, and n represents the effect of the irradiated normal liver volume on $TD_{50}(1)$. The difference in $TD_{50}(1)$ indicates that radiation tolerance of the liver is reduced in patients with primary liver cancer vs. liver metastases. The n value close to 1 indicates a large volume effect of the liver complication and a strong correlation between RILD and the mean dose received by the normal liver.

Using these parameters, the effective volume, V_{eff}, was calculated for each patient, and the prescription dose, which leads to an NTCP smaller than an acceptable limit, was determined using the Lyman model. The V_{eff} is defined as the normal liver volume, which, if irradiated uniformly to a dose, would be associated with the same NTCP as the nonuniform dose distribution actually delivered with the same dose. It can be calculated using (differential) dose–volume histogram (DVH) by $V_{eff} = \sum_i v_i (D_i/D)^{1/n}$, where v_i and D_i are the percentage volume and dose in the ith dose bin of DVH. The correlations between V_{eff} and the prescription dose at 5% risk of RILD for a fractional

TABLE 20.2 Radiation-Induced Liver Disease (RILD) and Grade ≥3 Toxicities after Liver Radiation Therapy (RT)

Study	Patients (*n*) (CPS), Tumor Type	Dose, Dose/fx (Gy)	Dose (*D*)/Volume Constraints	RILD (*N*) (CPS) Definition	Grade ≥3 (Nonhepatic) GI Toxicity
Park 2002	158 117 (A), 41 (B) HCC	48.2 (25.2–59.4) 1.8/fx	$V_{50\%} < 25\%$, $D > 59.4$ Gy $50\% < V_{50\%} < 25\%$, $D = 45$–54 Gy $75\% < V_{50\%} < 50\%$, $D = 30.6$–41.4 Gy $V_{50\%} > 75\%$ no treatment	7 (A) 4 (B) Classic	Gastroduodenal ulcer (9), gastroenteritis (8)
Li 2003	45 No Child C HCC	50.4 1.8/fx	MDTNL <30 Gy D_{mean}(stomach/small intestine) <45 Gy D_{mean}(right kidney) <20 Gy D_{mean}(left kidney) <10 Gy D_{max}(stomach/small intestine) <54 Gy	9 (1 death) Classic and nonclassic	GI bleeding (3)
Wu 2004	94 43 (A) 51 (B) HCC	56 (48–60) 4–8 /fx	$V_{50\%} < 25\%$, $D = 60$ Gy @ 7.5/fx $50\% < V_{50\%} < 25\%$, $D = 54$ Gy @ 6 Gy/fx $75\% < V_{50\%} < 50\%$, $D = 48$ Gy @ 4 Gy/fx $V_{50\%} > 75\%$ no treatment	12 (4 death) N/A	Gastroduodenal ulcer (5)
Cheng 2004	89 68 (A) 21 (B) HCC	49.9 (36–66) 2/fx	NTPL <1/3, ICG-R15 ≤10% 40 Gy 1/3 < NTPL < 1/2, ICG-R15 ≤ 10% 50 Gy 1/3 ≤ NTPL ≤ 1/2, 10% < ICG-R15 ≤ 20% 40 Gy NTPL > 1/2, ICG-R15 ≤ 10% 60–66 Gy NTPL > 1/2, 10 < ICG-R15 ≤ 20% 50 Gy NTPL > 1/2, 20 < ICG-R15 ≤ 30% 40 Gy	10 (A) 7 (B) Classic and nonclassic	Not reported
Ben-Josef 2005	128 Child A PLC and metastatic disease (mets)	60.8 (40–90) 1.5/fx	NTCP < 10%–15% for liver Duodenum/stomach max <68 Spine cord max <37.5 If $V_{20} > 50\%$ for one kidney, other kidney $V_{18} < 10\%$	5 (1 death) Classic	30% in total with 6 GI ulcer and bleeding
Xu 2006	109 93 (A) 16 (B) PLC	54 (38–68) 4.6 (4–6)/fx	Prescription was decided by physicians' own judgments (trial and error)	17 (13 death) 8 (A) 9 (B) Classic and nonclassic	Not reported
Zhou 2007	50 48 (A) 2 (B) HCC	43 (30–54) 2/fx	MDTNL ≤30 Gy	2 (2 death, A) Classic and nonclassic	GI bleeding (1)
Kim 2007	105 85 (A) 20 (B) HCC	54 (44–58.5) 2/fx	Not reported	13 Classic	Not reported
Lee 2009	131 114 (A) 17 (B) HCC	45 ± 16.5 1.5–2.5/fx	$V_{50\%} < 25$, $D ≥ 59.4$ Gy $25 ≤ V_{50\%} ≤ 49$, $D = 45$–54 Gy $50 ≤ V_{50\%} ≤ 75$, $D = 30.6$–45 Gy $V_{50\%} > 75$, no treatment	13 9 (A) 4 (B) Classic and nonclassic	Not reported

$V_{50\%}$ = fraction of the nontumor liver treated with more than 50% of prescribed dose; NTPL = nontumor part of liver; ICG-R15 = indocyanine green retention rate at 15 minutes; MDTNL = mean dose to normal liver; CPS = Child-Pugh Score.

dose of 1.5 Gy twice a day were given by Dawson and Ten Haken (2005) and are shown in Figure 20.2.

The risk of RILD is significantly correlated with the degree of liver cirrhosis and baseline liver function. The patients treated in the University of Michigan study had a liver cirrhosis score of Child-Pugh A, while patients treated in Asia had liver cirrhosis scores of both Child-Pugh A and, less commonly, Child-Pugh B. The liver that had a score of Child-Pugh B was much more susceptible to the radiation damage than that of Child-Pugh A. The Lyman model parameters derived by Cheng et al. (2004) at 2 Gy per fraction for HCC were $n = 0.35$, $m = 0.30$, and $TD_{50}(1) = 48.9$ Gy for Child-Pugh A and $n = 0.23$, $m = 0.22$ and

$TD_{50}(1) = 38.7$ Gy for Child-Pugh B, demonstrating a reduced tolerance to the liver for Child-Pugh B patients. Note that this analysis included classic and nonclassic RILD as toxicity end points. In addition, the majority of Asian patients listed in Table 20.2 were hepatitis B or C carriers, which also presented with a statistically greater susceptibility to nonclassic RILD after 3D-CRT (Cheng et al. 2004).

One should be cautious when using parameters derived in one patient group to another group without taking into account the heterogeneity of patients, difference in dose fractionation schemes, different end points, different clinical etiology of liver disease, and so on. For example, the group at the University

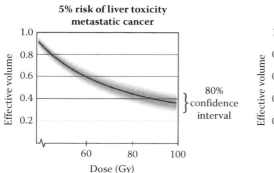
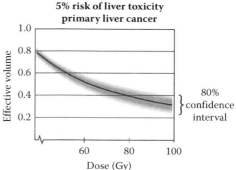

FIGURE 20.2 Dependence of the effective volume on the prescribed dose that leads to a normal tissue complication probability (NTCP) of 5% for a fractional dose of 1.5 Gy for metastatic liver cancer (left panel) and primary liver cancer (right panel). (Reprinted from Dawson, L. A., and R. K. Ten Haken. 2005. *Semin Radiat Oncol* 15:279–83. With permission.)

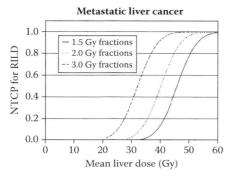
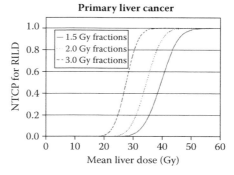

FIGURE 20.3 Normal tissue complication probability (NTCP) as a function of mean dose for a fractional dose of 1.5, 2, and 3 Gy for metastatic liver cancer (left panel) and primary liver cancer (right panel) as calculated using the linear quadratic (LQ) model. (Reprinted from Dawson, L. A., and R. K. Ten Haken. 2005. *Semin Radiat Oncol* 15:279–83. With permission.)

of Michigan found no RILD for a patient treated with 1.5 Gy/fraction twice a day if the MDTNL ≤31 Gy (Dawson et al. 2002). However, Zhou et al. (2007) reported two cases of RILD with a mean dose of 24.9 Gy and 23.3 Gy at 2 Gy/fraction, respectively. Both RILD patients were positive for hepatitis B virus and had liver cirrhosis with a Child-Pugh A score.

As expected, the disparity of liver response to radiation between Child-Pugh A and Child-Pugh B increases with the increasing dose per fraction. Xu et al. (2006) reported the Lyman model parameters extracted from the patients treated with a fractional dose of 4.6 Gy. These parameters were $n = 1.1$, $m = 0.28$, and $TD_{50}(1) = 40.5$ Gy for Child-Pugh A and $n = 0.7$, $m = 0.43$, and $TD_{50}(1) = 23$ Gy for Child-Pugh B.

How to use the Lyman model parameters obtained at a given dose fractionation scheme to calculate NTCP for another dose fractionation scheme has been an issue. Xu et al. (2006) reported that the Lyman model parameters derived by the Michigan group failed to predict the RILD incidence rate for their patients. However, it has been pointed out that the radiobiological effect of dose per fraction should be taken into account when applying the Lyman model parameters derived at 1.5 Gy per fraction to calculate NTCP for 4.6 Gy per fraction (Ten Haken and Dawson 2006). Dawson and Ten Haken (2005) used the LQ model to extrapolate their NTCP calculation to 2 and 3 Gy per fraction, as shown

in Figure 20.3, and found that this was a means to allocate and individualize the dose safely using a different dose per fraction. However, Tai, Erickson, and Li (2009) found that such an extrapolation using the LQ model overestimates NTCP for Child-Pugh A, when the fractional dose of interest is much different from 1.5 Gy per fraction. They proposed a new expression to calculate the normalized total dose (NTD) and extracted the Lyman model parameters for HCC by fitting the RILD data from multiple institutions. The fitting results for both Child-Pugh A and Child-Pugh B are shown in Figure 20.4. The data from Zhou et al. and Lee et al., although not used in the original fitting, are also shown in the figure for comparison. The extracted Lyman model parameters are $m = 0.36$, $TD_{50}(1) = 40.3$ Gy for Child-Pugh A at 1.5 Gy per fraction and $m = 0.41$, $TD_{50}(1) = 23.9$ Gy for Child-Pugh B at 4.6 Gy per fraction (n and α/β were fixed to be 1 and 2 Gy in the fitting). Using these parameters, the plots similar to Figure 20.3 can be generated for a few hypofractionated treatments for Child-Pugh A and Child-Pugh B, as shown in Figure 20.5.

Several comments should be made about Figures 20.3 and 20.5. The Lyman model cannot generate a relation between the mean dose and NTCP for various fractional doses without a DVH (Tome and Fenwick 2004). In order to obtain Figures 20.3 and 20.5, an approximation was made such that the doses in all voxels were corrected with the same fractional dose, d, which

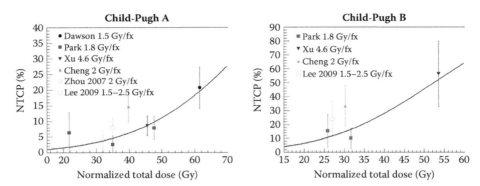

FIGURE 20.4 Normal tissue complication probability (NTCP) data plotted as a function of normalized total dose (NTD) from hepatocellular carcinoma (HCC) patients of Child-Pugh A (left panel) and Child-Pugh B (right panel). NTD was calculated by $\left(\frac{\alpha/\beta+d+f\times N}{\alpha/\beta+d_{ref}+f\times N_{ref}}\right)D(d)$, where N is the number of fractions and f is a fitting parameter (0.156 and 0 for Child-Pugh A and B, respectively; Tai 2009). The subscript refers to the reference dose fraction scheme at which the Lyman model parameters were derived. (Adapted from Tai, A., B. Erickson, and X. A. Li. 2009. *Int J Radiat Oncol Biol Phys* 74:283–9. With permission.)

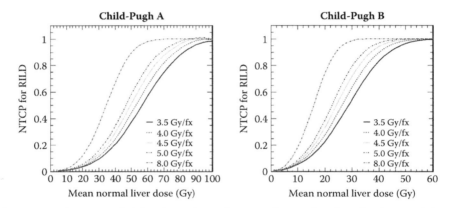

FIGURE 20.5 Normal tissue complication probability (NTCP) as a function of mean dose for a few fractional doses for hepatocellular carcinoma (HCC) patients with Child-Pugh A (left panel) and Child-Pugh B (right panel) as calculated using a modified linear quadratic model. (From Tai, A., B. Erickson, and X. A. Li. 2009. *Int J Radiat Oncol Biol Phys* 74:283–9. With permission.)

is an overestimation for a nonuniform dose distribution of the normal liver DVH. However, because of the great uncertainties in the determination of the Lyman model parameters (the gray bands in Figure 20.2 are an example of these uncertainties), an overestimation of NTCP in Figures 20.3 and 20.5 represents a conservative prediction of NTCP (Dawson, Lawrence, and Ten Haken 2004). An excellent discussion on the error analysis of the NTCP calculations can be found in the article by Jackson et al. (1995).

The Lyman model parameters extracted based on data of a certain cohort are population-averaged parameters. It will be more helpful if the patient-specific parameters correlated with the dose response of the normal liver can be collected during the course of RT before the onset of liver complication. Cao et al. (2007) studied the correlation between the radiation dose and the reduction of venous perfusion using the data of functional CT in an effort to eventually relate the change in venous perfusion to RILD for each individual patient. However, this method is not ready to be used in the clinic at this time, and the use

of venous perfusion changes as an indicator of subregional clinically relevant injury requires validation.

Understanding the dependence of patients' survival and RILD on dose fractionation schemes is helpful in designing an individualized treatment plan aimed at improving patient survival, while keeping RILD under an acceptable limit. The idea can be implemented into an inverse treatment planning system to find the best dose fractionation scheme for a given patient by minimizing a predefined cost function. However, in order to achieve this goal, more patient data, especially data from well-controlled, multi-institutional trials, are needed for better modeling and more reliable model parameters.

20.3 Management of Intra- and Interfractional Organ Motions

In order to be able to deliver tumoricidal radiation dose to liver tumors while keeping normal tissue complication under an acceptable limit, many methods, including the management of intra- and interfractional organ motion, have been applied for reducing the PTV.

20.3.1 Management of Respiration Motion

Respiratory motion management has considerable impact on the treatment planning and delivery for liver irradiation. The liver can move significantly up to 4 cm during free breathing (Keall et al. 2006). Without proper management, such motion would lead to target delineation errors, increased normal tissue dose, and inaccurate dose delivery. The organ motion can be categorized into two components—intrafractional and interfractional motion. The intrafractional motion refers to the motion during one fraction of radiation treatment, whereas the interfractional motion refers to the motion among different fractions. The simplest way to capture motion management is to measure the range of tumor respiratory motion in free breathing for each individual patient and to use this range to decide the individualized (often reduced) PTV margin. Ten Haken et al. (1997) found that a PTV margin reduction of 1–2 cm due to breathing control allowed an increase in the normal liver tolerance dose of 6–8 Gy.

The internal target volume (ITV), which covers the tumor volume during breathing cycles for a patient, can be generated from 4D-CT scans. 4D-CT acquires breathing-synchronized CT images for treatment planning and tumor motion assessment. In the 4D-CT scan, images are sorted into various phase or amplitude bins (usually 10 bins), and 3D images are reconstructed from images in the same bin (Pan et al. 2004). The drawback of 4D-CT for liver tumors is that the quality of tumor enhancement using intravenous contrast suffers from the extended time of the 4D-CT scan. The kV fluoroscopy and cine MR images are also useful for assessing tumor breathing motion and the MR image can be fused together with CT for the target delineation (Dawson, Eccles, and Craig 2006). Breathing coaching may be helpful during simulation and treatment for improving reproducibility of patient breathing (Linthout et al. 2009).

Ideally, tumors on the images of all breathing bins should be contoured to generate ITV by tracing these contours. But in practice, only two sets of images corresponding to the end of exhale (EE) and the end of inhale (EI) were contoured to generate ITV in most cases (Dawson, Eccles, and Craig 2006; Wagman et al. 2003). A recent study for liver treatment showed that the ITV generated from the contours on EE and the EI encompassed 94% of the ITV generated using all contours (Xi et al. 2009).

When the breathing motion is significant (>1 cm), the motion during the treatment should be managed in order to reduce the PTV margin, so that the conformal dosimetric goals can be achieved. Methods such as voluntary breath hold (Kimura et al. 2004), active breathing control (ABC; Eccles et al. 2006), abdominal compression (Wunderink et al. 2008; Heinzerling et al. 2008), gating (Wagman et al. 2003), and tumor tracking (Shirato et al. 2007) may be applied.

A commercial product is available for performing breath hold for liver immobilization during RT (active breathing coordinator, Elekta, Crawley, United Kingdom). This system was used for liver treatment by Wong et al. (1999) and Dawson et al. (2001). The system immobilizes organs through breath holds controlled by a therapist at a selected phase of the breathing cycle based on lung volume monitored by a remote computer. It includes a mouthpiece, a nose-plug, a balloon-valve, and so on. When triggered to inflate, the balloon-valve prevents airflow to and from the patient, and the treatment beam is turned on for a predetermined period. The synchronization between the breath-hold and beam-on time is controlled manually by therapists because the ABC system and linear accelerator (linac) are not electronically integrated.

At Princess Margaret Hospital in Toronto, Canada, an educational section is provided for the patient to be familiar with ABC prior to the treatment. At that time, the position and phase (EE vs. EI) of the breath hold as well as the length of comfortable and stable breath hold are determined for each patient. Before each breath hold, the patient is asked to take two deep breaths to help maintain a longer period of suspended ventilation. The EE phase is the preferred position for breath hold for liver treatment because it is the most stable and producible position for most of the patients. The average length of ABC breath-hold treatment was 12 seconds (range, 5–32 seconds; Eccles et al. 2006).

In gated radiation treatment, patients either breathe freely or follow some online breathing instruction while the treatment beams are delivered only in a preselected breathing window (gating window or duty-cycle). The beam on/off time is triggered by the breathing signal measured either with a reflective marker on the patient belly and infrared tracking camera as in the real-time position management system (Wagman et al. 2003) or with a pressure sensor under a belt around the patient's abdomen as in the Anzai system (Li, Stepaniak, and Gore 2006). For gated delivery, the respiration-triggered scan can also be taken at the EE phase for planning, while fluoroscopy may be used to estimate the residual motion of the liver (represented by the right diaphragm) inside the gating window (Wagman et al. 2003). The residual motion inside the gating window can also be estimated using 4D-CT by contouring images of the phases included in the gating window.

Wagman et al. (2003) showed a significant reduction in average tumor motion from 8.5 to 1.3 mm and from 5.2 to 1.3 mm with gating of the 30% duty-cycle at EE along the craniocaudal (CC) and anteroposterior (AP) directions, respectively (Wagman et al. 2003). Assuming that gating is used in conjunction with image guidance, the dosimetric benefit of gated vs. nongated delivery for a patient treated with 1.8 Gy per fraction for the metastatic liver tumor is shown in Figure 20.6. The gated plan was done at the EE phase with the residual motion measured using the three-phase images around EE. The mean dose was reduced from 33 to 29 Gy for the normal liver (liver volume minus CTV) and 29 to 22 Gy for the right kidney. Wagman et al. also reported that gated delivery enabled dose escalation of 21% on average.

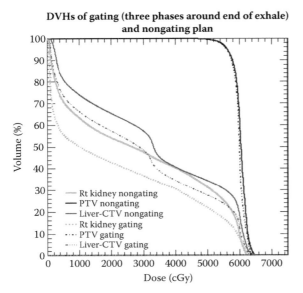

DVHs of gating (three phases around end of exhale) and nongating plan

Legend:
— Rt kidney nongating
— PTV nongating
— Liver-CTV nongating
···· Rt kidney gating
-·-· PTV gating
·-·- Liver-CTV gating

(x-axis: Dose (cGy), 0 to 7000; y-axis: Volume (%), 0 to 100)

FIGURE 20.6 Dose–volume histogram comparisons of gated and free-breathing plans for a liver patient.

Wunderink et al. (2008) reported that abdominal compression reduced the patient groups' median excursion of the gold fiducial markers implanted in the healthy liver tissue surrounding the tumor by 62% in the CC and 38% in the AP directions with respect to the median free-breathing excursions and the residual excursions reproduced well during the treatment course. The tumor-tracking technique for liver treatment is being investigated in a few research centers. However, many technical difficulties need to be solved before it can be widely applied for liver patients.

PTV can be generated from the ITV by including all the other margin components, such as interfractional setup variations and interfraction/intrafraction changes in tumor position and shape that may affect the tumor position during treatment. A study of couch shift data by Li et al. (2007) using tomotherapy megavoltage CT images for patients' daily repositioning revealed that a setup margin of about 8 mm was needed in the ITV-to-PTV expansion for irradiating abdominal targets if daily IGRT was not available. By studying the kV cone-beam CT (CBCT) data of non-breath-hold patients with or without the abdominal compression, Case et al. (2009) concluded that interfractional liver position variations relative to the vertebral bodies are more significant than the variations of the intrafraction liver position, with mean variations of 2 mm (90th percentile, 4.2), 3.5 mm (90th percentile, 7.3), and 2.3 mm (90th percentile, 4.7) for the interfractional variations and 1.3 mm (90th percentile, 2.9), 1.6 mm (90th percentile, 3.6), and 1.5 mm (90th percentile, 3.1) for the intrafractional variations along the CC, medial-lateral (ML), and AP directions, respectively. A similar conclusion was obtained by Eccles et al. (2006) for patients using ABC with an average interfraction and intrafraction CC reproducibility (sigma) of 3.4 mm (range, 1.5–7.9 mm) and 1.5 mm (range, 0.6–3.9 mm), respectively. If possible, each institution should study the variation of liver positions according to its own patient setup

technique, respiratory management method, image modality, and so on and estimate the necessary margin to generate PTV from ITV. For example, a minimum margin of 5 mm was used to account for the reproducibility of immobilization at Princess Margaret Hospital (Dawson, Eccles, and Craig 2006).

The method of respiratory management that should be used depends on many factors, including tumor motion amplitude, reproducibility of the tumor position, patients' tolerance, and communication concerns. For example, Dawson et al. reported that two-thirds of patients were suitable for treatment with ABC. Treatment time is prolonged with gating, which may not be tolerable for some patients. Abdominal compression may even increase tumor motion in some patients (Eccles, Patel, and Lockwood 2008). The data for each individual patient should be obtained during the initial simulation. Regardless of the motion management strategy that is used, shifts in liver position relative to bones can occur, motivating image guidance to reduce the residual geometric uncertainties during liver RT.

20.3.2 Image-Guided Treatment Delivery

For treatment delivery to the liver, patients are usually placed in an immobilization device in a supine position with their arms above their head and a leg support under their knees, replicating the exact positioning of the initial simulation. In addition to traditional immobilization devices like the alpha cradle and the Vac-Lok bag, immobilization equipment specially designed for SBRT (with abdominal compression included) including the Body-Pro-Lok system (CIVCO Medical Solutions, Kalona, Iowa) and the BodyFix system (Elekta, Stockholm, Sweden) are also available. With daily IGRT now being used for liver treatment, the function of the immobilization device is less essential for reproducibility of patient position, but it is used more to prevent patient movement during treatment.

The same breathing control method used during the initial simulation is applied for treatment delivery. Due to baseline shifts in liver position and the need for tight margins in treatment planning to reduce normal tissue toxicities (Case et al. 2009; Eccles et al. 2006), daily IGRT is becoming a standard for liver treatment. In some cases, more than one imaging modality and repeated image acquisitions may be used for monitoring the reproducibility of inter- or intrafractional motion (Lee, Brock, and Dawson 2008).

IGRT significantly reduces the PTV margin needed. A study by Dawson et al. (2005) using repeated orthogonal MV images for liver patients treated with ABC found that image guidance and repositioning reduced the population random setup errors (σ) from 6.5, 4.2, and 4.7 mm to 2.5, 2.8, and 2.9 mm and the population systematic setup errors (Σ) from 5.1, 3.4, and 3.1 mm to 1.4, 2, and 1.9 mm along the CC, ML, and AP directions, respectively.

Because most liver tumors are not visible without using contrast, it is impossible to directly register the treatment tumor position with its planning position. Surrogates such as the vertebral body, diaphragm, or liver are used for patient repositioning. The diaphragm may be used for registration in the CC direction, and vertebral body is used for registration along ML and AP directions (Dawson, Eccles, and Craig 2006). Alternatively, the whole liver or a portion of the liver can be used as a surrogate for the liver tumor when volumetric image guidance with sufficient soft-tissue contrast is available.

For patients treated with ABC at Princess Margaret Hospital, intrabreath-hold liver position stability was measured using online kV fluoroscopy at the treatment position before each fraction and MV beam's-eye-view cine loop images during treatment (Eccles et al. 2006). Intra- or interfractional liver position reproducibility were investigated at the time of treatment, before each treatment fraction, by comparing repeated AP MV verification images and then comparing them with the position on the AP digitally reconstructed radiograph produced in the planning system. In order to acquire 3D kV CBCT images of the liver patients in the ABC breath-hold position, an in-house "stop-and-go" CBCT was also developed (Hawkins et al. 2006). For patients treated without breath hold, the pre- and post-SBRT 4D kV CBCT scans were obtained for each fraction, and the images were sorted into various phase bins (Case et al. 2009). The intra- or interfractional variations of liver position were then calculated by comparing sorted images before and after treatment and then comparing the sorted images with corresponding phase-sorted 4D-CT images acquired for treatment planning during the initial simulation.

It is important to understand how the images of a moving target are acquired and reconstructed by each imaging modality to assure that the tumor or surrogate position in the treatment images actually reflects its corresponding position in the planning images. In gated delivery, for example, the tumor position acquired using the CBCT before treatment may not reflect its position in the gating window unless the CBCT images can be synchronized with patient breathing. In this case, online MV fluoroscopy can be used for the tumor position verification (Tai et al. 2010). In addition, the reproducibility of patient setup against the planned patient position also depends on the quality of image modalities used in IGRT.

20.4 Stereotactic Body Radiotherapy

The idea of SBRT is in line with the effort of delivering tumoricidal dose to ablate tumor cells. As shown in Figure 20.1, 3D-CRT for HCC with conventional fractionation schemes achieved a low 3-year survival (between 10% and 30%), while some select patients with small tumors treated using the hypofractionated proton or light-ion beams had much-improved 3-year survivals. This is not only true for proton or light-ion therapy. The data from the hypofractionated RT reported by Liang et al. (2005) and Wu, Liu, and Chen (2004) for patients with a tumor size < 5 cm also showed significantly higher survival rates ($p = .000$). In contrast, those patients with a tumor size <5 cm in the Seong et al. study (2003) treated with 1.8 Gy per fraction did not show much improved survival in comparison with the other patients in the study ($p = .047$). These data indicate that the clinical outcome may be significantly improved for some select patients with

small-size tumors if being treated with high doses in a few fractions. SBRT can be considered as the translation of intracranial stereotactic radiosurgery and stereotactic radiotherapy toward application for tumors in the chest and abdomen.

Most liver cancers currently treated with SBRT in one to five fractions were liver metastases with a limited number of lesions and a tumor size <6 cm. However, patients were treated with six fractions at Princess Margaret Hospital, and no restriction in the number of tumors or tumor size as long as dose limits to normal tissue were acceptable. The prescription dose was highly individualized based on the Lyman NTCP model (Dawson, Eccles, and Craig 2006). Multiple coplanar or noncoplanar beams should be used to comply with conformity requirements. The dose was usually prescribed to an isodose line such that a high percentage of PTV could be covered by the prescribed dose. This choice leads to higher doses inside PTV, which is considered to be reasonable if assuming that the center of the tumor may contain radioresistant hypoxic tumor cells. Table 20.3 lists data of SBRT for metastatic and primary liver cancer from multiple institutions.

RILD and grade ≥ 3 toxicities are rare in SBRT. However, Hoyer et al. (2006) and Méndez-Romero et al. (2006) reported that one patient died of liver failure in each group after SBRT. Two nonclassic RILD cases were also observed by Méndez-Romero et al. in which these two patients had less than 700 cc (638 cc and 639 cc) normal liver receiving 15 Gy or less. In addition to some GI toxicities, which were reported previously in conventional treatments, SBRT-related liver function and normal liver tissue changes have been reported. Lee, Kim, et al. (2009) reported that four metastatic liver patients (6%) had a decline in their liver function 3 months after SBRT from Child-Pugh A to B (3) and C (1). Seven (17%) experienced progression from Child-Pugh A to B for treatment of primary liver cancers (Tse et al. 2008). Kavanagh, Schefter, and Cardenes (2006) reported characteristic hypodensity in the first few months after liver SBRT and apparent atrophy of surrounding normal liver parenchyma after extended follow-up, but this was not correlated with clinically important toxicity. Severe skin toxicity was also seen when the hot spot of 48 Gy (16 Gy per fraction) was spread out to the skin. $V_{30} < 10$ cc was then suggested to prevent chest wall toxicity (Rusthoven et al. 2009). In addition, special caution is in order if targets are close to the mediastinum (Wulf et al. 2001).

The local control data from multiple institutions, listed in Table 20.3, vary from 57% to 92%. Patient heterogeneity, differences in dose fractionation scheme, treatment planning criteria, and dose delivery accuracy may all contribute to the observed variation. However, additional data from Wulf et al. (2006) with an escalated dose 3×12–12.5 Gy or 1×26 Gy revealed a higher local control of 100% in 1 year and 82% in 2 years in comparison to 76% in 1 year and 61% in 2 years when patients were treated with 3×10 Gy. The multivariate analysis showed that a higher dose is the only significant factor for achieving better tumor local control. Using the radiobiological parameters derived by Tai et al. (2008), the BEDs for 3×10 Gy and 3×12–12.5 Gy are 45 Gy and 62 Gy, respectively. Dependence of local control on

TABLE 20.3 Clinical Outcomes after Stereotactic Body Radiotherapy (SBRT). Patients Had Metastatic Liver Tumors Unless Specified in the Table

Study	Patients (n) Tumor Size Lesions (n)	Dose Scheme	Dose/Volume Constraints	LC (%)	SR(%)/MS	RILD, Grade ≥3 GI and Skin Toxicity
Herfarth 2001	37 (4 PLC) <6 cm 1–3 lesions	14–26 Gy One fraction	Esophagus <14Gy Stomach/small bowel <12 Gy Normal liver D_{30}< 12 Gy Normal liver D_{50}< 7 Gy	67 (18 m)		No major side effect
Wulf 2001	24 (1 CC) CTV = 50 cc (9–516) 1 lesion	30 Gy 3 fractions 6–12 days	Normal liver V_5< 50% Normal liver V_7< 30% All OAR <7 Gy	76 (1 y) 61 (2 y)	61 (1 y) 43 (2 y)	Esophagus ulceration (1) Pulmonary artery bleeding (1)
Katz 2005	N/A ≤6 cm 1–5 lesions	35–50 Gy 10 fractions	≥1000 cc of tumor-free liver V_{27}< 30% and V_{24}< 50% for tumor-free liver V_{10}< 10% one funct. kidney V_{18}< 33% two funct. kidneys Spine cord <34 Gy Small bowel/stoma. V_{37}< 1 cc		Pending	Pending
Schefter 2005	18 <6 cm 1–3 lesions	36–60 Gy 3 fractions 2 wks	Normal liver 700 cc < 15 Gy Right kidney V_{15}< 33% Total kidney V_{15}< 35% Cord <18 Gy Stomach/small bowel <30 Gy		N/A	No grade >3 toxicity
Méndez-Romero 2006	25(11 HCC) <7 cm 1–3 lesions	37.5 Gy 3 fractions[b] 5–6 days	Normal liver D_{33}< 21 Gy Bowel, duodenum, stomach D_{50}< 15 Gy Esophagus D_{5cc}< 21 Gy Cord <15 Gy Kidney D_{33}< 15 Gy	94 (1 y)[c] 82 (2 y)[c]	82 (1 y)[c] 54 (2 y)[c]	Classic RILD: 1 HCC patient of Child B (death with liver failure) Nonclassic RILD: 2 metastatic patients
Kavanagh 2006	36 <6 cm 1–3 mets	60 Gy 3 fractions 3–14 days	700 cc of normal liver ≤15 Gy Total kidney V_{15}< 35% Cord <18 Gy Stomach <30 Gy	93 (18 m) 92 (24 m)	30 (24 m)[d]	Subcutaneous tissue breakdown (1)
Hoyer 2006	44 <6 cm 1–4 mets[a]	45 Gy 3 fractions 5–8 days	Liver V_{10}≤ 30% Cord <18 Gy	86 (2 y)[e]	38 (2 y)[e] 22 (3 y)[e] 13 (5 y)[e]	One hepatic failure (death), colonic ulceration (1), duodenal ulceration (2), skin ulceration (2)
Katz 2007	69 Median (range) 2.7 cm (0.6–12) 2.5 (1–6)	48 (30–55) Gy 2–6 Gy/fx 2 wks	Normal liver ≥1000 cc D_{60}< 30 Gy (healthy livers) Liver D_{70}< 27 Gy (hepatitis/ cirrhosis)	76 (10 m) 57 (20 m)	37 (20 m) MS: 14.5 (m)	No grade >3 toxicity
Lee, Kim 2009[a]	68 75.2 cc (1.19–3090) no limit	41.8 (27.7–60) Gy by NTCP model 6 fractions 2 wks	More than 800 cc uninvolved liver bowel/stomach V_{30}< 0.5 cc[f] Cord+5 mm < 27 Gy 2/3 combined kidney <18 Gy 90% of one kidney <10 Gy	71 (1 y)	MS: 17.6 (m)	No dose-limiting toxicity, small bowel obstruction (1)

CC = cholangiocellular carcinoma; MS = median survival; CTV = clinical target volume; OAR = organ at risk; D_{50} = dose at 50% of volume; NTCP = normal tissue complication probability.

[a] With exception if the expected dose-volume relationship was satisfied.

[b] Three metastatic patients were treated at 30 Gy in 3 fractions in 5–6 days and another 3 patients with HCC ≥4 cm and cirrhosis were treated with 25 Gy in 5 fractions in 10 days.

[c] Local control rates of 1 and 2 years are 100% and 86% for the metastases group, respectively. The overall survival rates at 1 and 2 years were 85% and 62% for the metastases group and 75% and 40% for HCC patients.

[d] Reported in Rusthoven, K. E., B. D. Kavanagh, H. Cardenes, et al. 2009. *J Clin Oncol* 27:1572–8.

[e] Including 20 patients of other sites.

[f] Bowel/stomach V_{33}< 0.5 cc for the first 11 patients. One patient developed a grade 4 duodenal bleeding and grade 5 malignant small bowel obstruction 6 months after SBRT with 32.1 Gy and 33.1 Gy to 0.5 cc of the stomach and duodenum, respectively.

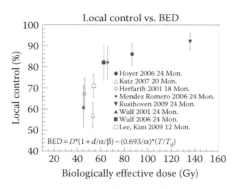
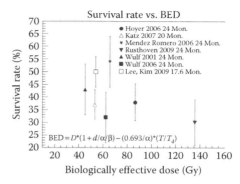

FIGURE 20.7 Local control as a function of biologically effective dose (BED; left panel) and survival rate as a function of BED (right panel) for stereotactic body radiotherapy (SBRT) treatments. Note that unlike the data points with solid symbols, those points with open symbols are not for 24 months due to lack of data at 24 months from these institutions. However, their reported data are also plotted for the purpose of completeness. The follow-up time is recorded from start of the treatment.

BED was plotted in Figure 20.7 (left panel), which seems to show a strong correlation between BED and local control. However, the patient survival rates do not show similar dependence on BED as shown in Figure 20.7 (right panel). The data by Wulf et al. with the increased BED did not lead to improved patient survival with 2-year survival rate of 43% vs. 32% before and after dose escalation. A similar result was seen in the recent publication by Rusthoven et al. (2009) with a median survival of 20.5 months and a 2-year survival rate of 30% even if BED in the treatment was as high as 136 Gy.

There is no clear dependence between the survival rate and BED. The low survival rate after SBRT for liver metastases may reflect poor prognosis and high risk of metastases in patients referred for SBRT. For example, in a study by Hoyer et al. (2006), 33% of patients had SBRT as second-line treatment after surgical resection or radiofrequency ablation and 52% patients received chemotherapy before SBRT. Similarly, in a study by Rusthoven et al. (2009), 69% of patients had received chemotherapy for metastatic disease before study enrollment, 45% had extrahepatic disease, and 51% had metastases from unfavorable primaries (metastases from colorectal and breast cancer are considered favorable primaries). The low survival rate may also lead to an artificially exaggerated local control because death and local failure are competing events.

Advantages of SBRT include (1) higher BED dose to the tumor, (2) low morbidity, (3) less anatomy variation during the course of treatment due to shorter treatment time, (4) patient convenience due to less hospital visits, (5) potentially less invasive alternative for patients not wanting surgery, and (6) effective use of medical resources due to shortened treatment fractions. Méndez-Romero et al. (2008) showed that quality of life had no significant change after SBRT.

Accuracy of dose delivery is the key to a successful SBRT. In order to achieve this goal, a few important components must be considered, including (1) secure patient immobilization for prolonged treatment sessions, (2) appropriate consideration for organ motion, (3) accurate delineation of target with proper margins, (4) optimal dose fractionation scheme for each individual patient, (5) highly conformal dose distribution, (6) precise patient repositioning based on daily image guidance, and (7) online monitoring of the tumor position. For SBRT, respiratory tumor motion control should be within 5 mm.

When gated treatment is used, patient tolerance should be considered because of prolonged treatment time. Prolonged treatment time may cause detrimental effects due to intrafraction radiation repair (Benedict et al. 1997; Fowler, Welsh, and Howard 2004). Total treatment times were less than 25 minutes for most patients at Princess Margaret Hospital (Hawkins et al. 2006) and no more than 45 minutes for those at the University of Colorado (Schefter et al. 2005).

Daily IGRT is a must for a precise SBRT delivery because of small margins, extremely conformal dose distribution, and large fractional dose. A summary of current IGRT strategy can be found in a study by Dawson and Jaffray (2007).

20.5 Remarks on Future Developments

Radiation treatment of liver cancer with curative intention has shown encouraging outcomes. Through the long journey from whole liver irradiation, 3D-CRT to SBRT, we have gained substantial knowledge of how liver tumors and normal liver respond to radiation. Technological advances allow delivery of high dose to the tumor, while still sparing the healthy tissue more precisely than ever before. These experiences and progresses warrant further randomized, well-designed, multi-institutional trials for treating liver tumors, such as those designed by the RTOG.

In this review, the dose-response relationships as a function of BED or NTD were studied using the data from multiple institutions, while BED and NTD were calculated with the model parameters derived by Tai et al. (Tai et al. 2008; Tai, Erickson, and Li 2009). Note that these parameters were obtained using median dosimetric quantities (prescribed dose, fractional dose, effective volume, and so on) instead of individual patient dosimetric data because they are not available in literature. In addition, the patient heterogeneity among these institutions would introduce bias to the dose-response relationships and therefore affect the extracted model parameters. Data from the RTOG trials will improve modeling for individualized, biologically based treatment planning.

SBRT has emerged as a promising modality of liver tumor treatment for some select patients. However, data are still lacking on long-term patient survival and toxicities after SBRT. We do not have a complete understanding on which patient groups would most likely benefit from SBRT because optimal dose fractionation schemes have not yet been established. Definitive evidence for the clinical advantages of this novel treatment modality over established treatments remains to be gathered.

There has been discussion as to where the LQ model is still valid for very high fractional dose (Guerrero and Li 2004; Park et al. 2008). In the trial at Princess Margaret Hospital with six fractions, a dose-V_{eff} relation similar to that in Figure 20.3 was derived based on the LQ model (Dawson, Eccles, and Craig 2006). The prescription dose was then calculated by setting a risk level of RILD to be 5%–20%. However, there was no dose-limiting toxicity observed from this study, which may imply an overestimate of RILD incidence (Tai, Erickson, and Li 2009) that may be attributed to both inaccuracy of the LQ model and the conservative method of converting the dose-V_{eff} relation to the higher dose per fraction. More clinical data will be helpful to clarify this important issue.

Online replanning, which takes 10 minutes, has been studied clinically for considering variation of prostate anatomy during treatment (Ahunbay et al. 2008). Substantial focal deformations in the liver were observed for some patients (Brock et al. 2008), which may be indicative of online adaptive replanning, especially SBRT. Alternatively, developing plans that are robust to residual uncertainties is another strategy for dealing with expected anatomic changes that may occur during RT.

In addition to the development of anatomical imaging technology, recent progress in functional and biological imaging may help in improving target delineation, identifying radioresistant regions, and monitoring response during treatment for an individual patient (Søvik, Malinen, and Olsen 2009). The information is helpful to develop subsequent adaptive treatment based on the data of tumor and normal tissue response collected during the course of a treatment (Eccles et al. 2009).

Respiratory motion management is a challenge to radiation treatment of the liver. Regardless of the technique used for breathing control, there are always uncertainties, even when using image guidance for baseline shifts in liver position. Radiation beams with very high dose rates may be developed to treat a moving target so that RT delivery can be completed within a couple of breathing cycles for reducing breathing variation and treatment time.

One reason that excellent tumor control does not lead to improved patient survival is intraheptic and/or extrahepatic tumor relapse soon after the local treatment of EBRT, providing rationale to perform a combination strategy of local modality and systemic therapies (Han, Lee, and Seong 2002). Right now, the first priority is to conduct well-designed, multi-institutional trials (Dawson 2009). Such trials may also include institutions from Asia because most of the HCC patients are in Asia. With well-controlled heterogeneity in prognostic factors, dose fractionation schemes, target and normal tissue dosimetric guidance, quality assurance strategies in planning and delivery, and consistent definition of dose response and toxicities, definitive knowledge regarding the efficacies of radiotherapy, suitable patient groups, and dose-outcome relations may be established. Such information may help to establish individualized adaptive radiotherapy, combined with other therapies, as a potentially less-invasive treatment option for liver cancer.

References

Ahunbay, E. E., C. Peng, G. P. Chen, et al. 2008. An on-line replanning scheme for interfractional variation. *Med Phys* 35:3607–15.

Altekruse, S. F., K. A. McGlynn, and M. E. Reichman. 2009. Hepatocellular carcinoma incidence, mortality, and survival trends in the United States from 1975 to 2005. *J Clin Oncol* 27:1485–91.

American Cancer Society. 2009. *Cancer Facts & Figures 2009*. Atlanta, GA: American Cancer Society.

Benedict, S. H., P. S. Lin, R. D. Zwicker, et al. 1997. The biological effectiveness of intermittent irradiation as a function of overall treatment time: Development of correction factors for linac-based stereotactic radiotherapy. *Int J Radiat Oncol Biol Phys* 37:765–9.

Ben-Josef, E., and T. S. Lawrence. 2005. Radiotherapy for unresectable hepatic malignancies. *Semin Radiat Oncol* 15:273–8.

Ben-Josef, E., D. Normolle, W. D. Ensminger, et al. 2005. Phase II trial of high-dose conformal radiation therapy with concurrent hepatic artery floxuridine for unresectable intrahepatic malignancies. *J Clin Oncol* 23:8739–47.

Brock, K. K., M. Hawkins, C. Eccles, et al. 2008. Improving image-guided target localization through deformable registration. IGRT 2008 special. *Acta Oncol* 47:1279–85.

Bush, D. A., D. J. Hillebrand, J. M. Slater, et al. 2004. High-dose proton beam radiotherapy of hepatocellular carcinoma: Preliminary results of a phase II trial. *Gastroenterology* 127:S189–93.

Cao, Y., J. F. Platt, I. R. Francis, et al. 2007. The prediction of radiation-induced liver dysfunction using a local dose and regional venous perfusion model. *Med Phys* 34:604–12.

Case, R. B., J. J. Sonke, D. J. Moseley, et al. 2009. Inter- and intra-fraction variability in liver position in non-breath-hold stereotactic body radiotherapy. *Int J Radiat Oncol Biol Phys* 75:302–8.

Cheng, J. C., V. P. Chuang, S. H. Cheng, et al. 2000. Local radiotherapy with or without transcatheter arterial chemoembolization for patients with unresectable hepatocellular carcinoma. *Int J Radiat Oncol Biol Phys* 47:435–42.

Cheng, J. C., J. K. Wu, P. C. Lee, et al. 2004. Biologic susceptibility of hepatocellular carcinoma patients treated with radiotherapy to radiation-induced liver disease. *Int J Radiat Oncol Biol Phys* 60:1502–09.

Chiba, T., K. Tokuuye, and Y. Matsuzaki. 2005. Proton beam therapy for hepatocellular carcinoma: A retrospective review of 162 patients. *Clin Cancer Res* 11:3799–805.

Cummings, L. C., J. D. Payes, and G. S. Cooper. 2007. Survival after hepatic resection in metastatic colorectal cancer: A population-based study. *Cancer* 109:718–26.

Dawson, L. A. 2009. Protons or photons for hepatocellular carcinoma? Let's move forward together. *Int J Radiat Oncol Biol Phys* 74:661–3.

Dawson, L. A., K. K. Brock, S. Kazanjian, et al. 2001. The reproducibility of organ position using active breathing control (ABC) during liver radiotherapy. *Int J Radiat Oncol Biol Phys* 51:1410–21.

Dawson, L. A., C. Eccles, J. P. Bissonnette, et al. 2005. Accuracy of daily image guidance for hypofractionated liver radiotherapy with active breathing control. *Int J Radiat Oncol Biol Phys* 62:1247–52.

Dawson, L. A., C. Eccles, and T. Craig. 2006. Individualized image guided iso-NTCP based liver cancer SBRT. *Acta Oncol* 45:856–64.

Dawson, L. A., and D. A. Jaffray. 2007. Advances in image-guided radiation therapy. *J Clin Oncol* 25:938–46.

Dawson, L. A., T. S. Lawrence, and R. K. Ten Haken. 2004. In response to Dr. Tome and Dr. Fenwick. *Int J Radiat Oncol Biol Phys* 58:1319–20.

Dawson, L. A., C. J. McGinn, D. Normolle, et al. 2000. Escalated focal liver radiation and concurrent hepatic artery fluoro-deoxyuridine for unresectable intrahepatic malignancies. *J Clin Oncol* 18:2210–8.

Dawson, L. A., D. Normolle, J. M. Balter, et al. 2002. Analysis of radiation-induced liver disease using the Lyman NTCP model. *Int J Radiat Oncol Biol Phys* 53:810–21.

Dawson, L. A., and R. K. Ten Haken. 2005. Partial volume tolerance of the liver to radiation. *Semin Radiat Oncol* 15:279–83.

Eccles, C., K. Brock, J. P. Bissonnette, et al. 2006. Reproducibility of liver position using active breathing coordinator for liver cancer radiotherapy. *Int J Radiat Oncol Biol Phys* 64:751–9.

Eccles, C. L., E. A. Haider, M. A. Haider, et al. 2009. Change in diffusion weighted MRI during liver cancer radiotherapy: Preliminary observations. *Acta Oncol* 48:1034–43.

Eccles, C. L., R. Patel, and G. Lockwood. 2008. Comparison of liver motion with and without abdominal compression using cine-MRI. *Int J Radiat Oncol Biol Phys* 72:S50.

Emami, B., J. Lyman, A. Brown, et al. 1991. Tolerance of normal tissue to therapeutic irradiation. *Int J Radiat Oncol Biol Phys* 21:109–22.

Ercolani, G., G. L. Grazi, M. Ravaioli, et al. 2003. Liver resection for hepatocellular carcinoma on cirrhosis: Univariate and multivariate analysis of risk factors for intrahepatic recurrence. *Ann Surg* 237:536–43.

Esnaola, N. F., N. Mirza, G. Y. Lauwers, et al. 2003. Comparison of clinicopathologic characteristics and outcomes after resection in patients with hepatocellular carcinoma treated in the United States, France, and Japan. *Ann Surg* 238:711–9.

Fong, Y., J. Fortner, R. L. Sun, et al. 1999. Clinical score for predicting recurrence after hepatic resection for metastatic colorectal cancer: Analysis of 1001 consecutive cases. *Ann Surg* 230:309–21.

Fowler, J. F. 1989. The linear-quadratic formula and progress in fractionated radiotherapy. *Br J Radiol* 62:679–94.

Fowler, J. F., J. S. Welsh, and S. P. Howard. 2004. Loss of biological effect in prolonged fraction delivery. *Int J Radiat Oncol Biol Phys* 59:242–9.

Fukumitsu, N., S. Sugahara, H. Nakayama, et al. 2009. A prospective study of hypofractionated proton beam therapy for patients with hepatocellular carcinoma. *Int J Radiat Oncol Biol Phys* 74:831–6.

Guerrero, M., and X. A. Li. 2004. Extending the linear-quadratic model for large fraction doses pertinent to stereotactic radiotherapy. *Phys Med Biol* 49:4825–35.

Guo, W. J., and E. X. Yu. 2000. Evaluation of combined therapy with chemoembolization and irradiation for large hepatocellular carcinoma. *Br J Radiol* 73:1091–7.

Han, K. H., J. T. Lee, and J. Seong. 2002. Treatment of non-resectable hepatocellular carcinoma. *J Gastroenterol Hepatol* 17:S424–7.

Hawkins, M. A., K. K. Brock, C. Eccles, et al. 2006. Assessment of residual error in liver position using kV cone-beam computed tomography for liver cancer high-precision radiation therapy. *Int J Radiat Oncol Biol Phys* 66:610–9.

Hawkins, M. A., and L. A. Dawson. 2006. Radiation therapy for hepatocellular carcinoma from palliation to cure. *Cancer* 106:1653–63.

Heinzerling, J. H., J. F. Anderson, L. Papiez, et al. 2008. Four-dimmensional computed tomography scan analysis of tumor and organ motion at varying levels of abdominal compression during stereotactic treatment of lung and liver. *Int J Radiat Oncol Biol Phys* 70:1571–8.

Herfarth, K. K., J. Debus, F. Lohr, et al. 2001. Stereotactic single-dose radiation therapy of liver tumors: Results of a phase I// II trial. *J Clin Oncol* 19:164–70.

Hoyer, M., H. Roed, A. Traberg Hansen, et al. 2006. Phase II study on stereotactic body radiotherapy of colorectal metastases. *Acta Oncol* 45:823–30.

Jackson, A., R. K. Ten Haken, J. M. Robertson, et al. 1995. Analysis of clinical complication data for radiation hepatitis using a parallel architecture model. *Int J Radiat Oncol Biol Phys* 31:883–91.

Kato, H., H. Tsujii, T. Miyamoto, et al. 2004. Results of the first prospective study of carbon ion radiotherapy for hepatocellular carcinoma with liver cirrhosis. *Int J Radiat Oncol Biol Phys* 59:1468–76.

Katz, A. W., M. Carey-Sampson, A. G. Muhs, et al. 2007. Hypofractionated stereotactic body radiation therapy (SBRT) for limited hepatic metastases. *Int J Radiat Oncol Biol Phys* 67:793–8.

Katz, A. W., L. A. Dawson, H. Elsaleh, et al. 2005. A phase I trial of highly conformal radiation therapy for patients with liver metastases. Radiation Therapy Oncology Group 0438.

Kavanagh, B. D., T. E. Schefter, and H. R. Cardenes. 2006. Interim analysis of a prospective phase I/II trial of SBRT for liver metastases. *Acta Oncol* 45:848–55.

Kawashima, M., J. Furuse, T. Nishio, et al. 2005. Phase II study of radiotherapy employing proton beam for hepatocellular carcinoma. *J Clin Oncol* 23:1839–46.

Keall, P. J., G. S. Mageras, J. M. Balter, et al. 2006. The management of respiratory motion in radiation oncology report of AAPM Task Group 76. *Med Phys* 33:3874–900.

Kim, T. H., D. Y. Kim, J. W. Park, et al. 2006. Three-dimensional conformal radiotherapy of unresectable hepatocellular carcinoma patients for whom transcatheter arterial chemoembolization was ineffective or unsuitable. *Am J Clin Oncol* 29:568–75.

Kim, T. H., D. Y. Kim, J. W. Park, et al. 2007. Dose-volumetric parameters predicting radiation-induced hepatic toxicity in unresectable hepatocellular carcinoma patients treated with three-dimensional conformal radiotherapy. *Int J Radiat Oncol Biol Phys* 67:225–31.

Kimura, T., Y. Hirokawa, Y. Murakami, et al. 2004. Reproducibility of organ position using voluntary breath-hold method with spirometer for extracranial stereotactic radiotherapy. *Int J Radiat Oncol Biol Phys* 60:1307–13.

Lawrence, T. S., L. M. Dworzanin, S. C. Walker-Andrews, et al. 1991. Treatment of cancers involving the liver and porta hepatis with external beam irradiation and intraarterial hepatic fluorodeoxyuridine. *Int J Radiat Oncol Biol Phys* 20:555–61.

Lawrence, T. S., J. M. Robertson, M. S. Anscher, et al. 1995. Hepatic toxicity resulting from cancer treatment. *Int J Radiat Oncol Biol Phys* 31:1237–48.

Lawrence, T. S., R. K. Ten Haken, M. L. Kessler, et al. 1992. The use of 3-D dose volume analysis to predict radiation hepatitis. *Int J Radiat Oncol Biol Phys* 23:781–8.

Lawrence, T. S., R. J. Tesser, and R. K. Ten Haken. 1990. An application of dose volume histograms to the treatment of intrahepatic malignancies with radiation therapy. *Int J Radiat Oncol Biol Phys* 19:1041–7.

Lee, M. T., K. K. Brock, and L. A. Dawson. 2008. Multimodality image-guided radiotherapy of the liver. *Imaging Decis MRI* 12:32–41.

Lee, M., J. Kim, R. Dinniwell, et al. 2009. Phase I study of individualized stereotactic body radiotherapy of liver metastases. *J Clin Oncol* 27:1585–91.

Lee, I. J., J. Seong, S. J. Shim, et al. 2009. Radiotherapeutic parameters predictive of liver complications induced by liver tumor radiotherapy. *Int J Radiat Oncol Biol Phys* 73:154–8.

Leibel, S. A., T. F. Pajak, V. Massullo, et al. 1987. A comparison of misonidazole sensitized radiation therapy to radiation therapy alone for the palliation of hepatic metastases: Results of a Radiation Therapy Oncology Group randomized prospective trial. *Int J Radiat Oncol Biol Phys* 13:1057–64.

Li, X. A., X. S. Qi, M. Pitterle, et al. 2007. Interfractional variations in patient setup and anatomic change assessed by daily computed tomography. *Int J Radiat Oncol Biol Phys* 68:581–91.

Li, X. A., C. Stepaniak, and E. Gore. 2006. Technical and dosimetric aspects of respiratory gating using a pressure-sensor motion monitoring system. *Med Phys* 33:145–54.

Li, B., J. Yu, L. Wang, et al. 2003. Study of local three-dimensional conformal radiotherapy combined with transcatheter arterial chemoembolization for patients with Stage III hepatocellular carcinoma. *Am J Clin Oncol* 26:e92–9.

Liang, S. X., X. D. Zhu, H. J. Lu, et al. 2005. Hypofractionated three-dimensional conformal radiation therapy for primary liver carcinoma. *Cancer* 103:2181–8.

Liang, S. X., X. D. Zhu, Z. Y. Xu, et al. 2006. Radiation-induced liver disease in three-dimensional conformal radiation therapy for primary liver carcinoma: The risk factors and hepatic radiation tolerance. *Int J Radiat Oncol Biol Phys* 65:426–34.

Linthout, N., S. Bral, I. Van de Vondel, et al. 2009. Treatment delivery time optimization of respiratory gated radiation therapy by application of audio-visual feedback. *Radiother Oncol* 91:330–5.

Liu, J. H., P. W. Chen, S. M. Asch, et al. 2004. Surgery for hepatocellular carcinoma: Does it improve survival? *Ann Surg Oncol* 11:298–303.

Liu, M. T., S. H. Li, T. C. Chu, et al. 2004. Three-dimensional conformal radiation therapy for unresectable hepatocellular carcinoma patients who had failed with or were unsuited for transcatheter arterial chemoembolization. *Jpn J Clin Oncol* 34:532–9.

McGinn, C. J., R. K. Ten Haken, W. D. Ensminger, et al. 1998. Treatment of intrahepatic cancers with radiation doses based on a normal tissue complication probability model. *J Clin Oncol* 16:2246–52.

Méndez-Romero, A., W. Wunderink, S. M. Hussain, et al. 2006. Stereotactic body radiation therapy for primary and metastatic liver tumors: A single institution phase I-II study. *Acta Oncol* 45:831–7.

Méndez-Romero, A., W. Wunderink, R. M. van Os, et al. 2008. Quality of life after stereotactic body radiation therapy for primary and metastatic liver tumors review. *Int J Radiat Oncol Biol Phys* 70:1447–52.

Mohiuddin, M., E. Chen, and N. Ahmad. 1996. Combined liver radiation and chemotherapy for palliation of hepatic metastases from colorectal cancer. *J Clin Oncol* 14:722–8.

Mornex, F., N. Girard, C. Beziat, et al. 2006. Feasibility and efficacy of high-dose three-dimensional-conformal radiotherapy in cirrhotic patients with small-size hepatocellular carcinoma non-eligible for curative therapies: Mature results of the French phase II RTF-1 trial. *Int J Radiat Oncol Biol Phys* 66:1152–8.

Pan, T., T. Lee, E. Rietzel, and G. T. Chen. 2004. 4D-CT imaging of a volume influenced by respiratory motion on multi-slice CT. *Med Phys* 31:333–40.

Park, W., D. H. Lim, S. W. Paik, et al. 2005. Local radiotherapy for patients with unresectable hepatocellular carcinoma. *Int J Radiat Oncol Biol Phys* 61:1143–50.

Park, C., L. Papiez, S. Zhang, et al. 2008. Universal survival curve and single fraction equivalent dose: Useful tools in understanding potency of ablative radiotherapy. *Int J Radiat Oncol Biol Phys* 70:847–52.

Park, H. C., J. Seong, K. H. Han, et al. 2002. Dose-response relationship in local radiotherapy for hepatocellular carcinoma. *Int J Radiat Oncol Biol Phys* 54:150–5.

Parkin, D. M., F. Bray, J. Ferlay, et al. 2005. Global cancer statistics, 2002. *CA Cancer J Clin* 55:74–108.

Robertson, J. M., T. S. Lawrence, L. M. Dworzanin, et al. 1993. The treatment of primary hepatobiliary cancers with conformal radiation therapy and regional chemotherapy. *J Clin Oncol* 11:1286–93.

Robertson, J. M., T. S. Lawrence, S. Walker, et al. 1995. The treatment of colorectal liver metastases with conformal radiation therapy and regional chemotherapy. *Int J Radiat Oncol Biol Phys* 32:445–50.

Russell, A. H., C. Clyde, T. H. Wasserman, et al. 1993. Accelerated hyperfractionated hepatic irradiation in the management of patients with liver metastases: Results of the RTOG dose escalating protocol. *Int J Radiat Oncol Biol Phys* 27:117–23.

Rusthoven, K. E., B. D. Kavanagh, H. Cardenes, et al. 2009. Mutiinstitutional phase I/II trial of stereotactic body radiation therapy for liver metastases. *J Clin Oncol* 27:1572–8.

Schefter, T. E., B. D. Kavanagh, R. D. Timmerman, et al. 2005. A phase I trial of stereotactic body radiation therapy (SBRT) for liver metastases. *Int J Radiat Oncol Biol Phys* 62:1371–8.

Seong, J., H. C. Park, K. H. Han, and C. Y. Chon. 2003. Clinical results and prognostic factors in radiotherapy for unresectable hepatocellular carcinoma: A retrospective study of 158 patients. *Int J Radiat Oncol Biol Phys* 55:329–36.

Shah, S. A., A. C. Wei, S. P. Cleary, et al. 2007. Prognosis and results after resection of very large (> or = 10 cm) hepatocellular carcinoma. *J Gastrointest Surg* 11:589–95.

Shirato, H., S. Shimizu, K. Kitamura, et al. 2007. Organ motion in image-guided radiotherapy: Lessons from real-time tumor-tracking radiotherapy. *Int J Clin Oncol* 12:8–16.

Søvik, Å., E. Malinen, and D. R. Olsen. 2009. Strategies for biologic image-guided dose escalation: A review. *Int J Radiat Oncol Biol Phys* 73:650–8.

Tai, A., J. D. Christensen, E. Gore, et al. 2010. Gated treatment delivery verification with on-line megavoltage fluoroscopy. *Int J Radiat Oncol Biol Phys* 76:1592–8.

Tai, A., B. Erickson, K. Khater, and X. A. Li. 2008. Estimate of radiobiological parameters from clinical data for biologically based treatment planning for liver irradiation. *Int J Radiat Oncol Biol Phys* 70:900–7.

Tai, A., B. Erickson, and X. A. Li. 2009. Extrapolation of normal tissue complication probability for different fractionations in liver irradiation. *Int J Radiat Oncol Biol Phys* 74:283–9.

Ten Haken, R. K., J. M. Balter, L. H. Marsh, et al. 1997. Potential benefits of eliminating planning target volume expansions for patient breathing in the treatment of liver tumors model. *Int J Radiat Oncol Biol Phys* 38:613–7.

Ten Haken, R. K., and L. A. Dawson. 2006. Prediction of radiation-induced liver disease by Lyman normal-tissue complication probability model in three-dimensional conformal radiation therapy for primary liver carcinoma: In regards to Xu, et al. *Int J Radiat Oncol Biol Phys* 66:1272.

Tome, W. A., and J. Fenwick. 2004. Analysis of radiation-induced liver disease using the Lyman NTCP model: In regards to Dawson et al. *Int J Radiat Oncol Biol Phys* 58:1318–9.

Tse, R., M. Hawkins, G. Lockwood, et al. 2008. Phase I study of individulized stereotactic body radiotherapy for hepatocellular carcinoma and intrahepatic cholangiocarcinoma. *J Clin Oncol* 26:657–64.

Wagman, R., E. Yorke, E. Ford, et al. 2003. Respiratory gating for liver tumors: Use in dose escalation. *Int J Radiat Oncol Biol Phys* 55:659–68.

Wang, M. H., Y. Ji, Z. C. Zeng, et al. 2010. Impact factor for microinvasion in patients with hepatocellular carcinoma: Possible application to the definition of clinical tumor volume. *Int J Radiat Oncol Biol Phys* 76:467–76.

Wei, A. C., P. D. Greig, D. Grant, et al. 2006. Survival after hepatic resection for colorectal metastases: A 10-year experience. *Ann Surg Oncol* 13:668–76.

Wong, J. W., M. B. Sharpe, D. A. Jaffray, et al. 1999. The use of active breathing control (ABC) to reduce margin for breathing motion. *Int J Radiat Oncol Biol Phys* 44:911–9.

Wu, D. H., L. Liu, and L. H. Chen. 2004. Therapeutic effects and prognostic factors in three-dimensional conformal radiotherapy combined with transcatheter arterial chemoembolization for hepatocellular carcinoma. *World J Gastroenterol* 10:2184–9.

Wulf, J., M. Guckenberger, U. Haedinger, et al. 2006. Stereotactic radiotherapy of primary liver cancer and hepatic metastases. *Acta Oncol* 45:838–47.

Wulf, J., U. Hadinger, U. Oppitz, et al. 2001. Stereotactic radiotherapy of targets in lung and liver. *Strahlenther Onkol* 177:645–55.

Wunderink, W., A. Méndez Romero, W. de Kruijf, et al. 2008. Reduction of respiratory liver tumor motion by abdominal compression in stereotactic body frame, analyzed by tracking fiducial markers implanted in liver. *Int J Radiat Oncol Biol Phys* 71:907–15.

Xi, M., M. Z. Liu, L. Zhang, et al. 2009. How many sets of 4DCT images are sufficient to determine internal target volume for liver radiotherapy? *Radiother Oncol* 92:255–9.

Xu, Z. Y., S. X. Liang, J. Zhu, et al. 2006 Prediction of radiation-induced liver disease by Lyman normal-tissue complication probability model in three-dimensional conformation radiation therapy for primary liver carcinoma. *Int J Radiat Oncol Biol Phys* 65:189–95.

Zeng, Z. C., Z. Y. Tang, J. Fan, et al. 2004. A comparison of chemoembolization combination with and without radiotherapy for unresectable hepatocellular carcinoma. *Cancer J* 10:307–16.

Zhou, Z. H., L. M. Liu, W. W. Chen, et al. 2007. Combined therapy of transcatheter arterial chemoembolization and three-dimensional conformal radiotherapy for hepatocellular carcinoma. *Br J Radiol* 80:194–201.

21

Adaptive Radiation Therapy for Prostate Cancer

C.-M. Charlie Ma
Fox Chase Cancer Center

21.1 Introduction

Radiation therapy is a local–regional therapy that has an established and major role in the management of prostate cancer. Modern radiation therapy is a complex process employing advanced imaging techniques in patient immobilization, target determination, critical structure delineation, treatment plan verification, target localization, and treatment assessment. Many clinical trials, including those carried out at Fox Chase Cancer Center (FCCC), have demonstrated that dose escalation with advanced radiation therapy techniques potentially increases the cure rate for prostate cancer, while keeping complicated risks at a reasonable level (Zelefsky et al. 1998; Hanks et al. 1998; Hanks 1999; Pollack et al. 2000; Pollack et al. 2002; Zietman et al. 2005; Peeters et al. 2006; Dearnaley et al. 2007).

New hypofractionation protocols aimed at escalated biological doses and a shorter treatment course may lead to improved local control, lower complications, more convenience for the patient, and reduced treatment costs (Lloyd-Davies, Collins, and Swan 1990; Amer et al. 2003; Yeoh et al. 2003; Fowler et al. 2003; Lukka et al. 2005; Pollack et al. 2006; Kupelian, Willoughby, Reddy, et al. 2007; Madsen et al. 2007; King et al. 2009). As dose levels are increased, the precise knowledge of target size, location, and the accuracy of dose delivery become crucial.

Imaging devices and techniques that are widely used for image guidance or adaptive therapy in radiation oncology have been reviewed in Chapters 9 through 14 of this book. This chapter will briefly review the treatment procedures of external-beam radiation therapy (EBRT) for prostate cancer based on the FCCC experience with a focus on the clinical implementation and practice of intensity-modulated radiation therapy (IMRT) and image-guided radiation therapy (IGRT) that have opened the door to adaptive therapy or individualized radiation therapy.

21.2 Image-Guided Treatment Procedures for Prostate Cancer

This section describes the treatment procedures of IGRT for prostate cancer based on experiences at FCCC, including patient immobilization, image acquisition, target and critical structure delineation, treatment planning, dose delivery, and quality assurance (QA). We have treated more than 3000 prostate patients with the IMRT technique incorporating a variety of imaging modalities for target and structure delineation and target localization for accurate dose delivery.

21.2.1 Treatment Simulation

21.2.1.1 Patient Immobilization

Advanced IGRT treatment techniques aimed at dose escalation and hypofractionation require adequate patient immobilization to facilitate multimodality imaging for treatment simulation and target localization for accurate dose delivery. The inter- and intrafractional organ motions that can be achieved with immobilization systems have been investigated in advance to establish realistic margins for planning purposes. Patients are immobilized and marked as closely as possible to the anticipated treatment isocenter. Devices to facilitate soft-tissue target localization before each treatment have been implemented to ensure the accuracy of isocenter placement (see Section 21.4.2).

All prostate patients are simulated in the supine position. Patient position reproducibility is achieved using a custom alpha cradle cast that extends from the mid-back to mid-thigh and the feet are positioned in a custom Plexiglas foot-holder (Figure 21.1). Patients are asked to have a comfortably full bladder and to empty the rectum using an enema prior to simulation. A low-residue diet the night before simulation is recommended to reduce gas. If at simulation the rectum is > 3 cm in width due to gas or stool, the patient is asked to try to expel the rectal contents. Our studies have demonstrated the benefits of an empty rectum during computed tomography (CT) simulation; it is relatively easy to achieve the rectal dose criteria for a large rectal volume during treatment planning, but the actual doses received by the rectum may be underestimated as the rectal volume decreases during subsequent treatments (Chen et al. 2010).

21.2.1.2 Imaging Parameters

IGRT has recently garnered growing interest, especially with the widespread use of IMRT. Image guidance is playing an increasingly important role in structure segmentation, target (treatment volume) determination, inter- and intrafractional target localization, and/or target redefinition. IMRT requires more precise information about the target and normal tissue structures for treatment planning. The use of contrast agents for the CT and registration of images from other modalities, such as magnetic resonance imaging (MRI), magnetic resonance spectroscopy (MRS), or positron emission tomography (PET), which provide additional anatomical as well as functional information, are often needed and may represent a change in typical practice. This information can help design heterogeneous dose distributions within the target to account for variations in tumor cell density and oxygenation conditions (i.e., IMRT for dose painting or sculpting). Image scans for treatment planning are performed with the patient in treatment position with the immobilization device. IMRT treatment planning requires 3D image data with more slices at a finer spacing than had been the norm previously, depending on the multileaf collimator (MLC) leaf width. For example, slice spacing of no more than 0.5 cm should be used for a MLC with 1-cm leaves, and finer spacing may be needed to generate digitally reconstructed radiographs (DRRs) of sufficient quality. The range of slice acquisition may also be expanded for noncoplanar beam arrangements.

CT scans are obtained in the treatment position using a flat-table insert on a GE LightSpeed 16-slice CT simulator (GE Medical Systems, Milwaukee, Wisconsin). Scans are acquired from approximately 2 cm above the top of the iliac crest to approximately mid-femur. This scan length will facilitate the use of noncoplanar beams when necessary. Scans in the region, beginning 2 cm above the femoral heads to the bottom of the ischial tuberosities, are acquired using a 3-mm slice thickness (1.25 mm for patients localized using the Calypso system, Calypso Medical, Seattle, Washington) and 3-mm table increment. All other regions are scanned to result in a 1-cm slice thickness. Retrograde urethrograms are not performed because all patients routinely undergo MRI for treatment planning where the prostate apex is well-defined.

All prostate patients undergo MRI on a GE 1.5 Tesla MR simulator (GE Medical Systems, Milwaukee, Wisconsin) located within the department, typically within 30 minutes before or after a CT scan. For patients localized with the Calypso system, the MR scan is performed before the implantation of the electromagnetic transponders because they interact with the magnetic field creating an artifact. MR scans are obtained without contrast media. CT images are loaded as the primary image data set and MR images are loaded as the secondary image data set. CT and MR images are first fused by dosimetrists according to bony anatomy using either chamfer matching or maximization of mutual information methods and then approved by oncologists to ensure the fusion accuracy of the prostate target.

Significant structure deformation may occur between CT and MRI due to variations in rectal or bladder fillings or weight changes for large intervals between image scans. All soft-tissue

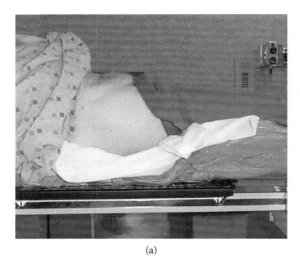

(a)

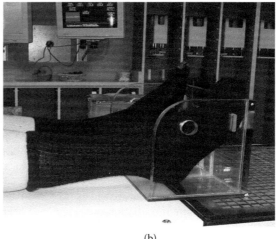

(b)

FIGURE 21.1 Patient setup with (a) a custom alpha cradle cast and (b) a custom polymethyl-methacrylate (PMMA) foot-holder.

structures are contoured based on either the CT or MRI at the discretion of the physician who evaluates the quality of the fusion. MRI is preferred for defining soft tissue in the pelvis, but difference in rectal or bladder fillings between scans may preclude using MRI. The external contour and bony structures are typically based on CT. Presently, we have the capability to perform MRI simulation without CT for prostate treatments (Chen, Price, Wang et al. 2004; Chen, Price, Nguyen et al. 2004).

21.2.2 Target and Structure Delineation

21.2.2.1 Clinical Target Volume Determination

Clinical target volume (CTV) includes prostate and proximal seminal vesicles. The proximal seminal vesicles (approximately 1 cm) are outlined separately to facilitate prostate localization using the B-mode acquisition and targeting (BAT) ultrasound system (Nomos Inc., Cleveland, Ohio). Because the junction between the prostate and proximal seminal vesicles is sometimes uncertain even with the aid of MR for planning, the full prescription dose is administered to the proximal seminal vesicles. For low-risk patients, the distal seminal vesicles are outlined as a separate structure as well, but no dose is prescribed; the ability to see the position of the distal seminal vesicles during the BAT ultrasound alignment process is useful.

MR images have much better soft-tissue contrast than CT and therefore can help identify the borders of the prostate more accurately. Sometimes, the levator ani muscles are hard to distinguish from the lateral aspect of the prostate and the base and apex of the prostate are also more difficult to discern on CT. If using CT images exclusively (as in some institutions where MRI is not available), contrast in the bladder may help distinguish the bladder–prostate interface; intravenous contrast is preferred because catheter placement distorts the anatomy. If contrast is used, heterogeneity corrections should not be applied to dose calculation because the contrast may change the CT numbers significantly. For imaging of the apex of the prostate, a retrograde urethrogram may be used; however, it (or any catheter) may alter the anatomical relationships such that it is no longer representative of what occurs during treatment.

Another way to approximate the apex of the prostate is to identify the penile bulb and then go superiorly about 1 cm (Plants et al. 2003). This approach will overestimate the position of the apex inferiorly, which may be appropriate, given that there is no prostate capsule inferiorly and there is tumor involvement of the apex about one-third of the time. It is important to note that there is a large degree of variation in the bulb apex distance and this method is not preferred. For these reasons, and if the transabdominal ultrasound localization is used, which is least accurate in the superior–inferior dimensions, we tend to overestimate the prostate apex position even when using MRI.

21.2.2.2 Planning Target Volume Determination

Planning target volume (PTV) for the prostate is 8 mm larger than the CTV (prostate and seminal vesicles) in all directions except posteriorly where the margin is typically 3–5 mm (Figure 21.2).

This margin is determined based on the target localization accuracy achievable using the localization techniques available at FCCC (cone-beam CT [CBCT]/on-board imaging [OBI] + gold seeds, CT-on-rails, BAT ultrasound, Calypso). This PTV is only used to facilitate the treatment optimization process using a treatment planning system. An *effective PTV* will be evaluated based on the actual target volume that received the prescription dose (see the plan acceptance criteria described in Section 21.2.3.4). This also defines an *effective margin*, that is, the distance between the CTV and the prescription isodose surface and is a minimum of 3 mm.

21.2.2.3 Critical Structure Delineation

The Radiation Therapy Oncology Group (RTOG) guidelines are used to delineate the rectal volume, which is contoured from the rectosigmoid flexure to the bottom of the ischial tuberosities. The rectal volume includes the entire rectal contents, and the rectal diameter is limited to less than 3 cm on any axial slice. The bladder is contoured in its entirety and is simulated and treated comfortably full. The femoral heads are contoured from the top of the femoral head to the level of the upper border of the lesser trochanter.

21.2.3 Treatment Planning

21.2.3.1 Dose Prescription

The typical dose prescribed to the prostate target, which includes both the prostate gland and the proximal seminal vesicles, is 78 Gy in 39 fractions (2 Gy/fraction) for low-risk prostate patients receiving standard IMRT and 80 Gy in 40 fractions for intermediate- and high-risk patients. The distal seminal vesicles and lymphatics are prescribed to 56 Gy (about 1.5 Gy/fraction) for high-risk patients. For postprostatectomy patients, the prostate bed is prescribed to 64–68 Gy (2 Gy/fraction). A

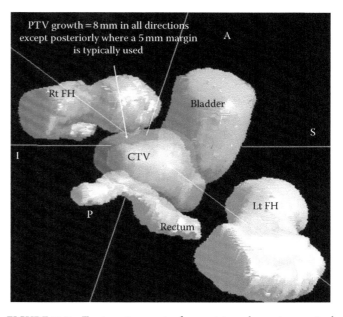

FIGURE 21.2 Treatment geometry for prostate and margins required to build the planning target volume (PTV).

hypofractionated IMRT protocol has been adopted at FCCC (Pollack et al. 2006), in which the prostate and the proximal seminal vesicles are prescribed to 70.2 Gy in 26 fractions (2.7 Gy/ fraction). This is equivalent to 84.4 Gy at 2 Gy/fraction, assuming a 1.5 Gy α/β ratio for the prostate. The distal seminal vesicles and lymphatics are given 50 Gy (about 1.9 Gy/fraction).

21.2.3.2 Treatment Planning Input Parameters

We have used the Corvus inverse planning system (Best Nomos, Sewickley, Pennsylvania) for prostate IMRT planning since 2000 and have accumulated extensive experience with this system. The plan acceptance criteria were also a moving target that reflected what we could achieve routinely based on the planning system and the techniques used (Price et al. 2003; Price et al. 2005; Price et al. 2006). The prescription input parameters that are used are listed in Table 21.1. The prostate and the proximal seminal vesicles are assigned a dose goal, a percentage of the volume that may be underdosed, as well as minimum and maximum doses to be delivered. Nontarget structures are assigned a dose limit, a percentage of the volume that may receive more than this limit, as well as minimum and maximum values (Pollack et al. 2005; Price et al. 2005).

The continuous annealing method is selected for optimization on the Corvus inverse planning system, which often results in high-quality plans with spiky intensity maps (Webb 1992). The CMS XiO and Varian Eclipse treatment planning systems are also used for IMRT prostate planning using the gradient algorithm (Webb 1997). Good-quality plans are generated using these planning systems with smoother intensity maps that require fewer monitor units (MUs) and thus shorter beam-on times. The intensity levels are typically set to 5. The minimum beamlet size is set to 1×1 cm for the Siemens accelerators with 1 cm width MLC leaves and 1×0.5 cm for the Varian accelerators with 0.5 cm width MLC leaves.

21.2.3.3 Beam Energies and Incident Directions

Prostate patients are treated using the IMRT technique, typically with 10 MV photon beams; lower energies increase the peripheral doses, especially for larger patients, while higher energies

generate more neutrons, leading to an increased risk of second cancers for younger patients. Treatment plans typically employ six to nine beam directions. The seven frequently used beam directions are 0°, 75°, 115°, 135°, 225°, 270°, and 330° (Figure 21.3). Beam directions are modified for individual patients to achieve optimal plan quality. In the interest of treatment quality, complexity, and delivery time, we typically begin with fewer beams and progress to nine beam directions and sometimes with noncoplanar beam arrangements. Simpler plans such as prostate-only or prostate + seminal vesicles typically result in fewer beam directions than with the addition of lymphatics.

21.2.3.4 Plan Acceptance Criteria

Our clinical acceptance criteria were established based on the treatment objectives, previous outcome data, and our planning

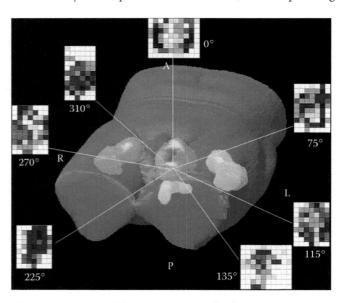

FIGURE 21.3 Typical beam incident angles for prostate intensity-modulated radiation therapy. The beam number and incident directions are adjusted to achieve a quality plan that meets our plan acceptance criteria. (Reprinted from Price, R., E. Horwitz, S. Feigenberg, and A. Pollack. 2005. Intact prostate cancer: Case study. In *Intensity Modulated Radiation Therapy: A Clinical Perspective*, ed. A. J. Mundt and J. C. Roeske. Hamilton, ON: BC Decker, Inc. With permission.)

TABLE 21.1 Treatment Planning Input Parameters for Prostate Intensity-Modulated Radiation Therapy

Target	Goal	Volume below Goal (%)	D_{min}	D_{max}
Prostate	78.0	5	72.2	81.3
Proximal SV	78.0	5	72.2	81.3
Distal SV	58.0	5	53.2	76.0
Critical Structure	Limit	Volume above Limit (%)	D_{min}	D_{max}
Tissue	78.0	0	0.0	76.0
Bladder	35.0	20	2.0	60.0
Right femoral head	40.0	30	10.0	45.0
Left femoral head	40.0	30	10.0	45.0
Rectum	15.0	10	5.0	76.0

SV = seminal vesicle.

Note: For low-risk patients, the prescription dose is 78 Gy to the prostate and proximal SVs. For intermediate- and high-risk, it is 80 Gy. High-risk patients also receive 56 Gy in 40 fractions to the regional lymph nodes.

TABLE 21.2 Treatment Plan Acceptance Criteria for Standard IMRT and Hypofractionated IMRT of Prostate Treatment

Standard IMRT		Hypofractionated IMRT	
PTV_{95}	≥78 Gy	PTV_{95}	≥70.2 Gy
R_{65Gy}	≤17%V	R_{50Gy}	≤17%V
R_{40Gy}	≤35%V	R_{31Gy}	≤35%V
B_{65Gy}	≤25%V	B_{50Gy}	≤25%V
B_{40Gy}	≤50%V	B_{31Gy}	≤50%V
FH_{50Gy}	≤10%V	FH_{40Gy}	≤10%V

Note: The treatment target is prescribed to 78 Gy in 39 fractions for standard IMRT and 70.2 Gy in 26 fractions for hypofractionated IMRT.

capabilities. The acceptance criteria listed in Table 21.2 are based on the dose–volume histogram (DVH). For standard IMRT, 95% of the prostate PTV (PTV_{95}) shall receive at least 100% of the prescription dose. In addition, 95% of the seminal vesicle PTV (SV PTV_{95}) shall receive at least 56 Gy. For the critical organs, no more than 17% of the rectum receives $\geq$ 65 Gy (R65) and no more than 35% receives $\geq$ 40 Gy (R40); no more than 25% of the bladder receives > 65 Gy (B65), and no more than 50% receives > 40 Gy (B40); and no more than 10% of the femoral heads receive > 50 Gy (FH50). Figure 21.4 shows the DVH curves for a typical prostate IMRT plan that meets our DVH criteria.

Other plan acceptance criteria are based on the isodose distributions. After the target dose is normalized to ensure 95% of the PTV receiving at least 100% of the prescription dose, we check the prescription isodose line slice by slice to ensure good target coverage and critical structure sparing. The prescription isodose surface defines an effective PTV, and the distance between the CTV and the prescription isodose surface becomes the effective margin. In general, the effective margin should not be < 3 mm posteriorly or < 6 mm in other directions on any individual CT slice. The 90% isodose line should not exceed one-half the diameter of the rectal contour and the 50% isodose line should fall within the rectal contour (following the Memorial experience) on any individual CT slice. These isodose constraints are nicely viewed on the sagittal dose distribution (Figure 21.5).

21.2.4 Treatment Delivery

A typical prostate treatment can be completed in 15–20 minutes depending on the localization techniques used. All prostate treatments at FCCC incorporate daily soft-tissue target localization using CBCT/OBI + gold seeds, BAT ultrasound, CT-on-rails, or Calypso. The estimated prostate localization uncertainty is < 3 mm using CBCT/OBI + gold seeds, Calypso, and CT-on-rails, while it is < 5 mm using BAT ultrasound (at 95% confidence level) based on extensive clinical measurements. CT-on-rails and BAT ultrasound are noninvasive and especially suitable for patients who cannot be implanted with gold seeds or Calypso transponders. The Calypso system is capable of monitoring the intrafractional prostate motion during treatment and thus allowing for real-time dose tracking or beam gating. More details on target localization will be provided in Section 21.4.

The treatment isocenter is chosen during the CT simulation process such that it falls within the prostate volume. The planning isocenter is forced to be identical to that depicted during CT simulation. This insures that any isocenter shifts occurring on a daily basis are the result of localization only and are an attempt to avoid treatment errors. Treatment portal images are not routinely taken although the original isocenter is checked on a weekly basis using a single set of orthogonal port films. This assures that the starting point used in the localization process is the same throughout the course of the treatment and allows us to minimize the required isocenter shifts.

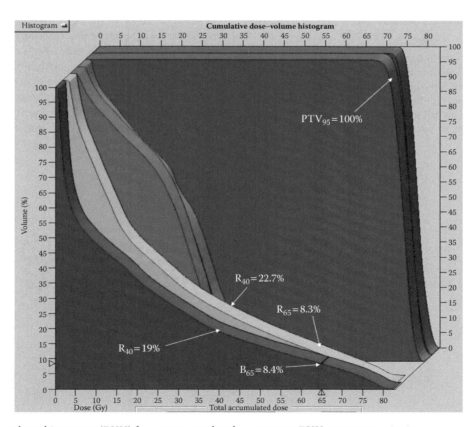

FIGURE 21.4 Dose–volume histograms (DVH) for a treatment plan that meets our DVH-acceptance criteria.

21.2.5 Quality Assurance

An extensive QA program has been established for IMRT with image guidance. Proper equipment was acquired to perform the QA tasks. The QA of IGRT generally consists of (1) commissioning and testing of the imaging, treatment planning, and delivery systems; (2) routine QA of the imaging and delivery systems; and (3) patient-specific dosimetry verification.

The first task is to ensure the integrity of the imaging, inverse planning, and IMRT delivery system. The second one is to ensure the normal operation of the imaging and beam delivery systems in addition to the daily, monthly, and annual QA protocols. For example, daily checks may be augmented to include a test of a dynamic treatment, including an output measurement that checks the stability of a programmed MLC gap width. The third task is to ensure an accurate and safe treatment of the patient using the IGRT technique. While IMRT produces conformal doses, the level of sophistication has been increased considerably. The goal of dosimetry validation is to verify that the correct dose distribution will be delivered to the patient (Ezzell et al.

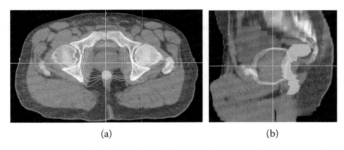

(a)   (b)

FIGURE 21.5 The (a) axial and (b) sagittal views of a treatment plan showing acceptable isodose distributions. The isodose lines are 100%, 90%, 80%, 70%, 60%, and 50% of the prescription dose. The 90% isodose line should not exceed half the diameter of the rectal contour and the 50% isodose line should fall within the rectal contour on any individual computed tomography slice.

2003; Ma 2007). This ensures that the plan has been properly computed and that the leaf sequence files and treatment parameters charted and/or stored in the record and verify system are correct and will be executable. To expedite IMRT delivery, an autosequencing delivery system is sometimes used. Such delivery systems (in different forms) are currently available from all major accelerator vendors and are routinely used at FCCC. Items that are validated before the first treatment include

- *MUs* (absolute dose at a point): Measurement in a phantom
- *MLC leaf sequences*: Measurement with film or a digital imaging device
- *Dose distribution*: Measurement with film or a 2/3D dosimeter array
- *Collision avoidance*: A "dry-run" before treatment

Figure 21.6 compares doses predicted by the treatment planning systems and those measured using an ionization chamber for more than 2000 IMRT patients. Our results showed that the differences were within 2% for 69% of the cases and within 4% for 97% of the cases. Differences of more than 3% were investigated individually to ensure the accuracy of the dose (distribution). So far, we have performed direct measurements on all of the IMRT plans. However, we are also developing reliable independent dose calculation methods for dosimetry QA (Ma et al. 2002; Ma et al. 2003; Ma et al. 2004; Luo et al. 2006; Fan et al. 2006).

Most of the treatment optimization systems use simple dose calculation algorithms to compute beamlet dose distributions used by the inverse planning process, which may introduce significant uncertainty in the optimized dose distribution due to the presence of heterogeneities, especially near tissue-air, tissue-lung, and tissue-bone interfaces. For example, the commonly used finite-size pencil beam algorithm can modify the beamlet dose distribution based on equivalent path length, but it does not handle dose perturbation due to electron transport at beam edges and near the interfaces. Simplified algorithms or models are also used to account for the effect of photon scatter in the treatment

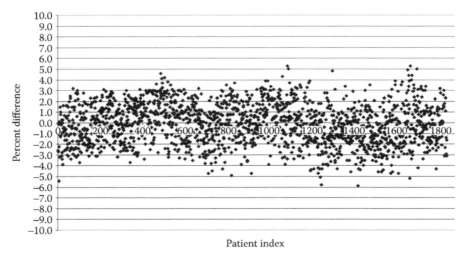

FIGURE 21.6 Percent dose differences between treatment planning dose calculation and ionization chamber measurements for the first 2000 intensity-modulated radiation therapy patients at Fox Chase Cancer Center (FCCC).

head and leakage through the collimators. Some treatment planning systems are implementing more accurate dose algorithms such as superposition convolution and Monte Carlo simulations for the final dose calculation after MLC leaf sequences have been determined in order to correct for the photon leakage effect. We have developed a Monte Carlo dose verification tool for IMRT QA, which may partially replace the phantom MU check and relative dose measurement with film.

21.3 Image Modalities for Target Determination

This section discusses the role of different imaging modalities in the target determination and critical structure delineation for advanced radiotherapy of prostate cancer employing dose escalation and hypofractionation schemes. In particular, we describe the use of CT and MRI for prostate treatment planning.

21.3.1 Image Fusion

Image fusion is an important step in IGRT treatment planning to reduce the overall geometric uncertainties of the target and critical structure volumes. Various advanced fusion software and tools are available on different commercial treatment planning systems that require adequate training for effective and efficient operation even for experienced planners. Image fusion procedures and policies have been established for different treatment sites to ensure quality and consistency within the department. As described in Section 21.2.1, patients are asked to have a half-full bladder and to empty the rectum using an enema prior to simulation. The CT and MR scans are performed within a 30-minute interval to minimize geometric variations between the two data sets. The CT and MR images can then be fused according to bony anatomy using either chamfer matching or maximization of mutual information methods. However, soft-tissue target volumes must be verified and approved by oncologists to ensure the fusion accuracy since significant structure deformation may occur between CT and MRI due to variations in rectal or bladder fillings or weight changes for large intervals between the image scans. Figure 21.7 shows the geometric uncertainties of soft-tissue structures after fusion based on bony anatomy; the CT and MR scans were performed 2 weeks apart, and the patient gained 15 pounds during this time.

21.3.2 Use of Multi-Image Modality for Structure Delineation

Multiple imaging modalities have been applied to target determination and critical structure delineation for prostate cancer radiotherapy to facilitate dose escalation and hypofractionation. CT has revolutionized radiotherapy by providing 3D patient anatomical information for conformal radiotherapy treatment planning and accurate dose calculation. MRI provides superior image quality for soft-tissue delineation over CT and is widely used for target and organ delineation in prostate radiotherapy

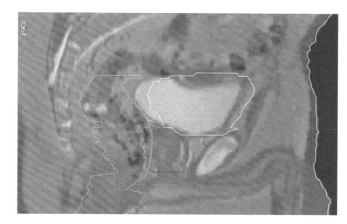

FIGURE 21.7 Screen capture of fused computed tomography–magnetic resonance (CT-MR) image showing significant soft-tissue mismatch after fusing CT and MR images based on bony structures only.

(Khoo et al. 2000; Tanner et al. 2000; Potter et al. 1992). There could be a 40% difference in prostate volume between CT and MRI (Rasch et al. 1999; Krempien et al. 2002).

Debois et al. (1999) showed that improved prostate and rectal volume delineation from MRI could lead to improvements in both target coverage and rectal sparing. Figure 21.8 shows the differences in prostate contours drawn based on CT and MRI. To reduce the geometric uncertainty of the target volume, the use of MRI for treatment planning of prostate cancer is desirable. MRS has been investigated at FCCC for treatment assessment and target determination for dose escalation (designing boosting volumes based on tumor cell density and/or hypoxia, see Section 21.3.3). MRS can analyze the chemical composition of prostate tissues that can be used to identify the volume of a prostate tumor and its stage, as well as contrast cancerous tissue to healthy prostate tissue; cancerous prostate tissue is found to be high in choline and low in citrate, whereas healthy prostate tissue is low in choline and high in citrate. Functional imaging, including PET and fMRI, has also been investigated for target determination for dose escalation and treatment assessment studies.

21.3.3 Use of Magnetic Resonance Imaging Alone for Prostate Intensity-Modulated Radiation Therapy

Fused MR-CT images have been widely accepted as a practical approach for both accurate anatomical delineation (using MRI data) and dose calculation (using CT data) and have been used routinely for prostate radiotherapy for more than 10 years at FCCC. We have also investigated using MRI alone for prostate radiotherapy treatment planning (i.e., MRI-based treatment planning).

The advantages of MRI-based treatment planning are several. The CT-MRI fusion process introduces additional errors due to the difficulties in coordinating the CT and MR images and significant discordance caused by soft-tissue deformation between image scans. When the Calypso system is used for target localization during treatment, the MR scan has to be performed before the implant of the radio transponders and the CT scan is usually 1 week or even longer after the implant. Significant organ deformation may occur for some patients. Furthermore, MRI-based treatment planning will avoid redundant CT imaging sessions, which in turn will avoid unnecessary radiation exposure to the patient. It also saves the patient, staff, and the machine time.

There are several perceived challenges of using MRI alone for radiotherapy planning: (1) the lack of electron density information that is needed for heterogeneity corrections in dose calculation, (2) image distortion that affects the patient's external contour determination and therefore introduces dose calculation error, and (3) lack of bony structures to derive effective DRR for patient setup. Many investigators have explored the efficacy of MRI-based treatment planning for radiotherapy (Michiels et al. 1994; Mizowaki et al. 2000; Guo 1998; Mah et al. 2002a; Mah et al. 2002b; Chen, Price, Wang et al. 2004; Chen, Price, Nguyen et al., 2004; Chen et al. 2010; Lee et al. 2003). Our studies have shown that it is possible to perform treatment planning dose calculation directly on MRI for prostate IMRT (Chen, Price, Wang et al. 2004; Chen, Price, Nguyen et al. 2004; Chen et al. 2010). Figure 21.9 shows IMRT dose distributions for a prostate patient calculated using MRI and CT with acceptable

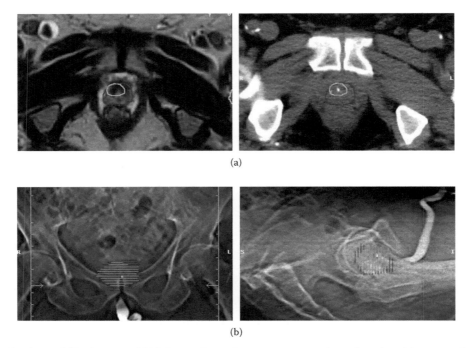

(a)

(b)

FIGURE 21.8 **(See color insert following page 204.)** Comparison of prostate contours drawn based on (a) computed tomography (CT) and magnetic resonance imaging (MRI) and (b) contour projections on DRRs: yellow—contours based on CT; blue—contours based on MRI.

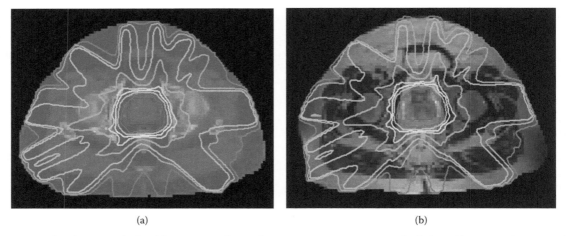

(a)   (b)

FIGURE 21.9 Dose distributions calculated by Monte Carlo simulations using patient geometry built from (a) computed tomography (CT) and (b) magnetic resonance imaging (MRI). The differences are clinically insignificant (within 2%/2 mm).

dosimetry accuracy. It is clear that before MRI-based patient geometry can be used for radiation therapy treatment planning, MR image distortions must be quantified and corrected. Currently, identified sites using MRI-based treatment planning clinically are for prostate and brain cancers.

21.4 Image Techniques for Target Localization

21.4.1 Patient Setup

Position verification will be a part of the initial plan validation. However, for subsequent IMRT treatments, patient setup and target localization will be the key to accurate dose delivery. The initial patient setup is to reproduce the patient treatment position using skin marks and bony anatomy to align the treatment fields with the planned fields. The latter is accomplished by comparing orthogonal films taken at simulation, DRRs from the planning system, and portal images from the treatment unit. When possible, portal images should be obtained for the fields used for treatment, and it is useful to have the MLC field boundary as apertures for the ports and compare it to corresponding DRRs from the planning system. Depending on the imaging system available, it may be possible to obtain a portal image of the modulated field superimposed on the patient's regional anatomy, but such images are often hard to interpret. One needs to consider how to acquire the maximum MLC field shape (based on a leaf sequence or an intensity map), how to verify the position of a slit collimator, or how to operate an electronic portal imaging system in the presence of dynamic fields.

Careful initial alignment by skin marks and bony structures will ensure the placement of the treatment isocenter in the proximity of the planned isocenter (within 5 mm) to facilitate soft-tissue target localization. The accuracy of this initial patient setup should be maintained to prevent large systematic or occasional human errors in the soft-tissue target localization process; our policies require that any isocenter shifts > 10 mm be confirmed by a physicist or a physician (it was 5 mm in the initial implementation period and relaxed after extensive experiences were gained). It is also necessary to ensure accurate initial patient setup when soft-tissue target localization systems become unavailable due to repair or other unexpected events.

21.4.2 Soft-Tissue Target Localization

Precise localization of the treatment target for prostate IMRT must rely on soft-tissue registration, not simply by aligning skin marks or bony structures. Specially designed imaging systems based on ultrasound or CT, implanted fiducial markers, and Calypso systems have been used for soft-tissue target localization at FCCC. Generally, patients are implanted with fiducial markers such as gold seeds or Calypso Beacons for accurate target localization. For those who are not eligible for or have declined such surgical insertions, ultrasound or CT-based localization techniques will be used.

21.4.2.1 Implanted Fiducial Markers

Surgically inserted fiducial markers have been routinely used for prostate localization in combination with an electronic portal imaging device (EPID), OBI, or other X-ray imaging devices. More than three gold markers are implanted with redundancy to detect prostate deformation or seed migration. At least 1 week later, the patient receives CT and MR scans according to our protocol. The marker locations can be clearly identified on the simulation CT and contoured. Prior to an IMRT treatment, orthogonal X-ray images are obtained to locate the fiducial markers. The fiducial markers can be manually aligned or the center of mass of the fiducial markers can be calculated and then aligned with that determined on the simulation CT using vendor-supplied or in-house localization software. Generally, translational shifts are applied to align the treatment isocenter with the planned isocenter using conventional treatment couches, while the CyberKnife robotic couch (Accuray Inc., Sunnyvale, California) also allows for rotational corrections based on the seed coordinates.

X-ray images with and without implanted radiographic markers have been implemented on the CyberKnife system (Accuray Inc.) for automatic target tracking in frameless stereotactic radiosurgery or radiotherapy for intracranial, spine, and lung tumor treatments (Murphy et al. 2001; Gerszten et al. 2003). These techniques have also been applied to prostate treatments for monitoring intrafractional prostate motion with implanted fiducial markers and for automatic target tracking during beam delivery. Applications for advanced prostate treatments (e.g., dose escalation and hypofractionation) using this technology are under investigation (King et al. 2003; Hara, Patel, and Pawlicki 2006; Fuller et al. 2007).

Fluoroscopy together with implanted radiopaque fiducials can also be used for real-time soft-tissue target motion monitoring. For this application, an X-ray image will be ideally taken in the beam's eye view during a treatment in order to adjust the field shape of the treatment beam. Special software will be needed to identify the variation of the target volume directly or to locate the fiducial markers to track the treatment isocenter in real time. Such systems are not available commercially at present.

The Calypso 4D localization system with implanted electromagnetic transponders (Calypso Beacons) can continuously monitor the location of the soft-tissue target, and it is commercially available for prostate treatments. The implanted electromagnetic transponders can be detected by the system at a frequency of 10 Hz and the 3D positions of the electromagnetic transponders and the target isocenter or the geometric center of the electromagnetic transponders can then be tracked. It provides continuous, real-time localization and monitoring of the prostate and allows for real-time treatment intervention (e.g., applying couch shifts) for 4D radiotherapy treatments (Balter et al. 2005; Kupelian, Willoughby, Mahadevan et al. 2007; Willoughby et al. 2006; Li et al. 2009; Li and Ma 2010).

We have evaluated the continuous motion tracking data for about 4000 fractions of 100 prostate patients. As shown in Figure 21.10, the majority of patients showed stable prostate

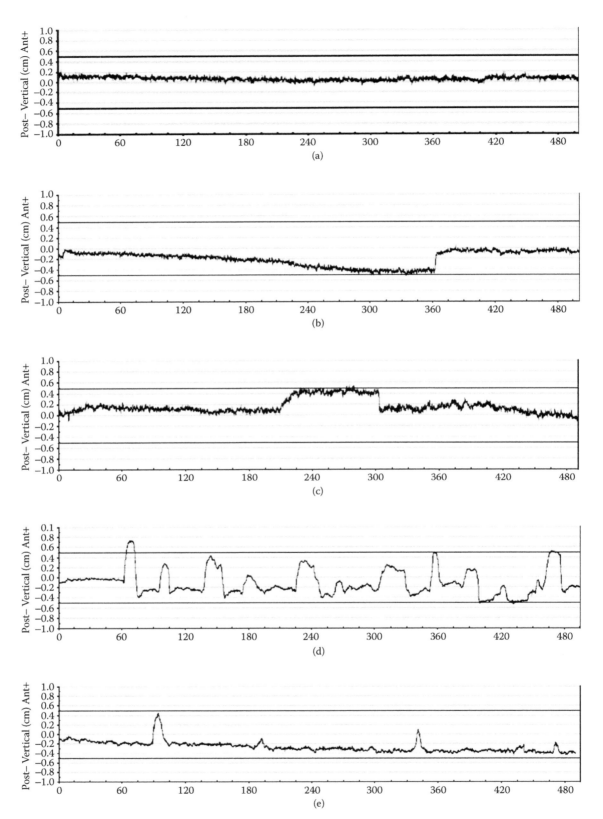

FIGURE 21.10 Examples of prostate motion observed with continuous tracking for 8 minutes: (a) stable target position, (b) continuous target drift, (c) transient excursion, (d) frequent excursions, and (e) mixed motion. The two solid straight lines indicate 5 mm threshold, and the patient position was changed at 360 seconds in (b) and at 300 seconds in (c).

positions, while some exhibited continuous drifts, transient excursions, frequent excursions, or mixed motion patterns. These have been observed by other investigators (Kupelian, Willoughby, Mahadevan et al. 2007; Willoughby et al. 2006). We found that a large number of significant (> 3 mm) prostate motions only happened in a small fraction of patients, indicating the potential improvement in radiotherapy margin management, for example, using smaller margins for the majority of patients while applying gating or tracking in beam delivery for this small group of patients. Besides the prostate displacement during the treatment, prostate rotation can also be detected. Figure 21.11 shows the histogram of the prostate rotation at the setup time for all the treatment fractions of the first 20 patients with a mean value of 6.4° and a standard deviation of 4.4° (Li and Ma 2010). Prostate rotational changes prior to and during IMRT treatment have generally been ignored, although in some cases parts of the target volume, for example, portions of the prostate apex, prostate base, and seminal vesicles may move out of the treatment field due to prostate rotation. We have also found that some large rotational changes reported by the Calypso system were caused by the migration of the radio transponders (Xu et al. 2009). Offline corrections (replanning) were performed for patients with significant Beacon migration.

The target localization accuracy is estimated at 2–3 mm (2σ) using the radiopaque markers and electromagnetic transponders, as reported by many investigators. Daily soft-tissue target localization based on implanted fiducial markers is easy to operate, accurate, and efficient especially with the use of vendor-supplied localization software. However, this technique also has some drawbacks due to its invasiveness and localization is based on the fiducial marker positions rather than the 3D treatment geometry. The use of X-ray imaging also adds radiation dose to the patient, which increases the risks of second cancers for younger patients.

21.4.2.2 Computed Tomography–Based Localization

CBCT has been widely adopted for target localization in recent years, which can provide 3D information of the patient geometry

under treatment conditions (AAPM 2009). CBCT that uses kilovoltage (kV) X-rays is superior in tissue absorption properties, but suffers more from photon scattering compared to megavoltage (MV) X-ray CBCT. Therefore, the image contrast of kV CBCT is compromised by the effect of photon scattering while MV CBCT is not affected by high-Z objects. The absorbed dose resulting from both kV and MV CBCT can be significant, for example, several cGy per scan depending on the patient size and imaging parameters used. The total dose from daily CBCT localization through an entire treatment course can be 1–3 Gy, which should be taken into account in treatment planning (Murphy et al. 2007). Due to the limitations in soft-tissue contrast, CBCT has been mainly used for bony anatomy alignment, for example, intracranial, head and neck, and spine. CBCT has been used alone for prostate localization by identifying prostate interfaces with the rectum and bladder or in combination with implanted fiducial markers. In particular, we have combined CBCT with the Calypso system to determine the migration of Calypso transponders (Xu et al. 2009).

While CBCT technology is being improved for better image quality, diagnostic CT scanners have been installed in the linear accelerator rooms for target localization due to their superior image quality (AAPM 2009; Ma and Paskalev 2006). While the installation of an independent diagnostic CT scanner in the treatment room is perceived to be costly and less efficient, it has a clear advantage in that it leverages all the development that has been invested in conventional CT technology over the past 20 years, leading to unquestioned image quality and clinical robustness. The geometrical accuracy, in combination with excellent image quality and less radiation dose (about 1 cGy per scan), provides a reliable platform for the clinical evaluation of the rapidly developing in-room CT technology and its impact on different treatment sites. FCCC was one of the first academic institutions to install a CT-on-rails system, in which the patient couch is shared by the linear accelerator and the CT scanner, and the CT scanning is accomplished by a movable CT gantry mounted on rails (Ma and Paskalev 2006).

Many studies and clinical experiences with the CT-on-rails systems have focused on the anatomical variation and soft-tissue target localization during the course of prostate treatment. Figure 21.12 shows the localization of the prostate using the Siemens CT-on-rails system and the volume targeting prototype software. Treatment planning contours and isocenters are imported via DICOM RT. The contours are aligned with the anatomy in the pretreatment scan. The machine isocenter (before the shift) is obtained using radiopaque markers. The sagittal view shows that the proximal seminal vesicles are used as a landmark for longitudinal alignment. The target localization procedure using the CT-on-rails system prior to a prostate IMRT treatment will add an extra 5 minutes to the total duration of the IMRT prostate treatment time.

Extensive studies on the accuracy of prostate localization have been carried out by investigators from the MD Anderson Cancer Center and FCCC, where vast experiences with the ultrasound-based daily prostate localization technique were

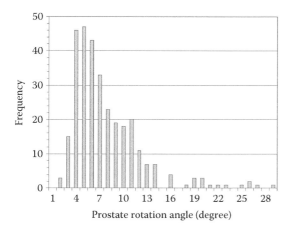

FIGURE 21.11 Prostate rotation as reported by the Calypso system based on the Beacon locations.

also accumulated. Their results showed that the systematic differences between CT-on-rails and ultrasound alignments were smaller than 1 mm, and the standard deviations of the random differences were in the range of 2–4 mm (Figure 21.13; Court et al. 2005; Paskalev et al. 2004; Paskalev, Feigenberg, Wang et al. 2005; Feigenberg et al. 2007). These studies demonstrated that CT-on-rails could be used as a gold standard to evaluate other alignment techniques. The CT-on-rails system was also

used to improve the localization of postprostatectomy patients in combination with the ultrasound localization technique (Paskalev, Feigenberg, Jacob et al. 2005). For such patients, the prostate bed is a target that is not associated with a given anatomical shape, and thus, target alignment is a very challenging task.

The ability to acquire 3D patient geometry prior to a radiotherapy treatment also allows for the evaluation of target coverage

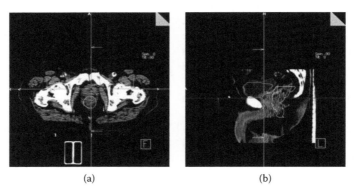

(a)   (b)

FIGURE 21.12 Soft-tissue target localization for prostate image-guided radiation therapy (IGRT) using the CT-on-rails system and volume targeting software ((a) axial view; (b) sagittal view). Organ contours drawn on simulation CT/MRI are used to match the patient anatomy on pretreatment CT to align the treatment isocenter with the planned isocenter. The sagittal view shows that the proximal seminal vesicles are used as a landmark for longitudinal alignment. (Reprinted from Feigenberg, S. J., K. Paskalev, S. Mcneeley, et al. 2007. *J Appl Clin Med Phys* 8:2268. With permission.)

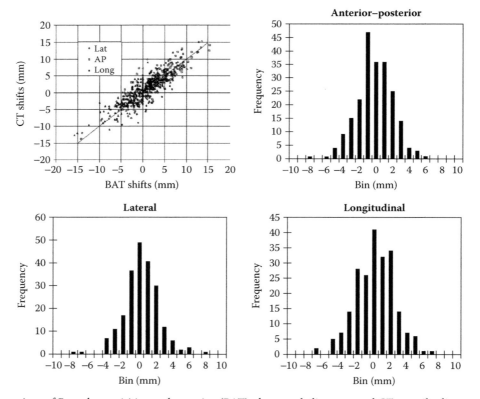

FIGURE 21.13 Comparison of B-mode acquisition and targeting (BAT) ultrasound alignment and CT-on-rails alignment for the same treatment fraction (top left) and histograms of the differences between ultrasound alignments and CT-on-rails alignment in anterior-posterior (AP), inferior–superior (longitudinal), and lateral directions, respectively, for the same treatment fraction. (Reprinted from Feigenberg, S. J., K. Paskalev, S. Mcneeley, et al. 2007. *J Appl Clin Med Phys* 8:2268. With permission.)

and critical structure sparing for advanced dose escalation and hypofractionation studies. An FCCC randomized clinical trial compares 76 Gy in 38 fractions (arm I: standard IMRT) to 70.2 Gy in 26 fractions (arm II: hypofractionated IMRT). The PTV margins in arms I and II were 5 and 3 mm posteriorly and 8 and 7 mm in all other dimensions, respectively. More details of the plan acceptance criteria are described in Section 21.2.3.4. The CT-on-rails system was used for target localization for both arms, and a retrospective study was performed on the target coverage and rectal sparing based on 98 CT-on-rails scans immediately after the dose delivery for 15 patients (Ma et al. 2009). The results showed that 7.1% of the treatment fractions exhibited compromised target coverage; the minimal PTV dose was less than 65 Gy for these fractions assuming that the CT-on-rails scans represented the actual patient geometry during the treatment.

Chen, Paskalev, and Zhu (2007) demonstrated the benefits of an empty rectum during CT simulation; it is relatively easier to achieve the rectal dose criteria for a large rectal volume during treatment planning, but the actual doses received by the rectum can be worse for some patients when the rectal volume becomes small in subsequent treatments (Figure 21.14). Other investigators have developed deformation registration methods for automated contouring to facilitate dose reconstruction or replanning for adaptive radiotherapy (Song et al. 2007; Gao et al. 2006; Godley et al. 2009).

21.4.2.3 Ultrasound-Based Localization

The BAT ultrasound system is an effective and efficient target relocation device for prostate treatment setup, which is excellent for localizing the prostate interfaces with the rectum and the bladder (Figure 21.15). It is noninvasive and sometimes may be the only option for patients who are not eligible for or have declined implantation of fiducial markers such as gold seeds

or Calypso transponders. The positioning accuracy is 3–5 mm (2σ) and an extra 3–5 minutes is added to the treatment time (Lattanzi et al. 2000; Feigenberg et al. 2007). The first two BAT systems were clinically implemented at FCCC and Cleveland Clinic.

A large number of prostate patients have been treated with the BAT or similar ultrasound soft-tissue localization systems, and extensive experiences have been accumulated. For example, soft-tissue contouring based on MRI during simulation has been found to improve the efficiency and accuracy in matching the prostate contours with the rectal and bladder interfaces shown on BAT. The ability to separate the proximal and distal seminal vesicles on MRI further helps reduce the localization uncertainty; more than 300 patients' scans randomly evaluated by the same physician showed a greater than 10% drop in substandard alignments (McNeeley et al. 2004).

The localization uncertainty for postprostatectomy patients could also be reduced by aligning BAT images with an ultrasound template, which was established prior to the first IMRT treatment by comparing the prostate bed geometry with a CT-on-rails system (Paskalev, Feigenberg, Jacob et al. 2005). The distribution of the differences between the template-assisted ultrasound alignments and control CT alignments are shown in Figure 21.16. These data were from the first 30 patients who underwent treatment of the prostate bed after the evaluation study was completed. The results show that a clinical margin of 8 mm fully covers the target for all 183 alignments but three, clearly demonstrating the improvement in alignment accuracy with the template-assisted ultrasound technique.

One concern with the transabdominal ultrasound system was that the abdominal pressure applied may displace the prostate position, thereby resulting in shifts that do not accurately reflect the true target position (McNeeley et al. 2002). We have

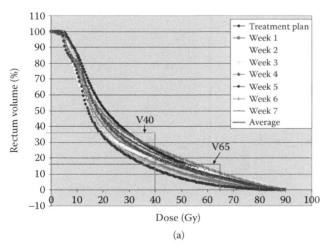

(a)

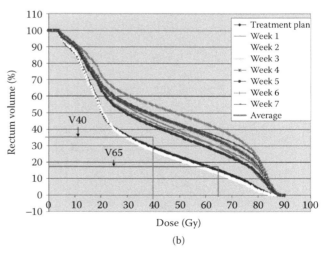

(b)

FIGURE 21.14 (See color insert following page 204.) Rectal dose–volume histograms (DVHs) based on the recomputed dose distributions using the CT-on-rails scans immediately after an intensity-modulated radiation therapy treatment. In (a) the original treatment plan was based on an empty rectum and in (b) the original treatment plan was based on a large rectum. The "average" DVH combines the doses received in all fractions. An empty rectum represented the most difficult scenario to plan, but resulted in better rectal dose distributions when rectal volume increases in subsequent treatments.

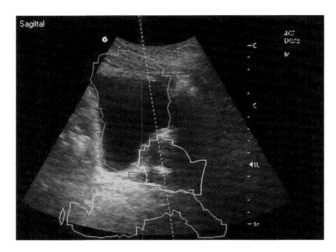

FIGURE 21.15 Soft-tissue target localization for prostate image-guided radiation therapy (IGRT) using the B-mode acquisition and targeting (BAT) ultrasound system. Organ contours drawn on simulation CT/MRI are used to match the patient anatomy on pretreatment ultrasound scan to align the treatment isocenter with the planned isocenter.

investigated this problem by scanning 12 patients with both light pressure (the least pressure necessary to capture an adequate image, i.e., normal conditions) and heavy pressure (the greatest pressure tolerated by the patient, rarely if never applied clinically). The amount of force applied and depth of probe penetration varied greatly depending on the patient's body habitus, abdominal wall muscle tone, and bladder filling. The same user performed all BAT scanning and the target alignment. Prostate position differences between images obtained with light and heavy pressures were determined using the scale printed on all the images by a single physician experienced with BAT alignment and blinded to the pressure used. The results showed that eight patients (67%) had no difference in alignment between light and heavy pressures. Three (25%) showed a 2 mm posterior displacement, and one showed a 3 mm anterior displacement with heavy pressure. We therefore concluded that extreme abdominal pressure did not alter the prostate position in the majority of patients, and the small shifts in the others are consistent with the BAT measurement uncertainty.

Due to the complexity of the prostate anatomy and the limitation of the existing technology, ultrasound-based soft-tissue

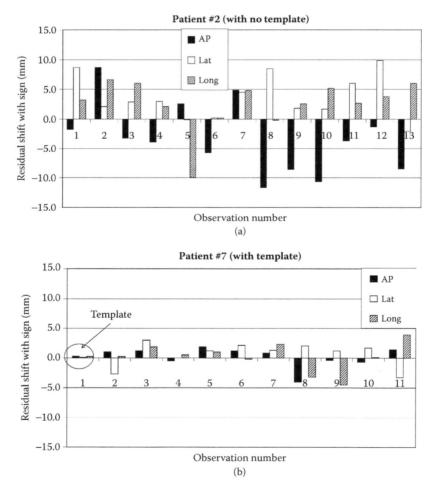

FIGURE 21.16 Daily residual isocenter shift based on CT-on-rails after B-mode acquisition and targeting (BAT) localization: (a) BAT alignment without a template and (b) BAT alignment with a template. (Reprinted from Paskalev, K., S. Feigenberg, R. Jacob, et al. 2005. *J Appl Clin Med Phys* 6:40–9. With permission.)

localization may appear to be more subjective than fiducial-based systems. Therefore, policies and procedures for clinical implementation, technical training, routine operation, and QA must be established to ensure the accuracy of ultrasound-based soft-tissue localization; > 1 cm localization errors have been reported by some investigators (Langen et al. 2003; Van Den Heuvel et al. 2003).

21.5 Future Trends and Challenges

This chapter reviewed the treatment procedures of EBRT for prostate cancer mainly based on the clinical implementation and practice of IMRT and IGRT at FCCC. As indicated in Sections 21.2 through 21.4, the procedural details for IMRT and image guidance or adaptive therapy depend on many factors, including equipment availability, treatment goals set, and imaging and delivery techniques used. Therefore, procedural differences are expected in other treatment centers where different treatment goals, strategies, and techniques have been explored. The accuracy and efficiency achievable in a routine clinic will depend on the implementation, training, experience, and above all, a good quality-control program with well-established policies and procedures for every aspect and each step of the treatment.

Image guidance and adaptive therapy procedures should be reviewed on a regular basis to ensure that the procedures are consistent with the initial design and to initiate appropriate changes if necessary. The review results and any new changes must be communicated between staff. The review of the elements of a patient's treatment can be integrated into the institutional chart rounds and quality control programs to verify that the entire IGRT or adaptive therapy procedure is operating correctly and desired treatment goals are achieved. Finally, a brief discussion on a few topics regarding prostate radiotherapy that may be of broad interest is given in Sections 21.5.1–21.5.3.

21.5.1 Dose Escalation

Despite the age-old question of whether to treat prostate cancer with radiotherapy, which has been debated among radiation oncologists between some European countries and North America, there has been continued effort in the determination of adequate dosing for prostate cancer. Extensive clinical research and large-scale randomized trials have been carried out in North American academic centers. The trends are unidirectional: to further escalate the target dose to achieve better local control.

There is clinical evidence that some tumor cells may survive the radiation doses employed in conventional radiation therapy of prostate cancer; a significant fraction of tumor cells remain morphologically viable as observed on postradiotherapy biopsies at FCCC and other academic centers. The fact that less morphologically viable tumor cells were found in patients treated with brachytherapy compared to EBRT may also suggest that more effective cell killing occurs in high-dose regions around the radiation sources or that dose rate or fractionation scheme plays a

significant role in different biochemical environments. The adequate doses to control prostate cancer are likely to be higher than those typically applied (e.g., 76–80 Gy in 2 Gy/fraction), at least for some patients with bulky diseases with or without hypoxic conditions. For EBRT, dose escalation can be achieved using the conventional fractionation scheme (1.8–2 Gy/fraction) with a higher total physical dose, a hypofractionation scheme (larger doses per fraction resulting in higher biologically equivalent doses) with a lower total physical dose, or the combination of the two (e.g., a standard treatment followed by a boost treatment). The availability of advanced image modalities and treatment techniques will enable us to explore different dose escalation and fractionation schemes and allow for the investigation of applications of multimodality treatments for prostate cancer, including EBRT, brachytherapy, surgery, chemotherapy, photodynamic therapy, high-intensity-focused ultrasound surgery, and so on.

21.5.2 Hypofractionation

Dose escalation in conventional fractionation is safe and straightforward (e.g., the Memorial Sloan Cancer Center increased the total prescription dose systematically from 68.4 to 86.4 Gy by increments of 5.4 Gy in consecutive groups of 3D-CRT patients). However, the treatment course has become prohibitively long and unacceptable for further dose escalation trials. Although the recent controversy on the low α/β ratio (1.5 Gy) for prostate (Brenner and Hall 1999) cannot be concluded due to the intrinsic limitations of the modeling process and the clinical outcome data, hypofractionation will still be a favorable approach if the α/β value is around 3 Gy.

Hypofractionation also offers other advantages over conventional fractionation, such as the reduction in treatment time and cost (when treatment is reimbursed based on the number of fractions). The FCCC hypofractionation trial prescribes 70.2 Gy in 26 fractions (equivalent to 84 Gy in 2 Gy/fraction assuming an α/β of 1.5 Gy), which is more convenient for patients (Pollack et al. 2006). Recent dose escalation or hypofractionation trials based on the CyberKnife system delivered either uniform or HDR-like dose distributions in five fractions; for example, a total prescription dose of 36.25 Gy is equivalent to 90.6 Gy in 2 Gy/fraction assuming an α/β of 1.5 Gy (King et al. 2009). Care must be taken to reduce the dosimetric uncertainty in image fusion, target delineation, dose calculation, patient setup, target localization, and dose delivery. The initial clinical outcome results are very encouraging (King et al. 2003; Hara, Patel, and Pawlicki 2006; Fuller et al. 2007).

21.5.3 Future of Adaptive Radiotherapy

Adaptive radiotherapy has attracted significant attention with the implementation of advanced dose escalation and hypofractionation treatment techniques, in which precise dose delivery to the treatment volume and critical structure avoidance become high priority. Image guidance plays an essential role in adaptive radiotherapy, although other technical issues also become critical to its development and clinical application. The ability to image radiation therapy patients in the treatment room has made it possible to address the problems of interfractional variations in patient anatomy and setup uncertainties (Ghilezan et al. 2004; Letourneau et al. 2005; Orton and Tome 2004; Yan et al. 2005).

Many strategies have been proposed for correcting interfractional variations, including offline adaptive correction schemes (Birkner et al. 2003; Rehbinder, Forsgren, and Lof 2004; Song et al. 2005) and online correction techniques (Yan et al. 2000; Court et al. 2005; Lee 2003; Mohan et al. 2005). Offline adaptive processes have been used to correct systematic positioning errors (Birkner et al. 2003; Rehbinder, Forsgren, and Lof 2004; Song et al. 2005; Yan et al. 2000) while offline replanning based on what has been delivered as reconstructed from treatment feedback has also been proposed (Birkner et al. 2003). For online correction, both patient repositioning (Ghilezan et al. 2004; Letourneau et al. 2005; Orton and Tome 2004; Yan et al. 2005; Birkner et al. 2003; Rehbinder, Forsgren, and Lof 2004; Ma et al. 2009) and plan adjustments (Yan et al. 2000; Court et al. 2005; Mohan et al. 2005) have been proposed.

Simple online modifications in treatment fields (e.g., small changes in rectal block positions or MLC leaf positions with corresponding minor MU adjustments) have been applied and/or investigated based on daily pretreatment image scans. Their impact on the treatment outcome may be very limited if conventional dose/fractionation and regular margins are used. Major online changes to treatment plans for individual fractions may require changes in policies and procedures that have both legal and technical implications. For example, if significant changes are found in the treatment geometry from a pretreatment image scan, should a therapist generate a new treatment plan and then treat? Should a physician be present at every fraction to approve the new treatment geometry (contours) and the new treatment plan for adaptive radiotherapy? Should the new treatment plan be reviewed and dosimetrically verified by a physicist?

Most technical issues for online adaptive therapy are related to time and are expected to be solved as technologies advance, for example, whether one can perform replanning and deliver the treatment in a short period so that the treatment geometry will not change significantly due to rectal or bladder filling between the image scan and the end of treatment. Manual target contouring will be impractical, while automated structure delineation based on deformation registration may be a good solution with the development of faster computers and computation methods. The treatment optimization process must be efficient and effective to produce quality plans that are acceptable for online adaptive radiotherapy.

A frequently encountered clinical ramification is to choose between efficiency and quality; simple and fast optimization and delivery methods usually result in inferior treatment plans. Our experiences have shown that typically two or more arcs are needed using the advanced RapidArc or volumetric arc therapy techniques to meet our clinical acceptance criteria for prostate treatment. New plan verification methodologies may be

developed based on the Monte Carlo dose calculation and a daily accelerator/MLC QA program to ensure the dosimetric accuracy of treatment planning and beam delivery for online treatment adaptation.

Adaptive radiotherapy is an advanced treatment technique that can facilitate the development of individualized radiation therapy. Population-based treatment strategies and planning parameters can be used to generate initial treatment plans for fractionated treatments. Treatment plans can be modified or regenerated offline or online when necessary based on the changes in patient geometry according to pretreatment images and the actual dose distributions received from previous treatment fractions. As we observed, based on our CT-on-rails and Calypso studies, a small fraction of patients may have large prostate movements and/or rectal or bladder volume variations. Treatment margins for these patients can be reassessed based on the daily image data (e.g., pre- and/or posttreatment CT and Calypso tracking information). Hypofractionated treatments, such as those employing a few fractions of large doses, may first benefit from online adaptive radiotherapy procedures. Adaptive radiotherapy can be expanded to combine with other treatment modalities and include additional (e.g., boost) treatments based on posttreatment image assessment.

Acknowledgments

I would like to thank my colleagues, Alan Pollack, Eric Horwitz, Mark Buyyounouski, Steve Feigenberg, Robert Price, Lili Chen, Jinsheng Li, Shawn McNeeley, and Kamen Paskalev for their contributions to this work and continuous discussions on the subject for many years. Comments and suggestions on the manuscript from Robert Price, Eric Horwitz, and Mark Buyyounouski are graciously acknowledged. Special thanks go to Lorraine Medoro for her excellent editorial work.

References

AAPM. 2009. *AAPM Report Number 104: The Role of In-Room kV X-Ray Imaging for Patient Setup and Target Localization.* College Park, MD: American Association of Physicists in Medicine.

Amer, A. M., J. Mott, R. I. Mackay, et al. 2003. Prediction of the benefits from dose-escalated hypofractionated intensity-modulated radiotherapy for prostate cancer. *Int J Radiat Oncol Biol Phys* 56:199–207.

Balter, J. M., J. N. Wright, L. J. Newell, et al. 2005. Accuracy of a wireless localization system for radiotherapy. *Int J Radiat Oncol Biol Phys* 61:933–7.

Birkner, M., D. Yan, M. Alber, J. Liang, and F. Nusslin. 2003. Adapting inverse planning to patient and organ geometrical variation: Algorithm and implementation. *Med Phys* 30:2822–31.

Brenner, D. J., and E. J. Hall. 1999. Fractionation and protraction for radiotherapy of prostate carcinoma. *Int J Radiat Oncol Biol Phys* 43:1095–101.

Chen, L., K. Paskalev, X. Xu, et al. 2010. Rectal dose variation during the course of image guided radiation therapy of prostate cancer. *Radiother Oncol* 95:198–202.

Chen, L., K. Paskalev, and J. Zhu. 2007. Image guided radiation therapy for prostate IMRT: daily rectal dose variations during the treatment course. In *Proceedings of the 15th International Conference on the Use of Computer in Radiation Therapy (ICCR)*, ed. J. -P. Bissonnette. Oakville, ON: Novel Digital Publishing.

Chen, L., R. A. Price Jr., T. B. Nguyen, et al. 2004. Dosimetric evaluation of MRI-based treatment planning for prostate cancer. *Phys Med Biol* 49:5157–70.

Chen, L., R. A. Price Jr., L. Wang, et al. 2004. MRI-based treatment planning for radiotherapy: Dosimetric verification for prostate IMRT. *Int J Radiat Oncol Biol Phys* 60:636–47.

Court, L. E., L. Dong, A. K. Lee, et al. 2005. An automatic CT-guided adaptive radiation therapy technique by online modification of multileaf collimator leaf positions for prostate cancer. *Int J Radiat Oncol Biol Phys* 62:154–63.

Dearnaley, D. P., M. R. Sydes, J. D. Graham, et al. 2007. Escalated-dose versus standard-dose conformal radiotherapy in prostate cancer: First results from the MRC RT01 randomised controlled trial. *Lancet Oncol* 8:475–87.

Debois, M., R. Oyen, F. Maes, et al. 1999. The contribution of magnetic resonance imaging to the three-dimensional treatment planning of localized prostate cancer. *Int J Radiat Oncol Biol Phys* 45:857–65.

Ezzell, G. A., J. M. Galvin, D. Low, et al. 2003. Guidance document on delivery, treatment planning, and clinical implementation of IMRT: Report of the IMRT Subcommittee of the AAPM Radiation Therapy Committee. *Med Phys* 30:2089–115.

Fan, J., J. Li, L. Chen, et al. 2006. A practical Monte Carlo MU verification tool for IMRT quality assurance. *Phys Med Biol* 51:2503–15.

Feigenberg, S. J., K. Paskalev, S. Mcneeley, et al. 2007. Comparing computed tomography localization with daily ultrasound during image-guided radiation therapy for the treatment of prostate cancer: A prospective evaluation. *J Appl Clin Med Phys* 8:2268.

Fowler, J. F., M. A. Ritter, R. J. Chappell, and D. J. Brenner. 2003. What hypofractionated protocols should be tested for prostate cancer? *Int J Radiat Oncol Biol Phys* 56:1093–104.

Fuller, D. B., C. Lee, and S. Hardy. 2007. Virtual HDRsm CyberKnife Prostate Treatment: Toward the development of non-invasive HDR dosimetry delivery and early clinical observations. *Int J Radiat Oncol Biol Phys* 69:S358–9.

Gao, S., L. Zhang, H. Wang, et al. 2006. A deformable image registration method to handle distended rectums in prostate cancer radiotherapy. *Med Phys* 33:3304–12.

Gerszten, P. C., C. Ozhasoglu, S. A. Burton, et al. 2003. Evaluation of CyberKnife frameless real-time image-guided stereotactic radiosurgery for spinal lesions. *Stereotact Funct Neurosurg* 81:84–9.

Ghilezan, M., D. Yan, J. Liang, et al. 2004. Online image-guided intensity-modulated radiotherapy for prostate cancer: How much improvement can we expect? A theoretical assessment of clinical benefits and potential dose escalation by improving precision and accuracy of radiation delivery. *Int J Radiat Oncol Biol Phys* 60:1602–10.

Godley, A., E. Ahunbay, C. Peng, and X. A. Li. 2009. Automated registration of large deformations for adaptive radiation therapy of prostate cancer. *Med Phys* 36:1433–41.

Guo, W. Y. 1998. Application of MR in stereotactic radiosurgery. *J Magn Reson Imaging* 8:415–20.

Hanks, G. E. 1999. Progress in 3D conformal radiation treatment of prostate cancer. *Acta Oncol* 38(Suppl. 13):69–74.

Hanks, G. E., A. L. Hanlon, T. E. Schultheiss, et al. 1998. Dose escalation with 3D conformal treatment: Five year outcomes, treatment optimization, and future directions. *Int J Radiat Oncol Biol Phys* 41:501–10.

Hara, W., D. Patel, and T. Pawlicki. 2006. Hypofractionated stereotactic radiotherapy for prostate cancer: Early results. *Int J Radiat Oncol Biol Phys* 66:S324–5.

Khoo, V. S., E. J. Adams, F. Saran, et al. 2000. A comparison of clinical target volumes determined by CT and MRI for the radiotherapy planning of base of skull meningiomas. *Int J Radiat Oncol Biol Phys* 46:1309–17.

King, C. R., J. D. Brooks, H. Gill, et al. 2009. Stereotactic body radiotherapy for localized prostate cancer: Interim results of a prospective phase II clinical trial. *Int J Radiat Oncol Biol Phys* 73:1043–8.

King, C. R., J. Lehmann, J. R. Adler, and J. Hai. 2003. CyberKnife radiotherapy for localized prostate cancer: Rationale and technical feasibility. *Technol Cancer Res Treat* 2:25–30.

Krempien, R. C., K. Schubert, D. Zierhut, et al. 2002. Open low-field magnetic resonance imaging in radiation therapy treatment planning. *Int J Radiat Oncol Biol Phys* 53:1350–60.

Kupelian, P., T. Willoughby, A. Mahadevan, et al. 2007. Multi-institutional clinical experience with the Calypso System in localization and continuous, real-time monitoring of the prostate gland during external radiotherapy. *Int J Radiat Oncol Biol Phys* 67:1088–98.

Kupelian, P. A., T. R. Willoughby, C. A. Reddy, E. A. Klein, and A. Mahadevan. 2007. Hypofractionated intensity-modulated radiotherapy (70 Gy at 2.5 Gy per fraction) for localized prostate cancer: Cleveland Clinic experience. *Int J Radiat Oncol Biol Phys* 68:1424–30.

Langen, K. M., J. Pouliot, C. Anezinos, et al. 2003. Evaluation of ultrasound-based prostate localization for image-guided radiotherapy. *Int J Radiat Oncol Biol Phys* 57:635–44.

Lattanzi, J., S. Mcneeley, S. Donnelly, et al. 2000. Ultrasound-based stereotactic guidance in prostate cancer—quantification of organ motion and set-up errors in external beam radiation therapy. *Comput Aided Surg* 5:289–95.

Lee, C. 2003. Supine vs. prone setup for prostate radiotherapy: A prospectively randomized study to assess dosimetric variability. *Int J Radiat Oncol Biol Phys* 57:S335–6.

Lee, Y. K., M. Bollet, G. Charles-Edwards, et al. 2003. Radiotherapy treatment planning of prostate cancer using magnetic resonance imaging alone. *Radiother Oncol* 66:203–16.

Letourneau, D., A. A. Martinez, D. Lockman, et al. 2005. Assessment of residual error for online cone-beam CT-guided treatment of prostate cancer patients. *Int J Radiat Oncol Biol Phys* 62:1239–46.

Li, J. S., L. Jin, A. Pollack, et al. 2009. Gains from real-time tracking of prostate motion during external beam radiation therapy. *Int J Radiat Oncol Biol Phys* 75:1613–20.

Li, J. S., and C. M. Ma. 2010. 4D radiation therapy for prostate cancer. *Int J Biomed Eng Technol*.

Lloyd-Davies, R. W., C. D. Collins, and A. V. Swan. 1990. Carcinoma of prostate treated by radical external beam radiotherapy using hypofractionation. Twenty-two years' experience (1962–1984). *Urology* 36:107–11.

Lukka, H., C. Hayter, J. A. Julian, et al. 2005. Randomized trial comparing two fractionation schedules for patients with localized prostate cancer. *J Clin Oncol* 23:6132–8.

Luo, W., J. Li, R. A. Price Jr., et al. 2006. Monte Carlo based IMRT dose verification using MLC log files and R/V outputs. *Med Phys* 33:2557–64.

Ma, C. M. 2007. Clinical implementation of IMRT. In *Handbook of Radiotherapy Physics (Theory and Practice)*, ed. W. P. Mayles, A. Nahum, and J. C. Rosenwald. New York: Taylor and Francis.

Ma, C. M., S. B. Jiang, T. Pawlicki, et al. 2003. A quality assurance phantom for IMRT dose verification. *Phys Med Biol* 48:561–72.

Ma, C. M., J. S. Li, T. Pawlicki, et al. 2002. A Monte Carlo dose calculation tool for radiotherapy treatment planning. *Phys Med Biol* 47:1671–89.

Ma, C. M., and K. Paskalev. 2006. In-room CT techniques for image-guided radiation therapy. *Med Dosim* 31:30–9.

Ma, C. M., R. A. Price Jr., J. S. Li, et al. 2004. Monitor unit calculation for Monte Carlo treatment planning. *Phys Med Biol* 49:1671–87.

Ma, C., G. Shan, W. Hu, et al. 2009. A dose-guided, volumetric target localization technique for prostate IGRT. *Int J Radiat Oncol Biol Phys* 75:S579.

Madsen, B. L., R. A. Hsi, H. T. Pham, et al. 2007. Stereotactic hypofractionated accurate radiotherapy of the prostate (SHARP), 33.5 Gy in five fractions for localized disease: First clinical trial results. *Int J Radiat Oncol Biol Phys* 67:1099–105.

Mah, D., M. Steckner, A. Hanlon, et al. 2002a. MRI simulation: Effect of gradient distortions on three-dimensional prostate cancer plans. *Int J Radiat Oncol Biol Phys* 53:757–65.

Mah, D., M. Steckner, E. Palacio, et al. 2002b. Characteristics and quality assurance of a dedicated open 0.23 T MRI for radiation therapy simulation. *Med Phys* 29:2541–7.

McNeeley, S., M. Buyyounouski, R. Price, E. Horwitz, and C. Ma. 2002. Prostate displacement during transabdominal ultrasound. *Med Phys* 29:1341.

McNeeley, S., A. Pollack, E. Horwitz, R. Price, and C. Ma. 2004. A transabdominal ultrasound alignment technique to improve alignment quality and consistency in the treatment of prostate carcinoma. *Med Phys* 31:1703.

Michiels, J., H. Bosmans, P. Pelgrims, et al. 1994. On the problem of geometric distortion in magnetic resonance images for stereotactic neurosurgery. *Magn Reson Imaging* 12:749–65.

Mizowaki, T., Y. Nagata, K. Okajima, et al. 2000. Reproducibility of geometric distortion in magnetic resonance imaging based on phantom studies. *Radiother Oncol* 57:237–42.

Mohan, R., X. Zhang, H. Wang, et al. 2005. Use of deformed intensity distributions for on-line modification of image-guided IMRT to account for interfractional anatomic changes. *Int J Radiat Oncol Biol Phys* 61:1258–66.

Murphy, M. J., J. Balter, S. Balter, et al. 2007. The management of imaging dose during image-guided radiotherapy: Report of the AAPM Task Group 75. *Med Phys* 34:4041–63.

Murphy, M. J., S. Chang, I. Gibbs, et al. 2001. Image-guided radiosurgery in the treatment of spinal metastases. *Neurosurg Focus* 11:e6.

Orton, N. P., and W. A. Tome. 2004. The impact of daily shifts on prostate IMRT dose distributions. *Med Phys* 31:2845–8.

Paskalev, K., S. Feigenberg, R. Jacob, et al. 2005. Target localization for post-prostatectomy patients using CT and ultrasound image guidance. *J Appl Clin Med Phys* 6:40–9.

Paskalev, K., S. Feigenberg, L. Wang, et al. 2005. A method for repositioning of stereotactic brain patients with the aid of real-time CT image guidance. *Phys Med Biol* 50:N201–7.

Paskalev, K., C. M. Ma, R. Jacob, et al. 2004. Daily target localization for prostate patients based on 3D image correlation. *Phys Med Biol* 49:931–9.

Peeters, S. T., W. D. Heemsbergen, P. C. Koper, et al. 2006. Dose-response in radiotherapy for localized prostate cancer: Results of the Dutch multicenter randomized phase III trial comparing 68 Gy of radiotherapy with 78 Gy. *J Clin Oncol* 24:1990–6.

Plants, B. A., D. T. Chen, J. B. Fiveash, et al. 2003. Bulb of penis as a marker for prostatic apex in external beam radiotherapy of prostate cancer. *Int J Radiat Oncol Biol Phys* 56:1079–84.

Pollack, A., A. L. Hanlon, E. M. Horwitz, et al. 2006. Dosimetry and preliminary acute toxicity in the first 100 men treated for prostate cancer on a randomized hypofractionation dose escalation trial. *Int J Radiat Oncol Biol Phys* 64:518–26.

Pollack, A., R. Price, L. Dong, S. Feigenberg, and E. Horwitz. 2005. Intact prostate cancer. In *Intensity Modulated Radiation Therapy: A Clinical Perspective*. ed. A. J. Mundt and J. C. Roeske. Hamilton, ON: BC Decker, Inc.

Pollack, A., G. K. Zagars, L. G. Smith, et al. 2000. Preliminary results of a randomized radiotherapy dose-escalation study comparing 70 Gy with 78 Gy for prostate cancer. *J Clin Oncol* 18:3904–11.

Pollack, A., G. K. Zagars, G. Starkschall, et al. 2002. Prostate cancer radiation dose response: Results of the M. D. Anderson phase III randomized trial. *Int J Radiat Oncol Biol Phys* 53:1097–105.

Potter, R., B. Heil, L. Schneider, et al. 1992. Sagittal and coronal planes from MRI for treatment planning in tumors of brain, head and neck: MRI assisted simulation. *Radiother Oncol* 23:127–30.

Price Jr., R. A., J. M. Hannoun-Levi, E. Horwitz, et al. 2006. Impact of pelvic nodal irradiation with intensity-modulated radiotherapy on treatment of prostate cancer. *Int J Radiat Oncol Biol Phys* 66:583–92.

Price, R., E. Horwitz, S. Feigenberg, and A. Pollack. 2005. Intact prostate cancer: Case study. In *Intensity Modulated Radiation Therapy: A Clinical Perspective,* ed. A. J. Mundt and J. C. Roeske. Hamilton, ON: BC Decker, Inc.

Price, R. A., S. Murphy, S. W. Mcneeley, et al. 2003. A method for increased dose conformity and segment reduction for SMLC delivered IMRT treatment of the prostate. *Int J Radiat Oncol Biol Phys* 57:843–52.

Price Jr., R. A., K. Paskalev, S. Mcneeley, and C. M. Ma. 2005. Elongated beamlets: A simple technique for segment and MU reduction for sMLC IMRT delivery on accelerators utilizing 5 mm leaf widths. *Phys Med Biol* 50:N235–42.

Rasch, C., I. Barillot, P. Remeijer, et al. 1999. Definition of the prostate in CT and MRI: A multi-observer study. *Int J Radiat Oncol Biol Phys* 43:57–66.

Rehbinder, H., C. Forsgren, and J. Lof. 2004. Adaptive radiation therapy for compensation of errors in patient setup and treatment delivery. *Med Phys* 31:3363–71.

Song, W., B. Schaly, G. Bauman, J. Battista, and J. Van Dyk. 2005. Image-guided adaptive radiation therapy (IGART): Radiobiological and dose escalation considerations for localized carcinoma of the prostate. *Med Phys* 32:2193–203.

Song, W. Y., E. Wong, G. S. Bauman, J. J. Battista, and J. Van Dyk. 2007. Dosimetric evaluation of daily rigid and non-rigid geometric correction strategies during on-line image-guided radiation therapy (IGRT) of prostate cancer. *Med Phys* 34:352–65.

Tanner, S. F., D. J. Finnigan, V. S. Khoo, et al. 2000. Radiotherapy planning of the pelvis using distortion corrected MR images: The removal of system distortions. *Phys Med Biol* 45:2117–32.

Van Den Heuvel, F., T. Powell, E. Seppi, et al. 2003. Independent verification of ultrasound based image-guided radiation treatment, using electronic portal imaging and implanted gold markers. *Med Phys* 30:2878–87.

Webb, S. 1992. Optimization by simulated annealing of three-dimensional, conformal treatment planning for radiation fields defined by a multileaf collimator: II. Inclusion of two-dimensional modulation of the x-ray intensity. *Phys Med Biol* 37:1689–704.

Webb, S. 1997. *The Physics of Conformal Radiotherapy: Advances in Technology,* Bristol, UK: IOP Publishing.

Willoughby, T. R., P. A. Kupelian, J. Pouliot, et al. 2006. Target localization and real-time tracking using the Calypso 4D localization system in patients with localized prostate cancer. *Int J Radiat Oncol Biol Phys* 65:528–34.

Xu, Q., J. Li, L. Chen, et al. 2009. Investigation of transponder migration and the effects on prostate localization: A cone beam CT study. *Int J Radiat Oncol Biol Phys* 75:S24.

Yan, D., D. Lockman, D. Brabbins, L. Tyburski, and A. Martinez. 2000. An off-line strategy for constructing a patient-specific planning target volume in adaptive treatment process for prostate cancer. *Int J Radiat Oncol Biol Phys* 48:289–302.

Yan, D., D. Lockman, A. Martinez, et al. 2005. Computed tomography guided management of interfractional patient variation. *Semin Radiat Oncol* 15:168–79.

Yeoh, E. E., R. J. Fraser, R. E. Mcgowan, et al. 2003. Evidence for efficacy without increased toxicity of hypofractionated radiotherapy for prostate carcinoma: Early results of a Phase III randomized trial. *Int J Radiat Oncol Biol Phys* 55:943–55.

Zelefsky, M. J., S. A. Leibel, P. B. Gaudin, et al. 1998. Dose escalation with three-dimensional conformal radiation therapy affects the outcome in prostate cancer. *Int J Radiat Oncol Biol Phys* 41:491–500.

Zietman, A. L., M. L. DeSilvio, J. D. Slater, et al. 2005. Comparison of conventional dose vs high-dose conformal radiation therapy in clinically localized adenocarcinoma of the prostate: A randomized controlled trial. *JAMA* 294:1233–9.

Adaptive Radiation Therapy for Gynecologic Cancers

Beth Erickson
Medical College of Wisconsin

Karen Lim
Princess Margaret Hospital

James Stewart
Princess Margaret Hospital

Eric Donnelly
Northwestern University

William Small
Northwestern University

Anthony Fyles
Princess Margaret Hospital

Michael Milosevic
Princess Margaret Hospital

22.1 Introduction

The adoption of advanced imaging tools in radiation oncology departments has brought with it the potential to improve the treatment of gynecologic cancers. Computed tomography (CT) simulators are now considered standard in most centers, and the number of centers with magnetic resonance imaging (MRI) and positron emission tomography (PET) capabilities is increasing. These tools can facilitate more intelligent external beam field design and customized brachytherapy dose distributions, which better conform dose to target tissues while excluding surrounding organs at risk (OARs). In addition, daily image-guided radiotherapy (IGRT) techniques can unveil any organ and target motion relative to the field, not only to confirm field placement but also possibly to facilitate tightening the radiation field around the target. These technologies have helped define the emergent field of adaptive radiation therapy (ART) in which the dose is adapted to individual anatomical considerations of each patient and, moreover, any anatomical changes that occur over the course of treatment.

22.2 External Beam Irradiation

Important in the implementation of ART for gynecologic cancers is a standard delineation of the target volumes, both in the postoperative and in the intact pelvis. The standard target volumes for pelvic irradiation in postoperative gynecologic malignancies include the parametrium, the upper vagina and vaginal cuff, and the pelvic lymph nodes. Traditional radiation techniques utilized to treat the pelvis consisted of either two (anteroposterior-posteroanterior [AP-PA]) or four (AP-PA, right and left laterals) fields. Historically, these fields were designed utilizing plain radiographs and bony landmarks for borders based on anatomical knowledge obtained during surgical trials (Greer et al. 1990). Conventionally, the superior field border for gynecologic malignancies has been placed at the level of the L4–L5 or L5–S1 interspace. If iliac or para-aortic nodal involvement was identified, the border was extended superiorly for additional coverage.

The lateral borders extended 1.5–2 cm beyond the widest portion of the bony, and the inferior border was placed at the lower edge of the obturator foramen or at least 3 cm below the lowest extent of vaginal disease. When there was mid-to-distal vaginal extension, the entire vagina, down to the level of the introitus, was included with distal vaginal involvement; inguinal nodal coverage was considered standard based on the anatomical drainage patterns. The anterior border on the lateral fields was designed to include the most anterior portion of the pubic symphysis, and the posterior border extended to include the sacral hollow (Perez 1997).

These traditional fields, however, can risk missing the pelvic target volumes if the entire extent of the clinical target does not fall within the borders based on the established bony landmark boundaries. Traditional field design did not rely on techniques to delineate nodal and primary target volumes (PTV) with their inherent individual anatomical variances. Studies evaluating the ability of standard pelvic fields to cover nodal volumes obtained using lymphangiograms and CT-simulation demonstrated that traditional fields do not provide optimal coverage in a significant proportion of patients (Chao and Lin 2002; Finlay et al. 2006). A study undertaken at Fox Chase Cancer Center, using lymphangiograms to delineate the clinical lymph node volumes, revealed that 45% of patients included in the study would have had inadequate nodal coverage based on standard field borders (Bonin et al. 1996).

22.3 Intensity-Modulated Radiotherapy

Motivated by a desire to improve target coverage, and potentially reduce treatment-induced toxicities, many centers have studied replacing the traditional parallel-opposed or four-field box external beam arrangement with advanced imaging and intensity-modulated radiotherapy (IMRT) techniques. IMRT treatments, utilizing CT-simulation and target delineation, have resulted in improved local control with similar late toxicity when compared with conventional four-field techniques (Roeske et al. 2000). In addition to improved coverage of target tissues, planning studies for cervical malignancies have suggested that IMRT reduces dose to the small bowel, rectum, bladder, and bone marrow while maintaining dose to target tissues (Roeske et al. 2000; Portelance et al. 2001; Lujan et al. 2003; Lim et al. 2008).

Early clinical outcomes are also encouraging and indicate a reduction in acute and late gastrointestinal (GI; Mundt et al. 2001; Mundt, Mell, and Roeske 2003) and acute hematological (Brixey et al. 2002) toxicities. Similar investigations for vulvar and vagina cancer, although limited in number, also demonstrate a reduction in dose to the small bowel, rectum, and bladder (Beriwal et al. 2006). Clinical results for a small cohort of patients with vulvar carcinoma suggest that IMRT treatments are well-tolerated, with little severe toxicity (Beriwal et al. 2006).

Patient selection plays an important role in maximizing the benefit of IMRT in this setting. The advantages of IMRT are likely to be most marked in patients at high risk for radiation toxicity (such as those with connective tissue disease) or where target volumes are very large (such as extended field irradiation). Patients requiring postoperative radiotherapy following hysterectomy may also benefit from IMRT because of the larger volume of small bowel in the pelvis and the advantage that IMRT offers to reduce small bowel dose while maintaining adequate target coverage.

A crucial prerequisite for the successful application of IMRT plans is a critical assessment of the anatomy of each individual patient. This assessment must be made on volumetric imaging to facilitate proper definition of customized treatment fields. Although CT imaging alone is adequate for routine treatment planning, MRI in addition to CT is strongly recommended if IMRT is being considered for patients with definitively treated cervical cancer. MRI offers substantially better soft-tissue definition of central pelvic structures around the cervix and the uterus than CT and aids in target delineation. Ideally, the MRI should be done immediately before or after the planning CT in order to minimize discrepancies in organ size and position due to factors such as bladder and rectal filling. With such imaging, the target tissues and OARs can then be contoured to direct the creation of an IMRT plan. In Sections 22.3.1 through 22.3.3, we discuss contouring methods for the nodal, postoperative, and intact uterus settings.

22.3.1 Contouring of the Female Pelvis: Nodal Targets

Directly detecting pelvic lymph nodes, whether normal or with metastases, on CT or MR images is particularly difficult, especially for smaller nodes. Published descriptions of lymph node locations with respect to MR and CT images for use with IMRT rely on expansions of the blood vessels to best balance the trade-off between contouring efficiency and nodal coverage.

Taylor et al. (2005), using an MRI after the administration of iron oxide nanoparticles to enhance lymph node contrast, recommended a 7-mm margin around the vessels to encompass >95% of the common iliac, internal iliac, medial and anterior external iliac, and obturator lymph node contours; except around the lateral external iliac vessels where a 10-mm margin was recommended. These volumes were further modified based on individual anatomical considerations, such as lateral coverage to the pelvic sidewall to create a surrogate target to cover 99% of the pelvic nodes in the patient cohort.

This volume is similar to that proposed by Dinniwell et al. (2009) in a similar study. This latter study recommended margins of 10 mm around the common and internal iliacs, 9 mm around the external iliac, and 12 mm around the distal para-aortics. This margin was further modified with a 12-mm expansion anterior to the sacrum and a 22-mm expansion medial to the pelvic sidewall. Although these margins initially appear larger than those proposed by Taylor et al., the authors note that the manual modifications of Taylor et al. increase the nodal volume to the point where the two margin recipes are generally coincident.

Chao and Lin's (2002) work using lymphangiograms to identify lymph node regions led to recommending a larger margin of 15–20 mm with further expansions to the pelvic sidewall and iliopsas muscles. The consensus guidelines developed for the Radiation Therapy Oncology Group (RTOG 0418) trial by Small et al. (2008) adopted a 7-mm margin around the vessels, based on the fact that lymphangiograms might overestimate the lymph node size.

22.3.2 Contouring of the Female Pelvis: Postoperative Endometrial and Cervical Cancer

The consensus guidelines by Small et al. (2008) included an atlas of the clinical target volumes (CTV) for postoperative radiotherapy of endometrial and cervical patients, especially when using IMRT (Figure 22.1). The committee concluded that the nodal CTV should include the common, external, and internal iliac lymph node regions. The upper 3 cm of the vagina and the paravaginal soft tissues lateral to the vagina should also be included as part of the target volume. It was recommended that in patients with involvement of the cervix, the presacral lymph nodes should also be included by contouring a 1.5-cm margin between the anterior border of the CTV and the anterior border of the vertebral bodies or sacrum and ending when the piriformis muscle becomes clearly visible. The superior extent of the CTV should start 7 mm below the L4–L5 interspace to allow for PTV expansion. In addition to the pelvic vasculature, the CTV should extend to include any adjacent visible or suspicious lymph nodes, lymphoceles, and applicable surgical clips. The volume should be modified to exclude bone, muscle, and bowel.

When creating the vaginal and parametrial volumes, a vaginal marker can be used to delineate the inferior extent of the vaginal cuff. At the superior portion of the vaginal cuff, the vaginal volume can be joined with the two nodal volumes making a single combined CTV. The vaginal and parametrial volume needs to account for internal organ motion through the creation of an internal target volume (ITV). The ITV is obtained by performing two separate treatment planning CT scans, one with a full bladder and the second with an empty bladder. The two scans are fused together, and the ITV, defined as the volume of the vagina in both the empty and full bladder CT scans, accounts for internal organ motion. The ITV is drawn after the full and empty bladder scans are fused together and should encompass the vagina and paravaginal soft tissues from both the scans.

Although patients should be treated with a full bladder to help push the small bowel up and out of the field, it is often difficult for patients to be able to maintain constant bladder filling throughout their treatment course. Thus, creation of an ITV accounts for vaginal cuff location despite the variations in bladder fullness. The inferior vaginal contours should end 3 cm below the vaginal marker or 1 cm above the bottom of the obturator foramen, whichever is most inferior. The lateral margin of the vaginal and parametrial CTV should be to the obturator internis muscle. Although specific recommendations were made to account for varying bladder volumes, the RTOG Consensus Committee offered no definitive proposals to account for varying rectal filling, other than the consideration of repeating the planning scan if excessive rectal distension was observed. Approaches to minimize rectal filling can include enema before treatment or

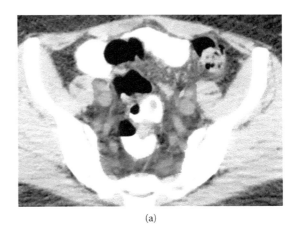

(a)

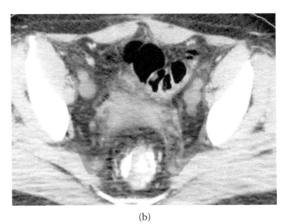

(b)

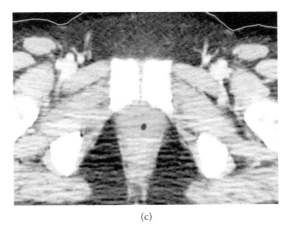

(c)

FIGURE 22.1 (See color insert following page 204.) (a) Superior clinical target volume (CTV)—upper external and internal iliac (red) and presacral (blue). (b) Middle CTV—external and internal iliac (red) and parametrial/vaginal (green). (c) Inferior CTV—vaginal.

evacuation of air with a catheter, but there is no consensus on this issue. These guidelines serve to define volumes that need to be included for effective treatment and to produce consistency between treating institutions for future comparisons.

As part of the RTOG 0418 protocol, the ability of the centers to comply with contouring guidelines outlined by Small et al. was assessed. When assessing the nodal CTV contours, only 41% were contoured per protocol. A minor deviation was noted in 38% of contours, and consisted primarily of contouring sacral nodes when there was no cervical involvement. A major deviation was identified in 22% of the cases. When the vaginal ITV was reviewed, 60% of the cases followed the guidelines. A minor deviation was found in 29% of the cases and a major deviation in only 12% (Jhingran et al. 2008, RTOG 0418). These initial results highlight the need for continued monitoring and teaching of target delineation, especially within the setting of protocols, so that accurate comparisons can be made.

To aid in imaging delineation, patients should have a radiopaque marker placed within the vagina abutting the vaginal cuff during the imaging scan to allow accurate visualization of the inferior extent of the cuff. Care must be taken to make certain the marker does not distend the vagina, and thus large obturators should not be utilized. A cotton-tipped applicator wrapped with CT-spots and a CT-spots BB marker at the top, covered with a condom, is an effective vaginal marker. Intravenous contrast material should be used to help define the vascular anatomy. Oral contrast should be given only if density corrections are done. A rectal tube can also be placed and a small amount of contrast injected to help delineate, but not distend, the rectum. Radiopaque markers can be placed at the introitus as well as the anal verge.

22.3.3 Contouring of the Female Pelvis: Intact Uterus

The inclusion of structures in the CTV for contouring in the intact uterus setting is largely extrapolated from the surgical management of cervix cancer (Wertheim 1912; Meigs and Dresser 1937; Querleu and Morrow 2008). The nodal CTV in this setting should include the relevant regional lymph nodes (common iliac, internal and external iliac, presacral) in addition to any suspiciously enlarged nodes. The details relating to contouring of the nodal CTV are identical to that of the postoperative setting previously described. The CTV for cervical cancer should include the gross tumor volume (GTV), cervix (if not already encompassed by the GTV), uterus, parametria, ovaries, and vaginal tissues (Figure 22.2). While radiological boundaries of some of these pelvic targets (such as parametrial tissues) have been poorly described in the past, recent efforts have been made to clarify this. Consensus guidelines for tumor CTV delineation are summarized next (Lim et al. 2009; Table 22.1).

The determination of an appropriate PTV margin around the CTV to account for inter- and intrafractional target motion in the setting of an intact uterus is complex and depends on several factors including setup variability, the dynamics of tumor regression and deformation, the effect of bowel and bladder filling, differences in movement dynamics between the primary tumor and nodal target volumes, and the availability of daily online soft-tissue imaging. The tumor and nodal CTVs are usually combined to form a single CTV for treatment planning. However, the primary tumor, cervix, and uterus may move substantially more, and in different ways, than the lymph nodes, which are in close proximity to the pelvic sidewall and vessels, and relatively immobile (Lim et al. 2008). It is important to recognize that simple translational shifts to compensate for tumor CTV displacement could compromise nodal CTV coverage in some circumstances.

CTV-PTV margin recommendations for conformal radiotherapy in patients with an intact cervix setting have ranged from 0.6 cm to 2.0 cm, depending on the methodology for assessing organ motion (Buchali et al. 1999; Kaatee et al. 2002; Chan et al. 2008; Taylor and Powell 2008; van de Bunt et al. 2008; Beadle et al.

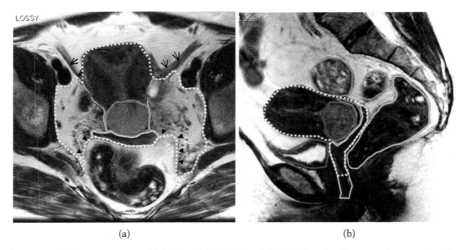

(a)   (b)

FIGURE 22.2 (See color insert following page 204.) (a) T2-weighted MR axial and (b) sagittal images of one patient demonstrating gross tumor volume (GTV; red), cervix (pink), uterus (blue), vagina (yellow), parametrium (green), bladder (purple), rectum (light blue), sigmoid (orange). Black arrow heads refer to uterosacral ligaments and mesorectal fascia. Open arrow heads refer to the broad ligament and top of the fallopian tube. Dashed white line represents the clinical target volume (CTV). (Reprinted from Lim, K., W. Small Jr., L. Portelance, et al. 2009. *Int J Radiat Oncol Biol Phys*. With permission.)

TABLE 22.1 Consensus Guidelines for Clinical Target Volume Delineation

Gross tumor volume (GTV)	Entire GTV (intermediate/high signal seen on T2-weighted MR images)
Cervix	Entire cervix (if not already included within GTV contour)
Uterus	Entire uterus
Parametrium	Entire parametrium including ovaries (include entire mesorectum if uterosacral ligament involved) Anteriorly: posterior wall of bladder or posterior border of external iliac vessel Posteriorly: uterosacral ligaments and mesorectal fascia Superiorly: top of fallopian tube/broad ligament. Depending on degree of uterus flexion, this may also form the anterior boundary of parametrial tissue Inferiorly: urogenital diaphragm Laterally: medial edge of internal obturator muscle/ischial ramus bilaterally
Vagina	Minimal or no vaginal extension: upper half of vagina Upper vaginal involvement: upper two-thirds of vagina Extensive vaginal involvement: entire vagina

2009). The use of IMRT without daily soft-tissue verification risks geographical target miss and should be avoided. Imaging devices integrated with the treatment unit, such as cone-beam CT (CBCT), can aid in this respect, but currently have limited soft-tissue contrast resolution. Fiducial markers inserted into the cervix prior to the start of treatment are also helpful particularly when used together with online volumetric imaging, but reflect the motion of only a part of the CTV and are subject to attrition and migration secondary to tumor regression and deformation.

In general, margins of 1.5–2 cm around the tumor CTV are recommended if daily soft-tissue verification is available during treatment. If bone matching alone is used, more generous margins are necessary, due to the uncertainty of tumor CTV position in relation to the nodal CTV. However, simple PTV margin class solutions may be inadequate because of large unpredictable motion that can occur from one fraction to the next and, if unrecognized, contribute to substantial target underdosing.

22.4 Dose–Volume Constraints

22.4.1 Postoperative Endometrial and Cervical Cancer

The majority of endometrial and early-stage cervical carcinomas are treated upfront with hysterectomy and appropriate surgical staging. Patients at risk for local recurrence, based on pathologic risk factors, may be considered for adjuvant radiation therapy, either in the form of external beam radiation therapy (EBRT) and/or brachytherapy. Postoperative adjuvant radiation therapy for endometrial and cervical cancer patients has been shown to significantly reduce the rate of local recurrences in several large randomized trials, especially in those patients with key risk factors (Keys et al. 2004; Sedlis et al. 1999; Creutzberg et al. 2000;

ASTEC 2009). Conventional whole pelvis radiation therapy (WPRT) homogeneously treats the majority of the true pelvis to the conventional dose of 45–50 Gy in 25–28 fractions. Within these treatment fields, large volumes of small bowel, rectum, and bladder are typically included. This is especially true after hysterectomy, when large portions of the small bowel fall into the high dose regions of the pelvis (Gallagher et al. 1986).

Small bowel irradiation can be associated with both acute and late morbidity. Typical acute GI symptoms include varying degrees of diarrhea, urgency, and/or abdominal cramping (Letschert et al. 1994). Several studies have shown that IMRT can reduce the volume of small bowel receiving the prescribed dose in both cervical and uterine cancers. Portelance et al. (2001) compared four-, seven-, and nine-field plans delivered by dynamic multileaf collimation with a four-field box technique and showed a 58%–67% reduction in the volume of small bowel irradiated to more than 45 Gy with IMRT.

Roeske et al. (2000) reported similar results in their study using radiation doses up to 45 Gy in 25 fractions to treat the proximal vagina, parametrial tissues, uterus, and pelvic nodes in 10 patients with cervical or uterine carcinoma. The study noted a 50% reduction in the volume of small bowel receiving more than 45 Gy with a nine-field Corvus plan (Nomos Corporation, Cranberry Township, PA), when compared with a four-field 3D conformal plan.

Mundt, Mell, and Roeske (2003) studied acute and chronic GI toxicity in postoperative gynecologic patients treated with IMRT compared with those treated with conventional WPRT. The study concluded that patients treated with IMRT had a lower rate of acute grade 2 GI toxicity than the WPRT-treated patients (60% vs. 91%; $p = .002$). The percentage of patients requiring no or only infrequent antidiarrheal medications was 75% in patients treated with IMRT, compared to 34% in patients treated with conventional WPRT. In this same group, patients treated with IMRT had a lower rate of chronic GI toxicity compared to patients treated with conventional WPRT (11.1% vs. 54%, respectively; $p = .02$; Mundt, Mell, and Roeske 2003). The current RTOG protocol guidelines recommend that the small bowel constraint be <30% of bowel receiving $\geq$40 Gy in the postoperative setting (Jhingran RTOG 0418).

Another normal tissue at risk for significant side effects in patients with gynecologic malignancies, especially in those patients receiving concurrent chemotherapy, is the bone marrow. In adults, the pelvis accounts for approximately 40% of the total body bone marrow reserve (Ellis 1961). For patients undergoing pelvic radiation therapy, a great percentage of this area will be within the treatment field. Hematologic toxicity can cause significant life-threatening problems during treatment, including a high risk of infection as white blood cell counts decrease. Anemia can also be a problem at presentation or during treatment. Several studies have documented the adverse impact on local control associated with decreased levels of hemoglobin in cervical patients (Brixey et al. 2002; Lujan, Roeske, and Mundt 2001). IMRT can help decrease the hematologic toxicity associated with WPRT.

Brixey et al. (2002) reported a reduction in hematological toxicity (grade 2 or greater white blood cell toxicity) observed in patients treated with a combination of chemotherapy and IMRT when compared with standard WPRT (31% vs. 60%). This reduction was attributed to a significant reduction in the volume of irradiated bone marrow, particularly within the iliac crests. Lujan et al. published IMRT plans that resulted in a 46% reduction in the pelvic bone marrow volume receiving 50% of the prescribed dose compared to conventional four-field plans (Lujan, Roeske, and Mundt 2001). The RTOG guidelines did not define bone marrow constraints, but the contouring guidelines did recommend modification of the CTV to exclude vertebral bodies and pelvic bone. A reduction in hematology toxicity can improve patient care by limiting breaks in treatment and potentially increasing the number of courses of chemotherapy that patients are able to tolerate.

In addition to the small bowel and hematologic toxicities observed in the treatment of gynecologic malignancies, the bladder and rectum are also susceptible to toxicity. Recent RTOG studies utilizing conventional radiation techniques reported high GI (35% grade 3 and 4) and genitourinary (9% grade 3 and 4) complications, highlighting the need to limit dose to these normal structures (Morris et al. 1999; Grigsby et al. 1998). Using IMRT in the treatment of gynecologic malignancies has the potential to not only avoid OAR, but deliver higher doses to areas at risk, including adjacent lymph node regions, without increasing toxicity. Similar to studies evaluating small bowel, IMRT has been shown to significantly limit the dose to both the rectum and bladder.

Portelance et al. (2001) demonstrated that conventional four field techniques had a significantly higher rectal volume receiving a higher dose compared to IMRT techniques: 46% for four-field vs. 6% and 3% for seven- and nine-fields, respectively. Likewise, conventional four-field techniques also showed a higher percentage of bladder volume receiving higher than the prescribed dose when compared to IMRT: 60% for four-field vs. 31% for seven-field and 27% for nine-field. Current RTOG guidelines suggest that <60% of the rectal volume receives ≥30 Gy and that <35% of the bladder volume receives ≥45 Gy to limit the toxicity to these normal structures.

The RTOG 0418 study was designed to assess the ability of the radiation oncologist to meet the aforementioned guidelines. The initial results showed that most deviations were minor, except for 16% major deviations, all of which were related to small bowel constraints. The percentages of minor deviations, however, were strikingly high with 67% found for the bladder, 76% for the rectum, 17% for the small bowel, and 33% for the femoral heads (Jhingran et al. 2008).

Overall, acute radiation-associated toxicity in the postoperative setting has been reported to occur in 65% of patients (Weiss et al. 1999).

Acute toxicity most commonly involves varying degrees of diarrhea, cramping, and abdominal discomfort. With the use of IMRT techniques, a significant amount of normal tissue can be avoided to help limit these toxicities. Late toxicities may arise months to years after radiation therapy, ranging from intermittent diarrhea and malabsorption to more severe toxicities including obstruction and fistulas. Although late toxicities are uncommon, grade 1 and 2 late toxicities have been reported in 18% of patients and grade 3 and 4 toxicities in 2%–11% (Letschert et al. 1994; Jereczek-Fossa et al. 1998; Weiss et al. 1999). Further follow-up is needed to assess the impact of IMRT in reducing late complications.

With the promise of more conformal treatments and ART come several issues of concern and caution. Careful attention needs to be given to the CTV used to generate these plans. Ahamad et al. (2005) studied a variety of different clinical tumor volumes to create IMRT plans. The study showed that the volume of normal tissue spared by IMRT relative to conventional techniques was exceedingly sensitive to small increases in the margin size utilized to create the planning target volumes. When a 5-mm margin increase was used around the CTV, the volume of small bowel spared 30 Gy or more was reduced by approximately 40%.

22.4.2 Dose–Volume Constraints: Intact Uterus

Target dose–volume constraints for definitive treatment of cervix cancer with IMRT have been varied. Within the published literature, dose prescriptions have ranged from 95%–100% of the PTV to receive 95%–107% of the prescribed dose (Portelance et al. 2001; Mundt et al. 2002; Ahmed et al. 2004; Georg et al. 2006; Gerszten et al. 2006; van de Bunt et al. 2006; Lim et al. 2008). Planning dose constraints for the OARs have not always been explicitly stated and have varied from center to center. Most of the contemporary normal tissue dose–volume data for pelvic organs has been derived from the prostate cancer literature (Fiorino et al. 2009). However, dose constraints appropriate for the treatment of prostate cancer may not be strictly applicable for cervix cancer patients, particularly as the target volumes are much larger and treatment consists of a combination of EBRT and brachytherapy.

Huang et al. (2007) have reported that in patients with an intact uterus receiving standard pelvic irradiation, the volume of bowel exposed to a low dose of radiation was more predictive of grade 2 or greater acute bowel toxicity. In contrast, Roeske et al. (2003) reported a correlation between grade 2 or greater acute bowel toxicity and the absolute volume of bowel receiving 45 Gy using an IMRT technique. However, their patient population included postoperative endometrial cancers as well as definitive cervix cancers. Differences in the definition of OARs, for the purposes of contouring (such as "individual loops of bowel" vs. "peritoneal cavity"), as well as inter- and intrafraction motion of these organs during treatment, make dosimetric correlations with toxicity outcomes difficult.

The dose–volume objectives in Table 22.2 have been adapted from the upcoming phase 2 RTOG 0918 cervix IMRT trial. They were derived from a careful review of the literature and experience in previous RTOG studies.

TABLE 22.2 Dose Constraints and Minor Deviations for Both Tumor Target and Organs at Risk

	Objectives	Minor Deviations
Primary target volumes (PTV)	≥98% volume receives ≥95% prescribed dose	≥98% volume receives ≥90% prescribed dose
	≤10% volume receives ≥105% prescribed dose	≤15% volume receives ≥105% prescribed dose
	≤5% volume receives ≥110% prescribed dose	≤10% volume receives ≥110% prescribed dose
	No volume receives ≥115% prescribed dose	
Bladder	≤40% volume receives >40 Gy	≤40% volume receives >30 Gy
	No volume receives 50 Gy	≤25% volume receives >45 Gy
		No volume receives 50 Gy
Rectum	≤60% volume receives >40 Gy	100% volume receives 45 Gy
		No volume receives 50 Gy
Bowel	≤30% volume receives >30 Gy	≤40% volume receives >30 Gy
		≤30% volume receives >40 Gy
		≤10% volume receives >45 Gy
Femoral heads	≤5% volume receives 45 Gy	≤25% volume receives >40 Gy
	No volume receives 50 Gy	≤15% volume receives >45 Gy
		No volume receives 50 Gy
Bone marrow	≤65% receives >20 Gy	≤75% receives >20 Gy
Genitalia/mons pubis	≤50% receives 20 Gy	≤50% receives 15 Gy
	No volume receives >15 Gy	No volume receives >25 Gy

22.5 Patient Positioning and Immobilization

It is imperative that patients are simulated in a reproducible position with the proper immobilization devices and normal organ opacification to enable successful delineation of target volumes and OAR and to account for possible motion during treatment. Patients with gynecologic malignancies are typically immobilized with devices that secure the pelvis and/or abdomen (Bentel 1992). The prone position has been shown to reduce small bowel dose in postoperative patients (where significant portions of bowel reside low in the pelvis) receiving either standard pelvic fields or IMRT (Gallagher et al. 1986; Adli et al. 2003; Weiss et al. 2003; Martin et al. 2005). It is not currently clear if this benefit also extends to the definitive setting where the uterus is present to help keep the small bowel out of the pelvis (Pinkawa et al. 2003; Georg et al. 2006).

22.5.1 Organ and Target Motion: Postoperative Endometrial and Cervical Cancer

The volume of normal tissue spared by IMRT relative to conventional techniques is heavily dependent on organ motion during treatment (Ahamad et al. 2005). This is of particular concern, as other studies have reported that normal tissues and target volumes can move during treatment (Weiss et al. 2007). The vaginal vault and central pelvic tissue locations are dependent on changes in rectal and bladder filling and can be found to have shifted relative to their original location based on static imaging studies. This motion increases the risk of missing areas containing subclinical disease when compared to standard techniques if very tight margins and conformal isodose curves are utilized around the target volumes. With the potential for motion and IMRT conforming dose distributions more precisely, assessment of this organ motion has become extremely important.

Both organ motion and patient movement can cause deficiencies in dose coverage during treatment. Although it is has been well-documented that the cervix and uterus have significant motion, fewer studies have looked at organ motion in gynecologic patients in the postoperative setting.

The nodal volume in the postoperative setting is felt to be more rigid and less prone to motion. Studies of rectal cancer looking at clinical tumor motion within the pelvis found minimal motion within the clinical nodal volumes. Nuyttens et al. (2002) noted that the superior CTV in rectal patients, where the volume was defined by the common and external/iliac vessels was relatively stable across all the patients. However, the vaginal cuff motion can be significant and needs to be accounted for during treatment planning.

Rusthoven et al. (2009) evaluated postoperative endometrial and cervical cancer patients to assess organ motion during IMRT. The study placed gold fiducial markers within the vaginal cuff to evaluate motion throughout the treatment by taking daily MVCTs. The average vaginal motion was only 2 mm, but the largest amount of motion was 18 mm. In total, 95% of the vaginal motion was found to be 7 mm or less. The study noted that the average vaginal motion was small, but in individual cases, significant motion was identified (Rusthoven et al. 2009). These findings highlight the need for specific instructions on bladder and rectal filling during treatment. The consensus guidelines adopted by the RTOG attempt to account for this inherent motion by creating an ITV. With the use of IMRT, both individual anatomy and motion can be accounted for.

22.5.2 Organ and Target Motion: Intact Uterus

The complex motion exhibited by pelvic organs during EBRT for the intact uterus has been well-documented by a number of investigators (Kerkhof et al. 2009; Ahmad et al. 2008; Chan et al. 2008; Taylor and Powell 2008; van de Bunt et al. 2008; Beadle et al. 2009).

In the case of cervix cancer, the motion is further complicated by concomitant tumor regression and deformation during treatment (Mayr et al. 2006; Van de Bunt et al. 2006; Chan et al. 2008; Lim et al. 2008). As a consequence, target motion can be unpredictable and highly individualized. These anatomical changes during treatment have raised concerns about the risk of target miss and OAR overdosing associated with IMRT plans (Randall and Ibott 2006). Simple PTV margin class solutions may be inadequate because of large unpredictable motion that can occur from one fraction to the next. Attempts to move beyond this necessitate daily IGRT with adaptive replanning to maintain target coverage, while simultaneously capitalizing on tumor regression during treatment to reduce dose to adjacent normal organs.

With an intact uterus, bladder filling may play a greater role in influencing target positioning during treatment. Large variations in target position due to changes in bladder filling between fractions are possible in these patients, and correlations between bladder filling and CTV motion have been noted by multiple investigators (Kerkhof et al. 2009; Chan et al. 2008; Taylor and Powell 2008; Van de Bunt et al. 2008; Beadle et al. 2009). In general, the relationship between uterine motion and bladder filling remains unclear with most studies showing only a weak correlation (Chan et al. 2008; Beadle et al. 2009). However, anecdotal evidence suggests that anteverted uteri may be more affected by bladder filling than upright or retroverted uteri (Kerkhof et al. 2009; Lim et al. 2008; Taylor and Powell 2008). While there is evidence that a full bladder helps limit the amount of bowel in the pelvis (Buchali et al. 1999; Georg et al. 2006), maintaining these bladder volumes is difficult, particularly as treatment progresses and patients develop radiation cystitis. Soft-tissue imaging during treatment has demonstrated that bladder volume is rarely consistent despite bladder filling instructions to the patients (Ahmad et al. 2008; Chan et al. 2008; Lim et al. 2008).

Rectal filling is also highly variable among patients from day to day, though the impact on target position has been less extensively studied (Chan et al. 2008; Taylor and Powell 2008; Van de Bunt et al. 2008). The issue of a bowel preparation during treatment in order to minimize rectal volume changes remains controversial for cervix cancer patients. Some groups recognize that it influences target position in prostate cancer, but the data is scarce in the setting of cervix cancer. Attempts to regulate rectal filling with the use of laxatives or enemas during treatment may result in greater gaseous distention.

The literature on vaginal motion during radiotherapy for the intact cervix is scarce. Taylor and Powell (2008) looked at one point within the vagina, located 2 cm below the cervix, and assessed motion relative to bladder and rectal filling on two consecutive MRIs. They reported mean upper vaginal motion of 2.6 mm (ranging up to 10 mm) in the AP direction and minimal lateral motion. They found significant correlation with rectal filling. Lim et al. (2008) also assessed vaginal motion, using surface meshes of the entire organ rather than a single representative point. Deformable soft-tissue modeling of this organ over a 5-week course of radiotherapy yielded motion estimates similar to Taylor's, with standard deviations of 4 mm and 3 mm in the AP and SI directions respectively.

The challenge of accommodating uterine motion during EBRT is particularly difficult. Investigations have documented interfraction point wise motion of the uterine fundus and cervix of up to 50 mm and 20 mm, respectively, in some patients (Kaatee et al. 2002; Huh et al. 2004; Chan et al. 2008; Taylor and Powell 2008). Over a treatment course, the uterus has even been observed rotating from an anteflexed to a retroflexed position (Huh et al. 2004). In addition, the potential for substantial cervix cancer regression and deformation during treatment (Beadle et al. 2009) further complicates the design of a population-based IMRT strategy. The dosimetric consequences of this motion and tumor regression in the setting of IMRT are currently being investigated (Kerkhof et al. 2008; Lim et al. 2008).

Figure 22.3 illustrates target and normal organ motion for a typical patient with cervix cancer over a 5-week course of EBRT (Lim et al. 2008). During treatment, the uterus rotates about the cervix and oscillates between a retroverted and nearly upright position. Bladder filling is also highly variable, despite explicit

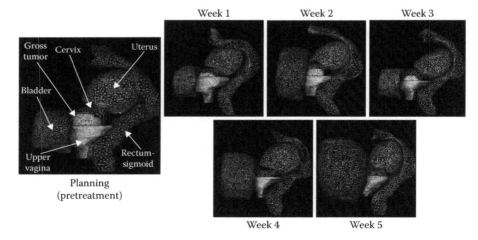

FIGURE 22.3 Organ motion and target regression over a 5-week radiotherapy treatment. (Reprinted from Stewart, J. K., S. H. Lim, et al. 2010. *Int J Radiat Oncol Biol Phys*. With permission.)

patient instructions in an attempt to maintain a consistent bladder volume. Finally, regression of the gross tumor is clearly evident as treatment progresses (Figure 22.3).

22.6 Adaptive Potential of Image-Guided Therapy

The highly deformable and individual nature of organ and tumor motion in the gynecological setting presents unique challenges for IMRT treatments. High-quality soft-tissue imaging during treatment will likely be necessary to facilitate PTV margin reduction and achieve the maximal dosimetric benefit (Kerkhof et al. 2008). Even with such imaging, the temporal lag between the imaging session and treatment can limit its relevance, especially in light of the demonstrated intrafractional uterine motion. Integrated volumetric imaging with CBCT or other similar modalities reduces this lag from imaging to treatment, although the current lack of soft-tissue contrast with this modality limits its efficacy. Such an integrated system may best be used clinically to reduce setup errors by aligning the bony pelvis or by verifying the bladder and rectal filling status at the start of treatment. The development of an integrated MR-accelerator treatment system would greatly improve soft-tissue contrast compared with CBCT and, in turn, greatly facilitate real-time imaging and adaptive replanning.

IGRT for cervix cancer may realize its greatest potential through the monitoring of treatment progress. Repeated imaging throughout treatment can, for example, identify patients with poor treatment prognosis due to minimal tumor regression (Mayr et al. 2009). Once recognized, more aggressive treatment options can be considered. In addition, repeated imaging facilitates dosimetric treatment evaluation after taking organ motion and target regression into account (Lim et al. 2008; Stewart et al. 2009). With this dosimetric knowledge, it is possible to determine if the target is underdosed or if the healthy organs are receiving more or less dose than planned. It would then be possible, for instance, to replan patients at strategic points during treatment to maintain the target dose or reduce dose to the rectum and other OARs as required to achieve a high likelihood of tumor control with a low risk of toxicity (Georg et al. 2009).

This approach is currently under investigation. The feasibility and future clinical impact of adaptive replanning will depend on the availability of real-time tools to facilitate high quality soft-tissue imaging, rapid recontouring of target volumes and OARs, deformable soft-tissue modeling and dose accumulation, rapid radiation treatment replanning, and appropriate quality assurance of new treatment plans.

22.7 Brachytherapy

Brachytherapy is perhaps one of the most conformal and adaptable approaches to deliver a dose to a target. It is a time-honored approach in the treatment of gynecologic cancers, in existence since the early 1900s. With a rich tradition of using point doses calculated from localization films rather than CT- or MR-based dosimetry, there has been some resistance in moving to 3D

volume-based brachytherapy. It has become increasingly obvious that orthogonal film-based dose distribution analysis, in which single or multiple reference points are chosen on films at the interface of the organs closest to the applicators and at select dose specification points, is inadequate for gynecologic brachytherapy. Single tumor reference points such as point A, chosen from localization films, do not give sufficient information about the dose distribution throughout the tumor volume. Nor do the reference points for the bladder and rectosigmoid accurately reflect the dose distribution within these organs.

Additionally, there is no recognition of the volume of tumor and normal tissues receiving these doses. The ICRU 38 report attempted to provide a guideline for the selection of universal reference points for bladder and rectum, but many times the doses in other parts of these organs are much higher than the ICRU 38 points (Datta et al. 2006). There is some doubt whether the dose at these reference points has any correlation with the complex 3D dose distributions within these organs and with the risk of complications (Erickson 2003). Furthermore, other chosen normal organ points are subjective and unreliable. In the high dose gradient area around the applicators, the dose at one point may be meaningless if it is not representative of or related to the high dose areas in the OARs.

This 2D approach is now frequently challenged with the advent of MRI and CT compatible applicators and digital imaging and communications in medicine (DICOM) image transfer used traditionally for external beam plans. Most radiation oncology departments have CT simulators and some are also acquiring MR scanners, making 3D image-guided brachytherapy a more common reality. Image-based brachytherapy enables defining both the disease and the OAR and then shaping the dose distribution to optimally cover the disease and exclude the normal tissues.

22.7.1 Computed Tomography-Based Brachytherapy

CT scans have been used for diagnostic and treatment planning purposes since the 1960s. CT-based treatment planning for intracavitary applications was piloted in the 1980s (Erickson 2000, 2003). In these formative years, the value of CT was limited by the artifacts produced by the metal applicators available to most brachytherapists. Procedures to minimize the effects of interference from the applicators and organ contrast agents included use of small volumes (20–30 cm³) of dilute contrast in the rectosigmoid and bladder, unshielded applicators, optimum window and level settings, and expanded CT numbers to visualize organ boundaries. Useful information, despite these limitations, included measurements of cervical diameter as well as rectovaginal septal and uterine wall thickness, uterine size and shape, tandem location in the uterus, uterine perforation, and applicator proximity to the bladder, rectosigmoid, and small bowel (Erickson 2000, 2003). CT imaging following applicator placement allows immediate confirmation of the position of the applicator relative to the cervix and adjacent OAR, both above and below the vaginal packing.

The greatest value of CT has been in gaining a better understanding of normal tissue doses in the organs close to the applicators (Figure 22.4). Point doses, isodose curves, and dose–volume histograms (DVH) can be generated and used to define treatment. There have been appreciable differences between the doses to the critical normal organs calculated from localization films vs. CT. CT-defined doses have been 1–5.4 times the point dose measurements taken from localization films (Erickson 2003). It is well-established that point dose measurements have underestimated normal tissue doses. What is not yet known is whether these higher doses are of clinical consequence. There seems to be a range of point doses that predict for complications when calculated from localization films.

There is emerging data that the volume receiving certain doses may be the key rather than a small high point dose somewhere in an organ. Nonetheless, CT often reveals that the highest dose to the bladder may well be the bladder horns draping off to one side of the applicator or the portion of the bladder at the intersection of the tandem and vaginal sources, above the vaginal packing and the ICRU 38 point. Likewise, the circuitous sigmoid may weave much closer to the applicator than the displaced rectum located near the ICRU 38 point. The sigmoid has typically been ignored given the difficulty in opacifying and localizing it on films, but

is often at greatest risk as it is above the rectal retractors and can weave very high around the uterine tandem or dip very low, close to the vaginal applicators. It is susceptible to late ulceration and stricture, and complications herein may be misinterpreted as rectal. These anatomical surprises explain why some patients with perfectly acceptable point doses will have unexpected late complications (Figure 22.5).

CT is excellent at "exposing" these organ locations (Shin et al. 2006). It should be considered the current standard, even if CT-compatible applicators are not available. CT-compatible applicators have enabled more useful imaging of the pelvis and CT-based dosimetry has become very easy to accomplish in most departments. CT simulation scout films can replace localization films. The normal organs can be contoured, and point doses, isodose curves, and DVH parameters can be generated to evaluate and potentially modify the implant dose distribution (Kim et al. 1997; Shin et al. 2006). Cumulative DVH are recommended for the OARs, whether using CT or MR. Outer-wall contours alone are acceptable for assessment of 0.1, 1, and 2 cm³ volumes, whereas inner and outer-wall contours are used for 5 and 10 cm³ volumes (Wachter-Gerstner et al. 2003a; Pötter et al. 2006). The smaller volumes are thought to most closely approximate the size of an ulcer or fistula. CT-based dosimetry should be done for every

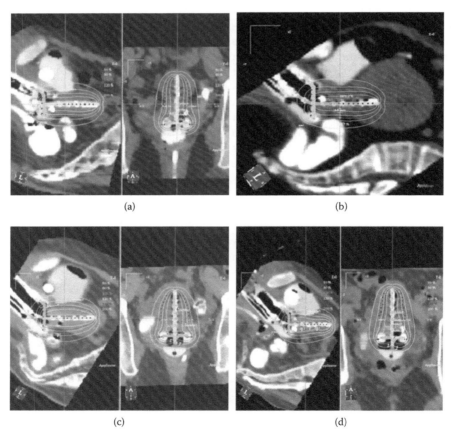

(a)　　　　　　　　　　(b)

(c)　　　　　　　　　　(d)

FIGURE 22.4　(**See color insert following page 204.**) (a) Sagittal and coronal CT plan with pear-shaped dose distribution and dose specification points; (b) Sagittal CT image revealing sigmoid above rectal retractor; (c) and (d) Sagittal and coronal CT scans in the same patient for two different fractions revealing change in position of organs at risk (OARs) relative to dose distribution.

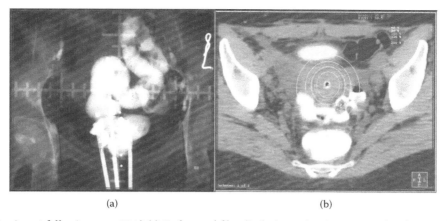

(a)   (b)

FIGURE 22.5   (See color insert following page 204.) (a) Orthogonal film displaying a circuitous sigmoid with point doses that are difficult to evaluate. (b) Axial CT relates the dose distribution to the adjacent sigmoid loop and allowed dose adjustment to prevent potential sigmoid toxicity.

fraction due to tumor shrinkage, the variable filling and position of the rectosigmoid and bladder, and the variable position and deformation of the uterus and vagina due to the applicator, packing, and adjacent normal organs (Shin et al. 2006).

22.7.2 Magnetic Resonance Imaging-Based Brachytherapy

With CT-based computerized dosimetry, rectosigmoid and bladder doses can be determined more accurately than with localization films. Even when using CT-compatible applicators, however, the boundaries between structures of interest are poorly defined. MR emerged in the 1980s as an alternative to CT. The value of MR in the imaging of gynecologic cancers lies in its multiplanar capability and superior soft-tissue resolution, compared to CT, enabling delineation of tumor within the cervix, uterus, and vagina as well as within the parametrial and vaginal tissues (Erickson 2003; Wachter-Gerstner et al. 2003a). Tumors of the cervix display moderately increased signals on T2-weighted images relative to normal cervical stroma, permitting definition of tumor volume. This is an advantage during brachytherapy as one can assess the proximity of the tumor to the applicator and the subsequent dose distribution throughout the tumor volume. The dose distribution can then be optimally conformed to the defined target volume, while accurately defining and limiting the dose to the adjacent normal OAR.

Tumor volume and dose response relationships have been increasingly important, but elusive, with CT because of the inability to define tumors and calculate the dose around or within them. The presence of dose gradients over the target volume, in combination with poor delineation of tumor boundaries on CT, has made dose specification to the tumor problematic and dose response data inconclusive.

A predictable evolution from CT to MR can occur. Such was the case at the Medical University of Vienna, where CT-based brachytherapy dominated from 1993 to 1998 and MR became available for treatment planning in 1998 (Pötter et al. 2000; Fellner et al. 2001; Wachter-Gerstner et al. 2003a; Wachter-

Gerstner et al. 2003b; Kirisits et al. 2005). From 1998 to 2002, an MR was done for only fraction 1 and if in subsequent applications, no substantial differences were seen with regard to implant geometry, GTV, or topography, the MR-based treatment used for fraction 1 was continued for all the subsequent insertions. In 2002, patient-specific MR-based treatment planning was systematically used for all four fractions. A comprehensive system was introduced that integrated all the dose/volume information available for the targets and OARs. The goal was to reach a prescribed dose of 85 Gy to the D90 with a V100 of at least 90%, while judiciously limiting the doses to the OARs. The GEC ESTRO guidelines formed the basis of this system (Kirisits et al. 2005).

22.7.3 GEC ESTRO Guidelines

The Gyn GEC ESTRO working group began to develop guidelines for recording and reporting 3D image-based treatment planning for cervix cancer brachytherapy in 2000. The guidelines were published in 2005, and they described a methodology using MR at the time of brachytherapy to define the GTV and CTV (Haie-Meder et al. 2005). The GTV at the time of brachytherapy was defined as residual tumor following external beam on clinical examination as well as the high signal regions on T2 FSE images in the cervix and paracervical tissues. High-risk CTV (HRCTV) included the GTV as well as the entire cervix and the extracervical tumor spread at the time of brachytherapy (Figure 22.6a). "High-risk volume" refers to tissues with a major risk of local recurrence because of residual macroscopic disease, which require a high dose of radiation, similar to that delivered traditionally to point A.

The intermediate risk CTV (IRCTV) was defined as encompassing the HRCTV with a margin of 5–15 mm, and refers to tissues carrying a significant microscopic tumor load. Doses of approximately 60 Gy are intended for this volume. With these different regions of risk defined according to physical examination and MR at the time of brachytherapy, dose–volume parameters were defined for the GTV, HRCTV, IRCTV, and OARs. For the rectum, contouring included the outer wall from the anorectal junction to the rectosigmoid flexure and the sigmoid contour continued alone until the sigmoid was approximately 2 cm from the uterus. The small bowel was contoured only if it was within 2 cm of the uterus. The outer contours of the bladder were also defined (Pötter et al. 2006; Kirisits et al. 2005). D100 and D90 as well as V100 were recommended for reporting as well as the minimum dose in the most irradiated tissue volume for 0.1, 1, and 2 cm³ of the OARs, contouring the outer walls only (Wachter-Gerstner et al. 2003b; Kirisits et al. 2005).

The radiobiological model equivalent dose (EQD2) is used to sum the external beam and HDR doses together over the course of treatment so that a cumulative biologically weighted dose is available. This allows for systematic evaluation of the doses to the targets and normal organs over the course of the treatment and for comparison between the centers. A number of international teams have adopted the GEC ESTRO guidelines

(Zwahlen et al. 2009; Lindegaard et al. 2008; Chargari et al. 2009; De Brabandere et al. 2008). As with CT, the sigmoid doses are often found to be higher than the rectal doses and recognition of this has prevented overdosage of the sigmoid in many series (Zwahlen et al. 2009; Lindegaard et al. 2008; De Brabandere et al. 2008). Lindegaard et al. (2008) found that it was much easier to meet the DVH parameters defined in the GEC ESTRO guidelines with MR-guided plans rather than library plans.

The goal of 3D-image-based brachytherapy is to define and improve target coverage while maximizing normal organ sparing. MR as an imaging tool improves soft-tissue resolution and has the capability of imaging in multiple planes in contrast to CT. The GEC ESTRO guidelines have been introduced and validated in Europe (Pötter et al. 2008). Dimopoulos et al. (2006) reviewed the pre- and postbrachytherapy insertion MR scans in 49 patients and used a scoring system to define the applicator,

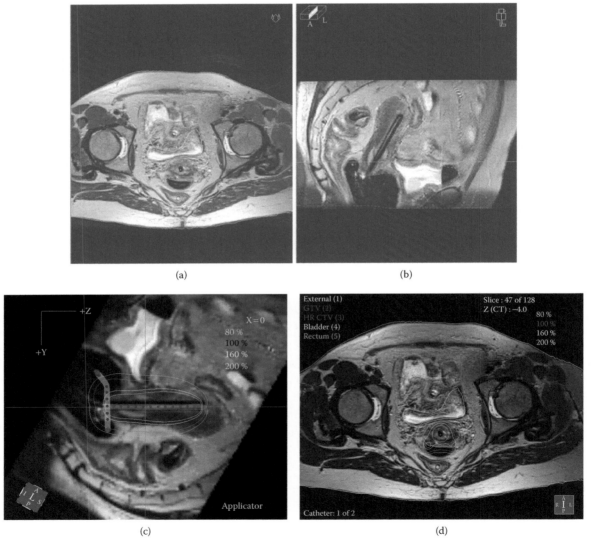

(a)　　　　　　　　　　　　　　(b)

(c)　　　　　　　　　　　　　　(d)

FIGURE 22.6　(See color insert following page 204.) (a) Axial and (b) sagittal MRI scans revealing contoured GEC-ESTRO GTV and high-risk clinical target volume (HRCTV). Isodose distribution displayed on this same patient in (c) sagittal and (d) axial planes.

targets, and OAR on these images. MR imaging was able to provide appropriate information for accurate definition of the applicator, GTV, CTV, and OAR consistently in the patients studied.

Inter- and intraobserver variation in contouring the HRCTV has also been a concern relative to reproducibility and consistency in following the guidelines. Lang et al. (2006) compared the contours and dosimetry from the three institutions in Europe that had developed and followed the GEC ESTRO guidelines and found reassuring similarities in the defined targets, tumor, and normal organ doses (Pötter et al. 2008). If the targets are not contoured appropriately, there is the potential to shape the dose distribution in accordance with these contours, while in actuality missing or erroneously expanding the target. The prescription isodose can be expanded by 5 mm with intracavitary techniques, but the addition of needles is necessary for further expansion up to 15 mm from point A (Pötter et al. 2008).

Petric et al. (2008) found that there was the least difference between the HRCTV contours of 13 patients by two study participants when contouring was done in the para-transverse plane, perpendicular to the cervical axis, due to the ability to have a circumferential view of the cervix. Unfortunately, most treatment planning systems will not directly plan on any plane other than the transverse plane. In addition, training and practice in contouring will lead to more consistency between participants. Such skills require a fundamental knowledge of MR pelvic anatomy (Dimopoulos, Lang et al. 2009; Pötter et al. 2008). Dimopoulos, Lang, et al. (2009) analyzed contouring disparities and found multiple reasons for lack of complete conformity, including differences in window and level settings, distinction between the GTV vs. fluid in the endocervical canal, incorporation of the clinical drawings and physical exam into the contour delineation, loss of the cervicouterine junction with advancing age, and use of unedited automargins for defining the IR CTV (Pötter et al. 2008).

Other potential pitfalls of MR-based dosimetry can include applicator reconstruction errors potentially compounded by MR-associated image distortion (Tanderup et al. 2008; Haack et al. 2009). The use of marker catheters in the applicators used to identify the first dwell positions in the tandem and vaginal applicators can lead to more accurate applicator reconstruction, as can the use of registration between radiographs and CT scans with the MR scans (Pötter et al. 2008). These MR markers can include saline, copper sulfate, gadolinium, glycerin, and oil. Gadolinium-soaked vaginal packing can help define the cervical tumor and any vaginal extensions of the tumor (Figure 22.7a).

The GEC ESTRO guidelines have now been adopted nationally and internationally with data rapidly accumulating. In 2007, Pötter et al. (2007, 2008) published a series on patients treated from 2001 to 2003 who had undergone systematic MR-based planning according to the GEC ESTRO guidelines, and compared them to those treated from 1998 to 2001, when CT-based planning was used. The MR-based approach led to an improvement in the outcome for patients with tumors >5 cm and a reduction in grade 3 and 4 morbidity. The GEC ESTRO DVH parameters were correlated with local control in one recent review.

Dimopoulos, DeVos, et al. (2009) investigated the impact of D90 and D100 for the HRCTV on local control, and if the D90 for the HRCTV was ≥87 Gy EQD2, pelvic control rates of >95% were achieved. There seems to be agreement among the international investigators using MR for image-based brachytherapy that this approach improves normal-tissue sparing and target coverage (Zwahlen et al. 2009; Lindegaard et al. 2008). Dose–volume constraints for the rectum, sigmoid, and bladder have yet to be validated (Pötter et al. 2008).

As experience has been gained in delineating volumes on MR scans, nuances to the techniques are becoming apparent. Certainly MR magnet strength may alter the image quality and change the ease of delineating the GTV and the HRCTV. Slice thickness as well as slice orientation can also vary. Pixelation of the images will vary by technique. There is very little published regarding the techniques used by individual institutions. As MR scanners have moved from diagnostic radiology to radiation oncology departments, more sequences and technical issues need to be addressed apart from the expertise of the diagnostic radiologists. The initial work in Vienna was done with a low strength magnet (0.2 T) and these images are now able to be compared to the newer higher strength magnets.

As initially done in Vienna (Kirisits et al. 2005), at the Medical College of Wisconsin, beginning in 2004, an MRI scan was obtained after the fraction 1 CT scan, in diagnostic radiology, on a 1.5-T magnet. A phased-array pelvic surface coil was strapped to the patient resulting in an increased signal-to-noise ratio that permitted acquisition of images with improved spatial resolution. To reduce time and cost, a streamlined MR protocol was used with no gadolinium and only T2 images were obtained in the axial, sagittal, and coronal planes with a 24–28-cm field of view, 5-mm slice thickness with a 1-mm gap, 16-kHz bandwidth, 256–521 × 256 matrix and NEX of 2. An echo train length of 8 was used for the FSE T2 sequences. The bladder catheter was left on the patient's legs during the scan with the 30 cm^3 of instilled contrast from the CT simulation. The patient remained sedated and monitored during the entire scan. Bowel motion was limited by having patients fast for 4–6 hours prior to the study and if there were no medical contraindications, an antiperistaltic agent (glucagon 1 mg IV or IM) was administered just before the examination.

In 2008, a 3-T MRI scanner was installed in the Department of Radiation Oncology, across from the CT simulator. Patients are now transferred from the insertion table directly to the MR cart, which then slides into the MR to become the MR table. Hollow plastic catheters (Best Medical part 536-30 and 536-35-inner catheter with filament) filled with saline are inserted into the tandem and ring and a phased-array pelvic surface coil is strapped to the patient. Images are acquired on a Siemens 3-T MAGNETOM Verio scanner (Siemens Healthcare, Erlangen, Germany) equipped with 45 mT/m (180 mT/m/ms) gradients. The body RF coil is used for signal transmission and commercial flex and spine RF array coils are used for signal reception. A scout series in the plane of the tandem and the ring is performed first to assess the position of the applicator relative to the cervix and vagina (Figure 22.7b). Nonfat-suppressed, axial, T2-weighted images are acquired using a 3D slab-selective SPACE (sampling perfection with application optimized contrasts using different flip angle evolutions) pulse sequence with the following parameters: field-of-view, 384 mm; matrix, 384 × 384; turbo factor, 93; bandwidth, 407 Hz/pixel; apparent echo time, 92 milliseconds; echo train duration, 422 milliseconds; repetition time, 2500 milliseconds; averages, 1.4; and slice thickness, 1 mm (Figure 22.6). These parameters are optimized to produce images with isotropic 1-mm cubic voxels, good signal-to-noise ratio and T2 contrast, and minimal respiratory artifacts. To reduce total acquisition time, partial Fourier encoding is used in the slice direction and parallel imaging is used in the phase encode direction. Acquisition times range from 12 to 16 minutes, depending on the required slice coverage. An intravenous injection of glucagon (1 mg GlucaGen, Novo Nordisk, Princeton, NJ) is administered to reduce effects of bowel motion that can occur during acquisition. After imaging is completed, the patient is carefully transferred on the MR slider board to a cart and returned to the brachytherapy suite for monitoring. Once the treatment fraction is delivered, the patient has the applicators removed while sedated and is then transferred to the recovery room for monitoring until fully stable and ready for discharge.

There are inherent challenges to obtaining an MR for every brachytherapy fraction, including cost, time, and treatment prolongation. There have been claims that the effort is not worth the

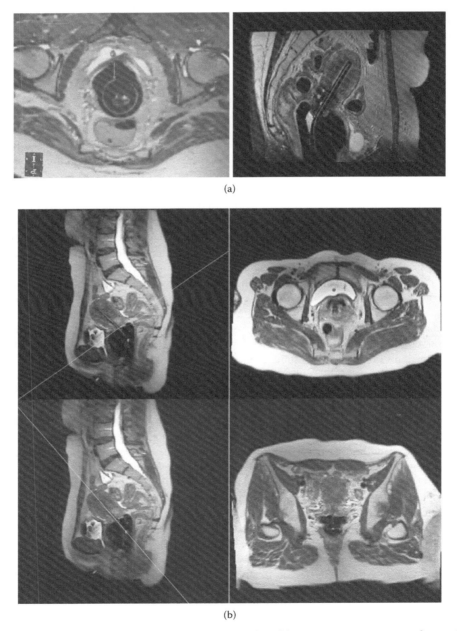

(a)

(b)

FIGURE 22.7   (a) Saline-filled catheter marks dwell position in ring and tandem. (b) 3-T MR Scout series to confirm applicator placement.

conformity achieved. To help answer the question of whether an MR should be done just for fraction 1 or for each fraction, Kirisits et al. (2006) compared the use of one treatment plan only vs. individualized treatment plans for each fraction in 14 patients. He concluded that the use of only one treatment plan for several applications resulted in higher doses to the target, presumably due to tumor shrinkage and the OAR. The sigmoid dose was increased relative to the bladder and the rectal doses. Use of alternative imaging modalities to complement the initial plan, such as CT, could help reduce the inherent uncertainties of the one plan approach. This approach of use of MR for fraction 1 and CT for the remaining insertions may be an economical solution for some individuals, and is currently allowed on the EMBRACE (a European study on MRI-guided brachytherapy in locally advanced cervical cancer) trial. CT is excellent for identifying the OAR. Chi et al. (2009) also found that customized planning using CT-based dosimetry for each HDR fraction, rather than using a single plan to quantify normal organ doses for only one fraction, led to a reduction in normal tissue doses. The other alternative would be to do one implant for more than one fraction, such as in Vienna, where patients have one applicator insertion for delivery of two fractions.

22.8 Future Developments

More data is needed on the use of IGRT to validate these complex technologies. Daily image-guided external beam can lead to fewer errors in treatment delivery and perhaps a decrease in field volume due to this inherent reliability. Results of IMRT for treatment of the postoperative pelvis seem to indicate a reduction in morbidity, and perhaps a worthy solution in patients with abundant pelvic small bowel or in those receiving chemotherapy in an attempt to spare bone marrow. Use of IMRT for the intact pelvis should be done with caution due to motion considerations. It is unclear if this will spare normal tissues due to the large margins needed to account for this motion.

Image-guided brachytherapy has enabled visualization and sparing of the OAR, and perhaps a methodology to give dose where it is most needed. Further follow-up will be needed to determine if local control can be improved. Future use of functional imaging such as MRS or PET may allow further dose escalation. Imaging what one needs to treat and not to treat is an important first step in adapting the radiation to the anatomy and improving the therapeutic ratio.

References

Adli, M., N. A. Mayr, H. S. Kaiser, et al. 2003. Does prone positioning reduce small bowel dose in pelvic radiation with intensity-modulated radiotherapy for gynecologic cancer? *Int J Radiat Oncol Biol Phys* 57:230–8.

Ahamad, A., W. D'Souza, M. Salehpour, et al. 2005. Intensity-modulated radiation therapy after hysterectomy: Comparison with conventional treatment and sensitivity of the normal-tissue-sparing effect to margin size. *Int J Radiat Oncol Biol Phys* 62:1117–24.

Ahmad, R., M. S. Hoogeman, S. Quint, et al. 2008. Inter-fraction bladder filling variations and time trends for cervical cancer patients assessed with a portable 3-dimensional ultrasound bladder scanner. *Radiother Oncol* 89:172–9.

Ahmed, R. S., R. Y. Kim, J. Duan, et al. 2004. IMRT dose escalation for positive para-aortic lymph nodes in patients with locally advanced cervical cancer while reducing dose to bone marrow and other organs at risk. *Int J Radiat Oncol Biol Phys* 60:505–12.

ASTEC/EN.5 writing committee on behalf of the ASTEC/EN.5 Study Group. 2009. Adjuvant external beam radiotherapy in the treatment of endometrial cancer: Pooled trial results, systemic review and meta-analysis. *Lancet* 373:137–46.

Beadle, B. M., A. Jhingran, M. Salehpour, et al. 2009. Cervix regression and motion during the course of external beam chemoradiation for cervical cancer. *Int J Radiat Oncol Biol Phys* 73:235–41.

Bentel, G. C. 1992. Treatment planning pelvis. In *Radiation Therapy Planning*. 2nd ed. 439–489. New York: McGraw-Hill.

Beriwal, S., D. E. Heron, H. Kim, et al. 2006. Intensity-modulated radiotherapy for the treatment of vulvar carcinoma: A comparative dosimetric study with early clinical outcome. *Int J Radiat Oncol Biol Phys* 64:1395–400.

Bonin, S. R., R. M. Lanciano, B. W. Corn, et al. 1996. Bony landmarks are not an adequate substitute for lymphangiography in defining pelvic lymph node location for the treatment of cervical cancer with radiotherapy. *Int J Radiat Oncol Biol Phys* 34:167–72.

Brixey, C. J., J. C. Roeske, A. E. Lujan, et al. 2002. Impact of intensity-modulated radiotherapy on acute hematologic toxicity in women with gynecologic malignancies. *Int J Radiat Oncol Biol Phys* 54:1388–96.

Buchali, A., S. Koswig, S. Dinges, et al. 1999. Impact of the filling status of the bladder and rectum on their integral dose distribution and the movement of the uterus in the treatment planning of gynaecological cancer. *Radiother Oncol* 52:29–34.

Chan, P., R. Dinniwell, M. A. Haider, et al. 2008. Inter- and intrafractional tumor and organ movement in patients with cervical cancer undergoing radiotherapy: A cinematic-MRI point-of-interest study. *Int J Radiat Oncol Biol Phys* 70:1507–15.

Chao, K. S., and M. Lin. 2002. Lymphangiogram-assisted lymph node target delineation for patients with gynecologic malignancies. *Int J Radiat Oncol Biol Phys* 54:1147–52.

Chargari, C., N. Magne, I. Dumas, et al. 2009. Physics contributions and clinical outcome with 3D-MRI-based pulsed-dose-rate intracavitary brachytherapy in cervical cancer patients. *Int J Radiat Oncol Biol Phys* 74:133–9.

Chi, A., M. Gao, J. Sinacore, et al. 2009. Single versus customized treatment planning for image-guided high-dose-rate brachytherapy for cervical cancer: Dosimetric comparison and predicting factor for organs at risk overdose with single plan approach. *Int J Radiat Oncol Biol Phys* 75:309–14.

Creutzberg, C. L., W. L. Van Putten, P. C. Koper, et al. 2000. Surgery and postoperative radiotherapy versus surgery alone for patients with stage 1 endometrial carcinoma: Multicenter randomized trial. PORTEC Study Group. *Lancet* 355:1404–11.

Datta, N. R., A. Srivastava, K. J. M. Das, et al. 2006. Comparative assessment of doses to tumor, rectum, and bladder as evaluated by orthogonal radiographs vs. computer enhanced computed tomography-based intracavitary brachytherapy in cervical cancer. *Brachytherapy* 5:223–29.

De Brabandere, M., A. G. Mousa, A. Nulens, et al. 2008. Potential of dose optimisation in MRI-based PDR brachytherapy of cervix carcinoma. *Radiother Oncol* 88:217–26.

Dimopoulos, J. C. A., V. DeVos, D. Berger, et al. 2009. Interobserver comparison of target delineation for MRI-assisted cervical cancer brachytherapy: Application of the GYN GEC-ESTRO recommendations. *Radiother Oncol* 91:166–72.

Dimopoulos, J., S. Lang, C. Kirisits, et al. 2009. Dose-volume histogram parameters and local tumor control in magnetic resonance image-guided cervical cancer brachytherapy. *Int J Radiat Oncol Biol Phys* 75:56–63.

Dimopoulos, J., G. Schard, D. Berger, et al. 2006. Systematic evaluation of MRI findings in different stages of treatment of cervical cancer: Potential of MRI on delineation of target, pathoanatomic structures, and organs at risk. *Int J Radiat Oncol Biol Phys* 64:1380–8.

Dinniwell, R., P. Chan, G. Czarnota, et al. 2009. Pelvic lymph node topography for radiotherapy treatment planning from ferumoxtran-10 contrast-enhanced magnetic resonance imaging. *Int J Radiat Oncol Biol Phys* 74:844–51.

Ellis, R. E. 1961. The distribution of active bone marrow in the adult. *Phys Med Biol* 5:255–63.

Erickson, B. 2000. Image guidance: CT guidance assists brachytherapy for gynecologic disease. *Diagn Imaging* 22:167–91.

Erickson, B. A. 2003. Presidential address. The sculpted pear: An unfinished brachytherapy tale. *Brachytherapy* 2:189–99.

Fellner, C., R. Pötter, T. H. Knocke, et al. 2001. Comparison of radiography- and computed tomography-based treatment planning in cervix cancer in brachytherapy with specific attention to some quality assurance aspects. *Radiother Oncol* 58:53–62.

Finlay, M. H., I. Ackerman, R. G. Tirona, et al. 2006. Use of CT simulation for treatment of cervical cancer to assess the adequacy of lymph node coverage of conventional pelvic fields based on bony landmarks. *Int J Radiat Oncol Biol Phys* 64:205–9.

Fiorino, C., R. Valdagni, T. Rancati, et al. 2009. Dose-volume effects for normal tissues in external radiotherapy: Pelvis. *Radiother Oncol* 93:153–67.

Gallagher, M. J., H. D. Brereton, R. A. Rostock, et al. 1986. A prospective study of treatment techniques to minimize the volume of pelvic small bowel with reduction of acute and late effects associated with pelvic irradiation. *Int J Radiat Oncol Biol Phys* 12:1565.

Georg, P., D. Georg, M. Hillbrand, et al. 2006. Factors influencing bowel sparing in intensity modulated whole pelvic radiotherapy for gynaecological malignancies. *Radiother Oncol* 80:19–26.

Georg, P., C. Kirisits, G. Goldner, et al. 2009. Correlation of dose-volume parameters, endoscopic and clinical rectal side effects in cervix cancer patients treated with definitive radiotherapy including MRI-based brachytherapy. *Radiother Oncol* 91:173–80.

Gerszten, K., K. Colonello, D. E. Heron, et al. 2006. Feasibility of concurrent cisplatin and extended field radiation therapy (EFRT) using intensity-modulated radiotherapy (IMRT) for carcinoma of the cervix. *Gynecol Oncol* 102:182–8.

Greer, B. E., W. J. Koh, D. C. Figge, et al. 1990. Gynecologic radiotherapy fields defined by intraoperative measurements. *Gynecol Oncol* 38:421–4.

Grigsby, P. W., J. D. Lu, D. G. Mutch, et al. 1998. Twice-daily fractionation of external irradiation with brachytherapy and chemotherapy in carcinoma of the cervix with positive para-aortic lymph nodes: Phase II study of the Radiation Therapy Oncology Group 92-10. *Int J Radiat Oncol Biol Phys* 41:817–22.

Haack, S., S. K. Nielsen, J. C. Lindegaard, et al. 2009. Applicator reconstruction in MRI 3D image-based dose planning of brachytherapy for cervical cancer. *Radiother Oncol* 91:187–93.

Haie-Meder, C., R. Pötter, E. Van Limbergen, et al. 2005. Recommendations from Gynaecological (GYN) GEC-ESTRO Working Group (I): Concepts and terms in 3D image based 3D treatment planning in cervix cancer brachytherapy with emphasis on MRI assessment of GTV and CTV. *Radiother Oncol* 74:235–45.

Huang, E. Y., C. C. Sung, S. F. Ko, et al. 2007. The different volume effects of small-bowel toxicity during pelvic irradiation between gynecologic patients with and without abdominal surgery: A prospective study with computed tomography-based dosimetry. *Int J Radiat Oncol Biol Phys* 69:732–9.

Huh, S. J., W. Park, Y. Han, et al. 2004. Interfractional variation in position of the uterus during radical radiotherapy for cervical cancer. *Radiother Oncol* 71:73–9.

Jereczek-Fossa, B., J. Jassem, R. Nowak, et al. 1998. Late complications after postoperative radiotherapy in endometrial cancer: Analysis of 317 consecutive cases with application of linear-quadratic model. *Int J Radiat Oncol Biol Phys* 41:329–38.

Jhingran, A., L. Portelance, B. Miller, et al. A phase II study of intensity modulated radiation therapy (IMRT) to the pelvis +/– chemotherapy for post-operative patients with either endometrial or cervical carcinoma. RTOG 0418.

Jhingran, A., K. Winter, B. E. Portelance, et. al. 2008. A phase II study of intensity modulated radiation therapy (IMRT) to the pelvis for post-operative patients with either endometrial or cervical carcinoma (RTOG 0418). ASTRO Abstract. *Int J Radiat Oncol Biol Phys* 72(1):S16–7.

Kaatee, R. S., M. J. Olofsen, M. D. Verstraate, et al. 2002. Detection of organ movement in cervix cancer patients using a fluoroscopic electronic portal imaging device and radiopaque markers. *Int J Radiat Oncol Biol Phys* 54:576–83.

Kerkhof, E. M., B. W. Raaymakers, U. A. van der Heide, et al. 2008. Online MRI guidance for healthy tissue sparing in patients with cervical cancer: An IMRT planning study. *Radiother Oncol* 88:241–9.

Kerkhof, E. M., R. W. van der Put, B. W. Raaymakers, et al. 2009. Intrafraction motion in patients with cervical cancer: The benefit of soft tissue registration using MRI. *Radiother Oncol* 93:115–21.

Keys, H. M., J. A. Roberts, V. L. Brunetto, et al. 2004. A phase III trial of surgery with or without adjunctive external pelvic radiation therapy in intermediate risk endometrial adenocarcinoma: A Gynecologic Oncology Group study. *Gynecol Oncol* 92:744–51.

Kim, R. Y., J. F. Caranto, P. N. Pareek, et al. 1997. Dynamics of pear-shaped dimensions and volume of intracavitary brachytherapy in cancer of the cervix: A desirable pear shape in the era of three-dimensional treatment planning. *Int J Radiat Oncol Biol Phys* 37:1193–9.

Kirisits, C., S. Lang, J. Dimopoulos, et al. 2006. Uncertainties when using only one MRI-based treatment plan for subsequent high-dose-rate tandem and ring applications in brachytherapy of cervix cancer. *Radiother Oncol* 81:269–75.

Kirisits, C., R. Pötter, S. Lang, et al. 2005. Dose and volume parameters for MRI-based treatment planning in intracavitary brachytherapy for cervical cancer. *Int J Radiat Oncol Biol Phys* 62:901–11.

Lang, S., A. Nulens, E. Briot, et al. 2006. Intercomparison of treatment concepts for MR image assisted brachytherapy of cervical carcinoma based on GYN GEC-ESTRO recommendations. *Radiother Oncol* 78:185–93.

Letschert, J. G., J. V. Lebesque, B. M. Aleman, et al. 1994. The volume effect in radiation-related late small bowel complications: Results of a clinical study of the EORTC Radiotherapy Cooperative Group in patients treated for rectal carcinoma. *Radiother Oncol* 32:116–23.

Lim, K. S. H., V. Kelly, J. Stewart, et al. 2008. Whole pelvis IMRT for cervix cancer: What gets missed and why? *Int J Radiat Oncol Biol Phys* 72:S112.

Lim, K., V. Kelly, J. Stewart, et al. 2009. Pelvic IMRT for cervix cancer: Is what you plan actually what you deliver? *Int J Radiat Oncol Biol Phys* 74(1):304–12.

Lim, K., W. Small Jr., L. Portelance, et al. 2009. Consensus guidelines for delineation of clinical target volume for intensity-modulated radiotherapy for the definitive treatment of cervix cancer. *Int J Radiat Oncol Biol Phys*. In press.

Lindegaard, J., K. Tanderup, S. K. Nielsen, et al. 2008. MRI-guided 3D optimization significantly improves DVH parameters of pulsed-dose-rate brachytherapy in locally advanced cervical cancer. *Int J Radiat Oncol Biol Phys* 71:756–64.

Lujan, A. E., A. J. Mundt, S. D. Yamada, et al. 2003. Intensity-modulated radiotherapy as a means of reducing dose to bone marrow in gynecologic patients receiving whole pelvic radiotherapy. *Int J Radiat Oncol Biol Phys* 57:516–21.

Lujan, A. E., J. C. Roeske, and A. J. Mundt. 2001. Intensity-modulated radiation therapy as a means of reducing dose to bone marrow in gynecologic patients receiving whole pelvis radiation therapy [Abstract]. *Int J Radiat Oncol Biol Phys* 51:220.

Martin, J., K. Fitzpatrick, G. Horan, et al. 2005. Treatment with a belly-board device significantly reduces the volume of small bowel irradiated and results in low acute toxicity in adjuvant radiotherapy for gynecologic cancer: Results of a prospective study. *Radiother Oncol* 74:267–74.

Mayr, N. A., J. Z. Wang, S. S. Lo, et al. 2010. Translating response during therapy into ultimate treatment outcome: A personalized 4-dimensional MRI tumor volumetric regression approach in cervical cancer. *Int J Radiat Oncol Biol Phys* 76(3):719–27.

Mayr, N. A., W. T. Yuh, T. Taoka, et al. 2006. Serial therapy-induced changes in tumor shape in cervical cancer and their impact on assessing tumor volume and treatment response. *Am J Roentgenol* 187:65–72.

Meigs, J. V., and R. Dresser. 1937. Carcinoma of the cervix treated by the roentgen ray and radium. *Ann Surg* 106:653–67.

Morris, M., P. J. Eifel, J. Lu, et al. 1999. Pelvic radiation with concurrent chemotherapy compared with pelvic and para-aortic radiation for high-risk cervical cancer. *N Engl J Med* 340:1137–43.

Mundt, A., L. Mell, J. C. Roeske, et al. 2003. Preliminary analysis of chronic gastrointestinal toxicity in gynecology patients treated with intensity-modulated whole pelvic radiation therapy. *Int J Radiat Oncol Biol Phys* 56:1354–60.

Mundt, A. J., J. C. Roeske, A. E. Lujan, et al. 2001. Initial clinical experience with intensity-modulated whole-pelvis radiation therapy in women with gynecologic malignancies. *Gynecol Oncol* 82:456–63.

Mundt, A. J., J. C. Roeske, A. E. Lujan, et al 2002. Intensity-modulated radiation therapy in gynecologic malignancies. *Med Dosim* 27:131–6.

Nuyttens, J. J., J. M. Robertson, D. Yan, et. al. 2002. The variability of clinical target volume for rectal cancer due to internal organ motion during adjuvant treatment. *Int J Radiat Oncol Biol Phys* 53:497–503.

Perez, C. A. 1997. Uterine Cervix. In *Principles and Practice of Radiation Oncology*. 3rd ed. ed. C. A. Perez and L. W. Brady, 1733–834. Philadelphia: Lippincott-Raven.

Petric, P., J. Dimopoulos, C. Kirisits, et al. 2008. Inter- and intraobserver variation in HR-CTV contouring: Intercomparison of transverse and paratransverse image orientation in 3D-MRI assisted cervix cancer brachytherapy. *Radiother Oncol* 89:164–71.

Pinkawa, M., B. Gagel, C. Demirel, et al. 2003. Dose-volume histogram evaluation of prone and supine patient position in external beam radiotherapy for cervical and endometrial cancer. *Radiother Oncol* 69:99–105.

Portelance, L., K. S. Chao, P. W. Grigsby, et al. 2001. Intensity-modulated radiation therapy (IMRT) reduces small bowel, rectum, and bladder doses in patients with cervical cancer receiving pelvic and para-aortic irradiation. *Int J Radiat Oncol Biol Phys* 51:261–6.

Pötter, R., J. Dimopoulos, P. Georg, et al. 2007. Clinical impact of MRI assisted dose–volume adaptation and dose escalation in brachytherapy of locally advanced cervix cancer. *Radiother Oncol* 83:148–55.

Pötter, R., C. Haie-Meder, E. Van Limbergen, et al. 2006. Recommendations from gynaecological (GYN) GEC ESTRO working group (II): Concepts and terms in 3D image-based treatment planning in cervix cancer brachytherapy-3D dose–volume parameters and aspects of 3D image-based anatomy, radiation physics, radiobiology. *Radiother Oncol* 78:67–77.

Pötter, R., C. Kirisits, E. Fidarova, et al. 2008. Present status and future of high-precision image guided adaptive brachytherapy for cervix carcinoma. *Acta Oncol* 47:1325–36.

Pötter, R., T. H. Knocke, C. Fellner, et al. 2000. Definitive radiotherapy based on HDR brachytherapy with iridium 192 in uterine cervix carcinoma: Report on the Vienna University Hospital findings (1993–1997) compared to the preceding period in the context of ICRU 38 recommendations. *Cancer/Radiother* 4:159–72.

Querleu, D., and C. P. Morrow. 2008. Classification of radical hysterectomy. *Lancet Oncol* 9:297–303.

Randall, M. D., and G. S. Ibott. 2006. Intensity-modulated radiation therapy for gynecologic cancers: Pitfalls, hazards, and cautions to be considered. *Semin Radiat Oncol* 16:138–43.

Roeske, J. C., D. Bonta, L. K. Mell, et al. 2003. A dosimetric analysis of acute gastrointestinal toxicity in women receiving intensity-modulated whole-pelvic radiation therapy. *Radiother Oncol* 69:201–7.

Roeske, J. C., A. Lujan, J. Rotmensch, et al. 2000. Intensity-modulated whole pelvic radiation therapy in patients with gynecologic malignancies. *Int J Radiat Oncol Biol Phys* 48:1613–21.

Rusthoven, C., K. Latifi, K. Javedan, et al. 2009. Assessment of organ motion in postoperative endometrial and cervical cancer patients treated with intensity modulated radiation therapy. *Int J Radiat Oncol Biol Phys* 75(3 Suppl):Abstract 2402.

Sedlis, A., B. N. Bundy, M. Z. Rotman, et al. 1999. A randomized trial of pelvic radiation therapy versus no further therapy in selected patients with stage IB carcinoma of the cervix after radical hysterectomy and pelvic lymphadenectomy: A Gynecologic Oncology Group study. *Gynecol Oncol* 73:177–83.

Shin, L. H., T. H. Kim, J. K. Cho, et al. 2006. CT-guided intracavitary radiotherapy for cervical cancer: Comparison of conventional point a plan with clinical target volume-based three-dimensional plan using dose-volume parameters. *Int J Radiat Oncol Biol Phys* 64:197–204.

Small Jr, W., L. K. Mell, P. Anderson, et al. 2008. Consensus guidelines for delineation of clinical target volume for intensity-modulated pelvic radiotherapy in postoperative treatment of endometrial and cervical cancer. *Int J Radiat Oncol Biol Phys* 71:428–34.

Stewart, J. K., S. H. Lim, et al. 2010. Automated weekly replanning for intensity-modulated radiotherapy of cervix cancer. *Int J Radiat Oncol Biol Phys* 78(2):350–8.

Tanderup, K., T. P. Hellebust, S. Lang, et al. 2008. Consequences of random and systematic reconstruction uncertainties in 3D image based brachytherapy in cervical cancer. *Radiother Oncol* 89:156–63.

Taylor, A., and M. E. Powell. 2008. An assessment of interfractional uterine and cervical motion: Implications for radiotherapy target volume definition in gynaecological cancer. *Radiother Oncol* 88:250–7.

Taylor, A., A. G. Rockall, R. H. Reznek, et al. 2005. Mapping pelvic lymph nodes: Guidelines for delineation in intensity-modulated radiotherapy. *Int J Radiat Oncol Biol Phys* 63:1604–12.

van de Bunt, L., I. M. Jurgenliemk-Schulz, G. A. de Kort, et al. 2008. Motion and deformation of the target volumes during IMRT for cervical cancer: What margins do we need? *Radiother Oncol* 88:233–40.

van de Bunt, L., U. A. van der Heide, M. Ketelaars, et al. 2006. Conventional, conformal, and intensity-modulated radiation therapy treatment planning of external beam radiotherapy for cervical cancer: The impact of tumor regression. *Int J Radiat Oncol Biol Phys* 64:189–96.

Wachter-Gerstner, N., S. Wachter, E. Reinstadler, et al. 2003a. Bladder and rectum dose defined from MRI based treatment planning for cervix cancer brachytherapy: Comparison of dose-volume histograms for organ contours and organ wall, comparison with ICRU rectum and bladder reference point. *Radiother Oncol* 68:269–76.

Wachter-Gerstner, N., S. Wachter, E. Reinstadler, et al. 2003b. The impact of sectional imaging on dose excalation in endocavitary HDR-brachytherapy of cervical cancer: Results of a prospective comparative trial. *Radiother Oncol* 68:51–9.

Weiss, E., P. Hirnle, H. Arnold-Bofinger, et al. 1999. Therapeutic outcome and relation of acute and late side effects in the adjuvant radiotherapy of endometrial carcinoma stage I and II. *Radiother Oncol* 53:37–44

Weiss, E., S. Richter, C. Hess, et al. 2003. Radiation therapy of the pelvic and paraaortic lymph nodes in cervical carcinoma: A prospective three-dimensional analysis of patient positioning and treatment technique. *Radiother Oncol* 68:41–9.

Weiss, E., K. Wijesooriya, S. V. Dill, et al. 2007. Tumor and normal motion in the thorax during respiration: Analysis of volumetric and positional variations using 4D CT. *Int J Radiat Oncol Biol Phys* 67:296–307.

Wertheim, E. 1912. The extended abdominal operation for carcinoma uteri (based on 500 operative cases). *Am J Obstet Dis Women Children* 66:169–232.

Zwahlen, D., J. Jezioranski, P. Chan, et al. 2009. Magnetic resonance imaging-guided intracavitary brachytherapy for cancer of the cervix. *Int J Radiat Oncol Biol Phys* 74:1157–64.

23

Adaptive Radiotherapy for Treatment of Soft-Tissue Sarcoma

Julie A. Bradley
Medical College of Wisconsin

Kristofer Kainz
Medical College of Wisconsin

X. Allen Li
Medical College of Wisconsin

Thomas F. DeLaney
Harvard Medical School

Dian Wang
Medical College of Wisconsin

23.1 Introduction

Soft-tissue sarcomas originate in the connective tissue of organs throughout the human body. Many histologies comprise the category of soft-tissue sarcoma. The natural history of the disease unites the various histology categories and anatomic locations. Soft-tissue sarcomas are commonly present as enlarging, painless masses. They most commonly develop in the extremities, followed by the trunk, retroperitoneum, and head and neck (O'Sullivan et al. 2007). Symptoms that develop are related to the anatomic site in which the sarcoma arises. Soft-tissue sarcomas develop locally within the confines of fascial planes and other natural anatomic barriers like bone. A pseudocapsule of normal tissue and reactive fibrosis appears to envelop the tumor. However, satellite tumor cells can spread beyond this false barrier (Enneking, Spanier, and Malawer 1981). This scattered extension around the primary tumor mandates that the extent of radiation and surgical margins encompass these areas of potential subclinical disease.

Soft-tissue sarcoma is an uncommon malignancy, affecting approximately 10,660 Americans annually (Jemal et al. 2009). The median age at diagnosis is 50–55 years. The workup includes a complete history including occupational history, family history, and personal history of previous radiation therapy and physical examination with attention to the tumor location, size, depth, and neurological function. Genetic syndromes, including Li Fraumeni, hereditary retinoblastoma, and neurofibromatosis type 1, carry an increased risk of certain types of soft-tissue sarcomas (Wong et al. 1997; Strong et al. 1992). Imaging includes magnetic resonance imaging (MRI) and computed tomography (CT) scan with contrast of the involved area as well as a chest X-ray or CT of the chest (Demas et al. 1998; Cormier and Pollack 2004). Currently, positron emission tomography (PET) does not have a standard role in evaluation of soft-tissue sarcomas

(Bastiaannet et al. 2004). An incisional biopsy or core needle biopsy is recommended for tissue diagnosis (Hoeber et al. 2001; Heslin et al. 1997). The approach to the biopsy incision is extremely important, as a biopsy that breaches an uninvolved compartment or results in a sizeable hematoma can alter patterns of tumor spread and require an increase in the radiation and/or surgical field (Anderson et al. 1999). Staging is based on the grade, tumor size, and superficial vs. deep location in addition to the presence of nodal or distant metastases. The French Federation of Cancer Centers Sarcoma Group (Trojani et al. 1984) and U.S. National Cancer Institutes (NCI) systems (Costa et al. 1984) are the most well-known grading systems. These systems use slightly different criteria to determine the tumor grade. The NCI system assigns grades 1–4 based on the histological type, cellularity, nuclear pleomorphism, degree of necrosis, and mitotic rate; the French system designates grades 1–3 based on differentiation, degree of necrosis, and mitotic rate. The International Union against Cancer (UICC) and American Joint Committee on Cancer (AJCC) utilize a four-tiered system. A two-tiered, high vs. low grade is also widely used. For the three-tiered systems, a low grade is equivalent to grade 1 and a high grade is equivalent to grades 2 and 3. Grades 1 and 2 are low grade and grades 3 and 4 are high grade in the four-tiered system (O'Sullivan et al. 2007). Staging of sarcomas is unique in that grade is incorporated in the assignment of stage I, II, or III disease.

Histologies include undifferentiated pleomorphic sarcoma, liposarcoma, leiomyosarcoma, myxofibrosarcoma, synovial sarcoma, peripheral nerve sheath tumors, rhabdomyosarcoma, and clear cell sarcoma (Stoeckle et al. 2001). Dermatofibrosarcoma protuberans, gastrointestinal stromal tumors, and angiosarcomas are not included in the UICC/AJCC staging system. Approximately 30% of sarcomas develop due to abnormalities in fusion genes (O'Sullivan et al. 2007). Synovial sarcomas can display a t(X;18)(p11;q11) translocation involving the *SSX1* or *SSX2* and *SYT* genes. The classical myxoid variant of liposarcoma carries the t(12;16)(q13;p11) translocation due to TLS-CHOP fusion. Alveolar rhabdomyosarcoma can contain the t(2;13)(q35;q14) translocation affecting the *PAX3* and *FKHR* genes. Other common mutations include p53 and MDM2 amplification.

Five-year survival ranges from 90% for stage I tumors to 10%–20% for stage IV tumors (Stojadinovic et al. 2002). For all stages and sites combined, 5-year overall survival is 50%–60% (O'Sullivan et al. 2007). Lymph node involvement and metastatic disease are uncommon at diagnosis. Overall, approximately 4% of people present with lymph node metastases. Spread to the lymph nodes is associated more commonly with certain histologies including clear cell, epitheloid, rhabdomyosarcoma, angiosarcoma, and synovial cell sarcoma (Fong et al. 1993).

Distant metastases are found in 10% of people at diagnosis. Sarcomas typically spread to the lungs first, followed by bone, liver, skin, and brain (Vezeridis, Moore, and Karakousis 1983). Liposarcoma, particularly the myxoid variant, has a predilection for soft-tissue sites (O'Sullivan et al. 2007).

Surgical resection is the mainstay of treatment, but a multimodality approach is often needed, particularly for high-grade tumors. In one study by Yang et al. (1998), 91 patients with extremity sarcomas were treated with limb-sparing surgery and postoperative chemotherapy for high-grade tumors. The patients were then randomized to radiation or observation. For 50 patients with low-grade extremity sarcomas, surgical resection was performed followed by radiation therapy vs. observation. Local control for both low- and high-grade tumors was improved with radiation, with local control at 10 years of 98% for patients treated with radiation therapy compared to 70% for patients on the observation arm ($p = .0001$). Overall survival did not differ between the two groups.

The Sarcoma Meta-Analysis Collaboration reviewed 14 randomized trials of doxorubicin-based chemotherapy in the adjuvant setting after local treatment for soft-tissue sarcomas (Tierney et al. 1997). Chemotherapy significantly decreased local recurrence and distant recurrence and improved overall recurrence-free survival. Chemotherapy failed to significantly improve overall survival, except for the subset of patients with extremity sarcomas. In this group, a 7% overall survival benefit occurred at 10 years. Criticisms of this study include the lack of central pathology review, inclusion of patients with low- and unknown-grade tumors, and inclusion of inactive chemotherapy drugs in the treatment regimens.

The phase II trial of the Radiation Therapy Oncology Group (RTOG 9514) evaluated neoadjuvant chemoradiation followed by surgical resection and three cycles of postoperative chemotherapy for extremity sarcoma (Kraybill et al. 2006). The chemotherapy regimen consisted of mesna, doxorubicin, ifosfamide, and dacarbazine with split-course radiation to a total dose of 44 Gy. A 27% complete pathologic response was observed, and a 62% partial pathologic response was also observed. However, grade 4 toxicity measured 83%, and three treatment-related deaths occurred. Therefore, this regimen requires modification before it can be considered as a standard treatment option.

The role of chemotherapy for soft-tissue sarcoma continues to be defined and remains controversial. RT alone yields inferior results, but should be considered for unresectable tumors. A retrospective review of 112 patients with soft-tissue sarcomas found a 5-year actuarial local control rate of 45%, disease-free survival of 24%, and overall survival 35% (Kepka et al. 2005). Tumor size, radiation dose, and stage influenced these outcomes.

Favorable prognostic factors include a low grade, small tumor size, superficial location, histology, and negative surgical margins (Ray and McGinn 2008). Relapse typically occurs early, with an average time to metastatic disease of 1 year (Pisters, Leung, et al. 1996 and Pisters, Harrison, et al. 1996). The risk of distant metastases at 5 years varies drastically with tumor grade and size.

23.2 Radiation Therapy and Image Guidance

Radiation therapy begins with a CT and/or MRI simulation to obtain images from which the radiation delivery is designed. Positioning of the patient at the time of simulation must be well-contemplated and executed, with particular attention on reproducibility and rotational motion. The primary and drain incisions should be wired at the time of simulation for postoperative cases. The planning scan should extend 10 cm above and below the primary tumor. The gross tumor volume (GTV) and clinical target volume (CTV) contours are drawn with respect to bone and fascial planes that separate the compartments.

Because MRI better delineates soft tissue compared to CT, MRI can be used to more accurately determine the radiation target through fusion of the MRI with the planning CT images, particularly for preoperative treatments. Ideally, the patient's position for the planning CT is replicated for the MRI to optimize the quality of the fused images. A detailed consensus for GTV, CTV, and planning target volume (PTV) margins has not been established. The Sarcoma Working Group of RTOG is currently developing the consensus report for sarcoma target definitions. General guidelines for radiotherapy of extremity sarcoma are as follows. For preoperative radiation therapy, the GTV includes the encapsulated tumor. This volume is best identified on postgadolinium T1-weighted MRI. The CTV includes the gross tumor and microscopic disease, designated CTV1. The T2-weighted image evinces suspicious edema, which harbors the satellite cells and should be included in the CTV (White et al. 2005). CTV2 includes CTV1 and 4–5 cm longitudinal margins and 1–2 cm radial margins. An additional volume of CTV2 plus 1 cm, the PTV, is then drawn to account for setup error and organ motion (van Herk 2004). Approximately 50 Gy delivered at 180–200 cGy per fraction is a standard preoperative dose. A 10–16-Gy postoperative boost (either external beam radiotherapy, brachytherapy, or intraoperative radiotherapy) may be delivered for positive margins. For postoperative radiation therapy, CTV1 includes the tumor bed, scar, and drainage sites with bolus plus 5–7 cm longitudinal margins and 2–3 cm radial margins. CTV1 is treated to approximately 45 Gy at 1.8 Gy per fraction. CTV2 consists of the tumor bed and surgical scar with 2-cm margins treated to 50.4–54 Gy. CTV3 includes the tumor bed with a 2-cm uniform margin typically treated to a total dose of 60–66 Gy. The PTV margin is an additional 1 cm. For unresectable tumors, doses in excess of 63 Gy at a median of 2 Gy per fraction were shown to improve tumor control and survival (Kepka et al. 2005).

Other factors that one should consider during the planning process include sparing skin to decrease lymphedema, avoiding total joint irradiation, and adequately dosing the superficial aspect of the tumor or tumor bed. Critical structures vary based on primary site. Toxicities associated with radiation therapy include skin reactions including desquamation, wound complications including delayed wound healing or need for surgical intervention, fibrosis, edema, decreased joint range of motion, bone fracture, and secondary malignancy (Cormier and Pollack 2004).

Image-guided radiotherapy (IGRT) is a technique that permits precise tumor localization on a periodic basis (Dawson and Jaffray 2007). This precision ensures coverage of the target volumes, including GTV and CTV (Mackie et al. 2003). It also allows for smaller PTV margins, as the setup error and organ motion are accounted for prior to treatment (Yan et al. 2005; Mageras and Mechalakos 2007). With IGRT, pretreatment images are acquired to evaluate the tumor location on a treatment day relative to the tumor location on the day of CT simulation and radiation planning. These images are obtained on a frequent basis (e.g., daily, several times per week). The contours drawn on the planning CT scan are coregistered with the pretreatment CT scan.

Because the radiation therapy is designed based on the contours drawn on the simulation CT, the goal is to meticulously align the treatment CT scan with the contours. If the tumor is misaligned, the patient and/or couch are adjusted to shift the tumor into its original location. This adjustment ensures that the planned dose is delivered to the tumor and that the surrounding normal structures have not migrated into the high-dose regions. However, use of IGRT can prolong the interval the patient is on the couch and therefore the time the treatment machine is occupied. In addition, IGRT increases the cost of treatment. Frequent imaging with X-rays can also result in a significant dose of additional radiation; this dose should be appraised in the treatment plan with appropriate adjustments performed as expedient (Chen, Sharp, and Mori 2009).

Several imaging modalities are available. Traditionally, portal images are acquired, usually on a weekly basis. These kilovoltage (kV) or megavoltage (MV) X-ray images provide a two-dimensional (2D) projection that display bony anatomy. However, alignments based on bony anatomy do not necessarily reflect proper alignment of soft tissues (Ippolito et al. 2008). Therefore, other imaging techniques have been studied, including implanted fiducials and ultrasound. More recent technologies

include MV fan-beam CT (tomotherapy), conventional kV fan-beam CT configured on rails, kV or MV cone-beam CT (CBCT), and three-dimensional (3D) laser surface imaging. The CT-on-rails is installed inside a linear accelerator room and shares the same treatment couch with the accelerator. The gantry of the CT scanner advances, while the couch remains stationary, thereby reducing patient movement and potential error (Kuriyama et al. 2003).

CBCT is performed through the use of a gantry-mounted MV or kV source and detector. The source executes a partial or full gantry rotation to acquire hundreds of projection images with subsequent conversion to CT-like axial images (Yamada, Lovelock, and Bilsky 2007). 3D reconstruction of the surface anatomy is created utilizing photogrammetry, and the patient is aligned based on the resultant surface model (Bert et al. 2006). This technique does not use ionizing radiation. The advantages of CT compared to X-ray portal images include acquisition of a 3D image, improved soft-tissue contrast, and ease of comparison with planning CT (Mackie et al. 2003).

IGRT may provide the most benefit for patients treated with intensity-modulated radiation therapy (IMRT), which typically renders highly conformal dose distributions with steep dose gradients (Mackie et al. 2003). With IGRT, one can safely apply smaller margins for setup uncertainty, and these smaller margins may diminish normal tissue toxicity and/or optimize radiation dose to the target (Mackie et al. 2003; Yan et al. 2005; Mageras and Mechalakos 2007). Smaller margins may be particularly advantageous in soft-tissue sarcoma through reduction in overall field size, which is postulated to reduce toxicity. The risk of geometric miss and normal tissue damage due to the steep dose gradients and the narrow margins can be abated with IGRT (Li et al. 2007).

Dramatic interfractional changes (increases and decreases) in tumor volume occur during radiation therapy of large soft-tissue sarcomas (Li et al. 2007). As an example, Figure 23.1a depicts the daily variation in tumor shape, position, and volume. These changes can result in significant underdosing of tumor targets and/or overdosing of adjacent normal tissues. For the case presented in Figure 23.1a, the significant tumor volume changes resulted in dramatic underdosing of the PTV (Figure 23.1b). The PTV coverage dropped from 95% on day 0 to 30% on day 31. This change in tumor coverage clearly indicates the need of image-guided adaptive treatment with replanning that can potentially improve tumor control and/or toxicity. By performing adaptive therapy, the intended dose can consistently be delivered to the target and/or the planned normal tissue sparing can be maintained.

23.3 Image-Guided Treatment Delivery

23.3.1 Head and Neck

Within the head and neck region, a plethora of vital structures exist in a confined space. For this reason, IMRT is often used in the treatment of head and neck malignancies, most notably for salivary gland preservation (Braam et al. 2006; Hsiung

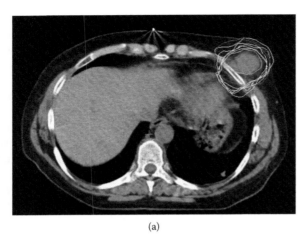
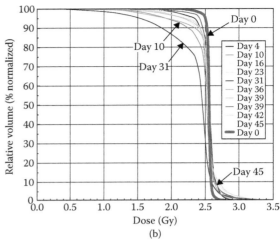

(a)   (b)

FIGURE 23.1 (See color insert following page 204.) Interfractional variations in clinical target volume (CTV) for a chest wall sarcoma. (a) A series of daily CTVs (yellow) contoured from daily megavoltage computed tomography (MVCT) scans. The planning CTV (green) and primary target volume (PTV; red) are overlaid on the axial KVCT image. The CTV in the middle of the treatment nearly doubled in volume compared to the planning CTV (day 0). (b) A series of verification DVHs for CTVs based on daily MVCTs (using tomotherapy-planned adaptive software). The thicker red line represents the dose-volume histogram (DVH) for the original/planned CTV.

et al. 2006). Surgery remains the mainstay of treatment for most head and neck sarcomas, but either preoperative or postoperative radiation is indicated for high-grade sarcomas, large tumors, and close or positive surgical margins to improve local control (Potter and Sturgis 2003; Shellenberger and Sturgis 2009). Preoperative radiotherapy typically includes smaller fields and lower doses compared to postoperative radiotherapy (O'Sullivan et al. 1999). In addition, the radiation is not delivered to a hypoxic postsurgical site (Tyldesley et al. 1997) and irradiation of skin grafts is avoided (O'Sullivan, Gullane, et al. 2003). A prospective series (O'Sullivan, Gullane, et al. 2003) demonstrated a 20% incidence of wound complications in patients who received preoperative radiotherapy for head and neck sarcoma. This incidence compares favorably to the 35% incidence of postoperative wound complications in the randomized NCIC study for extremity soft-tissue sarcoma (O'Sullivan et al. 2002).

For radiation therapy, patients are usually immobilized with a thermal plastic face mask. While this immobilization device improves the reproducibility of the patient's daily position (Gilbeau et al. 2001), other factors can still contribute to misalignment. Zhang et al. investigated setup uncertainties for 14 patients with head and neck cancer using CT-on-rails (Zhang et al. 2006). Three bony landmarks were selected as regions of interest (ROI) (C2 vertebral body, C6 vertebral body, and palatine process of the maxilla) and separately registered from the pretreatment CT to the corresponding position in the planning CT. The palatine process of the maxilla demonstrated stability in the right-left direction, but motion in the superior-inferior direction due to "in plane rotations and variations in jaw positions" (Zhang et al. 2006, p. 1559). Of the three ROIs, the C6 vertebral body demonstrated the largest motion in the right-left direction. On average, the ROIs rotated less than one degree for each plane. However, the ample range of displacement and rotation

illustrates that for an individual patient, the setup error can have a considerable effect on the precision of daily positioning.

Schubert et al. evaluated translational and rotational setup variations for four anatomic treatment sites, including 1179 brain and head and neck cancer patients, with daily MVCT (Schubert et al. 2009). About 5.9% of head and neck patients had a rotational setup error of ≥3 degrees despite the use of an immobilization mask. A similar study investigated rotational motion for eight patients with head and neck cancer who experienced "substantial weight loss or tumor shrinkage" (Ezzell et al. 2007, p. 3233). They concluded that rotational shifts may be misinterpreted as translational shifts on orthogonal portal images, resulting in setup error in 3D. Another study has demonstrated that rotational setup error may translate into a clinically relevant discrepancy in target coverage for tumors with certain characteristics, specifically "nonspherical, elongated target volumes and sharp dose gradients close to organs at risk" (Guckenberger et al. 2006, p. 940).

Hong et al. (2005) prospectively enrolled 10 patients in a study to evaluate daily head and neck setup accuracy in radiation therapy. The patients were immobilized with a conventional thermoplastic mask with baseplate fixation to the treatment couch. They were then aligned using three-point laser alignment and weekly port films. Prior to each treatment, setup accuracy was assessed by measuring three Cartesian coordinates (craniocaudal, anteroposterior, right-left) and three rotational coordinates (couch, spin, and tilt); the positional errors were applied to 10 disparate head and neck IMRT plans. The mean absolute setup error in a single dimension measured 3.33 mm. Utilizing analyses of the six degrees of freedom, the mean composite setup deviation from isocenter measured 6.97 mm. When these setup measurements were applied to IMRT plans, the dose-volume histogram (DVH) demonstrated PTV underdosing, a decrease in equivalent uniform dose (EUD) for tumor coverage, and an increase in EUD for the parotid glands. The authors concluded that "Our paradigm is to use daily verification of inter and intrafractional positional accuracy when treating patients with H&N IMRT" (Hong et al. 2005, p. 787).

Barker et al. (2004) obtained CT scans three times weekly for 14 patients throughout their course of head and neck radiotherapy. GTV and normal structures including parotid glands, mandible, external contour, and spinal canal were delineated, and changes in volume and position were assessed in relation to bony anatomy. The GTV of both tumor and lymph nodes decreased in volume at a rate of 1.8% per treatment day, and the center of mass varied with time as tumor shrinkage was asymmetric. The parotid glands also decreased in volume and shifted medially (median positional change, 3.1 mm) corresponding with patient weight loss. The authors concluded that these measurable anatomic changes over the radiotherapy course could impact dosimetry, particularly in more conformal radiotherapy techniques.

Zeidan et al. (2007) evaluated the setup error for 24 patients undergoing treatment for head and neck malignancy with daily image-guidance retrospectively using several image-guided protocols that differed in the percentage of treatment fractions utilizing MVCT. The protocols included the absence of MVCT imaging, imaging for the first one, three, five, or seven fractions, weekly imaging, imaging for the first five fractions followed by weekly imaging, and imaging every other day. For the protocols in which imaging was employed in ≥9% of treatment fractions, systematic error averaged over all the fractions was determined to be <1.5 mm in all directions. Weekly image guidance resulted in 33% of treatments with ≥5 mm setup error. The systematic error further decreased to ~0.5 mm in all directions

when imaging frequency escalated to 50% of treatment fractions. However, 11% of all treatments still resulted in ≥ 5 mm misalignment error. The authors determined that daily imaging results in the greatest reduction in systematic error, but less frequent imaging still successfully reduces setup error, albeit to a lesser degree. Therefore, they concluded that imaging frequency should be selected with consideration of treatment margins and proximity to critical structures.

Head and neck sarcomas comprise 5%–9% of all adult soft-tissue sarcomas (Hsiung et al. 2006) and 1%–11% of all head and neck malignancies (Patel, Shaha, and Shah 2001). Although sarcomas of the head and neck are of a different cell origin, they are subject to comparable inherent difficulties encountered with radiating squamous cell carcinomas in the head and neck region. No randomized trials have been conducted to provide level I evidence regarding the benefit of IGRT in head and neck cancers, but studies such as these indicate that a potential for missing the target exists if imaging is not used to account for positional variations in setup during the course of treatment. A myofibroblastic sarcoma arising in the right mandible was treated postoperatively with IMRT using tomotherapy, as depicted in Figure 23.2.

23.3.2 Orbit

Within the head and neck region, the orbit is a particularly challenging site to treat. In a case of primary sarcoma of the orbit shown in Figure 23.3, the patient was treated with daily IG-IMRT. The goal was to decrease the tumor size prior to surgical resection. Extrapolating from the Canadian NCIC study (O'Sullivan et al. 2002) for soft-tissue sarcoma of the extremities, the lower dose required for preoperative radiation and the

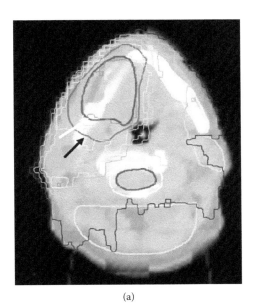
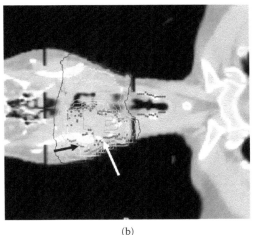

(a)   (b)

FIGURE 23.2 (See color insert following page 204.) Primary myofibroblastic sarcoma of the right mandible. (a) Axial view and (b) coronal view of the patient depicting clinical target volume (CTV1; tumor bed), CTV2 (operative bed), and isodose lines (blue 18 Gy, cyan 30 Gy, green 42 Gy, light green 48 Gy, orange 54 Gy, dark orange 57 Gy, red 60 Gy). In both (a) and (b), the black arrow indicates CTV (purple contour) and the white arrow indicates primary target volume (PTV; red contour).

reduced treatment volume is anticipated to decrease toxicity to the bone, soft tissue, and nerves in the radiation field. A prospective study by O'Sullivan, Gullane, et al. (2003) also supports the use of preoperative radiotherapy in head and neck sarcoma with a 2-year actuarial local relapse-free rate of 80%.

In addition, if the tumor demonstrates a favorable response, the patient may be spared an enucleation. The radiation dose for this large tumor would still likely result in blindness of the treated eye, but the patient's esthetics would improve. For a smaller orbital tumor, sparing of the lens to preserve vision is particularly important. Other critical structures include the optic nerves, optic chiasm, lacrimal gland, bony orbit, pituitary gland, and frontal and temporal lobes.

23.3.3 Retroperitoneum

Patients with retroperitoneal sarcomas often remain asymptomatic until the tumor has become quite large, and even at that time symptoms can be nonspecific. Complete surgical resection constitutes the only curative option (Stoeckle et al. 2001). Radiation is used with the intent of reducing the risk of local failure. A randomized trial comparing surgical resection and postoperative radiation therapy vs. surgery alone has not been conducted, but some retrospective data support a benefit with postoperative radiation (Stoeckle et al. 2001; Zlotecki et al. 2005; Catton et al. 1994). However, there are significant challenges to delivering high-dose irradiation postoperatively for retroperitoneal sarcoma.

One difficulty arises because bowel and/or other critical organs almost always migrate into the surgical bed, into the region of maximal radiation dose, once the tumor has been excised. Another difficult challenge for postoperative radiation consists of identification of the target volume(s), as the tumor has been removed and the anatomical region has been disturbed.

In addition, a relatively high dose is required to eradicate microscopic disease in the tumor bed. With preoperative radiation therapy, the tumor displaces bowel and other structures away from the radiation field, although they still remain in close proximity. White et al. (2007) reported the use of a saline-filled tissue expander placed in the abdomen prior to preoperative radiation therapy. The purpose of the tissue expander is to further displace small bowel out of the radiation field in order to decrease bowel toxicity. In preoperative radiotherapy, the target volume is more conspicuously demarcated and the radiation fields are smaller as they do not need to encompass an entire operative site (Lawenda and Johnstone 2008).

Researchers have endeavored to determine the value of preoperative radiation therapy. The American College of Surgeons Oncology Group opened a phase III trial (Z9031) in which patients were randomized to preoperative radiation therapy and en bloc surgical resection vs. surgical resection alone. However, the study was prematurely closed due to low accrual; therefore, no randomized data are available to direct therapeutic recommendations. Retrospective reviews and prospective, nonrandomized studies offer promising results in terms of local control and acceptable toxicity associated with preoperative radiotherapy (Zeidan et al. 2007; Jones et al. 2002; Pawlik et al. 2006).

In the patients treated preoperatively (when the target is well-defined), IMRT can reduce dose to selected structures through conformal avoidance of OAR. Koshy et al. (2003) compared 3D-CRT with IMRT for 10 patients with retroperitoneal sarcoma and one patient with inguinal sarcoma. Nine patients were treated with preoperative radiation and two patients underwent postoperative radiation. With IMRT, the small bowel received a mean dose of 27 Gy, compared to 36 Gy with 3D-CRT. In addition, the volume of small bowel receiving > 30 Gy decreased from 63.5% with 3D-CRT to 43.1% with IMRT ($p = .043$). The volume of the tumor receiving the prescribed dose improved

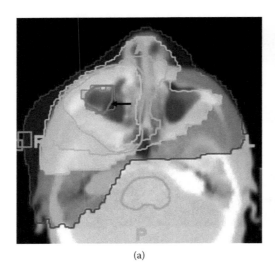

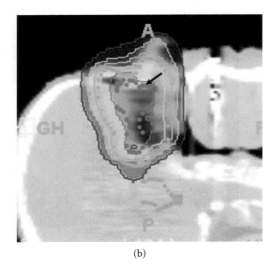

(a)   (b)

FIGURE 23.3   (See color insert following page 204.) Primary sarcoma of the right orbit. (a) Axial view and (b) sagittal view of the patient depicting primary target volume (PTV) and isodose regions (dark blue 10 Gy, light blue 20 Gy, green 30 Gy, yellow 40 Gy, red 50 Gy). The patient was treated preoperatively with tomotherapy. In both (a) and (b), the black arrow indicates PTV (red contour).

with IMRT: the volume that received 95% percent of the dose was determined to be 95.3% with 3D-CRT compared to 98.6% with IMRT. The authors concluded that sparing the normal structures provides the opportunity to investigate preoperative dose escalation, which has the potential to enhance local control.

Tzeng et al. (2006) prospectively enrolled 16 patients with retroperitoneal sarcoma to assess selective dose escalation using preoperative radiation. While 45 Gy was delivered in 28 fractions to a PTV encompassing the entire tumor, the volume considered at high risk for positive margins received a boost to 57.5 Gy. Treatment was well-tolerated, with 25% of patients experiencing nausea and emesis during radiation therapy and no patients with severe postoperative toxicity. The 2-year actuarial local control rate was 80%.

IGRT improves consistency of daily targeting, thereby permitting margin reduction and the prospect for normal-tissue spring. It could also assist in facilitation of dose escalation. Further studies are imperative to substantiate these hypotheses. Figure 23.4 demonstrates the daily pretreatment image for a patient with recurrent retroperitoneal sarcoma abutting the right kidney that was treated preoperatively using tomotherapy. Note the absence of even 14 Gy to the majority of the kidney despite its close proximity to the PTV (Figure 23.4a). The image-guided repositioning based on the registration of bony anatomy and soft tissue ensures coverage of the target and avoidance of the critical structures (e.g., kidney and cord; Figure 23.4b).

23.3.4 Paraspinal

Paraspinal sarcomas may abut or encompass the spinal cord region. Radiation delivery is designed to minimize the dose to the spinal cord. If the patient is not properly aligned, even a difference of several millimeters may result in a higher dose than

anticipated to the spinal cord. As the spinal cord dose escalates, so does the risk of spinal cord myelopathy. Even a small length of the spinal cord treated to a high dose can result in severe sequela for the patient (Hall and Giaccia 2005). IGRT allows continual assessment of the position of the spinal cord and the tumor. In addition to pretreatment adjustments to shift the tumor into the planned field, tumor response can be assessed. In tumors with a significant response, the pressure on the spinal cord may be released, resulting in a change in position of the spinal cord and the tumor. IGRT may elucidate such changes in the tumor and OAR; adaptive planning may be necessary to safely and effectively complete radiation treatment.

Hansen et al. (2006) reported a case in which MV CBCT was utilized in treatment of a patient with high-grade sarcoma after surgical resection with hardware placement. A hypothetical IMRT plan was compared to the original IMRT plan. Without daily shifts based on the pretreatment imaging, the dose to 0.1 cubic centimeter of the spinal cord increased by 9.4 Gy and the doses to the CTV1 and CTV2 were decreased by 4 and 4.8 Gy, respectively.

Terezakis et al. (2007) conducted a retrospective review of 27 patients with paraspinal tumors (partially resected or unresectable). Eighteen patients had a diagnosis of sarcoma. The median dose prescribed to the PTV was 66 Gy (range, 5396–7080 cGy) at 180 or 200 cGy per fraction. Seven patients (26%) experienced an in-field recurrence at a median of 9.4 months. No patient developed radiation-induced myelopathy. Cord edema detected with MRI only manifested in patients with local progression. About 84% of patients who initially presented with pain reported improvement in or stability of their pain. Almost two-thirds (62.5%) of patients discontinued narcotics after completing radiation therapy. Around 89% of patients experienced a stable or improved performance status. Only patients with

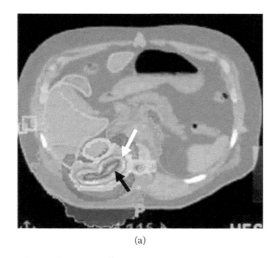

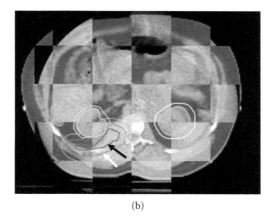

(a)   (b)

FIGURE 23.4 (See color insert following page 204.) Recurrent retroperitoneal sarcoma (axial view). (a) Tomotherapy plan depicting gross tumor volume (GTV; indicated by black arrow), primary target volume (PTV; indicated by white arrow), and isodose regions (light blue 14 Gy, dark blue 25 Gy, cyan 35 Gy, orange 45 Gy, pink 50 Gy). Note the absence of even 14 Gy to the majority of the kidney despite its close proximity to the PTV. (b) Daily megavoltage computed tomography (MVCT) superimposed on planning CT scan. Black arrow indicates GTV (red contour) and white arrow indicates PTV (pink contour). Right kidney is contoured in cyan; left kidney, in green.

disease progression suffered a decline in motor or sensory function. The authors concluded that high doses of photon radiation therapy can be safely delivered using IGRT.

Among the cases presented in the MVCT-guided shift and margin-calculation study by Li et al. (2007) was that of five patients with sarcoma proximal to the cord. These patients were simulated using kVCT and treated using helical tomotherapy. Prior to each fraction, a fan-beam MVCT image encompassing the PTV was acquired. The primary purpose of the pretreatment MVCT image was to refine patient positioning by comparison of anatomic landmarks within the pretreatment MVCT and the planning kVCT; as will be discussed in Section 23.4, the pretreatment MVCT scans also helped to identify dramatic changes to the target and region-at-risk geometry that may require a revision to the treatment plan. The optimal alignment of the MVCT and the plan kVCT determined the additional lateral, longitudinal, and vertical shifts that were required for the patient beyond the initial external marker-based setup in order to ensure coverage to the target volume. From the systematic (μ) and random (σ) uncertainties calculated among the MVCT-guided daily shifts, a CTV-to-PTV expansion margin $\Sigma = 2|\mu| + 0.7\sigma$ was calculated. In this way, an estimate was made of the CTV-to-PTV expansion margin that would be required, in the absence of MVCT-based image guidance, to ensure 95% coverage of the target volume for paraspinal sarcoma treatment (van Herk 2004). Among the paraspinal cases, 7.6 mm was calculated to be the overall CTV-to-PTV expansion margin, obtained by adding in quadrature the margins determined along the three translational directions. The primary contribution to this overall uncertainty was the systematic uncertainty in setup along the lateral direction. One possible explanation for this is that the anterior skin mark that is used for initial patient setup along the lateral direction may not be rigidly positioned relative to the spine. As the vertebrae are easily distinguishable in both MVCT and kVCT images, it is easy to reposition the patient along the lateral direction using axial images.

Given that the penumbra from the high-dose region of a typical IMRT plan is around 3 mm, and given that paraspinal targets often require prescribed doses in excess of the cord $TD_{5/5}$ dose of 45 Gy, the availability of IGRT to reduce the CTV-to-PTV treatment margin by up to 8 mm makes it possible to boost the paraspinal tumor dose to levels necessary for control, while still sparing the cord.

23.3.5 Extremity

Limb-salvage therapy with a combination of radiotherapy and surgery has replaced extremity amputation as the standard of care for management of soft-tissue sarcomas of the extremities (Rosenberg et al. 1982). The sequence of these two modalities and the role of chemotherapy remain controversial. The benefits of preoperative radiation therapy include a lower radiation dose, a smaller radiation field, and potential decrease in tumor size prior to resection (Nielsen et al. 1991; Sadoski et al. 1993).

A challenge of safely and effectively delivering radiation therapy to extremity sarcomas is the proper immobilization of the patient that allows consistent reproducibility of position. Even with the extremity stabilized at points above and below the tumor, the potential for rotational motion remains. This challenge is often compounded by large field size (O'Sullivan, Ward, et al. 2003).

IMRT enhances dose conformity to the tumor and allows for improved sparing of selected normal-tissue structures (Jaffray 2005).

Hong et al. (2004) reported a comparison of 3D conformal radiotherapy (3D-CRT) plans with IMRT plans for 10 patients with soft-tissue sarcoma of the thigh. The $V_{100\%Rx}$ for the femur decreased at an average of 57%, from $44.7 \pm 16.8\%$ with 3D-CRT to $18.6 \pm 9.2\%$ with IMRT ($p < .01$). In addition, the volume of surrounding soft tissues (excluding PTV and bone) receiving prescription dose (63 Gy) decreased on average by 78%, from 997 ± 660 cm^3 with 3D-CRT to 201 ± 144 cm^3 with IMRT ($p < .01$) The variation in mean dose to the surrounding soft tissues was not statistically significant ($p = .92$). A reduction of 45% in the volume of the skin receiving the prescription dose of 63 Gy was observed, from 115 ± 40 cm^3 with 3D-CRT to 61 ± 20 cm^3 with IMRT ($p < .01$). Griffin et al. (2007) performed a dosimetric evaluation of conformal radiation therapy compared to IMRT regarding dose to the planned surgical skin flaps with the rationale that decreased dose to the skin could translate into lower rates of wound-healing complications. IMRT resulted in a significantly lower mean dose to the planned skin flaps (40.12 Gy with conformal vs. 26.71 Gy with IMRT, $p = .0008$) and significantly decreased the V_{30} for the planned skin flaps (83.4% with conformal vs. 34% with IMRT, $p = .0001$). This study was not designed to evaluate the projected outcome on postoperative wound healing, but it demonstrated the target conformality afforded by IMRT.

The target dose conformity and improved OAR sparing compels the correction of uncertainties in patient setup. Li et al. (2007) analyzed MVCT-guided daily translational shifts that included 10 patients with extremity tumors. From the systematic and random uncertainties of these shifts, the CTV expansion required to ensure $\geq 95\%$ coverage to the PTV in the absence of image guidance was calculated and determined to be 7.9 mm. However, shifts as large as 13.8 mm were observed in order to achieve optimal positioning prior to execution of treatment for these extremity tumors. The magnitude of these shifts exceeded the calculated PTV margin; without image guidance to verify the tumor position, treatment delivery would have resulted in suboptimal dose distribution.

Daily confirmation of tumor location with reduction in field size related to better understanding of CTV has the potential to enhance outcome. Conservative management with surgery and radiotherapy yields >90% local control (Pisters, Leung, et al. 1996; Yang et al. 1998); however, morbidity such as fibrosis, edema, and imperfect limb function detracts from patients' quality of life (O'Sullivan, Gullane, et al. 2003). Acute wound complications developed in 35% of patients in the preoperative arm and 17% of patients in the postoperative arm ($p = .01$) in a prospective phase III randomized trial comparing preoperative with postoperative radiotherapy (O'Sullivan et al. 2002). Davis et al. (2005) reported late toxicity for 129 patients, 73 treated with preoperative radiation therapy and 56 treated with postoperative radiation therapy. About 31.5% of patients in the preoperative arm suffered grade 2 or greater fibrosis, while 48.2% of patients in the postoperative arm were affected. Edema occurred in 15.1% of preoperative patients and 23.3% of postoperative patients; 17.8% of preoperative patients and 23.2% of postoperative patients developed joint stiffness. A larger field size correlated with higher rates of fibrosis and joint stiffness. The dosimetric comparison of Hong et al. (2004) suggests a potential for decreased toxicity with IMRT given the reduced volume of normal structures receiving the prescribed dose. IGRT may also abate toxicity through application of smaller margins, which curtails the amount of normal tissue irradiated to a high dose.

An ongoing RTOG 0630 study led by Wang et al. aims to evaluate the hypothesis that more conservative CTV and IGRT can decrease morbidity for soft-tissue sarcomas of the extremity. This phase II study is designed to evaluate the effect of more conservative CTV and daily IG-IMRT on late radiation morbidity and patterns of failure.

In this study, all patients receive radiation therapy prior to surgical resection. The radiation technique consists of either 3D-CRT or IMRT. Cohort A consists of patients receiving (1) neoadjuvant and/or adjuvant chemotherapy and radiation therapy to a total dose of 50 Gy in 25 daily fractions or (2) concurrent or interdigitated chemotherapy and radiation therapy to a total dose of 44 Gy in 22 daily fractions. Cohort B consists of patients receiving single modality radiation therapy to a total dose of 50 Gy in 25 daily fractions. Surgical resection is performed 4–8 weeks after the completion of preoperative therapy. For patients with positive surgical margins, a postoperative radiation boost is required. The boost can be delivered via external beam radiation, brachytherapy, or intraoperative radiation therapy. The study design permits chemotherapy for patients with deep tumors ≥8 cm in size and intermediate-to-high-grade histology.

This study mandates daily pretreatment imaging. Imaging can be performed using 3D techniques (fan or CBCT) or 2D techniques (electronic kV or MV X-ray portal images). If orthogonal images are used, placement of fudicial markers under ultrasound or CT guidance is advised. The goal for alignment is an error ≤3 mm. The recommended CTV includes GTV, suspicious edema defined by MRI T2 images, and 2–3 cm margins in the longitudinal directions or end of the compartment plus 1 cm. The radial margin should be 1–1.5 cm or anatomic barrier. The PTV includes CTV with a 0.5-cm margin as described in the RTOG 0630 protocol (www.rtog.org). In contrast, the standard margins include a CTV that extends 5–8 cm longitudinally and 1.5–2 cm radially. The standard PTV consists of the CTV expanded by least 1 cm.

The primary objective of the RTOG 0630 protocol is to evaluate lymphedema, fibrosis, and joint stiffness 2 years after the initiation of treatment. The frequency of wound complications will also be compared to non-IGRT rates. Given the reduction in irradiated volumes in this study, a decrease in late toxicity is expected. The patterns of failure will also be assessed to ensure that the smaller margins do not compromise tumor eradication.

23.4 Adaptive Strategies

A review of pretreatment images may identify dramatic changes to the patient anatomy that would require a modification to the treatment plan in order to maintain region-at-risk avoidance as well as target volume coverage. An example presented here is a patient with a chondrosarcoma who was treated using helical tomotherapy (33 fractions) with daily acquisition of pretreatment MVCT images. Figure 23.5 depicts the MVCT scan acquired in the middle of the treatment course, with the planning-kVCT's spinal cord contours superimposed upon the MVCT image. The curvature of the spine had significantly changed from what was apparent on the plan kVCT image, and the misalignment could not be corrected using either translation or rotation alone. A 50.4-Gy target volume for this patient encompassed the cord, and a 59.4-Gy tumor bed was situated just posterior to the cord. Figure 23.6

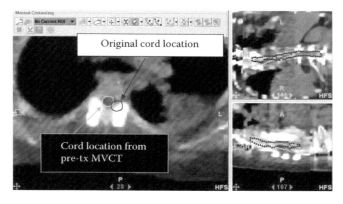

FIGURE 23.5 A pretreatment megavoltage computed tomography (MVCT) scan acquired for a patient with chondrosarcoma treated with tomotherapy, with the plan kilovoltage computed tomography (kVCT)-based cord contours superimposed upon the image. The curvature of the spine could not be corrected using translation or rotation alone.

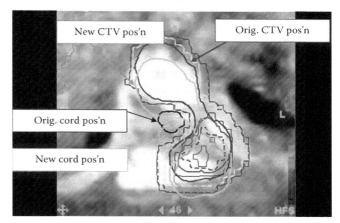

FIGURE 23.6 (See color insert following page 204.) The target-volume and cord contours are superimposed upon the pretreatment megavoltage computed tomography (MVCT) image corresponding to their original locations (according to the planning kilovoltage computed tomography (kVCT) and to their apparent locations on the MVCT image. Note that the new location of the cord overlaps with the original position of the 50.4-Gy CTV.

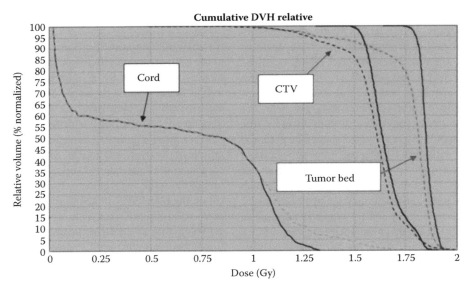

FIGURE 23.7 Cumulative single-fraction dose-volume histograms for the cord, 50.4-Gy clinical target volume (CTV), and 59.4-Gy tumor bed, according to the original kilovoltage computed tomography (kVCT)-based treatment plan (solid lines) and the dose recalculation based on the mid-treatment megavoltage computed tomography (MVCT) image (dashed lines). Note that for the MVCT-based dose calculation (indicative of the patient anatomy on the day of acquisition), not only does the dose to the target volumes decrease, but also the maximum point dose to the cord increases. Over a full treatment course, this increased dose to the cord would exceed cord-dose tolerance.

depicts the mid-treatment MVCT image with the MVCT-based and kVCT-based target-volume contours superimposed upon it. Note that the cord now impinges upon the high-dose region that corresponds to the original CTV location. Had this fraction been administered, the cord would have received the dose that was intended for the CTV. The pretreatment MVCT image can also be used as the basis for a recalculation of the dose distribution within the patient resulting from the original planned tomotherapy IMRT beam. Isodose distributions are displayed in Figure 23.6, and single-fraction DVHs comparing the MVCT-based recalculation with the original kVCT-based plan are presented in Figure 23.7. Had the fraction been administered, the maximum point dose to the cord would increase from 1.33 Gy per fraction (44 Gy over a full 33-fraction treatment) to 1.75 Gy per fraction (58 Gy over 33 fractions, exceeding tolerance).

Since the cord and the paraspinal targets were identifiable within the MVCT images, as shown in Figures 23.5 and 23.6, it was possible to modify the target volume and cord contours and to use them and the MVCT data set to generate an adaptive treatment plan to restore target coverage and cord sparing. Isodose distributions from the adaptive plan (assuming a full 33-fraction treatment) are shown in Figure 23.8. The conformity and uniformity of the target volume-dose coverage is favorable, as is the sparing of the cord from these high doses. The DVHs for the adaptive plan are depicted in Figure 23.9. Coverage of the 50.4-Gy and 59.4-Gy target volumes were restored to their prescribed coverage of $V_{100\%Rx} \geq 95\%$, and the maximum point dose to the cord was reduced to approximately 41 Gy over a full 33-fraction treatment course. Thus, for a case in which significant changes to the spine curvature arose in mid-treatment,

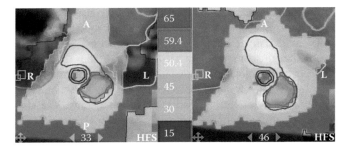

FIGURE 23.8 (**See color insert following page 204.**) Isodose distributions for an adaptive tomotherapy plan using a megavoltage computed tomography (MVCT) scan acquired in the middle of the treatment course, and the modified target volume and cord contours based upon the MVCT image. Dose levels in Gy are included.

MVCT-based image guidance was an essential component in identifying the need for an adaptive plan and also served as a suitable basis for the development of an adaptive plan that maintained the cord dose below tolerance.

23.5 Summary

Adaptive radiation therapy, assisted by IGRT, is effective in addressing the large interfractional setup errors and anatomic variations during treatment for soft-tissue sarcoma. Clinical trials similar to RTOG 0630 are necessary to quantify the benefit associated with IGRT for the various anatomic locations in which soft-tissue sarcomas arise. The advantages must then be compared to the disadvantages of IGRT, which include prolongation of the interval the patient is on the treatment couch,

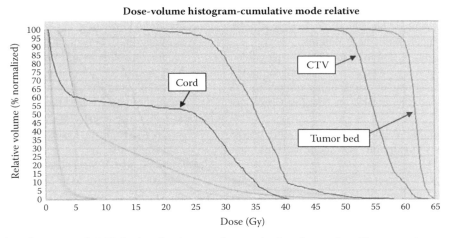

FIGURE 23.9 Dose-volume histograms (DVH) for the adaptive tomotherapy plan based upon the mid-treatment megavoltage computed tomography (MVCT) image. Target-volume coverage and sparing of the cord from high dose were restored with adaptive radiotherapy.

accounting for the radiation dose from imaging in the treatment plan, and economic cost (Chen, Sharp, and Mori 2009). The benefit of IGRT may be more pronounced in this era in which IMRT is frequently used. IGRT has the potential to improve precision in treating the target, thereby diminishing suboptimal target dosing and augmenting normal tissue sparing. We await the results of clinical trials to prove the efficacy of IGRT.

References

Anderson, M. W., H. T. Temple, R. G. Dussault, et al. 1999. Compartmental anatomy: Relevance to staging and biopsy of musculoskeletal tumors. *Am J Roentgenol* 173:1663–71.

Barker Jr., J. L., A. S. Garden, K. K. Ang, et al. 2004. Quantification of volumetric and geometric changes occurring during fractionated radiotherapy for head-and-neck cancer using an integrated CT/linear accelerator system. *Int J Radiat Oncol Biol Phys* 59:960–70.

Bastiaannet, E., H. Groen, P. L. Jager, et al. 2004. The value of FDG-PET in the detection, grading and response to therapy of soft tissue and bone sarcomas: A systematic review and meta-analysis. *Cancer Treat Rev* 30:83–101.

Bert, C., K. G. Metheany, K. P. Doppke, et al. 2006. Clinical experience with a 3D surface patient setup system for alignment of partial-breast irradiation patients. *Int J Radiat Oncol Biol Phys* 64:1265–74.

Braam, P. M., C. H. Terhaard, J. M. Roesink, et al. 2006. Intensity-modulated radiotherapy significantly reduces xerostomia compared with conventional radiotherapy. *Int J Radiat Oncol Biol Phys* 66:975–80.

Catton, C. N., B. O'Sullivan, C. Kotwall, et al. 1994. Outcome and prognosis in retroperitoneal soft tissue sarcoma. *Int J Radiat Oncol Biol Phys* 29:1005–10.

Chen, G., G. Sharp, and S. Mori. 2009. A review of image-guided radiotherapy. *Radiol Phys Technol* 2:1–12.

Cormier, J. N., and R. E. Pollack. 2004. Soft tissue sarcomas. *CA Cancer J Clin* 54:94–109.

Costa, J., R. A. Wesley, E. Glatstein, et al. 1984. The grading of soft tissue sarcomas: Results of clinicohistopathologic correlation in a series of 163 cases. *Cancer* 53:530–41.

Davis, A. M., B. O'Sullivan, R. Turcotte, et al. 2005. Late radiation morbidity following randomization to preoperative versus postoperative radiotherapy in extremity soft tissue sarcoma. *Radiother Oncol* 75:48–53.

Dawson, L., and D. Jaffray. 2007. Advances in image-guided radiation therapy. *J Clin Oncol* 25:938–45.

Demas, B. E., R. T. Heelan, J. Lane, et al. 1988. Soft-tissue sarcomas of the extremities: Comparison of MR and CT in determining the extent of disease. *Am J Roentgenol* 150:615–20.

Enneking, W. F., S. S. Spanier, and M. M. Malawer. 1981. The effect of the anatomic setting on the results of surgical procedures for soft parts sarcoma of the thigh. *Cancer* 47:1005–22.

Ezzell, L. C., E. K. Hansen, J. M. Quivey, and P. Xia. 2007. Detection of treatment setup errors between two CT scans for patients with head and neck cancer. *Med Phys* 34:3233–42.

Fong, Y., D. G. Coit, J. M. Woodruff, et al. 1993. Lymph node metastasis from soft tissue sarcoma in adults. Analysis of data from a prospective database of 1772 sarcoma patients. *Ann Surg* 217:72–7.

Gilbeau, L., M. Octave-Prignot, T. Loncol, et al. 2001. Comparison of setup accuracy of three different thermoplastic masks for the treatment of brain and head and neck tumors. *Radiother Oncol* 58:155–62.

Griffin, A. M., C. I. Euler, M. B. Sharpe, et al. 2007. Radiation planning comparison for superficial tissue avoidance in radiotherapy for soft tissue sarcoma of the lower extremity. *Int J Radiat Oncol Biol Phys* 67:847–56.

Guckenberger, M., J. Meyer, D. Vordermark, et al. 2006. Magnitude and clinical relevance of translational and rotational patient setup errors: A cone-beam CT study. *Int J Radiat Oncol Biol Phys* 65:934–42.

Hall, E., and A. Giaccia. 2005. Clinical response of normal tissues. In *Radiobiology for the Radiologist,* ed. E. Hall, 327–48. Philadelphia: Lippincott Williams & Wilkins.

Hansen, E. K., D. A. Larson, M. Aubin, et al. 2006. Image-guided radiotherapy using megavoltage cone-beam computed tomography for treatment of paraspinous tumors in the presence of orthopedic hardware. *Int J Radiat Oncol Biol Phys* 66:323–26.

Heslin, M. J., J. J. Lewis, J. M. Woodruff, et al. 1997. Core needle biopsy for diagnosis of extremity soft tissue sarcoma. *Ann Surg Oncol* 4:425–31.

Hoeber, I., A. J. Spillane, C. Fisher, et al. 2001. Accuracy of biopsy techniques for limb and limb girdle soft tissue tumors. *Ann Surg Oncol* 8:80–7.

Hong, L., K. M. Alektiar, M. Hunt, et al. 2004. Intensity-modulated radiotherapy for soft tissue sarcoma of the thigh. *Int J Radiat Oncol Biol Phys* 59:752–59.

Hong, T. S., W. A. Tome, R. J. Chappell, et al. 2005. The impact of daily setup variations on head-and-neck intensity-modulated radiation therapy. *Int J Radiol Oncol Biol Phys* 61:779–88.

Hsiung, C. Y., H. M. Ting, H. Y. Huang, et al. 2006. Parotid-sparing intensity-modulated radiotherapy (IMRT) for nasopharyngeal carcinoma: Preserved parotid function after IMRT on quantitative salivary scintigraphy and comparison with historical data after conventional radiotherapy. *Int J Radiat Oncol Biol Phys* 66:454–61.

Ippolito, E., I. Mertens, K. Haustermans, et al. 2008. IGRT in rectal cancer. *Acta Oncol* 47:1317–24.

Jaffray, D. A. 2005. Emergent technologies for 3-dimensional image-guided radiation delivery. *Semin Radiat Oncol* 15:208–16.

Jemal, A., R. Siegel, E. Ward, et al. 2009. Cancer statistics, 2009. *CA Cancer J Clin* 59:225.

Jones, J. J., C. N. Catton, B. O'Sullivan, et al. 2002. Initial results of a trial of preoperative external beam radiation therapy and postoperative brachytherapy for retroperitoneal sarcoma. *Ann Surg Oncol* 9:346–54.

Kepka, L., T. F. DeLaney, H. D. Suit, et al. 2005. Results of radiation therapy for unresected soft-tissue sarcomas. *Int J Radiat Oncol Biol Phys* 63:852–59.

Koshy, M., J. C. Landry, J. D. Lawson, et al. 2003. Intensity modulated radiation therapy for retroperitoneal sarcoma: A case for dose escalation and organ at risk toxicity reduction. *Sarcoma* 7:137–48.

Kraybill, W. G., J. Harris, I. J. Spiro, et al. 2006. Phase II study of neoadjuvant chemotherapy and radiation therapy in the management of high risk, high grade, soft-tissue sarcomas of the extremities and body wall: Radiation Therapy Oncology Group Trial 9514. *J Clin Oncol* 24:619–25.

Kuriyama, K., H. Onishi, N. Sano, et al. 2003. A new irradiation unit constructed of self-moving gantry-CT and linac. *Int J Radiat Oncol Biol Phys* 55:428–35.

Lawenda, B., and P. Johnstone. 2008. Retroperitoneum. In *Perez and Brady's Principles and Practice of Radiation Oncology,* ed. E. Halperin, C. Perez and L. Brady, 1708–15. Philadelphia: Lippincott Williams and Wilkins.

Li, X. A., X. S. Qi, M. Pitterle, et al. 2007. Interfractional variations in patient setup and anatomic change assessed by daily computed tomography. *Int J Radiat Oncol Biol Phys* 68:581–91.

Mackie, T. R., J. Kapatoes, K. Ruchala, et al. 2003. Image guidance for precise conformal radiotherapy. *Int J Radiat Oncol Biol Phys* 56:89–105.

Mageras, G. S., and J. Mechalakos. 2007. Planning in the IGRT context: Closing the loop. *Semin Radiat Oncol* 17:268–77.

Nielsen, O. S., B. Cummings, B. O'Sullivan, et al. 1991. Preoperative and postoperative irradiation of soft tissue sarcomas: Effect of radiation field size. *Int J Radiat Oncol Biol Phys* 21:1595–99.

O'Sullivan, B., P. Chung, C. Euler, et al. 2007. Soft tissue saracoma. In *Clinical Radiation Oncology,* ed. L. L. Gunderson and J. E. Tepper, 1519–49. Philadelphia: Churchill Livingstone.

O'Sullivan, B., A. M. Davis, R. Turcotte, et al. 2002. Preoperative versus postoperative radiotherapy in soft-tissue sarcoma of the limbs: A randomized trial. *Lancet* 359:2235–41.

O'Sullivan, B., B. Gullane, J. Irish, et al. 2003. Preoperative radiotherapy for adult head and neck soft tissue sarcoma: Assessment of wound complication rates and cancer outcome in a prospective series. *World J Surg* 27:875–83.

O'Sullivan, B., I. Ward, T. Haycocks, et al. 2003. Techniques to modulate radiotherapy toxicity and outcome in soft tissue sarcoma. *Curr Treat Options Oncol* 6:453–64.

O'Sullivan, B., J. Wylie, C. Catton, et al. 1999. The local management of soft tissue sarcoma. *Semin Radiat Oncol* 9:328–48.

Patel, S. G., A. R. Shaha, and J. P. Shah. 2001. Soft tissue sarcomas of the head and neck: An update. *Am J Otolaryngol* 22:2–18.

Pawlik, T. M., P. W. T. Pisters, L. Mikula, et al. 2006. Long-term results of two prospective trials of preoperative external beam radiotherapy for localized intermediate- or high-grade retroperitoneal soft tissue sarcoma. *Ann Surg Oncol* 13:508–17.

Pisters, P. W., L. B. Harrison, D. H. Leung, et al. 1996. Long-term results of a prospective randomized trial of adjuvant brachytherapy in soft tissue sarcoma. *J Clin Oncol* 14:859–68.

Pisters, P. W., D. H. Leung, J. Woodruff, et al. 1996. Analysis of prognostic factors in 1,041 patients with localized soft tissue sarcomas of the extremities. *J Clin Oncol* 14:1679–89.

Potter, B. O., and E. M. Sturgis. 2003. Sarcomas of the head and neck. *Surg Oncol Clin N Am* 12:379–417.

Ray, M. E., and C. J. McGinn. 2008. Soft tissue sarcomas (excluding retroperitoneum). In *Perez and Brady's Principles and Practice of Radiation Oncology,* ed. E. Halperin, C. Perez and L. Brady, 1808–21. Philadelphia: Lippincott Williams and Wilkins.

Rosenberg, S. A., J. E. Tepper, E. Glatstein, et al. 1982. The treatment of soft tissue sarcomas of the extremities: Prospective randomized evaluations of (1) limb-sparing surgery plus radiation therapy compared with amputation and (2) the role of adjuvant chemotherapy. *Ann Surg* 196:305–15.

Sadoski, C., H. D. Suit, A. Rosenberg, et al. 1993. Preoperative radiation, surgical margins, and local control of extremity sarcomas of soft tissues. *J Surg Oncol* 52:223–30.

Schubert, L., D. Westerly, W. A. Tome, et al. 2009. A comprehensive assessment by tumor site of patient setup using daily MVCT imaging from more then 3,800 helical tomotherapy treatments. *Int J Radiol Oncol Biol Phys* 73:1260–69.

Shellenberger, T. D., and E. M. Sturgis. 2009. Sarcomas of the head and neck region. *Curr Oncol Rep* 11:135–42.

Stoeckle, E., J. M. Coindre, S. Bonvalot, et al. 2001. Prognostic factors in retroperitoneal sarcoma: A multivariate analysis of a series of 165 patients of the French Cancer Center Federation Sarcoma Group. *Cancer* 92:359–68.

Stojadinovic, A., D. H. Leung, P. Allen, et al. 2002. Primary adult soft tissue sarcoma: Time-dependent influence of prognostic variables. *J Clin Oncol* 20:4344–52.

Strong, L. C., W. R. Williams, and M. A. Tainsky. 1992. The Li-Fraumeni syndrome: From clinical epidemiology to molecular genetics. *Am J Epidemiol* 135:190–99.

Terezakis, S. A., D. M. Lovelock, M. H. Bilsky, et al. 2007. Image-guided intensity-modulated photon radiotherapy using multifractionated regimen to paraspinal chordomas and rare sarcomas. *Int J Radiat Oncol Biol Phys* 69:1502–08.

Tierney, J. F., L. A. Stewart, M. K. B. Parmar, et al. 1997. Adjuvant chemotherapy for localised resectable soft-tissue sarcoma of adults: Meta-analysis of individual data. *Lancet* 350:1647–54.

Trojani, M., G. Contessa, J. M. Coindre, et al. 1984. Soft-tissue sarcomas of adults; study of pathologic prognostic variables and definition of a histopathological grading system. *Int J Cancer* 33:37–42.

Tyldesley, S., K. Fryer, A. Minchinton, et al. 1997. Effects of debulking surgery on radiosensitivity, oxygen tension and kinetics in a mouse tumour model. *Clin Invest Med* 20:S83.

Tzeng, C. W. D., J. B. Fiveash, R. A. Popple, et al. 2006. Preoperative radiation therapy with selective dose escalation to the margin at risk for retroperitoneal sarcoma. *Cancer* 107:371–79.

van Herk, M. 2004. Errors and margins in radiotherapy. *Semin Radiat Oncol* 14:52–64.

Vezeridis, M. P., R. Moore, and C. P. Karakousis. 1983. Metastatic patterns in soft-tissue sarcomas. *Arch Surg* 118:915–18.

White, J. S., D. Biberdorf, L. M. DiFrancesco, et al. 2007. Use of tissue expanders and pre-operative external beam radiotherapy in the treatment of retroperitoneal sarcoma. *Ann Surg Oncol* 14:583–90.

White, L. M., J. S. Wunder, R. S. Bell, et al. 2005. Histologic assessment of peritumoral edema in soft tissue sarcoma. *Int J Radiat Oncol Biol Phys* 61:1439–45.

Wong, F. L., J. D. Boice Jr., D. H. Abramson, et al. 1997. Cancer incidence after retinoblastoma: Radiation dose and sarcoma risk. *JAMA* 278:1262–67.

Yamada, Y., D. M. Lovelock, and M. H. Bilsky. 2007. A review of image-guided intensity-modulated radiotherapy for spinal tumors. *Neurosurgery* 61:226–35.

Yan, D., D. Lockman, A. Martinez, et al. 2005. Computed tomography guided management of interfractional patient variation. *Semin Radiat Oncol* 15:168–79.

Yang, J. C., A. E. Chang, A. R. Baker, et al. 1998. Randomized prospective study of the benefit of adjuvant radiation therapy in the treatment of soft tissue sarcomas of the extremity. *J Clin Oncol* 16:197–203.

Zeidan, O. A., K. M. Langen, S. L. Meeks, et al. 2007. Evaluation of image-guidance protocols in the treatment of head and neck cancers. *Int J Radiat Oncol Biol Phys* 67:670–77.

Zhang, L., A. S. Garden, J. Lo, et al. 2006. Multiple regions-of-interest analysis of setup uncertainties for head-and-neck cancer radiotherapy. *Int J Radiat Oncol Biol Phys* 64:1559–69.

Zlotecki, R. A., T. S. Katz, C. G. Morris, et al. 2005. Adjuvant radiation therapy for resectable retroperitoneal soft tissue sarcoma: The University of Florida experience. *Am J Clin Oncol* 28:310–16.

Subject Index

383

T - #0246 - 111024 - C0 - 280/210/20 - PB - 9780367577001 - Gloss Lamination